Kurt Gottfried
Tung-Mow Yan

Quantum Mechanics: Fundamentals

Second Edition

With 75 Figures

Springer

Kurt Gottfried
Laboratory for Elementary Particle Physics
Cornell University
Ithaca, NY 14853
USA
kg13@cornell.edu

Tung-Mow Yan
Laboratory for Elementary Particle Physics
Cornell University
Ithaca, NY 14853
USA
ty18@cornell.edu

Series Editors
R. Stephen Berry
Department of Chemistry
University of Chicago
Chicago, IL 60637
USA

Joseph L. Birman
Department of Physics
City College of CUNY
New York, NY 10031
USA

Mark P. Silverman
Department of Physics
Trinity College
Hartford, CT 06106
USA

H. Eugene Stanley
Center for Polymer Studies
Physics Department
Boston University
Boston, MA 02215
USA

Mikhail Voloshin
Theoretical Physics Institute
Tate Laboratory of Physics
The University of Minnesota
Minneapolis, MN 55455
USA

Cover illustration: The arrangement of the experiment by Y-Ho. Kim, R. Yu, S.P. Kulik, Y. Shih and MO. Scully described in chapter 12.

Library of Congress Cataloging-in-Publication Data
Gottfried, Kurt.
Quantum mechanics : fundamentals.—2nd ed. Kurt Gottfried, Tung-Mow Yan.
p. cm. — (Graduate texts in contemporary physics)
Includes bibliographical references and index.
ISBN 978-0-387-22023-9
1. Quantum theory. I. Yan, Tung-Mow. II. Title. III. Series.
QC174.12 .G68 2003
530.12—dc21 printed on acid-free paper. 2002030571

ISBN 978-0-387-22023-9 ISBN 978-0-387-21623-2 (eBook)
DOI 10.1007/978-0-387-21623-2

Originally published by Springer-Verlag New York, Inc. in 2003
Softcover reprint of the hardcover 2nd edition 2003

9 8 7 6 5 4 3 2 1 SPIN 10959909

springeronline.com

Graduate Texts in Contemporary Physics

Series Editors:

R. Stephen Berry
Joseph L. Birman
Mark P. Silverman
H. Eugene Stanley
Mikhail Voloshin

Springer Science+Business Media, LLC

Graduate Texts in Contemporary Physics

S.T. Ali, J.P. Antoine, and J.P. Gazeau: **Coherent States, Wavelets and Their Generalizations**

A. Auerbach: **Interacting Electrons and Quantum Magnetism**

T.S. Chow: **Mesoscopic Physics of Complex Materials**

B. Felsager: **Geometry, Particles, and Fields**

P. Di Francesco, P. Mathieu, and D. Sénéchal: **Conformal Field Theories**

A. Gonis and W.H. Butler: **Multiple Scattering in Solids**

K. Gottfried and T-M. Yan: **Quantum Mechanics: Fundamentals, 2nd Edition**

K.T. Hecht: **Quantum Mechanics**

J.H. Hinken: **Superconductor Electronics: Fundamentals and Microwave Applications**

J. Hladik: **Spinors in Physics**

Yu.M. Ivanchenko and A.A. Lisyansky: **Physics of Critical Fluctuations**

M. Kaku: **Introduction to Superstrings and M-Theory, 2nd Edition**

M. Kaku: **Strings, Conformal Fields, and M-Theory, 2nd Edition**

H.V. Klapdor (ed.): **Neutrinos**

R.L. Liboff (ed): **Kinetic Theory: Classical, Quantum, and Relativistic Descriptions, 3rd Edition**

J.W. Lynn (ed.): **High-Temperature Superconductivity**

H.J. Metcalf and P. van der Straten: **Laser Cooling and Trapping**

R.N. Mohapatra: **Unification and Supersymmetry: The Frontiers of Quark-Lepton Physics, 3rd Edition**

R.G. Newton: **Quantum Physics: A Text for Graduate Students**

H. Oberhummer: **Nuclei in the Cosmos**

G.D.J. Phillies: **Elementary Lectures in Statistical Mechanics**

R.E. Prange and S.M. Girvin (eds.): **The Quantum Hall Effect**

S.R.A. Salinas: **Introduction to Statistical Physics**

B.M. Smirnov: **Clusters and Small Particles: In Gases and Plasmas**

(continued after index)

To Sorel and Caroline

Preface

Quantum mechanics was already an old and solidly established subject when the first edition of this book appeared in 1966. The context in which a graduate text on quantum mechanics is studied today has changed a good deal, however. In 1966, most entering physics graduate students had a quite limited exposure to quantum mechanics in the form of wave mechanics. Today the standard undergraduate curriculum contains a large dose of elementary quantum mechanics, and often introduces the abstract formalism due to Dirac. Back then, the study of the foundations by theorists and experimenters was close to dormant, and very few courses spent any time whatever on this topic. At that very time, however, John Bell's famous theorem broke the ice, and there has been a great flowering ever since, especially in the laboratory thanks to the development of quantum optics, and more recently because of the interest in quantum computing. And back then, the Feynman path integral was seen by most as a very imaginative but rather useless formulation of quantum mechanics, whereas it now plays a large role in statistical physics and quantum field theory, especially in computational work.

For these and other reasons, this book is not just a revision of the 1966 edition. It has been rewritten throughout, is differently organized, and goes into greater depth on many topics that were in the old edition. It uses Dirac notation from the outset, pays considerable attention to the interpretation of quantum mechanics and to related experiments. Many topics that did not appear in the 1966 edition are treated: the path integral, semiclassical quantum mechanics, motion in a magnetic field, the S matrix and inelastic collisions, radiation and scattering of light, identical particle systems and the Dirac equation.

We thank Thomas von Foerster and Robert Lieberman for creating the opportunity to publish with Springer. Bogomil Gerganov devoted a great deal of skill and unending attention to the preparation of the manuscript, supported with equal zeal in the last stage by Kerryann Foley. Rhea Garen and Harald Pfeiffer have done the art work and calculations displayed in the figures. We thank the authors of experimental papers whose results are shown, and the American Institute of Physics for the photographs from the Emilio Segré Visual Archive.

Kurt Gottfried has here, at last, the opportunity to thank colleagues who have, either in writing or face-to-face, shared their knowledge, patiently answered questions and corrected misconception: in particular Aage Bohr, Michael Berry, Roy Glauber, Daniel Greenberger, Martin Gutzwiller, Eric Heller, Michael Horne, Ben Mottelson, Michael Nauenberg, Abner Shimony, Ole Ulfbeck and Anton Zeilinger, and the late John Bell, Rudolf Peierls and Donald Yennie. He is deeply indebted to David Mermin and Yuri Orlov for innumerable enlightening and stimulating discussions over many years.

We both wish to take this occasion to thank and remember those who first taught us quantum mechanics: David Jackson, Julian Schwinger and Victor Weisskopf.

Kurt Gottfried
Tung-Mow Yan
Ithaca, NY
January 2003

Road Signs

This book treats considerably more material than fits into a standard *two-semester* course. Furthermore, the rigid organization of this (or any) book does not map in a simply-connected manner onto a good course. We therefore offer some guidance based on our experience with what we have found suitable for the *first* semester. Personal judgments and interests should be used to select material for the second semester, or for self-instruction. Various selections have been taught by us in the first semester. The emphasis has been on physical phenomena, with the general theory and approximation techniques intermingled with applications. The last time the semester was confined to general theory as motivated and illustrated by various one-body problems:

- The portions of the first three chapter with which students are not yet familiar, but leaving almost all of § 2.6 - § 2.9 on propagators, the path integral and semiclassical quantum mechanics for later.
- Low Dimensional Systems: spectroscopy in two-level systems, the harmonic oscillator and motion in a magnetic field.
- Hydrogenic atoms: fine and hyperfine structure, and the Zeeman and Stark effects, i.e., most of chapter 5, leaving aside Pauli's solution (§ 5.2).
- Symmetries: the rotation group (§ 7.4), some consequences of rotational symmetry (part of § 7.5); and tensor operators (§ 7.6).
- Elastic Scattering: general theory (§ 8.1 – § 8.2), approximation methods (§ 8.3), and scattering of particles with spin (§ 8.5).

References. Complete citations are given in Endnotes and the text, except for the following which recur often and are denoted by abbreviations:

Bethe and Salpeter: *Quantum Mechanics of One- and Two-Electron Atoms,* H.A. Bethe and E.E. Salpeter, Springer-Verlag (1957)

Jackson: *Classical Electrodynamics*, J.D. Jackson, 3rd. ed., Wiley (1999)

LLQM: Quantum Mechanics, L.D. Landau and E.M. Lifshitz, 3rd. ed., Pergamon (1977)

WZ: *Quantum Theory and Measurement*, J.A. Wheeler and W.H. Zurek (eds.), Princeton (1982)

Equation numbering. Equations are numbered sequentially in each chapter, and those in different chapters are cited as $(n.m)$, where n is the number of the other chapter.

Errata. Corrections should be sent to both kg13@cornell.edu and ty18@cornell.edu. They will be posted on http://www.lepp.cornell.edu/books/QM-I/

Contents

1

Fundamental Concepts

Readers of this book are assumed to have some knowledge of the fundamental concepts that underlie the elaborate mathematical structure of quantum mechanics. Nevertheless, a recapitulation of these basic ideas is in order here. Introductory treatments take the beginner to base camp, and for good reason do not raise issues that can be postponed safely until demanding climbing is about to start. That is to come, however. For that reason, this review of the fundamentals will include some discussion of the role of relativity in quantum mechanics, why consistency imposes non-classical conditions on the electromagnetic field, and fundamental features of quantum mechanics that are only revealed by multi-particle states.

1.1 Complementarity and Uncertainty

With the wisdom of hindsight, today one can say that certain qualitative features of nature, apparent well before Planck and Einstein invented quantization, posed profound enigmas for the conceptual structure of classical physics. One example will suffice to make this point.

According to the equipartition theorem of classical statistical mechanics, the specific heat of a gas is proportion to the number of degrees of freedom N_f of the individual molecules that compose the gas. If the gas is noble, with apparently structureless constituents, the straightforward count would give $N_f = 3$ for the translational degrees of freedom, which agrees with experiment at all but low temperatures. But even a 19th-century skeptic could ask why this is credible, for if the atoms are structureless, how can they display an excitation spectrum — can a structureless piano play a sonata? Furthermore, according to classical statistical mechanics, the degrees of freedom that generate the excitation spectrum must be included at all temperatures in calculating the specific heat, and this would destroy the agreement with experiment.

The classical *Weltanschauung* leads to whole array of profound puzzles. To give just one other example, it cannot explain why the properties of a sample of an element are identical to that of *any other* sample of the same element, irrespective of their prior chemical and physical history — whether one sample had been extracted from one compound and another from a different compound, by totally different methods. Such puzzles, recognized by only a handful before 1900, were removed by the hypothesis that energy is quantized. With this assumption, it no longer matters that the molecules have electronic, nuclear, and even subnuclear modes of excitation if the gas in which the molecule finds itself is at room temperature, for then the naive count of N_f is correct because the internal degrees of freedom are effectively frozen out. Should the molecule be at the center of the sun, however, that count would be deeply wrong because then it would dissociate into its electrons and nuclei, and these separate constituents would react with the surrounding matter.

(a) Complementarity

That quantization of energy is incompatible with classical mechanics was obvious and distressing to Planck from the start. In the following quarter century, it became increasingly clear that quantization in any guise could not be grafted onto classical mechanics or electrodynamics — that a fundamentally new theory was necessary because classical physics could not even provide a language for describing a host of atomic phenomena in purely qualitative terms. This theory — quantum mechanics and quantum electrodynamics — would have to lead logically from basic assumptions to quantization of such quantities as energy and angular momentum, and also incorporate atomic phenomena with continuous energy spectra, such as collisions. And the new theory would have to satisfy Bohr's *correspondence principle* — loosely speaking, reduce to classical mechanics and electrodynamics in the limit of large quantum numbers. How experiment and theory combined to give birth to quantum mechanics and electrodynamics is a fascinating story that is beyond the scope of this book; brief remarks about this history and subsequent developments, and some references to the original literature and to historical studies, can be found at the end of this chapter, and in chapter 13.

The study of the interaction between light and matter provided most of the crucial clues in the historical development of the quantum theory, and still offers a fairly direct path to central concepts of quantum mechanics and electrodynamics. To be specific, consider the Compton effect, which emerges from the scattering of γ-rays by electrons. Classical electrodynamics would describe the incident γ-ray as a nearly monochromatic wave packet of mean wave vector $\boldsymbol{k}$, and claim that its energy and momentum are proportional to the square of the field strength.[1] This field would force the electrons to oscillate with frequency $\omega = ck$; and the acceleration of the charge would produce an outgoing ("scattered") electromagnetic wave having the *same* frequency and wavelength as the incident radiation, and an angular distribution that is roughly dipolar. After the collision, classical physics would say that the electron would acquire a momentum along $\boldsymbol{k}$ if it had been at rest initially.

The experiments of Compton and his followers gave results in striking disagreement with these predictions. Most, but not all, of their results could be accounted for by accepting Einstein's photon concept: that the electromagnetic wave is to be viewed as an assembly of massless particles, *photons*, each carrying an energy and momentum given by

$$E = \hbar\omega\,, \quad \boldsymbol{p} = \hbar\boldsymbol{k}\,, \qquad \hbar \simeq 1.054 \times 10^{-27}\,\text{erg sec}\,, \tag{1}$$

where $\hbar$ is Planck's constant. By treating the process as a collision between a massless photons and an electron of mass m obeying the relativistic laws of energy and momentum conservation for *particles*, Compton could account for his observation

[1] The magnitude k of the wave vector is related to the wavelength λ by $k = 2\pi/\lambda$, and its direction $\boldsymbol{k}/k = \hat{\boldsymbol{k}}$ is normal to the wavefront. We usually use the reduced wave length $\lambdabar = \lambda/2\pi$; also $\hat{\boldsymbol{a}}$ always stands for a unit vector.

that the scattered γ-rays had a wavelength λ' longer than that of the incident ray,

$$\lambda' = \lambda + 2\lambda_C \sin^2 \tfrac{1}{2}\theta \tag{2}$$

where

$$\lambda_C = \hbar/mc \tag{3}$$

is the electron's Compton wavelength, 3.86×10^{-11} cm, and θ the laboratory scattering angle. Furthermore, Bothe and Geiger confirmed that the scattered electrons came off in coincidence at an angle correlated with that of the scattered photon as expected from these conservation laws.

From the classical perspective, the relations in Eq. 1 are a scandalous liaison between unrelated concepts — that of particle and wave. This is but one example of Bohr's *complementarity principle*, which asserts that to attain an understanding of phenomena in the quantum realm it is necessary to engage two or more concepts that are mutually inconsistent from the classical viewpoint. In Eq. 1 the complementary concepts constitute *wave-particle duality.* As we shall learn in chapter 10, when electrodynamics is reformulated as a quantum theory, this rather vague formulation of complementarity becomes much sharper: the γ-rays produced by Compton's radioactive source are states of the electromagnetic field with a definite number of photons, but in such states the field strengths do not have definite values. The states produced by an ideal laser, on the other hand, have well-defined field strengths, but not a definite number of photons. Here the complementary concepts are those of photon and field strengths; one or the other can be given a sharp meaning, but not both simultaneously.

In the Compton effect, the relation between wavelength and scattering angle is correctly given by (2), and the conservation of energy and momentum gives the observed angle at which the electron emerges once the photon's angle θ is specified, but that does not say *which* value of θ will be observed when the next radioactive decay produces a γ-ray. It is difficult, at best, to imagine anything the experimenter could do to specify further the collision in the hope of getting an in-principle prediction of the value of θ, e.g., by measuring the impact parameter between the photon and the electron, as one can do in a collision between billiard balls. This suggests that the only *empirically reproducible* data the experimenter can hope to find is the *probability distribution* for the scattering angle θ. Einstein's original "naive" photon concept has no means for calculating such a probability distribution. This can, it turns out, be done successfully with quantum electrodynamics.

The Compton effect also serves as an illustration of the correspondence principle. For incident light of low enough frequency or long enough wavelength, the scattered light has the same frequency and wavelength as the incident light. According to (2), here "long enough" means wavelengths long compared to the Compton wavelength, $\lambda \gg (\hbar/mc)$. When this inequality is satisfied, the angular distribution calculated from quantum electrodynamics reduces to the classical Thomson cross section.

A large body of evidence attesting to the wave nature of light had been accumulated in the century preceding the discovery of the photoelectric and Compton effects. How can the phenomena of interference and diffraction be reconciled with these corpuscular manifestations of light?

To examine this question, consider the diffraction of a plane wave incident normally on a transmission grating of parallel slits, each of width d, separated by D, with $D \gg d$. According to classical optics, the angular separation $\Delta\alpha$ between the maxima in the interference pattern, and the angular width $\delta\alpha$ of each maximum, are

$$\Delta\alpha \approx (\lambda/D) , \qquad \delta\alpha \approx (d/D)\Delta\alpha \ll \Delta\alpha , \tag{4}$$

where λ is the wavelength. As detector of the transmitted light a photographic emulsion will do. If the light intensity is very low, and the plate is exposed for a sufficiently short time, a few dark spots will be seen, each due to a photochemical reaction triggered by a single photon. As the length of exposure is increased, the number of spots will increase and they will acquire a density distribution that approaches that of the classically predicted interference pattern. In short, the wave picture gives the correct probability for where a photon is detected.

If we accept the proposition that photons have corpuscular properties, this would presumably imply that the "size" of a photon is microscopic and small compared to a single slit. Then it is only natural to ask what would be observed if the plane wave that covers the whole grating were replaced by sequential exposures that illuminate just one slit at a time. According to classical optics, however, this would produce a succession of diffraction patterns from individual slit, which have an angular width λ/d, i.e., cover many of the maxima in the diffraction pattern when the whole grating is illuminated.

This consideration, and others of a similar vein, lead to an important conclusion:

> *In any setup that allows light to traverse different paths, these paths can either be combined coherently to form an interference pattern in which case the experiment cannot reveal which path a photon follows, or the apparatus can be modified to determine which path is followed but this destroys the interference pattern.*

This is yet another illustration of complementarity: an arrangement designed to manifest one of the properties of a phenomenon expected from the classical viewpoint contains the possibility of observing at least some of the other classical properties. In this instance, the complementary properties are the wave property of interference and the particle property of path.

The elegant experiment sketched in Fig. 1.1 further elucidates the complementarity of the concepts of path and interference. Here two trapped and effectively stationary Hg^+ ions are illuminated by a laser beam. The photons that comprise this beam are scattered by the ions and then detected. Because the probability of scattering is small, the experiment is in effect a succession of individual photon collisions by one or the other of the two ions. If, for a moment, we assume the ions to be structureless, and that there is no such thing as polarization, the scattered light would have the same angular distribution as in the classical Young two hole interference experiment. The ions are, however, spin $\frac{1}{2}$ objects. Furthermore, the laser produces linearly polarized light, and the experiment can select scattered photons of various linear polarizations. As a consequence, by judicious choices of the polarizations of the incident and scattered photons, the events can be separated into two categories:

1. those in which there is *no possibility* that either ion underwent a spin flip ;
2. those in which one of the ions *must* have undergone a spin flip.

In case 1, it is impossible to determine which of the two ions was responsible for scattering, so the path taken by the photon is unknown in principle. In case 2, on the

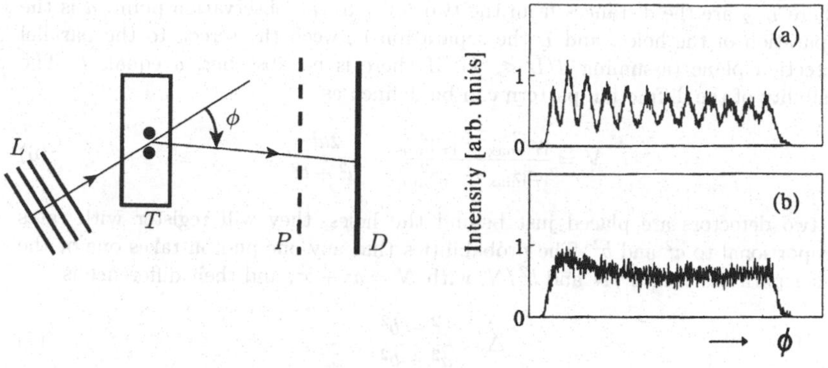

FIG. 1.1. Sketch of an interference experiment. L is a linearly polarized laser beam, T a trap that holds two Hg^+ ions effectively at rest, P a polarizer, and D a detections screen. (a) Events in which spin flip is excluded and the two possible photon paths are indistinguishable. (b) Events in which one of the ions must have had a spin flip which, in principle, determines the photon path. From U. Eichmann, J.C. Berquist, J.J. Bollinger, J.M. Gilligan, W.M. Itano, D.J. Wineland and M.G. Raizen, *Phys. Rev. Lett.* **70**, 2359 (1993). See also W.M. Itano et al., *Phys. Rev. A* **57**, 4176 (1998).

other hand, an examination of the spins of the ions after the photon has scattered would select which one caused the scattering and thus determine the photon's path. Hence the events in category 1 should show the Young interference pattern, while those in category 2 should not. The data confirming this is shown in Fig. 1.1.

Note well that in this experiment there is no instrument that actually measures the spin of the ions after scattering. This fact establishes two important points:

- The question of whether there is interference is settled solely by whether it is impossible or possible to determine which path is taken by a photon. Whether or not the determination is actually made does not matter.

- The "destruction" of the interference pattern arising from collisions where one ion has undergone a spin flip cannot be ascribed to some "irreducible disturbance caused by measurements" carried out on these photons. The collisions which do not cause a spin flip, and do produce the interference pattern, disturb the photons as much or as little as those in which there is a spin flip. To repeat the preceding point, all that counts is whether or not evidence exists that can reveal the photon's path.

For clarity's sake, the absence or presence of interference was just stated as a crystal clear distinction. The full story is that the visibility of the interference pattern diminishes as the confidence grows with which the photon path becomes knowable. For this purpose, consider diffraction by two holes in an opaque screen, with one of the holes being partially absorbing.[1] The amplitude and intensity are then

$$\varphi = ae^{ikL_1} + be^{ikL_2} , \quad |\varphi|^2 = a^2 + b^2 + 2ab\cos[(d/L)kx] , \tag{5}$$

[1] D.M. Greenberger and A. YaSin, *Phys. Lett. A* **128**, 391 (1988).

where $L_{1,2}$ are the distances from the two holes to the observation point, d is the separation of the holes, and L the separation between the screen to the parallel detection plane (assuming $d/L \ll 1$). If there is no absorber, a equals b. The visibility of the diffraction pattern can be defined as

$$V = \frac{|\varphi|^2_{\text{max}} - |\varphi|^2_{\text{min}}}{|\varphi|^2_{\text{max}} + |\varphi|^2_{\text{min}}} = \frac{2ab}{a^2+b^2} \,. \tag{6}$$

If two detectors are placed just behind the holes, they will register with rates proportional to a^2 and b^2. The probabilities that any one photon takes one or the other path are then a^2/N and b^2/N, with $N = a^2 + b^2$, and their difference is

$$\Delta = \frac{a^2 - b^2}{a^2 + b^2} \,. \tag{7}$$

Thus there is a smooth transition as $|a/b|$ departs from $a = b$, when the two paths are equally likely ($\Delta = 0$) and the visibility has its maximum value of $V = 1$, to zero visibility as $|b/a| \to 0$ when it is certain that every photon takes one of the two paths, i.e., $|\Delta| = 1$. Moreover,

$$V^2 + \Delta^2 = 1 \,, \tag{8}$$

so there is a relationship between the degrees to which the complementary wave-like and particle-like features can be in evidence simultaneously. As V is linear in b, a substantial diffraction pattern survives even when $b^2 \ll a^2$ and it is reasonably safe to bet on the path of the next photon.

A similar conclusion applies to the famous debate in which Einstein proposed to determine which of two slits in a plate each particle traverses in contributing to a diffraction pattern by measuring the recoil of the plate, which Bohr showed is incorrect because the uncertainty principle must also be applied to the plate (see the discussion following Eq. 17). A more elaborate analysis shows that when measurement of the plate's recoil provides incomplete knowledge of which slit is traversed there is also a visible diffraction pattern.[1]

(b) The Uncertainty Principle

The uncertainty principle emerges when the Einstein relations for the photon's momentum and energy are combined with the assumptions that the energy density of classical electrodynamics gives the probability of detecting individual photons in a given space-time volume. The classical electric field $\boldsymbol{E}(\boldsymbol{r}, t)$ in empty space satisfies

$$\left(\nabla^2 - \frac{1}{c^2}\frac{\partial^2}{\partial t^2}\right)\boldsymbol{E} = 0 \,, \quad \boldsymbol{\nabla}\cdot\boldsymbol{E} = 0 \,, \tag{9}$$

with boundary conditions appropriate to the apparatus in question; so does the magnetic field. The most general solution of (9) is

$$\boldsymbol{E}(\boldsymbol{r}, t) = \int d\boldsymbol{k}\, e^{i(\boldsymbol{k}\cdot\boldsymbol{r} - \omega t)}\, \boldsymbol{a}(\boldsymbol{k}) \,, \tag{10}$$

[1] W.K. Wooters and W.H. Zurek, *Phys. Rev. D* **19**, 473 (1979); *WZ*.

where $\omega = ck$ and $\boldsymbol{k} \cdot \boldsymbol{a} = 0$. If the photon is detected in some region with sides Δx_i $(i = 1, 2, 3)$, then the electromagnetic energy and momentum densities are nonzero in that region, and fall off quickly outside it. This requires $\boldsymbol{E}(\boldsymbol{r}, t)$ to have the same character. The theory of Fourier integrals then tells us that the size of the region in $\boldsymbol{k}$-space in which the Fourier amplitude $\boldsymbol{a}(\boldsymbol{k})$ is substantial is related to the size of the spatial region by

$$\Delta x_i \Delta k_j \gtrsim \delta_{ij} \, . \tag{11}$$

Furthermore, the time Δt that the packet takes to pass any point is related to the dispersion in frequency by

$$\Delta\omega \Delta t \gtrsim 1 \, . \tag{12}$$

The Einstein relations then recast these inequalities for classical wave packet propagation into inequalities involving dispersions in photon momentum and energy,[1]

$$\Delta x_i \Delta p_j \gtrsim \hbar \delta_{ij}, \qquad \Delta E \Delta t \gtrsim \hbar \, . \tag{13}$$

These are the *Heisenberg uncertainty relations*, first formulated by him for non-relativistic particles, not photons. However, *once the ability to determine any object's momentum and energy is restricted by the uncertainty relations, those of all other objects with which it can, in principle, interact must also satisfy such restriction,* for if they did not, those other objects could be used to "defeat" the uncertainty principle.

It must be emphasized that momentum conservation must be used to reach this last conclusion, and plays a central role in what follows. *That energy and momentum are conserved by any isolated systems on an event-by-event basis, and not just as an average over a large number of events, is, as we shall see, a consequence of the fundamental principles of quantum mechanics.* It is supported by an enormous body of evidence, beginning with the experiments on Compton scattering done before the invention of quantum mechanics, and most strikingly by experiments that observe the recoil of electrons, nuclei, nucleons and other particles when they emit, absorb or scatter neutrinos.

If the uncertainty relations also hold for massive particles, that would suggest that they too display wave-particle duality. This supposition is confirmed by diffraction experiments first done with electrons, then with neutrons, and more recently with atoms. An especially striking illustration of the universality of wave-particle duality is given in Fig. 1.2, which shows the diffraction pattern of electrons scattered from a standing light wave. These all demonstrate that a *de Broglie wavelength*

$$\lambda_{\text{deB}} = h/p \tag{14}$$

is to be ascribed to a particle of momentum p. Here, again, our tale is anachronistic, because de Broglie's conjecture was the opening breakthrough in the invention of quantum mechanics.

The uncertainty relations are still another manifestation of complementarity, for they stipulate that in a description of a system using one classical attribute, say,

---

[1]The momentum-coordinate uncertainty relation is an unambiguous inequality that follows from the formalism of quantum mechanics, whereas the energy-time relation is more qualitative and subtle, a distinction that this argument fails to reveal. This issue will be discussed in §2.4(d).

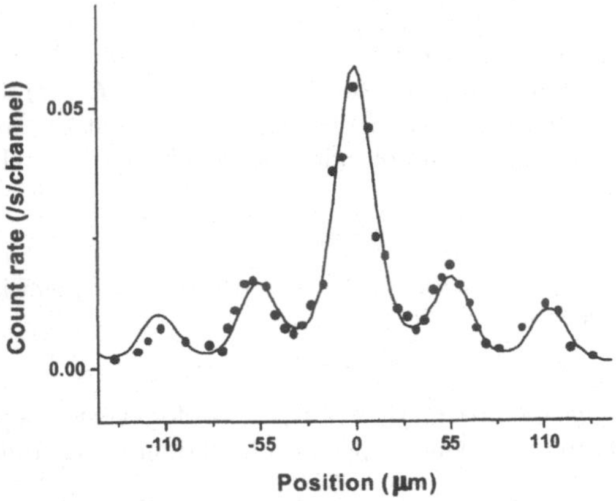

FIG. 1.2. Diffraction pattern produced by scattering electrons from the standing light wave created by two opposed lasers. From D.L. Freimund, K. Aflatooni and H. Batelaan, reprinted by permission from *Nature* **413**, 142 (2001) ©Macmillan Magazines, Ltd. The solid curve is based on the original theory by P.L. Kapitza and P.A.M. Dirac (*Proc. Camb. Phil. Soc.* **29**, 297 (1933)).

position, one surrenders the possibility of specifying the complementary classical attribute, momentum.

It so happens that Heisenberg derived the uncertainty relations by also using the Compton effect, but in the complementary sense — by analyzing to what extent the position and momentum of an electron can be determined. For this purpose he devised a famous thought experiment, the γ-ray microscope. His argument will not be repeated here. But one implication of the argument, recognized afterward by Pauli, is important and often overlooked.

The uncertainty relations do not say whether it is possible to determine one member of the pair x_i and p_i to arbitrary accuracy by surrendering all knowledge of the other. To increase the resolving power of the microscope, the wavelength $\lambda\!\!\!^{-}$ of the scattered light that is to form an image of the electron must be shortened. But according to (2), the shortest wavelength attainable is of order the Compton wavelength $\hbar/mc$, unless the scattering angle θ approaches zero, in which circumstance the microscope would be swamped by the incident light. Furthermore, the wavelength varies across the microscope's aperture by an amount of order $\hbar/mc$, and this again limits the resolution to that same degree. The conclusion, therefore, is that a massive particle's position can only be determined to an accuracy of order its Compton wavelength $\lambda\!\!\!^{-}_C$, no matter how poorly its momentum is known. From today's perspective this is not surprising, because for incident photon frequencies above $2mc^2/\hbar$, or wavelengths short compared to $\lambda\!\!\!^{-}_C$, pair creation becomes possible, and the notion that a measurement is being performed on a one-particle system breaks down. This remark is our first indication that the concepts of non-relativistic quantum mechanics must undergo far-reaching modifications when relativistic effects become important.

This conclusion does not mean that a massive particle's position can never be specified in quantum mechanics. The Compton wavelength is very short compared to all lengths that characterize non-relativistic quantum states because λ_C vanishes in the nonrelativistic limit $c \to \infty$. Put in a dimensionless manner, the ratio of the Compton to de Broglie wavelengths is

$$\frac{\lambda_C}{\lambda_{deB}} = \frac{p}{mc} = \frac{v}{c}, \tag{15}$$

and therefore the inability to specify a massive particle's position to arbitrary accuracy is acceptable in nonrelativistic quantum mechanics.

By the same token, however, position is an ill-defined concept for the massless photon. A photon's position cannot be specified to better than the mean wavelength of the wave packet that describes its propagation. This distinction is reflected in the formalism. In nonrelativistic quantum mechanics, it is possible to define a probability distribution in coordinate space, and in momentum space.[1] In contrast, quantum electrodynamics allows an unambiguous definition of the momentum distribution for photons, but not of a photon probability distribution in coordinate space.

Several questions remain unsettled about the degree to which coordinates or momenta of massive particles can be determined.

First, can a definite time be assigned to a position determination? The uncertainty in time is the interval δt taken by the scattered photon of wavelength λ' to pass a point, which is given by

$$\delta t \approx \lambda'/c \gtrsim \hbar/mc^2 . \tag{16}$$

This is the time required for light to traverse a Compton wavelength, and is negligible in nonrelativistic physics.

Second, is a position determination reproducible? After a time τ the particle struck by the γ-ray will have moved a distance $\tau(\boldsymbol{p} + \Delta\boldsymbol{p})/m$, and this stochastic drift can be reduced to any desired degree by repeating the measurement quickly enough. So, in the nonrelativistic regime, position determinations are, in effect, arbitrarily accurate and reproducible.

Third, can momenta be determined to arbitrary accuracy? To determine the momentum, a particle's position is measured at two points $\boldsymbol{r}_1$ and $\boldsymbol{r}_2$ with inaccuracies $\Delta\boldsymbol{r}_1$ and $\Delta\boldsymbol{r}_2$ sufficiently large to produce negligible uncertainty of the momentum, but small compared to the separation $L = |\boldsymbol{r}_1 - \boldsymbol{r}_2|$. By increasing L the accuracy can be refined to any desired degree, in contrast with the case of position determinations. The setup produces states that continue to have a well-defined momentum, because an almost free particle has a nearly constant momentum. On the other hand, a free particle state that has been determined to have a well-defined position has a large uncertainty in momentum, and as a consequence will lose its localization in a time that shrinks as the accuracy of the position determination increases.

Fourth, how quickly can a momentum determination be made? The preceding determination only produces a state of well-defined momentum after the long time interval Lm/p. But it can be used to prepare a target of low momentum particles for subsequent Compton scattering, and then a selection of the appropriate scattered photon will prepare a particle with the desired momentum. Because there is no need to form an image with the scattered photons in this case, the wavelength can be chosen so as to make the preparation time $\delta t \approx \lambda/c$ arbitrarily short.

---

[1]Because of the uncertainty principle, it is not possible, however, to define a joint probability distribution in coordinates and momenta (i.e., in phase space). In this connection, see §2.2(f).

These considerations justify the assumption of nonrelativistic quantum mechanics that coordinate and momentum space probability distributions exist at a definite instant. It should be understood, however, that knowledge of the momentum distribution does not determine the coordinate distribution, or vice versa, for that would fly in the face of the uncertainty principle. To determine these distributions one must know the Schrödinger wave function, in either coordinate or momentum space; the wave function, however, cannot be computed from either or even both probability distributions except in very special cases.

Finally, we must still confirm that the uncertainty principle is a universal constraint, as it was assumed to be in extending it to massive particles from its first appearance for photons in our account.

Consider, first, a simple position measurement, in which a particle moving along the z-direction is passed through a perpendicular slit of width d, which determines the x-coordinate to within an uncertainty $\Delta x = d$. Diffraction will cause the beam to spread by an angle

$$\sin\theta \simeq \frac{\lambda}{d} \simeq \frac{\hbar}{p_z d}, \tag{17}$$

which produces an uncertainty Δp_x of momentum in the x-direction of order $p_z \sin\theta$. This conforms with the uncertainty principle, i.e., $d\Delta p_x \sim \hbar$. But the assumption that momentum conservation is strictly valid would imply that the uncertainty principle could be "defeated" if the screen's momentum were precisely known. The latter precision is limited, however, by the requirement that the slit's position must be known with a precision much better than d if the previous claim about the determination of the particle's position is to stand, which implies that the screen has a momentum uncertainty much greater than $\hbar/d$. This greatly exceeds the uncertainty Δp_x of the particle's momentum along x, which is of order $\hbar/d$, and therefore precludes a circumvention of the uncertainty principle. In a nutshell, consistency demands that the apparatus also obeys the uncertainty principle. (The screen's mass is macroscopic, so its momentum uncertainty will produce negligible motion, and not undermine the position measurement.)

And now we can close the circle, so to say, by asking whether there are any restrictions on the accuracy of electromagnetic quantities beyond those on photon position and momentum, from which we began. There must be such restrictions because, if classical electrodynamics were left untouched, the position and momentum of a charged particle could be determined from the electric and magnetic field emanating from that particle. If those fields could be determined simultaneously with arbitrary precision, that knowledge could, once again, be used to "defeat" the uncertainty principle. Hence the uncertainty relations for the sources of the fields must also impose uncertainty relations on the fields.

The argument that leads to these relations, due to Bohr and Rosenfeld, is subtle and will be sketched in §10.2 after the quantum theory of the electromagnetic field has been developed. In brief, they noted that field strengths cannot be measured at mathematically sharp space-time points, but must be done with test charges of finite spatial extent V whose response to the fields is observed over a finite time interval T, because the response of point test bodies would be infinitely large as they would be sensitive to arbitrarily high frequency modes of the field. Hence, realistic measurements determine an average over the space-time volume $\Omega = VT$ of, say, the electric field, $\bar{\boldsymbol{E}}(\Omega)$, where Ω can be as small as one pleases, but must be finite.

The uncertainty in momentum and position of a charged test body will cause it to radiate uncertain electromagnetic fields which will confuse field measurements by another test body in another space-time region, and thereby produce an uncertainty in the fields in the latter region.

One example of such a field uncertainty relation is that for uncertainties in space-time averaged electric field components:

$$\Delta \bar{E}_x(\Omega_1)\Delta \bar{E}_y(\Omega_2) \geq \left| \frac{\hbar}{8\pi\Omega_1\Omega_2} \int_{\Omega_1} d\Omega_1 \int_{\Omega_2} d\Omega_2 \frac{\partial^2}{\partial x_1 \partial y_2} \frac{\delta(t - r/c)}{r} \right| . \tag{18}$$

The integral runs only over the portions of the two regions, Ω_1 and Ω_2, that can be connected by light signals, because the delta-function vanishes unless the distance r and time difference t between the two space-time integration points constitute a light-like interval.

This elegant result has two features that should be noted. First, the uncertainty product does not depend on the charge or mass of the test bodies, but depends only on the natural constants and on the regions in which the fields are measured, and of course it vanishes as $\hbar \to 0$. Thus this is an important illustration of the universality of the uncertainty principle. Second, if the two regions cannot be connected by light signals, then there is no restriction on the accuracy to which the two fields in question can be determined, because no measurement in one region can then produce an effect in the other region. This is an example of the relativistic causality principle, which plays a central role in quantum field theory.

A last, important, word. Precisely the same result, Eq. 18, will be derived in §10.2 from the "canonical" quantum theory of the electromagnetic field without resort to intuitive arguments involving test bodies, or order-of-magnitude estimates. Such concordance between heuristic and formal arguments should always be demanded of a basic theory.

1.2 Superposition

Quantum mechanics is a strictly linear theory, and all experimental data are consistent with this assumption. In the history of physics, it is the first and only basic theory with this property. Newton's equations are linear only in special circumstances, such as harmonic motion. Maxwell's equations are linear in empty space, but lose that property when charges are present. The theory of gravitation, general relativity, is inherently nonlinear. All this does not mean that quantum mechanics does not lead to nonlinear effects. Obviously it must, or it could not even reduce to classical mechanics and electrodynamics in the appropriate limit. So what does the opening sentence then mean?

(a) The Superposition Principle

The Schrödinger equation, for any system no matter how complicated, has the form

$$\left(H - i\hbar\frac{\partial}{\partial t} \right) \Psi(t) = 0 , \tag{19}$$

where H is the Hamiltonian and $\Psi(t)$ the wave function. The specifics of the system, including such essentials as the number and types of degrees of freedom it possesses, and whether it is to be treated in a Lorentz-invariant manner by quantum field theory, are buried in the form of H, which will then be reflected in Ψ. But all such "details" have no bearing on the following fact: if $\Psi_1(t)$ and $\Psi_2(t)$ are any two solutions of (19), and $c_{1,2}$ are arbitrary complex numbers, then the linear superposition

$$\Psi(t) = c_1\Psi_1(t) + c_2\Psi_2(t) \tag{20}$$

is also a solution of the Schrödinger equation (19).

This fundamental statement is called the *superposition principle.* In important respects it is the most profound principle of quantum mechanics, and it is often the root cause of those features of quantum mechanics that are most enigmatic from the perspective of our everyday experience. In this section we examine some of the surprising implications of the superposition principle.

The discussion of the preceding section already exploited superposition repeatedly — for example, in connection with the diffraction grating, and in other situations where electromagnetic or de Broglie waves were superposed coherently. These are all examples that have a more or less direct counterpart to phenomena in classical optics. The full import of the superposition principle can only be appreciated, however, by considering phenomena that have no classical counterpart.

The simplest examples of these deeply non-classical features are provided by two-particle states. This was first revealed by Heisenberg's treatment of the helium spectrum in 1926, in which, among other things, he showed that what would appear to be a magnetic interaction between electrons is really due to two-particle interference; this topic will be taken up in §6.2. A fuller appreciation of the astonishing properties of many-particle states, and their implications for the foundations of quantum mechanics, arose from a 1935 paper by Einstein, Podolsky and Rosen, cited henceforth as EPR. This paper and related topics are treated at length in chapter 12.

(b) Two-Particle States

All the examples considered in §1.1 involved superposition of waves in three-dimensional coordinate space, as in classical optics or in acoustics. The Schrödinger equation deals with waves in a three-dimensional space only in the special instance of one-particle systems, however, in which case $|\Psi(\boldsymbol{r},t)|^2 d^3r$ is the probability for finding one particle in an infinitesimal interval about the point $\boldsymbol{r}$ in everyday space. But if quantum mechanics is to supercede classical mechanics, it must, for an N-body system, specify probability distributions in $3N$ coordinates; that is, it must deal with waves not in three-dimensional coordinate space but in $3N$-dimensional configuration space. This is evident in the simplest situation, a system of N free particles of mass m, in which case the Schrödinger equation is

$$-\frac{\hbar^2}{2m}\nabla_N^2\Psi(\boldsymbol{r}_1\ldots\boldsymbol{r}_N;t) = i\hbar\frac{\partial}{\partial t}\Psi(\boldsymbol{r}_1\ldots\boldsymbol{r}_N;t) \ , \tag{21}$$

where ∇_N^2 is the Laplacian in $3N$ dimensions,

$$\nabla_N^2 \equiv \sum_{i=1}^{N}\left(\frac{\partial^2}{\partial x_i^2} + \frac{\partial^2}{\partial y_i^2} + \frac{\partial^2}{\partial z_i^2}\right) \ . \tag{22}$$

The multi-dimensional wavelike character of quantum states, which has no counterpart in classical wave phenomena, gives the concept of coherence a more subtle and richer meaning in quantum mechanics than it has in classical optics, and leads to phenomena that are astonishing and counterintuitive from a classical perspective.

It suffices for now to consider two-body systems without mutual interactions, the generalization to $N > 2$ being rather straightforward. The Hamiltonian is then

$$H = H_1 + H_2 \,, \tag{23}$$

where H_i is the Hamiltonian of body i. Each body (e.g., a many electron atom) may have internal degrees of freedom which interact with each other. All that matters is that there be no interaction between the two bodies. Two distinct types of Schrödinger equations then come into play. The complete two-body equation

$$(H_1 + H_2)\Psi(q_1q_2;t) = i\hbar\frac{\partial}{\partial t}\Psi(q_1q_2;t), \tag{24}$$

and the one-body equations

$$H_i\psi_{a_i}(q_i;t) = i\hbar\frac{\partial}{\partial t}\psi_{a_i}(q_i;t) \,, \tag{25}$$

where q_i stands for *all* the coordinates of body i, including those needed to describe its internal motions, and a_i is a label that distinguishes between the different solutions of this last equation.

Because H_1 does not depend on the coordinates of particle 2, and vice versa, H acting on the product of one-body wave functions results in

$$(H_1 + H_2)\psi_{a_1}(q_1;t)\psi_{a_2}(q_2;t) = \psi_{a_2}(q_2;t)\Big(H_1\psi_{a_1}(q_1;t)\Big) + \psi_{a_1}(q_1;t)\Big(H_2\psi_{a_2}(q_2;t)\Big),$$

and therefore

$$\left(H - i\hbar\frac{\partial}{\partial t}\right)\psi_{a_1}(q_1;t)\psi_{a_2}(q_2;t) = 0 \,. \tag{26}$$

Hence a product of one-body wave functions, $\psi_{a_1}(q_1;t)\psi_{a_2}(q_2;t)$, is a solution of the two-body Schrödinger equation provided there is no interaction between them.[1] As the two-body equation (24) is linear, the superposition of such products,

$$\Psi(q_1q_2;t) = \sum_{a_1a_2} c_{a_1a_2}\psi_{a_1}(q_1;t)\psi_{a_2}(q_2;t) \,, \tag{27}$$

where the $c_{a_1a_2}$ are arbitrary complex constants, is again a solution of the Schrödinger equation for a non-interacting two-body system. It is essential to note that in the preceding sentence "constants" means just that — *independent of time.*

If there is an interaction, the Hamiltonian will have an additional term involving the coordinates of both particles, $H_{12}(q_1q_2)$, and a product of one-body wave

[1]In the language of partial differential equations, this is just the method of separation of variables. Put in the more abstract form used here, it is not only more transparent, but it is illustrative of quantum mechanical thinking. That is, if A_1 and A_2 are any two operators that commute with each other (as do H_1 and H_2 because they act on different variables), then eigenstates of $A_1 + A_2$ are products of separate eigenstates of A_1 and A_2. Those who find this illuminating comment inscrutable should ignore it for now; it will be explained in detail in §2.1.

functions is then no longer a solution, nor are linear combinations with constant coefficients. For now we are only concerned with interactions of finite range, however, in which case the simple solutions (27) remain valid in many circumstances. That is, if the coordinates q_i are separated into a center-of-mass position $\mathbf{r}_i$, and internal coordinates, then in such systems H_{12} tends rapidly to zero as the separation $R = |\mathbf{r}_1 - \mathbf{r}_2|$ becomes large compared to an appropriate microscopic length scale, e.g., the Bohr radius $\sim 10^{-8}$ cm if the bodies are neutral atoms. At such separations, therefore, $H = H_1 + H_2$, and Eq. 27 is a solution wherever R is large compared to that scale. Furthermore, if the bodies are not involved in any processes sufficiently energetic to excite their internal motions, then in such an energy regime they are in effect "elementary" particles, and their wave functions can be treated as if they depend only on the center-of-mass coordinates $\mathbf{r}_i$. We will, therefore, often use the terms *particle*, *system* and *body* indiscriminately.

(c) Two-Particle Interferometry

Two-particle states can display interference effects if both particles are detected in coincidence while not showing conventional one-particle interference patterns. This is the phenomenon we shall now examine with the help of a simple and yet quite realistic *Gedanken* (or thought) experiment. Careful study of this experiment proves to be a sound investment for it reveals features that are central to both the formulation and interpretation of quantum mechanics.

Consider two particles a and b described by a wave function $\Psi(\mathbf{r}_a \mathbf{r}_b; t)$. From Ψ we can form the joint probability distribution

$$P_{ab}(\mathbf{r}_a \mathbf{r}_b; t) = |\Psi(\mathbf{r}_a \mathbf{r}_b; t)|^2 . \tag{28}$$

This is the probability of detecting a at $\mathbf{r}_a$ and b at $\mathbf{r}_b$ *in coincidence*. We can also form the one-particle probability distributions, e.g., that for detecting a when b is *not* observed at all:

$$P_a(\mathbf{r}_a; t) = \int d^3r_b \, P_{ab}(\mathbf{r}_a \mathbf{r}_b; t) . \tag{29}$$

Our purpose is to establish the following assertion:

> In any experimental setup that allows the two particles to traverse different paths, and in which *it is possible, in principle, to determine the path taken by one particle by some observation on the other,* neither particle will, by itself, display an interference pattern (i.e., in P_a or in P_b), but there may be an interference pattern in the a-b coincidence rate P_{ab}, i.e, in the correlation of positions for a and b. On the other hand, if the setup is such that *no observation on one particle can, in principle, determine the path of the other*, then either particle by itself, or both, may display an interference pattern (i.e., in P_a and/or P_b).

The words "in principle" are critical here. They allude, as we shall see, to the fact that *whether an observation is or is not made does not matter — what matters is whether such an observation is possible at all.*[1]

[1] Note also the similarity to the statement on p. 4 regarding one-particle interference effects.

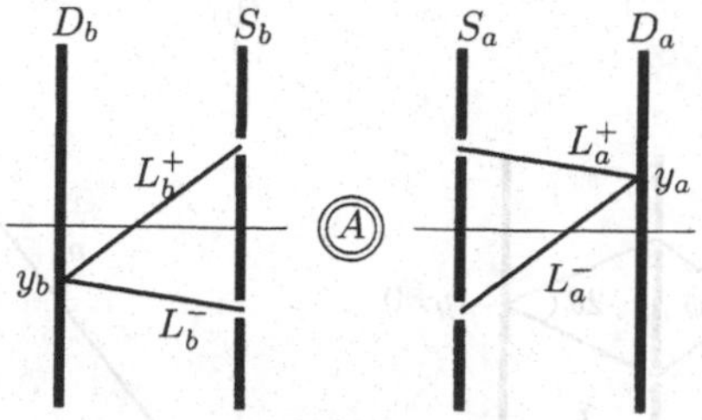

FIG. 1.3. Two-particle interferometer. The free particle A undergoes the two-body decay $A \to a+b$, with a passing through one of the two pin holes on the right and then detected on the right-hand screen, while b does the same on the left.

The state that will be analyzed to establish these contentions describes the particles produced in a decay process $A \to a + b$, and the experimental setup is the two-particle interferometer shown in Fig. 1.3. This consists of two parallel opaque screens S_a and S_b, each pierced by two pinholes symmetrically placed about the axis normal to the screens, and two parallel detection screens D_a and D_b sensitive only to a and b, respectively. The detectors record the coordinates of particles striking them in coincidence, i.e., determine the joint probability distribution P_{ab} of Eq. 28. This thought experiment is not far-fetched because there are several real-life examples of A. One is positronium, the bound electron-positron system, whose ground state annihilates into two photons; another is the neutral kaon K^0, a particle that decays into two π mesons.

In the process $A \to a + b$ momentum is conserved, and so the decay products (daughters) will go in exactly opposite directions if A was at rest. Then if a passes through one of the two holes on the right, b must pass through the diametrically opposed hole on the left, and therefore a determination of the path of one determines that of the other. On the other hand, if A were at rest its position would be totally uncertain; it could be anywhere with respect to the interferometer. Conversely, if A is at the exact center of the setup, its momentum would be totally uncertain, there would be no correlation between the directions of a and b, and observation on one daughter would not determine the path of the other.

Hence A's vertical localization s must exceed some lower limit to assure that the daughters can only pass through one pair of diametrically opposed pinholes. This limit is set by the uncertainty principle and momentum conservation. The former states that A's momentum uncertainty satisfies $\Delta p_A \gtrsim \hbar/s$; the latter that the spread in angles Θ between the daughters' momenta (see Fig. 1.4(b)) is given by $\Theta \simeq (\Delta p_A/\hbar k) \gtrsim (1/sk)$, where we assume that the energy release in the decay is large enough so that both daughters have momenta of approximately the same magnitude $\hbar k$. But if the source A is to only illuminate one or the other of the opposed holes, Θ must be much smaller than ϕ, the angle subtended by the two pinholes on one screen as seen from A. Consequently, the condition on A's localization is

$$s \gg \frac{1}{k\phi} . \tag{30}$$

We now turn to the daughters' wave function $\Psi_{\text{out}}(\boldsymbol{r}_a\boldsymbol{r}_b)$ outside the screens S_a and S_b. In this region they do not interact with each other, and therefore Ψ_{out} must

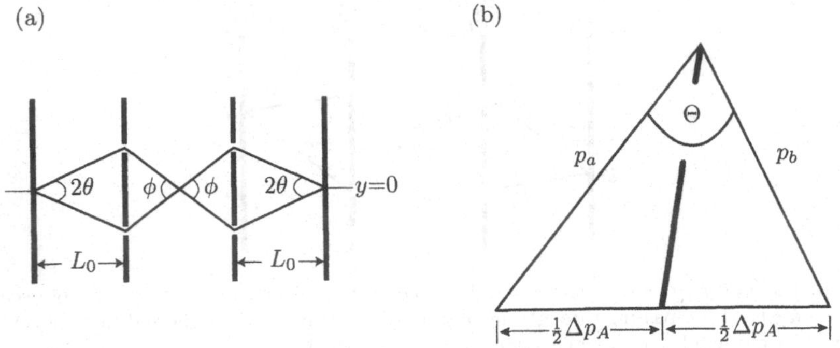

FIG. 1.4. (a) Angles pertaining to the two-particle interferometer. (b) Kinematics in the process $A \to a + b$.

be a linear combination of products of one-particle wave functions of the form (27), with each particle's wave function being a spherical wave $\psi(r) = e^{ikr}/r$ emanating from one of the pinholes. For example, one such term is $\psi(L_a^+)\psi(L_b^-)$, for the case where a emerges from the upper hole on the right and b from the lower one on the left, with the distances $L_{a,b}^{\pm}$ being defined in Fig. 1.3.

In general, Ψ_{out} is a linear combination of four such products with four arbitrary coefficients determined by matching $\Psi_{\text{out}}(\boldsymbol{r}_a\boldsymbol{r}_b)$ to the interior wave function $\Psi_{\text{in}}(\boldsymbol{r}_a\boldsymbol{r}_b)$ at the pinholes. Nothing essential is lost, and considerable simplification is gained, by considering the case where A is in a spherically symmetric wave packet. In this case the initial state A is invariant under reflection through the $y = 0$ plane perpendicular to the interferometer, which is equivalent to a rotation through π about an axis perpendicular to the interferometer's symmetry axis. Assuming that the interaction responsible for the process $A \to a + b$ is invariant under rotations, the daughters' state $\Psi_{\text{in}}(\boldsymbol{r}_a\boldsymbol{r}_b)$ must have the same reflection symmetry, and as a consequence the values of this wave function at the four pinholes are given by just two complex numbers:

$$\Psi_{\text{in}}(\boldsymbol{r}_a^+\boldsymbol{r}_b^-) = \Psi_{\text{in}}(\boldsymbol{r}_a^-\boldsymbol{r}_b^+) = \alpha\,, \qquad \Psi_{\text{in}}(\boldsymbol{r}_a^+\boldsymbol{r}_b^+) = \Psi_{\text{in}}(\boldsymbol{r}_a^-\boldsymbol{r}_b^-) = \beta\,, \tag{31}$$

where $\boldsymbol{r}_{a,b}^{\pm}$ are the positions of the pinholes.

These coefficients have a simple meaning: $|\alpha|^2$ is the probability for a and b to pass through diametrically opposed holes, i.e., for A to have undergone back-to-back decay, whereas $|\beta|^2$ is the probability for both to pass through either the two upper or two lower holes. Clearly, $|\beta/\alpha|^2 \ll 1$ if the initial state of A satisfies the source size condition (Eq. 30).

The outside wave function evaluated at the detectors, when (31) holds, is

$$\Psi_{\text{out}} \doteq \alpha\left(e^{ikL_a^+}e^{ikL_b^-} + e^{ikL_a^-}e^{ikL_b^+}\right) + \beta\left(e^{ikL_a^+}e^{ikL_b^+} + e^{ikL_a^-}e^{ikL_b^-}\right). \tag{32}$$

Here the distance L_0 from the screens to the detectors is assumed to be much larger than that between the holes on either screen, so that the denominators in e^{ikr}/r can all be replaced by L_0, which has been absorbed into an irrelevant overall factor, and $\doteq$ means equal apart from such a factor. In this geometry (the Fraunhoffer diffraction limit), the various lengths can be approximated by

$$L_a^{\pm} \simeq X \mp \theta y_a\,, \qquad L_b^{\pm} \simeq X \mp \theta y_b\,, \tag{33}$$

where X is the same length in all four cases, while θ and the y coordinates are defined in Fig. 1.4. With these small-angle approximations, (32) simplifies to

$$\Psi_{\text{out}}(y_a y_b) \doteq \alpha \cos[k\theta(y_a - y_b)] + \beta \cos[k\theta(y_a + y_b)]\,. \tag{34}$$

The novel two-particle interference phenomena require, as we will now see, that the decay be dominantly back-to-back, that is, $|\beta/\alpha|^2 \simeq 0$. The joint probability distribution for detecting a at y_a and b at y_b in coincidence is then

$$P_{ab}(y_a y_b) \doteq |\cos[k\theta(y_a - y_b)]|^2\,. \tag{35}$$

This asserts that *the coincidence rate will display an interference pattern in the variable* $|y_a - y_b|$*, the "distance," so to say, between locations on the widely separated detection screens.*

In striking contrast, *the distributions of locations of particles on the individual detection screens show no interference pattern.* This is so because, according to (29), the probability for detecting a at y_a regardless of where b struck the other detector, is

$$P_a(y_a) = \frac{1}{2Y}\int_{-Y}^{Y} dy_b\, P(y_a, y_b) = \text{const.} + O(1/Y)\,. \tag{36}$$

This distribution is independent of a's position, a result that requires the size $2Y$ of the detection region for b to be large enough to yield no information about b's position, i.e., $Y \gg (1/k\theta)$, the distance between interference fringes.

Quantum optics experiments that confirm the remarkable effect just described have been done, though for technical reasons, not by using photon pairs produced in a momentum-conserving decay such as just discussed. The results of an outstanding example are shown in Fig. 1.5.

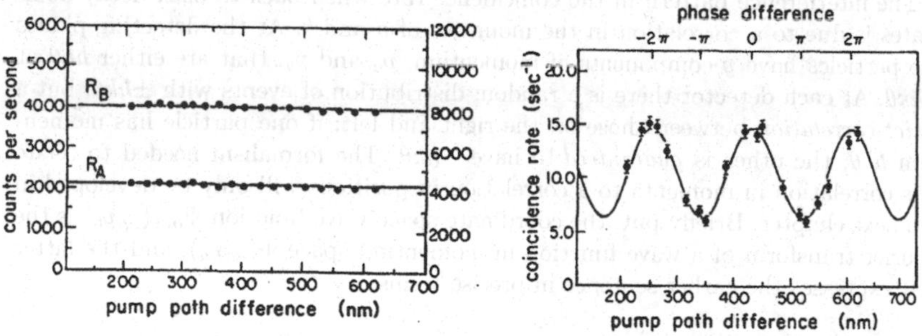

FIG. 1.5. Results of a two photon interference experiment which is effectively equivalent to the setup in Fig. 1.3 (L. Mandel, *Rev. Mod. Phys.* **71**, S274 (1999)). The horizontal scales are the counterpart of $y_a - y_b$. On the left are the singles rates at the two detectors D_a and D_b, on the right the coincidence rate, with the solid line being the theoretical expectation.

The existence of interference effect in the coincidence rate when there is none in the individual rates is a quantum mechanical phenomenon. Hence it is important to understand why this initially surprising result is, in retrospect, not surprising!

The absence of an interference pattern in the individual rates is due to the possibility of determining the path that both particles took by an observation on just one. To accomplish this one can, for example, replace b's position detector D_b by a device that determines b's momentum along y as it arrives at the left-hand detection plane. This would determine which hole in S_b it traversed, and because of the back-to-back decay which hole in the other screen was traversed by a. Hence there can be no interference pattern in the locations of a's position, for that would require a coherent addition of amplitudes from the two holes on the right side, which is excluded by the knowledge of which hole a traversed. Of course, the same conclusion holds for b.

The preceding paragraph might lead one to suspect either an error or a swindle, because it talks of an apparatus that measures position *and* momentum of one and the same object, which is impossible if the uncertainty principle is valid. The suspicion is unfounded, however. The experimental setup measures momentum *or* position, with the choice of which measurement is actually made being on an event-by-event basis, and if one wishes decided only *after* the decay $A \rightarrow a+b$ has occurred while the daughters are still *en route*. It is this feature of two-particle states that lies at the heart of EPR experiments, which will be discussed shortly.

To shed further light on what has just been discussed, consider what happens if the back-to-back decay does not dominate, and β is not negligible. In particular, consider the case where the source size s is small enough so that the two holes on either side are illuminated equally, i.e., $\alpha = \beta$. Then (34) becomes

$$\Psi_{\text{out}}(y_a y_b) \doteq \cos(k\theta y_a) \cdot \cos(k\theta y_b) , \tag{37}$$

which describes *independent* diffraction patterns on each of the two detection screens. There are no correlations in either coordinates or momenta, the reason being that determining which hole was traversed by one particle does not determine which path the other took.

The interference pattern in the coincidence rate when back-to-back decay dominates is due to a correlation in the momenta of a and b. At the detecting planes the particles have y-components of momentum, p_a and p_b, that are either $\hbar k\theta$ or $-\hbar k\theta$. At each detector there is a random distribution of events with $\pm\hbar k\theta$, but a *strict correlation* between those on the right and left: if one particle has momentum $\hbar k\theta$, the other is *guaranteed* to have $-\hbar k\theta$. The formalism needed to relate this correlation in momenta to a correlation in positions will only be developed in the next chapter. Briefly put, the coordinate space wave function $\Psi_{\text{out}}(y_a y_b)$ is the Fourier transform of a wave function in momentum space $\Phi(p_a p_b)$, and the latter just expresses the earlier sentence in precise terms:

$$\Phi \doteq \delta(p_a - \hbar k\theta)\delta(p_b + \hbar k\theta) + \delta(p_a + \hbar k\theta)\delta(p_b - \hbar k\theta) , \tag{38}$$

where $\delta(p)$ is the Dirac delta function, which is the statement that the momentum is precisely p. When this expression is Fourier transformed, it immediately leads to $\cos[k\theta(y_a - y_b)]$, in agreement with (32) when $\beta = 0$.

(d) EPR Correlations

The phenomena revealed by the two-particle interferometer have remarkable features that Einstein called "spooky action-at-a-distance," and which were first discussed by Einstein, Rosen and Podolsky (EPR).

There is, of course, nothing novel about using a particular measurement on one particle to determine a property of another that is, at the time, arbitrarily far away. If it is known that a missile with known linear and angular momenta will separate into two pieces, it suffices to determine these momenta of one piece to determine those of the other. Here, however, there are remarkable features that are totally foreign to classical physics. The wave-like correlation in position between particles that are arbitrarily far apart is such a feature — a feature that is complementary (in Bohr's sense) to the sharp correlations in their momenta, Eq. 38. While this momentum correlation seems very natural from a classical perspective, it must not be forgotten that the observed momentum distribution $|\Phi(p_a p_b)|^2$ is a coherent superposition of the two classically understandable terms in (38).

It is even more remarkable that a *free choice* between determining the position or momentum of b assigns the distant particle a, *without disturbing it,* to distinct categories of events that form distinct diffraction patterns. Furthermore, this choice can be made while the particles are on the way to the detectors. This feature is not evident with arbitrarily small pinholes, because such an idealized aperture produces no angular variation in intensity. With a finite aperture of width $2d$, the one-particle distribution is no longer uniform, as it is Eq. 36, but becomes

$$P_a(y_a) \doteq \left|\frac{\sin k\epsilon(\rho - y_a)}{k\epsilon(\rho - y_a)}\right|^2 + \left|\frac{\sin k\epsilon(\rho + y_a)}{k\epsilon(\rho + y_a)}\right|^2 , \tag{39}$$

where $\epsilon = d/L_0, \rho = l + \frac{1}{2}\varphi L_0$, and $2l$ is the distance between the pinholes. This is the incoherent sum of two conventional one-hole diffraction patterns with centers separated by 2ρ.

This EPR feature can be demonstrated by the following experiment. On the right, the apparatus always measures the coordinate of a; and on the left, the apparatus switches at random between measurements of position or momentum of b. All measurements are done in coincidence. This does not require communication between the widely separated laboratories, because there can be a protocol to insert only one specimen of A at regularly spaced intervals sufficiently long to insure that both laboratories carry out observations on the same $a + b$ specimen. After the run is over, the list of choices that were made in observing b is transmitted to the distant laboratory, where the data on the coordinate of a are separated into three sets: (1) those where $p_b = \hbar k\theta$, (2) those where $p_b = -\hbar k\theta$, and (3) those where y_b was measured. No human intervention is necessary after the apparatus is set to work; the collection, transmission and processing of the data can be fully automated.

The predictions of quantum mechanics are that

- set (1) will display the diffraction pattern of the upper hole alone (the first term of Eq. 39);
- set (2), the diffraction pattern of the lower hole (the second term);
- set (3), the oscillating correlation function $|\cos[k\theta(y_a - y_b)]|^2$.

While this particular experiment has not been done, enough experiments of the EPR variety have been so that there is no reason to doubt that what has just been claimed would be confirmed.

This setup provides no means for instantaneous signaling between the left-hand and right-hand laboratories. The observations on a in themselves give no information about what is learned about b in coincidence. No correlation can be elicited until the list of observation by one arm of the experiment are combined with the data from the other arm.

The interference patterns displayed, or not displayed, by the two-particle wave function are, to underscore it yet again, determined by what the experimenter can in principle do, and not by what may actually be done. This is best brought out by the *delayed choice* property — by the freedom to decide whether the momentum or coordinate is to be determined just before the particles are actually detected. What counts is whether the option to make this choice exists, and not whether the option is exercised. If the wave function is to pass the delayed choice test, it must possess a sufficient richness of properties to cope with a measurement that will be done in the future. It must, in this example, be able to display itself in a wave-like or particle-like guise, or in various combinations of these guises, depending, so to say, on what question will be asked of it.

The two cases discussed in detail, the one where there are independent one-particle diffraction patterns and no interference effects in the coincidence rate, and the other where there is an interference effect in the coincidence rate but no one-particle diffraction patterns, are limiting cases of the general situation described by Eq. 32 for arbitrary values of β/α. The quality that changes along this continuum is the degree of confidence with which it possible to determine the path of one particle by an observation on the other.[1] If $\beta = 0$, it is known for sure which path a took from a momentum measurement on b, but as β increases this becomes progressively less certain, and the two possible paths become equally probable when $\beta = \alpha$. The smooth transition from two-particle to one-particle interference goes hand-in-hand with this decrease of knowledge attainable *in principle.*

1.3 The Discovery of Quantum Mechanics

Quantum mechanics is among the great intellectual achievements of the 20th century, and how this came about is interesting in itself. The following only offers the briefest of sketches.

The history separates naturally into two eras: 1900–1925, the development of the "Old Quantum Theory"; and 1925 to circa 1935, in which quantum mechanics and electrodynamics were discovered and their principal features elucidated.

In 1900 Max Planck discovered that he could only account for the spectrum of thermal radiation, which was in violent contradiction with classical electrodynamics, by assuming that the material sources of radiation have a discrete ("quantized") energy spectrum. That this entailed a grave contradiction with classical physics was

[1]This complementarity between one-particle and two-particle interference, which is related to the confidence of knowledge about paths, can be phrased quantitatively as a relationship, similar to (8), between the visibility of the corresponding diffraction patterns; see G. Jaeger, A. Shimony and L. Vaidman, *Phys. Rev. A* **51**, 54 (1995).

clear to Planck and troubled him. In 1905, Albert Einstein produced a much deeper departure from classical concepts by applying quantization to the thermodynamics of the electromagnetic field itself, introduced what was later to be called the photon, and predicted correctly the basic feature of the photo-electric effect. Although this and later Einstein papers on the quantum theory had a great influence, his idea that light has corpuscular aspects was only widely accepted after the discovery of the Compton effect in 1923. The next major step came in 1913 with Niels Bohr's quantum theory of the hydrogen spectrum based on Ernest Rutherford's model of the atom. This led to a great advance in the understanding of atomic structure and spectra, as can be seen in the massive 1924 edition of Arnold Sommerfeld's *Atombau und Spektrallinien*. The culminating theoretical advances of this first period were Wolfgang Pauli's exclusion principle and the discovery of electron spin[2] by Samuel Goudsmit and George Uhlenbeck, both in 1925.

Experimental physics played an indispensable role throughout this first period, of course. The 19th-century studies of thermal ("black body") radiation gave the first clearcut demonstrations that classical physics suffers from fatal deficiencies. In 1914 the Franck-Hertz experiment used inelastic electron scattering to establish that atoms did indeed have quantized excitation energies; in 1921 the Stern-Gerlach experiment demonstrated the quantization of angular momentum; and to repeat, in 1923 Arthur Compton performed his fundamental experiment on the scattering of γ rays. These were, of course, only the highest peaks of an imposing mountain range that held many riches that were to prove invaluable in what was to come.

The practitioners of the Old Theory understood full well that a logical structure was lacking — that it was not really a theory, but a jumble of folklore and recipes with many striking successes and deep insights undermined by perplexing failures.[1]

The discovery of quantum mechanics arose from two distinct and seemingly unrelated streams of ideas: wave mechanics and matrix mechanics. Wave mechanics started first in 1923–24 with Louis de Broglie's suggestion that massive particles have wave-like properties — that wave-particle duality does not only apply to photons. This idea did not take on a powerful form until Erwin Schrödinger discovered his wave equation in 1926. Before then matrix mechanics had been initiated by Werner Heisenberg in the summer of 1925 and was already well established by year's end.

Heisenberg sought to focus on what is actually observable and to discard notions, such as electronic orbits in atoms, which he argued are not. He therefore replaced the classical coordinates and momenta of a charged particle by arrays of observable radiative transition amplitudes. Then he showed that these arrays obey a noncommutative algebra whose rules were constrained by empirically confirmed knowledge about radiative transitions that had been developed with the Old Theory, such as the Kramers-Heisenberg dispersion formula (see §10.7). In keeping with Bohr's correspondence principle, which states that the quantum theory must reduce to classical theory in the limit of large quantum numbers, Heisenberg assumed that his arrays obeyed equations of motion that had the same form as those of classical

---

[2]Electron spin was also proposed at the same time by R. de L. Kronig, but not published because Pauli convinced him the idea was nonsensical. It is said that the Nobel Prize was never awarded for electron spin because of this.

[1]In this connection, Hans Bethe recalled that he had a great advantage over older people in learning quantum mechanics in 1926 because he did not know the Old Quantum Theory!

mechanics. It should be said that this ground-breaking paper, which Heisenberg in private referred to as "fabricating quantum mechanics," is very difficult to follow, in contrast to most of the others that will now be referred to.

That Heisenberg's arrays are matrices was quickly recognized by Max Born, who with Pascual Jordan then derived the canonical (q, p) commutation rule. This was followed, in November 1925, by the Born, Jordan and Heisenberg *magnum opus* which gave a nearly complete formalism: conservation laws, angular momentum, canonical transformations, some stationary state perturbation theory, and a first stab at quantization of the electromagnetic field. Independently, and based only on Heisenberg's first paper, Paul Dirac in the same month presented a development of almost comparable sweep that postulated a correspondence between classical Poisson brackets and commutation rules. Finally, in January 1926 Pauli published his *tour de force* matrix mechanics solution of the Kepler problem, including both the Stark and Zeeman effects (see §5.2 and §5.4(c)).

Schrödinger called his papers "Quantization as an Eigenvalue Problem." His approach was inspired by de Broglie's suggestion and the analogy between geometrical optics and classical point mechanics that William R. Hamilton had created in the 1830s. Schrödinger's first paper, submitted in January 1926, put forward the time-independent wave equation. His point of departure was to replace Hamilton's characteristic function W by $\hbar \log \psi$ in the Hamilton-Jacobi equation (see §2.8(a)). From this he formulated a variational principle which led to the wave equation, and he then showed that its eigenvalues in the case of a Coulomb field reproduce the Bohr spectrum. In the following half year Schrödinger published a series of papers which also produced the complete theory. Particularly noteworthy was his demonstration that his "wave mechanics" is mathematically equivalent to "matrix mechanics."

Among the many successful early applications of quantum mechanics three merit special attention. First, in 1926, Heisenberg's successfully analyzed the He spectrum, which had baffled the Old Theory. In solving this problem he discovered that the Pauli principle requires antisymmetrical wave functions[1] (something which he could only do with Schrödinger's theory!), and that the correlations resulting therefrom produce large electrostatic level splittings that look as if they are magnetic in origin (see §6.2). Second, also in 1926, Born used Schrödinger's equation to solve the first scattering problem, a phenomenon that the Old Theory could not even formulate, and in doing so introduced the interpretation of $|\psi(\boldsymbol{r})|^2$ as a probability distribution. And third, in 1927, Dirac devised second quantization to quantize the radiation field and showed how that accounted for photons and radiative transitions.

Although Heisenberg set out to restrict the new theory to observable quantities, this proved to be elusive and the physical meaning of the mathematical formalism has been controversial to this day. The "Copenhagen interpretation" of quantum mechanics was developed in that city by Bohr and Heisenberg working both in collaboration and independently. The famous outcome was Bohr's principle of complementarity and Heisenberg's uncertainty principle, both published in 1927. Another and related early landmark was J. von Neumann's *Mathematische Grundlagen der Quantenmechanik*, Springer (1932). Einstein's critique of the Copenhagen interpre-

---

[1] That Bose and Fermi statistics require, respectively, symmetric and antisymmetric wave functions was discovered at the same time by Dirac.

tation culminated in the 1935 paper with Boris Podolsky and Nathan Rosen (see §12.1).

In the period 1927–1934 relativistic quantum theory was developed, and this will be sketched in chapter 13.

Enrico Fermi, Werner Heisenberg and Wolfgang Pauli
Photograph by F.D. Rasetti

James Franck and Max Born (with Born's son Gustav)

Arthur H. Compton

P.A.M. Dirac
Photo by A. Börtzells Tryckeri

Erwin Schrödinger
Photo by Francis Simon

Otto Stern
All pictures courtesy of the AIP Emilio Segrè Visual Archives

1.4 Problems

1. Derive Eq. 39 for the two-particle interference pattern produced by finite apertures.

Endnotes

For an instructive discussion of various neutron interference experiments, see D.M. Greenberger, *Rev. Mod. Phys.* **55**, 875 (1983). Remarkable coherence phenomena also arise in one-particle states describing particles possessing "internal" quantum numbers that have no classical counterpart. The most famous and extensively explored examples arise from neutral K mesons, and in particular, the phenomenon of "regeneration." In this case there is, however, an clear analogy with coherence phenomena in classical optics – for example, the propagation of polarized light through a magnetized medium. Neutral K meson phenomena are discussed in all books on particle physics; the analogy with propagation in a magnetized medium is in K. Gottfried and V.F. Weisskopf, *Concepts of Particle Physics,* Vol. I, Oxford University Press (1984); p. 152.

§1.2 is drawn from D.M Greenberger, M.A. Horne and A Zeilinger, *Physics Today,* pp. 22–29, August 1993, and from K. Gottfried, *Am. J. Phys.* **68**, 143 (2000). These articles cite additional literature.

The key original papers by Heisenberg, Born, Jordan, Dirac and Pauli, translated into English as necessary, on what began as matrix mechanics, together with invaluable comments, appear in B.L. van der Waerden (ed.), *Sources of Quantum Mechanics,* North-Holland (1967), Dover (1968). English translations of Schrödinger's papers are in E. Schrödinger, *Collected Papers on Wave Mechanics,* Blackie and Son (1928). Dirac's paper on the quantization of the electromagnetic field is reproduced in J. Schwinger (ed.), *Quantum Electrodynamics,* Dover (1958). English translations of Bohr's and Heisenberg's first papers on complementarity and uncertainty appear in J.A. Wheeler and W.H. Zurek (eds.), *Quantum Theory and Measurement,* Princeton University Press (1983), and so does the article by Bohr and Rosenfeld on the measurement of electromagnetic field strengths. The English translation of von Neumann's book was published by Princeton University Press in 1955.

For technical historical monographs, see E.T. Whittaker, *A History of the Theories of Aether and Electricity,* Vol. 2, Thomas Nelson and Sons (1953); M. Jammer, *The Conceptual Development of Quantum Mechanics,* McGraw-Hill (1966); and S.S. Schweber, *QED and the Men who Made it,* Princeton University Press (1994). For summaries more detailed than given here see A. Pais, *Inward Bound,* Oxford University Press (1986); S. Weinberg, *The Quantum Theory of Fields,* Vol. 1, chapter 1, Cambridge University Press (1995); and H.A. Bethe in *More Things in Heaven and Earth,* B. Bederson (ed.), Springer (1999), or *Rev. Mod. Phys.* **71**, S1 (1999). The historical development of quantum mechanics can also be inferred from the extensive citations in W. Pauli, *Die allgemeine Prinzipien der Wellenmechanik,* Encyclopedia of Physics, Vol. 5, Springer (1955), which is identical to *Handbuch der Physik,* Vol. 24, Part 1, Springer (1933). For the best eyewitness account see W. Pauli, *Scientific Correspondence,* Vol. 1, A. Hermann, K. v. Meyenn and V.F. Weisskopf (eds.), Springer-Verlag (1979).

Original papers on the Old Quantum Theory appear in D. ter Haar, *The Old Quantum Theory,* Pergamon (1967); those most directly relevant to the discovery of matrix mechanics are in van der Waerden (*loc. cit.*). A collection of excerpts from landmark papers, with commentaries, appear in *100 Years of Planck's Quantum,* I. Duck and E.C.G. Sudarshan, World Scientific (2000). For the role of black body radiation see T. Kuhn, *Blackbody Theory and the Quantum Discontinuity,* Oxford University Press (1978); A. Pais, *Subtle is the Lord ...,* Oxford University Press (1982); and *Einstein's Miraculous Year,* J. Stechel (ed.), Princeton University Press (1998), pp. 165-177. A recent discussion of developments at the turn of the last century, with references, appears in R.D. Purrington, *Physics in the Nineteenth Century,* Rutgers University Press (1997), chapter 8.

2
The Formal Framework

The concepts described impressionistically in chapter 1 must be cast into a well-defined mathematical formalism if one is to do theoretical physics. This chapter is devoted to a first cut at this task. Much of the material, such as that on symmetries and on the interpretation of quantum mechanics, will be reconsidered in greater depth in later chapters.

As in chapter 1, we assume that this is not the reader's first encounter with this material. Hence the presentation is occasionally terse, but as readers will have various levels of prior knowledge, it also seeks to be reasonably self-contained. One important point should be made at the outset. The level of mathematical rigor will be typical of the bulk of the theoretical physics literature — slovenly. Mathematicians view sloppy rigor as an oxymoron. Readers who share this view should consult the bibliography at the end of this chapter.

Any physical theory employs concepts more primitive than those on which the theory sheds light. In classical mechanics these primitive concepts are time, and points in three-dimensional Euclidean space. In terms of these, Newton's equations give unequivocal definitions of such concepts as momentum and force, which had only been qualitative and ambiguous notions before then. Maxwell's equations and the Lorentz force law, in conjunction with Newton's equations, define the concepts of electric and magnetic fields in terms of the motion of test charges.

There is no corresponding clearcut path from classical to quantum mechanics. In many circumstances, though certainly not all, classical physics tells us how to construct the appropriate Schrödinger equation that describes a particular phenomenon. But the statistical interpretation of quantum mechanics is not implied by the Schrödinger equation itself.[1] For this and other reasons the interpretation of quantum mechanics is still controversial despite its unblemished empirical success. In this chapter we therefore adopt a brashly pragmatic attitude to many issues concerning the interpretation of quantum mechanics, in the belief these matters are best pondered at length after combat experience, as they will be in chapter 12.

2.1 The Formal Language: Hilbert Space

The quantum state is a fundamentally abstract concept. For brevity, name it Ψ. Abstract concepts are also important in classical physics, but as mathematical and intuitive idealizations of concrete things. That is not so of the quantum state.

[1] This can be put as a fable. If Newton were to be shown Maxwell's equations and the Lorentz force law, he could figure out what is meant by the unfamiliar symbols $\boldsymbol{E}$ and $\boldsymbol{B}$, but if Maxwell was handed Schrödinger's equation he could not decipher the meaning of ψ. For an elaboration, see K. Gottfried, *Nature* **405**, 533 (2000).

The fact that any Ψ pertaining to an N-particle system can be given a numerical representation as a Schrödinger wave function $\psi(\boldsymbol{r}_1 \dots \boldsymbol{r}_N)$, whose variables are in good old configuration space, should not be read as implying that ψ is any more concrete than is Ψ. This warning is not just an ideological prejudice; if it is ignored one is led to seemingly plausible conceptions of reality that must be abandoned, as we will see in chapter 12.

In classical mechanics, the state of an N-particle system is specified by a point $X = (q_1 \dots q_{3N}; p_1 \dots p_{3N})$ in $6N$ dimensional phase space Ω_{qp}, where the q_j and p_j are the canonical coordinates and momenta. The equations of motion then specify how X moves about in Ω_{qp}. The states of all systems with $3N$ degrees of freedom inhabit this same space Ω_{qp}, with systems that have different forces and masses moving along different families of trajectories. Given X, all properties of the system are completely specified.

The quantum state Ψ differs from X in two ways: (i) it "lives" in an abstract space (Hilbert space) that has no connection to the "real" world; (ii) given a mathematical expression for Ψ, quantum mechanics provides unambiguous statistical predictions about the outcomes of all observations on the system that Ψ describes. There are analogies, on the other hand. The quantum states of all systems having the same degrees of freedom live in the same Hilbert space. And just as X gives a complete description in classical mechanics, so does Ψ in quantum mechanics, though of course the meaning of "complete" is very different.

Because quantum mechanics makes only statistical predictions in most circumstances, there is a tendency to overstate quantum mechanical uncertainty. In important ways, quantum mechanics provides more knowledge than does classical mechanics, a point that is often given short shrift. For example, if a monoatomic gas is in an environment that cannot impart enough energy to lift its atoms out of their ground state, we know all about it, so to say, no matter whether it is in a laboratory or in intergalactic space. That all these atoms have precisely the same ground state cannot be understood in classical mechanics.

The quantum state can be given a numerical form in an infinity of ways — by functions in the configuration space $\mathfrak{C}$, or in momentum space, to name just two. There is an analogy here to classical mechanics, where a description in terms of intuitively appealing variables can be translated into an infinity of others by canonical transformations. But phase space as a whole cannot be used to describe quantum mechanical probability distributions, because the uncertainty principle rules out simultaneous precise knowledge of coordinates and conjugate momenta.

Furthermore, wave-particle duality implies that Ψ cannot, over time, be strictly confined to any subspace $\mathfrak{C}$. At some instant it may be so confined, but thereafter it will spread, with finite speed in a quantum theory that complies with relativity, but instantaneously in nonrelativistic quantum mechanics. The whole setting in which the system can exist, in principle, is involved in defining Ψ. Restated with poetic license, a classical golfer whose ball leaves the tee flying toward a gap between two trees cares not one whit about their size, or about other trees that it will never encounter, whereas a quantum golfer does — "will never encounter" has no clear meaning for the quantum golfer.

(a) Hilbert Space

The elements of the language appropriate for describing quantum states in the configuration space $\mathfrak{C}$ are *complex* "wave" functions $\psi(q_1 \dots q_{3N})$. Any such function — whether or not it is a solution of some Schrödinger equation — can be specified uniquely in terms of a complete orthonormal set of functions. For the moment, let's restrict the range of all the coordinates to be finite, $-\frac{1}{2}L \leq q_j \leq \frac{1}{2}L$, with L larger than any physical dimension of interest, so that the volume of configuration space is L^{3N}. In this case such complete sets are denumerable: $u_a(q_1 \dots q_{3N})$, $a = 1, 2, \dots$. The term *complete* means that any ψ can be expressed as a linear combination of the u_a,

$$\psi(q_1 \dots q_{3N}) = \sum_{a=1}^{\infty} c_a u_a(q_1 \dots q_{3N}) \, . \qquad (1)$$

The term *orthonormal* means that

$$\int dq_1 \dots dq_{3N} \, u_a^*(q_1 \dots q_{3N}) u_{a'}(q_1 \dots q_{3N}) \equiv \int (dq) \, u_a^*(q) u_{a'}(q) = \delta_{aa'} \, , \qquad (2)$$

which also defines a convenient shorthand. As a consequence, the expansion coefficients c_a are

$$c_a = \int (dq) \, u_a^*(q) \psi(q) \, . \qquad (3)$$

Substitution into the original expansion for ψ then produces the completeness relation for the set $\{u_a\}$:

$$\sum_a u_a(q) u_a^*(q') = \delta^{3N}(q - q') \, , \qquad (4)$$

where now the shorthand means

$$\delta^{3N}(q - q') \equiv \prod_{j=1}^{3N} \delta(q_j - q_j') \, . \qquad (5)$$

In these equations $\delta_{aa'}$ is the Kronecker delta, i.e., 1 or 0 according to whether a and a' are equal or not, and $\delta(x - x')$ the one-dimensional Dirac delta function, defined by the property

$$f(x) = \int \delta(x - x') f(x') dx'$$

for any function f that is continuous at x.

The set of square integrable functions $\{u_a(q)\}$ constitute a *basis* in an infinite-dimensional complex vector space, called a *Hilbert space* $\mathfrak{H}$. It is the space that plays a role akin to that of phase space in classical mechanics — the stage on which quantum mechanics performs.

The functions $u_a(q)$ on the configuration space $\mathfrak{C}$ do not form a unique description of this particular basis. Consider, for example, the Fourier transform of u_a:

$$v_a(p_1 \dots p_{3N}) \equiv v_a(p) = \prod_{j=1}^{3N} \int_{-\frac{1}{2}L}^{\frac{1}{2}L} \frac{dq_j}{\sqrt{L}} \, e^{-ip_j q_j/\hbar} \, u_a(q_1 \dots q_{3N}) \, , \qquad (6)$$

where

$$p_j/\hbar = 2n_j\pi/L, \quad n_j = 0, \pm 1, \pm 2, \dots \ . \tag{7}$$

so that the functions $\Pi_j \exp(ip_jq_j/\hbar)$ satisfy periodic boundary conditions on the surface of $\mathfrak{C}$. At the moment, $\hbar$ is merely any constant having the dimension of action so as to make $p/\hbar$ an inverse length and to give p the dimension of momentum. The physical content required to make it Planck's constant has not yet entered the story.

Eq. 6 says that the set of functions $\{v_a(p)\}$ in momentum space provides a "representation" of the *same* basis in $\mathfrak{H}$ as does the set $\{u_a(q)\}$ in configuration space. This statement can be rephrased in a fruitful manner by noting that the plane waves form a complete orthonormal basis which is distinct from the one formed by the u_a (or equivalently, the v_a). That is, define

$$\phi_{p_1 \dots p_{3N}}(q_1 \dots q_{3N}) \equiv \phi_p(q) = \prod_{j=1}^{3N} \frac{e^{iq_jp_j/\hbar}}{\sqrt{L}} \ ; \tag{8}$$

then, as just claimed,

$$\int (dq)\, \phi_p^*(q)\phi_{p'}(q) = \prod_j \delta_{p_jp'_j} \equiv \delta_{pp'}^{3N} \ , \tag{9}$$

$$\sum_{p_1 \dots p_{3N}} \phi_p(q)\phi_p^*(q') = \delta^{3N}(q-q') \ . \tag{10}$$

In terms of this notation, Eq. 6 is

$$v_a(p) = \int (dq)\, \phi_p^*(q)u_a(q) \ , \tag{11}$$

and conversely, because of (9),

$$u_a(q) = \sum_{\{p\}} \phi_p(q)v_a(p) \ . \tag{12}$$

(b) Dirac's Notation

These somewhat unsightly expressions follow with seductive ease from a notation of great power due to Dirac. Notation is no laughing matter as Newton learned from his competitor Leibniz.

The basic idea, long familiar to mathematicians, is borrowed from vector algebra. Consider a real three-dimensional Euclidean space $\mathfrak{E}_3$. Any two vectors $(\boldsymbol{v}, \boldsymbol{w})$ can be described by an infinity of bases, and in particular, by any one of an infinity of distinct triads of mutually perpendicular unit vectors. Given such a basis, $\boldsymbol{v}$ and $\boldsymbol{w}$ can be specified by two triplets of real numbers, their projections onto the basis triad. However — and this is the crux of the matter, the geometrical meaning of $\boldsymbol{v}$ and $\boldsymbol{w}$, and of their scalar and vector product, are basis-independent: in all bases $\boldsymbol{v} \cdot \boldsymbol{w}$ is the same real number, and $\boldsymbol{v} \times \boldsymbol{w}$ is a vector perpendicular to both $\boldsymbol{v}$ and $\boldsymbol{w}$ and of a length that is basis-independent.

The generalization of these concepts appropriate to quantum mechanics is a d-dimensional complex vectors space — a Hilbert space if d is infinite; usually the space will be denoted by $\mathfrak{H}$ in either case. In Dirac's nomenclature, the vectors in this space are called *kets*, and denoted by the symbol $|\ \rangle$. The kets are *not* numerical functions, just as $\boldsymbol{v}$ is not a triplet of real numbers. As in Euclidean vector algebra, the kets are defined by their formal properties.

Let $(\alpha, \beta, \ldots)$ be complex numbers and $(|\xi\rangle, |\eta\rangle, \ldots)$ be kets. By the definition of a complex vector space, the product $\alpha|\xi\rangle$ is again a vector in the space, and so is the sum

$$|\zeta\rangle = \alpha|\xi\rangle + \beta|\eta\rangle \ . \tag{13}$$

Because the scalars in the space (i.e., the 'field' over which it is defined, in mathematical parlance) are complex numbers, scalar products cannot be defined directly between pairs of kets, however. It is necessary, first, to associate a dual vector to every ket in a one-to-one manner, called a *bra*, denoted by the symbol $\langle\ |$; and second, to define the scalar products as being between bras and kets.[1] Let $\langle\xi|$ and $\langle\eta|$ be the bras dual to the indicated kets. Their scalar product is then denoted by the bra-ket $\langle\xi|\eta\rangle$, a complex number having the property

$$\langle\xi|\eta\rangle = \langle\eta|\xi\rangle^* \ , \tag{14}$$

so that the *norm* $\langle\xi|\xi\rangle$ of any ket is real, and by definition positive.[2]

In view of (14), the bra corresponding to $|\zeta\rangle$ as defined in (13) is

$$\langle\zeta| = \alpha^*\langle\xi| + \beta^*\langle\eta| \ . \tag{15}$$

Furthermore, if $|\omega\rangle$ is any ket, then the scalar product between $|\omega\rangle$ and $|\zeta\rangle$ satisfies the linear relationship

$$\langle\omega|\zeta\rangle = \alpha\langle\omega|\xi\rangle + \beta\langle\omega|\eta\rangle \ . \tag{16}$$

Ordinary Euclidean vectors satisfy the inequality $|\boldsymbol{v}\cdot\boldsymbol{w}| \leq vw$. There is a generalization of this property in $\mathfrak{H}$, the *Schwarz inequality,*

$$|\langle\xi|\eta\rangle|^2 \leq \langle\xi|\xi\rangle\langle\eta|\eta\rangle \ , \tag{17}$$

with equality being attained only if $|\xi\rangle$ and $|\eta\rangle$ are proportional to each other, or collinear. The proof of Eq. 17 is left for a problem.

A basis in $\mathfrak{H}$ is a set of kets $\{|k\rangle\}$, $k = 1, 2, \ldots$, that spans the space in the sense that any ket can be expressed as a linear combination of the basis kets. With but few exceptions, we will use orthonormal bases, that is, bases such that

$$\langle k|k'\rangle = \delta_{kk'} \ . \tag{18}$$

If $|\omega\rangle$ is an arbitrary ket, it can be expressed as a linear combination of basis kets:

$$|\omega\rangle = \sum_k c_k|k\rangle \ . \tag{19}$$

Because of (18),

$$c_k = \langle k|\omega\rangle \ , \tag{20}$$

and therefore

$$|\omega\rangle = \sum_k |k\rangle\langle k|\omega\rangle \ . \tag{21}$$

[1] There is an analogy here to covariant and contravariant vectors in a non-Euclidean space, such as Minkowski space, where both types of vectors are needed to form Lorentz invariants.

[2] In the mathematics literature the norm is often defined as the positive square root of $\langle\xi|\xi\rangle$.

(c) Operators

This last expression gives the first hint of the power of Dirac's notation. To bring this out, consider the concept of a *subspace* of $\mathfrak{H}$, spanned by a subset of the basis vectors, and denoted by $\mathfrak{H}_S$. This subspace may be finite or infinite dimensional. Its complement $\mathfrak{H}_C$ is the remainder of $\mathfrak{H}$, so that

$$\mathfrak{H} = \mathfrak{H}_S \oplus \mathfrak{H}_C \ . \tag{22}$$

Take, in particular, the 1-dimensional subspace $\mathfrak{H}_k$ spanned by the single basis ket $|k\rangle$, and *define the projection operator* P_k *onto* $\mathcal{H}_k$ by

$$P_k|\omega\rangle = c_k|k\rangle = |k\rangle\langle k|\omega\rangle \ . \tag{23}$$

In this book we shall only indicate that an object P *is an operator by the special notation* $\hat{P}$ *when this is not obvious from the context.*

The next step is motivated by (23), namely, the introduction of what could be called an outer product between bras and kets as the definition of a *linear operator*, i.e., as an object that maps kets into linear combination of other kets (and of course bras into bras). To that end, write the right-hand side of (23) as $(|k\rangle\langle k|)|\omega\rangle$, and the projection operator as

$$P_k \equiv |k\rangle\langle k| \ . \tag{24}$$

The sum of the projection operators on *all* the basis kets is a projection on the entire space $\mathfrak{H}$, and when acting on any ket will leave it unchanged. This sum, therefore, is just the identity operator, which we simply call 1. (When it seems necessary to emphasize that it is an operator, it will be denoted by **1**.) That is,

$$1 = \sum_k P_k = \sum_k |k\rangle\langle k| \ . \tag{25}$$

This is the statement that the basis spans the whole space, and is called the completeness relation. By judicious use of this form of 1, a considerable portion of what in the wave-mechanical formalism is often time and thought consuming "arithmetic" is reduced to flying by autopilot.

To see that this is so, consider

$$|\omega\rangle = 1|\omega\rangle = \left(\sum_k |k\rangle\langle k|\right)\cdot|\omega\rangle = \sum_k |k\rangle\langle k|\omega\rangle \ , \tag{26}$$

demonstrating how effortlessly Eq. 21 emerges. So does the basic identity satisfied by the projection operators on an orthonormal set:

$$P_k P_{k'} = \delta_{kk'} P_k \ , \tag{27}$$

because $P_k P_{k'} = |k\rangle\langle k|k'\rangle\langle k'|$.

Projection operators onto multi-dimensional subspaces can also be defined by generalizing (24). Let $\mathcal{K}$ designate a set of kets $|k_i\rangle, i = 1, \dots, n$, which span the subspace $\mathfrak{H}_\mathcal{K}$. The projection operator onto this subspace is then

$$P_\mathcal{K} = \sum_{i=1}^{n} |k_i\rangle\langle k_i| \ , \qquad P_\mathcal{K}^2 = P_\mathcal{K} \ . \tag{28}$$

Let $\mathcal{K}'$ be some other subspace *orthogonal* to $\mathcal{K}$, i.e., having no basis kets in common with $\mathcal{K}$, and define $P_{\mathcal{K}'}$ as in (28). Then

$$P_{\mathcal{K}} P_{\mathcal{K}'} = 0 \,, \tag{29}$$

in obvious generalization of (27).

Next, define two operators $X = |a\rangle\langle b|$ and $Y = |c\rangle\langle d|$, where the bras and kets are arbitrary. The product of an operator with with any ket or bra are then defined by generalizing the foregoing:

$$X|\xi\rangle = (|a\rangle\langle b|)|\xi\rangle = |a\rangle\langle b|\xi\rangle \,, \tag{30}$$

$$\langle\xi|X = \langle\xi|(|a\rangle\langle b|) = \langle\xi|a\rangle\langle b| \,. \tag{31}$$

The product of X and Y is defined as

$$XY = (|a\rangle\langle b|)(|c\rangle\langle d|) = |a\rangle(\langle b|c\rangle)\langle d| \,, \tag{32}$$

i.e., the operator $|a\rangle\langle d|$ multiplied by the number $\langle b|c\rangle$, and best written *without* the parentheses as

$$XY = |a\rangle\langle b|c\rangle\langle d| \,. \tag{33}$$

Properties of operators and their representation in a particular basis now follow easily. Thus if A is a linear operator, its action on a basis ket $|k\rangle$, i.e., $A|k\rangle$, is first written as $1 \cdot A|k\rangle$, and therefore

$$A|k\rangle = \sum_{k'} |k'\rangle\langle k'|A|k\rangle \,. \tag{34}$$

The complex numbers $\langle k'|A|k\rangle$ in this linear combination of kets are called the *matrix elements* of A in the k representation. The term *matrix* is appropriate because it describes the linear combination of kets that is produced by the application of two operators in succession:

$$BA|k\rangle = \sum_{k} |k'\rangle\langle k'|B \cdot 1 \cdot A|k\rangle = \sum_{k'k''} |k'\rangle\langle k'|B|k''\rangle\langle k''|A|k\rangle \,. \tag{35}$$

This is the law for matrix multiplication:

$$\langle k|BA|k'\rangle = \sum_{k''} \langle k|B|k''\rangle\langle k''|A|k'\rangle \,. \tag{36}$$

Furthermore, any operator can be written in terms of its matrix elements by writing $A = 1 \cdot A \cdot 1$ and using Eq. 25:

$$A = \sum_{kk'} |k\rangle\langle k|A|k'\rangle\langle k'| \,. \tag{37}$$

The diagonal matrix element $\langle k|A|k\rangle$ is called the expectation value of A in $|k\rangle$, for reasons to be explained in §2.2(b).

Two numbers that are global characteristics of operators appear frequently. The first is the *trace*, defined as the sum of the diagonal matrix elements, the second is the *determinant* of the matrix elements,

$$\mathrm{Tr}\, A = \sum_{k} \langle k|A|k\rangle, \qquad \det A = \det\,\{\langle k|A|k'\rangle\} \,. \tag{38}$$

These numbers would be of limited value were they dependent on the basis used to define the matrix elements. That they are not will emerge shortly. A caution: the trace and determinant of many operators important in quantum mechanics do not exist because the expressions that define them fail to converge when the Hilbert space in question is infinite dimensional.

From any operator A we can construct two other operators which may, or may not, differ from A. The first is the *transpose*, the second the complex conjugate:

$$A^T = \sum_{kk'} |k\rangle\langle k'|A|k\rangle\langle k'|, \qquad A^* = \sum_{kk'} |k\rangle\langle k|A|k'\rangle^* \langle k'| \,. \tag{39}$$

When $A = A^T$, the operator is called symmetric. Transposition combined with complex conjugation produces the *Hermitian adjoint* $A^\dagger$ of A:

$$A^\dagger = (A^*)^T = \sum_{kk'} |k\rangle\langle k'|A|k\rangle^* \langle k'| \,. \tag{40}$$

If $A = A^\dagger$, it is an *Hermitian operator.* Such operators play a central role in quantum mechanics. Note that

$$\langle k|A|k'\rangle^* = \langle k'|A^\dagger|k\rangle \,. \tag{41}$$

Hence if $A|\psi\rangle = |\psi'\rangle$,

$$\langle\psi'| = \langle\psi|A^\dagger \,. \tag{42}$$

It follows from these definitions that

$$(AB)^\dagger = B^\dagger A^\dagger \,. \tag{43}$$

Hence if A and B are Hermitian, their product is only Hermitian if they commute. The *commutator* is defined as

$$[A, B] = AB - BA \,; \tag{44}$$

if A and B are Hermitian, and the commutator is not zero, it is *anti-Hermitian,* the latter meaning an operator C with the property $C^\dagger = -C$. By the same token, the *anticommutator*, defined as

$$\{A, B\} = AB + BA \,, \tag{45}$$

is Hermitian if A and B are. Operators of any type satisfy the Jacobi identity

$$[[A, B], C] + [[C, A], B] + [[B, C], A] = 0 \,. \tag{46}$$

Any operator A can be decomposed into its Hermitian and anti-Hermitian components, A_1 and A_2:

$$A_1 = \tfrac{1}{2}(A + A^\dagger) \,, \quad A_2 = \tfrac{1}{2}(A - A^\dagger) \,. \tag{47}$$

A *positive operator* is defined by the requirement that its expectation value in any ket be real and positive. Hence any positive operator is Hermitian. If A is an arbitrary operator, then both $AA^\dagger$ and $A^\dagger A$ are positive operators, but they are only identical if A_1 commutes with A_2 because

$$AA^\dagger - A^\dagger A = 2[A_2, A_1] \,. \tag{48}$$

(d) Unitary Transformations

If an operator U satisfies

$$U^\dagger U = UU^\dagger = 1 , \tag{49}$$

it is said to be *unitary.* In terms of the k basis, this reads

$$\sum_{k'} \langle k|U|k'\rangle\langle k''|U|k'\rangle^* = \delta_{kk''}, \quad \text{or} \quad \sum_{k'} U_{kk'}U^*_{k''k'} = \delta_{kk''} , \tag{50}$$

which is the familiar definition of a unitary matrix in the notation $U_{kk'} = \langle k|U|k'\rangle$.

Unitary operators are of great importance in quantum mechanics because they describe symmetry operations and time evolution. This is intimately connected with the fact that they describe the relationship between distinct bases in $\mathfrak{H}$. Let $|r\rangle, |r'\rangle, \ldots$ be another orthonormal basis that differs from the k basis. The relations between them are

$$|r\rangle = \sum_k |k\rangle\langle k|r\rangle, \qquad |k\rangle = \sum_r |r\rangle\langle r|k\rangle . \tag{51}$$

The scalar products $\langle k|r\rangle$ are often called *transformation functions* because they specify how one basis is transformed into the other. They are the elements of a unitary matrix because

$$\sum_k \langle r|k\rangle\langle k|r'\rangle = \delta_{rr'}, \qquad \sum_r \langle k|r\rangle\langle r|k'\rangle = \delta_{kk'} , \tag{52}$$

which has the same form as (50) with $U_{rk} = \langle r|k\rangle$.

When the bases related by a unitary transformation are denumerable, the unitary operator can be written explicitly in terms of the two sets of basis vectors. Let $|a_1\rangle \ldots |a_i\rangle \ldots$ be one basis ordered in a specific manner, and $|b_1\rangle \ldots |b_i\rangle \ldots$ be the other basis ordered so that $|a_i\rangle$ is mapped into $|b_i\rangle$; define

$$U_{ba}|a_i\rangle = |b_i\rangle . \tag{53}$$

Then

$$U_{ba} = \sum_i |b_i\rangle\langle a_i| , \tag{54}$$

and

$$U^\dagger_{ba} = U_{ab}. \tag{55}$$

Furthermore, if $\{|c_i\rangle\}$ is a third basis, then

$$U_{bc}U_{ca} = U_{ba} . \tag{56}$$

This known as the group property of sequential unitary transformation.

The combination of operators $UAU^\dagger$, where U is unitary, is called a *unitary transformation* of A. For any pair of operators, the trace and determinant of their product satisfy the following identities:

$$\text{Tr } AB = \text{Tr } BA, \quad \det AB = (\det A)(\det B) . \tag{57}$$

Hence

$$\text{Tr } UAU^\dagger = \text{Tr } A, \quad \det UAU^\dagger = \det A . \tag{58}$$

The latter follows from (57) and

$$|\det U| = 1 \ , \tag{59}$$

because $\det U^\dagger = (\det U)^*$, so that

$$\det UAU^\dagger = (\det U)(\det A)(\det U^\dagger) = |\det U|^2 \det A = \det A \ . \tag{60}$$

Hence the trace and determinant of any operator are invariant under unitary transformations, or as stated earlier, under a change of basis.

The projection operator $P_\mathcal{K}$ onto a subspace $\mathfrak{H}_\mathcal{K}$, defined in (28), is also invariant under unitary transformations of the basis in $\mathfrak{H}_\mathcal{K}$. This is self-evident because $P_\mathcal{K}$ is the unit operator in $\mathfrak{H}_\mathcal{K}$.

There are many ways of expressing a unitary operator U in terms of a Hermitian operator. The most common by far is

$$U = e^{iQ} = \sum_{n=0}^{\infty} \frac{(iQ)^n}{n!} \ , \tag{61}$$

where Q is Hermitian, so that $U^\dagger = e^{-iQ}$ and $UU^\dagger = 1$. This power series arises in a multitude of situations, especially in perturbation theory and in the description of continuous symmetries. Another useful form can be

$$U = \frac{1 + iK}{1 - iK} \ , \tag{62}$$

where now K is Hermitian.

The abstract language of kets and bras can now be related to the concrete language with which we started — that of the basis functions $u_a(q)$ in configuration space, their counterparts $v_a(p)$ in momentum space, and the Fourier transform involving $\phi_p(q)$ that relates these functions. To that end, recall first Eq. 1, and introduce the arbitrary ket $|\psi\rangle$, and a complete orthonormal set $|a\rangle, |a'\rangle, \ldots$, so that

$$|\psi\rangle = \sum_a |a\rangle\langle a|\psi\rangle \ . \tag{63}$$

On comparing with (1), we have $c_a = \langle a|\psi\rangle$.

Next, introduce a *nondenumerable* set of kets $|q_1 \ldots q_{3N}\rangle$, where all the labels take on all values on the real interval $(\frac{1}{2}L, -\frac{1}{2}L)$, which is orthonormal in the sense that

$$\langle q_1 \ldots q_{3N}|q'_1 \ldots q'_{3N}\rangle = \delta^{3N}(q - q') \ , \tag{64}$$

and complete

$$1 = \int_{-\frac{1}{2}L}^{\frac{1}{2}L} dq_1 \ldots \int_{-\frac{1}{2}L}^{\frac{1}{2}L} dq_{3N} \ |q_1 \ldots q_{3N}\rangle\langle q_1 \ldots q_{3N}| \ , \tag{65}$$

where δ^{3N} was defined in (5). Obviously, deep issues of convergence and the like are being ignored, in keeping with our rough-and-ready attitudes. That being said, the linear relation between complex functions (Eq. 1) is simply the scalar product of (63) with the bra $\langle q_1 \ldots q_{3N}|$:

$$\langle q_1 \ldots q_{3N}|\psi\rangle = \sum_a \langle q_1 \ldots q_{3N}|a\rangle\langle a|\psi\rangle \ . \tag{66}$$

In short, ordinary functions such as $u_a(q)$ are scalar products $\langle q|a\rangle$ in the shorthand defined in Eq. 2, and in particular

$$u_a(q_1 \dots q_{3N}) = \langle q_1 \dots q_{3N}|a\rangle \ . \tag{67}$$

The representation of these *same* relationships in momentum space follows in the same way, by means of momentum space bras and kets satisfying

$$\langle p_1 \dots p_{3N}|p_1' \dots p_{3N}'\rangle = \delta_{pp'}^{3N} \ , \tag{68}$$

$$1 = \sum_{\{p\}} |p_1 \dots p_{3N}\rangle\langle p_1 \dots p_{3N}| \ . \tag{69}$$

Here the p's take on the discrete set of values (7), so the basis is denumerable, and Kronecker as compared to Dirac deltas appear in the orthogonality relation. Equations (11) and (12) are then simply

$$\langle p|a\rangle = \int (dq)\langle p|q\rangle q\langle a\rangle, \quad \langle q|a\rangle = \sum_p \langle q|p\rangle\langle p|a\rangle \ , \tag{70}$$

which identify the product of plane waves $\phi_p(q)$ of Eq. 8 as the elements of the unitary transformation from the q to the p basis:

$$\langle q_1 \dots q_{3N}|p_1 \dots p_{3N}\rangle = \prod_{j=1}^{3N} \frac{e^{iq_jp_j/\hbar}}{\sqrt{L}} \ . \tag{71}$$

This last expression brings out an important if rather obvious point: *the Hilbert spaces for the separate degrees of freedom are independent, and the space for the whole system is a direct product of such spaces:*

$$\mathfrak{H} = \mathfrak{H}_1 \otimes \mathfrak{H}_2 \dots \otimes \mathfrak{H}_{3N} \ . \tag{72}$$

The kets in the q and p bases are also products, e.g.,

$$|p_1p_2 \dots p_{3N}\rangle = |p_1\rangle \otimes |p_2\rangle \dots \otimes |p_{3N}\rangle \ . \tag{73}$$

Not all kets in $\mathfrak{H}$ are of this product form, however! An arbitrary ket $|\psi\rangle$ will be such a product only if just one term appears in the sum of Eq. 1, and the function u_a is itself a product.

There is an asymmetry in the preceding definitions of the p and q bases: the former is denumerable and the latter is not. The q basis can also be made denumerable by replacing the spatial continuum $(-\frac{1}{2}L, \frac{1}{2}L)$ by a discrete lattice with a spacing much smaller than any dimension of physical interest. That is often done in numerical work, and sometimes in purely theoretical discussions. Conversely, the discrete momentum space can be replaced by a continuum by taking the limit $L \to \infty$. That we will often do. To see how, it suffices to look at just one degree of freedom, i.e., one factor in (72), and to replace the plane wave in (71) as follows:

$$\frac{e^{iqp/\hbar}}{\sqrt{L}} \longrightarrow \frac{e^{iqp/\hbar}}{\sqrt{2\pi\hbar}} \equiv \varphi_p(q) \ . \tag{74}$$

These functions are normalized to delta functions:

$$\int_{-\infty}^{\infty} dp\, \varphi_p(q)\varphi_p^*(q') = \delta(q-q'), \qquad \int_{-\infty}^{\infty} dq\, \varphi_p^*(q)\varphi_{p'}(q) = \delta(p-p') , \tag{75}$$

so that with these definitions the $q-p$ transformation function is

$$\langle q|p\rangle_\infty = \varphi_p(q) , \tag{76}$$

and

$$1 = \int_{-\infty}^{\infty} dp\, |p\rangle_\infty\langle p|, \qquad \langle p|p'\rangle_\infty = \delta(p-p') . \tag{77}$$

The suffix ∞ will not be shown again because it should be obvious whether the discrete or continuous basis is in use.

(e) Eigenvalues and Eigenvectors

Finally, we summarize crucial properties of Hermitian operators. The following are well-known facts about d-dimensional complex vector spaces $\mathfrak{C}_d$, provided d is finite:

1. Any *single* Hermitian operator A on $\mathfrak{C}_d$ can be diagonalized by a unitary transformation.

2. The elements of this diagonalized form are real and are called the *eigenvalues* of A, designated by $a_1, \dots a_d$. They need not all be different; sets having the same value are called *degenerate*. The set of all eigenvalues is called the *spectrum* of A.

3. The eigenvalues are the roots of *the secular equation*

$$\det\,(A - a\mathbf{1}) = 0 , \tag{78}$$

   i.e., the roots of an algebraic equation of degree d.

4. The basis vectors $|1\rangle \dots |d\rangle$ that diagonalize A are called its *eigenvectors or eigenkets,* and satisfy

$$A|n\rangle = a_n|n\rangle . \tag{79}$$

   Hence

$$A = \sum_n |n\rangle a_n \langle n| . \tag{80}$$

   This is called the *spectral decomposition* of the Hermitian operator A. If there are degeneracies, *all* eigenvectors within degenerate subspaces must be included in the sum for (80) to be valid.

5. Eigenvectors with different eigenvalues are orthogonal. Those that belong to a degenerate subspace will not be orthogonal automatically, but orthogonal linear combinations can always be built from them.

6. If A_i, where $i = 1, \dots, K$, is a set of K *commuting* Hermitian operators, these operators can be diagonalized simultaneously, with eigenvalues $\{a_n^{(i)}\}$. The eigenvectors satisfy

$$(A_i - a_n^{(i)})|a_n^{(1)}a_n^{(2)} \dots a_n^{(K)}\rangle = 0 , \tag{81}$$

where the eigenvectors are now designated by the simultaneous eigenvalues. Any two eigenvectors are orthogonal if any of their eigenvalues differ.

7. If a pair of Hermitian matrices do *not* commute, they *cannot* be diagonalized simultaneously.

8. If $|a\rangle$ is an eigenket of A with eigenvalue a, and B is any operator (in general, not Hermitian) such that

$$[A,\ B] = \lambda B\,, \qquad \text{then} \quad B|a\rangle = \text{const.}|\lambda + a\rangle\ . \tag{82}$$

It is no small matter to establish under what circumstances these statements hold in an infinite-dimensional Hilbert space $\mathfrak{H}$, especially when the space is spanned by a nondenumerable basis, as is assumed in writing Eq. 77. For example, a force law that is singular at zero separation may produce no difficulty in classical mechanics because only a subset of classical orbits can reach the singularity, whereas in quantum mechanics it may be impossible to elude the singularity. In that case some of the statements just listed are false. While bearing this in mind, we shall assume that they are correct until disaster strikes, and march into the jungle like a naive tourist.

2.2 States and Probabilities

In the phenomena to which quantum mechanics applies, only statistical data are experimentally reproducible in most, thought not all, circumstances. It is fitting, therefore, that quantum mechanics only makes statistical predictions about most, though not all, experimental outcomes.

> *It is a central claim of quantum mechanics that the statistical character of quantum mechanics is irreducible — that there are no underlying "hidden variables" which behave in a deterministic manner, and that this statistical character is not an expression of ignorance about such hidden substructure.*

Here the word "claim" is not to be read as hypothesis. At one time it was possible to leave it for the future to decide whether the existence of such hidden substructure is a plausible hypothesis, but that option only survives now in a non-local form we find unacceptable. It must be stressed that here "hidden substructure" does *not* mean further degrees of freedom that are still invisible at presently available energies, but degrees of freedom that are at work in familiar phenomena but not seen, and which supposedly account for the properties of quantum states that are perplexing from a classical perspective.[1] Hidden variable embellishments of quantum mechanics, as we shall see in chapter 12, are in conflict with experiment unless the additional variables interact instantaneously over arbitrary distances and thereby violate the relativistic principle of causality. As we learned from two-particle interferometry in §1.2, the quantum state can have both wave-like and particle-like

[1]Further *quantum* degrees of freedom, such as nucleons "inside" nuclei but ignored in atomic physics, or speculative constituents of leptons, are irrelevant to this issue.

correlations between distant non-interacting subsystems, and it has this property without violating causality, an amazing performance that classically motivated depictions cannot duplicate.

(a) Quantum States

The quantum state, as represented by the Schrödinger wave function $\Psi(\boldsymbol{r}_a \boldsymbol{r}_b)$, already appeared in our treatment of two-particle interferometry. Whether Ψ refers to an individual pair, or to an ensemble of such pairs, is a question that was not raised. Given the statistical nature of phenomena in the quantum realm, it seems natural to associate the quantum state with an ensemble of identical systems $\mathcal{S}$, with the probabilities the frequencies with which various specific properties of $\mathcal{S}$ occur. That is a widely held view, which we shared for many years, though as believers in the ensemble interpretation we were in the habit of speaking and thinking in the singular — *the* atom, *the* electron, etc.

Consider, however, the statement which closed §1.2: Ψ *must be able to display itself in a wave-like or particle-like guise, or in various combinations of these guises, depending, so to say, on what question will be asked of it in the future.* Recall, furthermore, that the EPR feature arises when the choice is made, *on an event-by-event basis and at random,* between determining the position or the momentum of one or the other particle belonging to a single pair. Thus an *individual* pair displays a richness of distinct properties which makes it rather contrived to think of the quantum state as being nothing beyond a description of an ensemble of pairs. Of course, this is not a proof that the ensemble interpretation is invalid. That said, *we will assume that the quantum state is associated with individual systems.* We should emphasize, moreover, that the predictions of observable effects are not contingent on whether one adheres to one or the other of these interpretations of the quantum state.

What then is meant by probability? In this interpretation, the quantum state of an individual system is endowed with *potentiality,* a term due to Heisenberg.[1] Potentiality has no numerical attribute; it is a primitive concept that quantum mechanics does not define, just as Euclidean geometry does not define the concept of point. Potentiality stands for the various properties that an individual system $\mathcal{S}$ has the potential for displaying in response to experimental tests, with probability being a property associated with the frequency of outcomes of tests on copies of $\mathcal{S}$.

We can now posit the three basic postulates of quantum kinematics; dynamics will be formulated in the following sections.

> **1.** *The most complete possible description of the state of any physical system $\mathcal{S}$ at any instant is provided by some particular vector $|\psi\rangle$ in the Hilbert space $\mathfrak{H}$ appropriate to the system. Every linear combination of such state vectors represents a possible physical state of $\mathcal{S}$.*

The last sentence is the *superposition principle.* If $\mathcal{S}$ is a system composed of N constituents which have no spin, and whose internal degrees of freedom cannot be excited in the energy regime of interest, then $\mathfrak{H}$ is the space defined in Eq. 72.

When $\mathcal{S}$ can be described by one vector, it is in a *pure state.* Should no single ket suffice, $\mathcal{S}$ is in a *mixed state,* or *mixture.* Introductory expositions of quantum

---

[1]Some authors use the term *propensity* in place of potentiality.

mechanics tend to focus almost exclusively on pure states, with the rationale that any mixed state can be expressed in terms of sets of pure states. While true, this does not do justice to the importance of mixtures.

The space $\mathfrak{H}$ specific to the system in question describes the full variety of pure states of $\mathcal{S}$. Properties that distinguish between states are needed, and this is provided by:

> **2.** *The physically meaningful entities of classical mechanics, such as momentum, energy, position and the like, are represented by Hermitian operators.*

Following Dirac, such operators will be called *observables.*[1] Observables that have no classical counterpart must also be anticipated, of course.

Let A be an observable. As it is Hermitian, it can be diagonalized and expressed by its spectral decomposition:

$$A = \sum_a |a\rangle a \langle a| \ . \tag{83}$$

Here $(a, a', \ldots)$ are the real eigenvalues of A, and $|a\rangle, \ldots$ the corresponding eigenkets.[2]

The last postulate defines how probabilities make their appearance:

> **3.** *A set of N identically prepared replicas of a system $\mathcal{S}$ described by the pure state $|\psi\rangle$, when subjected to a measurement designed to display the physical quantity represented by the observable A, will in each individual case display one of the values $(a, a', \ldots)$, and as $N \to \infty$ will do so with the probabilities $p_\psi(a), p_\psi(a'), \ldots$, where*
>
> $$p_\psi(a) = |\langle a|\psi\rangle|^2 \ . \tag{84}$$

This is the definition of probability in terms of the frequency of specific outcomes in a sequence of identical tests on copies of $\mathcal{S}$.

Eq. 84 is a consistent definition of a probability distribution: the real positive numbers $|\langle a|\psi\rangle|^2$ obey the sum rule

$$\sum_a |\langle a|\psi\rangle|^2 = \langle\psi|\psi\rangle = 1 \ , \tag{85}$$

where by convention all kets have unit norm; that is,

$$p_\psi(a) \geq 0, \qquad \sum_a p_\psi(a) = 1 \ . \tag{86}$$

---

[1]Dirac also used the terms c-number for ordinary numbers (in allusion to classical quantities), though possibly complex, and q-numbers for operators. Though not much used today, we sometimes return to this usage.

[2]As written, (83) refers to a discrete spectrum; should the spectrum be partially or wholly continuous, the sum is to be understood as including the appropriate integral, and some Kronecker deltas may be Dirac delta functions. This is a trivial complication at the formal level of this discussion, and will be ignored by writing all formulas as if the spectra in question are discrete.

The mean value $\langle A\rangle_\psi$ of the measurement results on a set of replicas of $\mathcal{S}$ in the state $|\psi\rangle$ is, by definition,

$$\langle A\rangle_\psi = \sum_a a\, p_\psi(a) \,, \tag{87}$$

which is the following diagonal matrix element of the observable A:

$$\langle\psi|A|\psi\rangle = \sum_a a|\langle a|\psi\rangle|^2 \,. \tag{88}$$

Such mean values are called *expectation values* in quantum mechanics. Should $|\psi\rangle$ actually be the eigenket $|a\rangle$, then

$$\langle a|A|a\rangle = a \,. \tag{89}$$

Note well that (87) implies that *the only values that an observable can display are its eigenvalues.*

The probability $p_\psi(a)$ can also be expressed as an expectation value, and in that form it generalizes readily to mixed states. This is so because $p = \langle\psi|a\rangle\langle a|\psi\rangle$, whence

$$p_\psi(a) = \langle\psi|P_a|\psi\rangle \,, \tag{90}$$

where $P_a = |a\rangle\langle a|$ is the projection operator onto $|a\rangle$.

In (87) the state $|\psi\rangle$ is arbitrary. Should $|\psi\rangle$ actually be one of the eigenkets of A, say $|a_0\rangle$, the distribution is sharp, $p_{a_0}(a) = \delta_{a_0 a}$, i.e., there is no dispersion in the values displayed when the state is one of the eigenstates of A; in this case the state is guaranteed to always display one particular eigenvalue.

Thus far the probabilities associated with a pure state $|\psi\rangle$ have been expressed in terms of the eigenvalues of an observable — presumably an observable having a useful physical significance. But the definition (84) generalizes as follows. Let $|\phi\rangle$ be an arbitrary state, which together with $|\phi'\rangle, |\phi''\rangle, \ldots$ forms a complete orthonormal set, but which may not diagonalize any operator that has a useful physical significance. Then by the argument that led to (86), *the probability that a system $\mathcal{S}$ in the state $|\psi\rangle$ will be found to be in the arbitrary state $|\phi\rangle$ is*

$$p_\psi(\phi) = |\langle\phi|\psi\rangle|^2 \,. \tag{91}$$

For this reason, the scalar product between states is called a *probability amplitude.*

It is natural to take it for granted that a particular value a displayed by a particular measurement of an observable A simply reveals a pre-existing value possessed by that individual specimen $\mathcal{S}$, just as your head has a definite circumference before the tape measure is unfurled in a statistical study of the egos of physicists. This entirely sensible supposition is not valid, however:

> *Values cannot be ascribed to observables prior to measurement; such values are only the outcomes of measurement.*

The common-sense inference that measurements reveal pre-existing values leads to implications that are contradicted by experiment, and are also incompatible with the Hilbert space structure of quantum mechanics. This conclusion is not obvious, and was not established firmly until some three decades after the discovery of quantum mechanics. The reasoning and experiments that lead to it will be given in chapter 12, but the conclusion is stated here to abort seductive misconceptions. One further point should, however, be made here. Because observables have no values prior to measurement, in the ensemble interpretation of the state it is illegitimate to distinguish between the ensemble's members, for doing so would be tantamount to introducing hidden variables. This can be taken as another motivation for defining the quantum state to be a description of an individual system as compared to an ensemble.

(b) Measurement Outcomes

From what has just been said, it is clear that measurement plays a central role in relating the mathematical formalism to observable facts. If quantum mechanics purports to be a complete description of physical phenomena, it had better also describe the measurement process itself. It turns out, however, that the quantum theory of measurement is far from straightforward. For that reason we also postpone its discussion to chapter 12. Until then we take a lowbrow approach, and offer the following example as a general purpose measurement paradigm.

Let $|\psi\rangle$ be the state of a photon of a particular frequency emitted in the direction $\boldsymbol{k}$ from an atom that is in a magnetic field pointing along $\boldsymbol{n}$. Imagine a measurement apparatus $\mathcal{M}_{\rm circ}$ that deflects any such photon into one of two directions $\boldsymbol{q}_{R,L}$ depending on whether its circular polarization is right or left-handed, and two photomultipliers that determine which of the two deflections a particular photon experienced. For arbitrary directions $\boldsymbol{k}$ and $\boldsymbol{n}$, it is impossible to predict what the next photon does, all that can be predicted is the probability for one as compared to the other deflection.

The application of the rule (91) is then as follows. Let $|\boldsymbol{k}_{R,L}\rangle$ be the states of a right- or left-circularly polarized photon propagating in the direction $\boldsymbol{k}$. Then $|\langle\boldsymbol{k}_{R,L}|\psi\rangle|^2$ are the probabilities that individual photons will be deflected in the directions $\boldsymbol{q}_{R,L}$. But any number of other questions can be addressed to $|\psi\rangle$. For example, the apparatus $\mathcal{M}_{\rm circ}$ can be replaced by another, $\mathcal{M}_{\rm lin}$, that measures linear polarization along two orthogonal directions $\boldsymbol{u}_{1,2}$. If $|\boldsymbol{k}_{1,2}\rangle$ are photon states with these polarizations, the corresponding probabilities are $|\langle\boldsymbol{k}_{1,2}|\psi\rangle|^2$.

Thus Eq. 91 for the probability takes for granted the existence in principle of an apparatus $\mathcal{M}_\phi$ that is able, when presented with a system S in the state $|\psi\rangle$, to assign it to the category that is in the state $|\phi\rangle$.

In general, a system will have several observables that all commute with each other; such observables are said to be *compatible*. It suffices to consider the case of two, A and B. There then exist simultaneous eigenkets $\{|ab\rangle\}$ of A and B with eigenvalues $(a, a', \ldots; b, b', \ldots)$. Let $f(x, y)$ be a function of two variables. Because A and B commute, the action of $f(A, B)$ on the simultaneous eigenkets is then

$$f(A,B)|ab\rangle = f(a,b)|ab\rangle \ . \tag{92}$$

Hence the expectation value of the operator f is

$$\langle\psi|f(A,B)|\psi\rangle = \sum_{ab} f(a,b)|\langle ab|\psi\rangle|^2 \ , \tag{93}$$

and by the preceding argument

$$p_\psi(ab) = |\langle ab|\psi\rangle|^2 \tag{94}$$

is the *joint probability distribution* that the observables A *and* B will display the two values a *and* b.

Note that $|\langle ab|\psi\rangle|^2$ *is not a conditional probability,* i.e., it is not the probability for the occurrence of b given that a has definitely occurred. Rather, it gives the probabilities for the occurrence of both a and b, conditional only on the state being $|\psi\rangle$. The conditional probability for b given a is

$$p_\psi(a|b) = \frac{p_\psi(ab)}{p_\psi(a)} . \tag{95}$$

Its distinction from the joint distribution $p_\psi(ab)$ is underscored by the identities

$$\sum_b p_\psi(a|b) = 1, \qquad \sum_b p_\psi(ab) = p_\psi(a) . \tag{96}$$

The most famous and familiar example of what has just been said is provided by the Schrödinger wave function. In Dirac's language it is the scalar product of $|\psi\rangle$ with a simultaneous eigenket $|q_1 \ldots q_{3N}\rangle$ of all the coordinate operators,

$$\psi(q_1 \ldots q_{3N}) \equiv \langle q_1 \ldots q_{3N}|\psi\rangle . \tag{97}$$

The *Born interpretation* of $|\psi(q_1 \ldots q_{3N})|^2$ as a joint probability distribution of the coordinates in an N-particle state is a special case of the meaning ascribed to scalar products quite generally.

Now to a situation that has no classical counterpart, the case of *incompatible observables, i.e., observables that do not commute.* Call such a pair P and Q, where these names are merely suggestive; they need not be momentum or coordinate. There are then eigenkets of P and Q, satisfying

$$P|p\rangle = p|p\rangle, \qquad Q|q\rangle = q|q\rangle , \tag{98}$$

but simultaneous eigenkets of P *and* Q *do not exist.* While probability distributions for the eigenvalues of P *or* of Q are of the same form as before, namely

$$p_\psi(p) = |\langle p|\psi\rangle|^2, \qquad p_\psi(q) = |\langle q|\psi\rangle|^2 , \tag{99}$$

it is not possible to define a joint distribution for the two eigenvalue sets $\{p\}$ and $\{q\}$.

To see why, try the following: calculate the probability that $|\psi\rangle$ will display a particular eigenvalue p of P, then the probability that this eigenstate of P will display the eigenvalue q of Q, and sum over *all* values of p in the expectation that this will give the probability $p_\psi(q)$ that $|\psi\rangle$ will display the value q when there is no knowledge about p. That is, form

$$\sum_p p_p(q) p_\psi(p) = \sum_p \left[|\langle q|p\rangle|^2 \times |\langle p|\psi\rangle|^2 \right] . \tag{100}$$

Compare this to

$$p_\psi(q) = |\langle q|\psi\rangle|^2 = \left| \sum_p \langle q|p\rangle\langle p|\psi\rangle \right|^2 . \tag{101}$$

The naive, or classical, way of composing probabilities (Eq. 100) disagrees with that of quantum mechanics (Eq. 101). In the latter the two probability amplitudes for $\psi \to p$ and $p \to q$ — both complex numbers — are *first* multiplied, *then* summed over the intermediate variable p, and it is the absolute square of this sum that gives the probability for q in the state $|\psi\rangle$, — a far cry from "common-sense" multiplication of probabilities! When the probabilities are combined as in (100), the phases of the probability amplitudes do not matter, whereas in the *coherent superposition* (101) these phases are crucial.

What is the reason for this striking difference? It is not that the familiar composition law of Eq. 100 is wrong, but that it corresponds to a sequence of measurements that differs from the one that determines $p_\psi(q)$.

In the situation summarized by (100), there is first a measurement apparatus $\mathcal{M}_P$ that assigns each specimen in the state $|\psi\rangle$ to the set that possess the eigenvalue p of P, as a consequence of which these individuals are *known* to be in the state $|p\rangle$; subsequently, each system that is in the state $|p\rangle$ is passed into a second apparatus $\mathcal{M}_Q$ which is able to ask "do you belong in $|q\rangle$," with the answer being "perhaps, with probability $|\langle q|p\rangle|^2$." The first segregation produces a set with the property p having the fractional population $p_\psi(p)$, so that the fraction of the systems that were in $|\psi\rangle$ that survive both filtrations is $p_p(q)p_\psi(p)$. The sum on p in (100) only results in a *recoverable* loss of knowledge, as in a demographic study that sums over cities when stating what fraction of a country's urban population ponders the meaning of quantum mechanics. That is to say, the same result $\sum_p p_p(q)p_\psi(p)$ is found by discarding all but one of the sets produced by $\mathcal{M}_P$ in the first measurement, then using this set of individuals as the input for $\mathcal{M}_Q$, and repeating this experiment for all the other sets produced by $\mathcal{M}_P$. The resemblance to demography is very poor, of course: as P and Q are incompatible, the second measurement $\mathcal{M}_Q$ produces individuals $|q\rangle$ which no longer have a definite value of p. This is confirmed by passing a set of outputs $|q\rangle$ from $\mathcal{M}_Q$ through the device $\mathcal{M}_P$, because the latter will have as its output sets $|p\rangle$ with a variety of values p.

The measurement that gives the result (101) is one in which a system in the state $|\psi\rangle$ is sent directly to $\mathcal{M}_Q$; the states $|p\rangle$ that appear in this equation are in essence a mathematical artifact, for any complete set gives the same result. The probability $p_\psi(q)$ determined in this way cannot be diagnosed or dissected to find $p_p(\psi)$. In contrast, if some other complete set $|a\rangle$ is used in (100), it will give a different result.

To summarize, both probability calculations make perfectly good sense, but they correspond to very different physical processes if P and Q are incompatible observables.

A final point must still be made about pure states. Probabilities involve only the *absolute values* of scalar products between vectors in a Hilbert space, while expectation values are *bilinear* in them. Therefore, replacing any ket $|\psi\rangle$ by $e^{i\alpha}|\psi\rangle$, where α is real, does not alter expectation values or probabilities. Thus it is not really correct to say that pure physical states corresponds to a single ket in a Hilbert space. Rather, the set of kets $e^{i\alpha}|\psi\rangle$ describes one physical state. Such a set is called a ray, and therefore *the state space of quantum mechanics is really a ray space, not a Hilbert space.*

It therefore seems unjustified to require various statements in quantum mechanics to be invariant under unitary transformations, because it is not relations between kets but between rays that have physical significance. Thanks to a theorem by Wigner (§7.1), this headache rarely arises, because the theorem proves that the only physically significant transformations that cannot be represented by unitary transformations on kets is time reversal.

(c) Mixtures and the Density Matrix

Although pure states abound in text books and research papers, systems in the real world are rarely in pure states. For example, the beam produced by an accelerator is hardly a coherent superposition of momentum eigenstates, as it would have to be if it were to be represented by a pure state. A more common example is that of a collection of atoms in thermodynamic equilibrium.

Consider the latter. Here the key observable is the Hamiltonian H with energy eigenvalues $\{E\}$. There will, in general, be other observables compatible with H, e.g., angular momentum, designated collectively by A, that together specify a basis $\{|Ea\rangle\}$ in $\mathfrak{H}$. At temperature T these states are populated in accordance with the Boltzmann probability distribution,

$$p_T(E) = e^{-E/kT}/Z \ , \tag{102}$$

where k is Boltzmann's constant and Z a normalization factor called the partition function. Requiring the probabilities to sum to one then fixes Z:

$$Z = \sum_{E,a} e^{-E/kT} \ . \tag{103}$$

In this and similar sums, all states belonging to a given energy eigenvalue must be included.

The expectation value of some observable Q, which need not be compatible with the the Hamiltonian, is then its expectation value in the various states $|Ea\rangle$ weighted by the thermal probability distribution,

$$\langle Q\rangle_T = \sum_{E,a} p_T(E)\, \langle Ea|Q|Ea\rangle \ . \tag{104}$$

This expectation value is thus the result of two very different averages: that due to the statistical distribution of eigenvalues of Q in the pure states $|Ea\rangle$, and that due to the probability that such a pure state occurs in the thermal ensemble:

$$\langle Q\rangle_T = \sum_{E,a}\sum_{q} q\, p_T(E)\, |\langle q|Ea\rangle|^2 \ . \tag{105}$$

By introducing an operator ρ called the *density matrix*, expectation values such as $\langle Q\rangle_T$ can be written in a form that at first sight is opaque but which turns out to be very powerful.[1] The nomenclature "matrix" is somewhat unfortunate, because ρ is an operator, and many authors call it the statistical or density operator. But the more common usage is so long established that, as with many other terms in physics, it is best to accept it as an indelible part of our cultural heritage, and to not balk at phrases like "the matrix elements of the density matrix are ...."

[1] The density matrix was introduced by Landau and developed into its prominent role by von Neumann.

The density matrix describing a thermal equilibrium ensemble is defined as

$$\rho_T = \sum_{E,a} |Ea\rangle p_T(E) \langle Ea| = \sum_{E,a} p_T(E) P_{Ea} , \tag{106}$$

i.e., a sum of projection operators onto the basis $\{|Ea\rangle\}$ weighted by the Boltzmann distribution. Because the probabilities $p_T(E)$ are normalized to 1,

$$\mathrm{Tr}\, \rho_T = 1 . \tag{107}$$

In terms of ρ_T, the expectation value of Q then has the compact and very useful form

$$\langle Q \rangle_T = \mathrm{Tr}\, \rho_T Q ; \tag{108}$$

This formulation allows us to give a basis-independent distinction between pure states and mixtures. Again, let $|\psi\rangle$ be some pure state, and

$$P_\psi = |\psi\rangle\langle\psi| \tag{109}$$

its associated projection operator. Then the expectation value in the pure state $|\psi\rangle$ is

$$\langle\psi|Q|\psi\rangle = \sum_{a,a'} \langle a|\psi\rangle\langle\psi|a'\rangle\langle a'|Q|a\rangle = \mathrm{Tr}\, P_\psi Q \equiv \mathrm{Tr}\, \rho_\psi Q . \tag{110}$$

That is to say, *in the case of a pure state, the density matrix is a projection operator onto the state.* Consequently, the density matrix of a pure state is characterized by

$$(\rho_\psi)^2 = \rho_\psi , \qquad \mathrm{Tr}\, (\rho_\psi)^2 = 1 . \tag{111}$$

The latter, as promised, is a statement that does not depend on the basis because the trace is invariant under unitary transformations.

Consider now the square of the thermal density matrix (Eq. 106):

$$(\rho_T)^2 = \sum_{E,a} [p_T(E)]^2 P_{Ea} . \tag{112}$$

Hence

$$\mathrm{Tr}\, (\rho_T)^2 = \sum_{E,a} [p_T(E)]^2 < 1 . \tag{113}$$

The sum is only one if the temperature is strictly zero, so that only the ground state is occupied, i.e., the case of a pure state.

The example of the thermal distribution is illustrative, but holds for any set of probabilities. The general situation is summarized as follows:

> *A state is pure if its density matrix ρ is a projection operator, and it is a mixture if it is not. The two cases are characterized by the invariant condition*
>
> $$\mathrm{Tr}\, \rho^2 \leq 1 , \tag{114}$$
>
> *with the equality only holding if the state is pure.*

Thus far the density matrix has been written in terms of projection operators. Like all operators, it can be written in any basis, and defined without reference to a particular basis — namely, as an observable whose eigenvalues $(p_1, p_2, \ldots)$ satisfy

$$0 \le p_i \le 1, \qquad \sum_i p_i = 1 \ . \tag{115}$$

The state is pure if, and only if, all eigenvalues p_i of ρ but one vanish.

Let $|a_i\rangle$ be the orthonormal basis that diagonalizes ρ, so that

$$\rho = \sum_i |a_i\rangle p_i \langle a_i| \ . \tag{116}$$

The expectation value of an observable Q is then given by (104),

$$\langle Q\rangle = \sum_i p_i \langle a_i|Q|a_i\rangle \ . \tag{117}$$

in general, therefore, the expectation value of an observable Q in a state ρ, whether pure or mixed, can be written in the invariant form

$$\langle Q\rangle = \operatorname{Tr} \rho\, Q \ . \tag{118}$$

The expression that gives the probability of a finding a state $|\phi\rangle$ in a pure state (Eq. 90) can be extended to the case of a mixture, namely,

$$p_\phi(\rho) = \operatorname{Tr} \rho\, P_\phi = \langle\phi|\rho|\phi\rangle \ . \tag{119}$$

In terms of the basis of (116), this reads

$$p_\phi(\rho) = \sum_i p_i |\langle\phi|a_i\rangle|^2 \ , \tag{120}$$

as it must.

The most important measure of the departure from purity is provided by *the von Neumann entropy S*. For any state ρ, it is defined as

$$S = -k \operatorname{Tr} \rho \ln \rho \ , \tag{121}$$

where k is Boltzmann's constant. When ρ is the Boltzmann distribution, S is the entropy of statistical thermodynamics (see Prob. 3); in terms of the probabilities p_i it has the familiar form

$$S = -k \sum_i p_i \ln p_i \ . \tag{122}$$

For a pure state, where only one $p_i = 1$ and the others vanish, $S = 0$. Furthermore $S \ge 0$ because $0 \le p_i \le 1$. This leads to the question of whether S has a maximum value. To answer this, one varies the probabilities in S under the constraint that they sum to 1. This is done by introducing a Lagrange multiplier λ,

$$\delta \sum_i p_i (\ln p_i + \lambda) = 0 \ , \tag{123}$$

or

$$\sum_i \delta p_i \, (\ln p_i + 1 + \lambda) = 0 \; . \tag{124}$$

The variations δp_i in (124) are now independent, and therefore $\ln p_i + 1 + \lambda = 0$, i.e., the p_i that maximize S do not depend on i. But they must sum to 1, hence this whole argument only makes sense as it stands if the Hilbert space $\mathfrak{H}_d$ has a finite dimension d, so that $p_i = 1/d$.

The entropy therefore satisfies the inequalities

$$0 \leq S \leq k \; \ln d \; , \tag{125}$$

and the density matrix that maximizes S is

$$\rho_{\max} = \frac{1}{d} \sum_i |a_i\rangle\langle a_i| \; . \tag{126}$$

At first sight, $\{|a_i\rangle\}$ is a particular basis that diagonalizes ρ. But the sum in (126) is just the unit operator, so

$$\rho_{\max} = \frac{1}{d} \; . \tag{127}$$

In short, the mixture in which the entropy is maximal is the one in which all states, in any basis, are populated with equal probability. Put another way, $S = 0$ when there is maximal knowledge in that the state is pure, and $S = S_{\max}$ when the available states are populated at random.

We must still address the following basic question: *Given an unlimited set of replicas of a system $\mathcal{S}$ in an unknown state ρ, known to reside in a d-dimensional Hilbert space $\mathfrak{H}_d$, what measurements must be carried out to determine this state?* The unknown density matrix (whether pure or mixed) is a d-dimensional Hermitian matrix of unit trace is specified by $d^2 - 1$ real parameters. According to Eq. 37 and Eq. 41, in terms of an the arbitrary basis $\{|a_i\rangle\}$ in $\mathfrak{H}_d$, the operator ρ has the form

$$\rho = \sum_{ij} |a_i\rangle r_{ij} \langle a_j| \; , \qquad r_{ij} = r_{ji}^* \; . \tag{128}$$

An arbitrary observable in $\mathfrak{H}_d$ is a linear combination, with real coefficients, of the d^2 Hermitian operators

$$X_{ij} = \tfrac{1}{2}\{|a_i\rangle\langle a_j| + |a_j\rangle\langle a_i|\} \; , \quad Y_{ij} = \tfrac{1}{2} i\{|a_i\rangle\langle a_j| - |a_j\rangle\langle a_i|\} \; . \tag{129}$$

Of these, the combination $\sum_i X_{ii} = 1$ is trivial. Because $\mathrm{Tr}\, \rho \, |a_i\rangle\langle a_j| = r_{ji}$,

$$\mathrm{Tr}\, \rho \, X_{ij} = \mathrm{Re}\, r_{ij} \; , \quad \mathrm{Tr}\, \rho \, Y_{ij} = \mathrm{Im}\, r_{ij} \; . \tag{130}$$

In consequence, when faced with an arbitrary unknown state in $\mathfrak{H}_d$, to identify the state unambiguously the expectation values of a complete set of observables must be measured, where by 'complete' is meant a set that allows the evaluation of all the $(d^2 - 1)$ nontrivial expectation values appearing in (130). A concrete example should help to make this inscrutable enumeration understandable. Take the case where $\mathcal{S}$ is an atom with angular momentum 1, whose states live in a 3-dimensional Hilbert space. According to what has just been learned, the expectation values of eight observables must be measured. As shown in §3.3, they are the three independent components of the atom's magnetic moment vector and the five independent components of its quadrupole tensor (a Cartesian tensor of rank 2).

(d) Entangled States

There is a common and serious misconception that mixtures only arise when pure states are "mixed" by the environment, such as a temperature bath, or by some apparatus, such as an accelerator. Not so: *If a composite system is in a pure state, its subsystems are in general in mixed states.* This is the context in which mixtures are often important in discussions of the interpretation of quantum mechanics, and also in many other contexts.

To see how mixtures arise from pure states, consider a system composed of two subsystems with coordinates q_1 and q_2. Let $|\Psi\rangle$ be an arbitrary pure state of the system, with wave function $\Psi(q_1q_2)$, so that its density matrix is

$$\langle q_1q_2|\rho|q_1'q_2'\rangle = \Psi(q_1q_2)\Psi^*(q_1'q_2') \, . \tag{131}$$

Let A_1 be an observable of the subsystem 1; that is, insofar as subsystem 2 is concerned it is the unit operator:

$$\langle q_1q_2|A_1|q_1'q_2'\rangle = \langle q_1|A_1|q_1'\rangle \, \delta(q_2 - q_2') \, , \tag{132}$$

where, to make contact with the wave-function language clearer the spectra are taken to be continuous. Hence the expectation value of A_1 in Ψ is

$$\langle A_1\rangle_\Psi = \int dq_1\, dq_1'\, dq_2\; \Psi^*(q_1q_2)\langle q_1|A_1|q_1'\rangle\Psi(q_1'q_2) \, . \tag{133}$$

This expectation value can be written in terms of a *reduced density matrix* ρ_1 describing only the first subsystem, and defined as

$$\langle q_1|\rho_1|q_1'\rangle = \int dq_2\; \langle q_1q_2|\Psi\rangle\langle\Psi|q_1'q_2\rangle \; = \int dq_2\; \langle q_1q_2|\rho|q_1'q_2\rangle \, . \tag{134}$$

Then Eq. 133 is

$$\langle A_1\rangle_\Psi = \int dq_1\, dq_1'\; \langle q_1|A_1|q_1'\rangle\langle q_1'|\rho_1|q_1\rangle = \mathrm{Tr}\, A_1\rho_1 \, . \tag{135}$$

That is to say, for all expectation values pertaining only to subsystem 1, the density matrix *is* ρ_1.

We will now show that *a subsystem is not in a pure state when the whole system that contains it is in a pure state* Ψ *unless* Ψ *has the form of a product.* In terms of our example, consider

$$\Psi(q_1q_2) = c_1u_1(q_1)v_1(q_2) + c_2u_2(q_1)v_2(q_2) \, , \quad |c_1|^2 + |c_2|^2 = 1 \tag{136}$$

where u_i and v_j are orthonormal,

$$\int dq_1\, u_i^*(q_1)u_j(q_1) = \delta_{ij}, \qquad \int dq_2\, v_i^*(q_2)v_j(q_2) = \delta_{ij} \, . \tag{137}$$

When neither coefficient in (136) vanishes, Ψ is called an *entangled state,*[1] i.e., one that *cannot* be written as a simple product

$$\Psi(q_1q_2) = \varphi(q_1)\chi(q_2) \, . \tag{138}$$

[1] Entangled states first appeared in Heisenberg's theory of He (see §6.2), but their full significance was first recognized by Schrödinger, and the term is due to him.

That (136) cannot be written in the form (138) will now be proven.

Thanks to (137), it follows from (136) that

$$\langle q_1|\rho_1|q_1'\rangle = |c_1|^2 u_1(q_1)u_1^*(q_1') + |c_2|^2 u_2(q_1)u_2^*(q_1') , \tag{139}$$

i.e., a density matrix for system 1 with probabilities $|c_i|^2$ for being in the state with wave function $u_i(q_1)$. That ρ_1 does not describe a pure state of *subsystem 1* is clear, because

$$\langle q_1|(\rho_1)^2|q_1'\rangle = |c_1|^4 u_1(q_1)u_1^*(q_1') + |c_2|^4 u_1(q_1)u_1^*(q_1') , \tag{140}$$

so that

$$\mathrm{Tr}\,(\rho_1)^2 = |c_1|^4 + |c_2|^4 < 1 . \tag{141}$$

Thus ρ_1 is not pure, and therefore cannot be represented by any single state in the Hilbert space of system 1, unless one of the coefficients $c_{1,2}$ vanishes, in which case Ψ has the product form, i.e., is not entangled, *QED*.

This result generalizes to composites of more than two subsystems and to more complicated states than Eq. 136. Namely, the density matrix of a subsystem ρ_s can only be that of a pure state if the density matrix ρ of the whole system is of the form $\rho_s \otimes \rho_R$, where ρ_s is itself pure and ρ_R is the density matrix of the remainder. If ρ has this product form all joint distributions will be products with one factor pertaining to the subsystem s and the other factor pertaining to the remainder.

Entangled states have correlations which are very weird from any classical perspective. This property is of crucial importance in such disparate topics as the spectrum of He and magnetism (§6.2), and in Bell's theorem (§12.3). The two-body probability distribution associated with the state (136) is

$$p(q_1q_2) = |c_1|^2|u_1(q_1)|^2|v_1(q_2)|^2 + |c_2|^2|u_2(q_1)|^2|v_2(q_2)|^2 + I_2(q_1q_2), \tag{142}$$

$$I_2(q_1q_2) = 2\,\mathrm{Re}\,\{c_1c_2^*\, u_1(q_1)u_2^*(q_1)v_1(q_2)v_2^*(q_2)\} . \tag{143}$$

The first two terms are mundane, but the interference term I_2 is very strange. For example, if the wave functions u_i and v_i are free particle wave packets describing particles that are running off to large distances in various directions, the interference term describes correlations even though the particles do not interact and are far apart.

The strange two-body interference term I_2 does not survive in the probability distribution for one subsystem, e.g., in $\langle q_1|\rho_1|q_1\rangle$, but this is only the case when the entangled state is composed of orthogonal wave functions. And even when they are orthogonal, the two-body interference term is crucial in situations such as those just mentioned.

Finally, we ask how and whether two pure states can be distinguished. For this purpose, consider a pure entangled state for a two-body system (a, b) of the type provided by the interferometer in §1.2:

$$\Phi(q_aq_b) = N[\varphi_1(q_a)\chi_1(q_b) + \varphi_2(q_a)\chi_2(q_b)] . \tag{144}$$

Here φ_n and χ_n are arbitrary wave functions for a and b, respectively, which in general are *not* orthogonal; N is the normalization factor. The probability distribution associated with Φ has the two-body interference term

$$I_2(q_aq_b) = 2N^2\mathrm{Re}\,\{\varphi_1(q_a)\varphi_2^*(q_a)\chi_1(q_b)\chi_2^*(q_b)\} . \tag{145}$$

The correlations between a and b when they are widely separated are contained in I_2. If, however, no measurement whatever is performed on b, the probability distribution for a is that of a mixture, with the one-body interference term

$$I_1(q_a) = 2N^2 \mathrm{Re}\,\{V\varphi_1(q_a)\varphi_2^*(q_a)\}\,, \tag{146}$$

where

$$V = \int dq_b\, \chi_1(q_b)\chi_2^*(q_b)\,. \tag{147}$$

Hence a by itself will only show an interference pattern if the states $\chi_{1,2}$ of the other body b are *not* orthogonal. This last statement is illustrated by the interferometer of §1.2: the two states of b that arrive at the detector D_b are orthogonal because they have differing momenta along the y direction, so in accordance with the preceding statement a does not display an interference pattern.

A somewhat different experimental setup than the one of §1.2 will elucidate the condition required for V not to vanish (see Fig. 2.1). Assume that the particles a

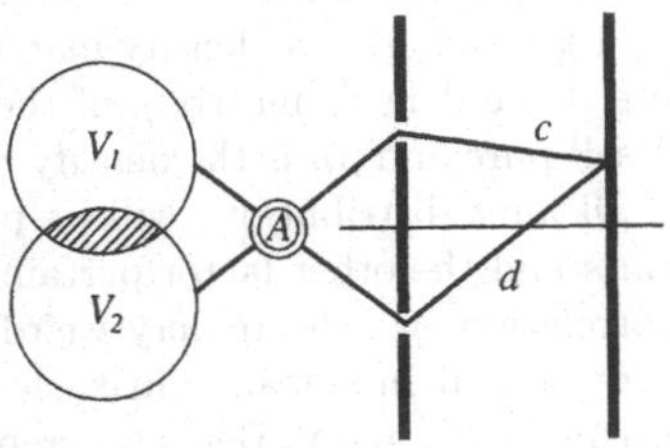

FIG. 2.1. A modification of the two particle interferometer of §1.2, in which the parent A decays into charged daughters (e.g., $K^0 \to \pi^+ + \pi^-$), with one decay product captured into somewhat overlapping traps on the left, while the other passes through one of two holes on the right on its way to the detector.

and b are charged, as in the decay $K \to \pi^+ + \pi^-$, and that the interferometer is replaced by an apparatus in which the left-hand screen and detector are replaced by magnetic traps that capture b into one of the states $\chi_{1,2}(q_b)$ when a has passed through either the upper or lower right-hand hole. Then a will show an interference pattern provided there is a spatial overlap between the two trapped states of b, i.e., provided they are *not* orthogonal. The significance of this overlap can be stated in two, at first sight quite different, ways: (i) on the mathematical side, that the states of b are not orthogonal; (ii) on the physical side, that the states of b do not unambiguously determine the path of a.

The general lesson to be drawn from these examples is that a pair of states, if they are orthogonal, can give a yes-no answer as to whether an observable has a particular value, whereas the answer becomes increasingly ambiguous as their overlap increases. For that reason, the *visibility* $|V|^2$ *of the interference pattern displayed by a alone is a measure of the confidence with which an observation on b determines the state of a.*

(e) The Wigner Distribution

Quantum states, whether pure or mixed, can be cast into a form that is astonishingly reminiscent of phase space distributions in classical statistical mechanics.

Consider a one-dimensional system with density matrix $\langle q|\rho|q'\rangle$ in the coordinate representation; the generalization to more degrees of freedom is trivial and will be stated at the end. *The Wigner distribution* is then defined by

$$f(q,p) = \int_{-\infty}^{\infty} ds\, e^{-ips/\hbar}\, \langle q + \tfrac{1}{2}s|\rho|q - \tfrac{1}{2}s\rangle \,. \tag{148}$$

Wigner's function $f(q,p)$ has three properties that coincide with those of a classical phase space distribution. First, as ρ is Hermitian, f is real. Second, integration of f over momentum gives the probability distribution of the coordinate:

$$\int \frac{dp}{2\pi\hbar}\, f(q,p) = \langle q|\rho|q\rangle \,. \tag{149}$$

And third, integration over the coordinate gives the probability distribution of the momentum:

$$\int \frac{dq}{2\pi\hbar}\, f(q,p) = \langle p|\rho|p\rangle \,. \tag{150}$$

Despite these marvelous properties, $f(q,p)$ is *not* a probability distribution because it is not positive definite for arbitrary states. This is hardly astonishing because the incompatibility of coordinate and momentum, or put equivalently, the uncertainty principle, implies that joint probability distributions of q and p cannot exist except under special circumstances. For example, as we will learn in §4.2, the ground state is the only energy eigenstate of the harmonic oscillator that has a positive definite Wigner distribution.

Observables can also be cast into a form due to Weyl that give them the appearance of functions in phase space. Let A be any observable, and define a function $\mathcal{A}(q,p)$ by requiring that it produce the correct expectation value when averaged over the Wigner distribution:

$$\int \frac{dq\, dp}{2\pi\hbar}\, \mathcal{A}(q,p)\, f(q,p) = \mathrm{Tr}\, A\rho \,. \tag{151}$$

It is left to the reader to show that $\mathcal{A}$ has precisely the same relationship to A as does f to ρ:

$$\mathcal{A}(q,p) = \int ds\, e^{-ips/\hbar}\, \langle q + \tfrac{1}{2}s|A|q - \tfrac{1}{2}s\rangle \,. \tag{152}$$

Two important special cases are observables which are diagonal in the coordinate or momentum representation, such as a potential V or a kinetic energy K, respectively. That is, define

$$\langle q|V|q'\rangle = \delta(q-q')\, V(q) \,, \qquad \langle p|K|p'\rangle = \delta(p-p')\, K(p) \,. \tag{153}$$

Then, according to (152), their "phase space" counterparts are

$$\mathcal{V}(q,p) = V(q) \,, \qquad \mathcal{K}(q,p) = K(p) \,. \tag{154}$$

This, of course, is required by (149) and (150).

The generalization to a N particles is simply

$$dq \to dq_1 \ldots dq_{3N} \,, \qquad \frac{dp}{2\pi\hbar} \to \frac{dp_1}{2\pi\hbar} \cdots \frac{dp_{3N}}{2\pi\hbar} \,, \tag{155}$$

and so forth. The element of integration in phase space for degree of freedom i is therefore $dq_i dp_i/h$, where $h = 2\pi\hbar$ is Planck's original constant.

2.3 Canonical Quantization

The first route from classical to quantum mechanics, discovered by Heisenberg, Born, Jordan and Dirac in 1925, is not the only one, but it is the most direct and succinct, and will now be summarized. Its point of departure is classical mechanics in the Hamiltonian format, that is, using canonical coordinates and momenta satisfying first order equations of motion. The leap to quantum mechanics is the replacement of these numbers by Hermitian operators acting on the infinite dimensional Hilbert space $\mathfrak{H}$ of the preceding discussion, and the specification of commutation rules between these operators.

(a) The Canonical Commutation Rules

The system of interest will, to begin with, again be that of N particles in three spatial dimensions treated as if they had no internal structure, i.e., a system with $3N$ degrees of freedom. For each degree of freedom Hermitian coordinate and momentum *operators* q_i and p_i $(i = 1, \ldots, 3N)$ are introduced. Eigenvalues of these operators will be designated by primes: q_i' and p_i'.

These operators are postulated to obey the following *canonical commutation rules*:

$$[q_i, p_j] = i\hbar\, \delta_{ij} \ , \tag{156}$$

$$[q_i, q_j] = 0, \qquad [p_i, p_j] = 0 \ . \tag{157}$$

The q-p commutator is the heart of the matter, and is the point of entry for $\hbar$.

Shortly we will see that symmetry considerations almost suffice to determine the commutation rules, though of course not $\hbar$. For now, we take (156) and (157) for granted and analyze their consequences.

Because all the q's commute, they can be diagonalized simultaneously; the same goes for the p's. On the other hand, q_i and p_j cannot be diagonalized simultaneously if $i = j$. Two sets of simultaneous eigenkets are thus

$$\begin{aligned} |q_1' \ldots q_{3N}'\rangle &\equiv |q_1'\rangle \otimes |q_2'\rangle \ldots \otimes |q_{3N}'\rangle \ , \\ |p_1' \ldots p_{3N}'\rangle &\equiv |p_1'\rangle \otimes |p_2'\rangle \ldots \otimes |p_{3N}'\rangle \ . \end{aligned} \tag{158}$$

In view of this factorization, it suffices to examine just one degree of freedom, with coordinate and momentum operators q and p satisfying

$$[q, p] = i\hbar \ . \tag{159}$$

Repeated use of the commutation rule (159) shows that

$$[q, p^n] = in\hbar p^{n-1}, \qquad [p, q^n] = -in\hbar q^{n-1} \ ; \tag{160}$$

apart from a factor, these right-hand sides have the form of derivatives of p^n and q^n. Motivated by this observation, consider any function $G(p)$ of the operator p that has a power series expansion, such as an exponential of p, and $F(q)$ any function of the operator q of the same type. Eq. 160 then generalizes to the important commutation rules

$$[q, G(p)] = i\hbar \frac{\partial G(p)}{\partial p} \ , \qquad [p, F(q)] = \frac{\hbar}{i} \frac{\partial F(q)}{\partial q} \ . \tag{161}$$

For functions of all the coordinates or momenta, these rules clearly are

$$[q_i, G(p_1 \dots p_{3N})] = i\hbar \frac{\partial G}{\partial p_i} , \qquad [p_i, F(q_1 \dots q_{3N})] = \frac{\hbar}{i} \frac{\partial F}{\partial q_i} . \tag{162}$$

The commutation rules Eq. 161, which are direct consequences of the canonical commutators, lead to an expected conclusion, which is nonetheless important: the spectra of q and p are both continuous from $-\infty$ to ∞. To demonstrate this, introduce the unitary operator

$$T(a) = e^{-iap/\hbar} , \tag{163}$$

where a is any real number having the dimension of length. $T(a)$ is unitary because p is Hermitian. Then, because of (161),

$$qT(a) = T(a)q + i\hbar\partial T/\partial p = T(a)(q + a) . \tag{164}$$

Let $|q'\rangle$ be some eigenket of q

$$q|q'\rangle = q'|q'\rangle , \tag{165}$$

and consider $qT(a)|q'\rangle$, which is

$$qT(a)|q'\rangle = (q' + a)T(a)|q'\rangle , \tag{166}$$

due to (164). Hence $T(a)|q'\rangle$ is an eigenket of q with eigenvalue $q' + a$. But a was arbitrary, and therefore all real numbers are eigenvalues of q.

This argument shows that the unitary operator $T(a)$ produces a *spatial translation* through the distance a. That is, by multiplying (164) by $T^\dagger$,

$$T^\dagger(a)qT(a) = q + a , \tag{167}$$

which is the unitary transformation of the coordinate operator corresponding to a spatial translation. Because T is unitary, it preserves norms, and therefore (166) implies

$$T(a)|q'\rangle = |q' + a\rangle . \tag{168}$$

This recognition that $T(a)$ produces a translation of the coordinate operators q_i and their eigenkets implies that *symmetry considerations largely determine the canonical commutation rules:* first, the fact that translations along different directions commute requires $[p_i, p_j] = 0$; and second, requiring the coordinates to obey (167) requires $[q_i, p_j] = i\delta_{ij}C$. The last commutation rule, $[q_i, q_j] = 0$, is based on the separate assumption that all coordinate components can be simultaneously specified. While C has the dimension of action, experiment is needed to confirm that C agrees with, for example, the value of $\hbar$ found from fitting the cosmic background radiation with the Planck distribution.

The same argument can be applied to p by use of the unitary operator

$$K(k) = e^{iqk/\hbar} , \tag{169}$$

where k is now any real number having the dimension of momentum. The result is that if $|p'\rangle$ is an eigenket of p with the indicated eigenvalue, so is $|p' + k\rangle$ for all k. The unitary operator $K(k)$ produces a translation in momentum space by k:

$$K^\dagger(k)pK(k) = p + k , \qquad K(k)|p'\rangle = |p' + k\rangle . \tag{170}$$

Translations in momentum space are often referred to as boosts.

Because of their continuous spectra, the coordinate and momentum eigenkets are normalized to delta functions:

$$\langle q'|q''\rangle = \delta(q' - q'') , \qquad \langle p'|p''\rangle = \delta(p' - p'') . \tag{171}$$

An important observation regarding the role of time, due to Pauli, is a consequence of the continuous spectra for coordinates and momenta. Were the goal to extend what has been said thus far to a relativistic theory, a first attempt might be to put time on the same footing as the spatial coordinates by generalizing the commutation rule to one between 4-vectors for position and momentum. This would imply the time-energy commutator

$$[\hat{t}, H] = -i\hbar , \tag{172}$$

where $\hat{t}$ is a Hermitian operator whose eigenvalues are time, H is the energy operator (the Hamiltonian), and the sign is chosen on the supposition that the coordinates and momenta are 4-vectors. But as we just saw, if $\hat{t}$ is to have a continuous spectrum like the coordinates, then so must H; i.e., there could be no lower bound to energies and no bound states with discrete energies! So this path to a relativistic theory is a disaster at the start. Indeed, what is done in relativistic quantum mechanics is to demote the spatial coordinates from operators to the same status as time, as 4-vector *parameters* that label observables such as the electromagnetic field operators to be discussed in chapter 10. That the coordinates are not eigenvalues of operators in relativistic quantum mechanics is consistent with the fact, pointed out in §1.1, that the position of a particle of mass m cannot be determined to an accuracy better than the Compton wave length $\hbar/mc$.

(b) Schrödinger Wave Functions

Schrödinger's originally distinct formulation of quantum mechanics in terms of wave functions and differential operators follows from canonical quantization by writing all operator and kets in the coordinate representation. In particular, the Schrödinger wave function is the scalar product $\langle q_1' \dots |\psi\rangle$.

The key is the transformation function between the coordinate and momentum representations for one degree of freedom, $\langle q'|p'\rangle$, that is to say, the wave function when $|\psi\rangle$ is actually $|p'\rangle$. Let $|0_q\rangle$ and $|0_p\rangle$ be the eigenkets of q and p with eigenvalue 0. Then by use of the unitary operators T and K,

$$\begin{aligned}\langle q'|p'\rangle = \langle 0_q|T^\dagger(q')|p'\rangle = e^{iq'p'/\hbar}\langle 0_q|p'\rangle \\ = e^{iq'p'/\hbar}\langle 0_q|K(p')|0_p\rangle = e^{iq'p'/\hbar}\langle 0_q|0_p\rangle .\end{aligned} \tag{173}$$

The constant $\langle 0_q|0_p\rangle$ is determined, up to an arbitrary phase, by requiring

$$\int dp' \, \langle q'|p'\rangle\langle p'|q''\rangle = \delta(q' - q'') , \tag{174}$$

and by recalling the Fourier representation of the delta function:

$$\int \frac{dq'}{2\pi\hbar} e^{ip'q'/\hbar} = \delta(p') . \tag{175}$$

The phase is then chosen so as to produce

$$\langle q'|p'\rangle = \frac{1}{\sqrt{2\pi\hbar}} e^{ip'q'/\hbar} = \varphi_{p'}(q') , \tag{176}$$

to return to the notation of Eq. 74. Hence for the N-particle system the wave function when all particles are in momentum eigenstates is a product of plane waves:

$$\langle q_1' \dots |p_1' \dots\rangle = (2\pi\hbar)^{-3N/2} \prod_{i=1}^{3N} e^{ip_i q_i/\hbar} . \tag{177}$$

Configuration and momentum space wave functions are defined as scalar products of $|\psi\rangle$ with coordinate and momentum eigenkets:

$$\psi(q') = \langle q'|\psi\rangle, \qquad \phi(p') = \langle p'|\psi\rangle . \tag{178}$$

Hence

$$\phi(p') = \int dq' \langle p'|q'\rangle\langle q'|\psi\rangle , \tag{179}$$

which is just the Fourier transform of $\psi(q')$ because of (176):

$$\phi(p') = \int \frac{dq'}{\sqrt{2\pi\hbar}} e^{-ip'q'/\hbar} \psi(q') ; \tag{180}$$

its inverse is

$$\psi(q') = \int \frac{dp'}{\sqrt{2\pi\hbar}} e^{ip'q'/\hbar} \phi(p') . \tag{181}$$

The probability distributions in configuration and momentum space are then $|\psi(q')|^2$ and $|\phi(p')|^2$. Given one it is *not* possible to construct the other. This is so because the probabilities only depend on the modulus of the wave functions, and phases that depend on either q' or p' must be known for the transformation from one representation to the other. The density matrix in any one representation does, of course, suffice, to determine that in any the other. In this context,

$$\langle q'|\rho|q''\rangle = \psi(q')\psi^*(q'') , \qquad \langle p'|\rho|p''\rangle = \phi(p')\phi^*(p'') , \tag{182}$$

and therefore

$$\langle p'|\rho|p''\rangle = \int \frac{dq'dq''}{2\pi\hbar} e^{-i(p'q'-p''q'')/\hbar} \langle q'|\rho|q''\rangle . \tag{183}$$

In short, to compute the momentum distribution $\langle p'|\rho|p'\rangle$ one must know the off-diagonal elements of ρ in the coordinate representation.

This example illustrates a general property of quantum mechanical probability distributions: *the probability distribution for a complete set of compatible observables does not determine the probability distribution for an incompatible observable.*

To determine the action of the momentum operator on configuration space wave functions the matrix elements of p in the q-representation is needed:

$$\langle q'|p^n|q''\rangle = \int dp' \langle q'|p'\rangle(p')^n\langle p'|q''\rangle = \int \frac{dq'}{2\pi\hbar}(p')^n \, e^{ip'(q'-q'')/\hbar} . \tag{184}$$

This is the nth derivative of a delta function:

$$\delta^{(n)}(x) = \frac{d^n}{dx^n}\delta(x) = \int dk\,(ik)^n\,e^{ikx}\;, \tag{185}$$

which has the property

$$\frac{d^n f(x)}{dx^n} = \int dx'\,\delta^{(n)}(x-x')f(x') \tag{186}$$

for a function that is sufficiently continuous at x. Hence (184) is

$$\langle q'|p^n|q''\rangle = (\hbar/i)^n\,\delta^{(n)}(q'-q'')\;, \tag{187}$$

and therefore

$$\langle q'|p^n|\psi\rangle = \left(\frac{\hbar}{i}\frac{\partial}{\partial q'}\right)^n\,\psi(q')\;. \tag{188}$$

The same argument gives the corresponding result for the action of coordinate operators on wave functions in momentum space:

$$\langle p'|q^n|\psi\rangle = \left(i\hbar\frac{\partial}{\partial p'}\right)^n\,\phi(p')\;. \tag{189}$$

The generalization of these formulas to more than one degree of freedom is straightforward and will not be spelled out. There is, however, one aspect of the extension to many degrees of freedom that merits attention: the displacement of *all* the coordinates. To construct the unitary operator that does this, we must group the three coordinates and momenta of each particle into vector operators: $\boldsymbol{x}_1 = (q_1, q_2, q_3)$, $\boldsymbol{x}_2 = (q_4, q_5, q_6)$, etc., and similarly with the momenta, which we now enumerate as $\boldsymbol{x}_n$, $\boldsymbol{p}_n$, with $n = 1, \ldots, N$. The displacement by the 3-vector $\boldsymbol{a}$ is to be produced by a unitary operator $T(\boldsymbol{a})$ that generalizes $T(a)$, as given by (163), so that (167) is replaced by

$$T^\dagger(\boldsymbol{a})\boldsymbol{x}_n T(\boldsymbol{a}) = \boldsymbol{x}_n + \boldsymbol{a}\;, \tag{190}$$

with $\boldsymbol{a}$ the same for all n. Because all individual momentum components commute with each other, T is simply a product of the operators T for each degree of freedom with the component of $\boldsymbol{a}$ appropriate to the momentum component in question, i.e.,

$$T(\boldsymbol{a}) = \prod_{n=1}^{N} \exp(-i\boldsymbol{p}_n\cdot\boldsymbol{a}/\hbar)\;. \tag{191}$$

But

$$\boldsymbol{P} = \sum_{n=1}^{N} \boldsymbol{p}_n \tag{192}$$

is the total momentum. Thus the spatial displacement operator for the whole system can written in terms of the total momentum as

$$T(\boldsymbol{a}) = e^{-i\boldsymbol{P}\cdot\boldsymbol{a}/\hbar}\;. \tag{193}$$

(c) Uncertainty Relations

The Heisenberg uncertainty relation for coordinates and momenta is but one example of such relations that hold for all pairs of incompatible observables.

To derive such relations, a precise definition of the uncertainty ΔA of an observable A is needed. The convenient and physically reasonable choice is the root-mean-square (rms) dispersion ΔA, defined as

$$\Delta A = \sqrt{\langle (A - \langle A \rangle)^2 \rangle} = \sqrt{\langle A^2 \rangle - \langle A \rangle^2}\,, \tag{194}$$

where $\langle \ldots \rangle$ is the expectation value in an arbitrary ket $|\phi\rangle$, i.e.,

$$(\Delta A)^2 = \sum_a (a - \langle A \rangle)^2 \, |\langle a|\phi\rangle|^2\,, \tag{195}$$

which is the second moment of the probability distribution $|\langle a|\phi\rangle|^2$.

Let B be an observable that does not commute with A. The goal is to find a lower bound for $\Delta A\,\Delta B$ as the precise statement of the uncertainty principle for arbitrary observables. For that purpose it proves to be convenient to define

$$\bar{A} = A - \langle A \rangle, \qquad \bar{B} = B - \langle B \rangle\,, \tag{196}$$

$$\bar{A}|\phi\rangle = |\phi_A\rangle, \qquad \bar{B}|\phi\rangle = |\phi_B\rangle\,. \tag{197}$$

Then

$$(\Delta A\,\Delta B)^2 = \langle \phi_A|\phi_A\rangle \langle \phi_B|\phi_B\rangle\,, \tag{198}$$

which, due to the Schwarz inequality (Eq. 17) gives

$$(\Delta A\,\Delta B)^2 \geq |\langle \bar{A}\bar{B}\rangle|^2\,. \tag{199}$$

Now

$$\bar{A}\bar{B} = \tfrac{1}{2}[A, B] + \tfrac{1}{2}\{\bar{A}, \bar{B}\} = \tfrac{1}{2}iC + \tfrac{1}{2}\{\bar{A}, \bar{B}\}\,, \tag{200}$$

where C and the anticommutator are both Hermitian. Hence

$$(\Delta A\,\Delta B))^2 \geq \tfrac{1}{4}|\langle\{\bar{A}, \bar{B}\}\rangle + i\langle C\rangle|^2\,, \tag{201}$$

and as both expectation values are real,

$$(\Delta A\,\Delta B))^2 \geq \tfrac{1}{4}\langle\{\bar{A}, \bar{B}\}\rangle^2 + \tfrac{1}{4}\langle C\rangle^2\,, \tag{202}$$

As shown by Prob. 6, when $|\phi\rangle$ is chosen to make $|\phi_A\rangle$ proportional to $|\phi_B\rangle$, the first term in (202) vanishes; therefore

$$\Delta A\,\Delta B \geq \tfrac{1}{2}|\langle [A, B]\rangle|\,. \tag{203}$$

This is *the general form of Heisenberg's uncertainty relation.* For the canonical variables, it reads

$$\Delta p_i\,\Delta q_j \geq \tfrac{1}{2}\hbar\,\delta_{ij}\,, \tag{204}$$

which now gives a precise lower bound in place of the order-of-magnitude estimate of §1.1.

Because time cannot be the eigenvalue of an operator, it is now clear that the time-energy uncertainty relation has a quite different meaning from that for quantities represented by operators. It will, therefore, be discussed separately in §2.4.

2.4 The Equations of Motion

What might be called kinematics is all that has been considered thus far. How states evolve in time will now be addressed.

(a) The Schrödinger Picture

The basic assumption will be that time evolution is represented by a unitary transformation parametrized by a continuous parameter t . If the superposition principle is taken to be fundamental, this assumption is almost inescapable, because superposition requires time evolution to be a linear transformation of vectors in the Hilbert space, i.e., if $|\psi;0\rangle$ is some state of a system at $t=0$, then at a later time $|\psi;t\rangle = L_t|\psi;0\rangle$, where L_t is a linear operator.

Consider some *time-independent* observable, A, with the spectrum $(a, a', \ldots)$. Its expectation value as a function of time will be

$$\langle\psi;t|A|\psi;t\rangle = \sum_a a|\langle\psi;t|a\rangle|^2 \; . \tag{205}$$

In general, the probabilities $|\langle\psi;t|a\rangle|^2$ for the various eigenvalues will change with time, but by hypothesis, not the eigenvalues themselves. On the other hand, rewriting (205) in terms of the operator L_t gives

$$\langle\psi;0|L_t^\dagger A L_t|\psi;0\rangle = \sum_{a_t} a_t|\langle\psi;0|a_t\rangle|^2 \; , \tag{206}$$

where $\{|a_t\rangle\}$ are the eigenkets of $L_t^\dagger A L_t$ and $\{a_t\}$ its eigenvalues. As just said, however, the spectrum must not change with time, whereas the probabilities can. But any unitary transformation of a Hermitian operator leave its spectrum invariant, and it is this fact that justifies the assumption that L_t is a unitary operator. Accepting this assumption means that there is a unitary operator $U(t', t)$ that manufactures $|\psi;t'\rangle$ from the state at time t:

$$|\psi;t'\rangle = U(t', t)|\psi;t\rangle \; . \tag{207}$$

Consider first a system that is isolated from external disturbances, so that the origin of time has no physical significance. When this is so, U simplifies greatly because it can only depend on time differences:

$$|\psi;t'\rangle = U(t'-t)|\psi;t\rangle \; . \tag{208}$$

These unitary operators must satisfy the following composition law:

$$U(t_2)U(t_1) = U(t_2+t_1) \; . \tag{209}$$

The crucial point here is that the composite is a function of the sum t_1+t_2, and therefore

$$U(t) = [U(t/N)]^N \; . \tag{210}$$

Now as $\delta t \to 0$, $U(\delta t) \to 1$, and therefore $U(\delta t) = 1 - i\Delta(\delta t)$, where $\Delta(\delta t)$ must be an infinitesimal Hermitian operator so as to make U unitary to first order in Δ.

The composition law (209) implies that $\Delta(\delta t_1) + \Delta(\delta t_2) = \Delta(\delta t_1 + \delta t_2)$, i.e., that $\Delta(\delta t)$ is proportional to δt. This fact is expressed by

$$U(\delta t) = 1 - i\delta t H/\hbar \,, \tag{211}$$

where $\hbar$ is introduced so that the operator H has the dimension of energy. It is *the Hamiltonian* of the system in question.

For finite time differences, (210) gives

$$U(t) = [U(t/N)]^N = \lim_{N\to\infty} \left(1 - \frac{1}{N}\frac{itH}{\hbar}\right)^N \,, \tag{212}$$

and therefore

$$U(t) = e^{-iHt/\hbar} \,. \tag{213}$$

An equivalent derivation is

$$U(t + \delta t) - U(t) = [U(\delta t) - 1]U(t) = -i(\delta t H/\hbar)U(t) \,, \tag{214}$$

so that

$$i\hbar\frac{\partial}{\partial t}U(t) = H\,U(t) \,, \tag{215}$$

which has the solution (213) because of the initial condition $U(0) = 1$. Because of its role in the unitary operator $U(t)$, the Hamiltonian is called the generator of translations in time.

The *Schrödinger equation* then follows from (207) and (215):

$$i\hbar\frac{\partial}{\partial t}|\psi; t\rangle = H|\psi; t\rangle \,. \tag{216}$$

The case of an eigenstate of H is of particular importance, i.e., a ket $|\psi_E; t\rangle$ that satisfies the *time-independent Schrödinger equation*

$$(H - E)|\psi_E; t\rangle = 0 \,. \tag{217}$$

The solution of (216) is then trivial:

$$|\psi_E; t\rangle = e^{-iEt/\hbar}|\psi_E; 0\rangle \,. \tag{218}$$

Energy eigenstates are called *stationary states* because they do not change in time, aside from a phase factor. Furthermore, the matrix elements of any time-independent observable A between stationary states also have a time dependence that is merely a phase:

$$\langle\psi_E; t|A|\psi_{E'}; t\rangle = e^{i(E-E')t/\hbar}\,\langle\psi_E; 0|A|\psi_{E'}; 0\rangle \,. \tag{219}$$

It is a mathematically trivial, but nonetheless important fact that for any operator A

$$\langle\psi_E|[A, H]|\psi_E\rangle = 0 \,; \tag{220}$$

in words, that the commutator of any observable with the Hamiltonian has zero expectation value in all stationary states.

Consider an N-particle system with a Hamiltonian that has the form familiar from classical mechanics when the forces are velocity-independent:

$$H = \sum_{n=1}^{N} \frac{\boldsymbol{p}_n^2}{2m_n} + V(\boldsymbol{x}_1 \ldots \boldsymbol{x}_N) \; . \tag{221}$$

Here $\boldsymbol{x}_n$ is the coordinate operator for particle n; its eigenvalues will be designated by $\boldsymbol{r}_n$. To put (216) into the coordinate representation, one takes the scalar product with a coordinate eigenket $|\boldsymbol{r}_1 \ldots \boldsymbol{r}_N\rangle$. For this purpose recall Eq. 188, which in the present notation is

$$\langle \boldsymbol{r}_1 \ldots | \boldsymbol{p}_n | \psi; t\rangle = \frac{\hbar}{i} \frac{\partial}{\partial \boldsymbol{r}_n} \psi(\boldsymbol{r}_1 \ldots; t) \; . \tag{222}$$

This then gives *the Schrödinger equation in the coordinate representation:*

$$i\hbar \frac{\partial}{\partial t} \psi(t) = \left(\sum_n \frac{1}{2m_n} \left(\frac{\hbar}{i} \frac{\partial}{\partial \boldsymbol{r}_n} \right)^2 + V(\boldsymbol{r}_1 \ldots \boldsymbol{r}_N) \right) \psi(t) \; . \tag{223}$$

If the classical Hamiltonian does not have the simple form (221), and contains expressions like qp, there is no unique recipe for turning it into a Hermitian operator. One obvious possibility is $qp \to \frac{1}{2}(qp + pq)$, but there are others. This ambiguity vanishes as $\hbar \to 0$, and by the same token classical mechanics does not always provide a clearcut rule for the move to quantum mechanics because anything that vanishes as $\hbar \to 0$ is invisible in the classical limit. It is remarkable that in so much of physics the unmodified classical Hamiltonian can be taken over successfully into quantum mechanics.

The unitary nature of time evolution leads to conservation laws and continuity equations involving the probability. These all stem from the following immediate consequence of Eq. 207: *the scalar product of any two solutions of the time-dependent Schrödinger equation is independent of time.*

The constancy in time of the norm $\langle \psi; t | \psi; t\rangle$ is perhaps the most important example. In terms of the coordinate space probability distribution,

$$w(\boldsymbol{r}_1, \ldots; t) \equiv |\psi(\boldsymbol{r}_1, \ldots; t)|^2 \; , \tag{224}$$

the constancy of the norm is

$$\frac{\partial}{\partial t} \int d^3r_1 \ldots d^3r_N \; w(\boldsymbol{r}_1, \ldots; t) = 0 \; . \tag{225}$$

If the interactions are diagonal in the coordinate representation, a much stronger statement holds in any infinitesimal region of configuration space, namely,

$$\frac{\partial}{\partial t} w(\boldsymbol{r}_1 \ldots; t) + \sum_{n=1}^{N} \frac{\partial}{\partial \boldsymbol{r}_n} \cdot \boldsymbol{i}_n(\boldsymbol{r}_1 \ldots; t) = 0 \; , \tag{226}$$

where

$$\boldsymbol{i}_n(\boldsymbol{r}_1 \ldots) = \frac{\hbar}{2im_n} \left(\psi^* \frac{\partial \psi}{\partial \boldsymbol{r}_n} - \psi \frac{\partial \psi^*}{\partial \boldsymbol{r}_n} \right) \; . \tag{227}$$

This is a continuity equation in the $3N$-dimensional configuration space, *not* in everyday 3-space, $\mathfrak{E}_3$; the N-tuple of 3-vectors $\boldsymbol{i}_n$ is needed to describe flow across a hypersurface in configuration space, and reflects the fact that the Schrödinger equation is a wave equation in configuration space, and only a conventional wave equation in the case of one particle. The global conservation law (Eq. 225) follows from the local law (226), if the wave function falls off sufficiently at large distances to have a finite norm.

In the important case of a stationary state, i.e., when $\partial w/\partial t = 0$, the continuity equation reduces to the condition that the current is divergence free:

$$\sum_n \frac{\partial}{\partial \boldsymbol{r}_n} \cdot \boldsymbol{i}_n = 0 \,. \tag{228}$$

To derive (226), note that (216) implies

$$-i\hbar\partial_t \langle\psi; t| = \langle\psi; t|H \tag{229}$$

because H must be Hermitian if probability is to be conserved. Using the shorthand $i\hbar\partial_t\langle \boldsymbol{r}|\psi\rangle = \langle \boldsymbol{r}|H|\psi\rangle$, etc.,

$$i\hbar\partial_t|\psi(\boldsymbol{r})|^2 = \langle\psi|\boldsymbol{r}\rangle\langle\boldsymbol{r}|H|\psi\rangle - \langle\psi|H|\boldsymbol{r}\rangle\langle\boldsymbol{r}|\psi\rangle \,. \tag{230}$$

The potential energy V does not contribute to the right-hand side because it is assumed to be diagonal in the coordinate representation; the kinetic energy K gives:

$$\psi^*(\boldsymbol{r})\langle\boldsymbol{r}|K|\psi\rangle - \langle\psi|K|\boldsymbol{r}\rangle\psi(\boldsymbol{r}) = -\sum_n \frac{\hbar^2}{2m_n}\frac{\partial}{\partial\boldsymbol{r}_n}\cdot\left(\psi^*\frac{\partial}{\partial\boldsymbol{r}_n}\psi - \psi\frac{\partial}{\partial\boldsymbol{r}_n}\psi^*\right), \tag{231}$$

which establishes (226).

If the particles in question are electrically charged, expressions for conventional charge and current densities, ρ and $\boldsymbol{j}$, are often needed. These are

$$\rho(\boldsymbol{r},t) = \sum_n e_n \int d^3r_1 \ldots d^3r_N \; \delta(\boldsymbol{r}-\boldsymbol{r}_n)\,|\psi(\boldsymbol{r}_1\ldots;t)|^2 \,, \tag{232}$$

$$\boldsymbol{j}(\boldsymbol{r},t) = \sum_n e_n \int d^3r_1 \ldots d^3r_N \; \delta(\boldsymbol{r}-\boldsymbol{r}_n)\,\boldsymbol{i}_n(\boldsymbol{r}_1\ldots;t) \,. \tag{233}$$

The proof that this density and current satisfy the conventional continuity equation in $\mathfrak{E}_3$ is left as an exercise.

The constancy of the scalar product $\langle\psi_1;t|\psi_2;t\rangle$ for two solutions of one Schrödinger equation is also accompanied by a continuity equation when the interaction is local:

$$\frac{\partial}{\partial t}\psi_1^*\psi_2 + \sum_n \frac{\hbar}{2im_n}\frac{\partial}{\partial\boldsymbol{r}_n}\cdot\left(\psi_1^*\frac{\partial\psi_2}{\partial\boldsymbol{r}_n} - \psi_2\frac{\partial\psi_1^*}{\partial\boldsymbol{r}_n}\right) = 0 \,. \tag{234}$$

The proof is again left as an exercise.

Thus far we have dealt with the time development of pure states. Let $\{|a\rangle\}$ be a basis that diagonalizes the density matrix at $t=0$:

$$\rho(0) = \sum_a |a\rangle p_a \langle a| \,. \tag{235}$$

At later times, $|a\rangle \to \exp(-iHt/\hbar)|a\rangle$, and therefore,

$$\rho(t) = e^{-iHt/\hbar}\rho(0)\, e^{iHt/\hbar} \ . \tag{236}$$

Note that the probabilities p_a do not change with time. The equation of motion for the density matrix follows by differentiating (236):

$$i\hbar\frac{d}{dt}\rho(t) = [H, \rho(t)] \ . \tag{237}$$

This is an important equation, especially in statistical physics. Loosely speaking, it is the counterpart of Liouville's equation in classical mechanics, though of course the latter pertains to phase space.

Finally, consider a system exposed to time-dependent applied forces. The origin of time is then no longer arbitrary, and the unitary time evolution operator $U(t', t)$ depends on both of its arguments, not just the interval $t' - t$. These operators now have the composition law

$$U(t'', t')U(t', t) = U(t'', t) \ , \tag{238}$$

which states that evolving to t' followed by evolving to t'' is the same as evolving directly from t to t''. The first question is whether infinitesimal transformations can still be written in the form of Eq. 211. To answer this, specialize (238) to two infinitesimal time increments:

$$U(t + \delta_2 + \delta_1, t + \delta_1)U(t + \delta_1, t) = U(t + \delta_2 + \delta_1, t) \ . \tag{239}$$

But $U(t + \delta, t) = 1 - iF(t, \delta)$ for an infinitesimal δ, with F being an infinitesimal Hermitian operator, and so (239) requires $F(t, \delta_2) + F(t, \delta_1) = F(t, \delta_2 + \delta_1)$, i.e., that $F(t, \delta)$ be proportional to δ. Hence in this case

$$U(t + \delta t, t) = 1 - i\delta t\, H(t)/\hbar \ , \tag{240}$$

where $H(t)$ is the *time-dependent Hamiltonian*. Therefore

$$U(t + \delta t, t') = U(t, t') - i\delta t H(t)U(t, t')/\hbar \ , \tag{241}$$

and so the differential equation (215) holds in the following form

$$i\hbar\frac{\partial}{\partial t}U(t, t') = H(t)U(t, t') \ . \tag{242}$$

Hence the Schrödinger equation for a time-dependent Hamiltonian still has the form of Eq. 216, and the scalar product of any two of its solutions is time-independent.

While the differential equation for $U(t, t')$ seems to be identical to the one for a time-independent Hamiltonian, that would be a misperception: in general, it is no longer possible to integrate the equation even in the formal manner that previously led to $\exp(-iHt/\hbar)$ to obtain $U(t, t')$ for finite time differences because, in most situations of physical interest, $H(t)$ does not commute with $H(t')$ when $t \neq t'$. (Recall that what made the integration possible (indeed, trivial) in the time-independent case was that H then behaved like a number.) A major industry has been devoted for decades to grappling with this time-dependent situation, because even when the system is isolated and the complete Hamiltonian is time-independent the perverse notion of turning the problem into a time-dependent one proves to be very fruitful, as we will see in subsection (e).

(b) The Heisenberg Picture

In the preceding discussion of time dependence, the state vector or density matrix changed with time, while the observables did not (unless, of course, they are explicitly time-dependent like the Hamiltonian $H(t)$ in Eq. 242.) Because time evolution is unitary, another equivalent formulation, in which the observables move while the state stays put, is also available. The former, where the states move is called the *the Schrödinger picture,* and the latter where they do not but the observables move, *the Heisenberg picture.* The Heisenberg picture is better suited to bringing out fundamental features, such as symmetries and conservation laws, and it is indispensable in systems with many degrees of freedom, like those dealt with in quantum field theory and statistical physics.

Assume a time-independent Hamiltonian, so that $U(t) = \exp(-iHt/\hbar)$. The case where H depends on t is not different in principle, but of course more complicated. Let $\{|\psi_b;t\rangle\}$ be a complete set of solutions of the Schrödinger equation, and call their $t=0$ values $|\psi_b\rangle$. The matrix elements of a time-independent observable A in this moving basis are

$$\langle\psi_b;t|A|\psi_{b'};t\rangle = \langle\psi_b|e^{iHt/\hbar}Ae^{-iHt/\hbar}|\psi_{b'}\rangle \; . \tag{243}$$

This tells us how to put the burden of carrying the time-dependence on the observables:

- *In the Heisenberg picture, kets that describe the time evolution of pure states are fixed in the Hilbert space, and observables A that are time-independent in the Schrödinger picture are replaced by operators $A(t)$ that evolve with the unitary transformation*

$$A(t) = e^{iHt/\hbar}\, A\, e^{-iHt/\hbar} \; . \tag{244}$$

Observables in the Heisenberg picture therefore obey the equation of motion

$$i\hbar\frac{d}{dt}\, A(t) = [A(t), H] \; . \tag{245}$$

This equation has two immediate consequences of great importance:

- *Observables that commute with the Hamiltonian are constants of motion.*
- *Any one constant of motion can be diagonalized simultaneously with the Hamiltonian, i.e, they possess simultaneous eigenstates. The Hamiltonian and a set of constants of motion can be diagonalized simultaneously provided that all these constants of motion commute with each other.*

One should not be misled by the superficial similarity between the equation of motion for ρ in the Schrödinger picture (Eq. 237) and the Heisenberg picture equation (245). *The density matrix does not move in the Heisenberg picture,* because the kets that describe the pure states that define ρ (e.g., as in Eq. 235) are fixed in this picture.

When the Hamiltonian has the familiar form of Eq. 221, the commutation rules (162) immediately yield the following equations of motion for the canonical coordinates and momenta in the Heisenberg picture:

$$\frac{dq_i(t)}{dt} = \frac{\partial H}{\partial p_i} \, , \qquad \frac{dp_i(t)}{dt} = -\frac{\partial H}{\partial q_i} \; . \tag{246}$$

These have exactly the same form as Hamilton's classical equations of motion.

Of course, integration of the quantum mechanical equations is, in general, a far more difficult proposition because the dynamical variables do not commute with each other. This complication is partially absent, however, if the Hamiltonian has the conventional form (221) and V has only linear and quadratic terms in the coordinates, for then the equations of motion are linear:

$$\dot{q}_i = p_i/m_i, \qquad \dot{p}_i = a_i + \sum_j \omega_{ij} q_j \,. \tag{247}$$

Hence the expectation values of q_i and p_i in any state, whether pure or mixed, evolve exactly like those of classical mechanics for the important cases of no forces, a constant force, and harmonic motion. Of course, this does not mean that there is then no difference between classical and quantum mechanics.

It should be noted that a ket $|\alpha\rangle_S$ that is stationary in the Schrödinger picture will become the moving ket $|\alpha(t)\rangle_H = U^\dagger(t)|\alpha\rangle_S$ in the Heisenberg picture. Furthermore, if an observable $B_S(t)$ is explicitly time-dependent in the Schrödinger picture, its counterpart in the Heisenberg picture is still given by (244) but obeys the equation of motion

$$i\hbar \frac{d}{dt} B_H(t) = i\hbar \frac{\partial}{\partial t} B_H(t) + [B_H(t), H] \,. \tag{248}$$

Finally, should the Hamiltonian $H_S(t)$ itself be time-dependent in the Schrödinger picture, the transformation (244) is replaced by

$$A_H(t,t') = U^\dagger(t,t') A_S U(t,t') \,, \tag{249}$$

and this also applies to the Hamiltonian. Consequently the equation of motion (245) becomes

$$i\hbar \frac{d}{dt} A_H(t,t') = [A_H(t,t'), H_H(t,t')] \,. \tag{250}$$

This situation is illustrated by the interaction picture, which is to be treated in §2.4(e).

(c) Time Development of Expectation Values

The Heisenberg picture is well suited to addressing the question of how expectation values evolve in time quite generally. For the Hamiltonian (221), the Heisenberg equations of motion are

$$\frac{dq_i(t)}{dt} = \frac{1}{m_i} p_i(t) \,, \qquad \frac{dp_i(t)}{dt} = -\frac{\partial V}{\partial q_i} \equiv F_i \,, \tag{251}$$

where F_i is the force operator.

Let $\bar{q}_i(t)$, etc., denote the expectation values $\langle q_i(t)\rangle$ in an arbitrary state. Then

$$\frac{d\bar{p}_i(t)}{dt} = \langle F(q_1(t) \ldots, q_f(t))\rangle \,, \qquad \frac{d\bar{q}_i(t)}{dt} = \frac{1}{m_i} \bar{p}_i(t) \,. \tag{252}$$

These would be Newton's equations for the expectation values if the following replacement were valid:

$$\langle F\rangle \longrightarrow F(\bar{q}_1, \ldots, \bar{q}_f) \,, \tag{253}$$

where it is to be understood that the $\bar{q}_i$ are time-dependent. To see what this approximation would entail, expand the force operator about the expectation values of the coordinates:

$$F_i(q) = F_i(\bar{q}) + \sum_j (q_j - \bar{q}_j) \left.\frac{\partial F_i}{\partial q_j}\right|_{\bar{q}} + \frac{1}{2}\sum_{jk} (q_j - \bar{q}_j)(q_k - \bar{q}_k) \left.\frac{\partial^2 F_i}{\partial q_j \partial q_k}\right|_{\bar{q}} + \ldots \tag{254}$$

The linear term in the deviation from $\bar{q}_i$ drop out when the expectation value is taken, which then results in approximate equations for the expectation values of the momenta:

$$\frac{d\bar{p}_i}{dt} \simeq F_i(\bar{q}) + \frac{1}{2}\sum_{jk} \Delta_{jk}(t) \left.\frac{\partial^2 F_i}{\partial q_j \partial q_k}\right|_{\bar{q}} , \tag{255}$$

where

$$\Delta_{ij}(t) = \langle q_i(t) q_j(t) \rangle - \bar{q}_i(t)\bar{q}_j(t) . \tag{256}$$

Once again, we see that *there are no corrections unless the forces are nonlinear.*

The size of the correction term in (256) depends both on the state and the forces. No generally valid statement about its magnitude can be made. This can already be seen in the simplest case, that of a free particle. In this case (see Prob. 8)

$$\langle q_i^2(t) \rangle - \langle q_i^2(0) \rangle = \frac{t}{m_i} \langle q_i(0) p_i + p_i q_i(0) \rangle + \frac{t^2}{m_i^2} \langle p_i^2 \rangle , \tag{257}$$

where there is *no sum* over the index i, and the coordinate system is chosen so as to make $\bar{p}_i = \bar{q}_i(0) = 0$. Hence for large values of t,

$$\Delta q_i(t) \longrightarrow t\,\frac{\Delta p_i}{m_i} \gtrsim t\,\frac{\hbar}{m_i \Delta q_i(0)} , \tag{258}$$

where the latter is based on the uncertainty principle. The rate at which the quantum mechanical spreading grows is therefore inversely proportional to both the mass and the initial spread. It should be noted, however, that in this force-free case, the spreading is just that of a set of classical free particle trajectories with initial spreads $\Delta q_i(0)$, $\Delta p_i(0)$. In this case, the appearance of $\hbar$ in (258) is due entirely to the requirement that the initial spreads satisfy the uncertainty principle.

(d) Time-Energy Uncertainty

Because it is not possible to define a sensible operator that has time as its spectrum, the time-energy uncertainty relationship is not expressible as one unambiguous theorem to which all must agree, as is the case for uncertainty relationships between the canonical variables and other pairs of dynamical observables.

Nevertheless, given a dynamical system it is usually possible to define one or more operators that play the role of a clock. A vivid and instructive example is the motion of a particle of charge e and mass m in a uniform magnetic field $\boldsymbol{B}$, a problem that will be solved in detail in §4.3. The classical Hamiltonian for motion in the $x-y$ plane perpendicular to the field is

$$H = \tfrac{1}{2}(\dot{x}^2 + \dot{y}^2) . \tag{259}$$

This is a quadratic form, and the detailed quantum mechanical solutions are not needed for our purpose: by the theorem stated following Eq. 247, the mean values of the dynamical variables in any wave packet move classically, i.e., with constant angular velocity ω_c on circles, where $\omega_c = eB/mc$ is the cyclotron frequency. This system will therefore toll time with an accuracy ΔT proportional to the sharpness of the clock hand, which is the angular width $\Delta\theta$ of the packet. The natural definition is $\Delta T = \Delta\theta/\omega_c$, so that ΔT equals the period when $\Delta\theta = 2\pi$.

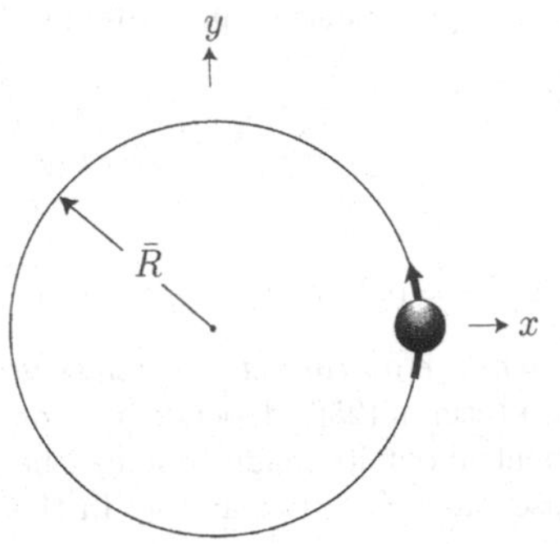

FIG. 2.2. A quantum clock formed by a charged particle wave packet moving in a homogeneous magnetic field perpendicular to the plane of the page.

A definition of $\Delta\theta$ in terms of the dynamical variables is needed which does not suffer from the singularities that afflict trigonometry, such as $\theta = \arctan y/x$. For that reason, we choose a coordinate system whose origin is at the center of the circle, and such that at the instant t of interest the packet is then passing across the x-axis (see Fig. 2.2). Angular positions within a packet are at that instant given by $y/\bar{R}$, where $\bar{R}$ is the r.m.s. radius of the circular orbit. We therefore define a "time" operator T, whose dispersion will be a measure of the clock's precision, by

$$T = y/\bar{R}\omega_c \,. \tag{260}$$

The general uncertainty relation for non-commuting observables (Eq. 203) when applied to such a non-stationary state, then gives

$$\Delta T\,\Delta E \geq \frac{1}{2\bar{R}\omega_c}\,|\langle[y,H]\rangle| \geq \frac{\hbar}{2\bar{R}\omega_c}\,\langle\dot{y}\rangle \tag{261}$$

because of (245). Continuing to take advantage of the classical solution, we have $\langle\dot{y}\rangle \simeq \bar{R}\omega_c$, where we have refrained from an equal sign because we are glossing over any difference between the mean and the r.m.s. radius. With this caveat, we have

$$\Delta T\,\Delta E \gtrsim \tfrac{1}{2}\hbar \,. \tag{262}$$

This is the time-energy uncertainty relation for this particular system with this reasonable, but neither unique nor perfectly precise, definition of the time uncertainty.

An arbitrary wave packet will change in angular size, and so the right-hand side of (261) is, in general, a function of time, a fact that is not reflected in the rough inequality (262). As we shall learn in §4.3(c), in the case of motion in a uniform magnetic field, there is a special set of packets, called coherent states, which do not change in shape, for which the uncertainty product is therefore constant, and which, furthermore, have the minimal time-energy uncertainty product $\sim \hbar$. This too is a special property of quadratic Hamiltonians.

To generalize from this example, consider an observable $Y(t)$ built from the dynamical variables of the system, which is to serve as the pointer of a dynamical "clock." Then, as in (261),

$$\Delta Y(t)\,\Delta E \geq \tfrac{1}{2}\,|\langle[Y(t),H]\rangle| \geq \tfrac{1}{2}\hbar\,\langle\dot{Y}(t)\rangle\,. \tag{263}$$

Now define

$$\Delta T(t) \equiv \frac{\Delta Y(t)}{\langle\dot{Y}(t)\rangle}\,. \tag{264}$$

Then

$$\Delta T(t)\,\Delta E \geq \tfrac{1}{2}\,\hbar\,. \tag{265}$$

This is a quite general time-energy uncertainty relationship; the quantity $\Delta T(t)$ is a time characteristic both of the system and the state in question, and ΔE is the spread in energy of the stationary states that are superposed to form the state. We have emphasized that, in general, both the numerator and denominator in Eq. 264 are functions of t, so ΔT is only a constant under special conditions. On the other hand, ΔE is time-independent if H is.

Instead of using some observable of the system to act as a clock, the issue can also be phrased as follows. We start at $t = 0$ with a system in a *non-stationary* state $|\Phi\rangle$, and ask for the probability that the evolving system is still in $|\Phi\rangle$ at a later time t:

$$P(t) = |\langle\Phi|e^{-iHt/\hbar}|\Phi\rangle|^2 = \left|\int_0^\infty dE\; e^{-iEt/\hbar}\,w_\Phi(E)\right|^2, \tag{266}$$

where

$$w_\Phi(E) = \sum_a |\langle Ea|\Phi\rangle|^2\,, \tag{267}$$

a being the eigenvalues of the observables other than H. There are various ways in which one can characterize the time dependence of $P(t)$, and for general energy distributions w_Φ the analysis involved is a subtle matter.[1] In analogy with (264), we can define the function

$$\tau(t) = \frac{P(t)}{dP(t)/dt}\,. \tag{268}$$

It can then be shown that

$$\langle\tau\rangle\,\Delta E \gtrsim \gamma\,\hbar\,, \tag{269}$$

where $\langle\tau\rangle$ is some useful time average of (268), $(\Delta E)^2 = \langle(H - \langle H\rangle)^2\rangle$, and γ is a number of order 1 that depends somewhat on w_Φ.

Clearly, there is only one circumstance under which the definition (268) gives a function $\tau(t)$ that is time-independent: $P(t) = e^{-t/\tau}$, *the exponential decay law*, with τ called the *lifetime* (not the half life $\tau_{1/2} = \tau \ln 2$). This form of $P(t)$ applies

[1] P. Pfeifer and J. Fröhlich, *Rev. Mod. Phys.* **67**, 759 (1995).

to very high accuracy in many phenomena, and we shall study some examples later in this volume. The spectral density $w_\Phi(E)$ that leads from (266) to the exponential decay law is

$$w_\Phi(E) = \frac{1}{\pi} \frac{\frac{1}{2}\Gamma}{(E - E_0)^2 + \frac{1}{4}\Gamma^2} \,. \tag{270}$$

When $\Gamma \ll E_0$, the integral in (266) is readily evaluated (Prob. 7), and gives

$$P(t) = e^{-\Gamma t/\hbar} \,, \tag{271}$$

or

$$\tau = \hbar/\Gamma \,. \tag{272}$$

Γ is a measure of the spread in energies of the distribution (270), and for that reason is called the *width* of the decaying state. Thus we see that the time-energy uncertainty relation is obeyed by the exponential decay law.

It should be noted that the exponential law is not a rigorous consequence of the distribution (270), and it is generally true that for times very short and very long compared to the lifetime there are departures from exponential decay. However, in the very important phenomena where the lifetime is far longer than the characteristic periods of the decaying system, such as beta decay or the radiative decay of excited states, the corrections to exponential decay are usually far too small to be observed.

(e) The Interaction Picture

Because the evolution of a system can be described either with stationary observables and moving state vectors, or the other way around, it is clear that descriptions in which both move are also available. Consider the case of a system whose Hamiltonian has the form $H_0 + V$, where both operators are time-independent, and H_0 is simple enough so that the time evolution it alone generates is known, but not that of the full Hamiltonian. An important example is scattering, with H_0 being the Hamiltonian of the free projectile and target, and V the interaction between them, without which there is no scattering. In this case the initial state, well before the collision, is a solution of the Schrödinger equation governed by H_0, but once the two objects come within range the evolution of the state is governed by $H_0 + V$. It can then be useful to remove the "trivial" time dependence due to H_0. The description that does this is called the *interaction picture*.

Let $|\psi; t\rangle_S$ be a state in the Schrödinger picture. Its counterpart in the interaction picture is then

$$|\psi; t\rangle_I = e^{iH_0 t/\hbar} |\psi; t\rangle_S \,; \tag{273}$$

were there no interaction V, the left-hand side would be time-independent, i.e., in the Heisenberg picture. To obtain the equation of motion for $|\psi; t\rangle_I$, one substitutes $|\psi; t\rangle_S$ from (273) into the Schrödinger equation, and multiplies from the left by $e^{iH_0 t/\hbar}$:

$$e^{iH_0 t/\hbar} (i\hbar\partial_t - H_0 - V)\, e^{-iH_0 t/\hbar} \, |\psi; t\rangle_I = 0 \,. \tag{274}$$

This then yields the sought-for equation:

$$\left(i\hbar \frac{\partial}{\partial t} - V_I(t) \right) |\psi; t\rangle_I = 0 \,, \tag{275}$$

where

$$V_I(t) = e^{iH_0t/\hbar}\, V\, e^{-iH_0t/\hbar} \tag{276}$$

is V in the interaction picture.

We have already learned how, at least in principle, to find the time evolution when the Hamiltonian is time-dependent: namely, by solving Eq. 242. That is to say, in the interaction picture the movement of states is given by

$$|\psi;t\rangle_I = U_I(t,t')|\psi;t'\rangle_I\ , \tag{277}$$

where U_I is the solution of

$$\left(i\hbar\frac{\partial}{\partial t} - V_I(t)\right) U_I(t,t') = 0\ , \tag{278}$$

with the initial condition $U_I(t,t) = 1$. This equation can only be solved directly if $[V_I(t), V_I(t')] = 0$ when $t \neq t'$. In this case, $V_I(t)$ is in effect a c-number, and

$$U_I(t,t') = \exp\left(-\frac{i}{\hbar}\int_{t'}^{t} d\tau\ V_I(\tau)\right)\ . \tag{279}$$

A complete solution is also known if the commutator $[V_I(t),\ V_I(t')]$ is a c-number.[1] In all but such exceptional cases, indeed, in almost all situations of real interest, no closed form solution is known. On the other hand, the solution in powers of V has been of enormous value. It is most easily derived form the integral equation that contains both the differential equation (278) and the initial condition:

$$U_I(t,t') = 1 - \frac{i}{\hbar}\int_{t'}^{t} d\tau\ V_I(\tau)\, U_I(\tau,t')\ . \tag{280}$$

By iteration this gives the Dyson expansion for the time evolution in the interaction picture:

$$U_I(t,t') = 1 - \frac{i}{\hbar}\int_{t'}^{t} d\tau\ V_I(\tau) + \left(\frac{i}{\hbar}\right)^2 \int_{t'}^{t} d\tau \int_{t'}^{\tau} d\tau'\, V_I(\tau)\, V_I(\tau') + \cdots\ . \tag{281}$$

2.5 Symmetries and Conservation Laws

Symmetries and conservation laws have a familiar relationship in classical physics. For a system whose Hamiltonian is invariant under both spatial translations and rotations, the linear and angular momenta are all constants of motion. This connection holds also in quantum mechanics, as will be shown shortly, though it has a somewhat different meaning because these six constants of motion do not all commute with each other. The other symmetry that plays a major role in classical physics concerns the relationship between inertial frames in relative motion, implemented by Galileo or Lorentz transformations, which are taken up in §7.3 and 13.2.

[1] J. Schwinger, *Phys. Rev.* **75**, 651 (1949).

The symmetries just mentioned are continuous. The discrete symmetries of space reflection and time reversal are also important in classical mechanics and electrodynamics, but in quantum mechanics their implications are stronger. Finally, while it is a common thing in classical mechanics to consider systems of particles that have identical masses and other attributes, their indistinguishability has no particular consequence in the classical framework, whereas in quantum mechanics indistinguishability has enormous ramifications, as we will learn in chapters 6 and 12.

(a) Symmetries and Unitary Transformations

It is important to have a clear understanding of what is, and is not, meant by a symmetry. Let F and F' be two inertial frames related by one or more of the following: a translation in space or time, a spatial rotation, a uniform relative motion, a reflection in space, a reversal in time. Consider a system S prepared in arbitrary states $|\Psi_\alpha\rangle, |\Psi_\beta\rangle, \ldots$ to certain specifications by observables attached to F, and the states $|\Psi'_\alpha\rangle, |\Psi'_\beta\rangle, \ldots$ of S satisfying precisely the same specifications by observables attached to F'. On the assumption that these frames are equivalent, all the probabilities relating the states prepared in F must be equal to the corresponding relations in F':

$$|\langle\Psi_\alpha|\Psi_\beta\rangle|^2 = |\langle\Psi'_\alpha|\Psi'_\beta\rangle|^2 . \qquad (282)$$

Were this not so, the probabilities could be used to distinguish between F and F'.

To carry this home, think of (282) as it applies to a one particle system, a state $|\Psi_{\boldsymbol{p}}\rangle$ of momentum $\boldsymbol{p}$ along the x-axis of F, another state $|\Psi_{lm}\rangle$ of total angular momentum l and projection m along the z-axis of F, the probability $|\langle\Psi_{\boldsymbol{p}}|\Psi_{lm}\rangle|^2$, and of the states having exactly these properties but with respect to a frame F' which is obtained by a rotation about the y-axis of F.

Note that according to the definitions used in (282), the condition of S described by the state $|\Psi'_\alpha\rangle$ differs from the condition described by $|\Psi_\alpha\rangle$ in that in the former S has been *moved* in some manner with respect to the latter as seen from the frame F; this is called the *active description.* There is another, equivalent, way of describing the situation: leaving the object fixed and defining $|\Psi'_\alpha\rangle$ to be state of this fixed object as observed from F'; this is called the passive description. *In this book we always use the active description.*

We stress that (282) is not a statement about interactions or dynamics — only about the properties of space and time. Eq. 282 must hold whatever the forces internal to or acting on the system S may be.

The probabilities in (282) will be equal if the corresponding probability amplitudes are equal apart from a phase factor,[1]

$$\langle\Psi_\alpha|\Psi_\beta\rangle = e^{i\lambda}\langle\Psi'_\alpha|\Psi'_\beta\rangle . \qquad (283)$$

The indifference of the physical consequences to the phase factor is a restatement of the fact that quantum mechanics lives, so to say, in a ray space and not a Hilbert space. If the physically relevant space were a Hilbert space, and an arbitrary phase were inadmissible, then demanding that (282) hold for all Hilbert space vectors would require the kets $|\Psi'_\alpha\rangle$ to be related to the kets $|\Psi_\alpha\rangle$ by a *unitary* transfor-

[1] The case of time reversal is more complicated; see §7.2.

mation, for it is unitary transformations that leave scalar products invariant in complex vector spaces.

One might well fear that the admissibility of the phase factor would greatly complicate matters, but that is not so. The previously mentioned theorem due to Wigner states that there is just one physically important symmetry, time reversal, that cannot be implemented with a unitary transformation. Until chapter 7 we will, therefore, always resort directly to unitary transformations without mentioning these complications.

(b) Spatial Translations

Spatial translations were discussed in §2.3(a), but they merit a closer look because, while being especially simple, they introduce concepts that are important in more complicated symmetries.

Recall, first, that the unitary operator for a spatial translation $\boldsymbol{a}$ is

$$T(\boldsymbol{a}) = e^{-i\boldsymbol{P}\cdot\boldsymbol{a}/\hbar} , \tag{284}$$

where $\boldsymbol{P}$ is the total momentum operator for the system in question, and $\boldsymbol{a}$ is a numerical 3-vector. From geometry we know that spatial translations along different directions commute with each other, and therefore we demand that

$$[P_i, P_j] = 0 . \tag{285}$$

Let $\boldsymbol{x}_n$ be the coordinate operator of particle n. By its very definition, a coordinate operator must have the following behavior under translation:

$$T^\dagger(\boldsymbol{a})\boldsymbol{x}_n T(\boldsymbol{a}) = \boldsymbol{x}_n + \boldsymbol{a} . \tag{286}$$

If $|\psi\rangle$ is any state, then

$$T(\boldsymbol{a})|\psi\rangle = |\psi;\, \boldsymbol{a}\rangle \tag{287}$$

is that state displaced through the distance $\boldsymbol{a}$. (In this connection, recall Eq. 167 and Eq. 168.)

The eigenvalues of $\boldsymbol{x}_n$ will, as before, be called $\boldsymbol{r}_n$. Then the coordinate eigenkets are transformed as follows:

$$T(\boldsymbol{a})|\boldsymbol{r}_1,\ldots\rangle = |\boldsymbol{r}_n + \boldsymbol{a},\ldots\rangle , \quad T^\dagger(\boldsymbol{a})|\boldsymbol{r}_1,\ldots\rangle = |\boldsymbol{r}_n - \boldsymbol{a},\ldots\rangle . \tag{288}$$

The original and spatially transformed wave function are, by definition,

$$\psi(\boldsymbol{r}_1,\ldots) = \langle \boldsymbol{r}_1,\ldots|\psi\rangle , \qquad \psi'(\boldsymbol{r}_1,\ldots) = \langle \boldsymbol{r}_1,\ldots|T(\boldsymbol{a})|\psi\rangle \tag{289}$$

Hence

$$\psi'(\boldsymbol{r}_1\ldots) = \psi(\boldsymbol{r}_1 - \boldsymbol{a},\ldots) . \tag{290}$$

Note carefully that the coordinates on the right-hand side of this relationship have been translated in the direction opposite to that of the state $|\psi\rangle$. Furthermore, from $T^\dagger T = 1$ it follows immediately that

$$\psi'(\boldsymbol{r}_1 + \boldsymbol{a}\ldots) = \psi(\boldsymbol{r}_1\ldots) . \tag{291}$$

Equations (291) and (290) are special cases of general results that holds for all unitary transformation. Namely, if P' is the image of the point P under any transformation τ, i.e., if $P' = \tau P$, and $|\psi\rangle$ and $|\psi'\rangle$ any state and its image under τ produced by the unitary operator $U(\tau)$, then the transformation laws for wave functions are

$$\psi'(P') = \psi(P) , \tag{292}$$

and

$$\psi'(P) = \psi(\tau^{-1}P) . \tag{293}$$

The geometrical explanation for this important and generic fact is illustrated below for the case of rotations. Now let $F(\boldsymbol{x}_n)$ be any observable constructed from coordinates. Then

$$T^{\dagger}(\boldsymbol{a})F(\boldsymbol{x}_n)T(\boldsymbol{a}) = F(\boldsymbol{x}_n + \boldsymbol{a}) . \tag{294}$$

In particular, for an infinitesimal translation,

$$F(\boldsymbol{x}_n + \delta\boldsymbol{a}) = F(\boldsymbol{x}_n) + \frac{i}{\hbar}\sum_i \delta a_i [P_i, F(\boldsymbol{x}_n)] . \tag{295}$$

Hence if a function of the coordinates is invariant under a translation along the ith direction, it commutes with that component of the total momentum.

The connection between the momentum conservation law and invariance under spatial translation now follows in close analogy to that of classical mechanics. Namely, *if the Hamiltonian is invariant under spatial translations, the total momentum commutes with the Hamiltonian and is therefore a constant of motion.* The last part of this statement is a special case of the general theorem that observables that commute with H are constants of motion. Equation (294) also tells us that if H is only invariant under translations along one direction (or in one plane), only the momentum along that direction (or in that plane) is a constant of motion.

(c) Symmetry Groups

By considering a sequence of translations, one is led quite naturally to the concept of a group of symmetries. Take the translation through $\boldsymbol{a}$ followed by $\boldsymbol{b}$:

$$T(\boldsymbol{b})T(\boldsymbol{a}) = T(\boldsymbol{a} + \boldsymbol{b}) . \tag{296}$$

Because the argument of the composite translation is the sum of $\boldsymbol{a}$ and $\boldsymbol{b}$, the order in which these translations are carried out does not matter; they commute. Of course, we already knew that because the various components of $\boldsymbol{P}$ commute with each other, but this tells us that $[P_i, P_j] = 0$ is a requirement set by Euclidean geometry itself, not dynamics. The special case $\boldsymbol{b} = -\boldsymbol{a}$, which just undoes the first translation, is also of interest:

$$T(\boldsymbol{a})T(-\boldsymbol{a}) = T(\boldsymbol{a})T^{\dagger}(\boldsymbol{a}) = T(0) = 1 , \tag{297}$$

where 1 expresses the fact that no change is produced.

The relations (296) and (297) show that *the operators $T(\boldsymbol{a})$ form an Abelian Lie group of unitary operators standing in one-to-one correspondence with the group of translation in the Euclidean 3-space $\mathfrak{E}_3$.*

This wallop of jargon will now be explained because it plays so large a role in the discussion of symmetries in quantum mechanics.

- A group $\mathfrak{G}$ is a finite or infinite set of elements $(g_1, g_2, \ldots)$ having a composition law for every pair of elements such that g_1g_2 is again an element of $\mathfrak{G}$; which is associative, i.e., $(g_1g_2)g_3 = g_1(g_2g_3)$; and with every element g_i having an inverse g_i^{-1} such that $g_ig_i^{-1}$ is the identity element I, i.e., $Ig_i = g_iI = g_i$ for all i.
- A group is Abelian if all its elements commute, i.e., $g_1g_2 = g_2g_1$.
- A group with an infinite set of elements is a Lie group if its elements can be uniquely specified by a set of *continuous* parameters $(z_1 \ldots z_r)$.

Two distinct but closely related groups are in play in the case of translations: both are Abelian Lie groups, parametrized by the components of a 3-vector $\boldsymbol{a}$. One is the group of translations in every-day Euclidean 3-space, $\mathfrak{E}_3$; it is Abelian because translations along different directions commute. The second is a group of unitary operators in Hilbert space; it is Abelian because the components of the total momentum $\boldsymbol{P}$ operator commute with each other. Their one-to-one correspondence is implemented by giving the elements of both groups the same parametrization and the same composition laws.

It would, therefore, be natural to say that the unitary operators T form a representation of the translation group in Hilbert space. But the term "representation" is, by tradition, reserved for matrices whose multiplication law stands in one-to-one correspondence with the algebra of the group. That is, if $\{|\xi\rangle\}$ is any basis in $\mathfrak{H}$, the matrices with elements $\langle\xi|T|\xi'\rangle$ form a representation of this group in that

$$\sum_{\xi''} \langle\xi|T(\boldsymbol{a})|\xi''\rangle\langle\xi''|T(\boldsymbol{b})|\xi'\rangle = \langle\xi|T(\boldsymbol{a}+\boldsymbol{b})|\xi'\rangle \,, \tag{298}$$

which is true by virtue of (296).

In equating sequences of transformation to products of unitary transformations, as we have done starting with (296), we have ignored the fact that multiplying the product by an arbitrary phase would not have any physical consequence because physical states are rays and not vectors. That is to say, if τ_i stands for the parameters specifying some particular transformation (such as $\boldsymbol{a}$ in the case of a translation), $\tau_2\tau_1$ for the parameters that specify the indicated sequence of transformations, and $U(\tau_i)$ the corresponding unitary operators, then in contrast to (296) or (298), we are free to write

$$U(\tau_2)U(\tau_1) = e^{i\lambda}\, U(\tau_2\tau_1) \,. \tag{299}$$

Once again, one could worry that ignoring the phase factor is illegitimate. However, it is only for Galileo transformations that this issue matters, as we shall see in §7.3; in all other cases the phases of the states can be chosen so that there is no phase factor in the product of unitary transformations.

Next, consider the generalization of the infinitesimal translation $\delta\boldsymbol{a}$,

$$T = 1 - i\delta\boldsymbol{a}\cdot\boldsymbol{P}/\hbar \,. \tag{300}$$

Quite generally, if a unitary operator $U(z_1 \ldots z_r)$ carries out a transformation belonging to a Lie group, then if the transformation is infinitesimal it has the form

$$U = 1 - i\sum_{l=1}^{r} \delta z_l\, G_l \,, \tag{301}$$

where the operators $\mathcal{G}_l$, which must be Hermitian for U to be unitary, are called the *generators* of the group $\mathfrak{G}$. In the case of the spatial translation group, the generators are the three components of the total momentum $\boldsymbol{P}$ (for convenience divided by $\hbar$ to give them the dimension of momentum). The algebra obeyed by the generators as defined by their commutators is called the *Lie algebra* of $\mathfrak{G}$. It is called Abelian if the generators commute, as they do for the translation group.

Because it is just Euclidean geometry that defines the properties of the translation group, the Lie algebra associated with this group cannot depend on quantum mechanics for its construction. To see this, let $f(x_1x_2x_3)$ be any function of the coordinates in $\mathfrak{E}_3$, taken now to be real numbers and not operators, and consider the infinitesimal translation $x_i \to x_i + \delta a_i$. Then

$$\delta f = f(x_i + \delta a_i) - f(x_i) = \sum_i \delta a_i \frac{\partial f}{\partial x_i}\,; \tag{302}$$

in this formulation the differential operators $\partial/\partial x_i$ are the commuting infinitesimal generators. Putting this into the form

$$\delta f = \frac{i}{\hbar}\sum_i \delta a_i \frac{\hbar}{i}\frac{\partial f}{x_i}\,, \tag{303}$$

shows how the momentum in the Schrödinger representation is related to translation in space.

Equation (302) is actually (295) in another guise, which can be seen by recalling Eq. 162 rewritten in the notation being used here:

$$[\boldsymbol{p}_n, F(\boldsymbol{x}_1 \ldots \boldsymbol{x}_N)] = \frac{\hbar}{i}\frac{\partial F}{\partial \boldsymbol{x}_n}\,, \qquad [\boldsymbol{x}_n, G(\boldsymbol{p}_1 \cdots \boldsymbol{p}_N)] = i\hbar\frac{\partial G}{\partial \boldsymbol{p}_n}\,; \tag{304}$$

thus

$$\delta F = \sum_n \delta \boldsymbol{a}\cdot\frac{\partial F}{\partial \boldsymbol{x}_n} = \frac{i}{\hbar}\sum_n [\delta\boldsymbol{a}\cdot\boldsymbol{p}_n, F]\,, \tag{305}$$

which is (295).

If the Lie algebra is non-Abelian, it is not straightforward to construct the unitary transformations for finite values of all the parameters. However, for the *subgroup* defined by setting all but one parameter, which we call z, to zero, it is straightforward to construct a finite transformation because only one generator, $\mathcal{G}$, is involved, so that it behaves like a number as nothing with which it does not commute is involved. Then the argument that led to the time translation operator (Eq. 213) carries through as it stands, and so for this subgroup

$$U(z) = e^{-iz\mathcal{G}}\,. \tag{306}$$

(d) Rotations

Many of the systems of interest in quantum mechanics are either exactly or approximately invariant under rotations. However, rotations about different axes do not commute, and therefore the rotation group is non-Abelian, which makes its analysis quite complicated. For these reasons a significant portion of this volume will be devoted to developing this analysis and applying it to various physical problems.

The parametrization of rotations in $\mathfrak{E}_3$ can be done in a variety of ways. Here we will specify a rotation R by the unit vector $\boldsymbol{n}$ along an axis of rotation, and an angle of rotation θ about that axis, with the convention that the rotation is counterclockwise when looking along $\boldsymbol{n}$ towards the origin, and $0 \leq \theta \leq 2\pi$.

As with translations, a great deal can be learned from infinitesimal transformations. An infinitesimal rotation will be parametrized by

$$\delta\boldsymbol{\omega} = \boldsymbol{n}\,\delta\theta\,. \tag{307}$$

Under this rotation, a vector $\boldsymbol{K}$ in $\mathfrak{E}_3$ transforms as follows:

$$\boldsymbol{K} \to \boldsymbol{K} + \delta\boldsymbol{\omega}\times\boldsymbol{K} \equiv \boldsymbol{K} + \delta\boldsymbol{K}\,, \tag{308}$$

or in terms of Cartesian components,

$$\delta K_i = \epsilon_{ijk}\delta\omega_j K_k\,, \tag{309}$$

where ϵ_{ijk} is the totally antisymmetric Levi-Civita tensor, with $\epsilon_{123} = 1$.

If $\boldsymbol{K}$ is written as a column 3-vector, rotations through any angle can be carried out with the help of the following 3×3 matrices:

$$I_1 = \begin{pmatrix} 0 & 0 & 0 \\ 0 & 0 & -i \\ 0 & i & 0 \end{pmatrix}, \quad I_2 = \begin{pmatrix} 0 & 0 & i \\ 0 & 0 & 0 \\ -i & 0 & 0 \end{pmatrix}, \quad I_3 = \begin{pmatrix} 0 & -i & 0 \\ i & 0 & 0 \\ 0 & 0 & 0 \end{pmatrix}. \tag{310}$$

A finite rotation of $\boldsymbol{K}$ about a single axis, say, about the axis 1 through the angle ϕ_1, is accomplished by

$$\boldsymbol{K} \to \boldsymbol{K}' = e^{-i\phi_1 I_1}\,\boldsymbol{K}\,. \tag{311}$$

Successive rotations of $\boldsymbol{K}$ about distinct axes do not commute, a fact that is captured in the commutation rule

$$[I_i, I_j] = i\epsilon_{ijk}\, I_k\,. \tag{312}$$

Quantum mechanics enters by assigning, to every rotation R in $\mathfrak{E}_3$, a unitary transformation $D(R)$ on the Hilbert space $\mathfrak{H}$ of the system of interest. To be concrete, one can think of $\boldsymbol{K}$ as a vector in everyday $\mathfrak{E}_3$ that identifies some point in an apparatus that is used to prepare or measure states of the system. Consistency requires the correspondence between rotations in $\mathfrak{E}_3$ and the unitary transformations in $\mathfrak{H}$ to be maintained when two (or more) rotations are carried out in succession. Let R_2R_1 be the rotation in $\mathfrak{E}_3$ that results from carrying out R_1 followed by R_2. The consistency requirement is then

$$D(R_2R_1) = D(R_2)\,D(R_1)\,. \tag{313}$$

As rotations in $\mathfrak{E}_3$ do not commute, it is clear that in general

$$D(R_1)D(R_2) \neq D(R_2)D(R_1)\,. \tag{314}$$

The exception is the set of successive rotations about one, and only one axis $\boldsymbol{n}$. The argument that led to the general result (306) implies that such rotations can be written in the form

$$D(R) = \exp(-i\theta\boldsymbol{n}\cdot\boldsymbol{J})\,, \tag{315}$$

where the Hermitian generator $\boldsymbol{n}\cdot\boldsymbol{J}$ is, by definition, *the component of angular momentum along the direction* $\boldsymbol{n}$.

The implications of the requirement (313) emerge from considering two successive infinitesimal rotations about two distinct axes, say, through $(\delta\phi_1, \delta\phi_2)$ about the axes 1 and 2, in both possible orders:

$$\boldsymbol{K}' = (1 - i\delta\phi_2 I_2 + \ldots)(1 - i\delta\phi_1 I_1 + \ldots)\,\boldsymbol{K}\,, \tag{316}$$

$$\boldsymbol{K}'' = (1 - i\delta\phi_1 I_1 + \ldots)(1 - i\delta\phi_2 I_2 + \ldots)\,\boldsymbol{K}\,. \tag{317}$$

These two rotated vectors are not identical:[1]

$$\boldsymbol{K}'' - \boldsymbol{K}' = \{-\delta\phi_1\delta\phi_2(I_1 I_2 - I_2 I_1) + \ldots\}\boldsymbol{K} = \{-i\delta\phi_1\delta\phi_2\, I_3 + O(\delta\phi^3)\}\boldsymbol{K}\,. \tag{318}$$

The correspondence to the unitary operators $D(R)$ must maintain this difference. In view of (315), they must thus satisfy

$$\lim_{\delta\phi_i\to 0}\left(e^{-i\delta\phi_1 J_1}\, e^{-i\delta\phi_2 J_2} - e^{-i\delta\phi_2 J_2}\, e^{-i\delta\phi_1 J_1}\right) = -i\delta\phi_1\delta\phi_2\, J_3 + O(\delta\phi^3)\,. \tag{319}$$

Hence $[J_1, J_2] = iJ_3$, and in general,

$$[J_i, J_j] = i\epsilon_{ijk}\, J_k\,. \tag{320}$$

This is *the angular momentum commutation rule.*

This derivation of the angular momentum commutation rule relied on no assumption concerning the dynamical variables of some system.[2] It therefore applies to all systems, whether they be composed of structureless particles with no internal degrees of freedom; or particles that have an intrinsic angular momentum, or spin, which is not a function of canonical coordinates and momenta; or the quantized electromagnetic field. Whatever the system may be, we have been speaking of rotating the system as a whole, and $\boldsymbol{J}$ is therefore the system's total angular momentum. (One can, of course, imagine rotations of some component of a system, and the same geometrical argument will then lead to the same commutation rule for the angular momentum of that component.)

Let A be any observable. Under the rotation R it undergoes the unitary transformation

$$A \to D^\dagger(R)\, A\, D(R)\,. \tag{321}$$

If R is infinitesimal,

$$A \to A + \delta A\,, \qquad \delta A = i\delta\theta\,[\boldsymbol{n}\cdot\boldsymbol{J}, A]\,. \tag{322}$$

Hence if an observable is invariant under rotations about $\boldsymbol{n}$*, it commutes with the corresponding component of angular momentum.* An observable that is invariant under rotations about all directions is called a *scalar* under rotations.

---

[1] The generalization of this result to arbitrary infinitesimal rotations is the subject of Prob. 10.

[2] While (320) follows from (312) by $I_i \to J_i$, this is only a mnemonic; I_i is a 3×3 matrix acting on garden variety 3-vectors whereas J_i is an operator in a Hilbert space that is only specified when the system in question is identified. As we shall see in §3.3, the I_i are the angular momentum operators in the 3-dimensional (or $j = 1$) representation, though not in the conventional form in which the 3-component is diagonal.

The angular momentum conservation law now follows as it does for linear momentum: *If the Hamiltonian is invariant under rotations about an axis, the component of the total angular momentum along that axis is a constant of motion.*

All sorts of quantities that are not scalars play important roles in physics. A general classification of non-invariant observables will be given in §7.6; here we only treat the most important and simplest class — *vector operators.* By definition, a set of three observables $(V_1, V_2, V_3) = \boldsymbol{V}$ is called a vector operator if it transforms under rotations in the same way as does the c-number vector $\boldsymbol{K}$ in (308):

$$\delta \boldsymbol{V} = i\delta\theta\,[\boldsymbol{n}\cdot\boldsymbol{J}, \boldsymbol{V}] = \delta\theta\,\boldsymbol{n}\times\boldsymbol{V}\,. \tag{323}$$

In short, the definitions (308) and (323) are completely general — whether the vector is the line from this word to your nose, or the momentum operator of a neutron in some nucleus. Or put another way, anything that transforms in this way *is* a vector.

The coefficients of n_i in (323) must be equal, and $\boldsymbol{V}$ must therefore obey the commutation rule[1]

$$[V_i, J_j] = i\epsilon_{ijk}\,V_k\,. \tag{324}$$

Any set of three observables that satisfy this commutation rule with the angular momentum constitute a vector operator. In particular, the angular momentum $\boldsymbol{J}$ is itself a vector operator, as comparison of (324) to (320) shows, and common sense requires.

It is instructive to verify that the unitary transformation (321) does properly transform a vector operator under finite rotations. For this purpose it suffices to consider a rotation about any one axis. Define

$$V_i(\theta) = e^{i\theta J_3} V_i e^{-i\theta J_3} \tag{325}$$

Then

$$\frac{dV_i(\theta)}{d\theta} \equiv \dot{V}(\theta) = ie^{i\theta J_3}[J_3, V_i]e^{-i\theta J_3}\,, \tag{326}$$

so $\dot{V}_1 = -V_2, \dot{V}_2 = V_1, \dot{V}_3 = 0$. Therefore $\ddot{V}_i = -V_i$ for $i = 1, 2$, which after integration gives the proper answer:

$$e^{i\theta J_3} V_1 e^{-i\theta J_3} = V_1\cos\theta - V_2\sin\theta\,, \quad e^{i\theta J_3} V_2 e^{-i\theta J_3} = V_2\cos\theta + V_1\sin\theta\,. \tag{327}$$

The total momentum $\boldsymbol{P}$ of a system, being a vector, does not commute with the total angular momentum; only components of $\boldsymbol{P}$ and $\boldsymbol{J}$ along the same direction commute.

The scalar product of two vector operators is invariant under rotation, and must therefore commute with $\boldsymbol{J}$, as is easily confirmed by applying (324) to $\boldsymbol{V}_1\cdot\boldsymbol{V}_2$. In particular, therefore, J^2 and P^2 both commute with $\boldsymbol{J}$, and also each other.

While all components of $\boldsymbol{P}$ and $\boldsymbol{J}$ are constants of motion if the Hamiltonian is invariant under translations and rotations, they cannot all be diagonalized simultaneously. Hence it is not possible to construct simultaneous eigenstates of all these six constants of motion. In addition to the rotational scalars P^2, J^2 and $\boldsymbol{P}\cdot\boldsymbol{J}$, one component of angular momentum, traditionally defined to be J_3, can be diagonalized simultaneously, and states can be designated by the associated eigenvalues.

---

[1] In the older literature this commutation rule is often written as $\boldsymbol{V}\times\boldsymbol{J} = i\boldsymbol{V}$.

As an illustration of this totally general discussion, consider a single particle with position and momentum operators $\boldsymbol{x}$ and $\boldsymbol{p}$. The *orbital angular momentum operator* $\boldsymbol{L}$ for this particle is then defined as

$$\boldsymbol{L} = (\boldsymbol{x}\times\boldsymbol{p})/\hbar\,, \tag{328}$$

or

$$L_i = \epsilon_{ijk}x_j p_k/\hbar\,. \tag{329}$$

The factor $1/\hbar$ in the definition of $\boldsymbol{L}$, and the absence of $\hbar$ in (320), have the consequence that *all angular momenta are dimensionless*, a convention that will be adhered to throughout. It should also be mentioned that the order of x_j and p_k in (329) does not matter because only commuting factors appear in L_i.

The commutation rule for the orbital angular momentum now follow from (329) when the canonical commutation rule $[x_i, p_j] = i\delta_{ij}\hbar$ is used:

$$[L_i, L_j] = i\epsilon_{ijk}\, L_k\,. \tag{330}$$

That this is identical in form to the general rule (320) is simply a consequence of the requirement that the rotations in $\mathfrak{E}_3$ are to be translated into unitary operators in one and the same way no matter what the system may be. Furthermore, from the canonical commutation rules

$$[x_i, L_j] = i\epsilon_{ijk}\, x_k\,, \qquad [p_i, L_j] = i\epsilon_{ijk}\, p_k\,. \tag{331}$$

Hence $\boldsymbol{x}$ and $\boldsymbol{p}$ are vector operators, again as common sense requires.

The role of the orbital angular momentum operator in rotations can also be seen in a slightly different light as follows. Let $\psi(\boldsymbol{r})$ be some wave function, where $\boldsymbol{r} = (r_1, r_2, r_3)$ is the eigenvalue of $\boldsymbol{x}$. Under a infinitesimal rotation about $\boldsymbol{n} = (0,0,1)$, the change in ψ is

$$\begin{aligned} \delta\psi(\boldsymbol{r}) &= \psi(r_1 - r_2\delta\theta, r_2 + r_1\delta\theta, r_3) - \psi(r_1, r_2, r_3) \\ &= -\delta\theta\left(r_2\frac{\partial}{\partial r_1} - r_1\frac{\partial}{\partial r_2}\right)\psi(\boldsymbol{r}) \\ &= \delta\theta\frac{i}{\hbar}(x_1p_2 - x_2p_1)\psi(\boldsymbol{r}) = i\delta\theta\, L_3\,\psi(\boldsymbol{r})\,, \end{aligned} \tag{332}$$

where in the last expressions the p_i and L_i have become differential operators. This too is required by consistency, because under a rotation R, the state $|\psi\rangle$ and the eigenket $|\boldsymbol{r}\rangle$ of $\boldsymbol{x}$ undergo the transformations

$$|\psi\rangle \to D(R)|\psi\rangle \equiv |\psi'\rangle\,, \qquad |\boldsymbol{r}\rangle \to D(R)|\boldsymbol{r}\rangle = |R\boldsymbol{r}\rangle\,, \tag{333}$$

where $R\boldsymbol{r}$ is the rotated image of $\boldsymbol{r}$. The wave function therefore transforms as follows:

$$\langle\boldsymbol{r}|\psi\rangle \to \langle\boldsymbol{r}|D(R)|\psi\rangle = \langle R^{-1}\boldsymbol{r}|\psi\rangle\,, \tag{334}$$

and therefore

$$\psi'(\boldsymbol{r}) = \psi(R^{-1}\boldsymbol{r})\,. \tag{335}$$

This illustrates the general relationship Eq. 293.

When R is the infinitesimal rotation of (332), this last equation becomes

$$\psi(\boldsymbol{r}) \to \psi'(\boldsymbol{r}) = (1 + i\delta\theta L_3)\,\psi(\boldsymbol{r})\,, \tag{336}$$

where L_3 is again the differential representation of the operator. The sign difference between (336) and $D = 1 - i\delta\theta L_3$ results from the fact that when an object (here the state $|\psi\rangle$) is rotated through R, the point that is then on the object at $\boldsymbol{r}$ was originally located at $R^{-1}\boldsymbol{r}$ in the fixed coordinate frame (see Fig. 2.3). Note that this last statement, and our whole treatment of rotations, has been in the active mode.

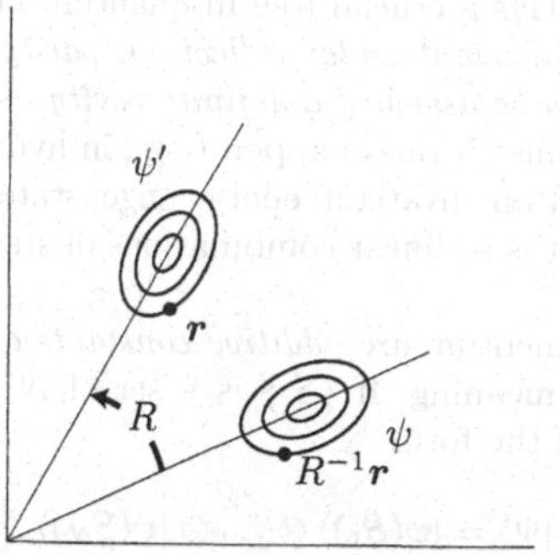

FIG. 2.3. The relationship between rotated states (Eq. 334) as an example of the general case (Eq. 293). The curves are contours of ψ = const.

(e) Space Reflection and Parity

Vectors that change sign under a reflection through the origin are called *polar vectors,* and those that do not change sign are called *axial vectors.* Coordinates, momenta and electric fields $\boldsymbol{E}$ are polar vectors, whereas angular momenta and magnetic fields $\boldsymbol{B}$ are axial vectors. A quantity like $\boldsymbol{E}\cdot\boldsymbol{B}$, which is invariant under rotations but changes sign under reflection, is called a *pseudoscalar.* It is sometimes convenient to reflect through a plane instead of the origin, but there is no fundamental distinction between these as they are related by a rotation of π about the normal to the reflection plane. Unless stated otherwise, the term "reflection" will mean through the origin.

Space reflection is implemented by a unitary operator I_s. By definition, it has the following effect on positions and momenta:[1]

$$I_s^\dagger \boldsymbol{x}_n I_s = -\boldsymbol{x}_n, \qquad I_s^\dagger \boldsymbol{p}_n I_s = -\boldsymbol{p}_n\,, \tag{337}$$

and therefore the opposite effect on angular momenta:

$$I_s^\dagger \boldsymbol{J} I_s = \boldsymbol{J}\,. \tag{338}$$

[1]The canonical commutation rule is, therefore, invariant under space reflection. Time reversal only changes the sign of $\boldsymbol{p}$, but not of $\boldsymbol{x}$, and therefore it would *not* leave the canonical commutation rule invariant if it were implemented by a unitary transformation. For this, and other related reasons, a non-unitary transformation is involved in time reversal, as will be discussed in §7.2.

Let $|\psi\rangle$ be some state. It may or may not be an eigenstate of I_s. However, if it is, and has eigenvalue i_s, i.e., if

$$I_s|\psi\rangle = i_s|\psi\rangle \ , \tag{339}$$

then because $(I_s)^2 = 1$, it follows that

$$i_s = \pm 1 \ . \tag{340}$$

This eigenvalue (or quantum number) is called the *parity*. It has no counterpart in classical mechanics, but plays a crucial role in quantum mechanics.

If the Hamiltonian is invariant under reflection, parity is a constant of motion and energy eigenstates can be assigned a definite parity — even or odd. The words "can be" appear here because it does happen (e.g., in hydrogen!) that even though the Hamiltonian is reflection invariant energy eigenstates of different parity are degenerate, and when that is so linear combinations of states of differing parity are also energy eigenstates.

Linear and angular momentum are *additive constants of motion.* This statement has the following simple meaning: If $\{S_i\}$ is a set of N non-interacting systems, then this set has states of the form

$$|\Psi\rangle = |\psi(S_1)\rangle \otimes \ldots \otimes |\psi(S_N)\rangle \, , \tag{341}$$

and if $|\psi(S_i)\rangle$ are eigenstates of, say, momentum with eigenvalue $\boldsymbol{P}_i$, then $|\Psi\rangle$ is an eigenstate of momentum with an eigenvalue that is the sum of those of the constituents. This statement holds also for any one component of angular momentum, and for the energy.

By contrast, *parity is a multiplicative quantum number.* For the state defined in (341), in an obvious notation

$$i_s = \prod_k i_s(S_k) \ . \tag{342}$$

(f) Gauge Invariance

Gauge invariance is a symmetry that arises in classical electrodynamics, but it has flowered into a theme of central importance in the generalizations of electrodynamics to the other fundamental interactions of physics.

In electrodynamics it is often advantageous to replace the electric and magnetic field strengths by the scalar and vector potentials $\phi(\boldsymbol{r},t)$ and $\boldsymbol{A}(\boldsymbol{r},t)$:

$$\boldsymbol{E}(\boldsymbol{r},t) = -\boldsymbol{\nabla}\phi(\boldsymbol{r},t) - \frac{1}{c}\frac{\partial}{\partial t}\boldsymbol{A}(\boldsymbol{r},t) \ , \tag{343}$$

$$\boldsymbol{B}(\boldsymbol{r},t) = \boldsymbol{\nabla} \times \boldsymbol{A}(\boldsymbol{r},t) \ . \tag{344}$$

The field strengths are left invariant under the gauge transformation

$$\boldsymbol{A}(\boldsymbol{r},t) \to \boldsymbol{A}'(\boldsymbol{r},t) = \boldsymbol{A}(\boldsymbol{r},t) + \boldsymbol{\nabla}\chi(\boldsymbol{r},t) \ , \tag{345}$$

$$\phi(\boldsymbol{r},t) \to \phi'(\boldsymbol{r},t) = \phi(\boldsymbol{r},t) - \frac{1}{c}\frac{\partial}{\partial t}\chi(\boldsymbol{r},t) \ , \tag{346}$$

where $\chi(\boldsymbol{r},t)$ is an arbitrary (smooth) function. In short, the field strengths have an invariant physical meaning, whereas potentials that differ only by gauge transformations represent the same physical situation in classical physics.

The classical Hamiltonian for a system of charged particles is turned into the operator H in Schrödinger's equation by the standard recipe

$$\boldsymbol{p}_n \to \frac{\hbar}{i}\frac{\partial}{\partial \boldsymbol{r}_n} , \tag{347}$$

so that

$$H = \sum_{n=1}^{N} \left\{ \frac{1}{2m_n} \left(\frac{\hbar}{i}\frac{\partial}{\partial \boldsymbol{r}_n} - \frac{e_n}{c}\boldsymbol{A}(\boldsymbol{r}_n,t) \right)^2 + e_n \phi(\boldsymbol{r}_n,t) \right\} + V , \tag{348}$$

where V is the electrostatic energy due to the interaction between these particles, and $\phi(\boldsymbol{r},t)$ the potential describing the applied electric field. To make Schrödinger's equation invariant under the gauge transformation on the potentials, the wave function must also change to compensate. Because of the spatial derivatives in H, and the time derivative in Schrödinger's equation, this is accomplished by changing the phase of the wave function as follows:

$$\psi(\boldsymbol{r}_1,\ldots,\boldsymbol{r}_N,t) \to \psi'(\boldsymbol{r}_1,\ldots,\boldsymbol{r}_N,t) = \prod_{n=1}^{N} \exp\left(\frac{ie_n}{\hbar c}\chi(\boldsymbol{r}_n,t) \right) \psi(\boldsymbol{r}_1,\ldots,\boldsymbol{r}_N,t) . \tag{349}$$

In quantum mechanics, therefore, the term *gauge transformation* refers to the combined transformations on ϕ, $\boldsymbol{A}$ and ψ. There is a fundamental difference between classical and quantum physics, however: in the quantum case, the potentials themselves can be of physical significance in situations where the electromagnetic field is in a topologically non-trivial space, such as the region outside a tube (see §4.3(d)).

The change of phase (349) is a unitary transformation in the Hilbert space, but the change of the potentials, which are not operators here, is not a unitary transformation. Nevertheless, the gauge transformation form an Abelian group: two successive transformations are again a gauge transformation, their order does not matter, and every transformation has an inverse.

The operator

$$\boldsymbol{v}_n = \frac{1}{m_n}\left(\frac{\hbar}{i}\frac{\partial}{\partial \boldsymbol{r}_n} - \frac{e_n}{c}\boldsymbol{A}(\boldsymbol{r}_n,t) \right) \tag{350}$$

represents the velocity of particle n; this is confirmed by computing the time derivative of the coordinate operator. For that reason it is to be expected that the probability current is not (227) when the system interacts with an electromagnetic field, but is rather

$$\boldsymbol{i}_n(\boldsymbol{r}_1,\ldots;t) = \frac{1}{2}\{\psi^* \boldsymbol{v}_n \psi + \psi(\boldsymbol{v}_n\psi)^*\} . \tag{351}$$

The probability density remains $|\psi|^2$. That (351) is correct is confirmed by showing that it satisfies the continuity equation (226) as a consequence of the Schrödinger equation when (348) is the Hamiltonian.

2.6 Propagators and Green's Functions

The two initial formulations of quantum mechanics, stemming from Heisenberg and Schrödinger, have now been described. The path integral, a later equivalent formulation due to Feynman, can be motivated by the insights revealed by the interference experiments discussed in chapter 1. For that reason, the path integral can be viewed as having a more intimate connection to the conceptual essentials of quantum mechanics. One can therefore take the position that quantum mechanics is defined by the path integral, which is how it was introduced originally by Feynman. We, however, will derive it from Schrödinger's equation in the following section. This section prepares the ground for this task, and in so doing it develops some powerful tools that are important in their own right.

The Schrödinger equation, like any linear partial differential equation, can be turned into an integral equation that states the boundary conditions up front. For the time-dependent equation, the kernel of the integral equation will be called the propagator, while for the time-independent equation it will be called Green's function. Readers should however note that many authors use the latter name for both. As is to be expected, these two kernels are each others Fourier transforms.

(a) Propagators

Schrödinger's differential equation is of first order in time, so an initial condition must be specified if it is to manufacture a definite solution. The initial condition can be made organic to the formulation if the differential equation is recast as an integral equation. For this purpose recall Eq. 207,

$$|\psi;t\rangle = U(t,t')|\psi;t'\rangle \,. \tag{352}$$

In terms of this unitary operator, the Schrödinger equation is

$$[i\hbar\partial_t - H(t)]U(t,t') = 0 \,, \tag{353}$$

which holds also for a time-dependent Hamiltonian. Let $\boldsymbol{r} \equiv (\boldsymbol{r}_1, \ldots, \boldsymbol{r}_N)$ be a point in the configuration space $\mathfrak{C}$ of the system, and $|\boldsymbol{r}\rangle$ be a simultaneous eigenket of all the coordinate operators. Then

$$\psi(\boldsymbol{r};t) = \int d\boldsymbol{r}' \, \langle\boldsymbol{r}|U(t,t')|\boldsymbol{r}'\rangle \, \psi(\boldsymbol{r}';t') \,, \tag{354}$$

where $d\boldsymbol{r} \equiv d^3r_1 \ldots d^3r_N$. This equation moves the state both into the future and the past.

An equation that singles out the future is obtained by introducing the *propagator* K, defined as the function

$$K(\boldsymbol{r}t, \boldsymbol{r}'t') = \langle\boldsymbol{r}|U(t,t')|\boldsymbol{r}'\rangle \, \theta(t-t') \,, \tag{355}$$

where $\theta(x)$ is the unit step (or Heaviside) function:

$$\theta(x) = \begin{cases} 1 & x > 0 \\ 0 & x < 0 \end{cases} \quad ; \qquad \frac{d}{dx}\theta(x) = \delta(x) \,. \tag{356}$$

Because $U(t,t') \to 1$ as $t \to t'$ from above,

$$\lim_{t\to t'} K(\boldsymbol{r}t, \boldsymbol{r}'t') = \delta(\boldsymbol{r} - \boldsymbol{r}') , \tag{357}$$

where the right-hand side is the $3N$-fold delta function.

As is clear from its definition, *the propagator is a probability amplitude — the amplitude for finding the system at some point* $\boldsymbol{r}$ *in configuration space at time* t *given that it was originally at* $\boldsymbol{r}'$ at time t'. For that reason the following notation is also often used:

$$K(\boldsymbol{r}t, \boldsymbol{r}'t') = \langle \boldsymbol{r}; t | \boldsymbol{r}'; t' \rangle , \qquad (t \geq t') . \tag{358}$$

The propagator satisfies an inhomogeneous counterpart of the Schrödinger equation:

$$[i\hbar\partial_t - H(t)]K(\boldsymbol{r}t, \boldsymbol{r}'t') = i\hbar\delta(t - t')\,\delta(\boldsymbol{r} - \boldsymbol{r}') , \tag{359}$$

where HK is shorthand for the operator H acting on the variables $\boldsymbol{r}$, e.g., by differentiation in the case of the kinetic energy. Equation (359) follows from (353), (355), and (356):

$$i\hbar\partial_t K(\boldsymbol{r}t; \boldsymbol{r}'t') = HK(\boldsymbol{r}t; \boldsymbol{r}'t') + \langle \boldsymbol{r}|U(t,t')|\boldsymbol{r}'\rangle\, i\hbar\delta(t - t') , \tag{360}$$

and $\delta(t - t')U(t,t') = \delta(t - t')$. As K satisfies the inhomogeneous equation (359) with a unit source, it is frequently called Green's function for the time-dependent Schrödinger equation, but to avoid confusion the term *Green's function* will be reserved here for a closely related time-independent object to be defined shortly.

The group property $U(t_3,t_2)U(t_2,t_1) = U(t_3,t_1)$ implies the following important composition law for the propagator:

$$K(\boldsymbol{r}t, \boldsymbol{r}t') = \int d\boldsymbol{r}''\, K(\boldsymbol{r}t, \boldsymbol{r}''t'')\, K(\boldsymbol{r}''t'', \boldsymbol{r}'t') , \quad (t > t'' > t') . \tag{361}$$

Another property of K is important — its dimension. Let $[Q]$ be the dimension of the quantity Q. A one-dimensional delta function $\delta(x)$ has dimension $[1/x]$, so the right side of (359) has dimension $L^{-d}T^{-1}\hbar$, where d is the dimension of $\mathfrak{C}$; the dimension of the left side is (energy)$\cdot[K]$, and therefore

$$[K] = L^{-d} . \tag{362}$$

(b) Green's Functions

When the Hamiltonian is time-independent, $U = e^{-iH(t-t')/\hbar}$, and the propagator then depends only on the interval $t - t'$. The preceding equations then simplify considerably. The propagator is now written as

$$K(\boldsymbol{r}, \boldsymbol{r}'; t) = \langle \boldsymbol{r}|e^{-iHt/\hbar}|\boldsymbol{r}'\rangle\, \theta(t) . \tag{363}$$

Let $\{\psi_{n\nu}(\boldsymbol{r})\}$ be a complete set of eigenfunctions of H, with energy eigenvalues E_n, and ν the additional quantum numbers beyond energy in the case of degeneracy:

$$\sum_{n\nu} \psi_{n\nu}(\boldsymbol{r})\psi^*_{n\nu}(\boldsymbol{r}') = \delta(\boldsymbol{r} - \boldsymbol{r}') . \tag{364}$$

Then

$$K(\boldsymbol{r},\boldsymbol{r}';t)=\theta(t)\sum_{n,\nu}e^{-iE_nt/\hbar}\psi_{n\nu}(\boldsymbol{r})\psi^*_{n\nu}(\boldsymbol{r}')\,. \tag{365}$$

Equation (365) says that the propagator for a time-independent Hamiltonian is a Fourier series in time with the frequency spectrum $\{E_n/\hbar\}$. For this and other reasons, the Fourier transform of the propagator often plays an important role. To construct this transform, consider, first, the function defined by

$$f(E)=\int_C dz\frac{e^{-izt/\hbar}}{z-E}\,, \tag{366}$$

where E is real, $z=\xi+i\eta$, and the contour C traverses $-\infty<\xi<\infty$ just above the real axis. The integrand has a simple pole on the real axis at $z=E$, and is evaluated with Cauchy's theorem by closing the contour in the half-plane in which the exponential vanishes as $z\to\infty$. Let

$$e(z)=e^{-izt/\hbar}=e^{-i\xi t/\hbar}e^{\eta t/\hbar}\,. \tag{367}$$

If $t<0$, $e(z)\to 0$ in the upper half-plane, and after being closed the contour contains no singularities so the integral vanishes. But if $t>0$, $e(z)\to 0$ in the lower half-plane, in which case the closed contour includes the pole at $z=E$ and the integral equals $-2\pi ie^{-iEt/\hbar}$. In short,

$$f(E)=-2\pi ie^{-iEt/\hbar}\theta(t)\,. \tag{368}$$

When applied to (365),

$$K(\boldsymbol{r},\boldsymbol{r}';t)=\frac{i}{2\pi}\int_C dz\,e^{-izt/\hbar}\sum_{n,\nu}\frac{\psi_{n\nu}(\boldsymbol{r})\psi^*_{n\nu}(\boldsymbol{r}')}{z-E_n}\,. \tag{369}$$

Green's function for the time-independent Schrödinger equation is now defined as the integrand of this Fourier integral when $z=E+i\epsilon$:

$$G(\boldsymbol{r},\boldsymbol{r}';E)=\sum_{n,\nu}\frac{\psi_{n\nu}(\boldsymbol{r})\psi^*_{n\nu}(\boldsymbol{r}')}{E-E_n+i\epsilon}\,, \tag{370}$$

where ϵ is a *positive* infinitesimal. G has this name because it satisfies the inhomogeneous Schrödinger equation with a unit source term,

$$(E-H)G(\boldsymbol{r},\boldsymbol{r}';E)=\delta(\boldsymbol{r}-\boldsymbol{r}')\,, \tag{371}$$

which follows by applying $(E-H)$ to (370). According to (370), when E is extended to the complex plane, *Green's function has simple poles at the eigenvalues of the Hamiltonian,*[1] *with residues R_n whose spatial dependence give the corresponding eigenfunctions*:

$$R_n=\sum_{\nu}\psi_{n\nu}(\boldsymbol{r})\psi^*_{n\nu}(\boldsymbol{r}')\,. \tag{372}$$

[1] If the Hamiltonian has a spectrum that is partially or wholly continuous, then the poles coalesce into branch cuts. This will be elucidated in §4.4 and 8.2.

The inverse to (369) gives Green's function in terms of the propagator:

$$G(\boldsymbol{r},\boldsymbol{r}';E)=\frac{1}{i\hbar}\int_0^\infty dt\, e^{izt/\hbar}\,K(\boldsymbol{r},\boldsymbol{r};t)\,, \qquad z=E+i\epsilon\,. \tag{373}$$

The integral runs only over positive times because, by definition, this propagator K vanishes for $t<0$, and for that reason is called "causal" or "retarded." Propagators with other boundary conditions in time can also be defined.

The expressions (365) and (370) tell us that once the propagator, or equivalently, Green's function, is known, the energy spectrum and stationary state wave functions are also known, at least implicitly. In that sense, the propagator (or Green's function) provides a complete solution to the quantum mechanics of a system governed by a time-independent Hamiltonian. It should, therefore, come as no surprise that exact analytic expressions for these quantities are only known for a very small number of systems. On the other hand, because the propagator and Green's function are so rich in content, approximate expressions for them are often of much greater use than approximations to individual states.

(c) The Free Particle Propagator and Green's Function

The propagator and Green's function for free particles are important in their own right, and their evaluation is instructive.

The Hamiltonian of a free particle system is a sum of commuting terms, so the unitary operator U is a product of terms not only for each particle but also for each component of momentum. Hence it suffices to evaluate K for one particle moving in one dimension:

$$K(x,x';t)=\langle x|\exp(-i\hat{p}^2t/2m\hbar)|x'\rangle\,, \tag{374}$$

where $\hat{p}$ is the momentum operator. The description of a free particle must be translation invariant, so K can only depend on $x-x'$, and x' can be set to zero. Introducing the complete set of momentum eigenfunction of (176) gives

$$K(x;t)=\theta(t)\int_{-\infty}^{\infty}\frac{dp}{2\pi\hbar}\exp[i(px-p^2t/2m)/\hbar]\,. \tag{375}$$

As always, it pays to use dimensionless variables. Specific values of t and p define the classical characteristic distance $l=pt/m$, but in view of the de Broglie relation $p=\hbar/l$, this translates into a relation between length and time alone:

$$l=\sqrt{\hbar t/m}\,. \tag{376}$$

This prompts the introduction of the dimensionless length and momentum variables $\xi=x/l$ and $\eta=lp/\hbar$, in terms of which (375) is

$$K(x;t)=\theta(t)\,\frac{1}{2\pi l}\int_{-\infty}^{\infty}d\eta\,e^{i(\xi\eta-\frac{1}{2}\eta^2)}=\theta(t)\,\frac{1}{2\pi l}\,e^{i\xi^2/2}\int_{-\infty}^{\infty}d\eta\,e^{-i\eta^2/2}\,, \tag{377}$$

where a shift in η gave the last result. The dependence on x and t is now explicit, and the dimension is as expected from (362).

Integrals of the type appearing in (377) are called Gaussian integrals. Define

$$I(a)=\int_{-\infty}^{\infty}dx\,e^{-ax^2/2}=\sqrt{\frac{2\pi}{a}} \tag{378}$$

for a real and positive. For complex $a = |a|e^{i\theta}$, the integral is defined by analytic continuation provided $-\frac{1}{2}\pi \leq \theta \leq \frac{1}{2}\pi$:

$$I(a) = \sqrt{2\pi/|a|}\, e^{-i\theta/2} . \tag{379}$$

In (377), a is pure imaginary; integrals of this type are called Fresnel integrals, but we will use the name Gaussian for all integrals of this type. When $a = \pm i\lambda$, $\theta = \pm\frac{1}{2}\pi$,

$$\int_{-\infty}^{\infty} dx\, e^{\mp i\lambda x^2/2} = \sqrt{\frac{2\pi}{|\lambda|}} \begin{cases} 1/\sqrt{i} \\ \sqrt{i} \end{cases} . \tag{380}$$

This dependence on the sign of Im a must not be forgotten when the phase of the quantity being calculated matters.

The final result for the free particle propagator in one dimension is therefore

$$K(x;t) = \theta(t) \sqrt{\frac{m}{2\pi i\hbar t}}\, e^{ix^2 m/2\hbar t} . \tag{381}$$

As already stated, for N particles of various masses in three dimensions, the propagator is just a product of such factors:

$$K(\boldsymbol{r};t) = \theta(t)\, (2\pi i\hbar t)^{-3N/2} \prod_{n=1}^{N} (m_n)^{3/2} \exp\left(i\frac{r_n^2 m_n}{2\hbar t}\right) . \tag{382}$$

The form of the free particle Green's function does not have such a trivial dependence on the dimension d of the configuration space $\mathfrak{C}$. The reason is that Green's function is a wave emanating from a point source with the symmetry of the Laplacian in a Euclidean space of dimension d: it is a cylindrical wave when $d = 2$, a spherical wave when $d = 3$, etc.[1]

Consider the important case of a single particle in three dimensions. It turns out to be most convenient to not use the general definition (371), but rather Green's function for the Helmholtz equation,

$$(\nabla^2 + k^2)G_0(\boldsymbol{r} - \boldsymbol{r}'; k) = \delta(\boldsymbol{r} - \boldsymbol{r}') . \tag{383}$$

Here translation invariance is exploited, and the energy is written as $\hbar^2 k^2/2m$. Define the Fourier representation of G_0 as

$$G_0(\boldsymbol{r};k) = \int \frac{d^3q}{(2\pi)^3}\, g_k(\boldsymbol{q})\, e^{i\boldsymbol{q}\cdot\boldsymbol{r}} . \tag{384}$$

Then (383) requires

$$(k^2 - q^2)g_k(\boldsymbol{q}) = 1 . \tag{385}$$

Without thought one might put $g = (k^2 - q^2)^{-1}$, but what then happens at $k^2 = q^2$? The answer is provided by (370): the singularity is to be averted by $k^2 \to k^2 + i\epsilon$, where ϵ is a *positive* infinitesimal. Hence

$$G_0(\boldsymbol{r};k) = \int \frac{d^3q}{(2\pi)^3}\, \frac{e^{i\boldsymbol{q}\cdot\boldsymbol{r}}}{k^2 - q^2 + i\epsilon} . \tag{386}$$

---

[1] Green's functions for the Helmholtz equation in an arbitrary number of dimensions are derived in A. Sommerfeld, *Partial Differential Equations in Physics*, Academic Press (1964), pp. 232–234.

Integrating over the orientation of $\boldsymbol{q}$, and *defining* k as positive, gives

$$\begin{aligned} G_0(\boldsymbol{r};k) &= \frac{1}{2\pi^2}\int_0^\infty q^2dq\,\frac{\sin qr}{qr}\,\frac{1}{k^2-q^2+i\epsilon} \\ &= \frac{1}{8\pi^2ri}\int_{-\infty}^{\infty} q\,dq\,\frac{e^{iqr}-e^{-iqr}}{(k+i\epsilon-q)(k+i\epsilon+q)}\,, \end{aligned} \tag{387}$$

where $r=|\boldsymbol{r}|\geq 0$. Now extend q to the complex plane; then $e^{\pm iqr}\to 0$ in the upper/lower half-planes of q, and the integral can, for the separate terms, be closed in these half-planes. The poles at $q=k\pm i\epsilon$ then give the final result

$$G_0(r;k) = -\frac{1}{4\pi}\frac{e^{ikr}}{r}\,, \qquad (d=3)\,. \tag{388}$$

As expected, this is a spherically symmetric wave centered at the origin. That the result could only depend on $|\boldsymbol{r}|$ was, of course, evident at the outset because the Laplacian is spherically symmetric. For the same reason g_k depends only on $|\boldsymbol{q}|$.

The fact that $G_0(r;k)$ is an *outgoing* wave merits some discussion. If the evaluation of (387) is repeated with $\epsilon\to-\epsilon$, the result is

$$G_0^{\text{adv}}(r;k) = -\frac{1}{4\pi}\frac{e^{-ikr}}{r}\,, \tag{389}$$

which is a spherical wave *converging* onto the origin. This change in the sign of ϵ, when traced back to (366), there amounts to $z\to\xi-i\epsilon$, and in (368) to $\theta(t)\to\theta(-t)$! In short, the incoming wave Green's function (389) is related to propagation into the past, not the future, and for that reason this is called the *advanced* Green's function.

Green's function for a free particle in one dimension will also be useful:

$$G_0(x,x';k) = \frac{1}{2ki}\left[\theta(x-x')e^{ik(x-x')}+\theta(x'-x)e^{ik(x'-x)}\right] \tag{390}$$

$$= \frac{1}{2ik}e^{ik|x-x'|}\,, \qquad (d=1)\,. \tag{391}$$

The derivation is left as an exercise.

(d) Perturbation Theory

One of the most important applications of propagators is to the development of *perturbation theory*. Consider the situation in which the Hamiltonian H has the form H_0+V, with the problem defined by H_0 being solvable, and V is in some sense small. Then it is natural to seek expressions for various quantities that depend on the "perturbation" V in terms of what is already known about H_0 by expanding in powers of V. What conditions H_0 and V must satisfy for such an expansion to be valid is often, indeed almost always, a sophisticated question, but for now assume that ignorance is bliss.

For the purpose of treating V as a perturbation, write[1] (359) as

$$[i\hbar\partial_{t_1}-H_0]K(\boldsymbol{r}_1t_1,\boldsymbol{r}_2t_2) = V(\boldsymbol{r}_1t_1)K(\boldsymbol{r}_1t_1,\boldsymbol{r}_2t_2)+i\hbar\delta(\boldsymbol{r}_1-\boldsymbol{r}_2)\delta(t_1-t_2)\,. \tag{392}$$

[1] If V is not diagonal in the coordinate representation, there is an obvious generalization requiring a further integration on the right-hand side.

Let K_0 be the propagator for the unperturbed problem:

$$[i\hbar\partial_{t_1} - H_0]K_0(\boldsymbol{r}_1t_1, \boldsymbol{r}_2t_2) = i\hbar\delta(\boldsymbol{r}_1 - \boldsymbol{r}_2)\delta(t_1 - t_2) . \tag{393}$$

Then Eq. 392 implies

$$K(\boldsymbol{r}_1t, \boldsymbol{r}_2t_2) = K_0(\boldsymbol{r}_1t_1, \boldsymbol{r}_2t_2) + \frac{1}{i\hbar}\int d\boldsymbol{r}_3dt_3\, K_0(\boldsymbol{r}_1t_1, \boldsymbol{r}_3t_3)V(\boldsymbol{r}_3t_3)K(\boldsymbol{r}_3t_3, \boldsymbol{r}_2t_2) , \tag{394}$$

or in an obvious and convenient shorthand,

$$K(1,2) = K_0(1,2) + \frac{1}{i\hbar}\int d3\; K_0(1,3)V(3)K(3,2) . \tag{395}$$

To confirm this, apply $(i\hbar\partial_{t_1} - H_0)$ to (394) and use (393).

Equation (395) can be iterated by substituting the full expression for K into the integral:

$$\begin{aligned} K(1,2) = K_0(1,2) &+ \frac{1}{i\hbar}\int d3\; K_0(1,3)V(3)K_0(3,2) \\ &+ \frac{1}{(i\hbar)^2}\int d3\, d4\; K_0(1,3)V(3)K_0(3;4)V(4)K(4,2) . \end{aligned} \tag{396}$$

This is still exact. The iteration can be repeated until the desired power of V is reached. The expansion in powers of V is thus

$$\begin{aligned} K(1,2) = K_0(1,2) &+ \frac{1}{i\hbar}\int d3\; K_0(1,3)V(3)K_0(3,2) \\ &+ \frac{1}{(i\hbar)^2}\int d3\, d4\; K_0(1,3)V(3)K_0(3,4)V(4)K_0(4,2) \ldots , \end{aligned} \tag{397}$$

or in a still more compact notation:

$$K = K_0\left(1 + \sum_{n=1}^{\infty}(i\hbar)^{-n}(VK_0)^n\right) . \tag{398}$$

This series is basic to time-dependent perturbation theory, and is used in a broad range of phenomena in many branches of physics. Here and as in quantum field theory, it leads to the famous Feynman diagrams, which are graphical mnemonics that greatly simplify the evaluation of the terms in the series.

A very important perturbation expansion in powers of V also exists for Green's function. This expansion will now be developed without paying heed to whether it is legitimate, for that is a complicated issue which will be taken up in the context of specific problems. One point can be made right now, however. Replace V by λV, where λ is a parameter that can be set to 1 in the end. The naive perturbation expansions assume that the quantities of interest, such as Green's function, the energy, the wave function, etc., are analytic functions of the complex variable λ inside the unit circle. Often a simple physical argument demonstrates that this condition is not met, and sometimes a sophisticated consideration will lead to such a conclusion. But in a host of important problems it is simply not known whether a power series expansion is valid, and it is still used because it seems to work, or nothing more sophisticated is practical.

The derivation of the perturbation series is facilitated by introducing an operator $\mathcal{G}$ called *the resolvent.* It is motivated by Eq. 370:

$$\mathcal{G}(z) = \sum_{n\nu} |n\nu\rangle \frac{1}{z - E_n} \langle n\nu| = \sum_n \frac{P_n}{z - E_n} \equiv \frac{1}{z - H} \,. \tag{399}$$

Here P_n is the projection operator onto *all* states of energy E_n, and the last expression is a very convenient form, as we shall soon see. The resolvent is well-defined as long as z is not on the real axis because the spectrum of the Hermitian operator H is real. As is clear from (370), Green's function is just the set of all matrix elements of the resolvent in the coordinate representation when $z = E + i\epsilon$:

$$G(\boldsymbol{r}, \boldsymbol{r}'; E) = \langle \boldsymbol{r}|\mathcal{G}(E + i\epsilon)|\boldsymbol{r}'\rangle \,. \tag{400}$$

In terms of $\mathcal{G}$, the inhomogeneous Schrödinger equation (371) is simply

$$(z - H)\mathcal{G}(z) = 1 \,. \tag{401}$$

The perturbation expansion follows from an identity whose derivation is left as an exercise. For any two operators A and B,

$$\frac{1}{A - B} = \frac{1}{A} + \frac{1}{A} B \frac{1}{A - B} = \frac{1}{A} + \frac{1}{A - B} B \frac{1}{A} \,. \tag{402}$$

These expressions only make sense, of course, if the denominators have no vanishing matrix elements; as before, this is accomplished by generalizing the energy to a complex variable. Now define $\mathcal{G}_0$, the "unperturbed" counterpart of $\mathcal{G}$, by

$$\mathcal{G}_0(z) = \frac{1}{z - H_0} \,. \tag{403}$$

The matrix elements of this resolvent form Green's function for the Schrödinger equation belonging to the unperturbed Hamiltonian H_0. Setting $A = z - H_0$ and $B = V$ in (402) thus gives

$$\mathcal{G} = \mathcal{G}_0 + \mathcal{G}_0 V \mathcal{G} \,, \tag{404}$$

which is shorthand for the integral equation

$$G(\boldsymbol{r}, \boldsymbol{r}') = G_0(\boldsymbol{r}, \boldsymbol{r}') + \int d\boldsymbol{s}\, G_0(\boldsymbol{r}, \boldsymbol{s}) V(\boldsymbol{s}) G(\boldsymbol{s}, \boldsymbol{r}') \,, \tag{405}$$

where E has been suppressed, and all the coordinates range over the whole of configuration space. The expansion of the exact resolvent in powers of the perturbation V again ensues by iteration:

$$\mathcal{G} = \mathcal{G}_0 + \mathcal{G}_0 V (\mathcal{G}_0 + \mathcal{G}_0 V \mathcal{G}) \,, \tag{406}$$

which is exact. Repeated iteration, and the assumption that terms beyond some given order are negligible, then yields the perturbation series:

$$\mathcal{G} = \mathcal{G}_0 + \mathcal{G}_0 V \mathcal{G}_0 + \mathcal{G}_0 V \mathcal{G}_0 V \mathcal{G}_0 + \ldots \,. \tag{407}$$

The expansion of Green's function in powers of the perturbation is often called the *Born series.* Only the first term in the expansion was used by Born in the paper that gave both the first treatment of a scattering problem in quantum mechanics, and (in a footnote) the *Born interpretation* of the wave function as a probability amplitude.

2.7 The Path Integral

The path integral is widely used in quantum field theory and statistical mechanics, and has also proven to be a powerful tool in numerical computations. On the other hand, in the simpler problems with which we deal in this volume, it must be said that the path integral is rarely a more powerful computational tool than those that emerge from the older formulations of quantum mechanics, or even competitive with them. (It should also be said that when ambiguities arise in the path integral formulation, and they do, these are ultimately resolved by comparing with canonical quantization or the Schrödinger equation.) Nevertheless, because of the powerful concepts on which it is based, and the widespread use just mentioned, acquaintance with the path integral is indispensable.

(a) The Feynman Path Integral

For systems composed of particles (as compared to dynamical fields), Feynman's path integral expresses the propagator as a coherent sum of an infinite number of amplitudes over *all* paths in configuration space, and not just those dictated by the classical equations of motion. As the classical limit is approached, the paths in the immediate neighborhood of a true classical path become ever more dominant. The path integral thus yield important insights into the connection between classical and quantum mechanics, quite apart from its other virtues. This connection will be developed in detail in §2.8.

This formulation of quantum mechanics has the remarkable feature of not using operators in Hilbert space. Rather, it constructs probability amplitudes *ab initio* from the classical concept of paths in configuration space, albeit paths that are not constrained by classical mechanics. As we will be deriving the path integral from the older versions of quantum mechanics, we will start from the Hilbert space formulation.

Consider, first, the simplest situation, a single particle in one dimension, with the Hamiltonian

$$H = \frac{\hat{p}^2}{2m} + V(\hat{x}, t) \equiv T(\hat{p}) + V(\hat{x}) , \tag{408}$$

where the notation $\hat{p}$, etc., emphasizes that these objects are operators. The generalization to interacting particles in three dimensions is straightforward, though velocity-dependent interactions, such as those with electromagnetic fields, involve some complications.

The object of interest is the propagator,

$$K(b, a) = \langle x_b | U(t_b, t_a) | x_a \rangle \, \theta(t_b - t_a) ; \qquad a \equiv (x_a t_a) , \; b \equiv (x_b t_b) . \tag{409}$$

The path integral representation for K is obtained by:

1. breaking the evolution from a to b into a large sequence of κ small forward steps in time of duration τ by means of the composition law (238) for U;

2. evaluating each small step explicitly;

3. showing that these steps sum to the form $\sum_P \exp(iS/\hbar)$, where S is the classical action for some path P composed of linear segments from a to b;

4. taking the limit, to be called Lim, defined by

$$\mathsf{Lim} : \tau \to 0, \quad \kappa \to \infty, \quad \kappa\tau = t_b - t_a \,, \tag{410}$$

with $t_b - t_a$ held fixed.

The sequence of steps is a product of κ unitary transformations:

$$K(b,a) = \langle x_b | U(t_b, t_b - \tau) \ldots U(t_a + 2\tau, t_a + \tau) U(t_a + \tau, t_a) | x_a \rangle \,. \tag{411}$$

Define

$$t_a \equiv t_0, \; t_k = t_0 + k\tau, \; t_b \equiv t_\kappa; \qquad x_a \equiv x_0, \; x_b \equiv x_\kappa \,. \tag{412}$$

On introducing $\kappa - 1$ sets of intermediate states $\{|x_k\rangle\}$, (411) becomes

$$\begin{aligned} K(b,a) = \int_{-\infty}^{\infty} & dx_{\kappa-1} \ldots dx_2 dx_1 \, \langle x_\kappa | e^{-iH(t_{\kappa-1})\tau/\hbar} | x_{\kappa-1} \rangle \ldots \\ & \ldots \langle x_2 | e^{-iH(t_1)\tau/\hbar} | x_1 \rangle \langle x_1 | e^{-iH(t_0)\tau/\hbar} | x_0 \rangle \,. \end{aligned} \tag{413}$$

With $\tau \to 0$ anticipated, the time argument of H in any step can be taken anywhere in the interval, and the lower end was chosen. Unless V has a discontinuous time dependence, this is legitimate, but not when V is velocity-dependent.

It is tempting to replace $e^{-iH\tau/\hbar}$ by $1 - iH\tau/\hbar$. This would, however, spoil the unitary property as expressed in the composition law (238), and thereby abandon the superposition principle. An approximation valid as $\tau \to 0$ is thus needed which maintains unitarity. In addition, as T and V are diagonal in the momentum and coordinate representations, respectively, this approximation should, ideally, replace $e^{-iH\tau/\hbar}$ by a product of unitary operators that are diagonal in these incompatible representations. This wish list is met by the Baker-Campbell-Hausdorff (BCH) theorem[1]

$$e^A e^B = \exp\left(A + B + \frac{1}{2}[A,B] + \frac{1}{12}\left([A,[A,B]] - [B,[A,B]]\right) \ldots \right) \,, \tag{414}$$

where A and B are operators in Hilbert space, and $\ldots$ alludes to multiple commutators of ever higher order. In our case, both A and B are proportional to τ, so the terms beyond $A + B$ are of order τ^2 or smaller, and permit the approximation

$$e^{-iH\tau/\hbar} \simeq e^{-iT\tau/\hbar} \, e^{-iV\tau/\hbar} \,. \tag{415}$$

Hence

$$\begin{aligned} \langle x_{k+1} | e^{-iH(t_k)\tau/\hbar} | x_k \rangle &= \langle x_{k+1} | e^{-iT\tau/\hbar} | x_k \rangle \, e^{-iV(x_k t_k)\tau/\hbar} \\ &= \sqrt{\frac{m}{2\pi i\hbar\tau}} \exp\left[\frac{i}{\hbar} \left(\frac{(x_{k+1} - x_k)^2 m}{2\tau} - \tau V(x_k t_k) \right) \right] \,, \end{aligned} \tag{416}$$

[1] The simplest special case is when $[A,B]$ commutes with both A and B, so that all terms beyond $[A,B]$ vanish. It has many applications, and the proof is the subject of Prob. 13.

where the free particle propagator (381) was used in the second step. Returning to (413) gives

$$K(b,a) = \mathsf{Lim}\left(\frac{m}{2\pi i\hbar\tau}\right)^{\frac{1}{2}\kappa}\int_{-\infty}^{\infty} dx_{\kappa-1}\dots dx_1 \times \exp\left[\frac{i\tau}{\hbar}\sum_{k=0}^{\kappa-1}\left(\frac{(x_{k+1}-x_k)^2 m}{2\tau^2} - V(x_k t_k)\right)\right]. \quad (417)$$

Imagine now a path $x(t)$ composed of linear segments from a to b passing through $(x_1,\dots,x_{\kappa-1})$ at $(t_1,\dots,t_{\kappa-1})$, as in the *integrand* of (417), and depicted in Fig. 2.4. The ratio $(x_{k+1}-x_k)/\tau$, as $\tau \to 0$, is the velocity $\dot{x}(t_k)$ in this step. Therefore

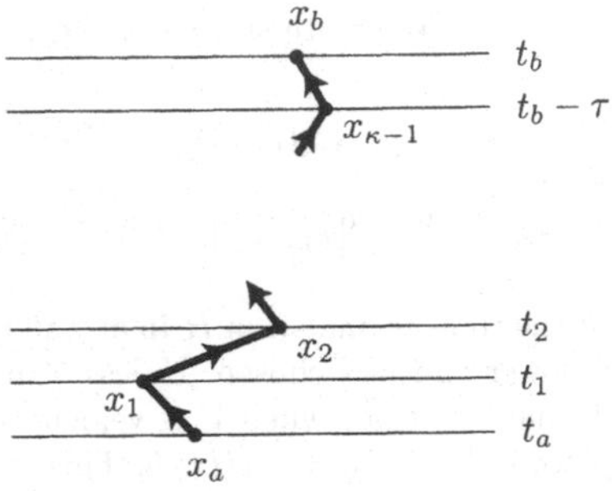

FIG. 2.4. Coordinates involved in the path integral (Eq. 417).

$$\lim_{\tau\to 0}\left(\frac{(x_{k+1}-x_k)^2 m}{2\tau^2} - V(x_k t_k)\right) = \frac{1}{2}m[\dot{x}(t_k)]^2 - V(x(t_k)t_k) = L(x(t_k),\dot{x}(t_k),t_k)\,, \quad (418)$$

i.e., *the Lagrangian* of the system at (x_k, t_k). The sum on k in (417) is therefore a discrete approximation to the time integral of the Lagrangian from a to b over the particular path $x(t) \equiv (x_a t_a;\dots;x_k t_k;\dots;x_b t_b)$,

$$\mathsf{Lim}\sum_{k=0}^{\kappa-1} L(t_0 + k\tau) = \int_{t_a}^{t_b} dt'\, L(x(t'),\dot{x}(t'),t') \equiv S_{ba}[x(t)]\,, \quad (419)$$

where $F[x(t)]$ denotes a functional of $x(t)$. *The functional $S_{ba}[x(t)]$ is the classical action for motion along the arbitrary path $x(t)$.* Hence (417) is

$$K(b,a) = \mathsf{Lim}\left(\frac{m}{2\pi i\hbar\tau}\right)^{\frac{1}{2}\kappa}\int_{-\infty}^{\infty} dx_{\kappa-1}\cdots\int_{-\infty}^{\infty}\int_{-\infty}^{\infty} dx_1\, \exp\left(\frac{i}{\hbar}S_{ba}[x(t)]\right). \quad (420)$$

In this expression the path $x(t)$ is unrestricted except at the and points, and is not just one selected by the classical equation of motion, i.e., not just a path that minimizes the action. In the limit $\tau \to 0$, the integrals over the intermediate points $(x_1,\dots,x_{\kappa-1})$ therefore include all paths from a to b. The last two sentences, and Eq. 420, *define* what is meant by the path integral. Following Feynman, it is customary to write this last equation as

$$K(b,a) = \int_a^b \mathcal{D}(x(t)) \, \exp\left(\frac{i}{\hbar} S_{ba}[x(t)]\right) , \tag{421}$$

where the meaning of the differential, or measure, $\mathcal{D}(x(t))$, and the integration symbol, are to be understood by referring back to (420).

The argument just given pays even less heed to mathematical discipline than is customary in theoretical physics, for it is reasonable to fear that here there could be fatal booby-traps because the set of all paths surely contains exotica that will not submit to routine integration. One is banking on the hope that that such paths will produce sufficiently rapid oscillations in $e^{iS/\hbar}$ to not contribute in the limit. This is not unreasonable, because the phase in (417) oscillates rapidly if the path is sufficiently erratic to violate

$$|x_{k+1} - x_k| \lesssim \sqrt{\hbar\tau/m} \, . \tag{422}$$

For a fixed time-slice parameter τ, a sufficiently large mass will suppress paths that are far from the differentiable classical path. (Indeed, the formal classical limit $\hbar \to 0$ forces the path to be smooth, a fact that will be explored in more detail in the following section.) And for any mass, there is a τ sufficiently small to suppress paths that jump about in space by an amount that violates (422). This argument does not pretend to be mathematically satisfactory; readers seeking such a treatment should consult the literature cited at the end of this chapter.

The generalization to more particles in three dimensions in the absence of velocity-dependent forces is straightforward; $x(t)$ is simply replaced by a path in the configuration space of the system. The evaluation of such path integrals is anything but straightforward, of course.

When there are velocity-dependent forces, the discretization of the evolution $a \to b$ involves a subtle point. This is illustrated by the important example of a particle of charge e in the presence of a magnetic field with vector potential $\boldsymbol{A}(\boldsymbol{r})$. (Time-dependent magnetic and electric fields do not produce any further problems.) The classical Lagrangian for this system is

$$L = \frac{1}{2} m[\dot{\boldsymbol{r}}(t)]^2 - V(\boldsymbol{r}) + \frac{e}{c} \dot{\boldsymbol{r}} \cdot \boldsymbol{A}(\boldsymbol{r}) \, . \tag{423}$$

The propagator $K(b,a)$ must satisfy a condition imposed by gauge invariance. Because of the definition (354) and (355) of the propagator, and the gauge transformation (349) of wave function, this condition is

$$K(\boldsymbol{r}t, \boldsymbol{r}'t') \to e^{ie\chi(\boldsymbol{r})/\hbar c} K(\boldsymbol{r}t, \boldsymbol{r}'t') \, e^{-ie\chi(\boldsymbol{r}')/\hbar c} \, . \tag{424}$$

The generalization of (420) when the Lagrangian is (423) must conform with this. It turns out that in slicing the time evolution the vector potential must be evaluated at the midpoint between $\boldsymbol{r}_k$ and $\boldsymbol{r}_{k+1}$ to meet this requirement.

(b) The Free-Particle Path Integral

Although we already know the propagator for free particles (recall Eq. 381), it is instructive to go through the much more demanding task of finding it by means of the path integral. For that matter, two separate derivations will be given, the first of which actually avoids doing the path integral.

As we already learned, when there are no interactions nothing is gained by treating more than one particle in more than one dimension. Let $x(t)$ be an arbitrary path, $Q(t)$ the classical path from (x_a, t_a) to (x_b, t_b), and

$$y(t) = x(t) - Q(t) \tag{425}$$

be the deviation from the classical path. In the variable y, all paths start and end at $y = 0$ because all paths $x(t)$ start at a and end at b.

First-order deviations from the classical path leave the action unchanged, and therefore the action for the arbitrary path $x(t)$ is

$$S[Q(t) + y(t)] = \tfrac{1}{2}m \int_{t_a}^{t_b} dt\, [\dot{Q}^2(t) + \dot{y}^2(t)]\,. \tag{426}$$

The first term is the classical action,

$$S_{\rm cl}(b,a) = \int_{t_a}^{t_b} dt\, \tfrac{1}{2}m\dot{Q}^2(t) = \frac{m}{2}\frac{(x_b - x_a)^2}{t_b - t_a}\,. \tag{427}$$

This action could also be expressed as $\frac{1}{2}p(x_b - x_a)$, but this would not make sense in this quantum mechanical context because the momentum and position cannot be specified simultaneously. The action must be expressed in terms of the initial and final coordinates and the elapsed time, as required by the definition of the propagator.

The separation of arbitrary paths into the classical path and the departure therefrom thus results in

$$K(b,a) = F(t_b - t_a)\exp\left(\frac{i}{\hbar}S_{\rm cl}(b,a)\right)\,, \tag{428}$$

where $F(t_b - t_a)$ is the integral over all paths from $y = 0$ back to $y = 0$ during the interval $t_b - t_a$:

$$F(t_b - t_a) = \int_{t_a, y=0}^{t_b, y=0} \mathcal{D}(y(t))\exp\left(\frac{i}{\hbar}\int_{t_a}^{t_b} dt\, \tfrac{1}{2}m\dot{y}^2(t)\right)\,. \tag{429}$$

Evaluation of the path integral can in this case (and some others) be evaded by a trick due to Feynman. It exploits the composition law (361), the fact that F is the propagator from the origin to the origin, and (428):

$$F(t) = \int_{-\infty}^{\infty} dy\, K(0,t;y,t')K(y,t';0,0) \tag{430}$$

$$= F(t-t')F(t')\int_{-\infty}^{\infty} dy\, \exp\left[iy^2\left(\frac{1}{t-t'} + \frac{1}{t'}\right)\frac{m}{2\hbar}\right]$$

$$= F(t-t')F(t')\,(2\pi i\hbar/m)^{\frac{1}{2}}\sqrt{t'(t-t')/t}\,. \tag{431}$$

Therefore

$$F(t_b - t_a) = \sqrt{\frac{m}{2\pi i\hbar(t_b - t_a)}}\,. \tag{432}$$

When combined with (428) the known result, Eq. 381, is recovered:

$$K(b,a) = \sqrt{\frac{m}{2\pi i\hbar(t_b - t_a)}}\,\exp\left(\frac{im(x_b - x_a)^2}{2\hbar(t_b - t_a)}\right). \tag{433}$$

A brute-force evaluation of the free-particle path integral is more instructive, however, because it hints at what must be done in more difficult problems, or when the path integral is computed numerically in problems where no analytic approach is known or practical.

From the definition (417) et seq. of the path integral, (429) is

$$F(t_b - t_a) = \mathsf{Lim}\left(\frac{m}{2\pi i\hbar\tau}\right)^{\frac{1}{2}\kappa} \int_{-\infty}^{\infty} dy_{\kappa-1}\ldots dy_1 \times \exp\left(\frac{im}{2\hbar\tau}\sum_{k=0}^{\kappa-1}(y_{k+1} - y_k)^2\right). \tag{434}$$

Introduce dimensionless variables, i.e.,

$$\eta_\kappa = y_k\sqrt{m/\hbar\tau}, \tag{435}$$

so that

$$F(t_b - t_a) = \mathsf{Lim}\left(\frac{m}{\hbar\tau}\right)^{\frac{1}{2}} \left(\frac{1}{2\pi i}\right)^{\frac{1}{2}\kappa} \int_{-\infty}^{\infty} d\eta_{\kappa-1}\ldots d\eta_1 \times \exp\left(\frac{i}{2}\sum_{k=0}^{\kappa-1}(\eta_{k+1} - \eta_k)^2\right). \tag{436}$$

The integral is now a pure number — the complete dependence on $\hbar$ and m is in the factor $\sqrt{m/\hbar}$, a fact that will be important when we examine the classical limit in the next section.

The argument of the exponential in (436) is a quadratic form, which, after it is diagonalized, reduces F to a product of Gaussian integrals. In detail,

$$\sum_{k=0}^{\kappa-1}(\eta_{k+1} - \eta_k)^2 = 2(\eta_1^2 + \eta_2^2 + \ldots + \eta_{\kappa-1}^2 - \eta_1\eta_2 - \eta_2\eta_3 - \ldots \eta_{\kappa-2}\eta_{\kappa-1}) \tag{437}$$

because $y_0 = y_\kappa = 0$. Next, define the symmetric matrix

$$\Lambda = \begin{pmatrix} 2 & -1 & 0 & \cdots & & \\ -1 & 2 & -1 & & & \\ 0 & -1 & 2 & & & \\ & & & \ddots & & \\ & & & & 2 & -1 \\ & & & & -1 & 2 \end{pmatrix}, \tag{438}$$

in terms of which

$$\sum_{k=0}^{\kappa-1}(\eta_{k+1} - \eta_k)^2 = \sum_k \zeta_k^2\lambda_k\,, \tag{439}$$

where $\{\zeta_k\}$ and $\{\lambda_k\}$ are the eigenvectors and eigenvalues of Λ. The new element of integration is $d\zeta_1 \dots d\zeta_{\kappa-1}$ because the transformation from $\{\eta_k\}$ to $\{\zeta_k\}$ is orthogonal. All λ_k are positive, and therefore (380) gives

$$F(t) = \mathsf{Lim} \sqrt{\frac{m}{2\pi i\hbar\tau}} \prod_{k=1}^{\kappa-1} \int_{-\infty}^{\infty} \frac{d\zeta_k}{\sqrt{2\pi i}} \, e^{\frac{1}{2}i\zeta_k^2\lambda_k} \tag{440}$$

$$= \mathsf{Lim} \sqrt{\frac{m}{2\pi i\hbar\tau \, \det\Lambda}} \,, \tag{441}$$

where

$$\det\Lambda = \prod_k \lambda_k \tag{442}$$

is the determinant of Λ. By computing $\det\Lambda$ for small values of κ, one quickly concludes that[1]

$$\det\Lambda = \kappa + 1 \,, \tag{443}$$

and Eq. 432 reappears when Lim is taken because $\kappa\tau = t_b - t_a$.

The heuristic argument made about the paths that dominate when there is a potential [recall Eq. 422, etc.] is manifestly correct in the free-particle case. From (440), and the fact that the eigenvalues λ_k are of order one, it follows that the spatial dispersion of the paths about the classical straight line is of order $|\zeta| \simeq 1$, or $|\delta y| \simeq \sqrt{\hbar\tau/m}$, in agreement with (422).

2.8 Semiclassical Quantum Mechanics

Quantum mechanics must somehow incorporate classical mechanics in the limit $\hbar \to 0$. Other examples of such relationships are common in physics. For example, Newton's equations of motion arise directly from their relativistic generalization by taking the limit $c \to \infty$ in the latter. The relation between classical and quantum mechanics is far less straightforward, however. At the formal level, the classical equations of motion do not arise out of Schrödinger's or Heisenberg's equations by simply setting $\hbar \to 0$. The mathematical reason is that that the probability amplitudes of quantum mechanics are highly singular as $\hbar \to 0$, as is evident in the form $\sum[\exp(iS/\hbar)]$ of the path integral, which reveals an essential singularity at $\hbar = 0$. From a conceptual viewpoint there is a much deeper chasm between classical and quantum mechanics than that between relativistic and nonrelativistic classical mechanics. The complexity of the mathematical connection between classical and quantum mechanics seems to reflect this profound conceptual separation.

The link to classical mechanics is more direct in the path integral than in the Hilbert space formulation, because the former is based on the classical concept of paths in configuration space. This link relies on the formulation of classical mechanics due to Hamilton and Jacobi, and for that reason we begin, in subsection (a), with a sketch of Hamilton-Jacobi theory.[2] As the derivation of the semiclassical

[1] For a proof of (443) that generalizes to more difficult cases, see Prob. 14.

[2] Indeed, Schrödinger's discovery of his wave equation stemmed from this connection; the Hamilton-Jacobi equation is the first equation in his first paper.

propagator involves some intricate analysis, we then present, in subsection (b), a heuristic semiclassical approximation to the Schrödinger wave function which is intuitively attractive but of limited validity. A description of the semiclassical propagator is then given in subsection (c), with detailed derivations postponed to subsection (d).

(a) Hamilton-Jacobi Theory

In classical mechanics, all possible motions of a system are determined by its *Lagrangian*, $L(q(t), \dot{q}(t), t)$. Here $q(t)$ is a shorthand for a point $(q_1(t), \ldots, q_f(t))$ in the configuration space $\mathfrak{C}_f$, and f the number of degrees of freedom. For an N-particle system, the relation to the notation of the earlier sections is $(q_1, q_2, q_3) = \boldsymbol{r}_1$, $m_{1,2,3} = m_1$, etc. In this section q_i and the conjugate momenta p_i are always c-numbers, never operators, whether classical or quantum mechanics is under discussion.

The *classical action* is the integral of L along a trajectory *allowed* by the equations of motion, from a configuration $q_a = (q_1(t_a), \ldots, q_f(t_a))$ at time t_a to another point q_b in $\mathfrak{C}_f$ at time t_b :

$$S(q_b t_b; q_a t_a) \equiv S(b, a) = \int_{t_a}^{t_b} dt\, L(q(t), \dot{q}(t), t) \,. \tag{444}$$

This is not to be confused with the action appearing in the path integral, which includes all paths, whether or not permitted classically.

The most familiar way of selecting a particular classical trajectory is to specify the initial q_i and p_i , but this does not translate into quantum mechanics because of the uncertainty principle. A classical formulation does exist which evades this problem, namely, that based on $S(b,a)$, for it selects a particular trajectory by specifying the initial and final configurations (q_a, t_a) and (q_b, t_b), which corresponds to the specification of a particular propagator $K(b, a)$.

As $S(b, a)$ is only a function of the initial and final coordinates, it remains to find the corresponding momenta. For that purpose, consider arbitrary variations δq:

$$\delta S(b, a) = \int_{t_a}^{t_b} dt \left(\frac{\partial L}{\partial q_i} - \frac{d}{dt}\frac{\partial L}{\partial \dot{q}_i} \right) \delta q_i \; + \; \frac{\partial L}{\partial \dot{q}_i}\delta q_i \bigg|_{t_a}^{t_b} \,. \tag{445}$$

The integral vanishes thanks to Lagrange's equations. Because $p_i = \partial L/\partial \dot{q}_i$, the end points contribute

$$\delta S(b, a) = p_{b,i}\, \delta q_i(t_b) - p_{a,i}\, \delta q_i(t_a) \,, \tag{446}$$

and therefore

$$p_{b,i} = \frac{\partial}{\partial q_{b,i}} S(b, a) = \; p_{b,i}(q_b t_b; q_a t_a) \,, \tag{447}$$

$$p_{a,i} = -\frac{\partial}{\partial q_{a,i}} S(b, a) = \; p_{a,i}(q_b t_b; q_a t_a) \,. \tag{448}$$

Note that these momenta are functions of the initial and final configurations.

The change of the action when the time at end point b varies is

$$\frac{dS}{dt_b} = L(t_b) = \frac{\partial S}{\partial t_b} + \frac{\partial S}{\partial q_{b,i}}\dot{q}_{b,i} = \frac{\partial S}{\partial t_b} + p_{b,i}\dot{q}_{b,i}\,; \tag{449}$$

at b, the relation between the Hamiltonian and Lagrangian is

$$H(t_b) = p_{b,i}\,\dot{q}_{b,i} - L(t_b)\,. \tag{450}$$

Therefore

$$\frac{\partial S}{\partial t_b} + H(t_b) = 0\,. \tag{451}$$

For now, we fix the initial configuration (q_a, t_a), and concentrate on the variation with the variables (q_b, t_b), called simply (q, t). The *Hamilton-Jacobi equation* is nothing but (451) after the momenta in $H(q, p, t)$ are eliminated by use of (447):

$$\frac{\partial S}{\partial t} + H\left(q_1, \ldots, q_f; \frac{\partial S}{\partial q_1}, \ldots, \frac{\partial S}{\partial q_f}; t\right) = 0\,. \tag{452}$$

As an important example, consider a system of charged particles exposed to an electromagnetic field, and having mutual interactions that are not velocity-dependent. The Hamiltonian and Lagrangian are

$$H = \sum_i \frac{1}{2m_i}\left[p_i - \Phi_i(t)\right]^2 + V(q_i, \ldots, q_f; t)\,, \qquad \Phi_i(t) \equiv \frac{e_i}{c}A_i(t)\,, \tag{453}$$

$$L = \sum_i \left(\tfrac{1}{2}m_i\,\dot{q}_i^2 + \dot{q}_i\Phi_i(t)\right) - V\,, \tag{454}$$

where $A_i(t)$ is a component of the vector potential evaluated at the position particle i. The potential V describes the interactions between the particles, and any applied electric field. According to (452), for this system the Hamilton-Jacobi equation is

$$\frac{\partial S}{\partial t} + \sum_i \frac{1}{2m_i}\left(\frac{\partial S}{\partial q_i} - \Phi_i\right)^2 + V = 0\,. \tag{455}$$

The velocities are found from $\dot{q}_i = \partial H/\partial p_i$ and (447):

$$\dot{q}_i = \frac{1}{m_i}\left(\frac{\partial S}{\partial q_i} - \Phi_i\right)\,. \tag{456}$$

The Hamilton-Jacobi equation thus offers a description of dynamics that is very different from those given by Hamilton's or Lagrange's equations of motion. The latters' multitudes of ordinary differential equations are replaced by just *one* partial differential equation. This equation is of first order in time, has $f+1$ independent variables, $(q_1, \ldots, q_f)$ and t, and one dependent variable S. This does smell like Schrödinger's equation in configuration space: the same order in time, the same independent variables, and one dependent variable, the wave function ψ. Of course, there is one obvious and crucial difference: Schrödinger's equation is linear in ψ (the superposition principle!), whereas the Hamilton-Jacobi equation is nonlinear in S.

There is also a remarkable analogy between the solution $S(b,a)$ of the Hamilton-Jacobi equation and the propagator $K(b,a)$: both handle whole families of trajectories in one fell swoop. Given initial data at time t_0 as a continuous function $S(q,t_0)$ in $\mathfrak{C}_f$, the partial differential equation (452) will manufacture $S(q,t)$ in the future.

Hamilton gave an elegant geometrical depiction of how the trajectories of particles emerge from this continuum formulation of point mechanics by drawing an analogy to the way in which the rays of geometrical optics emerge from wave optics. For our purpose, it suffices to consider a time-independent Hamiltonian, equal masses and no vector potential. Then

$$S(q,t) = W(q) - Et\,, \tag{457}$$

where E is the energy, and (455) simplifies to

$$|\boldsymbol{\nabla} S| = |\boldsymbol{\nabla} W| = \sqrt{2m(E-V)}\,, \tag{458}$$

where $\boldsymbol{\nabla}$ is the gradient in f-dimensional Euclidean space, i.e., the configuration space $\mathfrak{C}_f$.

Consider, first, the family of surfaces $W(q) = \text{const.}$, which coincide with the surfaces of constant $S(q,t)$ at $t=0$. As time increases, the surfaces of constant S move in obedience to (457). Take a particular surface $W(q) = W_0$; the surface of S that originally satisfied $S = W_0$ must move by time t to coincide with the surface $W(q) = W_0 + Et$. In short, to visualize the motion of S it suffices to visualize the family $\mathcal{F}$ of surfaces of constant $W(q)$ (see Fig. 2.5).

Furthermore, given one member $\mathcal{F}_0$ (say $W(q) = W_0$) of $\mathcal{F}$, all others can be constructed from it as follows. Eq. 458 specifies the gradient of W throughout $\mathfrak{C}_f$; erect a vector $\boldsymbol{u}$ normal to $\mathcal{F}_0$ of length ds at each point, where

$$ds = \frac{dW}{\sqrt{2m[E - V(q)]}} \tag{459}$$

Then the locus of the end points of these vectors $\boldsymbol{u}$ is the surface on which $W(q) = W_0 + dW$. The continuation of this process constructs the whole family $\mathcal{F}$.

The motion of a surface of constant $S(q,t)$ is found from $dW = Edt$; such a surface therefore moves normal to itself in the direction given by the sign of E and with a velocity of magnitude

$$\frac{ds}{dt} = \frac{E}{\sqrt{2m[E - V(q)]}}\,. \tag{460}$$

If we think of the surfaces $S = \text{const.}$ as wave-fronts, then ds/dt is analogous to the phase velocity.

Finally, because

$$p_i = \partial W(q)/\partial q_i\,, \tag{461}$$

the family of curves normal to the family of surfaces $\mathcal{F}$ are the trajectories; i.e., the set of paths in $\mathfrak{C}_f$ along any one of which a system point $q(t)$ moves in accordance with the laws of mechanics. The magnitude of the velocity of any particular motion as it passes through q is not ds/dt, however, but $\sqrt{2(E-V)/m}$ according to (461) and (458). Therefore, in its motion along a trajectory, a moving point $q(t)$ does not stay on one surface $S=$ const.

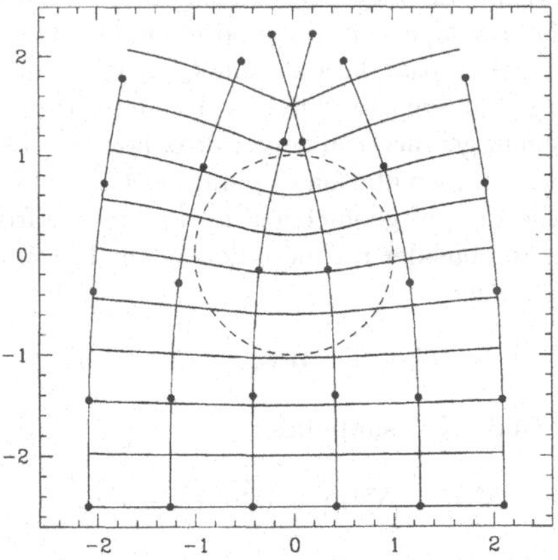

FIG. 2.5. Solution of the Hamilton-Jacobi equation in two dimensions for the potential $V(x,y) = -(1+x^2+y^2)^{-1}$, on the dashed circle $V = -\frac{1}{2}$. The dots are the positions of particles at $t = T, 2T, 3T, 4T$, which started at $t = 0$ with equal velocities in the y-direction at $y = -2.5$, and form trajectories that are everywhere normal to the surfaces of constant W, as shown. Note that these positions at any time after $t = 0$ do not stay on a single surface of constant W (see discussion following Eq. 460). The surfaces $W =$ const. develop cusps when trajectories cross, which is of importance in the construction of the semiclassical propagator (see Eq. 480 et seq.).

(b) The Semiclassical Wave Function

The earliest papers by de Broglie and Schrödinger made it evident that $\hbar$ appears explicitly in quantum mechanical amplitudes in the form $\exp(i\Theta/\hbar)$. This raises the question of whether there are circumstances under which Θ does not depend on $\hbar$, and if so, whether Θ then has a meaning in classical mechanics. This question is answered by the *Ansatz* due to Brillouin and Wentzel,

$$\psi(q,t) = e^{i\Theta(q,t)/\hbar} . \tag{462}$$

For the Hamiltonian (453), the Schrödinger equation is

$$\left(\sum_i \frac{1}{2m_i}\left(\frac{\hbar}{i}\frac{\partial}{\partial q_i} - \Phi_i\right)^2 + V(q)\right)\psi = i\hbar\frac{\partial\psi}{\partial t} . \tag{463}$$

A short calculation shows that

$$\left(\frac{\hbar}{i}\frac{\partial}{\partial q_i} - \Phi_i\right)^2 e^{i\Theta/\hbar} = \left[\left(\frac{\partial\Theta}{\partial q_i} - \Phi_i\right)^2 + \frac{\hbar}{i}\left(\frac{\partial^2\Theta}{\partial q_i^2} - \frac{\partial\Phi_i}{\partial q_i}\right)\right] e^{i\Theta/\hbar} . \tag{464}$$

In the gauge $\nabla\cdot\boldsymbol{A} = 0$, $\partial\Phi_i/\partial q_i = 0$. Having made no approximation, the Schrödinger equation thus requires Θ to satisfy

$$\frac{\partial\Theta}{\partial t} + \sum_i \frac{1}{2m_i}\left(\frac{\partial\Theta}{\partial q_i} - \Phi_i\right)^2 + V(q) = -\frac{\hbar}{i}\sum_i \frac{1}{2m_i}\frac{\partial^2\Theta}{\partial q_i^2} . \tag{465}$$

The right- and left-hand sides of (465) are very different: $\hbar$, i, and the second derivative of Θ appear only on the right. If the right side is dropped, (465) becomes the Hamilton-Jacobi equation (455). Roughly speaking, therefore,

$$\psi \sim \exp\left(\frac{i}{\hbar} \times \text{classical action}\right) \qquad \hbar \to 0\,, \tag{466}$$

which is the idea underlying the path integral.

It is now natural to consider a systematic expansion of Θ in powers of $\hbar$:

$$\Theta = S + \frac{\hbar}{i} S_1 + \left(\frac{\hbar}{i}\right)^2 S_2 + \ldots\,, \tag{467}$$

where $S(q,t)$ is a solution of the Hamilton-Jacobi equation. According to (465), S will be a good approximation to Θ provided

$$|\partial S/\partial q_i|^2 \gg \hbar\,|\partial^2 S/\partial q_j^2|\,. \tag{468}$$

This inequality can be stated in a more meaningful way by considering the one-dimensional case, and defining the *local de Broglie wavelength* $\lambda(q)$ as

$$\frac{\hbar}{\lambda(q)} \equiv p(q) = \frac{\partial S}{\partial q}\,. \tag{469}$$

Then (468) becomes

$$\left|\frac{\partial \lambda}{\partial q}\right| \ll 1\,. \tag{470}$$

That is, *the phase of the wave function is given by the classical action if the local wavelength $\lambda(q)$ changes but little in a distance of order the wavelength.*

An equation for the next term, S_1, is found by substituting (467) into the Schrödinger equation, and equating powers of $\hbar$:

$$\frac{\partial S_1}{\partial t} + \sum_k \frac{1}{2m_k}\left[2\left(\frac{\partial S}{\partial q_k} - \Phi_k\right)\frac{\partial S_1}{\partial q_k} + \frac{\partial^2 S}{\partial q_k^2}\right] = 0\,. \tag{471}$$

This will produce a real function S_1 from the real action S in the regions of configuration space available to classical motions. When the higher-order terms, S_2, etc., are dropped, ψ is approximated by the *semiclassical wave function*[1]

$$\Psi_{\rm sc}(q,t) = e^{S_1(q,t)}\, e^{iS(q,t)/\hbar} = \sqrt{D(q,t)}\, e^{iS(q,t)/\hbar}\,. \tag{472}$$

In this approximation, the spatial probability distribution is

$$D(q,t) = |\Psi_{\rm sc}(q,t)|^2 = e^{2S_1(q,t)}\,, \tag{473}$$

which is independent of $\hbar$. The remarkable fact that the next-to-leading order term, $D(q,t)$, is $\hbar$-independent would seem to imply that it too is determined by classical mechanics. As we shall in the next subsection, that is indeed so.

[1]Many authors refer to (472), and even the propagator $K_{\rm sc}$, as the WKB approximation. This is historically incorrect; we will only apply the name WKB to the stationary state one-dimensional case (see §4.5).

The relationship between $D(q,t)$ and $S(q,t)$ is opaque when expressed by (471), but will become clear when it is rewritten as follows:

$$\frac{\partial}{\partial t} e^{2S_1} + \sum_k \frac{\partial}{\partial q_k} \left[\frac{1}{m_k} \left(\frac{\partial S}{\partial q_k} - \Phi_k \right) e^{2S_1} \right] = 0 . \tag{474}$$

The Schrödinger current i_k belonging to the wave function Ψ_{sc} is

$$i_k = \frac{1}{2m_k} \Psi_{sc}^* \left(\frac{\hbar}{i} \frac{\partial}{\partial q_k} - \Phi_k \right) \Psi_{sc} + \text{c.c.} = \frac{D}{m_k} \left(\frac{\partial S}{\partial q_k} - \Phi_k \right) . \tag{475}$$

Recalling (456) then gives the following intuitively appealing expression for the current,

$$i_k(q,t) = D(q,t)\, \dot{q}_k(t) , \tag{476}$$

where $\dot{q}_k$ is the velocity of a particle at the position $q_k(t)$ as it moves in accordance with the classical equations of motion. Eq. 474 is therefore

$$\frac{\partial D}{\partial t} + \sum_k \frac{\partial}{\partial q_k} (D\, \dot{q}_k) = 0 , \tag{477}$$

i.e., *the continuity equation for a classical fluid of density* $D(q,t)$ *and local velocity* $\dot{q}(t)$ *moving through configuration space.* In short, (477) is the approximation to the quantum mechanical continuity equation for the flow of probability when a solution of the Schrödinger equation is replaced by the semiclassical wave function Ψ_{sc}.

(c) The Semiclassical Propagator

The semiclassical wave function $\Psi_{sc}(q,t)$ has intuitively pleasing properties. Nevertheless, given a wave function $\psi(q)$ at $t=0$, it is not clear what conditions this initial data must satisfy if it is to be evolved in time in the form of Ψ_{sc}. For example, if $\psi(q)$ is real, then the implication that $S(q,t=0)=0$ renders the Hamilton-Jacobi equation useless, and yet that is the equation that should supposedly be solved to evolve the semiclassical wave function. In this example, some more or less ad hoc method for assigning phases to such a real input would have to be devised.

This question does not arise, and compliance with the superposition principle is assured, if the the wave function is evolved with a properly constructed semiclassical propagator $K_{sc}(b,a)$, so that time development is given by

$$\Psi(q;t) = \int d^f q'\, K_{sc}(q,t;q',t')\, \Psi(q';t') . \tag{478}$$

The derivation of this approximate propagator is quite intricate, and therefore we first give the expression for K_{sc} and discuss its properties.

In §2.7(b) we found that the propagator for a free particle in one dimension is

$$K_0(b,a) = \sqrt{\frac{m}{2\pi i\hbar(t_b - t_a)}}\, \exp\left(\frac{i}{\hbar} S^0(b,a) \right) , \tag{479}$$

where S^0 is *the classical action for a free particle.* Eq. 479 is exact. It turns out that this is a special case of a general but approximate result:

- In the limit $\hbar \to 0$, the propagator for an arbitrary system of particles has the form

$$K_{sc}(b,a) = \sum_{\mathcal{P}_r} A_r(b,a) \exp\left\{ i\left(\frac{1}{\hbar} S_r(b,a) - \frac{\beta_r \pi}{2}\right)\right\}, \tag{480}$$

where $S_r(b,a)$ is the classical action for the system in question when it follows the classically allowed path $\mathcal{P}_r$. In general, more than one classical path may connect b to a.

- The amplitude factor is known once the classical action $S_r(b,a)$ is known:[1]

$$A_r(b,a) = \left(\frac{i}{2\pi\hbar}\right)^{f/2} \sqrt{\det \frac{\partial^2 S_r(b,a)}{\partial q_{a,i} \partial q_{b,j}}} \,. \tag{481}$$

The determinant in (481) is called the Van Vleck determinant $D_r(b,a)$. For each path, D_r and S_r satisfy the continuity equation (477).

- The integer β_r is called the Maslov index. It too is determined by the classical trajectories, and in particular, by the number of trajectory crossings encountered on the various paths. The $\hbar$-independent phase factor $\exp(-i\beta_r\pi/2)$ is essential, for if it is overlooked the wave function generated by (478) will not be continuous. (In this connection, recall Fig. 2.5.)

As it stands, K_{sc} can only produce an approximate wave function in classically allowed regions, so it cannot describe quantum tunneling. Techniques for analytically continuing the propagator into classically forbidden regions exist, but this is a difficult topic which we will not treat. A semiclassical technique for handling tunneling problems in one dimension is presented in §4.5, however.

The amplitude factor (481) is a classical quantity, so it should have a classical meaning. Indeed it does: $|A(b,a)|^2$ is the probability $P_{cl}(q_b t_b; q_a t_a)$ for finding the configuration q_b at t_b for an ensemble of systems moving classically and known to have been in the configuration q_a at t_a. When an initial configuration $(q_a t_a)$ is specified, the uncertainty principle demands all momenta to be equally probable at that time. The normalization of this white momentum spectrum is given by (177):

$$|\langle p_a(t_a)|q_a(t)\rangle|^2 d^f p_a = \prod_i \left|\frac{e^{ip_{a,i}q_{a,i}/\hbar}}{(2\pi\hbar)^{1/2}}\right|^2 dp_{a,i} = \frac{d^f p_a}{(2\pi\hbar)^f} \,. \tag{482}$$

By assumption, each member of the ensemble moves classically from the initial condition $(q_a p_a)$, and will be in the interval $d^f q_b$ at t_b with a probability

$$P_{cl}(q_b t_b; q_a t_a) d^f q_b = \frac{d^f p_a}{(2\pi\hbar)^f} \,. \tag{483}$$

In computing $d^f p_a / d^f q_b$, bear in mind that when the initial and final coordinates are specified, as they are here, the $p_{a,i}$ are functions of these coordinates and times. The ratio of differentials is therefore the Jacobian

$$\frac{d^f p_a}{d^f q_b} = \frac{\partial(p_{a,1}(b,a), \ldots, p_{a,f}(b,a))}{\partial(q_{b,1}, \ldots, q_{b,f})} \,, \tag{484}$$

[1] Note that because of (427) this agrees with (479).

where $p_{a,i}(b,a) = p_{a,i}(q_b t_b; q_a t_a)$. Using (448) for these initial momenta gives

$$\frac{\partial(p_{a,1}(b,a),\ldots,p_{a,f}(b,a))}{\partial(q_{b_b,1},\ldots,q_{b,f})} = (-1)^f \det \frac{\partial^2 S(b,a)}{\partial q_{b,i}\,\partial q_{a,j}} \,. \tag{485}$$

The quantum and classical probabilities, $|K(b,a)|^2$ and $P_{\rm cl}(b,a)$, therefore agree when the propagator is approximated by $K_{\rm sc}$ and there is only one path connecting b to a. If there is more than one, quantum mechanical interference terms will survive as $\hbar \to 0$ unless an average over rapidly oscillating terms is imposed by the resolution of an experiment.

The following subsection describes the steps that lead to $K_{\rm sc}(b,a)$ from the path integral: (i) The action for arbitrary paths is expanded about the classical path, and it is shown that only quadratic departures survive in the limit $\hbar \to 0$. The path integrals containing these quadratic terms form the factor $A_r(b,a)$. (ii) If the paths connecting a to b do not cross, an evaluation of the path integral expressing A_r can be sidestepped by requiring $K_{\rm sc}$ to satisfy the composition law for unitary transformations. (iii) If there are paths that cross, the path integral for the amplitude factors must be evaluated, which yields the phase shifts shown in (480).

Because only quadratic departures from the classical path survive the limit $\hbar \to 0$, the semiclassical approximation $K_{\rm sc}(b,a)$ is the exact propagator for all Hamiltonians that do not contain terms higher than quadratic in the canonical coordinates and momenta. Consequently, the classical action determines the propagators exactly for systems of arbitrary many harmonically coupled degrees of freedom; for particles subjected to constant forces; and for charged particles in a uniform magnetic field.

(d) Derivations

Our first objective is to find the $\hbar \to 0$ behavior of the path integral representing the propagator. The essential features can be seen in one dimension. As in §2.7(b), let $x(t)$ be an arbitrary path, $Q(t)$ the path prescribed by classical mechanics, $x(t) = Q(t) + y(t)$, and $S[x(t)]$ the action for the arbitrary path. Assume also that only one classical path exists for this particular case $a \to b$. As $Q(t)$ minimizes S,

$$S[Q(t)+y(t)] = S[Q(t)] + \int_{t_a}^{t_b} dt \left(\tfrac{1}{2}y^2 \frac{\delta^2 S}{\delta x^2} + \tfrac{1}{2}\dot{y}^2 \frac{\delta^2 S}{\delta \dot{y}^2} + y\dot{y}\,\frac{\delta^2 S}{\delta y \delta \dot{y}} \right) + \ldots \tag{486}$$

$$= S(b,a) + \int_{t_a}^{t_b} dt\, [a(t)\dot{y}^2 + b(t)y^2 + c(t)y\dot{y}] + \ldots \,, \tag{487}$$

where $+\ldots$ stands for terms of order y^3 and higher. In the free-particle case, $a = \frac{1}{2}m$ and all the other terms vanish, which is why the approximate answer is exact in this instance.

In consequence of (487), the propagator is

$$K(b,a) = \mathcal{A}(b,a)\, e^{iS(b,a)/\hbar} \,, \tag{488}$$

where $S(b,a)$ is the action for the classical trajectory, and $\mathcal{A}$ is the path integral

$$\mathcal{A}(b,a) = \int_{t_a, y=0}^{t_b, y=0} \mathcal{D}(y(t)) \exp\left(\frac{i}{\hbar} \int_{t_a}^{t_b} dt\, [a(t)\dot{y}^2 + b(t)y^2 + c(t)y\dot{y}] + \ldots \right). \tag{489}$$

This should be compared to the corresponding expression for the free particle (Eq. 429). If the classical equation of motion allows several paths $Q_r(t)$ from a to b, then a sum on r is required, and the departures from each $Q_r(t)$ must be treated as in (489).

When the path integral is first approximated by κ steps of duration τ, as in Eq. 434, and dimensionless coordinates $\eta_\kappa = y_\kappa\sqrt{m/\hbar\tau}$ are introduced, it takes the form

$$\mathcal{A}(b,a) = \mathrm{Lim}\sqrt{m/\hbar\tau}\int_{-\infty}^{\infty} d^{\kappa-1}\eta \\ \times \exp\left(f_2(\eta) + (\hbar\tau/m)^{\frac{1}{2}} f_3(\eta) + (\hbar\tau/m) f_4(\eta) + \ldots\right) , \tag{490}$$

where $f_n(\eta)$ comes from the terms of order y^n in (487), and is a polynomial of degree n in the variables $(\eta_1, \ldots, \eta_{\kappa-1})$. Hence all terms in the Lagrangian that are of higher order than quadratic are removed by the limit $\hbar \to 0$:

$$A(b,a) = \sqrt{m/\hbar}\ \mathrm{Lim}\,\tau^{-\frac{1}{2}} \int_{-\infty}^{\infty} d^{\kappa-1}\eta\ \exp[f_2(\eta)] . \tag{491}$$

The evaluation of the path integral (491) can be sidestepped when there is only one path connecting (a,b) by requiring the propagator to satisfy

$$K_{\mathrm{sc}}(x't';xt) = \int d\bar{x}\ K_{\mathrm{sc}}(x't';\bar{x}\bar{t})K_{\mathrm{sc}}(\bar{x}\bar{t};xt) \tag{492}$$

$$= \int d\bar{x}\ A(x';\bar{x})A(\bar{x};x)\exp\left(\frac{i}{\hbar}[S(x';\bar{x}) + S(\bar{x};x)]\right) , \tag{493}$$

which must hold for all $t < \bar{t} < t'$. (In Eq. 493 the time arguments have been suppressed.) The amplitude factors have a spatial variation that does not depend on $\hbar$. In contrast, the actions in the exponent are very large compared to $\hbar$ in the semiclassical limit, so the exponential varies rapidly as a function of $\bar{x}$. The integral can therefore be evaluated by the stationary phase method. The relevant formula is

$$\int dx\, g(x)e^{if(x)/\hbar} \simeq \sum_\alpha e^{if(x_\alpha)/\hbar}\int dx\, g(x)\exp\left[\tfrac{1}{2}i(x-x_\alpha)^2 f''(x_\alpha)/\hbar\right] \\ \simeq \sum_\alpha g(x_\alpha)\sqrt{\frac{2\pi i\hbar}{f''(x_\alpha)}}\, e^{if(x_\alpha)/\hbar} , \tag{494}$$

where x_α are the stationary phase points, $f'(x_\alpha) = 0$. According to (447) and (448), the stationary phase condition is

$$\frac{\partial}{\partial\bar{x}}[S(x';\bar{x}) + S(\bar{x};x)] = -p_0(x';\bar{x}) + p_0(\bar{x};x) = 0 , \tag{495}$$

where p_0 are momenta at the intermediate time $\bar{t}$. Eq. 495 has a simple meaning: a stationary phase point $x_0(\bar{t})$ is one where the momentum arriving and leaving that point are equal, so that the whole trajectory from (xt) to $(x't')$ is continuous at $\bar{t}$ (see Fig. 2.6).

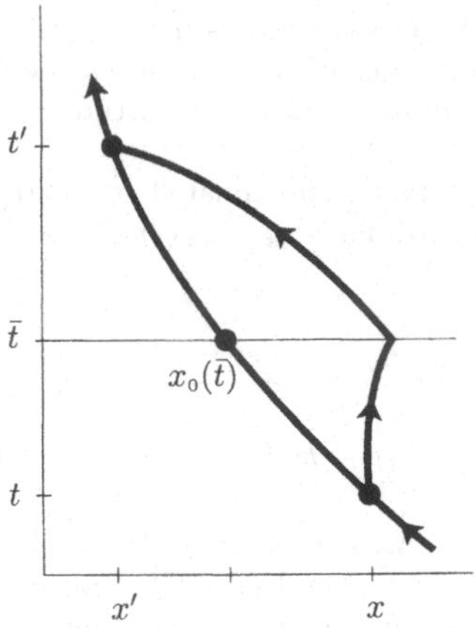

FIG. 2.6. The semiclassical propagators will only satisfy the composition law required by unitarity if the two trajectories join smoothly, which occurs at the stationary phase point $x_0(\bar{t})$.

This is required to permit the statement

$$S(x't; x_0\bar{t}) + S(x_0\bar{t}; xt) = S(x't'; xt) , \tag{496}$$

which must hold if (492) is to hold. In one dimension and for one given path there is only one point at any $\bar{t}$ at which the entering and leaving momenta are equal, and thus only one stationary phase point. Equation (493), in view of (494), requires the amplitude factors to satisfy

$$A(x'; x) = \sqrt{2\pi i\hbar}\, A(x'; x_0)A(x_0; x) \left(\frac{\partial^2}{\partial \bar{x}^2} [S(x'; \bar{x}) + S(\bar{x}; x)]_{\bar{x}=x_0} \right)^{-\frac{1}{2}} . \tag{497}$$

If (481) is to satisfy (497), the following must hold:

$$\frac{\partial^2 S(x'; x)}{\partial x' \partial x} = -\frac{\partial^2 S(x'; x_0)}{\partial x' \partial x_0} \frac{\partial^2 S(x_0; x)}{\partial x_0 \partial x} \left(\frac{\partial^2}{\partial x_0^2} [S(x'; x_0) + S(x_0; x)] \right)^{-1} . \tag{498}$$

Now (495) is a relationship between (x, x', x_0) for given values of $(t, t', \bar{t})$. Differentiation of this relationship with respect to x with x' held fixed gives

$$\left(\frac{\partial}{\partial x} \right)_{x'} [p_0(x_0; x) - p_0(x'; x_0)]$$
$$= \frac{\partial p_0(x_0; x)}{\partial x} + \frac{\partial x_0}{\partial x} \frac{\partial}{\partial x_0} [p_0(x_0; x) - p_0(x'; x_0)] , \tag{499}$$

because $p_0(x'; x_0)$ does not depend explicitly on x. Therefore the last factor in (498) is

$$\frac{\partial}{\partial x_0} [p_0(x'; x_0) - p_0(x_0; x)] = \frac{\partial p_0(x_0; x)}{\partial x} \frac{\partial x}{\partial x_0} . \tag{500}$$

Furthermore, because of (496),

$$\frac{\partial^2 S(x';x_0)}{\partial x' \partial x_0} = \frac{\partial^2 S(x';x)}{\partial x' \partial x} \frac{\partial x}{\partial x_0}, \tag{501}$$

which with (500) establishes (498).

Two questions now arise: (i) How are more degrees of freedom to be handled? (ii) How are the phases of the propagators to be computed? The first question is answered by an exercise in Jacobian gymnastics, which can be found in the review by Berry and Mount.

The second question is much deeper, and we only offer a hint as to how it was first answered by Gutzwiller. Basically, the propagators for distinct classically allowed paths must be sown together to form a continuous function. If there are no paths between a and b that cross each other, the phases are those of the free particle propagator. Consider, however, the one-dimensional case where two paths connecting a to b cross at an intermediate point (x_c, t_c), which is called a point conjugate to a. The convolution (492) must hold for all times between t_a and t_b, and must, therefore, properly continue the phases generated by the two paths that converge on (x_c, t_c) from $t_c - \epsilon$ to $t_c + \epsilon$. This is done by returning to the path integral, in which only quadratic departures from classical paths survive when $\hbar \to 0$. The evaluation is then similar to that for free particles (see Eq. 436, etc.), though with a more complicated quadratic form in the fluctuations η_k, and hence a product of Gaussian integrals with different eigenvalues λ_k. In the free particle case, all the eigenvalues are positive, and give the same root of i; in the present case that is no longer true when there are conjugate points, and the Maslov index counts the number of eigenvalues that have the "wrong" sign and the extra phase factor as given by Eq. 380. An example of such a calculation can be found in §17 of Schulman's monograph. A powerful theory exists for multidimensional problems because the indices characterize the topology of the manifolds of orbits in configuration space.[1]

The proof that the Van Vleck determinant satisfies the continuity equation is sketched in Prob. 15.

2.9 Problems

1. Let A be a positive definite Hermitian operator. Show that for all $|u\rangle$ and $|v\rangle$,

$$|\langle u|A|v\rangle|^2 \geq \langle u|A|u\rangle\langle v|A|v\rangle .$$

Under what conditions does the equality hold?

2. Let $A(x)$ be an operator that depends on a continuous variable x. Define its derivative by

$$\frac{dA}{dx} \equiv A'(x) = \lim_{\epsilon \to 0} \frac{A(x+\epsilon) - A(x)}{\epsilon} .$$

If A has an inverse, show that

$$\frac{d}{dx} A^{-1} = -A^{-1} A' A^{-1} ;$$

and finally, if A and B both depend on x, that $d(AB)/dx = A'B + AB'$.

[1] V.I. Arnold, *Mathematical Methods of Classical Mechanics*, Springer (1980), pp. 442–445.

3. The object is to derive the density matrix $\rho_T = e^{-\beta H}/Z$ for a system in thermal equilibrium, where $\beta = 1/kT$. Show it results from maximizing the entropy for a fixed value of the internal energy $U = \mathrm{Tr}\,\rho_T H$.

4. Consider a system composed of two parts s and R, and let ρ be the density matrix for some state of the combined system. Show that for the subsystem to be in a pure state, ρ must have the form $P_s \otimes \rho_R$, where P_s is some projection operator onto the Hilbert space of s.

5. Confirm that Eq. 150 follows from Eq. 148. Show that the phase space representation of an arbitrary observable $\mathcal{A}(q,p)$ is given by Eq. 152.

6. Show that the uncertainty relation Eq. 203 becomes an equality when $|\phi\rangle$ is such that $|\phi_A\rangle$ and $|\phi_B\rangle$ are collinear, i.e., $|\phi_A\rangle = \lambda|\phi_B\rangle$, and $\lambda = \langle C\rangle/2i(\Delta B)^2$.

7. Show that the electrical charge and current densities, as given by (232) and (233), satisfy the continuity equation.

8. (a) By integrating the Heisenberg equations of motion for a free particle, derive Eq. 257. Show that this equation agrees with the evolution of the momentum distribution of a system of non-interacting particles in classical mechanics. Was this to be expected?

(b) Use Eq. 258 to compute the spreading of the following free particle wave packets confined at $t = 0$ to volumes of dimension R: (i) an electron with R the Bohr radius in hydrogen, and R the electron's Compton wavelength; (ii) an alpha particle for the same values of R, and the radius of the Pb nucleus; and (iii) and a golf ball with $R = 1\ \mu$m.

9. Show that the energy distribution of Eq. 270 leads to the exponential decay law if $\Gamma \ll E_0$.

10. Define $\Delta\boldsymbol{K} = (R-1)\boldsymbol{K}$ for the rotation R of any 3-vector $\boldsymbol{K}$, and let R_1 and R_2 be two infinitesimal rotations parametrized with $\delta\boldsymbol{\omega}_1$ and $\delta\boldsymbol{\omega}_2$. If $\Delta\boldsymbol{K}$ is the change induced in $\boldsymbol{K}$ by $R = R_2^{-1}R_1^{-1}R_2R_1$, show that to leading order

$$\Delta\boldsymbol{K} = -(\delta\boldsymbol{\omega}_1\times\delta\boldsymbol{\omega}_2)\times\boldsymbol{K}\ .$$

11. (a) Let $\exp(i\Xi)$ be the phase factor in Eq. 349. Show that under a gauge transformation

$$\left(\boldsymbol{p}_n - \frac{e_n}{c}\boldsymbol{A}_n\right)\psi \to e^{i\Xi}\left(\boldsymbol{p}_n - \frac{e_n}{c}\boldsymbol{A}_n\right)\psi\ ,$$

where $\boldsymbol{A}_n = \boldsymbol{A}(\boldsymbol{r}_n, t)$, and $\boldsymbol{A}'_n$ is given by (345). Show that the current as defined in (351) is gauge invariant, whereas (227) is not.

(b) Let $\boldsymbol{x}_n$ be the coordinate operator for a particle whose motion is governed by the Hamiltonian (348). Show that $\boldsymbol{v}_n = \dot{\boldsymbol{x}}_n$, where $\boldsymbol{v}_n$ is given by (350).

(c) Show that the probability current of Eq. 351 satisfies the continuity equation in configuration space (Eq. 226).

12. Derive Green's function for a free particle in one dimension (Eq. 390) by (i) evaluating the $d = 1$ counterpart of (386); and (ii) by directly solving the inhomogeneous Schrödinger equation.

13. This problem concerns the following important special case of the BCH theorem: If A and B are two operators that do not commute with each other but which both commute with $[A, B]$, they satisfy

$$e^{A+B} = e^A\, e^B\, e^{-\frac{1}{2}[A,B]}\ .$$

(a) To prove this, first show that $[B, e^{xA}] = e^{xA}[B, A]x$. Next, define $G(x) = e^{xA}e^{xB}$, and show that

$$\frac{dG}{dx} = (A + B + [A, B]x)G\,.$$

Integrate this to obtain the desired result.

(b) More generally, show that for arbitrary A and B

$$\lim_{\alpha,\beta\to 0} e^{\alpha A}\, e^{\beta B} = e^{\alpha A + \beta B + \frac{1}{2}\alpha\beta[A,B] + X}\,,$$

where X is of higher order in α, β.

14. Let D_κ be the determinant of the κ-dimensional matrix (438). Show that $D_\kappa = 2D_{\kappa-1} - D_{\kappa-2}$, and that (443) is the solution of this recursion relation.

15. To prove that the Van Vleck determinant D satisfies the continuity equation, show first that for any non-singular matrix A with elements a_{ij} that depend on the parameter z,

$$\frac{\partial D}{\partial z} = D \sum_{ij} (A^{-1})_{ji} \frac{\partial a_{ij}}{\partial z}\,.$$

Calling $a_{ij} = \partial^2 S/\partial q_i \partial q'_j$, show that the continuity equation follows in virtue of the Hamilton-Jacobi equation.

Endnotes

For mathematically respectable presentations of quantum mechanics, see T.F. Jordan, *Linear Operators for Quantum Mechanics,* Wiley (1969); W. Thirring, *A Course in Mathematical Physics, v.3, Quantum Mechanics of Atoms and Molecules*, Springer (1981).

The question of whether the quantum state should be associated with individual systems, or with ensemble of systems, and the related issue of how probabilities should be introduced into quantum mechanics, are discussed by W. Heisenberg, *Physics and Philosophy*, Harper (1962); and A. Shimony, *Search for a Naturalistic World View,* Vol. II, Cambridge University Press (1993); A. Shimony in *Epistemological and Experimental Perspectives on Quantum Physics,* D. Greenberger et al. (eds.), Kluwer (1999). This question has been analyzed in mathematical terms by E. Farhi, J. Goldstone and S. Gutmann, *Ann. Phys.* **192**, 368 (1989); and S. Gutmann, *Phys. Rev. A* **52**, 3560 (1995). They formulate the postulates in terms of the state $|\psi\rangle$ of an individual system $\mathcal{S}$; introduce the state $|\psi\rangle^\infty$ of an infinite number of copies of $\mathcal{S}$, i.e., $|\psi\rangle \otimes |\psi\rangle \otimes \ldots$, and a frequency operator $F(a_i)$ for the eigenvalue a_i of the observable A pertaining to an individual $\mathcal{S}$; and show that $|\psi\rangle^\infty$ is an eigenstate of $F(a_i)$ with eigenvalue $|\langle a_i|\psi\rangle|^2$.

Probabilities having the quantum mechanical interference properties are derived by Y.F. Orlov from the absence, in principle, of any algorithm that determines the outcome of measurements: *Phys. Rev. Lett.* **82**, 243 (1999); *Ann. Phys.* **234**, 245 (1994).

Because quantum states are in general not pure, and also because the state space is a ray space, a formulation that treats the density matrix as basic, rather than kets, has considerable attraction. For such a formulation see R. Balian, *Am. J. Phys.* **57**, 1019 (1989); and *From Microphysics to Macrophysics*, Vol. 1, Springer (1991), chapter 2.

The path integral is treated in a mathematically serious manner in B. Simon, *Functional Integration in Quantum Physics*, Academic Press (1979). The classic reference on path integrals is R.P. Feynman and A.R. Hibbs, *Quantum Mechanics and Path Integrals*, McGraw Hill (1965).

Our treatment of semiclassical quantum mechanics draws on M.V. Berry and K.E. Mount, *Rep. Prog. Phys.* **35**, 315-397 (1972); M.C. Gutzwiller, *Chaos in Classical and Quantum Mechanics*, Springer (1990); E.J. Heller, in *The Physics and Chemistry of Wave*

Packets, J.A. Yeazell and T. Uzer (eds.), Wiley (2000); W.H. Miller, *Adv. Chem. Phys.* **25**, 69-177 (1974); W. Pauli, *Pauli Lectures in Physics*, Vol. 6, C.P. Enz (ed.), MIT Press (1973); L.S. Schulman, *Techniques and Applications of Path Integration*, Wiley (1981); and H. Kleinert, *Path Integrals in Quantum Mechanics, Statistics and Polymer Physics*, 2nd. ed., World Scientific (1995). An extensive and remarkable bibliography on semiclassical quantum mechanics and related topics is provided by M.C. Gutzwiller, *Am. J. Phys.* **66**, 304 (1998). The connection between mechanics and geometrical optics is treated in M. Born and E. Wolf, *Principles of Optics*, Pergamon (1959).

3
Basic Tools

This chapter is devoted to a set of topics that arise repeatedly in developing quantum mechanics to the point where it can be applied to specific phenomena and systems. These basic tools should be at the disposal of the student as the need arises. Most readers will already be familiar with at least some of these, but we cannot know which. To maximize attention to the essence of the physics that will be treated in the remainder of this volume, we develop these tools *ab initio* here. Many readers will already know enough to skip this chapter to begin with, and may prefer to return to it should they find that their knowledge needs refreshment or elaboration.

The bulk of this chapter is devoted to the theory of angular momentum and the behavior of key observables under rotation. These are of great importance because quantum mechanics deals so often with systems that are, to a high degree of precision, isolated and therefore invariant under rotation. The quantum mechanics of free particles, and of the isolated two-body system, which are fundamental to the theory of collisions and spectra, respectively, are then developed. Finally, as in all of physics, exactly solvable problems that describe realistic situations are very rare, and therefore approximation methods play essential roles in applications. The last section provides an introduction to such methods.

3.1 Angular Momentum: The Spectrum

The entire theory of angular momentum, with all its elaborations, stems from the commutation rules derived in §2.5(c):

$$[J_i, J_j] = i\epsilon_{ijk} J_k \,, \qquad (i,j,k = 1,2,3) \,. \tag{1}$$

These, please recall, are imposed by the properties of rotations in Euclidean 3-space, $\mathfrak{E}_3$, and are therefore required to hold for the angular momentum of any dynamical system in quantum mechanics. It should be understood, therefore, that what follows in this section is general, and is not restricted to the orbital angular momentum of a single particle.[1]

Equation (1) implies that only one component of $\boldsymbol{J}$ can be diagonalized, and it is traditional to chose J_3 (or if you prefer, J_z) for this role. Furthermore, J^2 (i.e., $\boldsymbol{J}^2$), being invariant under rotations, commutes with all three J_i. Hence J^2 and J_3 are chosen as the compatible observables to be diagonalized simultaneously. The simultaneous eigenkets are called $|jm\rangle$, and the eigenvalues are defined by

$$J^2|jm\rangle = j(j+1)|jm\rangle, \qquad J_3|jm\rangle = m|jm\rangle \,. \tag{2}$$

---

[1]There is one important exception to this statement: states of zero mass, and in particular one photon states, require the separate treatment given in §10.1.

At this point all we know is that m is real because J_3 is Hermitian, and $j(j+1)$ is real and positive, the proof of the latter property being left as an exercise. The "peculiar" expression $j(j+1)$ is chosen in the wisdom of hindsight, and could for now be replaced by any symbol. The spectra of j and m are discrete, and so the orthonormality relations are

$$\langle jm|j'm'\rangle = \delta_{jj'}\delta_{mm'} \,. \tag{3}$$

Two non-Hermitian operators play a key role in the theory of angular momentum:

$$J_\pm = J_1 \pm iJ_2 \,; \tag{4}$$

clearly $J_+ = J_-^\dagger$. Short calculations establish the commutation rules among J_+, J_-, J_3,

$$[J_+, J_-] = 2J_3 \,, \tag{5}$$

$$[J_+, J_3] = -J_+ \,, \qquad [J_-, J_3] = J_- \,, \tag{6}$$

and the identities

$$J_+J_- = J^2 - J_3(J_3 - 1) \,, \tag{7}$$

$$J_-J_+ = J^2 - J_3(J_3 + 1) \,. \tag{8}$$

A constraint on m follows from

$$\langle jm|J_1^2 + J_2^2|jm\rangle = \langle jm|J^2 - J_3^2|jm\rangle = j(j+1) - m^2 \,. \tag{9}$$

But this must be positive, and therefore

$$m^2 \leq j(j+1) \,. \tag{10}$$

Thus m is bounded above and below for a given j; call these bounds $m_>$ and $m_<$.

Next, consider $J_\pm|jm\rangle$; this too is an eigenket of J^2 with eigenvalue $j(j+1)$ because the $J_\pm$ commute with J^2. To determine whether they are eigenkets of J_3, use (6):

$$J_3J_\pm|jm\rangle = (J_\pm J_3 \pm J_\pm)|jm\rangle = (m \pm 1)J_\pm|jm\rangle \,. \tag{11}$$

Thus $J_\pm|jm\rangle$ are eigenkets of J_3 with eigenvalues $m \pm 1$:

$$J_\pm|jm\rangle = a_\pm(jm)|jm \pm 1\rangle \,. \tag{12}$$

For this reason, $J_\pm$ are called angular momentum raising and lowering operators.

The constants $a_\pm$ are evaluated with the help of (7) and (8):

$$|a_\pm|^2 = \langle jm|J_\mp J_\pm|jm\rangle = \langle jm|J^2 - J_3(J_3 \pm 1)|jm\rangle \,, \tag{13}$$

and therefore

$$|a_\pm(jm)|^2 = j(j+1) - m(m \pm 1) \,. \tag{14}$$

The standard phase convention is to take $a_\pm$ as positive, which then gives the following important relationship:

$$J_\pm|jm\rangle = \sqrt{j(j+1) - m(m \pm 1)}\;|jm \pm 1\rangle \,. \tag{15}$$

The right-hand side of this last equation must vanish when m reaches its limiting values, for if it did not, states would be produced that did not satisfy the condition $m_< \leq m \leq m_>$. Hence

$$j(j+1) = m_>(m_> + 1) = m_<(m_< - 1) \; . \tag{16}$$

Thus $(m_> + m_<)(m_> - m_< + 1) = 0$, and since the second factor is positive, $m_> = -m_<$, which when combined with (16) gives

$$m_> = -m_< = j \; . \tag{17}$$

The allowed values of m, and via (17) of j, now follow because $a_+ = 0$ only if $m = m_>$, and $a_- = 0$ only if $m = m_<$. Assume m_0 is an allowed eigenvalue, k some integer, and consider $(J_+)^k|jm_0\rangle$. This is proportional to $|jm_0 + k\rangle$, which must for some k satisfy the condition $m_0 + k = m_>$. By the same argument but using $(J_-)^l|jm_0\rangle$, it must be that $m_0 - l = m_< = -m_>$, where l is another integer. Hence $k + l = 2m_> = 2j$ must be an integer.

This establishes the spectrum of the angular momentum operators:

- *The eigenvalues of* J^2 *are* $j(j+1)$, *where* $j = 0, \frac{1}{2}, 1, \frac{3}{2}, \ldots$;
- *Given* j, *there are* $2j+1$ *eigenstates of* J_3 *with eigenvalues* $m = j, j-1, \ldots, -j$.

The $2j+1$ dimensional complex vector space spanned by the states of a given value of j will be designated as $\mathfrak{H}_j$.

The whole space $\mathfrak{H}_j$ can be constructed once one of its members is known by repeated use of J_+ and/or J_-. In particular

$$|jm\rangle = \left[\frac{(j+m)!}{(2j)!(j-m)!}\right]^{\frac{1}{2}} (J_-)^{j-m}|jj\rangle = \left[\frac{(j-m)!}{(2j)!(j+m)!}\right]^{\frac{1}{2}} (J_+)^{j+m}|j-j\rangle \; . \tag{18}$$

The spectrum of J_3 implies that the angular momentum operators have two important algebraic properties that do not depend on the choice of basis. These are most easily seen by first stating them in a basis-dependent manner. The eigenvalues of J_3 are symmetrically distributed about $m = 0$, and therefore $\sum_m \langle jm|J_3|jm\rangle = 0$. But this is Tr J_3 in the (j, m) basis, and as the trace is invariant under any unitary transformation on the basis, Tr $J_3 = 0$. Next, recall the unitary transformation in $\mathfrak{H}_j$ corresponding to a rotation. From Eq. 2.315 in §2.5(c),

$$D^j(R) = e^{-i\theta \boldsymbol{n} \cdot \boldsymbol{J}} \; , \tag{19}$$

where the J_i are the $(2j+1)$ dimensional matrices. By an appropriate choice of $(\boldsymbol{n}, \theta)$, $D^\dagger J_3 D$ will turn J_3 into any desired direction, leaving its trace unchanged. Hence

$$\mathrm{Tr}\; \boldsymbol{J} = 0 \tag{20}$$

for the whole vector operator $\boldsymbol{J}$.

The second representation-independent property of $\boldsymbol{J}$ is a consequence of the Caley-Hamilton theorem: If M is a Hermitian $n \times n$ matrix with eigenvalues $\lambda_1, \ldots,$ λ_n, then $\prod_i (M - \lambda_i) = 0$. This equation is invariant under $M \to U^\dagger M U$. Now let

$M = J_3$, and carry out the unitary transformation that turns J_3 into another direction. As a consequence, *for all components of* $\boldsymbol{J}$,

$$\prod_{m=-j}^{j} (J_i - m) = 0 . \tag{21}$$

This means that in $\mathfrak{H}_j$ any function of the angular momentum operators, such as $D^j(R)$, is a polynomial of at most degree $2j$.

There is a fundamental distinction between the states that have integer and half-integer eigenvalues j. This is revealed by a rotation about the axis of the component of $\boldsymbol{J}$ chosen to be diagonal, i.e., J_3 in our convention:

$$e^{-i\theta J_3}|jm\rangle = e^{-im\theta}|jm\rangle . \tag{22}$$

For a complete rotation through $\theta = 2\pi$, this state will not change if j is an integer, but will change sign if j is half-integer. That is to say, *states of integer angular momentum are single-valued under rotation, whereas states of half-integer angular momentum are double-valued under rotation!*

Experiments have confirmed that the neutron, which has spin $\frac{1}{2}$, has double-valued wave functions.

Because kets (or wave functions) are not in themselves observable quantities, they need not be single-valued. On the other hand, a Hermitian operator A that purports to be an observable must be single-valued under rotation to insure that its expectation value $\langle\Psi|A|\Psi\rangle$ is single-valued in an arbitrary state. Because this is a bilinear form, expectation values of single-valued observables will be single-valued under rotation even if $|\Psi\rangle$ is double-valued.

Consider, however, a state that is a coherent superposition of single- and double-valued states, $|\Psi_+\rangle + |\Psi_-\rangle$, and the expectation value of some observable in this state:

$$\langle\Psi|A|\Psi\rangle = \langle\Psi_+|A|\Psi_+\rangle + \langle\Psi_-|A|\Psi_-\rangle + 2\text{Re}\,\langle\Psi_+|A|\Psi_-\rangle . \tag{23}$$

This will be double-valued whether or not the observable A is single-valued. *Hence coherent superpositions of integer and half-integer angular momentum eigenstates are mathematical constructs that do not describe actual physical states.* This is an example of a *superselection rule,* a restriction on what can be coherent whatever the system or its interactions may be.

3.2 Orbital Angular Momentum

The general discussion of the preceding section, which made no commitment to the nature of the system in question, will now be specialized to that of a single particle whose canonical variables are the position and momentum operators $\boldsymbol{x}$ and $\boldsymbol{p}$.

In this case the angular momentum is purely orbital, $\boldsymbol{L} = \boldsymbol{x}\times\boldsymbol{p}/\hbar$, and as shown in §2.5, the canonical commutation rules then lead to the generic commutation rules $[L_1, L_2] = iL_3$ and cyclic permutations. It is again convenient to define raising and lowering operators,

$$L_\pm = L_1 \pm iL_2 , \tag{24}$$

with commutation rules

$$[L_+, L_-] = 2L_3 , \quad [L_\pm, L_3] = \mp L_\pm . \tag{25}$$

Once again, simultaneous eigenstates $|lm\rangle$ of L^2 and L_3 are introduced, and their eigenvalues defined by

$$L^2|lm\rangle = l(l+1)|lm\rangle\,, \qquad L_3|lm\rangle = m|lm\rangle\,. \tag{26}$$

Thus far nothing has been said to differentiate the orbital angular momentum $\boldsymbol{L}$ from the general operator $\boldsymbol{J}$. This distinction is made by defining L_i to be a component of $\boldsymbol{x}\times\boldsymbol{p}/\hbar$, which means that the L_i can be represented by differential operators acting on one-particle Schrödinger wave functions. In short, whereas the general discussion of §3.1 applies, for example, to the scattering of light by a molecule, provided $\boldsymbol{J}$ is the sum of all the angular momenta in the problem — that of the electromagnetic field and the electrons and nuclei that form the molecule, what follows would only apply to one of the nuclei as long as it has spin zero.

The differential operators just mentioned are derived from the Schrödinger representation of the linear momenta, $(\hbar/i)\partial/\partial x_i$ by first representing the coordinate operator by spherical coordinates, $\boldsymbol{x} \to (r,\theta,\phi)$. Then an exercise in calculus produces the following relationships:

$$L_3 \to \frac{1}{i}\frac{\partial}{\partial\phi}\,, \tag{27}$$

$$L_\pm \to \pm e^{\pm i\phi}\left(\frac{\partial}{\partial\theta} \pm i\cot\theta\,\frac{\partial}{\partial\phi}\right)\,, \tag{28}$$

where $\to$ means that the abstract operator is replaced by the differential operator. One example will suffice to show the details of this relationship:

$$\langle\theta\phi|L_3|lm\rangle = \frac{1}{i}\frac{\partial}{\partial\phi}\langle\theta\phi|lm\rangle\,. \tag{29}$$

The wave function in this last expression is called a *spherical harmonic*:

$$Y_{lm}(\theta\phi) \equiv \langle\theta\phi|lm\rangle\,. \tag{30}$$

In view of (29), the eigenvalue equation for L_3 is

$$\left(\frac{1}{i}\frac{\partial}{\partial\phi} - m\right)Y_{lm}(\theta\phi) = 0\,, \tag{31}$$

and therefore

$$Y_{lm}(\theta\phi) = e^{im\phi}y_{lm}(\theta)\,. \tag{32}$$

The raising and lowering operators give zero when acting on the states with $m = l$ and $m = -l$, respectively, i.e., $L_\pm Y_{l,\pm l} = 0$. Equation (28) therefore gives the differential equation

$$\left(\frac{d}{d\theta} - l\cot\theta\right)y_{l,\pm l} = 0\,, \tag{33}$$

which has the solution

$$y_{l,\pm l} = c_l(\sin\theta)^l\,. \tag{34}$$

Equation (34) suffices to eliminate half-integer values of l as eigenvalues of orbital angular momentum. The simplest case is $l = \frac{1}{2}$, and as shown in Prob. 1, the functions $e^{\pm i\phi/2}\sqrt{\sin\theta}$ do not transform properly under rotation. That is, *orbital angular momenta can only have integer eigenvalues.*

All the the spherical harmonics for a given l can be computed from $y_{l,\pm l}$ by repeated use of $L_{\mp}$. For one such step, Eq. 15 already gives the answer:

$$L_{\pm} Y_{lm} = \sqrt{l(l+1) - m(m \pm 1)}\ Y_{lm\pm 1}\,. \tag{35}$$

As in (18), repeated use gives

$$Y_{lm} = \sqrt{\frac{(l+m)!}{(2l)!(l-m)!}}\,(L_-)^{l-m} Y_{ll}\,. \tag{36}$$

Eq. 35 is then expressed as a differential operator in θ by use of (28):

$$\begin{aligned} L_- e^{im\phi} f(\theta) &= -e^{i(m-1)\phi}\left(\frac{d}{d\theta} + m\cot\theta\right) f(\theta) \\ &= e^{i(m-1)\phi}(\sin\theta)^{l-m}\frac{d}{d\cos\theta}[f(\theta)\sin^m\theta]\,. \end{aligned} \tag{37}$$

This is again a function of the form $e^{i\lambda\theta}g(\theta)$. Once more, repetition leads to

$$L_-^k e^{im\phi} f(\theta) = e^{i(m-k)\phi}\sin^{k-m}\theta\left(\frac{d}{d\cos\theta}\right)^k f(\theta)\sin^m\theta\,. \tag{38}$$

All that remains is to determine the overall coefficient c_l, which is set by normalization and phase conventions. The spherical harmonics are normalized as follows:

$$\langle lm|l'm'\rangle = \delta_{ll'}\delta_{mm'} = \int_0^{2\pi} d\phi \int_{-1}^{1} d\cos\theta\ \ Y_{lm}(\theta\phi)Y^*_{l'm'}(\theta\phi)\,. \tag{39}$$

The case of $\langle ll|ll\rangle$ is simple and determines c_l:

$$1 = |c_l|^2 4\pi(2^l l!)^2[(2l+1)!]^{-1}\,, \tag{40}$$

and with the traditional phase convention this gives

$$c_l = (-1)^l \frac{1}{2^l l!}\sqrt{\frac{(2l+1)!}{4\pi}}\,. \tag{41}$$

Using (36) and (38) then gives the final result

$$Y_{lm}(\theta\phi) = \frac{(-1)^l}{2^l l!}\sqrt{\frac{2l+1}{4\pi}\,\frac{(l+m)!}{(l-m)!}}\,e^{im\phi}\sin^{-m}\theta\left(\frac{d}{d\cos\theta}\right)^{l-m}\sin^{2l}\theta\,. \tag{42}$$

The θ dependence of Y_{lm} is seen to be a trigonometric polynomial, and when $m = 0$ it reduce to the Legendre polynomial P_l:

$$Y_{l0}(\theta\phi) = \sqrt{\frac{2l+1}{4\pi}}\,P_l(\cos\theta)\,. \tag{43}$$

For $m > 0$, the spherical harmonics are proportional to associated Legendre functions P_l^m:

$$Y_{lm}(\theta\phi) = (-1)^m \sqrt{\frac{2l+1}{4\pi}\frac{(l+m)!}{(l-m)!}}\, e^{im\phi} P_l^m(\theta) \quad (m \geq 0)\ ; \tag{44}$$

for negative m one uses a relation that follows from (42), and which is also useful in other contexts:

$$Y_{lm}^* = (-1)^m Y_{l,-m}\,. \tag{45}$$

The spherical harmonics form a complete, single-valued, and orthonormal set on the unit sphere. The completeness relation is derived in Prob. 3, and reads

$$\sum_{l=0}^{\infty}\sum_{m=-l}^{l} Y_{lm}(\theta\phi)Y_{lm}^*(\theta'\phi') = \frac{\delta(\theta-\theta')\delta(\phi-\phi')}{\sin\theta}\,. \tag{46}$$

The spherical harmonics satisfy many relations, among which the *addition theorem*, familiar from electrostatics, is especially useful. Let $\boldsymbol{n}_1, \boldsymbol{n}_2$, be two unit vectors whose orientations are specified by the polar angles $(\theta_i\phi_i)$, then

$$P_l(\boldsymbol{n}_1\cdot\boldsymbol{n}_2) = \frac{4\pi}{2l+1}\sum_{m=-l}^{l} Y_{lm}(\theta_1\phi_1)Y_{lm}^*(\theta_2\phi_2)\,. \tag{47}$$

This identity will be derived by use of the rotation group in §7.4.

Because spherical harmonics are used so much, it is convenient to have a more compact notation. The angular variables, and the element of solid angle, will often be denoted as follows:

$$\Omega \equiv \boldsymbol{n} \equiv (\theta,\phi) \qquad d\Omega \equiv d\boldsymbol{n} \equiv \sin\theta d\theta d\phi = d\phi d(\cos\theta)\,. \tag{48}$$

Here $\boldsymbol{n}$ is a unit vector whose orientation is specified by the angles θ and ϕ. In this notation (39) is

$$\int d\Omega\, Y_{lm}(\Omega)Y_{l'm'}^*(\Omega) = \int d\boldsymbol{n}\, Y_{lm}(\boldsymbol{n})Y_{l'm'}^*(\boldsymbol{n}) = \delta_{ll'}\delta_{mm'}\,. \tag{49}$$

Explicit formulas for small l are in the Appendix, but those for $l = 1$ merit special attention. They are

$$Y_{1,\pm1} = \mp\sqrt{3/8\pi}\; e^{\pm i\phi}\sin\theta, \quad Y_{10} = \sqrt{3/4\pi}\;\cos\theta\,. \tag{50}$$

An important point is to be noted here and remembered: these three angular momentum eigenfunctions are nothing but linear combinations of the three Cartesian components of the position vector (x_1, x_2, x_3) :

$$(Y_{11}, Y_{10}, Y_{1,-1}) = \left(-\frac{x_1+ix_2}{\sqrt{2}}, x_3, \frac{x_1-ix_2}{\sqrt{2}}\right)\frac{1}{r}\sqrt{\frac{3}{4\pi}}\,. \tag{51}$$

The basis on the right is called the spherical basis. The inverse relation is

$$(x_1,x_2,x_3) = \left(-\frac{Y_{11}-Y_{1,-1}}{\sqrt{2}}, \frac{i(Y_{11}+Y_{1,-1})}{\sqrt{2}}, Y_{10}\right)\cdot r\sqrt{\frac{4\pi}{3}}\,, \tag{52}$$

and the connection between them can be written in terms of a matrix acting on column vectors:

$$\begin{pmatrix} x_1 \\ x_2 \\ x_3 \end{pmatrix} = W \begin{pmatrix} Y_{11} \\ Y_{10} \\ Y_{1,-1} \end{pmatrix} N\,, \qquad \begin{pmatrix} Y_{11} \\ Y_{10} \\ Y_{1,-1} \end{pmatrix} = W^\dagger \begin{pmatrix} x_1 \\ x_2 \\ x_3 \end{pmatrix} N^{-1}\,, \tag{53}$$

where $N = r\sqrt{4\pi/3}$, and W is the unitary matrix

$$W = \begin{pmatrix} -\sqrt{\frac{1}{2}} & 0 & \sqrt{\frac{1}{2}} \\ i\sqrt{\frac{1}{2}} & 0 & i\sqrt{\frac{1}{2}} \\ 0 & 1 & 0 \end{pmatrix}. \tag{54}$$

We of course know how the components of the vector (x_1, x_2, x_3), or for that matter any vector, transform under rotation. Equation (53) tells us that the three spherical harmonics Y_{1m} transform in essentially the same way. To be specific, let R be the matrix that produces the rotated coordinates x_i' when acting on the column vector formed by the x_i:

$$\begin{pmatrix} x_1' \\ x_2' \\ x_3' \end{pmatrix} = R \begin{pmatrix} x_1 \\ x_2 \\ x_3 \end{pmatrix}; \tag{55}$$

then

$$\begin{pmatrix} Y_{11}(\boldsymbol{n}') \\ Y_{10}(\boldsymbol{n}') \\ Y_{1,-1}(\boldsymbol{n}') \end{pmatrix} = W^\dagger R W \begin{pmatrix} Y_{11}(\boldsymbol{n}) \\ Y_{10}(\boldsymbol{n}) \\ Y_{1,-1}(\boldsymbol{n}) \end{pmatrix}, \tag{56}$$

and this must also apply to the three kets $|1m\rangle$.

3.3 Spin

The term "spin" is among the most abused items in the jargon of physics. It is used in so many different contexts that it is virtually impossible to avoid confusing the uninitiated, and sometimes even the initiated. Here, in this first discussion, we strive to avoid this pitfall, but in later chapters we will surely fail at times, if for no other reason than to avoid long-winded explanations of what is meant when that is hopefully clear from the context.

The concept of spin is all too often identified with angular momentum $\frac{1}{2}$. That is a misunderstanding. In nonrelativistic quantum mechanics, spin should be taken as the total angular momentum about the center-of-mass, whatever magnitude it may have, and whether the system is known to be composite or believed to be "elementary." Thus the spin $\frac{1}{2}$ charged leptons, e, μ and τ, and their associated neutrinos are, as far as we now know, structureless. The nucleons, which are well described as bound states of three structureless spin $\frac{1}{2}$ quarks, behave like elementary particles in all processes that impart less energy than several hundred MeV to them, and can in this regime be treated as if they were elementary spin $\frac{1}{2}$ particles. Nuclei composed of various numbers of protons and neutrons act as elementary particles of various spins in atomic and condensed matter physics. But there are particles now believed to be elementary with spin other than $\frac{1}{2}$, in particular the spin 1 particles W and Z involved in the weak interaction, which have masses of roughly 80 and

91 GeV/c^2, respectively, i.e., comparable to those of nuclei in the middle of the periodic table. Finally, the photon is also structureless and elementary to the best of our knowledge, but because it has zero mass, the concept of its center-of-mass is meaningless and its angular momentum states cannot be described by the the theory we are developing at this point; in particular, the one-photon state has spin 1 in the sense that the projection of its angular momentum along its direction of motion is either 1 or -1, but there is no projection of magnitude 0, something which is incomprehensible in nonrelativistic quantum mechanics, and which is related to absence of longitudinally polarized electromagnetic waves in vacuum.

In this section the two cases of spin $\frac{1}{2}$ and spin 1 will be analyzed, the first rather extensively because spin $\frac{1}{2}$ is important not only in angular momentum theory itself, but also because it provides a language that fully describes any two-state system, whether or not angular momentum has anything to do with the matter.

Before beginning this discussion, a remark about the behavior of spin under transformation between inertial frames is in order. If $\boldsymbol{J}$ is defined as the total angular momentum of a system of N massive particles about their center-of-mass in the frame where the total momentum vanishes, then $\boldsymbol{J}$ is invariant under a Galileo transformation to a frame moving with uniform velocity (see Prob. 4). This result does not hold for a Lorentz transformation; in fact, the concept of center-of-mass cannot be defined in relativistic mechanics. As a consequence, in nonrelativistic quantum mechanics the state of a particle of nonzero spin can be written in the form $|\text{momentum}\rangle \otimes |\text{spin}\rangle$ in any frame, whereas Lorentz transformations from the rest frame to frames where the momentum is comparable to mc produces a mingling of the spin and momentum dependence, as we shall see when we come to the Dirac equation.

(a) Spin $\frac{1}{2}$

When $j = \frac{1}{2}$, it is convenient to replace the 2×2 angular momentum matrices by the *Pauli matrices* σ_i:

$$J_i = \tfrac{1}{2}\sigma_i \,. \tag{57}$$

Their explicit forms, in the conventions of §3.1, are

$$\sigma_1 = \begin{pmatrix} 0 & 1 \\ 1 & 0 \end{pmatrix}, \quad \sigma_2 = \begin{pmatrix} 0 & -i \\ i & 0 \end{pmatrix}, \quad \sigma_3 = \begin{pmatrix} 1 & 0 \\ 0 & -1 \end{pmatrix}. \tag{58}$$

The raising and lowering operators $\sigma_\pm = (J_1 \pm iJ_2)$ are

$$\sigma_+ = \begin{pmatrix} 0 & 1 \\ 0 & 0 \end{pmatrix}, \quad \sigma_- = \begin{pmatrix} 0 & 0 \\ 1 & 0 \end{pmatrix}. \tag{59}$$

The eigenstates $j = \frac{1}{2}$ and $m = \pm\frac{1}{2}$ will be called $|\pm\rangle$, and they can, if one prefers, be represented by 2-component column vectors:

$$|+\rangle = \begin{pmatrix} 1 \\ 0 \end{pmatrix}, \quad |-\rangle = \begin{pmatrix} 0 \\ 1 \end{pmatrix}. \tag{60}$$

The actions of the Pauli matrices on this basis are

$$\sigma_3|\pm\rangle = \pm|\pm\rangle, \quad \sigma_\pm|\mp\rangle = |\pm\rangle, \quad \sigma_1|\pm\rangle = |\mp\rangle, \quad \sigma_2|\pm\rangle = \pm i\,|\mp\rangle \,. \tag{61}$$

The Pauli matrices must, of course, satisfy the angular momentum commutation rules:

$$[\sigma_i, \sigma_j] = 2i\epsilon_{ijk}\sigma_k \,. \tag{62}$$

But Eq. 21 states that there are simpler relations specific to $j = \frac{1}{2}$, and in particular $J_i^2 = \frac{1}{4}$, or

$$\sigma_i^2 = 1 \,. \tag{63}$$

These other relations, between quadratic and linear forms, are

$$\sigma_i\sigma_j + \sigma_j\sigma_i = 0 \qquad (i \neq j) \,, \tag{64}$$

$$\sigma_i\sigma_j = i\epsilon_{ijk}\sigma_k \quad (i \neq j) \,. \tag{65}$$

They can all be combined into

$$\sigma_i\sigma_j = i\epsilon_{ijk}\sigma_k + \delta_{ij} \,. \tag{66}$$

The universal commutation rule (62) is, in this instance, a consequence of this special rule.

A bit of advice. Whenever possible, avoid use of the explicit representation (58). Essentially all calculations can be carried out much more simply, and with far less likelihood of error, by use of the identities (63–66), and

$$\mathrm{Tr}\, \sigma_i = 0 \,, \tag{67}$$

which is Eq. 20 for $j = \frac{1}{2}$.

The importance of the spin $\frac{1}{2}$ angular momentum operators stems from the following theorem: *The Pauli matrices, and the 2×2 unit matrix, form a complete set of 2×2 matrices.* That is to say, an arbitrary 2×2 matrix M can be written as a linear combination of 1 and the three σ_i:

$$M = m_0\mathbf{1} + \sum_i m_i\sigma_i = \begin{pmatrix} m_0 + m_3 & m_1 - im_2 \\ m_1 + im_2 & m_0 - m_3 \end{pmatrix} . \tag{68}$$

For example, M will be Hermitian if the numbers $(m_0, \ldots, m_3)$ are real, because the σ_i are Hermitian. They can, however, be complex. In consequence, all observables, unitary operators, etc., in *any* two-dimensional Hilbert space can be expressed as a linear combination of 1 and the σ_i, whether or not that space refers to a spin $\frac{1}{2}$ system.

The theorem just stated in italics is really a restatement of Eq. 66, as illustrated by the following important identity which reduces a frequently encountered quadratic form in the σ_i to a linear form:

$$\sum_{ij}(\sigma_i A_i)(\sigma_j B_j) = \sum_i A_i B_i + i\sum_{ijk} \epsilon_{ijk} A_i B_j \sigma_k \,, \tag{69}$$

where the A_i and B_i are six quantities that commute with the Pauli matrices, but not necessarily with each other.

It is natural to write (69) in vector notation without further thought:

$$(\boldsymbol{\sigma}\cdot\boldsymbol{A})(\boldsymbol{\sigma}\cdot\boldsymbol{B}) = \boldsymbol{A}\cdot\boldsymbol{B} + i\boldsymbol{\sigma}\cdot(\boldsymbol{A}\times\boldsymbol{B}) \,. \tag{70}$$

There are important special cases of this identity. If the *components* of $\boldsymbol{A}$ commute among themselves

$$(\boldsymbol{\sigma}\cdot\boldsymbol{A})^2 = A^2\ . \tag{71}$$

If they do not, $\boldsymbol{A}\times\boldsymbol{A}$ need not vanish, e.g., when $\boldsymbol{A}$ is some other angular momentum $\boldsymbol{J}$, in which case

$$(\boldsymbol{\sigma}\cdot\boldsymbol{J})^2 = J^2 - \boldsymbol{\sigma}\cdot\boldsymbol{J}\ . \tag{72}$$

Finally, there is the simplest case,

$$(\boldsymbol{\sigma})^2 = 3\ . \tag{73}$$

Is there a deeper meaning to the vector notation, however? To answer this, consider the unitary rotation operator for $j=\frac{1}{2}$, which is just Eq. 19 with the replacement (57):

$$D^{\frac{1}{2}}(R) = e^{-\frac{1}{2}i\theta\boldsymbol{n}\cdot\boldsymbol{\sigma}}\ . \tag{74}$$

In §2.5(c) we learned that the angular momentum is a vector operator, and this can be confirmed for $\boldsymbol{\sigma}$ by using (74) for an infinitesimal rotation $\delta\boldsymbol{\omega} = \boldsymbol{n}\delta\theta$,

$$(1+\tfrac{1}{2}i\delta\boldsymbol{\omega}\cdot\boldsymbol{\sigma})\,\boldsymbol{\sigma}\,(1-\tfrac{1}{2}i\delta\boldsymbol{\omega}\cdot\boldsymbol{\sigma}) = \boldsymbol{\sigma} + \delta\boldsymbol{\omega}\times\boldsymbol{\sigma}\ , \tag{75}$$

which can be checked with the identities obeyed by the σ_i. As expected, $\boldsymbol{\sigma}$ *transforms like a vector under rotations generated by the angular momentum operator* $\frac{1}{2}\boldsymbol{\sigma}$.

Whether it is justified to think of $\boldsymbol{A}$ and $\boldsymbol{B}$ in (70) as vector operators depends on the situation. For example, if the they are the electric and magnetic fields $\boldsymbol{E}$ and $\boldsymbol{B}$ applied *externally* to a system being treated quantum mechanically, then a rotation of that system will not affect them, and the expression (70) will not be a scalar. On the other hand, if the expression of interest is $S = (\boldsymbol{\sigma}\cdot\boldsymbol{p})(\boldsymbol{\sigma}\cdot\boldsymbol{x})$, where $\boldsymbol{p}$ and $\boldsymbol{x}$ are the momentum and coordinate of a particle with spin $\frac{1}{2}$, these operators will also behave like vectors if the rotation is generated by the particle's *total* angular momentum $\boldsymbol{L}+\frac{1}{2}\boldsymbol{\sigma}$, and then S will be a scalar, i.e., $D^\dagger(R)SD(R) = S$, where

$$D(R) = e^{-i\theta\boldsymbol{n}\cdot(\boldsymbol{L}+\frac{1}{2}\boldsymbol{\sigma})} = e^{-i\theta\boldsymbol{n}\cdot\boldsymbol{L}}e^{-\frac{1}{2}i\theta\boldsymbol{n}\cdot\boldsymbol{\sigma}} = D^L(R)D^{\frac{1}{2}}(R)\ . \tag{76}$$

The factorization is allowed because the orbital and spin angular momenta commute with each other as they act in distinct subspaces of $\mathfrak{H}$. The spin factor will rotate $\boldsymbol{\sigma}$ as in (75), and the orbital factor will produce the *same* rotation of $\boldsymbol{x}$ and $\boldsymbol{p}$, leaving S invariant.

The unitary operator for a finite rotation can be evaluated explicitly with ease in the case of $j=\frac{1}{2}$ because any higher-order polynomial in the Pauli matrices can be reduced to a linear form. Let $\Theta = \frac{1}{2}\theta\boldsymbol{n}\cdot\boldsymbol{\sigma}$, so that $D(R) = e^{-i\Theta}$. Equation (70) implies that

$$\Theta^n = (\tfrac{1}{2}\theta)^n \quad (n \text{ even}), \qquad \Theta^n = \Theta(\tfrac{1}{2}\theta)^{n-1} \quad (n \text{ odd})\ , \tag{77}$$

and so

$$e^{-i\Theta} = \sum_{n=0,2,\ldots}\frac{(\frac{1}{2}i\theta)^n}{n!} - i\,\Theta\sum_{n=1,3,\ldots}\frac{(\frac{1}{2}i\theta)^{n-1}}{n!}\ , \tag{78}$$

Therefore the spin $\frac{1}{2}$ rotation matrix is

$$D^{\frac{1}{2}}(R) = e^{-\frac{1}{2}i\theta\boldsymbol{n}\cdot\boldsymbol{\sigma}} = \cos\tfrac{1}{2}\theta - i\boldsymbol{\sigma}\cdot\boldsymbol{n}\sin\tfrac{1}{2}\theta\ . \tag{79}$$

This *rotation matrix is double-valued* as it must be for any half-integer angular momentum. Let the orientation of $\boldsymbol{n}$ be given by the polar angles (ϑ, φ) (see Fig. 3.1). Then from (58)

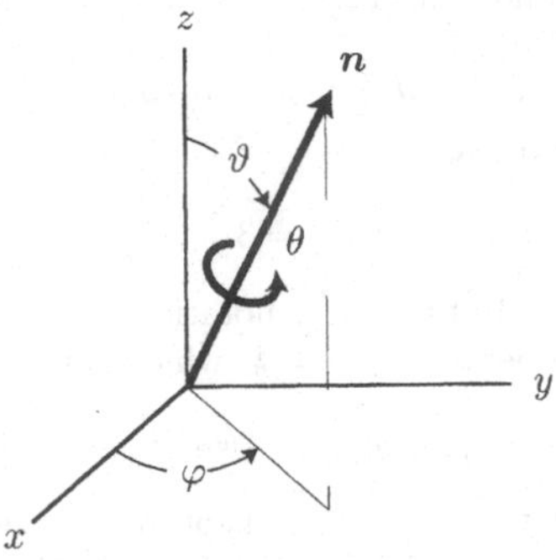

FIG. 3.1. Parametrizations of rotations: in terms of the axis $\boldsymbol{n}$ and angle θ of rotation; the angles $(\theta, \vartheta, \varphi)$ that specify the Caley-Klein parameters (see Eq. 81).

$$D^{\frac{1}{2}}(R) = \begin{pmatrix} a & b \\ -b^* & a^* \end{pmatrix}, \tag{80}$$

where the so-called Caley-Klein parameters are

$$a = \cos\tfrac{1}{2}\theta - i\sin\tfrac{1}{2}\theta\cos\vartheta\ , \quad b = -ie^{-i\varphi}\sin\tfrac{1}{2}\theta\sin\vartheta\ , \tag{81}$$

$$|a|^2 + |b|^2 = 1\ , \tag{82}$$

in virtue of which the matrix (80) is unitary. Indeed, the most general 2×2 unitary matrix has the form $e^{i\alpha}D^{\frac{1}{2}}(R)$, with α real, but as one can check in the equation that will follow, such an overall phase drops out of all observable quantities.

Consider now

$$D^\dagger(R)|+\rangle = a^*|+\rangle + b^*|-\rangle\ , \tag{83}$$

where $\frac{1}{2}$ is deleted from $D^{\frac{1}{2}}$, because in this subsection we only deal with $j = \frac{1}{2}$. An arbitrarily normalized ket in $\mathfrak{H}_{\frac{1}{2}}$ has the form (83), apart again from an overall phase. We now ask, is it an eigenstate of angular momentum? It certainly has $j = \frac{1}{2}$, as do all states in $\mathfrak{H}_{\frac{1}{2}}$, or better put, a rotation cannot change the eigenvalue of the scalar $\boldsymbol{J}^2$. But it also is an eigenstate of a particular component of $\boldsymbol{\sigma}$ with eigenvalue +1. To verify this, take $\sigma_3|+\rangle = |+\rangle$, from which

$$(D^\dagger\sigma_3 D)D^\dagger|+\rangle = D^\dagger|+\rangle\ , \tag{84}$$

and this states that the ket (83) is an eigenstate of $D^\dagger\sigma_3 D \equiv \tilde{\sigma}_3$ with eigenvalue +1, with $\tilde{\sigma}_3$ being the component of $\boldsymbol{\sigma}$ produced when σ_3 undergoes the rotation R.

To make the last statement more explicit and *inter alia* to show how to work with the representation invariant properties of Pauli matrices, let R be a rotation about $\boldsymbol{n} = (0,1,0)$, and $c = \cos\frac{1}{2}\theta$, $s = \sin\frac{1}{2}\theta$. Then

$$\begin{aligned}\tilde{\sigma}_3 &= (c + is\sigma_2)\sigma_3(c - is\sigma_2) = (c + is\sigma_2)^2\sigma_3 \\ &= (c^2 - s^2 + 2ics\sigma_2)\sigma_3 = \sigma_3\cos\theta - \sigma_1\sin\theta\ ,\end{aligned} \tag{85}$$

as promised. Note here how the half-angles, which do appear in the rotation of states, become "full" angles in the rotation of a vector operator, i.e., an observable. It is always thus, as it must be, or the observable would be double-valued. Note also that when $\theta \to 0$, Eq. 85 confirms (75).

The most general state in $\mathfrak{H}_{\frac{1}{2}}$ is described by a density matrix which must also be expressible in the form of Eq. 68:

$$\rho = \tfrac{1}{2}(1 + \boldsymbol{P} \cdot \boldsymbol{\sigma}) \,, \tag{86}$$

where $\boldsymbol{P}$ is a real numerical vector because ρ is Hermitian. As required, $\mathrm{Tr}\,\rho = 1$. If ρ is to be a density matrix, it must also satisfy $\mathrm{Tr}\,\rho^2 \leq 1$. From (71), $\mathrm{Tr}\,\rho^2 = \frac{1}{2}(1+P^2)$, and therefore

$$|\boldsymbol{P}| \leq 1 \,, \tag{87}$$

with $|\boldsymbol{P}| = 1$ being the case of a pure state. The vector $\boldsymbol{P}$ is called *the polarization*, because it gives the expectation value of $\boldsymbol{\sigma}$, i.e., of $2\boldsymbol{J}$:

$$\langle \boldsymbol{\sigma} \rangle = \mathrm{Tr}\,\rho\boldsymbol{\sigma} = \boldsymbol{P} \,. \tag{88}$$

The polarization also determines the expectation value of any vector operator in $\mathfrak{H}_{\frac{1}{2}}$, because such an operator must have the form $\boldsymbol{V} = \lambda\boldsymbol{\sigma}$, and therefore

$$\langle \boldsymbol{V} \rangle = \lambda \boldsymbol{P} \,. \tag{89}$$

The form of ρ is instructive when written in the representation that diagonalizes $\sigma_n = \boldsymbol{\sigma} \cdot \boldsymbol{P}$:

$$\rho = \begin{pmatrix} \frac{1}{2}(1+P) & 0 \\ 0 & \frac{1}{2}(1-P) \end{pmatrix} \,. \tag{90}$$

If $P = 1$, this is the density matrix of the pure state with eigenvalue $+1$ of σ_n. At the other extreme, $P = 0$, the state is a mixture that carries no directional information whatsoever; for any choice of two orthogonal spin $\frac{1}{2}$ states, it assigns probability 50% to each state. Because ρ is rotationally invariant in this case, any vector operator in $\mathfrak{H}_{\frac{1}{2}}$ must have a vanishing expectation value, in accord with (89).

What has just been established about the spin $\frac{1}{2}$ space leads to very important and general conclusions about the properties of any system of spin $\frac{1}{2}$. Because all observables in $\mathfrak{H}_{\frac{1}{2}}$ can be written as linear combinations of 1 and $\boldsymbol{\sigma}$, *in any state of a spin $\frac{1}{2}$ system the only possible observables must either be scalars or vectors under rotations.*[1] As a consequence, *if a system with total angular momentum $\frac{1}{2}$ carries charges and currents, its electromagnetic properties are fully described by a total charge and dipole moment;* higher multipoles vanish because in $\mathfrak{H}_{\frac{1}{2}}$ it is impossible to construct a second (or higher rank) tensor. Whether the dipole moment is electric or magnetic cannot be settled unless it is known whether space reflection and/or time reversal are symmetries.

(b) Spin 1

For $j = 1$ the state space $\mathfrak{H}_1$ is 3-dimensional. The angular momentum operators will be designated by $\boldsymbol{J} = (J_1, J_2, J_3)$, which can in this case be represented by

[1] This statement has no geometric meaning, of course, if the 2-dimensional space is being used to represent a system where the states are not related by rotation, e.g., a pair of states connected by radiative transition.

3×3 matrices. In addition to the universal angular momentum commutation rule, they satisfy a relation specific to $j = 1$ which, like Eq. 63, is a special case of Eq. 21:

$$J_i^3 = J_i \ . \tag{91}$$

The unitary rotation operator is again a special case of the generic form:

$$D^1(R) = e^{-i\theta \boldsymbol{n} \cdot \boldsymbol{J}} \ . \tag{92}$$

When expanded into power series, (91) produces a quadratic form in the operator $J_n = \boldsymbol{n} \cdot \boldsymbol{J}$,

$$D^1(R) = 1 - iJ_n \sin\theta - J_n^2(1 - \cos\theta) \ , \tag{93}$$

whose derivation is left as an exercise. The explicit form of this matrix can be found fairly easily by constructing the matrices J_i from the formulas in §3.1. We will not do this because it is not needed for now, and also because a powerful technique for calculating the rotation matrices for all values of j will be developed in §7.4.

A few remarks about $D^1(R)$ are appropriate here, nevertheless. First, because the states of $j = 1$ transform like a Euclidean 3-vector under rotations, it must be that $D^1(R)$ is the matrix that rotates such a vector. Second, the appearance of the quadratic operator J_n^2 means that when acting, for example, on the state $m = -1$, a rotation about any axis other than the 3-axis will produce a linear combination of states with all three values of m. That is to say, starting with any single ket in $\mathfrak{H}_1$, it is possible to reach all others by rotations. This was also true for $j = \frac{1}{2}$. It generalizes to all j, as can be seen from the theorem that followed from Eq. 21, because it stated that for spin j the rotation operator $D^j(R)$ is a polynomial in J_n of degree $2j$, which provides just enough raising and lowering operators $J_\pm$ to reach any state from any one of the kets $|jm\rangle$.

In the $j = \frac{1}{2}$ case, all operators in $\mathfrak{H}_{\frac{1}{2}}$ can be constructed as a linear combination of 1 and the three angular momentum operators. There is a counterpart of this important result for the case of $j = 1$. Any operator in $\mathfrak{H}_1$ is a 3×3 matrix, and can therefore be expressed as a linear combination of 9 linearly independent matrices, the most obvious set being those in which all elements but one vanish. This is not a convenient set, however, because it has no simple relation to rotations.

A set that has such a relationship can be built from the J_i. The first four are 1 and the J_i themselves, leaving five to be determined. Because of (91), these must be quadratic forms in the J_i. As always, rotations provide the clue: 1 is a scalar and $\boldsymbol{J}$ is a vector under rotations. The next more complicated objects insofar as rotations are concerned are second-rank tensors, such as the nine bilinears J_iJ_j. These are not fully independent of $(1, \boldsymbol{J})$, however. The antisymmetric combination $(J_iJ_j - J_jJ_i)$ equals $i\epsilon_{ijk}J_k$, and one symmetric combination is a constant: $\boldsymbol{J}^2 = 1(1+1) = 2$. Thus what is left are the symmetric combinations that do not contain $\boldsymbol{J}^2$, namely,

$$T_{ij} = \frac{1}{2}(J_iJ_j + J_jJ_i) - \frac{2}{3}\delta_{ij} \ , \tag{94}$$

which is a symmetric traceless tensor.

By what right has the term "tensor" been used here? Well, by the same right as "vector" was applied to $\boldsymbol{J}$. Under a rotation,

$$J_i \to D^1(R)^\dagger J_i D^1(R) = \sum_j a_{ij}(R) J_j \ , \tag{95}$$

where, as has now been demonstrated repeatedly, the coefficients $a_{ij}(R)$ produce precisely the linear combination of the three J_i that holds for any 3-vector when it undergoes the rotation R, whether it be an operator in some Hilbert space or a garden-variety vector. By the same token, a bilinear term in T_{ij} will transform as follows:

$$J_i J_j \to \left(D^1(R)^\dagger J_i D^1(R)\right)\left(D^1(R)^\dagger J_j D^1(R)\right) = \sum_{kl} a_{ik}(R) a_{jl}(R) J_k J_l \,, \qquad (96)$$

which is the transformation law of a second-rank tensor in $\mathfrak{E}_3$.

Any observable in the spin 1 space can now be written as some linear combination of 1, J_i, and T_{ij}. In particular, the density matrix has the form

$$\rho = \tfrac{1}{3}(1 + \boldsymbol{P}\cdot\boldsymbol{J} + W_{ij}T_{ij}) \,, \qquad (97)$$

where W_{ij} is a set of five real numbers.

The problem of specifying the state of a spin 1 system is clearly more complicated than the case of spin $\frac{1}{2}$. For example, such a state may have no spin polarization, $\boldsymbol{P} = 0$, but will not be spherically symmetric if $W_{ij} \neq 0$. In this particular example, no vector operator, such as dipole moment, would have an expectation value in this state, but a quadrupole moment would. Conversely, W_{ij} may vanish but not $\boldsymbol{P}$, in which case only vector operators have an expectations value. However, as it is evident from (97), *multipoles higher than a quadrupole must have zero expectation values in any state of a spin 1 system.* In §7.6 the preceding statement will be shown to be a special case of the powerful Wigner-Eckart theorem, but it is important not to use such a howitzer when gnats are the target. Addiction to mathematical overkill impairs the development of intuitive understanding — in this case, an understanding of what rotational symmetry does (and does not) imply in relatively simple situations, which would not be acquired if the whole machinery of the rotation group were put on the table at the slightest provocation.

Several important issues related to spin 1 are taken up in Prob. 5.

(c) Arbitrary Spins

The approach just used for spin $\frac{1}{2}$ and 1 can, of course, be extended to higher spins, but as spin 1 already indicated, it becomes progressively more clumsy. It is then more efficient to develop a more sophisticated analysis of the rotation group, as will be done in chapter 7.

There are, however, some facts we have already established which generalize effortlessly to all values of j. A complete set of matrices in the $2j + 1$-dimensional space $\mathfrak{H}_j$ must have $(2j+1)^2$ linearly independent members. Let J_i be the angular momentum matrices in $\mathfrak{H}_j$. Then this set consists of 1, the three J_i, the five components of T_{ij}, and so forth up to (irreducible) tensors of rank $2j$. The last part of this statement will be proven in §7.6.

In the majority of concrete applications, the observables of interest are either scalars, vectors, or second-rank tensors. As we saw in the spin $\frac{1}{2}$ and 1 cases, any vector operator in either space can be expressed as $V_i = \lambda_V J_i$, where λ_V is rotationally invariant, and not an operator on the states that form $\mathfrak{H}_j$, though it may be a horrendous operator involving other degrees of freedom. Hence all the matrix elements of V_i in any $\mathfrak{H}_j$ are given by

$$\langle jm|V_i|jm'\rangle = \lambda_V \langle jm|J_i|jm'\rangle \,. \qquad (98)$$

That is to say, the whole horde of matrix elements of the vector operator V_i in $\mathfrak{H}_j$ are given by the known matrix elements of J_i, apart from one common constant. This constant cannot be determined from rotational symmetry unless $V_i = J_i$. This is not a facetious remark. The coefficient is determined in this one case because the angular momenta obey nonlinear relations, their famous commutation rules. In contrast, the commutation rules between the angular momenta and a vector operator V_i are $[V_1,\ J_2] = iV_3$, etc., which are linear in the V_i, and therefore unchanged if V_i is multiplied by a factor that commutes with the J_i. So only if $V_i = J_i$ do the commutation rules determine the overall scale of the matrix elements, apart from phases.

The same argument applies to any symmetric traceless second-rank tensor operator X_{ij} in $\mathfrak{H}_j$: its matrix elements are

$$\langle jm|X_{ij}|jm'\rangle = \lambda_X \langle jm|\tfrac{1}{2}(J_iJ_j + J_jJ_i) - \tfrac{1}{3}j(j+1)\delta_{ij}|jm'\rangle \ , \tag{99}$$

where λ_X is a constant that depends only on the specifics of X_{ij}. In this case, the amount of information that symmetry provides is even more impressive: five $(2j+1)$-dimensional matrices are completely specified apart from one constant.

These relationships between matrix elements of vector and tensor operators and polynomials in the J_i only determine the matrix element *within* a subspace of a given j, because the angular momentum operators cannot change j. It is also possible to simplify greatly the matrix elements of vectors and tensors between states of differing j by means of rotational symmetry, but for that one needs the Wigner-Eckart theorem.

A final remark. One should not be misled into thinking that the rotation matrices $D^j(R)$ in $\mathfrak{H}_j$ constitute the most general unitary transformations in this $(2j+1)$-dimensional space. For every j they depend only on the three real parameters needed to specify a rotation in the three-dimensional space $\mathfrak{E}_3$, whereas the general unitary transformation in $\mathfrak{H}_j$ is specified by $(2j+1)^2$ real parameters. It is only in the case of $j = \frac{1}{2}$ that the rotations suffice to exhaust the unitary transformations in $\mathfrak{H}_j$ (aside from the overall phase factor mentioned after Eq. 82). A detailed discussion of the last point can be found in §7.4(b).

3.4 Free-Particle States

Free-particle wave functions are needed in many contexts: when particles are indeed free; as a complete basis of states to construct various operators, such as those for the electromagnetic field; to represent incoming and outgoing states in collisions; as a first approximation to the complete wave function in such processes; and in other circumstances that are best described as the occasion arises.

We first consider a particle of zero spin. The time-independent Schrödinger equation for a free particle of mass m is

$$\frac{p^2}{2m}\psi_E(\boldsymbol{r}) = E\psi_E(\boldsymbol{r}) \ . \tag{100}$$

As in §3.2, here we work in the coordinate representation, i.e., $\boldsymbol{p} \to \hbar\nabla/i$. One set of solutions of (100) are simultaneous eigenfunctions of the three components of momentum, with eigenvalue $\hbar\boldsymbol{k}$ and energy $E = (\hbar k)^2/2m$:

$$\varphi_{\boldsymbol{k}}(\boldsymbol{r}) = (1/2\pi)^{\frac{3}{2}} e^{i\boldsymbol{k}\cdot\boldsymbol{r}} . \tag{101}$$

The pre-factor has been chosen to produce the following orthogonality and completeness relations:

$$\int d^3r\, \varphi^*_{\boldsymbol{k}}(\boldsymbol{r})\varphi_{\boldsymbol{k}'}(\boldsymbol{r}) = \delta(\boldsymbol{k} - \boldsymbol{k}') , \tag{102}$$

$$\int d^3k\, \varphi_{\boldsymbol{k}}(\boldsymbol{r})\varphi^*_{\boldsymbol{k}}(\boldsymbol{r}') = \delta(\boldsymbol{r} - \boldsymbol{r}') . \tag{103}$$

But p^2 is invariant not only under translation, which has been taken advantage of in the states $\varphi_{\boldsymbol{k}}$ — it is also rotationally invariant. Hence another choice of compatible constants of motion is (p^2, L^2) and one component of $\boldsymbol{L}$, say L_3, with eigenstates identified by the eigenvalues (E, l, m). These are eigenfunctions of the Laplacian in spherical coordinates:

$$\nabla^2 = \frac{1}{r^2}\frac{\partial}{\partial r} r^2 \frac{\partial}{\partial r} - \frac{L^2}{r^2} , \tag{104}$$

where

$$L^2 = -\left(\frac{1}{\sin\theta}\frac{\partial}{\partial\theta}\sin\theta\frac{\partial}{\partial\theta} + \frac{1}{\sin^2\theta}\frac{\partial^2}{\partial\phi^2}\right) . \tag{105}$$

The wave functions then separates into a radial and angular part,

$$\langle \boldsymbol{r}|Elm\rangle = \psi^0_{Elm}(\boldsymbol{r}) = R_{El}(r)Y_{lm}(\theta\phi) , \tag{106}$$

where Y_{lm} is a spherical harmonic. The radial wave function therefore satisfies the ordinary differential equation

$$\left(\frac{\hbar^2}{2m}\left(-\frac{1}{r^2}\frac{d}{dr}r^2\frac{d}{dr} + \frac{l(l+1)}{r^2}\right) - E\right) R_{El}(r) = 0 . \tag{107}$$

As $E = (\hbar k)^2/2m$, the variable $\rho = kr$ turns this into the dimensionless equation

$$\left(\frac{1}{\rho^2}\frac{d}{d\rho}\rho^2\frac{d}{d\rho} - \frac{l(l+1)}{\rho^2} + 1\right) R_l(\rho) = 0 . \tag{108}$$

This is an equation of the Bessel type, the general solution having the form

$$R_l(\rho) = C_l Z_{l+\frac{1}{2}}(\rho)/\rho^{\frac{1}{2}} , \tag{109}$$

where Z_ν is a cylinder function. When $\nu = l + \frac{1}{2}$, these are finite combinations of elementary functions, and are called spherical cylinder functions. Like any second-order equation, (108) has two linearly independent solutions; furthermore, because it has a singular point at $\rho = 0$, these solutions are either regular (i.e., analytic) at $\rho = 0$, or irregular there. The regular solutions are the spherical Bessel functions:

$$j_l(\rho) = \sqrt{\frac{\pi}{2\rho}} J_{l+\frac{1}{2}}(\rho) . \tag{110}$$

They have the limiting forms

$$j_l(\rho) \longrightarrow \frac{1}{\rho}\sin(\rho - \tfrac{1}{2}l\pi) \quad (\rho \to \infty)\,, \tag{111}$$

$$j_l(\rho) \longrightarrow \frac{\rho^l}{(2l+1)!!} \qquad (\rho \to \infty)\,, \tag{112}$$

where $(2l+1)!! = (2l+1)(2l-1)\ldots 1$. The spherical Bessel functions form complete orthonormal sets on $(0, \infty)$:

$$\int_0^\infty y^2 dy\, j_l(xy) j_l(x'y) = \frac{\pi}{2x^2}\delta(x - x')\,. \tag{113}$$

The linearly independent irregular solutions are the spherical Neumann functions

$$n_l(\rho) = \sqrt{\frac{\pi}{2\rho}} N_{l+\frac{1}{2}}(\rho)\,, \tag{114}$$

whose limiting forms are

$$n_l(\rho) \longrightarrow \frac{1}{\rho}\cos(\rho - \tfrac{1}{2}l\pi) \qquad (\rho \to \infty)\,, \tag{115}$$

$$n_l(\rho) \longrightarrow -(2l-1)!!(1/\rho)^{l+1} \qquad (\rho \to 0)\,. \tag{116}$$

From these two, other important functions can be formed, the spherical Hankel functions:

$$h_l(\rho) = j_l(\rho) + i n_l(\rho)\,, \tag{117}$$

and their complex conjugates.[1] Their asymptotic form is

$$h_l(\rho) \longrightarrow \frac{1}{i^{l+1}\rho} e^{i\rho}\,. \tag{118}$$

The explicit forms for $l = 0$ are[2]

$$j_0(\rho) = \frac{1}{\rho}\sin\rho\,, \quad n_0(\rho) = -\frac{1}{\rho}\cos\rho\,, \quad h_0(\rho) = \frac{1}{i\rho}e^{i\rho}\,. \tag{119}$$

The complete free-particle wave functions are then

$$\psi^0_{Elm}(\mathbf{r}) = \sqrt{\frac{2}{\pi}\frac{k^2 dk}{dE}}\, j_l(kr)\, Y_{lm}(\theta\phi)\,, \tag{120}$$

where the first factor, in light of (113), is chosen to produce the following orthogonality relation:

$$\int d^3r\, [\psi^0_{Elm}(\mathbf{r})]^* \psi^0_{E'l'm'}(\mathbf{r}) = \delta_{ll'}\,\delta_{mm'}\,\delta(E - E')\,. \tag{121}$$

---

[1] Some authors use the notation $h_l^{(1)}$ for h_l, and $h_l^{(2)}$ for h_l^*.

[2] Further information about these functions is in the Appendix.

The functions (120) are also complete in virtue of (113) and the completeness property of the spherical harmonics (Eq. 46):

$$\sum_{lm} \int_0^\infty dE\, \psi^0_{Elm}(\boldsymbol{r})[\psi^0_{Elm}(\boldsymbol{r}')]^* = \delta(\boldsymbol{r} - \boldsymbol{r}') , \tag{122}$$

where

$$\delta(\boldsymbol{r} - \boldsymbol{r}') = \frac{\delta(r - r')\delta(\theta - \theta')\delta(\phi - \phi')}{r^2 \sin^2\theta} . \tag{123}$$

The unitary transformation from the linear to the angular momentum basis is of importance in its own right, and in many applications. As we will show, it is

$$\langle \boldsymbol{k}|E'lm\rangle = \delta(k - k')Y_{lm}(\hat{\boldsymbol{k}})i^{-l}\sqrt{\frac{dk}{dE}\frac{1}{k^2}} . \tag{124}$$

This asserts that the probability of finding the linear momentum in the direction $\hat{\boldsymbol{k}}$ in a state that has angular momentum (l, m) is proportional to $|Y_{lm}(\hat{\boldsymbol{k}})|^2$, a fact that has important applications in collision and decay processes.

To confirm (124), the integral

$$\int d^3r\, \varphi^*_{\boldsymbol{k}}(\boldsymbol{r})\psi^0_{E'lm}(\boldsymbol{r}) \tag{125}$$

must be evaluated. Note, however, that $\varphi_{\boldsymbol{k}}$ is invariant under rotations about $\boldsymbol{k}$, and if that direction is chosen to be the axis for the azimuthal angle ϕ, the integral must vanish unless $m = 0$. Put quantum mechanically, the states ψ^0 can be chosen as simultaneous eigenstates of L^2 and $\boldsymbol{L} \cdot \boldsymbol{p}$, so that

$$\varphi_{\boldsymbol{k}}(\boldsymbol{r}) = \sum_l c_l\, \psi^0_{El0}(\boldsymbol{r}) , \tag{126}$$

provided the 3-axis is chosen along $\boldsymbol{k}$.

The constant c_l is found by using

$$j_l(\rho) = \frac{1}{2i^l} \int_{-1}^{1} dx\, e^{ix\rho} P_l(x) , \tag{127}$$

the final result being

$$\varphi_{\boldsymbol{k}}(\boldsymbol{r}) = \sqrt{\frac{k^2 dk}{dE}} \sum_{l=0}^{\infty} \sqrt{\frac{2l+1}{4\pi}}\, i^l\, \psi^0_{El0}(\boldsymbol{r}) . \tag{128}$$

This relationship is frequently used in the equivalent forms

$$\begin{aligned} e^{ikr\cos\theta} &= \sum_l (2l+1) i^l j_l(kr) P_l(\cos\theta) \\ &= \sum_l \sqrt{4\pi(2l+1)}\, i^l j_l(kr) Y_{l0}(\theta) . \end{aligned} \tag{129}$$

Here θ is the angle between $\boldsymbol{r}$ and $\boldsymbol{k}$. By using the addition theorem (Eq. 47) this can be written as

$$e^{i\boldsymbol{k}\cdot\boldsymbol{r}} = 4\pi \sum_{lm} i^l j_l(kr) Y^*_{lm}(\hat{\boldsymbol{k}}) Y_{lm}(\hat{\boldsymbol{r}}) , \tag{130}$$

or equivalently, as

$$\varphi_{\boldsymbol{k}}(\boldsymbol{r}) = \sqrt{\frac{dE}{k^2 dk}} \sum_{lm} i^l Y_{lm}^*(\hat{\boldsymbol{k}}) \psi_{Elm}^0(\boldsymbol{r}) \,. \tag{131}$$

In these last two expressions the projection along which $\boldsymbol{L}$ has the eigenvalue m is arbitrary. The inverse to the last equation is

$$\psi_{Elm}^0(\boldsymbol{r}) = i^{-l} \sqrt{k^2 dk/dE} \int d\hat{\boldsymbol{k}}\; Y_{lm}(\hat{\boldsymbol{k}}) \varphi_{\boldsymbol{k}}(\boldsymbol{r}) \,. \tag{132}$$

The scalar product (124) is now an immediate consequence of this last relation.

The preceding treatment is based on momentum eigenstates traveling in any and all directions with all values of $|\boldsymbol{k}|$. Handling a continuous spectrum can sometimes be tricky, however, and some derivations may be easier and clearer if the momentum spectrum is discrete with a finite mesh size that is, in the end, taken to zero. Naturally, in an empty space of infinite extent all actually observable quantities (cross section, transition probabilities, and the like) must not depend on which spectrum is used.

The most convenient way of constructing a discrete momentum spectrum was already described in §2.1(d), namely, to require the wave functions to be periodic on the surface of a large cube of volume L^3. Plane waves satisfying this boundary condition have wave vectors $\boldsymbol{k}$ lying on the lattice

$$\boldsymbol{k} = (n_1, n_2, n_3) \cdot (2\pi/L) \,, \qquad n_i = 0, \pm 1, \pm 2, \ldots \,. \tag{133}$$

The momentum eigenfunctions are now

$$\phi_{\boldsymbol{k}}(\boldsymbol{r}) = L^{-3/2}\, e^{i\boldsymbol{k}\cdot\boldsymbol{r}} \,. \tag{134}$$

They again form a complete orthonormal set:

$$\frac{1}{L^3} \int d^3r\; e^{i(\boldsymbol{k}-\boldsymbol{k}')\cdot\boldsymbol{r}} = \delta_{\boldsymbol{k}\boldsymbol{k}'} \,, \tag{135}$$

$$\frac{1}{L^3} \sum_{\boldsymbol{k}} e^{i\boldsymbol{k}\cdot(\boldsymbol{r}-\boldsymbol{r}')} = \delta(\boldsymbol{r}-\boldsymbol{r}') \,. \tag{136}$$

Equation (136) is a sum over all the three integers in Eq. 133, and the Kronecker delta in (135) requires equality of all the three pairs of integers that define $\boldsymbol{k}$ and $\boldsymbol{k}'$. The relationship between periodic boundary conditions and a continuous spectrum of wave vectors is evidently given by the rules

$$\frac{1}{L^3} \sum_{\boldsymbol{k}} \Leftrightarrow \int \frac{d^3k}{(2\pi)^3} \,, \qquad \frac{L^3}{(2\pi)^3} \delta_{\boldsymbol{k}\boldsymbol{k}'} \Leftrightarrow \delta(\boldsymbol{k}-\boldsymbol{k}') \,, \tag{137}$$

in the limit $L \to \infty$.

Thus far the particles in question have been assumed to have no spin. Consider now the general case in which a particle has an angular momentum of magnitude s in a frame where it is at rest. This internal degree of freedom is describes by the observable $\boldsymbol{s}$, which satisfies the universal angular momentum commutation rule

$$[s_i, s_j] = i\epsilon_{ijk}\, s_k \,. \tag{138}$$

This operator commutes with the position and coordinate operators, and consequently with the orbital angular momentum,

$$[s_i, L_j] = 0 . \tag{139}$$

A complete set of observables now has four members. There are various ways of choosing these. The most obvious is any of the three discussed already (e.g., linear momentum), and one of the components of $\boldsymbol{s}$, say s_3, with eigenvalues $(s, s-1, \ldots, -s)$. Each of the states of the spin 0 system is therefore accompanied by $(2s+1)$ "spin states," $|s\rangle$, which span the $(2s+1)$-dimensional space $\mathfrak{H}_s$. As spin and linear momentum are independent degrees of freedom, an example of such an enlarged ket is then

$$|\boldsymbol{k}, s\rangle = |\boldsymbol{k}\rangle \otimes |s\rangle . \tag{140}$$

The corresponding wave function, often called a spinor, is written as

$$\langle \boldsymbol{r}|\boldsymbol{k}, s\rangle = \varphi_{\boldsymbol{k}}(\boldsymbol{r})\,|s\rangle . \tag{141}$$

This is a hybrid notation, to be sure, because the scalar product is partial in that the bra does not have a spin eigenvalue. Some might prefer to replace (141) by

$$\langle \boldsymbol{r}, s'|\boldsymbol{p}, s\rangle = \varphi_{\boldsymbol{k}}(\boldsymbol{r})\,\delta_{ss'} . \tag{142}$$

We will, almost always, use the mongrel notation of Eq. 141, because it is more descriptive than that of (142).

Sometimes there are disadvantages to the space-spin product form for states, however. In particular, the product of a state $|lm\rangle$ of orbital angular momentum with a spin state $|s\rangle$ is not, in general, an eigenstate of total angular momentum, which makes the product form inconvenient in many collision and decay problems. This handicap can be overcome, but it requires the more advanced understanding of rotations that will be developed in §7.5(b).

3.5 Addition of Angular Momenta

Many phenomena involve systems that have a total angular momentum that is the sum of angular momenta belonging to various of its subsystems. For example, an isolated atom has a conserved total angular momentum which is the sum of the nuclear and electronic angular momenta. The latter, in turn, is the sum of the angular momenta of the individual electrons, each of which is the sum of an orbital and spin angular momentum. Depending on the accuracy required for the problem at hand, all, some or none of these separate angular momenta are constants of motion. But whether they are or are not separate constants of motion, it is necessary to know how to add them up.

(a) General Results

Consider the addition of two angular momenta $\boldsymbol{J}_1$ and $\boldsymbol{J}_2$ that belong to distinct, independent degrees of freedom whose commutators therefore vanish, and having magnitudes j_1 and j_2. The state space in question, $\mathfrak{H}_{j_1 j_2}$, is spanned by products of eigenstates of the separate angular momenta,

$$|j_1 m_1\rangle \otimes |j_2 m_2\rangle \equiv |j_1 m_1 j_2 m_2\rangle\,, \tag{143}$$

and has the dimension

$$d = (2j_1 + 1)(2j_2 + 1)\ . \tag{144}$$

In the basis (143) the commuting observables are the operators $J_1^2, J_{1z}, J_2^2, J_{2z}$. From our knowledge of individual angular momenta, we know that there are no other angular momentum operators that commute with these four. (In this section the axis chosen for the diagonal component of angular momenta is called z instead of 3, to avoid confusion with subsystem labels.)

The total angular momentum operator is

$$\boldsymbol{J} = \boldsymbol{J}_1 + \boldsymbol{J}_2\,. \tag{145}$$

The objective is to construct simultaneous eigenstates $|jm\rangle$ of $\boldsymbol{J}^2$ and J_z. There must, however, be two additional operators that commute with J^2 and J_z, because the product basis (143) is specified by four commuting observables. An operator that commutes with J^2 and J_z must be invariant under rotations, and this identifies the missing ones as J_1^2 and J_2^2. The scalar $\boldsymbol{J}_1 \cdot \boldsymbol{J}_2$ could also be taken as a member of the commuting set, but it is a function of those already chosen because

$$\boldsymbol{J}_1 \cdot \boldsymbol{J}_2 = \tfrac{1}{2}(\boldsymbol{J}^2 - \boldsymbol{J}_1^2 - \boldsymbol{J}_2^2) = \tfrac{1}{2}[j(j+1) - j_1(j_1+1) - j_2(j_2+1)]\,. \tag{146}$$

The total angular momentum eigenstates are thus simultaneous eigenstates of the four operators (J_1^2, J_2^2, J^2, J_z.) They are written as $|j_1 j_2 jm\rangle$, and satisfy

$$J^2|j_1 j_2 jm\rangle = j(j+1)|j_1 j_2 jm\rangle\,, \tag{147}$$

$$J_z|j_1 j_2 jm\rangle = m|j_1 j_2 jm\rangle\,. \tag{148}$$

Because $\boldsymbol{J}$ satisfies the the generic angular momentum commutation rule, j must be an integer or half-integer, and for any given j the values of m are $j, j-1, \ldots, -j$. Further, because $J_z = J_{1z} + J_{2z}$, the product states (143) are eigenkets of J_z with eigenvalues

$$m = m_1 + m_2\,. \tag{149}$$

Hence if j_1 and j_2 are both integers, or both half-integers, the allowed values of m, and therefore of j, are integers, whereas if one of (j_1, j_2) is an integer while the other is not, then j is half-integer.

The problem remaining has two parts: (i) given j_1 and j_2, to find the spectrum of permitted values of j; and (ii) to determine the transformation between the product basis (Eq. 143) and that defined by (147) and (148).

The answer to the first question can be stated succinctly:

1. The total angular momentum has the values $j = j_1+j_2, j_1+j_2-1, \ldots, |j_1-j_2|$, for given values of j_1 and j_2, or briefly put, j satisfies a triangular inequality reminiscent of ordinary vector addition,

$$j_1 + j_2 \geq j \geq |j_1 - j_2|\,, \tag{150}$$

it being understood that the allowed values of j differ by integers.

2. There is only one sequence of $(2j+1)$ states, with $m = j, j-1, \ldots, -j$, for each allowed value of j.[1]

A check on this last assertion is

$$\sum_{j=|j_1-j_2|}^{j_1+j_2} (2j+1) = (2j_1+1)(2j_2+1) \,. \tag{151}$$

The proof of Eq. 150 for arbitrary j_1 and j_2, and the transformation from the product basis $|j_1 m_1\, j_2 m_2\rangle$ to the total angular momentum basis $|j_1 j_2\, jm\rangle$ will be given in subsection (c).

(b) Adding Spins $\frac{1}{2}$ and Unit Spins

Instead of tackling the general case straight away, we start with adding spin $\frac{1}{2}$, even if most readers will be familiar with this. Not only will this warm-up exercise clarify what is afoot in the general case, but the special case of spin $\frac{1}{2}$ is so important in applications that it is worthwhile to acquire familiarity with the details.

It will pay to use the compact notation of §3.3(a) here, with the individual spin states designated by $|\pm\rangle_{1,2}$ for spin projections $\pm\frac{1}{2}$.

The simplest case of all is the addition of two spin $\frac{1}{2}$'s. The space $\mathfrak{H}_{\frac{1}{2}\frac{1}{2}}$ is then four-dimensional, with the basis Eq. 143 the product kets $|\mu\rangle_1 \cdot |\nu\rangle_2$, where $\mu, \nu = \pm$, and $\otimes$ is written more modestly as "$\cdot$". The total angular momentum will be called $\boldsymbol{S}$ in this case, because that is the name it traditionally has in spectroscopic applications:

$$\boldsymbol{S} = \tfrac{1}{2}(\boldsymbol{\sigma}_1 + \boldsymbol{\sigma}_2) \,. \tag{152}$$

The eigenvalues of $\boldsymbol{S}^2$ are $S(S+1)$, and those of S_z are here called M_S. Because of (149), there are two states with $M_S = 0$, and one each of $M_S = \pm 1$, which is consistent with Eq. 150, according to which the allowed values of S are (i) an $S=1$ triplet with $M_S = 1, 0, -1$; and (ii) an $S = 0$ singlet.

The triplet is the "stretched" or "parallel" configuration. The states with maximum $|M_S|$, namely 1, are $|\pm\rangle_1 \cdot |\pm\rangle_2$. These have $S = 1$ because no larger values of $|M_S|$ exist. The one with $S = 1, M_S = 0$ is constructed from $M_S = \pm 1$ by using the famous identity (15)

$$J_\pm |jm\rangle = \sqrt{j(j+1) - m(m\pm 1)}\, |jm \pm 1\rangle \,, \tag{153}$$

which in this case is

$$S_\pm |1 M_S\rangle = \sqrt{2 - M_S(M_S \pm 1)}\, |1 M_S \pm 1\rangle \,. \tag{154}$$

The final result for the triplet is therefore

$$|11\rangle = |+\rangle_1 \cdot |+\rangle_2 \,, \tag{155}$$

$$|10\rangle = \frac{1}{\sqrt{2}}\Big(|+\rangle_1 \cdot |-\rangle_2 + |-\rangle_1 \cdot |+\rangle_2\Big) \,, \tag{156}$$

$$|1\, {-1}\rangle = |-\rangle_1 \cdot |-\rangle_2 \,. \tag{157}$$

[1]This may sound self-evident, but that is not quite so. In more complicated groups than the rotation group (e.g., unitary transformations in a 3-dimensional complex space), the analogue of angular momentum addition can produce more than one multiplet of a given dimension.

The remaining state is the $S=0$ singlet, which must be orthogonal to $|10\rangle$, i.e.,

$$|00\rangle = \frac{1}{\sqrt{2}}\Big(|+\rangle_1 \cdot |-\rangle_2 - |-\rangle_1 \cdot |+\rangle_2\Big)\,. \tag{158}$$

The symmetry of these states is often of great importance: *under interchange of the two individual spin projections, the $S=1$ spin triplet is symmetric, while the $S=0$ singlet is antisymmetric.* That all states within a multiplet of definite S have the same symmetry follows from the fact that they are related by the ladder operators $S_\pm$, which are symmetric under interchange because they are the sum of the individual spin operators.

The way in which states of definite total spin S transform under rotations is also of great significance. As always, the rotation operator is

$$D(R) = e^{-i\theta \boldsymbol{n}\cdot \boldsymbol{S}}\,. \tag{159}$$

Because the operators $\boldsymbol{S}$ cannot change the eigenvalue S, a rotation applied to a particular member of the basis $|SM_S\rangle$ will produce some linear combination of states with *same* S, i.e., no mixing between the singlet and members of the triplet. Hence the 4-dimensional space $\mathfrak{H}_{\frac{1}{2}\frac{1}{2}}$ decomposes into 3-dimensional and 1-dimensional subspaces $\mathfrak{H}_S$. This decomposition is rotationally invariant: under a rotation, the component of any ket that is in $\mathfrak{H}_S$ will be mapped into another ket in that subspace.

Operators that project onto the invariant subspaces $\mathfrak{H}_S$ are often handy. From (152), $\boldsymbol{S}^2 = S(S+1) = \frac{1}{2}(3 + \boldsymbol{\sigma}_1 \cdot \boldsymbol{\sigma}_2)$, so that

$$\boldsymbol{\sigma}_1 \cdot \boldsymbol{\sigma}_2 = 1 \;\; (S=1), \qquad \boldsymbol{\sigma}_1 \cdot \boldsymbol{\sigma}_2 = -3 \;\; (S=0)\,. \tag{160}$$

The projection operators onto the subspaces of definite S are therefore

$$P_s = \tfrac{1}{4}(1 - \boldsymbol{\sigma}_1 \cdot \boldsymbol{\sigma}_2) \qquad \text{(singlet)}\,, \tag{161}$$
$$P_t = \tfrac{1}{4}(3 + \boldsymbol{\sigma}_1 \cdot \boldsymbol{\sigma}_2) \qquad \text{(triplet)}\,. \tag{162}$$

As required, $P_s + P_t = 1$. It is also convenient to have an operator P_{12} that interchanges the two spins. In view of (160), and the fact that $S=1$ is symmetric while $S=0$ is antisymmetric,

$$P_{12} = \tfrac{1}{2}(1 + \boldsymbol{\sigma}_1 \cdot \boldsymbol{\sigma}_2)\,. \tag{163}$$

Note that all three projection operators are rotationally invariant, as they had better be because they must not discriminate among the three $S=1$ states.

Next, consider the addition of spin $\frac{1}{2}$ to an orbital angular momentum l:

$$\boldsymbol{J} = \boldsymbol{L} + \tfrac{1}{2}\boldsymbol{\sigma}\,. \tag{164}$$

The calculation does not change if l is replaced by an arbitrary half-integer j_1, but the most frequent application is to the addition of orbital angular momentum and spin.

Because $m = m_l \pm \frac{1}{2}$, the possible total angular momenta are now $j = l \pm \frac{1}{2}$, and at most two of the product states $|lm_l\rangle \cdot |\pm\rangle$ can contribute to any state $|jm\rangle$:

$$|jm\rangle = c_+|l\, m + \tfrac{1}{2}\rangle \cdot |-\rangle + c_-|l\, m - \tfrac{1}{2}\rangle \cdot |+\rangle\,. \tag{165}$$

The coefficients are found by requiring this to be an eigenstate of J^2, or using (146), of

$$\boldsymbol{L}\cdot\boldsymbol{\sigma} = \tfrac{1}{2}(L_+\sigma_- + L_+\sigma_-) + L_z\sigma_z = j(j+1) - l(l+1) - \tfrac{3}{4}\,. \qquad (166)$$

The operators $L_\pm\sigma_\mp$ acting on the right-hand side of (165) turn one state in the superposition into the other, while each is already an eigenstate of $L_z\sigma_z$. One relation then emerges between c_+ and c_-, which together with normalization determines the coefficients, apart from an overall phase. Those who have never done this calculation are encouraged to do so; the final results are given in Eq. 183.

As in the previous case, it can be useful to have operators that decompose the space $\mathfrak{H}_{l\frac{1}{2}}$ into its subspaces of definite $j = l \pm \frac{1}{2}$. These projection operators are

$$\Lambda_l^+ = \frac{l+1+\boldsymbol{\sigma}\cdot\boldsymbol{L}}{2l+1}\,, \qquad (j = l+\tfrac{1}{2})\,, \qquad (167)$$

$$\Lambda_l^- = \frac{l-\boldsymbol{\sigma}\cdot\boldsymbol{L}}{2l+1}\,, \qquad (j = l-\tfrac{1}{2})\,. \qquad (168)$$

Another special case that appears in many guises is the addition of $j_1 = 1$ and $j_2 = 1$, where the general theorem claims that the nine product states can be recast into eigenstates of $j = 0, 1, 2$. This can be done, with maximum obscurity, by using the powerful machinery to be developed in a moment. But it is a simple problem if one remembers, yet again, that $j = 1$ states transform like a 3-vector.

To that end, let A_i and B_i be the Cartesian components of any two 3-vectors. The 9-dimensional space of interest is spanned by the products A_iB_j. From these we can form a scalar I, a 3-vector $\boldsymbol{V}$, and a symmetric traceless tensor X_{ij}:

$$I = \boldsymbol{A}\cdot\boldsymbol{B}, \quad \boldsymbol{V} = \boldsymbol{A}\times\boldsymbol{B}, \quad X_{ij} = A_iB_j + A_jB_i - \tfrac{2}{3}I\delta_{ij}\,. \qquad (169)$$

These bilinear forms span the three subspaces $\mathfrak{H}_j$ with $j = 0, 1, 2$. As written here, the coefficients in these forms (e.g., ϵ_{ijk}) are only appropriate to the Cartesian basis. If it is necessary to know their counterparts in the spherical basis, where states have eigenvalues of J_z, the transformation of Eq. 54 must be applied, but in many circumstances it is not necessary to work in the spherical basis and then the forms of Eq. 169 are more convenient.

(c) Arbitrary Angular Momenta; Clebsch-Gordan Coefficients

Our first task is to confirm the earlier statements about the spectrum of J^2 for arbitrary j_1 and j_2. Because $m = m_1 + m_2$, the largest value of m is $j_1 + j_2 = m_{\max}$, and therefore the largest value of j is $j_1 + j_2 = j_{\max}$. Only one product ket has $m = m_{\max}$, so it is already an eigenstate of $j_{\max}$. There are two product states with $m_1 + m_2 = m_{\max} - 1$; one linear combination must belong to the multiplet with $j_{\max}$, and therefore the one orthogonal to it to $j = j_{\max} - 1$. Repetition of this argument seems to produce, at every step, a new eigenstate of J^2 with eigenvalues $j_{\max} - 2, j_{\max} - 3, \ldots$. This cannot go on forever, of course. We could have started with the lowest possible value of m, namely, $m_{\min} = -j_1 - j_2 = -m_{\max}$, and increased m one step at a time, adding a new multiplet of j at each step. This upward ladder in m will eventually bump into the downward ladder that started with $m = j_1 + j_2$. From Fig. 3.2 one infers that the degeneracy stops growing when $|m| = |j_1 - j_2|$ is reached, which establishes that the lowest value of j is indeed

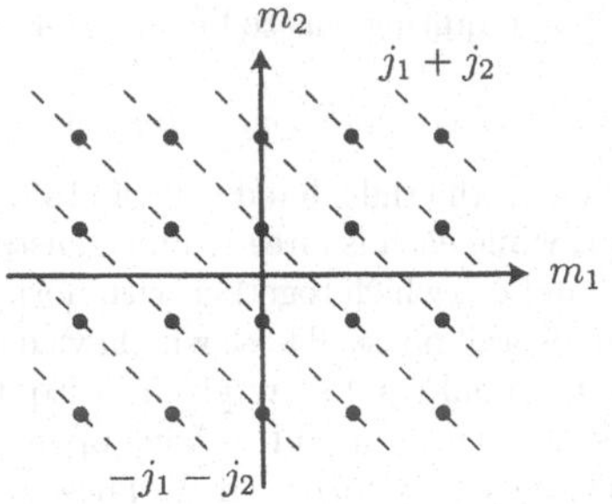

FIG. 3.2. Addition of angular momentum. The dots show the values of $m = m_1 + m_2$ for the case $j_1 = 2,\ j_2 = \frac{3}{2}$.

$|j_1 - j_2|$. That there is only one multiplet of a given j is also evident from this argument. *QED*

There is a standard group-theoretic shorthand for what has just been established: all angular momentum multiplets are designated by their multiplicity, e.g, **3** for $j = 1$. The space formed by all the products of two independent $j = 1$ multiplets is designated by $\mathbf{3} \otimes \mathbf{3}$. This 9-dimensional space, as we saw in (169), decomposes into states of total angular momentum 0, 1 and 2. This fact is written as

$$\mathbf{3} \otimes \mathbf{3} = \mathbf{1} \oplus \mathbf{3} \oplus \mathbf{5}\,. \tag{170}$$

In the same way, the fact that the product of two independent spin $\frac{1}{2}$ states is a linear combination of a singlet and a triplet is expressed as

$$\mathbf{2} \otimes \mathbf{2} = \mathbf{1} \oplus \mathbf{3}\,. \tag{171}$$

In general,

$$(\mathbf{2j_1 + 1}) \otimes (\mathbf{2j_2 + 1}) = (\mathbf{2}\,|\mathbf{j_1 + j_2}| + \mathbf{1}) \oplus \ldots \oplus (\mathbf{2}\,|\mathbf{j_1 - j_2}| + \mathbf{1})\,. \tag{172}$$

These silly looking identities are taken from the theory of groups more complicated than the three-dimensional rotation group. They express a concept more sophisticated than $2 \times 2 = 1 + 3$, and which we have already encountered in specific examples. Namely, the $(2j_1 + 1)(2j_2 + 1)$ dimensional space spanned by the product states $|j_1 m_1\rangle \otimes |j_2 m_2\rangle$ can be decomposed into rotationally invariant spaces $\mathfrak{H}_j$:

$$\mathfrak{H}_{j_1} \otimes \mathfrak{H}_{j_2} = \sum_{j=|j_1-j_2|}^{j_1+j_2} \mathfrak{H}_j \oplus\ . \tag{173}$$

One says that the space on the left-hand side is reducible into the *irreducible* spaces $\mathfrak{H}_j$ on the right side, where irreducibility refers to the fact that within *each* of the latter *any* ket will, under arbitrary rotations, become a linear combination of *all* the kets in that *same* space $\mathfrak{H}_j$. In terms of the example (169), this last statement is just the fact that under rotations any component of the tensor X_{ij} will become some linear combination of all five components, but not acquire any component of the vector $\boldsymbol{V}$, etc.

We must still learn how to construct the total angular momentum states $|j_1j_2jm\rangle$ in terms of the product states $|j_1m_1j_2m_2\rangle$:

$$|j_1j_2jm\rangle = \sum_{m_1m_2} |j_1m_1j_2m_2\rangle\langle j_1m_1j_2m_2|j_1j_2jm\rangle \,. \tag{174}$$

The elements in this unitary transformation are called *Clebsch-Gordan coefficients*, or CG coefficients for short, and will be abbreviated to $\langle j_1m_1j_2m_2|jm\rangle$; they are the generalization of $c_\pm$ in (165) to arbitrary (j_1, j_2). The inverse relation is

$$|j_1m_1j_2m_2\rangle = \sum_{jm} |j_1j_2jm\rangle\langle jm|j_1m_1j_2m_2\rangle \,. \tag{175}$$

Because the transformation between bases is unitary,

$$\sum_{jm} \langle j_1m_1j_2m_2|jm\rangle\langle jm|j_1m_1'j_2m_2'\rangle = \delta_{m_1m_1'}\delta_{m_2m_2'} \,, \tag{176}$$

$$\sum_{m_1m_2} \langle jm|j_1m_1j_2m_2\rangle\langle j_1m_1j_2m_2|j'm'\rangle = \delta_{jj'}\delta_{mm'} \,. \tag{177}$$

It is of course true that the sums on magnetic quantum numbers are restricted by $m = m_1 + m_2$, but if this is made explicit clumsier expressions result.

All CG coefficients can be computed from recursion relations that follow from applying $J_\mp$ to (174):

$$\begin{aligned} a_\mp(jm) &\sum_{m_1m_2} |m_1m_2\rangle\langle m_1m_2|j\,m\mp 1\rangle \\ &= \sum_{m_1m_2} \{a_\mp(j_1m_1)|m_1\mp 1\,m_2\rangle + a_\mp(j_2m_2)|m_1\,m_2\mp 1\rangle\}\langle m_1m_2|jm\rangle \,, \end{aligned} \tag{178}$$

where

$$a_\pm(jm) = \sqrt{j(j+1) - m(m\pm 1)} \,, \tag{179}$$

and (j_1, j_2) are suppressed where appropriate as they do not change. Taking the scalar product of (178) with $\langle m_1'm_2'|$ gives two recursion relationships:

$$a_\mp(jm)\langle m_1m_2|jm\mp 1\rangle = a_\pm(j_1m_1)\langle m_1\pm 1\,m_2|jm\rangle + a_\pm(j_2m_2)\langle m_1m_2\pm 1|jm\rangle \,. \tag{180}$$

When combined with the normalization condition these relations suffice to determine all CG coefficients apart from an arbitrary phase. Consider first the case $m = j$. Then from $J_+|jj\rangle = 0$ (i.e., the lower variant of (180)), and remembering that $m_1 + m_2 = m$, we have

$$a_-(j_1m_1)\langle m_1 - 1\,j+1-m_1|jj\rangle = -\,a_-(j_2\,j+1-m_1)\langle m_1\,j-m_1|jj\rangle \,. \tag{181}$$

Set $m_1 = j_1$, and call $\langle j_1\,j-j_1|jj\rangle = C_{j_1j}$; then (181) determines the coefficient for $m_1 = j_1 - 1$, and by iteration all the coefficients for lower values of m_1, in terms of C_{j_1j}. The latter is determined by

$$\sum_{m_1m_2} |\langle m_1m_2|jj\rangle|^2 = 1 \,, \tag{182}$$

and by requiring $C_{j_1 j}$ to be real and positive, because with $a_\pm(jm)$ being chosen to be real and positive, this has the consequence that all the coefficients will be real. The coefficients for $|jm\rangle$ with lower values of m are found by applying J_- enough times to $|jj\rangle$, using the upper variant of (178), which evaluates $\langle m_1 m_2 | j\, j-1\rangle$, etc., directly from $\langle m_1 m_2 | jj\rangle$.

To confirm one's understanding of this method it is best to work through an example, e.g., that of adding spin $\frac{1}{2}$ to an arbitrary j_1, which was already treated in connection with Eq. 165 though with a technique that does not suffice for more complicated cases. The result for $\langle j_1 \frac{1}{2} m_2 | jm\rangle \cdot \sqrt{2j_1+1}$ is

$$\begin{array}{ccc} & m_2 = \frac{1}{2} & m_2 = -\frac{1}{2} \\ j = j_1 + \frac{1}{2} & \sqrt{j_1 + m + \frac{1}{2}} & \sqrt{j_1 - m + \frac{1}{2}} \\ j = j_1 - \frac{1}{2} & -\sqrt{j_1 - m + \frac{1}{2}} & \sqrt{j_1 + m + \frac{1}{2}} \end{array} \quad . \tag{183}$$

Other tables of CG coefficients are in the Appendix.

In many applications it is necessary to interchange j_1 and j_2, or to carry out other changes of variables. While this can be done by using various identities satisfied by the CG coefficients, it is rather clumsy (and therefore error prone) because they are not defined in a manner that treats the three angular momenta symmetrically. A symmetric scheme, due to Wigner, is based on $\boldsymbol{J}_1 + \boldsymbol{J}_2 + \boldsymbol{J}_3 = 0$ instead of the asymmetric choice of Eq. 145. The resulting *Wigner 3-j symbol* is related to the CG coefficient by

$$\begin{pmatrix} j_1 & j_2 & j_3 \\ m_1 & m_2 & m_3 \end{pmatrix} = \frac{(-1)^{j_1 - j_2 - m_3}}{\sqrt{2j_3+1}} \langle j_1 m_1 j_2 m_2 | j_3 - m_3\rangle \,. \tag{184}$$

With this definition, the 3-j symbol is invariant under a cyclic permutation of of its columns, while for a noncyclic permutation

$$\begin{pmatrix} j_2 & j_1 & j_3 \\ m_2 & m_1 & m_3 \end{pmatrix} = (-1)^{j_1+j_2+j_3} \begin{pmatrix} j_1 & j_2 & j_3 \\ m_1 & m_2 & m_3 \end{pmatrix} . \tag{185}$$

Other symmetry relations satisfied by the 3-j symbols are listed in the Appendix.

(d) Matrix Elements of Vector Operators

Vector operators appear in an enormous number of phenomena, and often one must know their matrix elements between angular momentum eigenstates. The transformation properties of such states, and of vector operators, greatly simplify the evaluation of these matrix elements.

In §3.3 we learned that the matrix elements of the components V_i of any vector operator between states of *the same* angular momentum j are equal to the matrix elements of J_i apart from an overall constant λ_V. The operator J_i cannot connect states of *differing* j, however, whereas a vector operator, in general, does. These off-diagonal matrix elements are also determined by the angular momentum algebra — in effect by the geometry of Euclidean 3-space. This is another special case of the Wigner-Eckart theorem, and the one that probably sees more use than any other. The result is that these matrix elements are proportional to a CG coefficient, with the constant of proportionality essentially the same constant λ_V.

For the present purpose it is best to express $\boldsymbol{V}$ in the spherical basis V_μ ($\mu = \pm 1, 0$), as defined in Eq. 51:

$$V_{+1} = -\frac{1}{\sqrt{2}}(V_1 + iV_2), \quad V_0 = V_3, \quad V_{-1} = \frac{1}{\sqrt{2}}(V_1 - iV_2) \,. \tag{186}$$

The objective is the matrix elements $\langle j'm'|V_\mu|jm\rangle$. The first question is to find the values of (j, m) and (j', m') for which the matrix has non-vanishing elements. This can be answered by the algebraic exercise that gives the whole answer, and which follows immediately. A more efficient and insightful approach is to apply a rotation to $V_\mu|jm\rangle$. The first factor V_μ transforms exactly like the spherical harmonic $Y_{1\mu}$, i.e., the angular momentum state $|1\mu\rangle$, and therefore $V_\mu|jm\rangle$ transforms like $|1\mu\rangle \otimes |jm\rangle$. According to the general rule of Eq. 172, we are dealing with $\mathbf{3} \otimes \mathbf{(2j+1)}$, and therefore the only possible values are $j' = j-1, j, j+1$. By the same argument, $V_\mu|jm\rangle$ is an eigenstate of J_z with eigenvalue $m + \mu$. To summarize:

The matrix elements of a vector operator between angular momentum states, $\langle j'm'|V_\mu|jm\rangle$, vanish unless j and j' differ by 0 or 1, and $m' = m + \mu$.

These are the *angular momentum selection rules for vector operators*. They govern the great majority of radiation phenomena because the electric dipole moments of atoms and molecules are by far the most important operators involved in the emission, absorption and scattering of electromagnetic radiation.

To proceed further, the commutation rules $[V_i, J_k] = i\epsilon_{ijk}V_k$ are rewritten in terms of spherical components, which, in the notation of (179) and (186), then read

$$[J_z, V_\mu] = \mu V_\mu \,, \tag{187}$$

$$[J_\pm, V_\mu] = a_\pm(1\mu)\, V_{\mu\pm 1} \,. \tag{188}$$

Matrix elements are then taken between the states of interest. Equation (187) requires $m + \mu = m'$, and confirms one of the selection rules already stated. Equation (188) produces

$$a_\mp(j'm')\,\langle j'\,m' \mp 1|V_\mu|jm\rangle - \langle j'm'|V_\mu|j\,m \pm 1\rangle\, a_\pm(jm) = a_\pm(1\mu)\,\langle j'm'|V_{\mu\pm 1}|jm\rangle \,. \tag{189}$$

These are the recursion relations (180) for the CG coefficients $\langle jm1\mu|j'm'\rangle$, showing that the matrix element is proportional to this CG coefficient, and confirming the second part of the selection rule, $|j - j'| = 0, 1$. In this case, however, there is no normalization condition like (182), and so an overall constant is left undetermined, as we already know from Eq. 98 for the $j = j'$ elements.

The final result, in the traditional if somewhat occult notation, is then

$$\begin{aligned} \langle n'j'm'|V_\mu|njm\rangle &= (-1)^{j-1+m'}\langle n'j'||V||nj\rangle \begin{pmatrix} j & 1 & j' \\ m & \mu & -m' \end{pmatrix} , \\ &= \langle n'j'||V||nj\rangle \frac{\langle jm1\mu|j'm'\rangle}{\sqrt{2j'+1}} \,. \end{aligned} \tag{190}$$

The object $\langle n'j'||V||nj\rangle$ is called the reduced matrix element; when $j = j'$ it is proportional to the constant called λ_V. The symbols (n, n') refer to all other quantum numbers unrelated to rotational symmetry that specify the states in question.

To calculate the reduced matrix element, one matrix element must be evaluated explicitly, and then the others are given by the appropriate 3-j (or CG) coefficient. The following is an example of a reduced matrix element, the derivation being left as an exercise:

$$\langle n'j'||J||nj\rangle = \delta_{nn'}\delta_{jj'}\sqrt{j(j+1)(2j+1)}\,. \tag{191}$$

The scalar product of two vector operators is also expressible in terms of reduced matrix elements. Again, the proofs are left as exercises. For this purpose it is best to express the scalar product in terms of the spherical components of the vector operators:

$$S \equiv \boldsymbol{U}\cdot\boldsymbol{V} = \sum_{\mu=-1}^{1}(-1)^{\mu}U_\mu V_{-\mu}\,. \tag{192}$$

Then

$$\langle n'j'm'|S|njm\rangle = \frac{\delta_{jj'}\delta_{mm'}}{2j+1}\sum_{n''}\sum_{j''=|j'-1|}^{j'+1}(-1)^{j'-j''}\langle n'j'||U||n''j''\rangle\langle n''j''||V||nj\rangle\,. \tag{193}$$

In general, the sum on n'' extends over all states that have the angular momentum j''.

3.6 The Two-Body Problem

In nonrelativistic classical mechanics the motion of a two-body system with a central interaction separates into the free motion of the center of mass and the motion of a single fictitious particle in a central field. The same simplification holds in quantum mechanics. The problem is then reduced to finding the eigenvalues and eigenfunctions for the radial motion of one particle in a central field, because the angular eigenfunctions are already known from the theory of orbital angular momentum. In short, what is called "separation of variables" in the theory of differential equations is done by exploiting the symmetry of the problem.

(a) Center-of-Mass and Relative Motion

Let $(\boldsymbol{p}_n, \boldsymbol{x}_n)$ be the momentum and coordinate operators of the two particles, with masses m_n. The Hamiltonian is

$$\mathcal{H} = \frac{p_1^2}{2m_1} + \frac{p_2^2}{2m_2} + V(|\boldsymbol{x}_1 - \boldsymbol{x}_2|)\,. \tag{194}$$

The center-of-mass and relative coordinates are defined as

$$\boldsymbol{R} = (m_1\boldsymbol{x}_1 + m_2\boldsymbol{x}_2)/M, \qquad \boldsymbol{x} = \boldsymbol{x}_1 - \boldsymbol{x}_2\,, \tag{195}$$

and the total and relative momenta by

$$\boldsymbol{P} = \boldsymbol{p}_1 + \boldsymbol{p}_2, \qquad \boldsymbol{p} = (m_2\boldsymbol{p}_1 - m_1\boldsymbol{p}_2)/M\,, \tag{196}$$

where $M = m_1 + m_2$ is the total mass. The inverse relations are

$$\boldsymbol{x}_1 = \boldsymbol{R} + \frac{m_2}{M}\boldsymbol{x}\,, \qquad \boldsymbol{x}_2 = \boldsymbol{R} - \frac{m_1}{M}\boldsymbol{x}\,, \tag{197}$$

$$\boldsymbol{p}_1 = \frac{m_1}{M}\boldsymbol{P} + \boldsymbol{p}\,, \qquad \boldsymbol{p}_2 = \frac{m_2}{M}\boldsymbol{P} - \boldsymbol{p}\,. \tag{198}$$

In terms of these variables, the Hamiltonian is

$$\mathcal{H} = \frac{1}{2M}P^2 + H, \qquad H = \frac{1}{2\mu}p^2 + V(r)\,, \tag{199}$$

where the reduced mass μ is

$$\frac{1}{\mu} = \frac{1}{m_1} + \frac{1}{m_2}, \quad \text{or} \quad \mu = \frac{m_1 m_2}{M}\,. \tag{200}$$

The symbol H has been reserved for the relative motion because it is the non-trivial part of the Hamiltonian.

The new momenta and coordinates are independent, i.e., they commute with each other, $[R_i, p_j] = [x_i, P_j] = [p_i, P_j] = 0$, and they are canonical:

$$[R_i, P_j] = i\hbar\,\delta_{ij}\,, \qquad [x_i, p_j] = i\hbar\,\delta_{ij}\,. \tag{201}$$

The total angular momentum can also be split into independent operators referring to the center-of-mass and relative motion:

$$\hbar\boldsymbol{J} = \boldsymbol{x}_1 \times \boldsymbol{p}_1 + \boldsymbol{x}_2 \times \boldsymbol{p}_2 = \hbar\boldsymbol{L}_{\text{com}} + \hbar\boldsymbol{L}, \tag{202}$$

where

$$\hbar\boldsymbol{L}_{\text{com}} = \boldsymbol{R} \times \boldsymbol{P}, \qquad \hbar\boldsymbol{L} = \boldsymbol{x} \times \boldsymbol{p}\,. \tag{203}$$

The two parts of $\boldsymbol{J}$ commute with each other; $\boldsymbol{L}$ generates rotations in the space of the relative motion, but not in that of the center-of-mass motion, while $\boldsymbol{L}_{\text{com}}$ has the opposite role. For example,

$$[p_i, L_j] = i\epsilon_{ijk}p_k \qquad [x_i, L_j] = i\epsilon_{ijk}x_k\,, \tag{204}$$

as required if $\boldsymbol{x}$ and $\boldsymbol{p}$ are to be vector operators, but $[R_i, L_j]$, etc., vanish.

The constants of motion are therefore

$$H,\ P^2,\ \boldsymbol{P},\ \boldsymbol{L},\ \boldsymbol{L}_{\text{com}}\,, \tag{205}$$

just as in classical mechanics. They do not all commute with each other, of course. The system has six degrees of freedom, and in general six mutually compatible observables. Here "in general" alludes to the fact that in special cases there may be more, as in the Kepler problem (see §5.2). For the center-of-mass motion the choices of compatible observables are those for free-particle motion, while for the relative motion the compatible set is H, L^2 and one component of $\boldsymbol{L}$.

If one or both particles carry spin, there are spin operators $\boldsymbol{s}^{(1)}$ and/or $\boldsymbol{s}^{(2)}$ in addition to the observables listed in (205). When the interaction is a central force, as in (194), both spins are constants of motion, for each one of their components can be added to the compatible set of observables, and the states of the system are simply products of the states $|Elm\rangle$ to be constructed below and a factor $|s_3^{(1)}\rangle \otimes |s_3^{(2)}\rangle$. If the forces are spin-dependent, however, the individual spins will not be constants of motion, and the correct spin state will depend on the nature of the forces. As long as the system is isolated, the total angular momentum $\boldsymbol{J} = \boldsymbol{L} + \boldsymbol{s}^{(1)} + \boldsymbol{s}^{(2)}$ will be a constant of motion. The question of how spin enters the two-body problem will be treated at various points in later chapters — at first in the case of hydrogen in chapter 5. For now, we stay with the central force problem, and can set the spin issue aside.

(b) The Radial Schrödinger Equation: General Case

As H, $\boldsymbol{L}^2$ and L_3 are constants of motion, the relative motion has eigenfunctions of the same form as a free particle when the latter is described in spherical coordinates. To be specific, the eigenkets satisfy

$$\left(\frac{p^2}{2\mu} + V(x)\right)|Elm\rangle = E|Elm\rangle \ , \tag{206}$$

and the corresponding wave functions are

$$\langle \boldsymbol{r}|Elm\rangle = \psi_{Elm}(\boldsymbol{r}) = R_{El}(r)Y_{lm}(\theta\phi) \, , \tag{207}$$

where Y_{lm} is a spherical harmonic. The radial functions R_{El} are the solutions of

$$\left[\frac{\hbar^2}{2\mu}\left(-\frac{1}{r^2}\frac{d}{dr}r^2\frac{d}{dr} + \frac{l(l+1)}{r^2}\right) + V(r) - E\right] R_{El}(r) = 0 \, . \tag{208}$$

This becomes neater on substituting

$$E = -\hbar^2\beta^2/2\mu, \quad U(r) = 2\mu V(r)/\hbar^2 \ , \tag{209}$$

namely,

$$\left(\frac{1}{r^2}\frac{d}{dr}r^2\frac{d}{dr} - \frac{l(l+1)}{r^2} - U(r) - \beta^2\right) R_l(\beta; r) = 0 \, . \tag{210}$$

This is the form appropriate to a bound state, i.e., when $\beta^2 > 0$.

It often pays to state the problem in terms of the functions

$$u_l(\beta; r) = r R_l(\beta; r) \ . \tag{211}$$

Thanks to

$$\frac{1}{r^2}\frac{d}{dr}r^2\frac{dR}{dr} = R'' + \frac{2}{r}R' = \frac{1}{r}u'' \ , \tag{212}$$

this substitution turns (210) into a form that looks one-dimensional:

$$\left(\frac{d^2}{dr^2} - \frac{l(l+1)}{r^2} - U(r) - \beta^2\right) u_l(\beta; r) = 0 \, . \tag{213}$$

Below we will show that $R(r)$ does not diverge as $r \to 0$, and therefore u must be a solution of (213) that satisfies the requirement

$$u_l(r) = 0 \ . \tag{214}$$

Equation (213) can thus be thought of as a one-dimensional Schrödinger equation with the effective potential

$$U_l(r) = U(r) + \frac{l(l+1)}{r^2} \qquad (r > 0)\,, \tag{215}$$

and an infinite barrier at $r = 0$ that imposes the boundary condition (214). The term $l(l+1)/r^2$ is called the centrifugal potential, for obvious reasons; it produces the so-called centrifugal barrier at short distances.

For potentials that approach a constant as $r \to \infty$, it is natural to set this constant to zero. There are then unbound solutions for the whole continuum of energies $E \geq 0$. For outgoing states in the continuum the replacement

$$\beta \to -ik,\; k > 0, \quad \text{i.e.,} \quad E = (\hbar k)^2/2\mu \tag{216}$$

must be made, where $1/k$ is the wavelength $\lambda\!\!\!^{-}$ as $r \to \infty$. The continuum wave functions will be written as $R_l(k;r) = u_l(k;r)/r$, and these u_l must also vanish at $r = 0$.

The behavior of $V(r)$ as $r \to \infty$ provides a powerful distinction between different types of potentials. At first sight one might think that any potential that vanishes as $r \to \infty$ would have solutions that approach those of the free Schrödinger equation asymptotically. That is not so. The determining criterion is whether

$$\lim_{r\to\infty} rV(r) = 0\,, \tag{217}$$

that is, whether or not the potential falls off *faster* than a Coulomb field.

The analysis that follows is only concerned with the asymptotic form of bound-state wave functions, and thus holds for all potentials that have a specific form as $r \to \infty$, whatever their behavior at shorter distances may be. Consider, then, potentials having the asymptotic form

$$U_l(r) \sim \lambda/r^s\,, \tag{218}$$

where $\sim$ now means "as $r \to \infty$." If $l \neq 0$ and $s > 2$, the centrifugal term dominates asymptotically, and in view of (216) the wave function must tend to a linear combination of the free-particle solutions, $ah_l(i\beta r) + bh_l^*(i\beta r)$ (recall Eq. 118). But $h^*(i\beta r)$ for $\beta > 0$ diverges as $r \to \infty$, so $b = 0$, and therefore $R_l(\beta;r) \sim e^{-\beta r}/r$. The crucial point here is that the asymptotic form is completely set by the binding energy.

For $l = 0$ and all s, and for $l \neq 0$ but $s < 2$, what is needed is the asymptotic form of the solution of

$$\left(\frac{d^2}{dr^2} - \frac{\lambda}{r^s} - \beta^2\right) u(\beta;r) = 0\,. \tag{219}$$

Note that $U' \sim 0$ like a power if $s > 1$. To see what this implies consider first a state in the continuum, for which we would expect a slowly varying departure from the phase of a free-particle wave function, that is, $u \sim e^{ikr}\, e^{i\chi(r)}$, with χ slowly varying. For a bound state this translates into the Ansatz

$$u_l(r) \sim e^{-\beta r} e^{\chi(r)}\,, \tag{220}$$

where χ is again slowly varying. Inserting (220) into (219), and dropping terms in χ'' and $(\chi')^2$ gives

$$\frac{d\chi}{dr} = -\frac{\lambda}{2\beta r^s} , \tag{221}$$

which agrees with the assumption that χ varies slowly. The asymptotic form of $u(\beta; r)$ is therefore

$$u(r) \sim e^{-\beta r} \exp\left(\frac{\lambda}{2\beta(s-1)} r^{1-s}\right) , \tag{222}$$

that is, $u \sim e^{-\beta r}$ *provided* $s > 1$.

To summarize, for potentials that fall off faster than a Coulomb field, bound-state radial wave functions for any angular momentum have the universal asymptotic form of an exponential decrease determined solely by the binding energy:

$$u_l(\beta; r) \sim e^{-\beta r} , \quad \text{i.e.,} \quad R_l(\beta; r) \sim \frac{e^{-\beta r}}{r} . \tag{223}$$

On the other hand, for potentials that are Coulombic as $r \to \infty$, $\chi \sim -(\lambda/2\beta) \ln r$, so the asymptotic forms of the radial wave functions are

$$R_l(\beta; r) \sim \frac{e^{-\beta r}}{r} r^{-\lambda/2\beta} . \tag{224}$$

Because $\lambda < 0$ if the potential is attractive at large distances, this shows that the bound states fall off more slowly than those belonging to potentials that are more attractive than $1/r$ at large distances.[1]

The asymptotic form of the continuum states also depends on whether the condition (217) is met. When it is, the asymptotic form is that of the free-particle wave functions. For potentials with a $1/r$ tail the asymptotes of the continuum and bound-state wave functions are obtained from (224) by the substitution (216):

$$R_l(k; r) \sim \frac{1}{r} \exp\left[\pm i\left(kr - \frac{\lambda}{2k} \ln kr\right)\right] . \tag{225}$$

We must still determine when the condition $u(0) = 0$ (Eq. 214) is valid. Consider, first, potentials that are less singular than $1/r^2$ at the origin, i.e., potential satisfying

$$\lim_{r\to 0} r^2 V(r) = 0 . \tag{226}$$

Then on multiplying (213) by r^2 one has

$$r^2 u'' - l(l+1)u = 0 \tag{227}$$

in the immediate neighborhood of $r = 0$, and therefore

$$u_l(r) \sim r^{l+1} , \qquad r \to 0 . \tag{228}$$

This is the same behavior as that for free particles, as is obvious from (227).

---

[1] The binding energy does not determine the parameter β except in the case of a pure Coulomb potential at all distances.

For potentials that do not satisfy (226), there is a crucial difference between the attractive and repulsive case. An attraction $\propto r^{-2}$ strong enough to overcome the centrifugal barrier presents complicated issues.[1] More singular attractions are physically unacceptable because they produce an energy spectrum *without a lower bound*, and usually arise when unreliable approximations are made in deriving the potential, e.g., in taking the nonrelativistic limit of a relativistic theory. Singular repulsive potentials are a very different matter, and are widely used to model complicated interactions, such as those between neutral atoms at short distance. Nothing pathological happens in this case, but the wave functions vanish more rapidly as $r \to 0$ than in (228), so (214) remains valid.

(c) Bound-State Coulomb Wave Functions

The Schrödinger equation for the attractive Coulomb field (also called the Kepler problem), is

$$\left(\frac{\hbar^2}{2\mu}\nabla^2 + \frac{Ze^2}{4\pi r} + E\right)\psi = 0 \,. \tag{229}$$

The parameters μ and Z can be removed by introducing modified atomic units, where "modified" means appropriate not just to an electron moving about an infinitely heavy unit charge, but to arbitrary reduced mass μ and nuclear charge Ze. This involves choosing the following units of length and energy:

$$\text{length} : a = 4\pi\hbar^2/\mu Ze^2 = a_0(\mu/Zm_e) \,, \tag{230}$$

$$\text{energy} : W_0 = Ze^2/4\pi a = 2(Z^2 m_e/\mu)\text{Ry} \,, \tag{231}$$

where $a_0 = 0.52918 \cdot 10^{-8}$ cm is the Bohr radius, W_0 is *twice* the ionization energy of the ground state, and Ry the Rydberg, 13.606 eV. In hydrogen, $a \to a_0$ and $W_0 \to 2\text{Ry}$ if the proton mass is taken to be infinite. (For further discussion of these units see §5.1.)

Here we will only consider the bound-state solutions of (229). The continuum solutions ($E > 0$) describe scattering in Coulomb field, which is treated in §8.4.

When the radial coordinate and energy are expressed in these units, i.e.,

$$r \to r/a \,, \qquad E \to E/W_0 \equiv -\tfrac{1}{2}\kappa^2 \,, \tag{232}$$

the radial Schrödinger equation becomes

$$\left(\frac{d^2}{dr^2} - \frac{l(l+1)}{r^2} + \frac{2}{r} - \kappa^2\right) u_l = 0 \,. \tag{233}$$

This is related to (213) and (218) by $\lambda \to -2, \beta \to \kappa$. From (224) we thus know that u_l has the asymptotic form

$$u_l \sim r^{1/\kappa} e^{-\kappa r} \,. \tag{234}$$

A somewhat simpler equation emerges if the known limiting forms (228) and (234) are taken into account by setting

$$u_l = \rho^{l+1} e^{-\frac{1}{2}\rho} f_l(\rho) \,, \qquad \rho = 2\kappa r \,, \tag{235}$$

[1] See *LLQM*, §35.

where f_l is now the solution of

$$\rho f_l'' + (2l + 2 - \rho) f_l' + \left(\frac{1}{\kappa} - l - 1\right) f_l = 0 \,. \tag{236}$$

The equation that defines the confluent hypergeometric (or Kummer) function is

$$\left(z\frac{d^2}{dz^2} + (c - z)\frac{d}{dz} - a\right) {}_1F_1(a; c; z) = 0 \,, \tag{237}$$

where z is a complex variable, and a and c are complex constants. Hence

$$f_l(r) = \text{const.}\, {}_1F_1(l + 1 - \kappa^{-1}; 2l + 2; 2\kappa r) \,. \tag{238}$$

The solution that is regular everywhere in the finite z plane has the power series expansion

$${}_1F_1(a; c; z) = \sum_{n=0}^{\infty} \frac{(a)_n}{n!(c)_n}\, z^n \,. \tag{239}$$

where $(a)_n = a(a+1)\dots(a+n-1)$, and $(a)_0 = 1$. For Re $z \geq 0$ the asymptotic form of ${}_1F_1(a; c; z)$ is $\sim z^{a-c}e^z$, and as Re $z = \rho$, this will overwhelm the factor $e^{-\frac{1}{2}\rho}$ in (235) and yield a function u_l that diverges exponentially as $r \to \infty$. Bound states must have normalizable wave functions, and therefore only exist if the series (239) terminates, which requires a to be zero or a negative integer. In terms of the physically meaningful parameters, this condition is $\kappa^{-1} - l - 1 = 0, 1, 2, \dots$, or

$$\frac{1}{\kappa} = l + 1, l + 2, \dots, \,. \tag{240}$$

This is *the energy quantization condition for the Kepler problem*, as derived in Schrödinger's first paper on wave mechanics, for then, according to (232), the energy eigenvalues are

$$E_{l,k} = -\frac{1}{2(k+l)^2} \,, \quad l = 0, 1, \dots, \quad k = 1, 2, \dots \,. \tag{241}$$

This is Bohr's famous result,

$$E_n = -\frac{1}{2n^2} = -\frac{1}{n^2}\frac{Z^2 m_e}{\mu}\,\text{Ry} \,, \qquad n = 1, 2, \dots \,, \tag{242}$$

where

$$n = 1/\kappa \tag{243}$$

is the principal quantum number, and the first expression is in atomic units while the second is in conventional units. Hence, for a given value of n, all angular momentum eigenstates with

$$l = 0, 1, \dots, n - 1 \,, \tag{244}$$

have the same energy.

According to (235) and (238), the complete radial wave functions are then

$$R_{nl}(r) = C_{nl}\, r^l e^{-r/n}\, {}_1F_1(l + 1 - n; 2l + 2; 2r/n) \,, \tag{245}$$

where C_nl is the normalization constant. For the first two levels the radial functions are

$$\begin{aligned} R_{10} &= 2e^{-r} , \\ R_{20} &= \frac{1}{\sqrt{2}}(1 - \tfrac{1}{2}r)e^{-\frac{1}{2}r} , \\ R_{21} &= \frac{1}{2\sqrt{6}} re^{-\frac{1}{2}r} . \end{aligned} \tag{246}$$

They are normalized to

$$\int_0^\infty r^2 dr |R_{nl}(r)|^2 = 1 . \tag{247}$$

Expectation values of various powers of r are needed in many applications. In atomic units, the most important are

$$\begin{aligned} \langle r \rangle &= \tfrac{1}{2}[3n^2 - l(l+1)] , \\ \langle r^2 \rangle &= \tfrac{1}{2}n^2[5n^2 + 1 - 3l(l+1)] , \\ \langle r^{-1} \rangle &= 1/n^2 , \\ \langle r^{-2} \rangle &= [n^3(l + \tfrac{1}{2})]^{-1} , \\ \langle r^{-3} \rangle &= [n^3 l(l + \tfrac{1}{2})(l+1)]^{-1} . \end{aligned} \tag{248}$$

The bound-state wave functions in momentum space are also needed in some applications. They are

$$F_{nl}(p)Y_{lm}(\hat{\mathbf{p}}) = (1/2\pi)^{3/2} \int d^3p \; e^{-i\mathbf{p}\cdot\mathbf{r}} R_{nl}(r) Y_{lm}(\hat{\mathbf{r}}) , \tag{249}$$

where p is in atomic units, i.e., $\hbar/a$. The normalized radial functions for $n = 1$ and 2 are

$$\begin{aligned} F_{10} &= \sqrt{\frac{2}{\pi}} \frac{4}{(p^2+1)^2} , \\ F_{20} &= \frac{32}{\sqrt{\pi}} \frac{4p^2 - 1}{(4p^2+1)^3} , \\ F_{21} &= \frac{128}{\sqrt{3\pi}} \frac{p}{(4p^2+1)^3} . \end{aligned} \tag{250}$$

Much more information about Coulomb wave functions can be found in *LLQM*, §35 and 49, and in Bethe and Salpeter.

3.7 Basic Approximation Methods

In the last analysis, theoretical physics hardly ever provides a mathematically exact answer to realistic questions. There are a small number of situations where an exact solution of a basic equation gives a remarkably accurate account of the data, but eventually refinement of experimental technique or previously overlooked effects will require a more complicated theoretical description that no longer yields to an exact analytic solution. This was true from the start in celestial mechanics, and has also been true of quantum physics. Indeed, the evaluation of effects which are known to be small, or hoped to be small, has always occupied a large portion of the life of most physicists.

There is an enormous arsenal of approximation methods for solving quantum mechanical problems. Before the invention of powerful computers, great effort was devoted to analytic methods, and these are still of great importance. They are often, though certainly not always, the basis for numerical computations vastly more elaborate and precise than was originally imaginable, and they continue to be indispensable for gaining a qualitative and often even a quantitative understanding. These analytic approximation techniques also play an essential role in quantum mechanical thinking and talking about phenomena from the mundane to the astonishing.

This section describes the rudiments of perturbation theory, wherein it is assumed that the Hamiltonian is a sum of two terms,

$$H = H_0 + \lambda H_1 \; . \tag{251}$$

The unperturbed Hamiltonian H_0 is assumed to have a complete set of known eigenstates and eigenvalues. The matrix elements of the perturbation H_1 are at most of the same order of magnitude as those of H_0, and the dimensionless parameter λ is small enough so that, by hypothesis, the states and spectrum of H_0 form a reasonable starting point for the evaluation of the states and spectrum of H itself.[1]

When the perturbation is time-independent, the goal is to find the change in the eigenstates and eigenvalues due to H_1. This is called stationary-state perturbation theory. When the perturbation depends on time, the goal is to find the probability for transitions between the unperturbed states, and the technique for solving this problem is called time-dependent perturbation theory. The distinction between stationary-state and time-dependent perturbation theory is not, however, as straightforward as the preceding statements would seem to indicate. Often the Hamiltonian is time-independent, but the state of interest is not stationary. The most important examples of this circumstance are collision processes, where the initial state is a wave packet moving toward a scattering center which will, after impact, produce waves scattered in various directions. Another and closely related phenomenon of this type is the decay of a state. Collision and decay phenomena will not, however, be taken up here. They are of sufficient importance and difficulty to require separate treatment.

(a) Stationary-State Perturbation Theory

Let $\{|\alpha\rangle\}$ be a complete set of eigenstates of the unperturbed Hamiltonian H_0 with energy eigenvalues $\{E_\alpha\}$:

$$(H_0 - E_\alpha)|\alpha\rangle = 0 \; . \tag{252}$$

The goal is to find the eigenstates $\{|a\rangle\}$ and eigenvalues $\{E_a\}$ of the complete Hamiltonian H:

$$(H - E_a)|a\rangle = 0 \; . \tag{253}$$

---

[1]In later chapters (and in the literature) the parameter λ is often set to 1 in the results obtained here, with the understanding that H_1 is small itself in comparison to H_0. Here λ will be understood as being small, as this makes the approximations more transparent.

In principle, the E_a can be found by solving for the roots E of the secular equation in the unperturbed basis,

$$\det[\delta_{\alpha\beta}(E_\alpha - E) - \langle\alpha|H_1|\beta\rangle] = 0 \,. \tag{254}$$

However, this is rarely a practical method when the dimension of the Hilbert space is large, let alone infinite.

Consider, first, the simplest situation, where the spectrum of H_0 is *non-degenerate*. In this case, the sought-for states $|a\rangle$ can be related uniquely to the unperturbed states:

$$|a\rangle \to |\alpha\rangle \,, \quad E_a \to E_\alpha \,, \qquad \text{as } \lambda \to 0 \,. \tag{255}$$

The perturbed states are then expanded in terms of the unperturbed ones:

$$|a\rangle = c_\alpha|\alpha\rangle + \sum_{\beta\neq\alpha} d_\beta|\beta\rangle \,, \tag{256}$$

$$|c_\alpha|^2 + \sum_{\beta\neq\alpha} |d_\beta|^2 = 1 \,. \tag{257}$$

Perturbation theory evaluates the eigenvalues E_a, and the coefficients c_α and d_α, as power series in λ by requiring Eq. 256 to be a solution of the Schrödinger equation (253). That is, inserting (256) into $\langle\gamma|(H - E_a)|a\rangle = 0$, with $\gamma \neq \alpha$, gives

$$\lambda\langle\gamma|H_1|\alpha\rangle c_\alpha + d_\gamma(E_\gamma - E_a) + \lambda\sum_{\beta\neq\alpha} d_\beta\langle\gamma|H_1|\beta\rangle = 0 \,. \tag{258}$$

But $d_\beta \to 0$ as $\lambda \to 0$, so the last term in (258) is at most $O(\lambda^2)$, and therefore

$$d_\beta = \lambda c_\alpha \frac{\langle\beta|H_1|\alpha\rangle}{E_a - E_\beta} + O(\lambda^2) \,. \tag{259}$$

As $d_\beta = O(\lambda)$, (257) implies that

$$c_\alpha = 1 - O(\lambda^2) \,, \tag{260}$$

and therefore c_α can be replaced by 1 in (259). Furthermore, to leading order E_a in (259) can be replaced by the unperturbed eigenvalue E_α. Hence the final result for the perturbed eigenstate is the important formula

$$|a\rangle = |\alpha\rangle + \lambda\sum_{\beta\neq\alpha} |\beta\rangle \frac{\langle\beta|H_1|\alpha\rangle}{E_\alpha - E_\beta} + O(\lambda^2) \,. \tag{261}$$

Clearly, this formula requires that for all $|\beta\rangle$

$$\left|\lambda\frac{\langle\beta|H_1|\alpha\rangle}{E_\alpha - E_\beta}\right| \ll 1 \,. \tag{262}$$

Note that if the off-diagonal matrix elements of H_1 do not grow as the energy difference $|E_\alpha - E_\beta|$ increases, the more "distant" a state is from the state of interest the smaller its influence will be.

The perturbed energies follow from $\langle\alpha|H - E_a|a\rangle = 0$, or

$$0 = [E_\alpha - E_a + \lambda\langle\alpha|H_1|\alpha\rangle]c_\alpha + \lambda\sum_{\beta\neq\alpha} d_\beta\langle\alpha|H_1|\beta\rangle\ . \tag{263}$$

Using (260) and (261) gives the other important result of stationary-state perturbation theory:

$$E_a = E_\alpha + \lambda\langle\alpha|H_1|\alpha\rangle + \lambda^2\sum_{\beta\neq\alpha}\frac{|\langle\alpha|H_1|\beta\rangle|^2}{E_\alpha - E_\beta} + O(\lambda^3)\ . \tag{264}$$

To summarize, the leading contribution to the energy shift is just the expectation value of the perturbation in the unperturbed state. The second-order term involves the other unperturbed states, and in many situations this is actually the leading correction because the expectation value $\langle\alpha|H_1|\alpha\rangle$ vanishes by symmetry. It is noteworthy that this leading-order calculation gives the energy to second order in the perturbation but the state only to first order. If the expansion is carried to higher orders, this distinction between eigenvalues and eigenvectors continues to hold.

The validity of the perturbation expansion is a much more subtle matter than this naive approach might lead one to suppose. At bottom, the true eigenvalues and eigenstates are being expanded as a power series in the perturbation λH_1. This means that it has been tacitly assumed that these quantities are analytic functions in the complex variable λ in a circle of finite radius about the origin of the λ plane. That this can be a very stringent condition can be seen from the example of a harmonic oscillator perturbed by a quartic potential, i.e, the Hamiltonian

$$H_0 = \frac{p^2}{2m} + \frac{1}{2}kq^2 + \frac{1}{4}\lambda q^4\ . \tag{265}$$

For $\lambda > 0$, the unperturbed and perturbed potentials are not dramatically different; both tend to infinity as $|q| \to \infty$, and therefore have a discrete spectrum of bound states. One might, therefore, conclude that all would be well with perturbation theory, at least for low-lying states which do not extend into regions of space where the perturbation is substantial. That is not correct, however! For consider the case $\lambda < 0$, which must not be forgotten because of the analyticity requirement even though $\lambda > 0$ is only of actual interest. No matter how small $|\lambda|$ may be, the perturbation eventually overpowers the harmonic term and sends the potential to negative infinity (see Fig. 3.3). In such a potential there are no bound states whatsoever! Clearly, perturbation theory cannot cross the deep chasm that separates the bound-state spectrum of the simple harmonic oscillator from that of the continuous spectrum of the "perturbed" problem when $\lambda < 0$. In light of the analyticity requirement, perturbation theory must also suffer from some serious disease when $\lambda > 0$ even though the dramatic effect of the perturbation is then not manifest.

This is too early in this volume for a proper explanation of what use perturbation theory is in this, and other similar situations; a brief description must suffice for now. First take the $\lambda < 0$ case. For energies well below the potential maximum at $kq = |\lambda|q^3$, there are quasi-bound states with energies that are well approximated by the perturbation expansion, where "quasi" alludes to the fact that they are not really stationary because decay through the potential barrier, with a very long lifetime τ, is possible. Put another, and equivalent way, plane waves sent against this

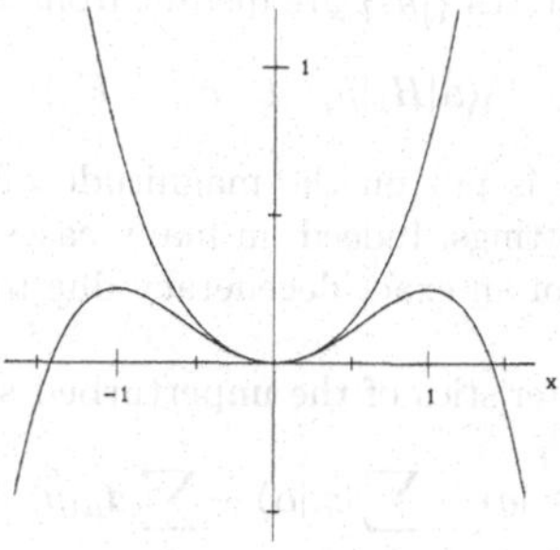

FIG. 3.3. The quartic potentials $\frac{1}{2}x^2 \pm \frac{1}{4}x^4$, showing that when the term x^4 is negative, the particle can escape to infinity by tunneling through the barrier.

double hump potential will have very narrow transmission and reflection resonances of width $\sim \hbar/\tau$ at the energies of these quasi-bound states. This phenomenon will be explained in detail in §4.4(c) in an exactly soluble model. In the "undramatic" $\lambda > 0$ case, the energies are of the form $E_a(\lambda) = A_a(\lambda) + S_a(\lambda)$, where $A_a(\lambda)$ is analytic in λ and has a power series expansion given by perturbation theory, while $S_a(\lambda)$ is exponentially small as $\lambda \to 0$ and is completely missed by the perturbation expansion.

(b) Degenerate-State Perturbation Theory

From (261) and (264) it is clear that even if the perturbation is small, it will produce large effects on unperturbed states that have nearby neighbors, i.e., if the unperturbed spectrum is degenerate or nearly degenerate. When this is the case, degenerate-state perturbation theory must be used.

Consider the spectrum of H_0 shown in Fig. 3.4, which contains a degenerate or nearly degenerate subspace $\mathfrak{D}$ spanned by the unperturbed states $\{|\alpha\rangle\}$. The

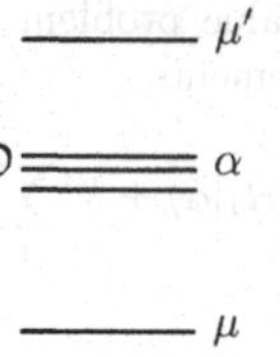

FIG. 3.4. Illustration of a situation in which degenerate-state perturbation theory is necessary; the levels are those of the unperturbed Hamiltonian, and $\mathfrak{D}$ is a nearly degenerate multiplet.

problem is to find the perturbed spectrum in $\mathfrak{D}$, in which case it is immaterial whether or not the states outside $\mathfrak{D}$ have degeneracies. All that matters is that these other unperturbed states $\{|\mu\rangle\}$ are distant from $\mathfrak{D}$ in the sense that

$$|\lambda\langle\alpha|H_1|\mu\rangle| \ll |E_\alpha - E_\mu| . \tag{266}$$

Within $\mathfrak{D}$, no constraint is put on the magnitude of matrix elements $\langle\alpha|H_1|\beta\rangle$ relative to the energy splittings. Indeed, in many cases of great importance, these splittings vanish because of an exact degeneracy due to a symmetry, such as rotational invariance.

In view of these characteristics of the unperturbed spectrum, (256) is amended to

$$|a\rangle = \sum_\alpha c_\alpha|\alpha\rangle + \sum_\mu d_\mu|\mu\rangle , \tag{267}$$

where the coefficients $\{c_\alpha\}$ are expected to be $O(1)$, whereas the $\{d_\mu\}$ are presumably $O(\lambda)$. Once again, exploit $(H - E_a)|a\rangle = 0$, i.e.,

$$\sum_\alpha c_\alpha(E_\alpha - E_a + \lambda H_1)|\alpha\rangle + \sum_\mu d_\mu(E_\mu - E_a + \lambda H_1)|\mu\rangle = 0 . \tag{268}$$

Projecting onto some state $|\beta\rangle$ in $\mathfrak{D}$ and another state $|\nu\rangle$ outside $\mathfrak{D}$ gives

$$c_\beta(E_\beta - E_a) + \lambda\sum_\alpha c_\alpha\langle\beta|H_1|\alpha\rangle + \lambda\sum_\mu d_\mu\langle\beta|H_1|\mu\rangle = 0 , \tag{269}$$

$$\lambda\sum_\alpha c_\alpha\langle\nu|H_1|\alpha\rangle + d_\nu(E_\nu - E_a) + \lambda\sum_\mu d_\mu\langle\nu|H_1|\mu\rangle = 0 . \tag{270}$$

The last term of (270) is $O(\lambda^2)$ and can be dropped, while E_a in $d_\nu(E_\nu - E_a)$ can be replaced by any of the E_α because their separation from each other and E_a is negligible in comparison to $E_\nu - E_\alpha$. Let $\bar{E}_D$ be some mean energy in $\mathfrak{D}$. Then

$$d_\nu = \frac{1}{\bar{E}_D - E_\nu}\lambda\sum_\alpha c_\alpha\langle\nu|H_1|\alpha\rangle , \tag{271}$$

and from (269)

$$c_\beta(E_\beta - E_a) + \sum_\alpha c_\alpha\left(\lambda\langle\beta|H_1|\alpha\rangle + \lambda^2\sum_\mu \frac{\langle\beta|H_1|\mu\rangle\langle\mu|H_1|\alpha\rangle}{\bar{E}_D - E_\mu}\right) = 0 . \tag{272}$$

This equation is just the eigenvalue problem in the subspace $\mathfrak{D}$ for the *effective Hamiltonian* H_{eff} with matrix elements

$$\langle\beta|H_{\text{eff}}|\alpha\rangle = \lambda\langle\beta|H_1|\alpha\rangle + \lambda^2\sum_\mu \frac{\langle\beta|H_1|\mu\rangle\langle\mu|H_1|\alpha\rangle}{\bar{E}_D - E_\mu} , \tag{273}$$

or in operator form,

$$H_{\text{eff}} = \lambda PH_1P + \lambda^2 PH_1\frac{1-P}{\bar{E} - H_0}H_1P , \tag{274}$$

where P is the projection operator onto $\mathfrak{D}$:

$$P = \sum_\alpha |\alpha\rangle\langle\alpha| . \tag{275}$$

The simplest situation occurs when H_1 only has matrix elements within $\mathfrak{D}$, and then the calculation merely amounts to diagonalizing the perturbation in this subspace. But the second-order term in $H_{\rm eff}$ has been kept because there are important cases where, because of symmetry, H_1 has no matrix elements within the degenerate subspace $\mathfrak{D}$, and then the degeneracy is lifted by distant states.

A simple, concrete eigenvalue problem will elucidate how stationary-state perturbation method works in different situations. Consider a 3-level system, i.e., a 3-dimensional Hilbert space, spanned by the eigenstates of an unperturbed Hamiltonian whose ground state is doubly degenerate with eigenstates $|1,2\rangle$, and excited state $|3\rangle$ at an energy Δ above the ground-state doublet. The perturbation only has matrix elements λM between the ground and excited states:

$$H_1 = \begin{pmatrix} 0 & 0 & \lambda M \\ 0 & 0 & \lambda M \\ \lambda M & \lambda M & \Delta \end{pmatrix} . \tag{276}$$

If $|\lambda M/\Delta| \ll 1$, the excited state will change little, but the ground-state doublet will split and the new eigenstates will differ dramatically from the unperturbed states. In this weak coupling situation, the new energies and eigenstates in the space spanned by $|1,2\rangle$ can be found from the effective Hamiltonian (273), which in this instance is

$$H_{\rm eff} = -\Delta x^2 \begin{pmatrix} 1 & 1 \\ 1 & 1 \end{pmatrix} , \qquad x \equiv \frac{\lambda M}{\Delta} . \tag{277}$$

The new energies and eigenstates are

$$E_S = -2\Delta x^2 , \qquad |E_S\rangle = \tfrac{1}{\sqrt{2}}\{|1\rangle + |2\rangle\} , \tag{278}$$

$$E_A = 0 , \qquad |E_A\rangle = \tfrac{1}{\sqrt{2}}\{|1\rangle - |2\rangle\} . \tag{279}$$

Note that the new ground state is the symmetric combination of the originally degenerate states; the other must then be antisymmetric to be orthogonal. This is a quite general feature of such problems.

It is easy to solve the eigenvalue problem posed by (276) exactly. One finds that there is again one eigenvalue of zero energy, which we will again call E_A ; this is clearly a quirk of the very simple Hamiltonian. The other two are

$$E_S = \tfrac{1}{2}\Delta\,[1 - \sqrt{1+8x^2}\,] \simeq -2\Delta x^2 , \tag{280}$$

$$E_+ = \tfrac{1}{2}\Delta\,[1 + \sqrt{1+8x^2}\,] \simeq \Delta + 2\Delta x^2 . \tag{281}$$

The last approximate expression is what non-degenerate perturbation theory gives to this order. Observe that the exact eigenvalues are not analytic functions of λ throughout the λ-plane, but have branch points at $\lambda = \pm i\Delta/\sqrt{8}M$, which is far from the origin if $M \ll \Delta$. In this example these singularities do not, however, invalidate perturbation theory as such. Another point worth noting is that the sum of the perturbed eigenvalues is Δ, as it must be because Tr $H = \Delta$; in more difficult problems this property of the spectrum can be used to simplify the calculation of the most recalcitrant eigenvalue.

The exact eigenstates are also interesting and informative. The unshifted state $|E_A\rangle$ is given by (279) for *all* values of λ. The other two states are

$$|E_{S,+}\rangle = \frac{1}{\sqrt{1+2(\lambda^2M^2/E^2)}}\,\frac{\lambda M}{E}\left(|1\rangle + |2\rangle + \frac{E}{\lambda M}|3\rangle\right) , \tag{282}$$

where E is the energy appropriate to either state. They are, of course, orthogonal to $|E_A\rangle$. In the perturbative regime, $|\lambda M/\Delta| \ll 1$, the top state is given by non-degenerate perturbation theory:

$$|E_+\rangle \simeq |E_3\rangle + x\{|E_1\rangle + |E_2\rangle\} . \tag{283}$$

The correction to (278) can be found from (282), or by requiring $|E_S\rangle$ to be orthogonal to $|E_+\rangle$:

$$|E_S\rangle \simeq \frac{1}{\sqrt{2}}\{|E_1\rangle + |E_2\rangle\} - \sqrt{2}x\,|E_3\rangle . \tag{284}$$

It is striking that all these states are either symmetric or antisymmetric under the interchange $|1\rangle \leftrightarrow |2\rangle$. This is as it must be, as can be seen by writing the Hamiltonian in the form

$$H = \Delta|3\rangle\langle 3| + \lambda M\{|1\rangle\langle 3| + |2\rangle\langle 3| + \text{h.c.}\} , \tag{285}$$

which displays its invariance under this interchange, and therefore requires all states to be either even or odd. The odd state $|E_A\rangle$ is unaffected by $|3\rangle$ because it is coupled symmetrically to $|1\rangle$ and $|2\rangle$.

(c) Time-Dependent Perturbation Theory

The problems we now turn to can, quite generally, be put as follows. The Hamiltonian of the system is

$$H = H_0 + V(t) , \tag{286}$$

where $V(t)$, called the perturbation, may be time-dependent (in the Schrödinger picture), but need not be. The state of interest, $|\Psi_i(t)\rangle$, is a solution of the complete Schrödinger equation

$$i\hbar\frac{\partial}{\partial t}|\Psi_i(t)\rangle = [H_0 + V(t)]|\Psi_i(t)\rangle . \tag{287}$$

This solution is to evolve out of a solution $|\Phi_i(t)\rangle$ of the *unperturbed* Schrödinger equation, i.e., to satisfy the initial condition

$$|\Psi_i(t)\rangle \to |\Phi_i(t)\rangle , \qquad t \to -\infty , \tag{288}$$

where $|\Phi(t)\rangle$ is a solution of

$$i\hbar\frac{\partial}{\partial t}|\Phi_i(t)\rangle = H_0|\Phi_i(t)\rangle . \tag{289}$$

Two concrete problems will make it clear what is up. The first type of problem is one where $V(t)$ is explicitly time-dependent. It is "turned on," say, at $t = 0$, and the initial state $|\Phi_i(t)\rangle$ is a stationary state of H_0. For $t > 0$ we wish to know the probability for the system to be in some other stationary state $|\Phi_f(t)\rangle$ of H_0. An example of this kind is an atom in its ground state which is subjected to an applied electromagnetic field.

The second example is the collision of a particle with a *time-independent* potential V of finite range. In this case the initial state $|\Phi_i(t)\rangle$ is a *free-particle* wave packet moving toward the region in which V acts. As soon as the packet overlaps with V, $|\Phi_i(t)\rangle$ ceases to be a solution of the complete Schrödinger equation, and waves will start propagating outward in various directions not contained in the incoming state. In due time these will reach a distant detector D_f which, in effect, asks for the probability that the state has evolved into another solution $|\Phi_f(t)\rangle$ of (289), because V has no influence at D_f. This *transition probability* is

$$P_{i\to f}(t) = |\langle\Phi_f(t)|\Psi_i(t)\rangle|^2 , \qquad t \to \infty , \tag{290}$$

where $|\Phi_f(t)\rangle$ is a packet with the momentum components accepted by D_f. Clearly, $P_{i\to f}$ is closely related to such observable quantities as scattering cross sections, but to make this connection important details remain to be settled. It is also common to introduce the term *transition amplitude* for

$$A_{i\to f} = \langle\Phi_f(t)|\Psi_i(t)\rangle . \tag{291}$$

Problems of the type just mentioned can be difficult to solve with adequate accuracy, and powerful methods exist for attacking them which exploit propagators, Green's functions, the interaction picture, and other sophisticated tools. Some of these will be introduced in later chapters. Here we only describe the most direct route to a simple but remarkably useful answer.

This crudest approximation is given by lowest-order time-dependent perturbation theory, which is based on the assumption the V is sufficiently weak to alter the initial state but little. This then allows the replacement

$$V(t)|\Psi(t)\rangle \to V(t)|\Phi(t)\rangle \tag{292}$$

in the exact Schrödinger equation (287), so that it then reads

$$\left(i\hbar\frac{\partial}{\partial t} - H_0\right)|\Psi_i(t)\rangle = V(t)|\Phi_i(t)\rangle . \tag{293}$$

To solve this equation, consider first a similar ordinary differential equation

$$\left(i\frac{d}{dt} - C\right)\psi(t) = s(t) , \tag{294}$$

which involves no operators, i.e., where C is a constant and $s(t)$ a numerical function. Assuming that $s(t) \to 0$ in the remote past, we seek a solution that satisfies the initial condition $\psi(t) \to \phi(t)$ as $t \to -\infty$, where $\phi(t)$ is a solution of (294) when the right-hand side vanishes. This solution is

$$\psi(t) = \phi(t) - i\int_{-\infty}^{t} dt'\, e^{-iC(t-t')}\, s(t') , \tag{295}$$

provided that $s(t) \to 0$ as $t \to -\infty$ sufficiently rapidly. We will not examine here what this requirement actually demands. As is to be expected from §3.6(b), in the case of three-dimensional potential scattering V must fall off faster at infinity than a Coulomb field. When V is an applied field that vanishes in the past, the condition is trivially satisfied.

This lesson can be applied immediately to (293) because H_0 commutes with itself and, in effect, acts like the number C. Thus

$$|\Psi_i(t)\rangle = |\Phi_i(t)\rangle - \frac{i}{\hbar}\int_{-\infty}^{t} dt'\, e^{-iH_0(t-t')/\hbar}\, V(t')|\Phi(t')\rangle \,. \tag{296}$$

The transition amplitude (Eq. 291) is therefore

$$A_{i\to f}(t) = \langle\Phi_f(t)|\Phi_i(t)\rangle - \frac{i}{\hbar}\int_{-\infty}^{t} dt'\, \langle\Phi_f(t)|e^{-iH_0(t-t')/\hbar}\, V(t')|\Phi_i(t')\rangle \,. \tag{297}$$

We now apply this result to problems of the first type. As an example, think of an atom exposed to a uniform time-dependent electric field $\boldsymbol{E}(t)$, in which case the perturbation $V(t)$ is $-E(t)d$, where d is the operator that corresponds to the component of the atom's electric dipole moment parallel to $\boldsymbol{E}$. Problems of this kind are due to perturbations of the form

$$V(t) = f(t)\, Q \,, \tag{298}$$

where $f(t)$ is a numerical function and Q some observable of the system. In this circumstance, the initial and final states are eigenstates of H_0, which we write as

$$|\Phi_{i,f}(t)\rangle = e^{-iE_{i,f}t/\hbar}\, |\Phi_{i,f}\rangle \,. \tag{299}$$

The transition probability (i.e., when $i \neq f$) is then

$$P_{i\to f}(t) = \frac{1}{\hbar^2}\, |\langle\Phi_f|Q|\Phi_i\rangle|^2\, |F(\omega_{fi}, t)|^2 \,, \tag{300}$$

where

$$F(\omega_{fi}, t) = \int_{-\infty}^{t} dt'\, f(t')\, e^{i\omega_{fi}t'} \,, \qquad \hbar\omega_{fi} = E_f - E_i \,. \tag{301}$$

Consider a periodic perturbation $f(t) = \sin\nu t$ that turns on at $t = 0$, in which case

$$F(\omega, t) = ie^{i\frac{1}{2}(\omega-\nu)t}\, \frac{\sin\frac{1}{2}(\omega-\nu)t}{\omega-\nu} - ie^{i\frac{1}{2}(\omega+\nu)t}\, \frac{\sin\frac{1}{2}(\omega+\nu)t}{\omega+\nu} \,. \tag{302}$$

The situation of primary interest is where the perturbation is resonant, i.e., has a frequency ν that is close to one of the excitation frequencies ω_{fi}. The first term in (302) is the one that resonates and dominates, the leading correction being its interference with the second term:

$$|F(\omega, t)|^2 \simeq \frac{\sin^2[\frac{1}{2}(\omega-\nu)t]}{(\omega-\nu)^2} - \frac{\sin[\frac{1}{2}(\omega-\nu)t]\sin\omega t\cos\omega t}{\omega(\omega-\nu)} \,, \qquad \omega \approx \nu \,. \tag{303}$$

At resonance, $\nu = \omega_{fi}$, the interference term is smaller than the resonant term by a factor $1/\omega_{fi}t$, i.e., it is miniscule because this is a characteristic atomic period divided by a macroscopic time. In the whole neighborhood of the resonance the transition probability is therefore

$$P_{i\to f}(t) = \frac{1}{\hbar^2}\, |\langle\Phi_f|Q|\Phi_i\rangle|^2\, \frac{\sin^2[\frac{1}{2}(\omega_{fi}-\nu)t]}{(\omega_{fi}-\nu)^2} \,. \tag{304}$$

This persistent oscillatory behavior is an artifact due to assuming that the perturbation is a perfect sinusoid forever. For example, the more realistic form

$$V(t) = Q\, e^{-t/\tau} \sin \nu t\ , \qquad (t \geq 0)\ , \tag{305}$$

gives a constant for the transition probability after the perturbation is over,

$$P_{i\to f} = \frac{1}{4\hbar^2} \frac{|\langle \Phi_f|Q|\Phi_i\rangle|^2}{(\omega_{fi} - \nu)^2 + (1/\tau)^2}\ , \qquad t \gg \tau\ , \tag{306}$$

which is a resonance of width $1/\tau$.

(d) The Golden Rule

We now turn to the second type of problem, in which the perturbation is time-independent. As already mentioned, one example of this is a collision. Another is the emission of light due to the interaction between a system's charged constituents with the electromagnetic field, which is alway present, but being weak in comparison to the forces that bind atoms, solids or nuclei, can be handled as a perturbation.

The conceptually simplest example is that of a particle colliding with a static potential V. A fully realistic description would use wave packets to describe the initial state, but is a somewhat complicated mathematical exercise, which is developed in §8.1(c). An ingenious argument due to Dirac leads more directly to the basic result in terms of the stationary states, (299), which in a collision problem are momentum eigenstates. Unless the scattering is in the exact forward direction, only the second term in (297) contributes, and the so the amplitude is

$$A_{i\to f}(t) = -\frac{i}{\hbar} \langle \Phi_f|V|\Phi_i\rangle \int_{-\infty}^{t} dt'\ e^{i\omega_{fi}t}\ . \tag{307}$$

We need this amplitude in the limit $t \to \infty$, but the integral integral does not exist as it stands. The trick is to assume that for some very large but finite time $-T$ in the past,

$$|\Psi_i(t)\rangle \to |\Phi_i(t)\rangle\ , \qquad t \to -T\ , \tag{308}$$

and to compute the amplitude at $t = T$:

$$A_{i\to f}(T) = -\frac{i}{\hbar} \langle \Phi_f|V|\Phi_i\rangle \int_{-T}^{T} dt\ e^{i\omega_{fi}t}\ , \tag{309}$$

which gives the transition probability

$$P_{i\to f}(T) = \frac{4}{\hbar^2} |\langle \Phi_f|V|\Phi_i\rangle|^2 \frac{\sin^2 \omega_{fi}T}{\omega_{fi}^2}\ . \tag{310}$$

The function $\sin^2 \omega T/\omega^2$ is shown in Fig. 3.5. As $\omega T \to \infty$, its peak height grows like T^2 and its width like $1/T$, so its area is $\sim T$. The exact value is

$$\int_{-\infty}^{\infty} \frac{\sin^2 \omega T}{\omega^2}\, d\omega = \pi T\ , \tag{311}$$

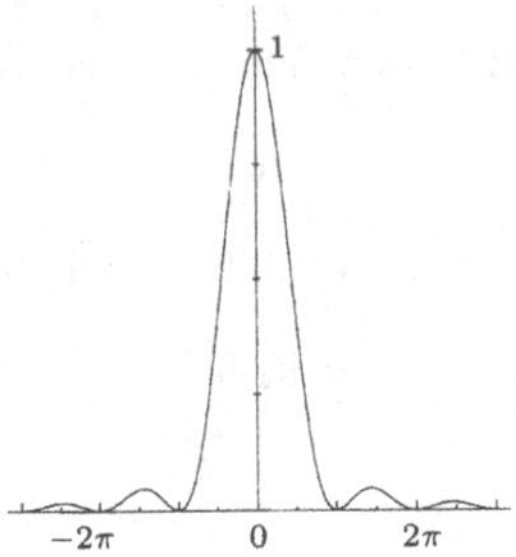

FIG. 3.5. The function $f(x) = \sin^2 x/x^2$ that occurs in time-dependent perturbation theory. Note that $T^2 f(\omega T)$ appears in (310) and (312).

or

$$\lim_{T\to\infty} \frac{\sin^2 \omega T}{\omega^2} = \pi T\, \delta(\omega) \,. \tag{312}$$

The transition probability therefore tends to the following result as $T \to \infty$:

$$P_{i\to f} = \frac{2\pi}{\hbar} \, |\langle \Phi_f|V|\Phi_i\rangle|^2 \, \delta(E_f - E_i) \cdot (2T) \,. \tag{313}$$

Thus this probability is proportional to the time interval $2T$ over which the interaction is active in virtue of the way it is idealized by (309). Equation (313) is therefore the consequence of the steady *transition rate*

$$\dot{P}_{i\to f} = \frac{2\pi}{\hbar} \, |\langle \Phi_f|V|\Phi_i\rangle|^2 \, \delta(E_f - E_i) \,. \tag{314}$$

This is Dirac's result; it is the most important result of time-dependent perturbation theory. As we shall learn in §9.1(c), its form holds to all orders in V. Fermi called (314) the *Golden Rule*, and it is often wrongly attributed to him. Note that (314) has the dimension (Energy/$\hbar$) if the states are normalized to unity, i.e., the dimension of 1/time, as required for a rate.

The transition rate is, among other things, an expression of *the law of energy conservation*, as it should be because it describes processes (in our example elastic scattering) due to time-independent interactions. When the interaction time $2T$ is finite, energy conservation is not perfect; indeed, as one sees from Fig. 3.5, the spread ΔE of values of $|E_f - E_i|$ is approximately

$$T\,\Delta E \gtrsim C\hbar \,, \tag{315}$$

where C is of order unity. This is an example of the time-energy uncertainty relation. In contrast to the uncertainty relations for incompatible observables, it does not have an unambiguous lower limit, something we already learned in §2.4(c).

Equation (314) gives the transition rate between states of perfectly sharp energies. In any actual experiment, the resolution is always finite, and the result observed is a sum over states in some interval $(E_f, E_f + dE_f)$. Let dN_f be the number of states in this interval. This number depends on the nature of the final states $|\Phi_f\rangle$ — whether they contain one or more particles. Here we take the simplest case, a one-particle final state; other cases will be treated in later chapters. To have states normalized to unity, we use plane waves satisfying periodic boundary conditions on the surface of a cube of volume L^3, as in §3.4. Then

$$dN_f = \left(\frac{L}{2\pi}\right)^3 d^3k_f \,, \tag{316}$$

and

$$\dot{P}_{i\to f}\, dN_f = \frac{2\pi}{\hbar}\, |\langle\Phi_f|V|\Phi_i\rangle|^2 \,\frac{L^3 d^3p_f}{(2\pi\hbar)^3 dE_f} dE_f\, \delta(E_f - E_i) \,. \tag{317}$$

The transition rate to the set F of all states in the interval $(E_f, E_f + dE_f)$ is then

$$\dot{P}_{i\to F} = \frac{2\pi}{\hbar} |\langle\Phi_f|V|\Phi_i\rangle|^2 \,\rho_f \,, \tag{318}$$

where

$$\rho_f = \frac{L^3 d^3p_f}{(2\pi\hbar)^3 dE_f} \tag{319}$$

is called *the density of final states.* Eq. 318 holds for final states of arbitrary complexity provided ρ_f is appropriately defined, and is often also called the Golden Rule; of course (319) only applies to the one-particle case. If the particle and/or target have spin, then the expressions just derived apply to transitions between states with specific spin projections.

(e) The Variational Principle

The time-independent Schrödinger equation can be derived by requiring a functional of the unknown state $|\psi\rangle$ to be stationary under arbitrary variations. This fact has many consequences. In particular, it offers a powerful tool for finding very accurate approximations to the energy eigenvalue of a system's ground state, and if there are constants of motion other than the energy, of the lowest energy eigenstate for each eigenvalue of these other constant of motion.

The *variational principle* states the following:

1. The functional
$$E[\psi] = \frac{\langle\psi|H|\psi\rangle}{\langle\psi|\psi\rangle} \tag{320}$$
is stationary with respect to arbitrary variations of $|\psi\rangle$ about any solution $|\psi_n\rangle$ of $(H - E_n)|\psi_n\rangle = 0$.

2. Let $|\psi_{0,\mu}\rangle$ be the *lowest* energy eigenstate, with definite eigenvalues μ of other exact constants of the motion should they exist (e.g., the lowest energy eigenstate of total angular momentum J and definite parity in a system that is invariant under rotations and reflection). Then if $|\Phi_\mu\rangle$ is an *arbitrary* state having this *same* set of other eigenvalues,
$$E[\Phi_\mu] \geq E[\psi_{0,\mu}] \,, \tag{321}$$
where the equality only holds if $|\Phi_\mu\rangle = |\psi_{0,\mu}\rangle$.

To prove that $E[\psi]$ is stationary about a solution of the Schrödinger equation, note that $|\psi\rangle$ and $\langle\psi|$ are to be treated as independent variables in carrying out the variation $\delta E[\psi]$ because wave functions such as $\langle q|\psi\rangle$ are complex. Varying only $\langle\psi|$ then gives

$$\delta E[\psi] = \frac{\langle\psi|H|\psi\rangle + \langle\delta\psi|H|\psi\rangle}{\langle\psi|\psi\rangle + \langle\delta\psi|\psi\rangle} - E[\psi] \tag{322}$$

$$= \frac{1}{\langle\psi|\psi\rangle}\Big(\langle\delta\psi|H|\psi\rangle - E[\psi]\,\langle\delta\psi|\psi\rangle\Big) + O(\delta\psi^2)\,. \tag{323}$$

But if $|\psi\rangle$ is the solution $|\psi_n\rangle$, then $E[\psi_n] = E_n$ and $\delta E[\psi] = 0$; *QED.*

To prove (321), let $|\psi_{n,\mu}\rangle$ be the complete set of exact energy eigenstates having the fixed set of other eigenvalues μ. Then

$$E[\Phi_\mu] - E[\psi_{0,\mu}] = \frac{\langle\Phi_\mu|H - E_{0,\mu}|\Phi_\mu\rangle}{\langle\Phi_\mu|\Phi_\mu\rangle} \tag{324}$$

$$= \sum_n (E_{n,\mu} - E_{0,\mu})\frac{|\langle\Phi_\mu|\psi_{n,\mu}\rangle|^2}{\langle\Phi_\mu|\Phi_\mu\rangle}\,. \tag{325}$$

This proves the theorem because $(E_{n,\mu} - E_{0,\mu}) \geq 0$, and the sum only vanishes if $|\Phi_\mu\rangle = |\psi_{0,\mu}\rangle$.

We shall apply the variational principle to several problems in later chapters, the first being the spectrum of Helium in §6.2. An instructive illustration is provided by the following one-dimensional Schrödinger equation:

$$\left(-\frac{1}{2}\frac{d^2}{dx^2} + \frac{1}{4x_0^2}(x^2 - x_0^2)^2\right)\psi = E\psi\,. \tag{326}$$

The potential is symmetric about $x = 0$, and has minima at $\pm x_0$. Because of this symmetry, the variational principle can be used to find approximate energies of both the lowest even and odd states.

Simple examples of trial functions of even and odd parity are

$$\Phi_\pm(x) = N[e^{-\alpha(x-x_0)^2} \pm e^{-\alpha(x+x_0)^2}]\,, \tag{327}$$

where N is the normalization factor. To find the approximate energies, one computes the expectation values $\langle K+V\rangle_\pm$, where K is the kinetic energy, and minimizes the resulting functions of the parameter α. This is an exercise left to the reader. The result, compared to accurate numerical solutions of the Schrödinger equation, is given in Fig. 3.6. As one sees, the variational principle gives remarkably good results, considering the degree of effort involved. Trial functions with more parameters give much better results, of course. In this problem with one degree of freedom such a more elaborate effort became archaic with the advent of modern computers, but the variational principle continues to be a powerful tool in combination with computers in a wide array of much more difficult problems.

3.8 Problems

1. The goal is to show explicitly that the spherical harmonics for half-integer angular momentum are inadmissible. In the simplest case of $l = \frac{1}{2}$, they are $u_\pm \propto e^{\pm i\phi/2}\sqrt{\sin\theta}$.

x_0	exact soln even	exact soln odd	one param family even	one param family odd	two param family even	two param family odd
1.0	0.397	1.122	0.531	1.128	0.399	1.124
2.0	0.572	0.688	0.718	0.738	0.587	0.694
3.0	0.675	0.677	0.717	0.717	0.685	0.685

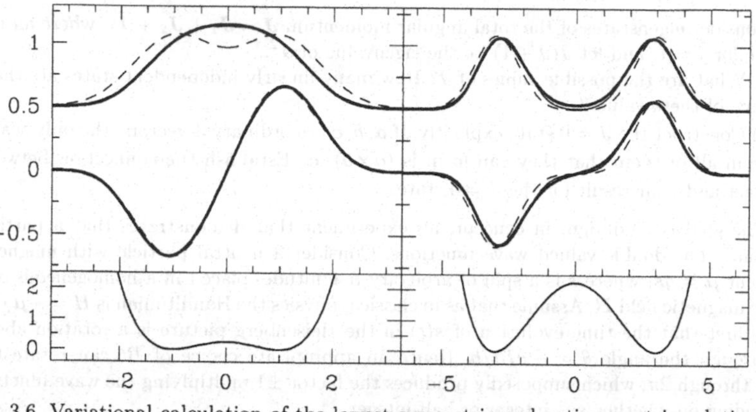

FIG. 3.6. Variational calculation of the lowest symmetric and antisymmetric solutions of the Schrödinger equation (326). The table shows the results for the one-parameter trial functions $\Phi_\pm$ of (327), and the trial function obtained from $\Phi_\pm$ when x_0 is replaced by a second variational parameter. The plot shows the results on the left for $x_0 = 1$ and on the right for $x_0 = 3$. The thick curves are the exact solutions, the dashed are the one-parameter solutions, and the barely discernible thin lines the two-parameter solutions. The even functions have been shifted upwards by 0.5 for clarity. The lower panel shows the potential for the two cases.

Apply rotation through $\frac{1}{2}\pi$ about the y-axis to u_+, and show it does not produce a linear combination of u_- and u_+. (A simpler but equivalent argument is to consider an infinitesimal rotation, i.e., to show that $L_- u_+$ does not do what it is supposed to.)

2. Confirm Eq. 28 by explicitly constructing $\langle \theta\phi | L_i | lm \rangle$.

3. Derive the completeness relation Eq. 46 from the multipole formula

$$\frac{1}{R} \equiv \frac{1}{|\boldsymbol{r} - \boldsymbol{r}'|} = \sum_l \frac{r_<^l}{r_>^{l+1}} P_l(\hat{\boldsymbol{r}} \cdot \hat{\boldsymbol{r}}'),$$

the fact that this is Green's function for Laplace's equation, and the addition theorem (47). (Hint: Write $(1/R)$ in terms of step functions.)

4. In nonrelativistic quantum mechanics the spin is treated as a Galileo invariant quantity; i.e., state vectors for a particle with spin are written as $|\boldsymbol{p}\rangle \otimes |sm_s\rangle$, with only the first factor changing under a transformation. Show that this is consistent by proving that the angular momentum about the center of mass of a system of particles that do not have spin is invariant under a Galileo transformation.

5. Show that the density matrix for a spin-1 object (Eq. 97) has unit trace, and find explicit expressions for $\boldsymbol{P}$ and W_{ij} analogous to Eq. 88.

6. Show that the projection operators P_s and P_t defined in (161) and (162) satisfy $P_s^2 = P_s, P_t^2 = P_t, P_s P_t = 0$.

7. The Wigner 3-j symbol is more symmetric than the CG coefficient because it is defined by summing three independent angular momenta $\boldsymbol{J}_i$ to zero. Let $|0\rangle$ be this state; show that

$$|0\rangle = \sum_{m_1 m_2 m_3} \begin{pmatrix} j_1 & j_2 & j_3 \\ m_1 & m_2 & m_3 \end{pmatrix} |j_1 m_1\rangle \otimes |j_2 m_2\rangle \otimes |j_3 m_3\rangle \,.$$

8. Consider eigenstates of the total angular momentum $\boldsymbol{J} = \boldsymbol{J}_1 + \boldsymbol{J}_2 + \boldsymbol{J}_3$, where all the $\boldsymbol{J}_i$ are for $j = 1$, and let $J(J+1)$ be the eigenvalue of $\boldsymbol{J}^2$.

(a) What are the possible values of J? How many linearly independent states are there for each of these values?

(b) Construct the $J = 0$ state explicitly. If $\boldsymbol{a}, \boldsymbol{b}, \boldsymbol{c}$ are ordinary 3-vectors, the only scalar linear in all of them that they can form is $(\boldsymbol{a}\times\boldsymbol{b})\cdot\boldsymbol{c}$. Establish the connection between this fact and your result for the $J = 0$ state.

9. The goal is to design, in concept, an experiment that demonstrates that a particle of spin $\frac{1}{2}$ has double-valued wave functions. Consider a neutral particle with magnetic moment $\boldsymbol{\mu} = g\boldsymbol{s}$, where $\boldsymbol{s}$ is a spin of arbitrary magnitude, placed in a homogeneous and static magnetic field $\boldsymbol{B}$. Assume that as in classical physics the Hamiltonian is $H = -\boldsymbol{\mu}\cdot\boldsymbol{B}$. Show first that the time evolution of $\boldsymbol{s}(t)$ in the Heisenberg picture is a rotation about $\boldsymbol{B}$ through the angle $\theta = -gBt/\hbar$. Hence an appropriate choice of Bt can rotate the state through 2π, which supposedly produces the factor ± 1 multiplying the wave function depending on whether s is integer or half-integer.

Allow a spin-polarized monochromatic beam to strike a screen with two holes, beyond which there are identical magnetic fields, but of opposite sign. Study the interference pattern produces, and show that a measurement of the intensity (i.e., no measurement of spin as such) for the cases $s = \frac{1}{2}$ and $s = 1$ leads to a confirmation of the different signatures after a rotation through 2π. In the case of cold neutrons at a temperature of 10° K, and a field of O.1 T, how long a path through the field is required to produce a rotation through 2π? Experiments of this type have been done; see S.A. Werner et al., *Phys. Rev. Lett.* **35**, 1053 (1975); and A.G. Klein and G.I. Opat, *ibid.* **37**, 238 (1976).

10. Consider a particle moving in the central potential $V(r) = Cr$. Introduce an appropriate dimensionless variable ρ in terms of which the radial Schrödinger equations is

$$\left(\frac{d^2}{d\rho^2} - \frac{l(l+1)}{\rho^2} - \rho + \lambda\right) u = 0 \,.$$

Show that the energies are related to the eigenvalues λ by $E = \lambda\,(\hbar^2 C^2/2m)^{\frac{1}{3}}$. The problem is to find the lowest eigenvalues for $l = 0, 1, 2$ by the variational method with a simple one-parameter trial function, and to compare the results with the exact values $\lambda = 2.338, 3.361, 4.248$. In choosing a trial function bear in mind the required behavior as $\rho \to 0$. Then try to calculate the energy of the first excited s state by choosing a trial function orthogonal to the ground-state trial function, and compare the result with the exact value $\lambda = 4.088$. Any comments?

11. Consider a two-body system with a central interaction $V(r)$, and let $\psi_n(r)$ be a bound s state eigenfunction. Show that

$$|\psi_n(0)|^2 = \frac{\mu}{2\pi}\int d^3r\, |\psi_n(r)|^2\, \frac{\partial V}{\partial r} \,.$$

Check this against the result for a Coulomb field, and use it to test the accuracy of the trial function in the preceding problem.

4

Low-Dimensional Systems

This chapter treats a number of problems concerning systems that either have a highly restricted number of degrees of freedom, or are described by a finite-dimensional Hilbert space. Problems of the first type are familiar from classical mechanics, but those of the second type have no classical counterpart whatever. All the systems of classical physics, whether they be particles or fields, have degrees of freedom with continuous values and require infinite-dimensional Hilbert spaces for their quantum mechanical description. For that reason, a full-fledged treatment of any phenomenon ultimately involves such a space. Nevertheless, finite-dimensional state spaces are of great importance, both in the description of actual experiments and in discussions of the foundations of quantum mechanics.

A familiar example illustrates the point. The energy spectrum of hydrogen in the absence of applied fields has an infinite number of bound states with negative energies $E_n \propto -n^{-2}$, where n is any positive integer, and a continuum of electron-nucleus scattering states with positive energies. The complete Hilbert space thus has the following decomposition,

$$\mathfrak{H} = \mathfrak{H}_1 \oplus \mathfrak{H}_2 \oplus \cdots \oplus \mathfrak{H}_{\text{cont}} ,$$

where $\mathfrak{H}_n$ contains the multiplet of bound states with energy E_n and has dimension n^2 if spins are ignored, while $\mathfrak{H}_{\text{cont}}$ contains all scattering states and is infinite-dimensional. Consider a common situation, a study of the the nth multiplet by means of applied static or time-dependent electromagnetic fields. If these fields are too weak or have frequencies too low to couple this multiplet significantly to others, which is often true, all the action takes place in $\mathfrak{H}_n$ to a very good approximation. And this is not peculiar to hydrogen. In spectroscopies of all varieties the focus is usually on a small subset of states. That there is an infinity of other states cannot be forgotten, of course, because phenomena such as spontaneous radiation would, in effect, also be forgotten. If, however, the coupling to these other states is weak, as it is for electromagnetism, the complete Hilbert space can be taken into account approximately after the finite subspace, whose states are strongly coupled, is fully understood.

This chapter begins with the simplest possible example, a system described by a two-dimensional Hilbert space. It is truly astonishing how many important phenomena are described with very high, or at least satisfactory, precision by this basic example. The remainder of the chapter deals with systems in one or two dimensions: the harmonic oscillator; the motion of a charged particle in a uniform magnetic field; and bound and scattering states in one dimensions.

4.1 Spectroscopy in Two-Level Systems

According to quantum mechanics, the two-dimensional complex vector space $\mathfrak{C}_2$ is the most austere stage on which *any* physical system can perform. In this space, any observable can display at most two (eigen)values, such as energies. But $\mathfrak{C}_2$ contains an infinite number of states — all the unit vectors, and therefore this simplest of all possible situations is far richer than a two-faced coin or an on-off switch, the simplest analogues in a classical world view. As a consequence, a network of quantum switches, or *qubits*, forms a circuit that is fundamentally different from a network of "classical" switches, and this has stimulated an enormous interest in quantum computing. This is a topic we will not be able to touch on, however. But other uses of the two-level system abound in quantum physics.

The most familiar example is a particle of spin $\frac{1}{2}$. One need not, however, be dealing with an actual spin $\frac{1}{2}$ object to exploit the machinery of $\mathfrak{C}_2$ provided by spin $\frac{1}{2}$. Consider two typical and important topics in spectroscopy: (a) two nearby energy levels $E_\pm(\lambda)$ which, as a function of some variable parameter λ, such as an applied magnetic field, come close to crossing or actually do so; (b) experiments that look for the resonant response of a system due to an applied oscillating field whose frequency ω is tunable. In both situations, the actual system under study will, in general, have an infinity of other energy eigenstates, but only the two already mentioned need be dealt with. In (a) the separation in energy to other states is so large that mixing is negligible with such distant states; in (b) mixing with distant states is negligible because for them the applied frequency is not resonant.

(a) Level Crossings

From examining spectroscopic data or energy level calculations, one sees that when an atom is exposed to a static applied electric or magnetic field whose strength is proportional to a parameter λ, certain pairs of energy levels may cross as λ is varied, while other pairs come close but barely avoid colliding. Our goal is to understand what determines whether or not a crossing occurs.

We need only concern ourselves with the neighborhood of a potential crossing point λ_0. If the levels are to cross, the energies will, in that region, have the form

$$E^0_\pm(\lambda) = (\lambda - \lambda_0)\epsilon_\pm + O((\lambda - \lambda_0)^2)\,, \tag{1}$$

apart from some overall constant. To incorporate the situation of interest, where the crossing is avoided, we first write the Hamiltonian in $\mathfrak{C}_2$ that has the intersecting eigenvalues (1):

$$H = \lambda\epsilon_+ \tfrac{1}{2}(1 + \sigma_3) + \lambda\epsilon_- \tfrac{1}{2}(1 - \sigma_3) \tag{2}$$

$$= \tfrac{1}{2}\lambda(\epsilon_+ + \epsilon_-) + \tfrac{1}{2}\lambda(\epsilon_+ - \epsilon_-)\,\sigma_3\,, \tag{3}$$

where $(\lambda - \lambda_0)$ was replaced by λ. The operators $\frac{1}{2}(1 \pm \sigma_3)$ project onto the eigenkets $|\pm\rangle$, i.e., "up" and "down" in the case of spin.

Recall that in §3.3(a) we learned that the most general Hermitian operator in $\mathfrak{C}_2$ is a linear combination of 1 and the three σ_i with real coefficients. We therefore generalize (3) to

$$H(\lambda) = \lambda B\sigma_3 + (a_1\sigma_1 + a_2\sigma_3)\,, \tag{4}$$

where the unit operator term in (3) is dropped as it can, if desired, be added to both eigenvalues afterwards. If a_1 or a_2 are not exactly zero, the off-diagonal terms proportional to $\sigma_{1,2}$ will produce mixing between the states $|+\rangle$ and $|-\rangle$, and prevent crossing. Thus crossing is very special in that the coefficients of σ_1 and σ_2 must either be *really* zero, or have simply been ignored.

Equation (4) can be written as

$$H(\lambda) = (\lambda^2 B^2 + a_1^2 + a_2^2)^{\frac{1}{2}} (\boldsymbol{\sigma} \cdot \boldsymbol{n}(\lambda)) , \tag{5}$$

where $\boldsymbol{n}(\lambda)$ is a real unit 3-vector. This Hamiltonian is diagonal in the basis $|\boldsymbol{n}_\pm(\lambda)\rangle$ of "spin up" and "down" states along $\boldsymbol{n}(\lambda)$, and has the eigenvalues

$$E_\pm(\lambda) = \pm\sqrt{\lambda^2 B^2 + a^2} , \qquad a^2 = a_1^2 + a_2^2 . \tag{6}$$

They are the two branches of the hyperbola with asymptotes $\pm\lambda B$ and intercepts $\pm a$ (see Fig. 4.1a). As claimed, crossing requires $a_1 = a_2 = 0$, and of course $\lambda B = 0$.

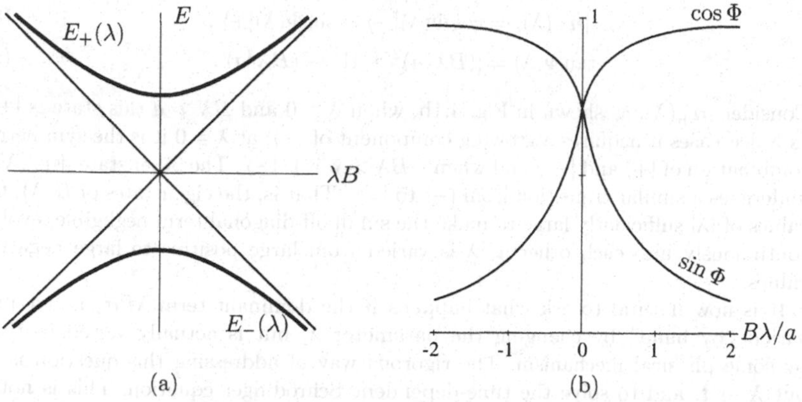

FIG. 4.1. In (a) the light lines are the eigenvalues of $\lambda B\sigma_3$, and the heavy lines those of the complete Hamiltonian, Eq. 4. In the adiabatic limit a state evolves along one or the other heavy lines, while in the sudden limit the evolution is along the light lines. The coefficients in the linear combinations (7) and (8) are shown in (b).

Under what circumstances, and with what confidence, can we know that the levels will cross, i.e., that $a = 0$? Take a concrete example, the Zeeman effect, in which an atom is exposed to a uniform magnetic field $\lambda\boldsymbol{B}$ with the interaction $-\lambda\boldsymbol{m}\cdot\boldsymbol{B}$, where $\boldsymbol{m}$ is the atom's magnetic moment operator. Because $\boldsymbol{m}$ is an axial vector operator, the atomic eigenstates in the presence of the field have a definite parity and projection M of angular momentum along $\boldsymbol{B}$. Hence, as λ is varied, levels of different parity and/or different M can cross, because there can be no matrix elements of the Hamiltonian between such states. Now consider a fly in this ointment: an imperfect (i.e., real) magnet which has a very small field $\boldsymbol{b}$ along a different direction. This can mix some of the crossing states, but not all, because the perturbation $-\boldsymbol{b}\cdot\boldsymbol{m}$ is also an axial vector operator, though along a different direction, so it can only mix states of the same parity and $M' = M \pm 1$.

If there is a stray electric field, it can mix levels of opposite parity, but again only if $M' = M \pm 1$. These stray effects will be imperceptible unless the precision with which energies are determined are comparable to the counterpart of $(a/\lambda B)^2$ in the generic Hamiltonian (4).

The examples just given are rather simple; more subtle and important examples of evaded crossings arise when the Hamiltonian of the system is treated approximately in the absence of an applied field. As an illustration, the approximation may treat orbital and spin angular momentum as separate constants of motion, whereas only their sum truly is. This may be fine on the whole, but as an applied field is varied it might allow crossing of levels that can actually mix when the heretofore ignored small terms in the underlying Hamiltonian are taken into account. *Therefore levels will only cross as a parameter is varied if these two states have different eigenvalues of an exact constant of motion.*

The eigenstates of $H(\lambda)$ have a very important property. It suffices to consider $a_2 = 0, a_1 \equiv a$, in which case $[H(\lambda) - E]|\Psi\rangle = 0$ results in

$$|\boldsymbol{n}_+(\lambda)\rangle = \cos\Phi(\lambda)|+\rangle + \sin\Phi(\lambda)|-\rangle\,, \tag{7}$$

$$|\boldsymbol{n}_-(\lambda)\rangle = \cos\Phi(\lambda)|-\rangle - \sin\Phi(\lambda)|+\rangle\,, \tag{8}$$

$$\tan\Phi(\lambda) = [(B\lambda/a)^2 + 1]^{\frac{1}{2}} - (B\lambda/a)\,. \tag{9}$$

Consider $|\boldsymbol{n}_+(\lambda)$; as shown in Fig. 4.1b, when $\lambda > 0$ and $B\lambda \gg a$ this state is $|+\rangle$; as λ decreases it acquires a growing component of $|-\rangle$; at $\lambda = 0$ it is the symmetric combination of $|+\rangle$ and $|-\rangle$; and when $-B\lambda \gg a$ it is $|-\rangle$. The other state, $|\boldsymbol{n}_-(\lambda)\rangle$, undergoes a similar transition from $|-\rangle$ to $|+\rangle$. That is, the eigenstates of $H(\lambda)$, for values of $|\lambda|$ sufficiently large to make the small off-diagonal term negligible, evolve continuously into each other as λ is varied from large positive to large negative values.

It is now natural to ask what happens if the dominant term $\lambda B\sigma_3$ is not just varied "by hand" by changing the parameter λ, but is actually varied in time by some physical mechanism. The rigorous way of addressing this question is to put $\lambda \to t$, and to solve the time-dependent Schrödinger equation. This is not a simple exercise, and is left for §7.7, but a physical argument gives the qualitative answer. For this purpose the characteristic frequency $1/\tau$, associated with the time τ the system spends passing through the influence of the perturbation, is to be compared to the characteristic frequency of the energy splitting, $2a/\hbar$. From Fig. 4.1a, $\frac{1}{2}B\tau \sim a$. If the latter frequency is small enough to satisfy $(1/\tau) \ll (2a/\hbar)$, transitions between the two branches in Fig. 4.1a are highly improbable, and the system will stay in the state in which it started, that is, remain on one or the other of the two branches $|\boldsymbol{n}_\pm(t)\rangle$.

In this limit

$$(4a^2/B) \gg \hbar\,, \tag{10}$$

the process is *adiabatic*: the system stays in an instantaneous eigenstate $|\Psi(t)\rangle$ of the "stationary"-state equation

$$[H(t) - E(t)]|\Psi(t)\rangle = 0\,, \tag{11}$$

in which time is treated as if it were just a parameter like λ. On the other hand, if passage is so sudden that $(4a^2/B) \ll \hbar$, the system is essentially oblivious to the off-diagonal perturbation and it stays put in one or the other of the eigenstates $|\pm\rangle$ of the unperturbed Hamiltonian. The pseudo-formal reasoning is that the perturbation produces the unitary evolution

$$U(\tfrac{1}{2}\tau, -\tfrac{1}{2}\tau) \sim 1 + \frac{i}{\hbar}\int_{-a/B}^{a/B} dt\, a\sigma_1 = 1 + i\frac{2a^2}{B\hbar}\,\sigma_1 \sim 1 \tag{12}$$

when the passage is sudden. This *sudden approximation* is valid for any perturbation that acts for a time that is very short compared to the system's characteristic period $\hbar/\Delta E$ in the neighborhood of the states of interest (provided, of course, that the perturbation does not have some pathological time dependence).

A remarkable illustration of this topic is provided by solar neutrinos.The thermonuclear processes in the sun's center produce only electron neutrinos ν_e, but other neutrinos having slightly different masses exist, and ν_e can transform into them by a term in the Hamiltonian like $a\sigma_1$, where a depends, among other things, on the density of matter. The mixing between ν_e and the other neutrino types will then be adiabatic or not depending on the rate with which this density changes as the neutrinos flow outward, a fact that can reveal information both about the sun and about neutrinos themselves.[1]

(b) Resonance Spectroscopy

Consider a system subjected to an external oscillating field, so that the complete *true* Hamiltonian is

$$\mathcal{H} = \mathcal{H}_0 + \mathcal{H}_1 \sin\omega t\,. \tag{13}$$

Let $|a\rangle$ and $|b\rangle$ be the two eigenstates of $\mathcal{H}_0$ of interest, and their splitting be $\hbar\Omega$:

$$\mathcal{H}_0|a\rangle = (E_0 - \tfrac{1}{2}\hbar\Omega)|a\rangle, \quad \mathcal{H}_0|b\rangle = (E_0 + \tfrac{1}{2}\hbar\Omega)|b\rangle\,. \tag{14}$$

The perturbation couples these states, the matrix element being

$$\langle a|\mathcal{H}_1|b\rangle = \hbar\lambda\,. \tag{15}$$

By appropriate choice of the arbitrary phases in the basis, λ can be made real. If the perturbation has a static part with an off-diagonal matrix element, this can be taken care of by redefining the basis, which is assumed to be done already. The factors of $\hbar$ have been inserted so as to give both Ω and λ the dimension of frequency.

In general, both $\mathcal{H}_0$ and $\mathcal{H}_1$ are operators in an infinite-dimensional Hilbert space. The restriction to the subspace $\mathfrak{C}_2$ spanned by $(|a\rangle, |b\rangle)$ permits the replacement of all operators by Pauli matrices, which form a complete set of operators in any such space. Taking E_0 as the zero of energy, the operator that impersonates $\mathcal{H}$ to perfection for the purpose at hand is then

$$H = \tfrac{1}{2}\hbar\Omega\sigma_3 + \hbar\lambda\sigma_2 \sin\omega t\,. \tag{16}$$

As a concrete example, consider a situation having nothing to do with spin an atom in an applied electric field $\boldsymbol{E}(t) = \boldsymbol{E}_0 \cos\omega t$. The corresponding vector potential is $\boldsymbol{A}(t) = -(c/\omega)\boldsymbol{E}_0 \sin\omega t$, and the interaction of the electrons with this field

[1] P. Pizzochero, *Phys. Rev. D* **36**, 2293 (1987).

is

$$\mathcal{H}_1 = \frac{e}{mc}\sum_{i=1}^{Z} \boldsymbol{p}_i \cdot \boldsymbol{A} + O(E_0^2) = -\frac{e}{m\omega}\sin\omega t \sum_i \boldsymbol{E}_0 \cdot \boldsymbol{p}_i \,. \tag{17}$$

If $\hbar\omega$ is equal to the energy difference between two atomic states $|a\rangle$ and $|b\rangle$, the applied field will produce resonant transitions between them, but not between any other states, and to high accuracy the other states can be ignored. The spin model will therefore describe this situation if the parameter λ is taken to be

$$\lambda = -\frac{e}{m\hbar\omega}\boldsymbol{E}_0 \cdot \langle a| \sum_i \boldsymbol{p} |b\rangle \,. \tag{18}$$

There is, of course, no relation between the direction in space of the electric field $\boldsymbol{E}$ and the orientation of the spin vector $\boldsymbol{\sigma}$ in the abstract space of the model.

Turning now to the solution of the model Hamiltonian (16), consider first the $\lambda = 0$ case — the unperturbed system. The solution of the Schrödinger equation is

$$|t\rangle = e^{-\frac{1}{2}i\Omega\sigma_3 t}|0\rangle \,, \tag{19}$$

where $|0\rangle$ is any linear combination of $|a\rangle$ and $|b\rangle$.

All observables in this $d = 2$ space can be expressed in terms of $\boldsymbol{\sigma}$. The expectation value of $\boldsymbol{\sigma}$ in $|t\rangle$ is

$$\boldsymbol{\mu}(t) \equiv \langle t|\boldsymbol{\sigma}|t\rangle = \langle 0|\boldsymbol{\sigma}(t)|0\rangle \,, \tag{20}$$

where $\boldsymbol{\sigma}(t)$ is the spin vector in the Heisenberg picture:

$$\boldsymbol{\sigma}(t) = e^{\frac{1}{2}i\Omega\sigma_3 t}\,\boldsymbol{\sigma}\, e^{-\frac{1}{2}i\Omega\sigma_3 t} \,. \tag{21}$$

If the system is actually a spin doublet, then $\boldsymbol{\mu}(t)$ is proportional to the magnetic moment.

Recall the following identity from §3.3(a):

$$\exp(-\tfrac{1}{2}i\vartheta \boldsymbol{n}\cdot\boldsymbol{\sigma}) = \cos\tfrac{1}{2}\vartheta - i\boldsymbol{n}\cdot\boldsymbol{\sigma}\,\sin\tfrac{1}{2}\vartheta \,, \tag{22}$$

where $\boldsymbol{n}$ is a unit vector. Then

$$\sigma_1(t) = e^{\frac{1}{2}i\Omega\sigma_3 t}\sigma_1\, e^{-\frac{1}{2}i\Omega\sigma_3 t} = \sigma_1 e^{-i\Omega\sigma_3 t} = \sigma_1\cos\Omega t - \sigma_2\sin\Omega t \,, \tag{23}$$

where the first step used $\sigma_3\sigma_1 = -\sigma_1\sigma_3$, and the second (22) and $\sigma_1\sigma_3 = -i\sigma_2$. The complete result for all the components is

$$\mu_1(t) = \mu_1(0)\cos\Omega t - \mu_2(0)\sin\Omega t \,, \tag{24}$$
$$\mu_2(t) = \mu_2(0)\cos\Omega t + \mu_1(0)\sin\Omega t \,, \tag{25}$$
$$\mu_3(t) = \mu_3(0) \,. \tag{26}$$

Thus $\boldsymbol{\mu}(t)$ precesses about the 3-axis with angular frequency Ω in the *counterclockwise* sense if $\Omega > 0$. This is precisely the motion of a magnetic moment in a static magnetic field along the 3-direction, with Ω the Larmor frequency in that analogy.

The perturbation $\sigma_1 \sin\omega t$ in (16) oscillates along the fixed direction $\boldsymbol{e}_1$. It can be viewed as a superposition of a clockwise and an anticlockwise rotating perturbation.

Only the latter component can resonate with the anticlockwise precession produced by the static portion of the Hamiltonian, and therefore it has a far larger effect than the clockwise rotating component if ω is in the vicinity of the resonance at $\omega = \Omega$. In the light of this observation, introduce unit vectors $\boldsymbol{e}_c$ and $\boldsymbol{e}_a$ rotating in the clockwise and anticlockwise sense:

$$\boldsymbol{e}_c = \boldsymbol{e}_1 \cos\omega t - \boldsymbol{e}_2 \sin\omega t\,, \tag{27}$$

$$\boldsymbol{e}_a = \boldsymbol{e}_1 \cos\omega t + \boldsymbol{e}_2 \sin\omega t\,. \tag{28}$$

The perturbing operator is then

$$\boldsymbol{e}_2 \cdot \boldsymbol{\sigma} \sin\omega t = \tfrac{1}{2}(\boldsymbol{e}_a - \boldsymbol{e}_c) \cdot \boldsymbol{\sigma}\,. \tag{29}$$

Because the anticlockwise term is the only one that can resonate, the clockwise term can be neglected. This is called the *rotating wave approximation*, and it is used to great advantage in a variety of fields, such as magnetic resonance and quantum optics. The approximate model Hamiltonian is now

$$\begin{aligned} H_{\mathrm{rw}} &= \tfrac{1}{2}\hbar\Omega\sigma_3 + \tfrac{1}{2}\hbar\lambda\, \boldsymbol{e}_a \cdot \boldsymbol{\sigma} \\ &= \tfrac{1}{2}\hbar\Omega\sigma_3 + \tfrac{1}{2}\hbar\lambda\, U(\omega t)\sigma_1 U^\dagger(\omega t)\,, \end{aligned} \tag{30}$$

where

$$U(\omega t) = e^{-\frac{1}{2}i\omega t\sigma_3}\,. \tag{31}$$

The Schrödinger equation in the rotating wave approximation is therefore

$$i\frac{\partial}{\partial t}|t\rangle = \left(\tfrac{1}{2}\Omega\sigma_3 + \tfrac{1}{2}\lambda U(\omega t)\sigma_1 U^\dagger(\omega t)\right)|t\rangle\,. \tag{32}$$

Because one of the two rotating perturbations has been dropped, (32) can now be turned into an equation with a time-independent Hamiltonian by replacing $|t\rangle$ by a state $|t\rangle_{\mathrm{r}}$ that is rotated about the 3-axis with angular frequency ω:

$$|t\rangle_{\mathrm{r}} \equiv U^\dagger(\omega t)|t\rangle\,. \tag{33}$$

Now for any unitary operator of the form $U = e^{-iQt}$

$$U^\dagger i\frac{\partial}{\partial t}U = i\frac{\partial}{\partial t} + Q\,. \tag{34}$$

Therefore, with $Q = \frac{1}{2}\omega\sigma_3$,

$$U^\dagger i\frac{\partial}{\partial t}|t\rangle = \left(U^\dagger i\frac{\partial}{\partial t}U\right)U^\dagger|t\rangle = \left(i\frac{\partial}{\partial t} + \tfrac{1}{2}\omega\sigma_3\right)|t\rangle_{\mathrm{r}}\,, \tag{35}$$

or

$$i\frac{\partial}{\partial t}|t\rangle_{\mathrm{r}} = [\tfrac{1}{2}(\Omega - \omega)\sigma_3 + \tfrac{1}{2}\lambda\sigma_1]|t\rangle_{\mathrm{r}}\,. \tag{36}$$

As expected, when $\omega = \Omega$ the precession due to the unperturbed Hamiltonian has been transformed away. The right-hand side of (36) is time-independent, and the equation can be integrated immediately. Returning to the original state via (33) then gives the complete answer:

$$|t\rangle = e^{-\frac{1}{2}i\omega t\sigma_3}\, e^{-\frac{1}{2}i[(\Omega-\omega)\sigma_3 + \lambda\sigma_1]t}|0\rangle\,. \tag{37}$$

The probability for the transition $a \to b$ is therefore

$$P_{ab}(t) = |\langle b|e^{-\frac{1}{2}i[(\Omega-\omega)\sigma_3+\lambda\sigma_1]t}|a\rangle|^2 , \tag{38}$$

where the first factor in (37) was dropped because it only produces a phase. To evaluate this, the operator is put into the form of (22). Clearly $\boldsymbol{n}$ is in the 1-3 plane at some angle φ to the 1-axis, where φ and ϑ are determined from

$$(\Omega - \omega)\sigma_3 + \lambda\sigma_1 = \vartheta \boldsymbol{n} \cdot \boldsymbol{\sigma} = \vartheta(\sigma_1 \cos\varphi + \sigma_3 \sin\varphi) . \tag{39}$$

That is,

$$\vartheta = \sqrt{(\Omega-\omega)^2 + \lambda^2} , \qquad \tan\varphi = (\Omega - \omega)/\lambda . \tag{40}$$

Only the term proportional to σ_1 can produce transitions, and therefore $P_{ab} = \sin^2(\frac{1}{2}\vartheta t)\cos^2\varphi$, or in terms of the physical parameters,

$$P_{ab}(t) = \frac{\lambda^2}{(\Omega-\omega)^2 + \lambda^2} \sin^2\left(\tfrac{1}{2}t\sqrt{(\Omega-\omega)^2+\lambda^2}\right) . \tag{41}$$

This is *Rabi's formula*; it displays the expected resonance at $\omega = \Omega$, with a width λ proportional to the strength of the oscillating perturbation.

It is instructive to compare this result with that found by treating the oscillating term as "small" in the sense of perturbation theory,

$$P_{ab}^0(t) = \left(\frac{\lambda}{\Omega - \omega}\right)^2 \sin^2[\tfrac{1}{2}(\Omega-\omega)t] . \tag{42}$$

By the very meaning of perturbation theory, this can only be taken seriously if the probability is small, i.e., if $\lambda \ll |\Omega - \omega|$, but this is violated at the resonant frequency no matter how weak the coupling parameter λ may be. The perturbation expansion replaces the unitary evolution operator by an operator that is not unitary, and it therefore can give absurd answers. In contrast, the rotating wave approximation replaces the exact unitary operator by the unitary operator in (37), and produces a transition probability that reaches one at resonance. What has been left out — the clockwise rotating terms, produces a shift in the resonant frequency of order λ^2/Ω, but the proof is rather difficult.[1] However, this shift is usually negligible.

In resonance experiment it is often illegitimate to treat the system as if were at zero temperature. An extension of the theory is then needed, and this also provides a good example of how the density matrix is used for such purposes.

The relative population of the two states at temperature T is

$$p_b/p_a = e^{-\hbar\Omega/kT} \equiv e^{-2\beta}, \tag{43}$$

and the density matrix of the system in thermal equilibrium is

$$\rho_T = \Lambda_a p_a + \Lambda_b p_b , \tag{44}$$

where $\Lambda_{a,b}$ are the projection operators onto the two states,

$$\Lambda_a = \tfrac{1}{2}(1 - \sigma_3) , \qquad \Lambda_b = \tfrac{1}{2}(1 + \sigma_3) . \tag{45}$$

[1] F. Bloch and A. Siegert, *Phys. Rev.* **57**, (1940); M.P. Silverman and F.M. Pipkin, *J. Phys. B* **5**, 1844 (1972).

The equilibrium density matrix is therefore

$$\rho_T = \tfrac{1}{2}(1 - \sigma_3 \tanh\beta)\ . \tag{46}$$

We may, for example, wish to know the probability $P_b(t)$ that $|b\rangle$ is occupied at time t if the oscillating field has been active since $t = 0$. This probability can be computed in the Schrödinger or Heisenberg picture; we use the latter. Then ρ_T remains (46), and

$$P_b(t) = \mathrm{Tr}\,\Lambda_b(t)\,\rho_T\ , \tag{47}$$

where $\Lambda_b(t) = U^\dagger(t)\Lambda_b U(t)$, and $U(t)$ is the unitary operator that evolves the states in the Schrödinger picture. Equation (47) is exact; it does not depend on the rotating wave approximation, though we shall soon resort to it. Hence

$$P_b(t) = \tfrac{1}{2} + \tfrac{1}{2}\mathrm{Tr}\,\sigma_3(t)[1 - \tfrac{1}{2}\sigma_3 \tanh\beta]\ , \tag{48}$$

where $\sigma_3(t)$ is in the Heisenberg picture. It is best to write the transformed Pauli matrix as

$$U^\dagger(t)\boldsymbol{\sigma}\,U(t) = \sum_{i=1}^{3} \boldsymbol{u}_i\,\sigma_i\,c_i(t)\ , \tag{49}$$

where the $\boldsymbol{u}_i$ are the Cartesian unit vectors. Recalling $\mathrm{Tr}\,\sigma_i = 0$, and the other properties of Pauli matrices, we see that (48) reduces to

$$P_b(t) = \tfrac{1}{2}[1 - c_3(t)\,\tanh\beta]\ . \tag{50}$$

To compute $c_3(t)$ we return to the rotating wave approximation, and replace the exact unitary operator in (49) by

$$U = e^{-\frac{1}{2}i\vartheta t\,\boldsymbol{\sigma}\cdot\boldsymbol{n}}\ , \tag{51}$$

where $\boldsymbol{n} = \boldsymbol{u}_1 \cos\varphi + \boldsymbol{u}_3 \sin\varphi$, with ϑ and φ given by (40). The following important identity, whose proof is left to Prob. 2, holds for the rotation of any vector operator, and in particular for $\boldsymbol{\sigma}$:

$$e^{\frac{1}{2}i\theta\boldsymbol{\sigma}\cdot\boldsymbol{n}}\,\boldsymbol{\sigma}\,e^{-\frac{1}{2}i\theta\boldsymbol{\sigma}\cdot\boldsymbol{n}} = \boldsymbol{\sigma}\,\cos\theta + \boldsymbol{n}\,(\boldsymbol{\sigma}\cdot\boldsymbol{n})\,(1 - \cos\theta) + (\boldsymbol{n}\times\boldsymbol{\sigma})\,\sin\theta\ . \tag{52}$$

For our $\boldsymbol{n}$, and with $\theta = t\vartheta$, this gives

$$c_3(t) = \cos\vartheta t + (1 - \cos\vartheta t)\sin^2\varphi\ . \tag{53}$$

The final result for the the occupation probability of the upper state is therefore

$$P_b(t) = \frac{1}{2}\left\{1 - \left[1 - \frac{2\lambda^2}{(\Omega-\omega)^2 + \lambda^2}\ \sin^2\left\{\tfrac{1}{2}t[(\Omega-\omega)^2 + \lambda^2]^{\frac{1}{2}}\right\}\right]\tanh(\hbar\Omega/2kT)\right\}\ . \tag{54}$$

Note that this reduces correctly to (41) when $T \to 0$.

4.2 The Harmonic Oscillator

The harmonic oscillator is ubiquitous in quantum physics because it often offers a powerful first approximation to departures from equilibrium. For example, molecular vibrational spectra are due to the harmonic oscillations of the nuclei about the fixed positions they would have if nuclear masses were infinite compared to the electron's. In a crystalline solid, again because the nuclei are so heavy, the departures from the static crystal are, to good approximation, simply harmonic. In both cases, these are imperfect approximations, and the imperfection can be important: a crystal could not conduct heat if the motions about equilibrium were truly harmonic. The most important example of harmonic motion is provided by the electromagnetic field, because Maxwell's equation in vacuum, when Fourier analyzed, turn into the equations of motion for an infinite set of uncoupled simple harmonic oscillators. In this case there is no approximation. As a consequence, the harmonic oscillator has a fundamental role in quantum electrodynamics. In short, the harmonic oscillator is not only an interesting and simple example of a dynamical system in its own right, but a tool of great importance in many complex phenomena.

(a) Equations of Motion

Let p and q be the canonical momentum and position *operators* of a particle of mass m moving in one dimension under the influence of a static linear restoring force, i.e., a potential energy proportional to q^2, of a strength such that the circular frequency is ω. The Hamiltonian is familiar from classical mechanics:

$$H = \frac{1}{2m}p^2 + \frac{1}{2}m\omega^2 q^2 \,. \tag{55}$$

The equations of motion in the Heisenberg picture are

$$\begin{aligned} i\hbar\dot{p}(t) &= [p(t), H] = -i\hbar m\omega^2 q(t) \,, \\ i\hbar\dot{q}(t) &= [q(t), H] = (i\hbar/m)p(t) \,. \end{aligned} \tag{56}$$

Because these equations are linear, the fact that the canonical variables do not commute with each other plays no role and the solutions have exactly the same form as in classical mechanics.

A simple argument reveals the appropriate dimensionless variables. Assume that low-lying states have coordinate and momentum uncertainties Δq and Δp whose product is of order $\hbar$, and that a crude approximation to their energy is

$$E \sim \frac{\hbar^2}{2m(\Delta q)^2} + \tfrac{1}{2}m\omega^2(\Delta q)^2 \,. \tag{57}$$

As a function of Δq, this has its minimum at

$$E_{\text{min}} = \hbar\omega \,, \qquad \Delta q = \sqrt{\hbar/m\omega} \,. \tag{58}$$

Hence the characteristic size of the low energy states is $\sqrt{\hbar/m\omega}$.

For that reason, we redefine the Hamiltonian, the canonical variables and the time in terms of the following dimensionless quantities:

$$(H/\hbar\omega) \to H \,; \quad q(m\omega/\hbar)^{\frac{1}{2}} \to q \,; \quad p(m\omega\hbar)^{-\frac{1}{2}} \to p \,; \quad \omega t \to t \,. \tag{59}$$

Then

$$H = \frac{1}{2}(p^2 + q^2) , \tag{60}$$

and

$$[q, p] = i . \tag{61}$$

These dimensionless variables will be used unless there is a statement to the contrary.

The solution of the problem and the interpretation of the results are greatly facilitated by introducing the following *non-Hermitian* operators:

$$a = \frac{1}{\sqrt{2}}(q + ip) , \qquad a^\dagger = \frac{1}{\sqrt{2}}(q - ip) , \tag{62}$$

$$q = \frac{1}{\sqrt{2}}(a + a^\dagger) , \qquad p = \frac{i}{\sqrt{2}}(a^\dagger - a) . \tag{63}$$

The operators a and $a^\dagger$ satisfy the commutation rule

$$[a, a^\dagger] = 1 \tag{64}$$

in virtue of (61). The Hamiltonian now reads

$$H = \tfrac{1}{2}(a^\dagger a + a a^\dagger) = a^\dagger a + \tfrac{1}{2} . \tag{65}$$

The Heisenberg equation of motion for any operator Q is

$$i\dot{Q}(t) = [Q(t), H] \tag{66}$$

in the units defined in (59); in particular, $\dot{a} = -ia$, and therefore

$$a(t) = e^{-it}a(0) , \qquad a^\dagger(t) = e^{it}a^\dagger(0) . \tag{67}$$

The canonical variables then develop in time exactly as they do in classical physics:

$$q(t) = q(0)\cos t + p(0)\sin t \tag{68}$$

$$p(t) = p(0)\cos t - q(0)\sin t . \tag{69}$$

There is a crucial difference in quantum mechanics, however: the operators at different times do not commute with each other. For example,

$$[q(t), q(0)] = \frac{-i\hbar}{m\omega}\sin\omega t , \tag{70}$$

where dimensional units were reintroduced to make it evident that the right-hand side vanishes when $\hbar \to 0$.

(b) Energy Eigenvalues and Eigenfunctions

Turning to the energy spectrum, define the important operator

$$N = a^\dagger a . \tag{71}$$

N is Hermitian and positive definite, i.e., $\langle N \rangle \geq 0$ in any state. The Hamiltonian is $N + \frac{1}{2}$, and the energy spectrum therefore has the lower bound $\frac{1}{2}$, or $\frac{1}{2}\hbar\omega$ in dimensional units. Let $|n\rangle$ be an eigenket of N with eigenvalue n:

$$(N - n)|n\rangle = 0 \,. \tag{72}$$

In view of what was just said

$$n \geq 0 \,. \tag{73}$$

Equation (64) leads directly to

$$[N, a] = -a \,; \qquad [N, a^\dagger] = a^\dagger \,. \tag{74}$$

Now recall the following simple but indispensable identity (cf. §2.1(e)):

$$[A, B] = \lambda B \ \ \& \ (A - \alpha)|\alpha\rangle = 0 \,, \quad \Longrightarrow \quad B|\alpha\rangle = \text{const.}|\alpha + \lambda\rangle \,. \tag{75}$$

Therefore

$$a^\dagger|n\rangle \propto |n+1\rangle \,, \qquad a|n\rangle \propto |n-1\rangle \,. \tag{76}$$

Furthermore, the norm of $a|n\rangle$ is $\langle n|a^\dagger a|n\rangle = n$ if $\langle n|n\rangle = 1$. Choosing the arbitrary phases as real therefore gives

$$a^\dagger|n\rangle = \sqrt{n+1}|n+1\rangle \,, \qquad a|n\rangle = \sqrt{n}|n-1\rangle \,. \tag{77}$$

The operator a lowers the eigenvalue n by one, and a^m lowers it by the integer m. Unless n is an integer, this will generate negative eigenvalues, in violation of (73). Furthermore, if 0 is an eigenvalue, $a|0\rangle = 0$ according to (77), and no negative eigenvalues can be reached if zero and the positive integers form the spectrum of N. For that reason N is called the *number operator.*

To summarize, *the spectrum of of the number operator $a^\dagger a = N$ is $n = 0, 1, \ldots,$ and the energy spectrum of the harmonic oscillator is*

$$E_n = n + \tfrac{1}{2} \,, \qquad \text{i.e.,} \ \ (n + \tfrac{1}{2})\hbar\omega \,, \ \ n = 0, 1, 2, \ldots \,. \tag{78}$$

The operators $a^\dagger$ and a raise and lower the energy by one unit (i.e., $\hbar\omega$), respectively, or put equivalently, they excite and de-excite by one unit. When the theory of the oscillator is applied to the electromagnetic field, this excitation or de-excitation takes the form of adding or subtracting a photon by emission or absorption, that is, creating or destroying a photon. In that context, therefore, $a^\dagger$ and a are called *creation and destruction operators*, respectively. This terminology will be used from now on, even though at this point it only means addition or removal of an excitation, and not of a particle.

The nonzero matrix elements of $a^\dagger$ and a can be read off from (77):

$$\langle n|a^\dagger|n'\rangle = \sqrt{n}\,\delta_{n,n'+1} \,, \qquad \langle n|a|n'\rangle = \sqrt{n+1}\,\delta_{n,n'-1} \,; \tag{79}$$

those for the canonical variables then follow from (63):

$$\langle n|q|n'\rangle = \frac{1}{\sqrt{2}}[\sqrt{n}\,\delta_{n,n'+1} + \sqrt{n+1}\,\delta_{n,n'-1}] \,, \tag{80}$$

$$\langle n|p|n'\rangle = \frac{i}{\sqrt{2}}[\sqrt{n}\,\delta_{n,n'+1} - \sqrt{n+1}\,\delta_{n,n'-1}] \,. \tag{81}$$

The expectation values of p^2 and q^2 follow from the observation that H is symmetric in q and p, so that in any state $\langle q^2\rangle = \langle p^2\rangle = \langle H\rangle$. In the basis of the energy eigenstates, q and p have no diagonal elements, and therefore

$$\Delta q = \sqrt{\langle q^2\rangle} = \Delta p = \sqrt{n+\tfrac{1}{2}}\,. \tag{82}$$

Thus the uncertainty product in the stationary states is

$$\Delta p \Delta q = n + \tfrac{1}{2}\,. \tag{83}$$

Hence in the ground state the minimum allowed by the uncertainty principle, $\frac{1}{2}\hbar$, is actually reached.

It is not necessary to solve the Schrödinger equation to find the wave functions; the properties of the creation and destruction operators produce them much more directly, just as the angular momentum operators $L_1 \pm iL_2$ lead more easily to the spherical harmonics than does Legendre's differential equation (see §3.2). Here the jump-off point is that $a|0\rangle = 0$. Let ξ be the eigenvalue of the dimensionless coordinate operator q, and define the wave function of the state with energy n in the usual way as

$$\varphi_n(\xi) = \langle \xi|n\rangle\,. \tag{84}$$

From (62),

$$0 = \langle \xi|a|0\rangle = \frac{1}{\sqrt{2}}\left(\xi + \frac{d}{d\xi}\right)\langle\xi|0\rangle\,. \tag{85}$$

Hence

$$\varphi_0(\xi) = \pi^{-\frac{1}{4}} e^{-\frac{1}{2}\xi^2}\,, \tag{86}$$

which is normalized to

$$\int_{-\infty}^{\infty} d\xi\, |\varphi_0(\xi)|^2 = 1\,. \tag{87}$$

In terms of the physical distance $x = \sqrt{\hbar/m\omega}\,\xi$, the wave function normalized to unity when integrated on x is

$$\psi_0(x) = \left(\frac{m\omega}{\pi\hbar}\right)^{1/4} \exp\left(-\frac{x^2}{2}\frac{m\omega}{\hbar}\right)\,. \tag{88}$$

The nth excited state can be constructed from the ground state by n applications of the creation operator, i.e., $(a^\dagger)^n|0\rangle = c_n|n\rangle$, with (77) determining c_n :

$$|n\rangle = \frac{1}{\sqrt{n!}}(a^\dagger)^n\,|0\rangle\,. \tag{89}$$

This identity should be memorized because it is used repeatedly. Here it gives an explicit formula for all the energy eigenfunctions:

$$\langle\xi|n\rangle = (n!)^{-\frac{1}{2}}\, 2^{-n/2}\langle\xi|(q-ip)^n|0\rangle\,, \tag{90}$$

and therefore

$$\varphi_n(\xi) = \frac{1}{\pi^{\frac{1}{4}}\sqrt{2^n n!}}\left(\xi - \frac{d}{d\xi}\right)^n e^{-\frac{1}{2}\xi^2} = \frac{1}{\pi^{\frac{1}{4}}\sqrt{2^n n!}}\, e^{\frac{1}{2}\xi^2}\left(-\frac{d}{d\xi}\right)^n e^{-\xi^2} \tag{91}$$

$$= \frac{1}{\pi^{\frac{1}{4}}\sqrt{2^n n!}}\, e^{-\frac{1}{2}\xi^2} H_n(\xi)\,, \tag{92}$$

where the $H_n(\xi)$ are the Hermite polynomials of degree n. The first five are

$$\begin{gathered} H_0 = 1, \quad H_1 = 2\xi, \quad H_2 = 4\xi^2 - 2, \quad H_3 = 8\xi^3 - 12\xi \ , \\ H_4 = 16\xi^4 - 48\xi^2 + 12 \ . \end{gathered} \tag{93}$$

The wave functions $\varphi_n(\xi)$ have an interesting and important property: they are even or odd under $\xi \to -\xi$ according to whether n is even or odd. The reason is that the oscillator's Hamiltonian is invariant under spatial reflection, and as a consequence the energy and parity can be specified simultaneously. The energy spectrum of the one-dimensional oscillator has no degeneracies, and therefore the energy eigenstates are automatically parity eigenstates. From (93), the parities are

$$i_s = (-1)^n \ . \tag{94}$$

It is "obvious" that the ground state is even, because only an even state can be free of nodes; a state with nodes has more curvature, and therefore more kinetic energy, than one that has none.[1]

(c) The Forced Oscillator

The behavior of an oscillator subjected to a time-dependent force is of importance in many contexts. When the oscillations are those of a small massive system, such as a molecule, the force can often be approximated as being constant over the dimensions of the unforced motions, and is described by adding the potential $f(t)q$ to the Hamiltonian. In the case of radiation, the unforced oscillations are those of the modes of the free electromagnetic field, while the sources responsible for emission and absorption of radiation add the term $\boldsymbol{j} \cdot \boldsymbol{A}$ to the Hamiltonian of the free field. After Fourier analysis, the latter also add a term of the form $f(t)q$ to each mode oscillator. The discussion that follows is therefore relevant to a wide range of phenomena.

The Hamiltonian, in the Schrödinger picture, is thus

$$H_F(t) = \tfrac{1}{2}(p^2 + q^2) + f(t)q \ . \tag{95}$$

The Heisenberg equation of motion is still (66), the Hamiltonian now being $H_F(t)$, with all the operators that appear in it in the Heisenberg picture, i.e., $q \to q(t)$, etc., and all equal time commutation rules the same as in the Schrödinger picture. Hence

$$i\dot{a}(t) = a(t) + \frac{1}{\sqrt{2}} f(t) \ . \tag{96}$$

From classical mechanics we know that H_F produces oscillations displaced from their unforced motions. Consider, therefore, the Ansatz

$$a(t) = a(0)e^{-it} + s(t) \ . \tag{97}$$

Substituting (97) into (96) gives

$$\left(i\frac{d}{dt} - 1 \right) s(t) = \frac{1}{\sqrt{2}} f(t) \ . \tag{98}$$

---

[1]That the number of nodes increases by one with each increase in the energy eigenvalue, as shown by (93), illustrates a general theorem about Sturm-Liouville differential equation.

Thus $s(t)$ is not an operator, and this is just a conventional differential equation, with the solution

$$s(t) = -\frac{i}{\sqrt{2}} e^{-it} \int_0^t dt'\, e^{it'} f(t') , \tag{99}$$

where the lower limit sets the initial condition $s(0) = 0$. The coordinate operator is therefore

$$q(t) = q_0(t) - \int_0^t dt' \sin(t - t')\, f(t') , \tag{100}$$

where $q_0(t)$ is the solution for the unforced oscillator (Eq. 68). This has the same form as in classical mechanics, as was to be expected in view of the linear equations of motion.

The evolution of states under the action of the applied force is determined by the unitary operator $U(t, t_S)$ that turns the Schrödinger picture operators into those of the Heisenberg picture, where t_S is the time when the two pictures coincide (cf. §2.4(b)). The initial condition $s(0) = 0$ is the choice $t_S = 0$, so call $U(t, 0)$ simply $U(t)$. It must produce (97),

$$U^\dagger(t) a U(t) = a(t) = a(0) e^{-it} + s(t) . \tag{101}$$

A spatial displacement through x is produced by the unitary operator $e^{-ipx/\hbar}$, i.e., is generated by the canonical variable conjugate to the coordinate. One might therefore guess that a displacement of a would be generated by the conjugate operator $a^\dagger$. But the latter is not Hermitian, and when exponentiated it does not produce a unitary operator. However, a slight generalization does, namely,

$$D(z) = \exp\left[z a^\dagger - z^* a\right] , \tag{102}$$

where z is a complex number. This is certainly unitary, and it will now be shown that it produces a displacement of a through z:

$$D^\dagger(z)\, a\, D(z) = a + z . \tag{103}$$

The first step is to use the BCH theorem in the case where $[A, B]$ commutes with both A and B:

$$e^A e^B = e^{A+B}\, e^{\frac{1}{2}[A,B]} . \tag{104}$$

Hence

$$D(z) = e^{\frac{1}{2}|z|^2}\, e^{-z^* a}\, e^{z a^\dagger} = e^{-\frac{1}{2}|z|^2}\, e^{z a^\dagger}\, e^{-z^* a} , \tag{105}$$

$$D^\dagger(z) = e^{\frac{1}{2}|z|^2}\, e^{z^* a}\, e^{-z a^\dagger} = e^{-\frac{1}{2}|z|^2}\, e^{-z a^\dagger}\, e^{z^* a} . \tag{106}$$

Now recall from §2.3(a) that if p and q satisfy the canonical commutation rules, then $[p, F(q)] = -i\hbar\, \partial F/\partial q$ for any function $F(q)$. The same argument leads to the analogous identities

$$[a, \Phi(a^\dagger)] = \partial\Phi/\partial a^\dagger , \qquad [a^\dagger, \Psi(a)] = -\partial\Psi/\partial a . \tag{107}$$

Consequently,

$$D^\dagger(z) a D(z) = a + e^{-z a^\dagger}[a, e^{z a^\dagger}] = a + z ; \tag{108}$$

QED. The hard part is now done. All that remains to get (97) is $a \to ae^{-it}$, i.e., the time dependence due to the unforced Hamiltonian H_0. The final result required for (101) is thus

$$U(t) = D(s(t))e^{-iH_0 t} . \tag{109}$$

On second thought, Eq. 101 only determines $U(t)$ up to a time-dependent numerical phase factor $e^{i\alpha(t)}$. That there is such a phase is shown in Prob. 3:

$$\alpha(t) = -\frac{1}{\sqrt{2}} \int_0^t dt' f(t') \,\mathrm{Re}\, s(t') , \tag{110}$$

i.e.,

$$U(t) = e^{i\alpha(t)} D(s(t))e^{-iH_0 t} . \tag{111}$$

As we will see, this phase factor drops out of all the results we will be considering in this volume, though it does have physical significance in quantum electrodynamics.

Any state $|\psi\rangle$ at $t = 0$ evolves into $U(t)|\psi\rangle$. The ground state $|0\rangle$ of H_0 is of particular interest in quantum electrodynamics, because $U(t)|0\rangle$ corresponds to the state that evolves out of the vacuum due to radiation from a time-varying current. To evaluate $U(t)|0\rangle$ it is most efficient to put $D(s)$ into a form where all the destruction operators are on the right, for they give zero when acting on $|0\rangle$. This is done by using (105):

$$U(t) = e^{i\alpha(t)} e^{-\frac{1}{2}|z|^2} e^{za^\dagger} e^{-z^* a} e^{-iH_0 t} , \qquad z \equiv s(t) . \tag{112}$$

Then $\exp(za^\dagger)$ is expanded in powers of $a^\dagger$, and (89) is used to obtain

$$|t;0\rangle = e^{i\alpha(t)} e^{-\frac{1}{2}it} e^{-\frac{1}{2}|z|^2} \sum_{n=0}^{\infty} \frac{z^n}{\sqrt{n!}} |n\rangle . \tag{113}$$

This is a coherent superposition of all the energy eigenstates. Among its many properties, two are especially important. The first is that (113) gives the amplitude $A_0(t)$ for finding the system in the ground state at time t if it was in that state originally. Aside from the unimportant phase $e^{i\alpha(t)}e^{-\frac{1}{2}it}$, this amplitude is

$$A_0(t) = \exp\left(-\frac{1}{4}\int_0^t dt_1 \int_0^t dt_2 \, e^{i(t_1-t_2)} f(t_1)f(t_2)\right) \tag{114}$$

$$= \exp\left(-\frac{1}{2}\int_0^t dt_1 \int_0^{t_1} dt_2 \, e^{i(t_1-t_2)} f(t_1)f(t_2)\right) . \tag{115}$$

The second, time-ordered form, Eq. 115, where $t_1 > t_2$, is given here to relate this result to later material.

The second important result is the probability at time t of finding the forced oscillator in the nth excited state given that it was in the ground state at $t = 0$:

$$p_n(t) = e^{-|s(t)|^2} \frac{|s(t)|^{2n}}{n!} . \tag{116}$$

This is a *Poisson distribution*, a fact of great significance in radiation theory.

(d) Coherent States

The displaced ground state (Eq. 113) belongs to a class of *coherent states* which play an important role in the quantum theory of light, and which therefore merit detailed examination.

A powerful formalism for handling coherent states will be developed shortly, but to clarify what is going on we first ask whether there is a solution $\psi(\xi, t)$ of the time-dependent Schrödinger equation for the *unforced* harmonic oscillator that has a coordinate space probability distribution with the shape $|\varphi_0(\xi)|$ of the ground state (Eq. 86) but which is centered about a moving point $y(t)$.

This sought-for wave function must, by assumption, have the form

$$\psi \propto \exp\{-\tfrac{1}{2}[\xi - y(t)]^2\}\, e^{i\chi(y,t)}\, e^{-\frac{1}{2}it} \tag{117}$$

where χ is real and the last phase factor merely moves the zero of energy to $\frac{1}{2}(\hbar\omega)$. With the benefit of hindsight, we make the Ansatz

$$\chi = \xi\, \dot{y}(t) - \phi(t)\,. \tag{118}$$

When used in (117) and substituted into the Schrödinger equation, this leads to the following equation for $y(t)$ and $\phi(t)$:

$$\dot{y}^2 + 2\xi(y + \ddot{y}) - y^2 = 2\dot{\phi}\,. \tag{119}$$

If $y(t)$ is assumed to be the coordinate of a classical oscillator (of unit mass and frequency), $\ddot{y} + y = 0$ and $\dot{y}^2 + y^2 =$ const., and (119) is solved provided the phase ϕ is

$$\phi(t) = \tfrac{1}{2}[\dot{y}(t)y(t) - \dot{y}(0)y(0)]\,. \tag{120}$$

The final result for ψ, apart from an arbitrary constant phase, is therefore

$$\psi(\xi, t) = \pi^{-\frac{1}{4}}\, e^{-\frac{1}{2}[\xi - y(t)]^2}\, e^{i\xi\dot{y}(t)}\, e^{-\frac{1}{2}i[\dot{y}(t)y(t) - \dot{y}(0)y(0)]}\, e^{-\frac{1}{2}it}\,, \tag{121}$$

where $y(t)$, the position of the packet's center, moves harmonically. *This is the wave function of the coherent state.* By design

$$|\psi(\xi, t)|^2 = |\varphi_0(\xi - y(t))|^2\,. \tag{122}$$

Now to *the coherent state formalism*, which is due to Glauber. Consider the *unforced* oscillator, and the set of states produced by acting on the ground state with the displacement operator $D(z)$:

$$|z\rangle = D(z)|0\rangle\,, \tag{123}$$

where z is now a *time-independent* complex parameter.

A better understanding of this displacement emerges when the exponent in (102) is expressed in terms of the coordinate and momentum operators. For this purpose, define the real constants (q_o, p_o) by

$$z = \frac{1}{\sqrt{2}}(q_o + ip_o)\,, \tag{124}$$

in terms of which[1]

$$D(z) \equiv D(q_0 p_0) = \exp\left[i(p_0 q - q_0 p)\right] . \tag{125}$$

This is *a displacement in phase space* through the distance q_0 and the momentum p_0. This is especially clear if the the BCH theorem is used to write

$$D(q_0 p_0) = e^{\frac{1}{2} i q_0 p_0}\, e^{-i q_0 p}\, e^{i p_0 q} = e^{-\frac{1}{2} i q_0 p_0}\, e^{i p_0 q}\, e^{-i q_0 p} , \tag{126}$$

from which

$$D^\dagger\, p\, D = p + p_0 , \qquad D^\dagger\, q\, D = q + q_0 . \tag{127}$$

It should be noted that the operator $D(z)$ produces the shift (127) *in any system, and is not specific to the harmonic oscillator.*

The meaning of the complex displacement through z is also apparent in the coordinate space wave function of the displaced ground state:

$$\langle \xi | D(z) | 0 \rangle = e^{-\frac{1}{2} i q_0 p_0} e^{i \xi p_0} \varphi_0(\xi - q_0) = \psi(\xi, 0) . \tag{128}$$

Proof of this result is left for a problem.

How does the displaced ground state $|z\rangle$ evolve in time? This is answered by evaluating

$$e^{-iH_0 t}|z\rangle = e^{-\frac{1}{2}it} \left(e^{-iH_0 t} D(z) e^{iH_0 t}\right)|0\rangle = e^{-\frac{1}{2}it} \exp[z a^\dagger e^{-it} - z^* a e^{it}]|0\rangle . \tag{129}$$

Note that here the time dependence of a and $a^\dagger$ is opposite to that in (67) because (129) is in the Schrödinger picture. The phase $e^{-it/2}$ can be dropped by redefining $\frac{1}{2}$ (i.e., $\frac{1}{2}\hbar\omega$) as the zero of energy. In view of (102), therefore,

$$e^{-iH_0 t}|z\rangle = |z e^{-it}\rangle ; \tag{130}$$

i.e., *time evolution merely replaces z by ze^{-it}*. According to (124), however,

$$z e^{-it} = \frac{1}{\sqrt{2}} [(q_0 \cos t + p_0 \sin t) + i(p_0 \cos t - q_0 \sin t)] . \tag{131}$$

Hence at time t the coherent state is the undistorted ground state displaced to the position and momentum a classical oscillator would have at that time were it to have been set in motion at $t = 0$ with the initial condition (q_0, p_0). We know this already from the discussion culminating in Eq. 121; in detail, $\langle \xi | z e^{-it} \rangle = \psi(\xi, t)$ apart from a inconsequential constant phase. That the coherent state does not develop any distortion is a special and remarkable property of the harmonic oscillator — of the circumstance that the Hamiltonian is a quadratic form, that is to say, of the linear equations of motion.

Important properties of coherent states follow immediately from the surprising fact that $|z\rangle$ is an eigenstate of the non-Hermitian operator a. From (108), $aD(z) = D(z)(z + a)$; when applied to $|0\rangle$ this gives

$$a|z\rangle = z|z\rangle . \tag{132}$$

[1]This form is known as the Weyl operator.

Hence $|z\rangle$ is an eigenstate of a with the complex eigenvalue z. Now let $\bar{A}(t)$ be the expectation value of any operator A in the coherent state. Then from (130) and (132)

$$\bar{a}(t) = ze^{-it} = \frac{1}{\sqrt{2}}(\bar{q}(t) + i\bar{p}(t)) \; . \tag{133}$$

The expectation value of the energy is therefore

$$\bar{H} = |z|^2 = \tfrac{1}{2}({p_0}^2 + {q_0}^2) \equiv E_0 \tag{134}$$

which is the energy of a classical oscillator when its canonical variables have the values (q_0, p_0). Thus

$$z = \sqrt{E_0}\, e^{i\varphi(t)} \; , \tag{135}$$

where φ is the phase of the motion. (That the lower bound on $\bar{H}$ is not $\frac{1}{2}\hbar\omega$ is due to our having redefined the ground-state energy as zero.)

The expansion of the coherent state in terms of energy eigenstates is given by (113),

$$|z\rangle = e^{-\frac{1}{2}|z|^2} \sum_{n=0}^{\infty} \frac{z^n}{\sqrt{n!}} |n\rangle \; . \tag{136}$$

Hence the probabilities are given by the Poisson distribution

$$p_n(z) = \frac{(E_0)^n}{n!}\, e^{-E_0} \; . \tag{137}$$

The Poisson distribution has properties that are of importance in many contexts. The mean value $\bar{N}$ of N is

$$\bar{N} = \sum_n n p_n = e^{-E_0} \frac{d}{dE_0} \sum_n \frac{{E_0}^n}{n!} = E_0 \; , \tag{138}$$

and therefore the distribution can be written as

$$p_n = \frac{\bar{N}^n}{n!} e^{-\bar{N}} \; , \quad \bar{N} = |z|^2 \; . \tag{139}$$

Furthermore, the trick used in (138) gives $\langle N(N-1)\rangle = \bar{N}^2$. Therefore

$$(\Delta N)^2 \equiv \langle N^2\rangle - \langle N\rangle^2 = \bar{N} \; , \tag{140}$$

or

$$\frac{\Delta N}{\bar{N}} = \frac{1}{\sqrt{\bar{N}}} \; , \tag{141}$$

and

$$\frac{\Delta H}{\bar{H}} = \frac{1}{\sqrt{\bar{H}}} \; . \tag{142}$$

Equation (142) means that when the amplitude of the oscillating wave packet is large compared to the size of the ground state, the energy eigenstates that contribute significantly are sharply peaked around the state with $n = E_0$.

The coherent states have some unusual properties because they are eigenstates of a non-Hermitian operator. Thus the adjoint of (132) is

$$\langle z|a^\dagger = \langle z|z^* \; . \tag{143}$$

Hence the relationship between bras and kets is unconventional; the eigenvalue of $a^\dagger$ acting on its "left-eigenvector" $\langle z|$ is z^*, not z, and for that reason some use the notation $\langle z^*|$. Non-Hermitian operators do not, in general, have orthogonal linearly independent eigenstates, and this is true of the coherent states. They do form a complete set, however. The scalar product between coherent states is

$$\langle z'|z\rangle = \langle D^\dagger(z')D(z)\rangle_0 = e^{-\frac{1}{2}(|z|^2+|z'|^2)}\,\langle e^{z'^*a}e^{za^\dagger}\rangle_0\,, \tag{144}$$

where the BCH theorem was used, and $\langle\ldots\rangle_0$ is the expectation value in $|0\rangle$. In the last expression the exponentials are expanded to produce the orthonormal eigenstates of N, which gives

$$\langle z'|z\rangle = e^{-\frac{1}{2}(|z|^2+|z'|^2)}\sum_n \frac{(z'^*z)^n}{n!} = e^{-\frac{1}{2}(|z|^2+|z'|^2)}e^{z'^*z}\,. \tag{145}$$

The absolute value of the scalar product is therefore

$$|\langle z'|z\rangle|^2 = e^{-|z-z'|^2}\,. \tag{146}$$

Thus coherent states with very different displacements in phase space ($|z-z'|\gg 1$) are nearly orthogonal.

Completeness is established by the following calculation:

$$\begin{aligned}\int_{-\infty}^{\infty} dq_o \int_{-\infty}^{\infty} dp_o\; D(z)|0\rangle\langle 0|D^\dagger(z) &= \int dq_o\, dp_o\; e^{-|z|^2}\, e^{za^\dagger}|0\rangle\langle 0|e^{z^*a} \\ &= \int dq_o\, dp_o\; e^{-|z|^2}\sum_{n,m}\frac{z^n(z^*)^m}{\sqrt{n!\,m!}}|n\rangle\langle m|\,;\end{aligned} \tag{147}$$

next set $q_o + ip_o = re^{i\phi}$, so that (147) becomes

$$\int_0^\infty r dr \int_0^{2\pi} d\phi\; e^{-\frac{1}{2}r^2}\sum_{n,m}\frac{(r/\sqrt{2})^{n+m}}{\sqrt{n!\,m!}}e^{i(n-m)\phi}|n\rangle\langle m|\,. \tag{148}$$

The angular integral vanishes unless $n=m$, and the radial integral reduces (148) to $2\pi\sum_n |n\rangle\langle n|$. The completeness relation is thus

$$1 = \frac{1}{\pi}\int d^2z\; |z\rangle\langle z|\,, \tag{149}$$

where

$$d^2z \equiv d(\mathrm{Re}\, z)\cdot d(\mathrm{Im}\, z)\,, \tag{150}$$

i.e., (149) is an integration over the whole complex plane, or equivalently, over the whole phase space plane (q_o, p_o). By combining (145) and (149) we see that the coherent states are not linearly independent.

(e) Wigner Distributions

The states of the harmonic oscillator provide instructive examples of the Wigner distribution (recall §2.2(d)).

We use the dimensionless coordinate and momentum eigenvalues ξ and $\eta = p(m\omega\hbar)^{-\frac{1}{2}}$. The Wigner distribution of the ground state is then

$$f_0(\xi,\eta) = \pi^{-\frac{1}{2}} \int ds\, e^{-i\eta s}\, e^{-\frac{1}{2}(\xi-\frac{1}{2}s)^2}\, e^{-\frac{1}{2}(\xi+\frac{1}{2}s)^2} . \tag{151}$$

This is an example of the Gaussian integral

$$\int_{-\infty}^{\infty} dx\, e^{-(\frac{1}{2}ax^2+ibx)} = (2\pi/a)^{\frac{1}{2}} e^{-b^2/2a} , \tag{152}$$

which holds provided $\mathrm{Re}\, a > 0$ for all complex b. Hence

$$f_0(\xi,\eta) = 2\, e^{-(\xi^2+\eta^2)} . \tag{153}$$

This is a positive definite distribution in phase space. It is normalized in accordance with the general format:

$$\int \frac{d\xi\, d\eta}{2\pi} f_0(\xi,\eta) = 1 . \tag{154}$$

The same calculation for the first excited state, as given by (92), gives

$$f_1(\xi,\eta) = (\xi^2+\eta^2-\tfrac{1}{2})\, e^{-(\xi^2+\eta^2)} \tag{155}$$

and this is *not positive definite!* Indeed, the nodes in the excited states produce nodes in their Wigner distributions, so only the ground state has the exceptional property of having a distribution that could be interpreted as a probability distribution in phase space.

Because the coherent states move in a manner that preserves the shape of the ground state, one might guess that they have a Wigner distribution that is also positive definite. In fact, the Wigner distributions of the coherent states have a remarkably simple form, as a straightforward calculation using (121) shows:

$$f_C(\xi,\eta;t) = f_0(\xi - y(t), \eta - \dot{y}(t)) . \tag{156}$$

This is the "phase space" distribution of the ground state centered on the phase space point $(y,\dot{y})$, which describes some motion of the classical oscillator.

Another interesting distribution is that of an oscillator in a thermal equilibrium state. The density matrix of such a state is

$$\rho_T = Z^{-1} \sum_n e^{-n\Omega} |n\rangle\langle n| , \tag{157}$$

where $\Omega = \hbar\omega/kT$, and Z is the partition function, defined for our purpose by requiring $\mathrm{Tr}\,\rho = 1$:

$$Z = \sum_n e^{-n\Omega} = \frac{1}{1-e^{-\Omega}} ; \tag{158}$$

here we have again set the ground-state energy to zero. As with so many properties of the harmonic oscillator, the sum on n required to evaluate the density matrix in the coordinate representation can be evaluated in closed form, and the same is true of the Fourier transform that turns it into the Wigner distribution. The result is

$$f_T(\xi,\eta) = 2\tanh(\tfrac{1}{2}\Omega)\exp[-(\xi^2+\eta^2)\tanh\tfrac{1}{2}\Omega] , \tag{159}$$

which reduces properly to (153) when $T \to 0$. Thus the thermal Wigner distribution is also positive definite for all temperatures, which is quite remarkable given that only the ground state has a positive definite distribution by itself.

(f) Propagator and Path Integral

The propagator for the harmonic oscillator can be evaluated by a considerable array of techniques. One might think this is just a pedantic reflection of the idealizations embodied in the linear equations of motion, and therefore not worthy of attention. Not so, however, because the oscillator is the point of departure for the theory of many complex phenomena.

First off, note that we already know the propagator for both the free and linearly forced oscillator! Consider Eq. 130; it states that

$$\langle z|e^{-iH_0t}|z'\rangle = \langle z|z'e^{-it}\rangle \,. \tag{160}$$

This scalar product is given by (145), and therefore the propagator of the free oscillator in the coherent state basis is

$$K(z,z';t) = e^{-\frac{1}{2}(|z|^2+|z'|^2)} \exp(z^* z' e^{-it})\, \theta(t) \,. \tag{161}$$

The extension to the forced oscillator is given in Prob. 8.

The Taylor expansion of the last exponential factor in (161) produces the Fourier representation of the propagator, with the frequencies giving the energy eigenvalues and the coefficients the energy eigenfunctions, as required by the universal formula Eq. 2.365 for a time-independent Hamiltonian:

$$K(z,z';t) = \sum_{n=0}^{\infty} u_n(z)\, u_n^*(z')\, e^{-int} \,. \tag{162}$$

The energies are now $n(\hbar\omega)$ because the ground-state energy has been redefined to being zero. The eigenfunctions in this representation are thus

$$u_n(z) = \frac{1}{\sqrt{n!\pi}}(z^*)^n e^{-\frac{1}{2}|z|^2} \,, \tag{163}$$

and they satisfy the orthonormality condition

$$\int d^2z\, u_n(z) u_m^*(z) = \delta_{n,m} \,. \tag{164}$$

The conventional and familiar wave functions in the coordinate basis (Eq. 91) are related to these by the transformation function (128).

The configuration space propagator for the oscillator can be evaluated by a host of path integral methods, and also less fancy techniques, which can be found in the bibliography. All are contingent on the quadratic nature of the Lagrangian, of course. For such Lagrangians the semiclassical propagator is exact, and as shown in §2.8(c), in one dimension it is

$$K(b,a) = \sqrt{-\frac{1}{2\pi i\hbar}\frac{\partial S_{\rm cl}(b,a)}{\partial x_b\, \partial x_a}}\, \exp\left(\frac{i}{\hbar} S_{\rm cl}(b,a)\right) \,, \tag{165}$$

were $S_{\rm cl}$ is the action for the classical path from $a = (x_a, t_a)$ to $b = (x_b, t_b)$.

The evaluation of the classical action, while only an exercise in classical mechanics, is instructive because the result must be expressed in a form appropriate to its use in the propagator, that is, in terms of specified initial and final coordinates and the elapsed time, not in terms of the initial coordinates and velocities.

The Lagrangian of interest is

$$L = \frac{1}{2}(\dot{Q}^2 - Q^2) - f(t)Q \,, \tag{166}$$

where $Q(t)$ is the classical coordinate in dimensionless form, the equation of motion being $\ddot{Q} + Q + f = 0$. Then

$$S_{\text{cl}} = \int_{t_a}^{t_b} dt\, L = \int_{t_a}^{t_b} dt \left(\frac{1}{2}\left[\frac{d}{dt}(Q\dot{Q}) - Q\ddot{Q} - Q^2 \right] - fQ \right) \tag{167}$$

$$= \frac{1}{2}[(Q\dot{Q})_{t_b} - (Q\dot{Q})_{t_a}] - \frac{1}{2}\int_{t_a}^{t_b} dt\, f(t)Q(t) \,. \tag{168}$$

In the unforced case, the solution of the equations of motion is

$$Q(t) = Q_a \cos(t - t_a) + \dot{Q}_a \sin(t - t_a) \,, \tag{169}$$

where $Q(t_a) = Q_a \,, \dot{Q}(t_a) = \dot{Q}_a$. The propagator requires the action for specified initial and final coordinates given the elapsed time $(t_b - t_a) = T$. Hence the initial and final velocities are

$$\dot{Q}_a = (Q_b - Q_a \cos T)/\sin T \,; \qquad \dot{Q}_b = (-Q_a + Q_b \cos T)/\sin T \,. \tag{170}$$

Equation (168) then gives the following result for the unforced case:

$$S_{\text{cl}}^{(0)}(b,a) = \frac{1}{\sin(t_b - t_a)} \left[\tfrac{1}{2}(Q_a^2 + Q_b^2)\cos(t_b - t_a) - Q_a Q_b\right] . \tag{171}$$

The rather long calculation for the driven oscillator is outlined in Prob. 9. The result is

$$\begin{aligned} S_{\text{cl}}(b,a) = \frac{m\omega}{\sin\omega T} \Bigg\{ & \tfrac{1}{2}(x_a^2 + x_b^2)\cos\omega T - x_a x_b \\ & + \frac{x_a}{m\omega}\int_{t_a}^{t_b} dt\, f(t)\sin\omega(t_b - t) + \frac{x_b}{m\omega}\int_{t_a}^{t_b} dt\, f(t)\sin\omega(t - t_a) \\ & - \frac{1}{m^2\omega^2}\int_{t_a}^{t_b} dt \int_{t_a}^{t} ds\, f(t)f(s)\, \sin\omega(t_b - t)\sin\omega(s - t_a) \Bigg\} \,, \end{aligned} \tag{172}$$

where the mass and frequency have been restored, with $f(t)$ now *redefined* by the Lagrangian $\frac{1}{2}m(\dot{x}^2 - \omega^2 x^2) - xf(t)$.

The complete result for the propagator then follows from (165):

$$K(b,a) = \sqrt{\frac{m\omega}{2\pi i\hbar \sin\omega T}} \exp\left(\frac{i}{\hbar} S_{\text{cl}}(b,a) \right) . \tag{173}$$

In the unforced case, the energy eigenfunctions in the coordinate representation can be found from this propagator; see Prob. 11.

4.3 Motion in a Magnetic Field

The motion of charged particles in electromagnetic fields is a central feature of many phenomena. In classical physics the problem of finding the trajectories can only be solved analytically for a very restricted set of field configurations, and it does not become easier in quantum mechanics. Here only the most important soluble situation will be treated, that of a spatially uniform static magnetic field. Nevertheless, this is a rich problem with many practical applications, especially in condensed matter physics. While classical physics offers an intuitive understanding of what quantum mechanics has to say about motion in a magnetic field, quantum mechanics predicts a startling effect that has no classical counterpart: under appropriate circumstances a charged particle can display interference phenomena produced by a magnetic field that it *never* encounters. This is the Aharonov-Bohm effect, which will also be described in this section.

(a) Equations of Motion and Energy Spectrum

The Hamiltonian for a particle of charge e and mass m in a static magnetic field described by the vector potential $\boldsymbol{A}$ is

$$H = \frac{1}{2m}\left(\boldsymbol{p} - \frac{e}{c}\boldsymbol{A}(\boldsymbol{q})\right)^2 . \tag{174}$$

Here $\boldsymbol{p}$ is the *canonical* momentum; the velocity is *not* $\boldsymbol{p}/m$, but rather

$$\boldsymbol{v} = \frac{1}{m}\left(\boldsymbol{p} - \frac{e}{c}\boldsymbol{A}\right) . \tag{175}$$

That is, from (174) and the canonical commutation rules, it follows that the velocity and coordinate are related as claimed in (175):

$$\boldsymbol{v}(t) = [\boldsymbol{q}(t), H]/i\hbar . \tag{176}$$

Consequently, the Hamiltonian is also

$$H = \tfrac{1}{2}m|\boldsymbol{v}|^2 . \tag{177}$$

Despite appearances, this is not the energy of a free particle as $\boldsymbol{v}$ does not obey the commutation rules of $\boldsymbol{p}/m$. The fundamental commutation rule of this problem follows from $[p_i, A_j] = (\hbar/i)\partial A_j/\partial q_i$:

$$[v_i, v_j] = i\frac{e\hbar}{cm^2}\epsilon_{ijk}B_k , \tag{178}$$

where B_k is a component of the magnetic field.

The velocity commutation rule (178) is gauge-independent — it contains only $\boldsymbol{B}$. Equation (177) is a gauge-invariant expression for the energy, and it is best to work with explicitly gauge invariant expressions when possible.

Equation (178) leads directly to the equation of motion for $\boldsymbol{v}$:

$$\dot{\boldsymbol{v}} = \frac{e}{2mc}[\boldsymbol{v}\times\boldsymbol{B} - \boldsymbol{B}\times\boldsymbol{v}] . \tag{179}$$

When $\boldsymbol{B}$ is not uniform,

$$[B_j, v_k] = \frac{1}{m}[B_j, p_k] = \frac{i\hbar}{m}\frac{\partial B_j}{\partial q_k}\,, \qquad (180)$$

and the equation of motion in the general case is therefore

$$\frac{d\boldsymbol{v}}{dt} = \frac{e}{mc}(\boldsymbol{v}\times\boldsymbol{B}) + i\frac{e\hbar}{2cm^2}\boldsymbol{\nabla}\times\boldsymbol{B}\,. \qquad (181)$$

The last term is absent in the classical case, but when the field is uniform Heisenberg's equation is identical to the classical equation of motion.

Consider now the case of a *uniform field,* namely, $\boldsymbol{B} = (0, 0, B)$, where $B > 0$. The only nonzero commutator among the velocity components is then

$$[v_x, v_y] = i\frac{e\hbar B}{cm^2}\,. \qquad (182)$$

In the light of this, it is best to write the Hamiltonian in terms of the Cartesian components of velocity:

$$H = \tfrac{1}{2}m(v_x^2 + v_y^2 + v_z^2)\,. \qquad (183)$$

Now $v_z = p_z/m$, and commutes with v_x and v_y: *the momentum parallel to the field is a constant of motion,* just as in classical physics. The motion along $\boldsymbol{B}$ is simply that of a free particle and can be set aside until needed.

For that reason we only concern ourselves with motion in the x-y plane perpendicular to $\boldsymbol{B}$, for which the Hamiltonian is

$$H = \tfrac{1}{2}m(v_x^2 + v_y^2)\,. \qquad (184)$$

The velocity components v_x and v_y do not commute with each other or with H. But according to (182), their commutator is a c-number. Consequently, *the Hamiltonian (184) for motion perpendicular to $\boldsymbol{B}$ is that of a simple harmonic oscillator!* To see this explicitly, define the operators

$$a = \sqrt{\frac{m}{2\hbar\omega_c}}\,(v_x + iv_y)\,, \quad a^\dagger = \sqrt{\frac{m}{2\hbar\omega_c}}\,(v_x - iv_y)\,, \qquad (185)$$

where ω_c is the *cyclotron frequency:*

$$\omega_c = \frac{eB}{mc}\,. \qquad (186)$$

A short calculation leads to the same commutation rule as for the operators of the same name in the theory of the harmonic oscillator, i.e.,

$$[a, a^\dagger] = 1\,. \qquad (187)$$

In terms of them, the Hamiltonian for motion in the plane transverse to $\boldsymbol{B}$ is

$$H = \hbar\omega_c(a^\dagger a + \tfrac{1}{2})\,. \qquad (188)$$

The energy spectrum for motion of a charged particle in a uniform magnetic field is therefore

$$E_n = \hbar\omega_c(n + \tfrac{1}{2})\,, \qquad n = 0, 1, \ldots\,. \qquad (189)$$

These are the *Landau levels.*

From the theory of the oscillator (see Eq. 59) it follows that the characteristic length in the problem is

$$l_B = \sqrt{\hbar/m\omega_c} = \sqrt{\hbar c/eB}\,. \tag{190}$$

This is called the *magnetic length.* For the electron[1]

$$\hbar\omega_c = 1.158 \times 10^{-4}\,\text{eV}(B/\text{T})\,, \quad l_B/a_0 = 4.85 \times 10^2(B/\text{T})\,, \tag{191}$$

where $a_0 = 4\pi\hbar^2/me^2$ is the Bohr radius and B/T means the magnetic field in units of tesla (T), or 10^4 gauss. The characteristic velocity of the state with energy E_n is therefore $(v/c)^2 \sim 5 \times 10^{-10}(B/\text{T})\,n$.

When the field is uniform, the equation of motion (181) simplifies to

$$\dot{v}_x = \omega_c v_y\,, \qquad \dot{v}_y = -\omega_c v_x\,. \tag{192}$$

The solution has the same appearance as in the classical case, i.e., for $e > 0$, clockwise circular motion with angular frequency ω_c:

$$x(t) = x_0 - \frac{1}{\omega_c}v_y(t)\,, \quad y(t) = y_0 + \frac{1}{\omega_c}v_x(t)\,. \tag{193}$$

The putative coordinates of the center, (x_0, y_0), are not c-numbers, however, because of the nonzero velocity commutator:

$$[x_0, y_0] = -il_B^2\,. \tag{194}$$

Another short calculation will confirm that x_0 and y_0 are constants of motion,

$$[H, x_0] = [H, y_0] = 0\,, \tag{195}$$

but because of (194), only one can be diagonalized simultaneously with the energy. In classical mechanics, the radius of the orbit is proportional to the energy. The same relation holds for the operators, namely

$$R^2 \equiv [x(t) - x_0]^2 + [y(t) - y_0]^2 = \frac{2}{m\omega_c^2}H\,. \tag{196}$$

Hence R^2 has a sharp value $(r^2)_n$ in an energy eigenstate,

$$(r^2)_n = (2n+1)l_B^2\,; \tag{197}$$

the radii of the stationary states therefore grow like $\sqrt{n}$.

(b) Eigenstates of Energy and Angular Momentum

It is clear that the Landau levels must be highly degenerate. The system is two-dimensional, and the Hilbert space must go over to that of a free particle in two

[1] In comparing this section with the literature, note that for the electron $e = -|e|$ in our expressions, whereas some authors set $e \to -e$ from the outset. Of course, in the numerical values for $\hbar\omega_c$ and l_B, e means $|e|$!

dimensions when $B \to 0$, which they would not if the states were really just those of a one-dimensional oscillator. The same conclusion can be drawn from the more intuitive remark that in the classical case the orbits for a given energy E are the circles of radius $\sqrt{2E/m\omega_c^2}$ with centers *anywhere* in the transverse plane. Indeed, this implies that the Landau levels must be infinitely degenerate. As we shall see, there are several distinct ways in which the these states can be differentiated. One is in terms of angular momentum, which will be developed now. It should be mentioned here that this richness of descriptions of the Landau levels plays a central role in the theory of the quantum Hall effect.[1]

To specify states uniquely, one quantum number beyond energy must be given. This quantum number, and the corresponding wave functions, are gauge-dependent to an astonishing degree. One can, as we shall do, choose a gauge in which L_z, the angular momentum along $\boldsymbol{B}$, and the energy commute. In this description *the subspace for a given energy is spanned by a discrete infinity of angular momentum eigenstates all centered about the point used to define the angular momentum.* In another popular choice (see Prob. 17), the energy and one linear combination of x_0 and y_0 is diagonalized simultaneously, and this leads to a *continuum* of states of a definite energy localized about all the points on a line in the x-y plane. Both of these descriptions make a subspace of the x-y plane exceptional, and then compensate for this by means of wave functions which are rather weird. Neither has a direct connection to the circular orbits of the classical theory, and for that reason they do not provide an easily visualizable description of the motion. To get a basis that does one must abandon the energy eigenstates in favor of time-dependent coherent states, which will be done in the next subsection.

A description in terms of angular momentum states requires the gauge to be symmetric about $\boldsymbol{B}$. Choose $\boldsymbol{A} = \frac{1}{2}(\boldsymbol{B}\times\boldsymbol{r})$, or

$$A_x = -\tfrac{1}{2}yB\,, \qquad A_y = \tfrac{1}{2}xB\,; \tag{198}$$

the velocity components are then

$$v_x = \frac{1}{m}p_x + \frac{\omega_c}{2}y\,, \qquad v_y = \frac{1}{m}p_y - \frac{\omega_c}{2}x\,. \tag{199}$$

The angular momentum along $\boldsymbol{B}$ (as always in units of $\hbar$) is

$$L_z = (xp_y - yp_x)/\hbar\,. \tag{200}$$

A bit of algebra then gives the Hamiltonian

$$H = \frac{1}{2m}(p_x^2 + p_y^2) + \frac{1}{8}m\omega_c^2(x^2 + y^2) - \tfrac{1}{2}\hbar\omega_c L_z\,. \tag{201}$$

As expected L_z, is a constant of motion.

The expression $H + \frac{1}{2}\hbar\omega_c L_z$ is the Hamiltonian of an isotropic two-dimensional oscillator of frequency $\frac{1}{2}\omega_c$. The spectrum of this oscillator is $\frac{1}{2}\hbar\omega_c(n_x + n_y + 1)$, where n_x and n_y are $0, 1, \ldots$. But n_x and n_y cannot be specified in a representation

[1]For an introduction to this topic, see R. Shankar, *Quantum Mechanics,* 2nd ed., Plenum (1994), pp. 587–592.

in which L_z is diagonal, though $n_x + n_y = n_\perp$ can be specified together with μ, the eigenvalue of L_z. This then gives the following formula for the energies:

$$E(n_\perp, \mu) = \tfrac{1}{2}\hbar\omega_c(n_\perp - \mu + 1) \, . \tag{202}$$

On the other hand, we know that the energies are given by the Landau formula (189), and furthermore, that the energy cannot depend on μ because all axes of rotation perpendicular to the x-y plane are physically equivalent when the field is uniform. Hence (202), while correct, is awkward.

In short, what is desired is a basis $|n, \mu\rangle$ where n is the quantum number in Landau's formula. Motivated by the theory of the harmonic oscillator, we seek operators that manufacture *all* these eigenstates from just one, the state with lowest energy ($n = 0$) and zero angular momentum, $\mu = 0$.

The operator $a^\dagger$ raises n by one. What is still needed are operators that change μ while leaving n fixed. Now recall from §3.5(d) that for any vector operator $\boldsymbol{V}$, the combination $V_x + iV_y$ raises the eigenvalue of L_z by one, while $V_x - iV_y$ lowers it by one. As the sought-for vector operators are not to change the energy, they must be constants of motion. Two such operators have already been identified: the center-of-the-circle coordinates (x_0, y_0). In the gauge (198),

$$x_0 = \frac{1}{2}x + \frac{1}{m\omega_c}p_y \, , \qquad y_0 = \frac{1}{2}y - \frac{1}{m\omega_c}p_x \, . \tag{203}$$

Define the dimensionless operators

$$Q_\pm = \frac{1}{\sqrt{2}\,l_B}(x_0 \pm iy_0) = \frac{1}{\sqrt{2}\,l_B}\left(\frac{x \pm iy}{2} \mp i\,\frac{p_x \pm ip_y}{m\omega_c}\right) \, . \tag{204}$$

When acting on the eigenstates of H and L_z, they therefore produce

$$Q_\pm|n, \mu\rangle = C_\pm|n, \mu \pm 1\rangle \, . \tag{205}$$

The coefficients are determined by requiring the states to have norm 1,

$$|C_+|^2 = \langle n, \mu|Q_- Q_+|n, \mu\rangle = |C_-|^2 + 1 \, , \tag{206}$$

where the last relation follows from (194), or equivalently,

$$[Q_-, Q_+] = 1 \, . \tag{207}$$

From (203), (201) and (194),

$$Q_- Q_+ = \frac{1}{\hbar\omega_c}H + L_z + \frac{1}{2} \, , \tag{208}$$

and therefore $|C_+|^2 = n + \mu + 1$. With an appropriate choice of the arbitrary phases we thus conclude that

$$Q_+|n, \mu\rangle = \sqrt{n + \mu + 1}\,|n, \mu + 1\rangle \, , \tag{209}$$

$$Q_-|n, \mu\rangle = \sqrt{n + \mu}\,|n, \mu - 1\rangle \, . \tag{210}$$

The last equation determines the angular momentum spectrum for every Landau level, because $Q_-|n, -n\rangle = 0$ implies

$$\mu = -n, -n + 1, -n + 2, \dots \, . \tag{211}$$

In words, *the angular momentum spectrum in the degenerate subspace of energy E_n is bounded from below, and takes on the integer values $\mu = -n, -n+1, \dots$.* This statement pertains to particles with positive charge; for a particle with negative charge, like the electron, the angular momentum is bounded from above at $\mu = n$.

The energy raising operator $a^\dagger$ is proportional to

$$v_x - iv_y = \frac{1}{m}(p_x - ip_y) + \frac{i\omega_c}{2}(x - iy) , \qquad (212)$$

and it therefore *lowers* the eigenvalue of L_z by one.[1] The action of a and $a^\dagger$ on the basis is therefore

$$a^\dagger|n, \mu\rangle = \sqrt{n+1}\,|n+1, \mu-1\rangle , \qquad a|n, \mu\rangle = \sqrt{n}\,|n-1, \mu+1\rangle , \qquad (213)$$

where the coefficients are the same as for the ordinary oscillator because the commutation rule is the same. Consequently,

$$\frac{1}{\sqrt{n!}}(a^\dagger)^n|0, 0\rangle = |n, -n\rangle . \qquad (214)$$

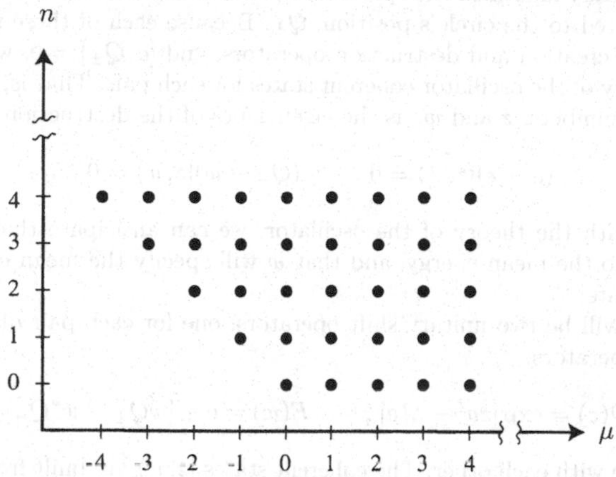

FIG. 4.2. The energy and angular momentum spectra for the motion of a charged particle in a uniform magnetic field; $E_n = \hbar\omega_c(n + \frac{1}{2})$, where ω_c is the cyclotron frequency.

The operators that produce the whole spectrum (see Fig. 4.2) from the state $|0, 0\rangle$ are then the direct extensions of those for the one-dimensional oscillator:

$$|n, \mu\rangle = \frac{1}{\sqrt{n!(n+\mu)!}}(Q_+)^{n+\mu}(a^\dagger)^n|0, 0\rangle . \qquad (215)$$

---

[1]Observe that $v_x - iv_y$ is a lowering operator for L_z even though v_x and v_y are linear combinations of canonical coordinates and momenta along different directions. For further discussion of this issue, see Prob. 15.

The wave functions can be found by the technique used for the oscillator (see Prob. 16). The one for $n = \mu = 0$ is

$$\Psi_{00}(\rho) = \frac{1}{\sqrt{2\pi}} e^{-\rho^2/4} , \tag{216}$$

where $\rho\, l_B$ is the radial distance in cylindrical coordinates. The other wave functions can then be generated from this one by differentiation. As already mentioned, *all these states single out the point* $\rho = 0$, *whereas there is no preferred point in the* x-y *plane.* The coherent states that will now be constructed do not suffer from this defect.

(c) Coherent States

As in the case of the one-dimensional oscillator, in this problem the coherent states are a complete but non-orthogonal set of time-dependent states, each of which describes the wave packet (216) moving along a classical trajectory *without distortion.* Four real parameters are needed to describe such a trajectory: the radius of the circle, or equivalently, the energy; the point on the circle where the particle is at $t = 0$; and the coordinates of the circle's center.

We have already identified the operators related to motion on the circle, a and $a^\dagger$; and those related to the circle's position, $Q_\pm$. Because each of these pairs satisfies the algebra of creation and destruction operators, and $[a, Q_\pm] = 0$, we can simply copy the theory of the oscillator coherent states for each pair. That is, we introduce two complex numbers, z and w, as the eigenvalues of the destruction operators:

$$(a - z)|z, w\rangle = 0 , \qquad (Q_- - w)|z, w\rangle = 0 . \tag{217}$$

By analogy with the theory of the oscillator, we can anticipate that $|z|^2$ will be proportional to the mean energy, and that w will specify the mean coordinates of the circle's center.

Now there will be two unitary shift operators, one for each pair of creation and destruction operators:

$$D(z) = \exp\left[z a^\dagger - z^* a\right] , \qquad F(w) = \exp\left[w Q_+ - w^* Q_-\right] ; \tag{218}$$

they commute with each other. The coherent states $|z, w\rangle$ are built from the energy and angular momentum eigenstate $|0, 0\rangle$ with eigenvalues $n = \mu = 0$ by replaying Eq. 123:

$$|w, z\rangle = F(w) D(z) |0, 0\rangle . \tag{219}$$

The time development of these states is almost identical to that of the oscillator; namely, $z \to z e^{-i\omega_c t}$ as before, but w does not change because $Q_\pm$ are constants of motion. Hence,

$$e^{-iHt}|z, w\rangle = |z e^{-i\omega_c t}, w\rangle , \tag{220}$$

where an overall and irrelevant phase $e^{-i\omega_c t/2}$ has again been dropped by defining zero to be the ground-state energy. Let $\bar{A}$ denote the expectation value of A in the state (220). According to (217) and (185),

$$\bar{a}(t) = z e^{-i\omega_c t} = \sqrt{m/2\hbar\omega_c}\, \left[\bar{v}_x(t) + i\bar{v}_y(t)\right] . \tag{221}$$

Setting $z = |z|e^{i\phi}$, we have

$$\bar{v}_x(t) = \sqrt{2\hbar\omega_c/m}\,|z|\cos(\omega_c t - \phi)\,, \quad \bar{v}_y(t) = -\sqrt{2\hbar\omega_c/m}\,|z|\sin(\omega_c t - \phi)\,. \tag{222}$$

The expectation value of the energy then follows from (184),

$$\bar{H} = \hbar\omega_c|z|^2\,, \tag{223}$$

and the mean square radius from (196),

$$\langle R^2 \rangle = 2l_B^2|z|^2\,. \tag{224}$$

The expectation value of the circle's center is found from

$$\bar{Q}_- = w = \frac{1}{\sqrt{2}\,l_B}\,(\bar{x}_0 - i\bar{y}_0)\,. \tag{225}$$

Combining this with (193) then gives

$$\bar{x}(t) = \sqrt{2}\,l_B[\mathrm{Re}\;w + |z|\sin\omega_c t\,] \tag{226}$$

$$\bar{y}(t) = \sqrt{2}\,l_B[-\mathrm{Im}\;w + |z|\cos\omega_c t\,]\,, \tag{227}$$

where the origin of t was chosen to eliminate the phase ϕ. As promised, this is clockwise motion on a circle whose radius is proportional to $|z|$, and whose central coordinates are proportional to the real and imaginary parts of w (see Fig. 4.3).

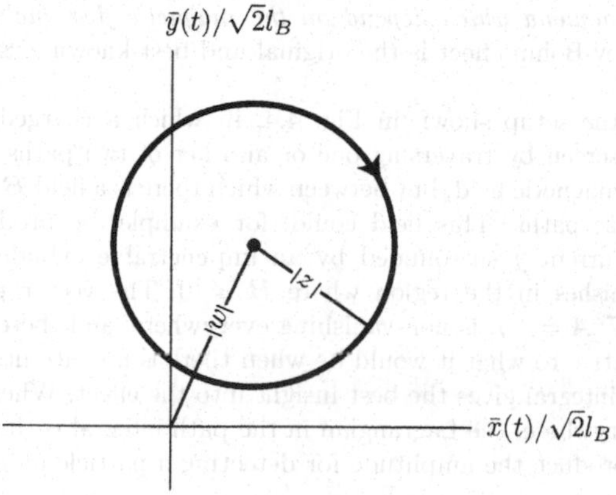

FIG. 4.3. The coherent state description of motion in a homogeneous magnetic field. These states are wave packets that move with constant angular frequency ω_c and without distortion on fixed circles. The complex parameter w specifies the position of the center of the circle, while $ze^{-i\omega_c t}$ gives the instantaneous mean position of the packet on such a circle. As intuition would indicate, the mean energy is proportional to $|z|^2$.

In the discussion of the time-energy uncertainty relation in §2.4(d), we used motion in a uniform magnetic field to illustrate how a system can serve as a clock. The coherent states form the ideal clock in this instance, because they always keep the minimum size l_B of the lowest state (216). Their angular uncertainty is thus $\Delta\theta \sim l_B/(\langle R^2\rangle)^{\frac{1}{2}} \sim (\hbar\omega_c/\bar{H})^{\frac{1}{2}}$. The energy uncertainty in a coherent state is $\sim (\hbar\omega_c\bar{H})^{\frac{1}{2}}$, and the time uncertainty, defined as before, is $\Delta T = \Delta\theta/\omega_c$. Consequently for the coherent states $\Delta T\,\Delta E \sim \hbar$, i.e, the time-energy uncertainty is also minimal for all of these states.

The scalar product and completeness relation satisfied by the coherent states in a magnetic field are the obvious extensions of those in §4.2(c):

$$|\langle z'w'|z,w\rangle|^2 = e^{-|z-z|^2}\, e^{-|w-w'|^2}\,, \tag{228}$$

$$1 = \frac{1}{\pi^2}\int d^2z \int d^2w\, |z,w\rangle\langle z,w|\,. \tag{229}$$

(d) The Aharonov-Bohm Effect

In classical electrodynamics, the motion of a charged particle is ultimately determined by the Lorentz force which is only a function of the field strengths at the instantaneous position of the particle, with the potentials being irrelevant. In quantum mechanics the behavior of the particle is determined by its wave function, which is the solution of a Schrödinger equation, and in it the potentials do appear. One might suppose that the gauge invariance of Schrödinger's equation must mean that the potentials themselves are merely a mathematical intermediary, and as in classical physics all observable effects only depend on field strengths "actually experienced" by the particle. That turns out to be false: *if the paths that combine to form the quantum mechanical amplitude do not cross any region in which there is a magnetic field, but surround a region in which there is a field, there are observable phenomena which depend on the magnetic flux enclosed by these paths.* The Aharonov-Bohm effect is the original and best known example of such a phenomenon.

Consider the setup shown in Fig. 4.4, in which a charged particle can reach a detection screen by traversing one or another of two paths on which it cannot encounter a magnetic field, but between which there is a field $\boldsymbol{B}$ confined to a region between these paths. This field could, for example, be produced by a solenoid carrying a current $\boldsymbol{j}$ surrounded by an impenetrable cylinder so that the wave function vanishes in the region where $\boldsymbol{B} \neq 0$. The vector potential, being the solution of $\nabla^2\boldsymbol{A} = -\boldsymbol{j}$, is non-vanishing everywhere, and therefore alters the wave function relative to what it would be when there is no current.

The path integral gives the best insight into the effect. When the magnetic field vanishes everywhere, the Lagrangian in the path integral is that for a free particle, and it will produce the amplitude for detecting a particle at D as the sum of two distinct terms,

$$\Psi(D) = \Psi_{P_1}(D) + \Psi_{P_2}(D)\,, \tag{230}$$

where Ψ_{P_i} is the contribution from the classical path P_i and its immediate neighbors. When the field is turned on, the Lagrangian changes by

$$\Delta L = \frac{e}{c}\,\boldsymbol{v}\cdot\boldsymbol{A}\,. \tag{231}$$

Each path then acquires an extra factor

$$\exp\left\{\frac{i}{\hbar}\int_{t_a}^{t_b} dt\,\Delta L\right\} = \exp\left\{\frac{ie}{\hbar c}\int_{t_a}^{t_b} dt\,\frac{d\boldsymbol{r}}{dt}\cdot\boldsymbol{A}\right\} = \exp\left\{\frac{ie}{\hbar c}\int_{P_i} d\boldsymbol{r}\cdot\boldsymbol{A}\right\}\,. \tag{232}$$

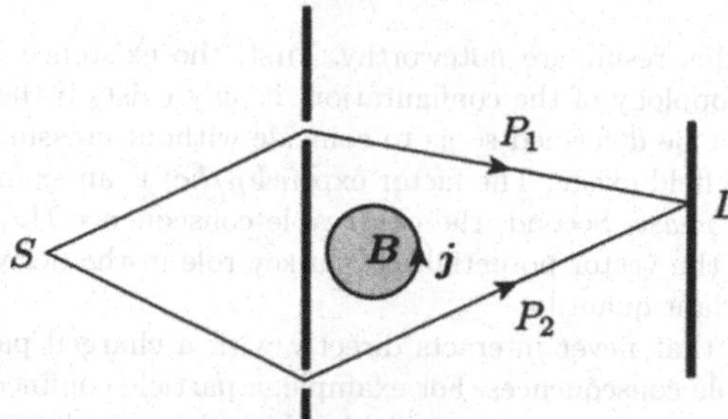

FIG. 4.4. Diffraction experiment displaying the Aharonov-Bohm effect; S is a source of particles, such as a hole in a screen.

Let P_i' be a neighbor to P_i; then the difference between the contribution of these paths is

$$\int_{P_i} d\boldsymbol{r}\cdot\boldsymbol{A} - \int_{P_i'} d\boldsymbol{r}\cdot\boldsymbol{A} = \oint d\boldsymbol{r}\cdot\boldsymbol{A} = \int_{S_i} d\boldsymbol{S}\cdot(\boldsymbol{\nabla}\times\boldsymbol{A}) \,, \tag{233}$$

where S_i is the surface whose boundary is formed by P_i and P_i'. Because $\boldsymbol{B} \equiv \boldsymbol{\nabla}\times\boldsymbol{A}$ vanishes in the neighborhood of both P_1 and P_2, all the paths in the neighborhood of P_1 give the same contribution to the amplitude, and the same holds separately for P_2.

On the other hand, there is a difference between the line integrals along P_1 and P_2:

$$\int_{P_2} d\boldsymbol{r}\cdot\boldsymbol{A} - \int_{P_1} d\boldsymbol{r}\cdot\boldsymbol{A} = \int_{S_{21}} d\boldsymbol{S}\cdot\boldsymbol{B} = \Phi_B \,, \tag{234}$$

where Φ_B is the magnetic flux crossing the surface S_{21} bounded by P_1 and P_2. This result holds for *any* pair of paths that enclose the region in which the magnetic field does not vanish. The path integral therefore produces a *relative* phase between the amplitudes that add coherently at D,

$$\begin{aligned} \Psi(D) &= C\left\{\Psi_{P_1}(D) + \Psi_{P_2}(D)\, e^{\frac{ie}{\hbar c}\oint d\boldsymbol{r}\cdot\boldsymbol{A}}\right\} & (235)\\ &= C\left\{\Psi_{P_1}(D) + e^{\frac{ie}{\hbar c}\Phi_B}\,\Psi_{P_2}(D)\right\} \,, & (236)\end{aligned}$$

where C is a common overall factor. By varying the strength of the magnetic field, and thereby the flux Φ_B, the relative phase of the two contributions to $\Psi(D)$ is varied. Therefore the interference pattern varies sinusoidally at a fixed point D as a function of Φ_B with a period

$$\Phi_0 = \frac{2\pi\hbar c}{|e|} = 4.14 \times 10^{-7}\ \text{Gauss cm}^2 \,, \tag{237}$$

which is the fundamental unit or quantum of magnetic flux. This is the Aharonov-Bohm effect.[1]

[1]For experimental confirmation, see A. Tonomura et al., *Phys. Rev. Lett.* **48**, 1443 (1982).

Two aspects of this result are noteworthy. First, the existence of the effect is contingent on the topology of the configuration: it only exists if the various paths that interfere cannot be deformed so as to coincide without crossing any region in which the magnetic field exists. The factor $\exp(ie\Phi_B/\hbar c)$ is an example of what is called a *topological phase*. Second, the observable consequence (Eq. 236) is gauge invariant; although the vector potential plays a key role in the derivation, the flux Φ_B is a gauge invariant quantity.

A magnetic field that never interacts directly with a charged particle can also have other observable consequences. For example, a particle confined to a multiply connected region that encloses magnetic fields which the particle cannot come into contact with, has energies that depend on the magnetic flux passing through the forbidden regions. The simplest example is the subject of Prob. 18.

4.4 Scattering in One Dimension

Only a small number of phenomena can be treated realistically by modeling them as a one-dimensional system, but such phenomena do exist. Of greater importance, the relative ease with which the one-dimensional Schrödinger equation can be solved makes it a very powerful pedagogical tool. Here we shall simplify more drastically still by only treating potentials that vanish everywhere except at isolated points. This has the virtue of reducing the Schrödinger equation to algebraic equations, which sidesteps a great deal of complicated and often uninformative mathematics while still providing sufficient meat for learning about such phenomena as bound states, transmission and reflection, resonance scattering and exponential decay, all of which arise in the real world.

(a) General Properties

If the potential is $V(x)$, it is convenient to write the time-independent Schrödinger equation as

$$\left(\frac{d^2}{dx^2}+k^2\right)\psi(x) \equiv \mathcal{L}_k\,\psi(x) = U(x)\psi(x)\,, \tag{238}$$

where $E=(\hbar^2k^2/2m)$ is the energy eigenvalue, $U=(2mV/\hbar^2)$, and it is assumed that $U(x)\to 0$ as $|x|\to\infty$. From §2.6 (Eq. 2.390) recall Green's function for the one-dimensional Helmholtz equation,

$$G_0(x,x';k)=\frac{1}{2ik}e^{ik|x-x'|}=\frac{1}{2ik}[\theta(x-x')e^{ik(x-x')}+\theta(x'-x)e^{-ik(x-x')}]\,, \tag{239}$$

which satisfies

$$\mathcal{L}_k\,G_0(x,x';k)=\delta(x-x')\,. \tag{240}$$

G_0 is therefore the inverse of the differential operator $\mathcal{L}_k$, and can be used to bring $\mathcal{L}_k$ to the right-hand side as follows:

$$\psi(x)=\int_{-\infty}^{\infty}dx'\,G_0(x,x';k)U(x')\psi(x')\,. \tag{241}$$

On applying $\mathcal{L}_k$ to this equation, Eq. 238 is recovered thanks to (240). A crucial possibility has been forgotten however: if $\varphi(x)$ is any solution of the free wave equation, $\mathcal{L}_k\,\varphi(x) = 0$, then $\varphi(x)$ can be added to the right-hand side of (241), because it will be eliminated when $\mathcal{L}_k$ acts. In general, therefore, Schrödinger's homogeneous differential equation can be rewritten as the inhomogeneous integral equation

$$\psi(x) = \varphi(x) + \int dx'\, G_0(x,x')U(x')\psi(x')\,, \tag{242}$$

where the eigenvalue parameter k in G_0, and the integration limits, have been suppressed.

The advantage of the integral over the differential equation is that it can explicitly incorporate the boundary conditions appropriate to the problem at hand. Thus G_0 insures that all waves scattered from the potential are outgoing, as is clear from (239), but of course other boundary conditions can be imposed by a different choice of Green's function.

It is essential to realize that the term $\varphi(x)$ in (242) is only allowed if the equation $\mathcal{L}_k\varphi = 0$ has a normalizable solution for the value of k in question, where in the continuous spectrum "normalizable" means that the integral of $|\varphi(x)|^2$ over any portion of the x-axis of length L does not grow faster than L. This requirement is met when $E > 0$ and k is real by linear combinations of e^{ikx} and e^{-ikx}. On the other hand, when $E < 0$, k is pure imaginary and all such $\varphi(x)$ grow exponentially as $x \to +\infty$ or $x \to -\infty$. Such functions $\varphi(x)$ are inadmissible and cannot appear in (242).

Consider, first, positive energies, and a state of momentum $\hbar k$ *incident from the left.* The appropriate form of (242) is then

$$\psi_{k,L}(x) = e^{ikx} + \int dx'\, G_0(x,x')U(x')\psi_{k,L}(x')\,, \tag{243}$$

where *k is positive.* That this integral equation imposes the correct boundary conditions is demonstrated by extracting the asymptotic form of the solution. In the integral x' is restricted to values where $U \neq 0$. Hence if U vanishes for $|x| > a$, then for such values of x the integral in (243), apart from the factor $1/2ik$, is

$$\theta(x-a)\,e^{ikx}\left(\int e^{-ikx'}U(x')\psi_{k,L}(x')\,dx'\right) + \theta(-a-x)\,e^{-ikx}\left(\int e^{ikx'}U(x')\psi_{k,L}(x')\,dx'\right). \tag{244}$$

The asymptotic forms of $\psi_{k,L}(x)$ are therefore

$$\psi_{k,L}(x) \sim e^{ikx} + R(k)\,e^{-ikx} \qquad (x \to -\infty)\,; \tag{245}$$

$$\sim T(k)\,e^{ikx} \qquad (x \to \infty)\,. \tag{246}$$

Thus $\psi_{k,L}$ has a reflected wave on the left of the potential and a transmitted wave to its right, as desired, *provided k is positive.* These asymptotic forms are also valid if U does not vanish identically outside some interval about $x = 0$, but decreases sufficiently rapidly as $|x| \to \infty$. We will not pause here to learn what "sufficiently" means for we shall do just that in the more realistic and important case of three-dimensional scattering in §8.2.

The derivation that led from (244) to the asymptotic forms also gives the following expressions for the transmission and reflection amplitudes $T(k)$ and $R(k)$:

$$T(k) = 1 + \frac{1}{2ik}\int dx\, e^{-ikx}\, U(x)\psi_{k,L}(x)\,, \tag{247}$$

$$R(k) = \frac{1}{2ik}\int dx\, e^{ikx}\, U(x)\psi_{k,L}(x)\,. \tag{248}$$

These are exact expressions, but they can only be evaluated if the exact wave function is known. Nevertheless, they are important and useful. First, they are the prototypes of similar expressions that appear repeatedly in the theory of collision phenomena. Second, they can be used to derive generally valid properties of scattering amplitudes (such as T and R) without explicit knowledge of the exact wave function. And third, they play a central role in approximation techniques. The simplest example of the latter is the Born approximation, in which the scattering amplitudes are expressed as power series in the potential. The leading Born approximation just is the replacement of the exact wave function by the unperturbed wave e^{ikx}:

$$T_0(k) = 1 + \frac{1}{2ik}\int dx\, U(x)\,, \tag{249}$$

$$R_0(k) = \frac{1}{2ik}\int dx\, e^{2ikx}\, Ux)\,. \tag{250}$$

Probability conservation requires the incident flux to equal the sum of the reflected and transmitted fluxes. These fluxes are $v|\psi_k(x)|^2$ apart from a common normalization factor for the various terms in the asymptotic expressions, and the velocity v is the same throughout because, by assumption, $V \to 0$ asymptotically. Hence the transmission and reflection amplitudes satisfy

$$|T(k)|^2 + |R(k)|^2 = 1\,. \tag{251}$$

Although this result is intuitively obvious, the proof is instructive because it is the simplest form of a very general argument in collision theory. The probability current is $i(x) = (\hbar/m)\text{Im}\; \psi^*(x)\psi'(x)$; for a stationary state in one dimension it satisfies $di(x)/dx = 0$. When this identity is integrated from $x = a$ to $x = -a$, it gives $i(a) = i(-a)$. Now let $a \to \infty$; then from (246), $i(a) = v|T|^2$. For $a \to -\infty$,

$$i(-a) = v\big[1 - |R|^2\big]\,, \tag{252}$$

which proves Eq. 251. The continuity equation also leads to the requirement

$$\text{Re}\; T(k)R^*(k) = 0\,, \tag{253}$$

which plays a significant role in the theory of the optical beam splitter (§10.5(a)). The proof is outlined in Prob. 19.

When the energy is negative, we write

$$E = -\frac{\hbar^2\alpha^2}{2m}\,, \qquad k \to i\alpha\,, \quad \alpha > 0\,, \tag{254}$$

where this choice of k is dictated by the requirement that G_0 be exponentially decaying, not growing, for if it were not the resulting wave function would not be normalizable. In this case, as already remarked, there is no admissible solution of the equation $\varphi'' - \alpha^2\varphi = 0$, and so the bound-state wave function must satisfy the integral equation

$$\psi(x) = -\frac{1}{2\alpha}\int dx'\, e^{-\alpha|x-x'|}\, U(x')\psi(x') . \tag{255}$$

There is a very significant difference between the inhomogeneous equation (243) and the homogeneous equation (255). This is most easily appreciated by supposing that they are to be solved numerically with the continuous x-axis approximated by a discrete set of points. Then the inhomogeneous linear integral equation (243) would become a system of *inhomogeneous* linear algebraic equations, and such a system always has a solution. On the other hand, the bound-state equation would become a *homogeneous* system of linear algebraic equations, which only has a non-trivial solution when the determinant vanishes. This vanishing only occurs, if at all, for a discrete set of values of E, which "explains" why the negative energy spectrum is discrete, whereas the positive energy spectrum is continuous.

According to (255), the asymptotic form of a bound state in one dimension is

$$\psi(x) \sim N\, e^{-\alpha|x|} , \qquad |x| \to \infty . \tag{256}$$

It is a quite general and important property of bound-state wave functions, true not only in one dimension, that their asymptotic form is completely determined by the binding energy. The proviso "quite general" alludes to the fact that in inferring (256) we assumed that $U(x)$ falls off sufficiently rapidly to allow the extraction of the factor $e^{-\alpha|x|}$ from the integral, which is not always permissible, an important exception being a Coulomb field in three dimensions (recall §3.6(b)).

The bound and scattering states form a complete orthonormal set, but some care must be taken in properly specifying the scattering states. Clearly, to be complete the set must have states whose incident components have both positive and negative momenta. But (245) and (246) are only sensible if $k > 0$, for if $k < 0$ these asymptotic forms have waves running in both directions to left of the interaction, whereas the correct asymptotic forms for a wave incident from the right are

$$\psi_{k,R}(x) \sim e^{-ikx} + R(k)\, e^{ikx} \qquad (x \to \infty) ; \tag{257}$$

$$\sim T(k)\, e^{-ikx} \qquad (x \to -\infty) , \tag{258}$$

provided again that k is positive. This wave function is the solution of

$$\psi_{k,R}(x) = e^{-ikx} + \int dx'\, G_0(x,x')U(x')\psi_{k,R}(x') . \tag{259}$$

The argument starting from (243) that led to expressions for T and R now gives expressions that can be combined with (247) and (248) to give

$$T(k) = 1 + \frac{1}{2ik}\int dx\, \varphi^*_{k,\sigma}(x)U(x)\psi_{k,\sigma}(x) , \tag{260}$$

$$R(k) = \int dx\, \varphi^*_{-k,\sigma}(x)U(x)\psi_{k,\sigma}(x) , \tag{261}$$

where $\sigma = (R, L)$, and $\varphi_{k,\sigma}(x)$ is the incident wave appropriate to the label σ.

Why did we not distinguish between states incident from the left and right when writing $T(k)$ and $R(k)$? The answer is that these amplitudes are the same for $\psi_{k,L}$ and $\psi_{k,R}$. When the potential is real (i.e., the Hamiltonian is Hermitian), time reversal is a symmetry, and this leads to the hardly surprising conclusion that wave packets incident from the left or right have the same amplitudes for transmission and reflection. Time reversal will be taken up in §7.2, and the proof of the preceding statement will then be a special case of more general properties of scattering amplitudes following from this symmetry. Incidentally, it should be understood that the independence of T and R on the direction of the incident state does *not* require a potential that is invariant under $x \to -x$. Time reversal and spatial reflection are distinct operations, and only invariance under the former matters here.

The orthonormality property of the continuum solutions is

$$\frac{1}{2\pi}\int dx\, \psi^*_{k,\sigma}(x)\psi_{k',\sigma'}(x) = \delta_{\sigma,\sigma'}\,\delta(k-k')\,. \tag{262}$$

Once again, this will not be demonstrated here, because it is a special case of a general result for scattering states best proved with a more powerful formalism (see §9.2). But one can understand why the scattering solutions have the same norm as the incident states, which is what (262) really means. In a time-dependent description, the state is initially a free-particle packet moving toward the potential before the reflected and transmitted packets even exist. Time evolution is unitary, and therefore preserves scalar products, whence the scalar product between states during and after scattering is the same as that between the incident states. Equation (262) is this statement in the limit where the incident states are momentum eigenstates.

The completeness relation is

$$\sum_b \psi_b(x)\psi_b(x') + \frac{1}{2\pi}\sum_\sigma \int_0^\infty dk\, \psi_{k,\sigma}(x)\psi^*_{k,\sigma}(x') = \delta(x-x')\,. \tag{263}$$

These orthonormality and completeness properties can be verified explicitly in the case of the simple potential to be treated now (see Prob. 21).

(b) The Delta-Function Potential

Consider an attractive potential that vanishes everywhere except at $x = 0$:

$$U(x) = -\gamma\delta(x)\,, \tag{264}$$

where γ is positive and has the dimension $(\text{length})^{-1}$. Then (243) becomes

$$\psi_{k,L}(x) = e^{ikx} - \gamma G_0(x,0)\,\psi_{k,L}(0)\,. \tag{265}$$

Setting $x = 0$, and using $G_0(0,0) = 1/2ik$, determines the only unknown, $\psi_{k,L}(0)$. The complete wave function is thus

$$\psi_{k,L}(x) = e^{ikx} - \frac{\gamma/2ik}{1+(\gamma/2ik)}\,[\theta(x)e^{ikx} + \theta(-x)e^{-ikx}]\,. \tag{266}$$

The reflection and transmission amplitudes are therefore

$$R(k) = -\frac{\gamma}{2ik+\gamma} \;, \qquad T(k) = \frac{2ik}{2ik+\gamma} \;, \tag{267}$$

and satisfy Eq. 251. A very short calculation will confirm that $\psi_{k,R}$, in which the incident state is e^{-ikx}, leads to the same reflection and transmission amplitudes, it being understood that k is positive throughout. Eq. 266 also shows that only even parity states are scattered, which is to be expected as odd states are oblivious to a zero range interaction.

Observe that $R \to -1$ and $T \to 0$ in *both* the attractive and repulsive cases as the strength $|\gamma|$ tends to infinity while k is held fixed. Hence either interaction reflects completely in this limit; indeed, $R = -1$ is just the statement that $\psi(0)$ vanishes when $|\gamma| \to \infty$, as it would for an infinitely high wall at $x = 0$. We will, therefore, often speak of the delta-function potential with $|\gamma| \ggg k$ as a nearly impenetrable wall even when it is an attractive potential.

The bound-state equation (255) is now

$$\psi(x) = \frac{\gamma}{2\alpha} e^{-\alpha|x|}\,\psi(0) \;, \tag{268}$$

and on setting $x = 0$ this determines the one and only bound-state eigenvalue:

$$\alpha = \tfrac{1}{2}\gamma \;, \qquad E_b = -\frac{\hbar^2\gamma^2}{8m} \tag{269}$$

Of course (268) has no solution when the potential is repulsive ($\gamma < 0$). The normalized bound-state wave function is

$$\psi_b(x) = \sqrt{\tfrac{1}{2}\gamma}\, e^{-\frac{1}{2}\gamma|x|} \;. \tag{270}$$

It is noteworthy that this potential $-\gamma\delta(x)$ has one (and only one) bound state no matter how small γ may be. That an arbitrarily weak attraction can bind a state is a special property of the one-dimensional Schrödinger equation that does not generalize to higher dimensions. This is connected to the fact that the density of momentum states goes like $p^{d-1}dp$ when the configuration space is d-dimensional.

Green's function for the Schrödinger equation with interaction can also be determined easily in this example. According to Eq. 2.405, it satisfies the integral equation

$$\begin{aligned} G(x,x') &= G_0(x,x') + \int dy\, G_0(x,y)U(y)G(y,x') \\ &= G_0(x,x') - \gamma G_0(x,0)G(0,x') \;. \end{aligned} \tag{271}$$

The solution is

$$G(x,x') = G_0(x,x') - \frac{k\gamma}{k + (\gamma/2i)} G_0(x,0)G_0(0,x') \;. \tag{272}$$

From our general discussion of Green's functions in §2.6(b), we know that G has poles at all energy eigenvalues and that the corresponding eigenfunctions appear in the residues. In this one-dimensional case, and with eigenvalues redefined by $w_n = 2mE_n/\hbar^2$, the eigenfunction expansion (2.370) becomes

$$G(x,x') = \sum_n \frac{\psi_n(x)\psi_n^*(x')}{k^2 - w_n + i\epsilon} . \tag{273}$$

Eq. 272 has a pole at $k = i\gamma/2$ in agreement with (269). To extract the wave functions, (272) has to be put into the form of (273):

$$G(x,x') = G_0(x,x') + \left(\frac{\gamma}{4k}\right)\frac{k+\frac{1}{2}i\gamma}{k^2+\frac{1}{4}\gamma^2}\, e^{ik|x|}\, e^{ik|x'|} . \tag{274}$$

The pole in the complex k^2 plane is at $-\frac{1}{4}\gamma^2 = w_0$ in the notation of (273), or at $k = i\gamma/2$, and its residue is

$$\tfrac{1}{2}\gamma\, e^{-\frac{1}{2}\gamma|x|}e^{-\frac{1}{2}\gamma|x'|} = \psi_b(x)\psi_b(x') , \tag{275}$$

in agreement with (270).

One should ask where the positive energy eigenvalues and their eigenfunctions appear in (274). The answer is that to extract this information from G, the latter has to be treated as a function of the complex variable E, or k^2 with the present convention, and therefore the factor γ/k must be written as $\gamma/\sqrt{k^2}$. The square root introduces a branch cut from $k^2 = 0$ to infinity, which is the positive energy spectrum. This more sophisticated point will be discussed in §8.2.

(c) Resonant Transmission and Reflection

The transmission and reflection amplitudes for the delta-function potential are featureless functions of the incident energy. Two (or more) strong delta-function potentials display far more interesting behavior because, at certain energies, they can almost trap a particle between them, and as a consequence the transmission and reflection probabilities have sharp resonances at such energies. Despite the unrealistic simplicity of the problem we are about to solve, it will reveal many features that appear in actual collision phenomena.

Consider then two delta-function potentials of equal strength separated by the distance $2a$:

$$U(x) = -\gamma[\delta(x+a) + \delta(x-a)] . \tag{276}$$

Important qualitative features of the solutions can be surmised without calculation. We know from (270) that the bound state of a single δ-function potential has a size of order $1/\gamma$, which means that when $a \gg (1/\gamma)$, each term in (276) will have a bound state that has an exponentially small overlap with the other. In this limit, therefore, there are two negative energy eigenvalues nearly degenerate at $E_b = -\hbar^2\gamma^2/8m$, and their splitting must be exponentially small as $a\gamma \to \infty$.

In thinking about the continuous spectrum, consider the situation where γ is very large compared to k. From the preceding example we know that waves impinging from the left or right will be strongly reflected at $x = -a$ or $x = a$, respectively. By the same token, a wave in the interior region $-a \le x \le a$ will be strongly reflected at $x = \pm a$. *Hence in the limit $\gamma \to \infty$, the interior and exterior regions are decoupled; a particle initially inside cannot get out, and if initially outside it cannot get in.*

In this limit the interior wave functions are very simple: they are free particle wave functions that vanish at $x = \pm a$, i.e., $\sin(n\pi x/a)$ and $\cos[(2n+1)\pi x/2a]$, where $n = 1, 2, 3, \ldots$. In short, in the $\gamma \to \infty$ limit, the energy spectrum contains the discrete levels

$$E_n^0 = \frac{\hbar^2 \pi^2}{8ma^2}\, n^2 \,, \qquad n = 1, 2, \ldots \,. \tag{277}$$

These discrete energy eigenvalues are, so to say, suspended in the continuum $0 \leq E < \infty$ *of free particle states that exist in the outside region* $|x| > a$ (see Fig. 4.5).

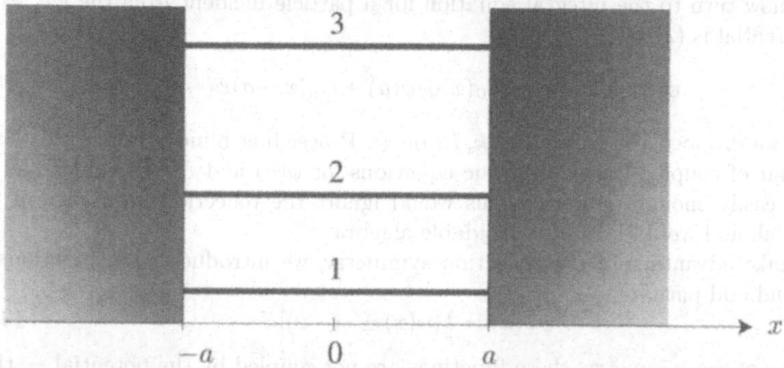

FIG. 4.5. The energy spectrum when the potential of Eq. 276 has infinite strength γ, and totally decouples the interior region $|x| < a$ from the exterior, so that the interior has a discrete spectrum "suspended" in the exterior continuum. When γ is finite, but no matter how small, the interior and exterior are coupled so that the discrete "bound" states then decay into the continuum states, while the continuum states display very sharp resonances when their energy is close to that of the discrete interior states.

This remarkable spectrum of discrete states suspended in a continuum changes dramatically when γ is finite. Indeed, from the amplitude T in the previous example we know that the enclosure's "walls" at $x = \pm a$ become penetrable when the energy is of order $\hbar^2\gamma^2/8m$, and therefore only the discrete inside states with

$$n \ll a\gamma/\pi \tag{278}$$

are nearly decoupled from the outside. The decoupling is imperfect, of course, even for states satisfying (278), because $T \neq 0$ for any finite γ and nonzero energy. In particular, according to (267) the lowest enclosed state with $k = \pi/2a$ has a probability for leakage to the exterior through either boundary of order

$$|T(\pi/2a)|^2 = \frac{1}{1 + (\gamma a/\pi)^2} \,. \tag{279}$$

Therefore the enclosed states are not really stationary; they are an especially simple example of decaying states, as we shall see.

The problem at hand therefore provides a toy model of a phenomenon that occurs in many branches of physics, such as alpha, beta, and gamma decay of nuclei, or of the decay of long-lived collective excitations in solids. The most familiar example is the radiative decay of atomic states, the analogy between this ubiquitous real-world phenomenon and our simple model being the following:

- the discrete atomic energy levels when the interaction between electrons and the electromagnetic field is ignored on the one hand, and the enclosed states between the walls on the other;
- the photon continuum, and the continuum of free particle outside states;
- the coupling of the electromagnetic field to the moving electrons, and leakage through the walls.

We now turn to the integral equation for a particle incident from the left when the potential is (276):

$$\psi(x) = e^{ikx} - \gamma[G_0(x,a)\psi(a) + G_0(x,-a)\psi(-a)] \,, \tag{280}$$

where we dropped the subscript (k, L) on ψ. Proceeding blindly from (280) leads to a pair of coupled linear algebraic equations for $\psi(a)$ and $\psi(-a)$, which can be solved easily enough. However, this would ignore the reflection invariance of the potential, and would lead into avoidable algebra.

To take advantage of the reflection symmetry, we introduce wave functions of even and odd parity:

$$\psi_\pm(x) = \tfrac{1}{2}[\psi(x) \pm \psi(-x)] \,. \tag{281}$$

Because of the symmetry, these functions are not coupled by the potential — they satisfy the separate "integral" equations

$$\psi_+(x) = \cos kx - \gamma[G_0(x,a) + G_0(x,-a)]\psi_+(a) \,, \tag{282}$$

$$\psi_-(x) = i\sin kx - \gamma[G_0(x,a) - G_0(x,-a)]\psi_-(a) \,. \tag{283}$$

This decoupling of simultaneous eigenstates of energy and a symmetry generator Λ (here spatial reflection) that commutes with the Hamiltonian is a consequence of the theorem that Λ is a constant of motion. That is, if the incident wave is a linear combination of e^{ikx} and e^{-ikx} that is odd, say, then the potential can only produce transmitted and reflected waves that are also purely odd. In this example the simplification brought by the symmetry is very modest, and ignorance of it will not produce a calamity. But such ignorance would inflict a high price in more complicated and realistic situations, such as scattering between particles with spin due to a rotationally invariant spin-dependent interaction.

The bound states satisfy the homogeneous counterparts of these equations, and after setting $k \to i\alpha_\pm$, they yield the eigenvalue conditions

$$\alpha_\pm - \tfrac{1}{2}\gamma\left(1 \pm e^{-2a\alpha_\pm}\right) = 0 \,. \tag{284}$$

As expected, when the separation is large ($a\gamma \gg 1$) there are two nearly degenerate bound states with binding energies close to that of a single potential, i.e., E_b of Eq. 269. Their splitting is found by setting $\alpha_\pm = \frac{1}{2}\gamma$ in the exponentially small term in (284), which results in

$$E_{b,\pm} = -|E_b|\left(1 \pm 2e^{-a\gamma}\right) \quad (\alpha \gg 1) \,. \tag{285}$$

Note that the symmetric state lies below the antisymmetric state, which is so because the latter has more curvature and therefore greater kinetic energy. Clearly, this is a generic property of such splittings.

The unknowns $\psi_\pm(a)$ in the scattering equations follow immediately from (282) and (283):

$$\psi_+(a) = \frac{\cos ka}{1+\gamma K_+}\,, \quad \psi_-(a) = \frac{i\sin ka}{1+\gamma K_-}\,, \quad K_\pm = \frac{1}{2ik}\left(1 \pm e^{2ika}\right)\,, \tag{286}$$

and the complete wave function is

$$\psi(x) = e^{ikx} - \frac{\gamma}{2ik}\left\{\left(e^{ik|x-a|} + e^{ik|x+a|}\right)\psi_+(a) + \left(e^{ik|x-a|} - e^{ik|x+a|}\right)\psi_-(a)\right\}. \tag{287}$$

The reflection amplitude can be read of from (287) as the coefficient of e^{-ikx} when $x < -a$:

$$R(k) = -\frac{\gamma}{ik}\left[\psi_+(a)\cos ka + i\psi_-(a)\sin ka\right] \tag{288}$$

$$= \frac{i\gamma}{k}\left(\frac{\cos^2 ka}{1+\gamma K_+} - \frac{\sin^2 ka}{1+\gamma K_-}\right). \tag{289}$$

It is prudent to confirm that this expression is correct in three known limits: (i) for $a = 0$ it simplifies to (267) with $\gamma \to 2\gamma$; (ii) for perfectly reflecting walls ($\gamma \to \infty$, k fixed), $R = -e^{-2ika}$ so as to produce $\psi(-a) = 0$; (iii) and $R \to 0$ for completely transparent walls ($k \to \infty$, γ fixed).

The transmission amplitude also follows from (287):

$$T(k) = 1 + \frac{i\gamma}{k}\left(\frac{\cos^2 ka}{1+\gamma K_+} + \frac{\sin^2 ka}{1+\gamma K_-}\right). \tag{290}$$

Some algebra will confirm that $|R|^2 + |T|^2 = 1$.

It will be convenient to introduce the dimensionless variables

$$\xi = ka\,, \qquad \lambda = 1/a\gamma\,. \tag{291}$$

Nearly impenetrable walls, or weak coupling between inside and outside, corresponds to small values of (k/γ), i.e., to

$$\lambda\xi \ll 1\,. \tag{292}$$

In terms of these variables the functions that appear in T and R are

$$F_+(\xi) = \frac{i\cos^2\xi}{\lambda\xi + \frac{1}{2}\sin 2\xi - i\cos^2\xi}\,, \qquad F_-(\xi) = \frac{i\sin^2\xi}{\lambda\xi - \frac{1}{2}\sin 2\xi - i\sin^2\xi}\,, \tag{293}$$

and the transmission and reflection amplitudes are

$$T(\xi) = 1 + F_+(\xi) + F_-(\xi)\,, \qquad R(\xi) = F_+(\xi) - F_-(\xi)\,. \tag{294}$$

It will be important to bear in mind that F_+ is due to even parity states and F_- to odd parity sates (see Eq. 288).

As already indicated, the transmission and reflection amplitudes are especially interesting when the potential is sufficiently strong to make the "walls" at $x = \pm a$ nearly impenetrable to states with momenta $k \ll \gamma$, or $\lambda\xi \ll 1$. Dropping $\lambda\xi$ in the denominators of $F_\pm$ gives

$$F_+(\xi) = -e^{-i\xi}\cos\xi\,, \qquad F_-(\xi) = -ie^{-i\xi}\sin\xi\,, \quad (\lambda\xi \to 0)\,, \tag{295}$$

and therefore $T = 0$, as expected. Dropping $\lambda\xi$ is only valid, however, if the other terms in the denominators of $F_\pm$ are large compared to $\lambda\xi$, and this is not true in F_+ if $\cos\xi \simeq 0$, or in F_- if $\sin\xi \simeq 0$. But $\cos\xi = 0$ and $\sin\xi = 0$ are just the conditions for a *perfectly* enclosed state of even and odd parity, respectively. *Thus without further analysis, we have learned that when the walls are nearly impenetrable, there is almost no transmission (and hence almost perfect reflection) unless the energy is very close to one of the energies that a perfectly enclosed state would have if the walls were truly impenetrable.*

The energy dependence of the amplitudes T and R, especially in the vicinity of the enclosed state energies, is most easily described when these functions are plotted in the complex plane. This is so because $F_\pm$ have the form

$$F_\pm(\xi) = \frac{iY_\pm}{X_\pm - iY_\pm} = \tfrac{1}{2}(e^{2i\delta_\pm} - 1) = i\delta_\pm \sin\delta_\pm \,, \tag{296}$$

$$\sin\delta_\pm(\xi) = \frac{Y_\pm}{X_\pm^2 + Y_\pm^2} \,, \tag{297}$$

$$X_\pm(\xi) = \lambda\xi \pm \tfrac{1}{2}\sin 2\xi \,, \quad Y_+(\xi) = \cos^2\xi \,, \quad Y_-(\xi) = \sin^2\xi \,. \tag{298}$$

Thus both $F_+(\xi)$ and $F_-(\xi)$ lie on circles of radius $\frac{1}{2}$ centered on the real axis at Re $F_\pm = -\frac{1}{2}$. It turns out that in three-dimensional elastic scattering, the amplitudes for each separate angular momentum state have exactly the form of Eq. 296 (see §8.1). The transmission and reflection amplitudes are also simple expressions in terms of the phases $\delta_\pm$:

$$T(\xi) = \tfrac{1}{2}(e^{2i\delta_+} + e^{2i\delta_-}) \,, \quad R(\xi) = \tfrac{1}{2}(e^{2i\delta_+} - e^{2i\delta_-}) \,. \tag{299}$$

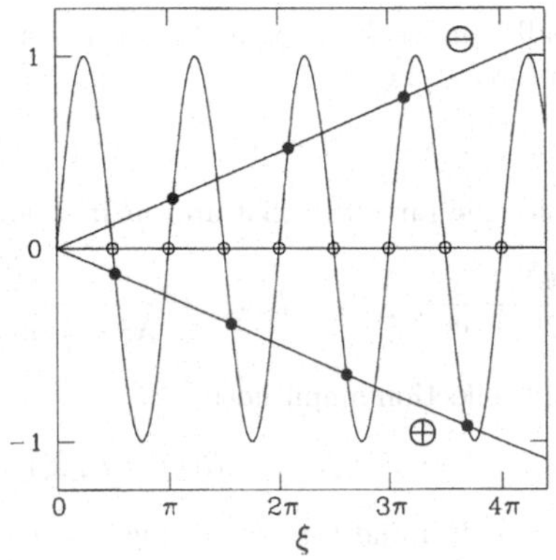

FIG. 4.6. The location of resonances when the strength parameter is $\lambda = 1/8\pi$, with $\xi = ka$. The heavy dots on the *lower* straight line ⊕ show the *even parity* roots of $X_+(\xi) = 0$ that are sharp resonances; the other unmarked intersections with $\sin 2\xi$ are not resonances. The upper straight line ⊖ gives the same information for odd parity states. The open circles on the axis form the discrete spectrum in the decoupling limit, $\lambda = 0$.

As we see, $|F_\pm|$ reach their maximum value 1 whenever $X_\pm = 0$. Let $\xi_n^\pm$ be the roots of $X_\pm(\xi) = 0$; their location is shown in Fig. 4.6. When $\lambda\pi \ll 1$ these roots are near $\frac{1}{2}n\pi$ for integer n such that $\frac{1}{2}n\lambda\pi \ll 1$. These roots fall into two very distinct categories:

1. There are values of $\xi_n^\pm$ of $X_\pm$ close to the energies of states with parity $\pm$ that would be perfectly enclosed in the limit $\lambda \to \infty$. These separate into two further categories:
   (a) Even parity roots, i.e., ξ_n^+, close to $\pi/2, 3\pi/2, \ldots$, so that $Y_+ = \cos\xi_n^+ \ll 1$, and therefore $|F_+| \simeq 1$ only in the immediate vicinity of the ξ_n^+. As we will see, these show up as narrow resonances in the transmission and reflection amplitudes.
   (b) Odd parity roots, i.e., ξ_n^-, close to $\pi, 2\pi, \ldots$, so that $Y_- = \sin\xi_n^- \ll 1$, and therefore $|F_-| \simeq 1$ only in the immediate vicinity of the ξ_n^-, which also show up as narrow resonances in $T(\xi)$ and $R(\xi$
2. There are other roots of $X_\pm = 0$ close to the energy of enclosed states, but of the "wrong" parity, i.e., $\mp$. At these roots $Y_\pm \simeq 1$, and therefore $F_\pm \simeq -1$ over a broad range of values of ξ in the region surrounding the root. These lead to rather featureless contributions to $T(\xi)$ and $R(\xi)$.

In short, sharp resonances are only associated with roots of $X_\pm = 0$ when the root lies close to the value that corresponds to the energy of an almost enclosed state of the *correct* parity $\pm$.

A more detailed understanding can be gained by examining the energy dependence in the neighborhood of an enclosed state energy. Consider, for example, a state of even parity. In this region the approximations $F_- \simeq -1$ can be made for this slowly varying functions, whereas the complete expression (296) must be kept for F_+. Consequently, according to (294),

$$T \simeq 1 + F_+ - 1 = F_+ \,, \qquad R \simeq F_+ + 1 \,, \tag{300}$$

and therefore

$$T(\xi) \simeq \frac{Y_+}{X_+ - iY_+} = \tfrac{1}{2}\left(e^{2i\delta_+} - 1\right) , \tag{301}$$

$$R(\xi) \simeq \frac{X_+}{X_+ - iY_+} = \tfrac{1}{2}\left(e^{2i\delta_+} + 1\right) . \tag{302}$$

Simplifying approximations can be made in the immediate proximity of the roots of ξ_n^+ when $\lambda\pi \ll 1$. The detailed calculations are left for Prob. 23, and the results will be quoted.[1] The roots are at

$$\xi_n^+ = \tfrac{1}{2}n\pi\left(1 + \lambda + O(\lambda^2)\right) , \qquad n \text{ odd} \,, \quad \tfrac{1}{2}n\pi\lambda \ll 1 \,. \tag{303}$$

Furthermore,

$$X_+(\xi) \simeq -(\xi - \xi_n^+) , \qquad \cos^2\xi \simeq (\tfrac{1}{2}n\pi\lambda)^2 \equiv \eta_n \,, \tag{304}$$

[1] It is actually somewhat simpler and also more insightful to analyze this problem by letting the variable ξ be complex, and to find the poles of $T(\xi)$ and $R(\xi)$. This approach is sketched in Prob. 23.

provided ξ lies close enough to ξ_n^+ to satisfy

$$|\xi - \xi_n^+| \lesssim O(\eta_n) \,. \tag{305}$$

To summarize, as long as ξ lies within the range specified by (305), and $(\frac{1}{2}n\pi\lambda) \ll 1$, the transmission and reflection amplitudes are given by

$$T(\xi) \simeq \frac{-i\eta_n}{\xi - \xi_n^+ + i\eta_n} \,, \qquad R(\xi) \simeq \frac{\xi - \xi_n^+}{\xi - \xi_n^+ + i\eta_n} \,. \tag{306}$$

The transmission probability has a Lorentzian shape:

$$|T(\xi)|^2 = \frac{\eta_n^2}{(\xi - \xi_n^+)^2 + \eta_n^2} \,. \tag{307}$$

When the requirement $\frac{1}{2}n\pi\lambda$ of (303) is met, the resonance is very sharp because $\eta_n \lll 1$. As the exact relationship $|T|^2 + |R|^2 = 1$ has been maintained by the approximations leading to (306), the reflection probability is an upside-down Lorentzian:

$$|R(\xi)|^2 = 1 - \frac{\eta_n^2}{(\xi - \xi_n^+)^2 + \eta_n^2} \,. \tag{308}$$

These approximate expressions only hold in the vicinity of resonances, i.e., for $|\xi - \xi_n^+|$ not exceeding a modest multiple of η_n. Fig. 4.7 shows the exact transmission and reflection probabilities over a broad range of energies; Fig. 4.8 shows more detailed information about one narrow resonance. The features shown here are characteristic of resonances quite generally.

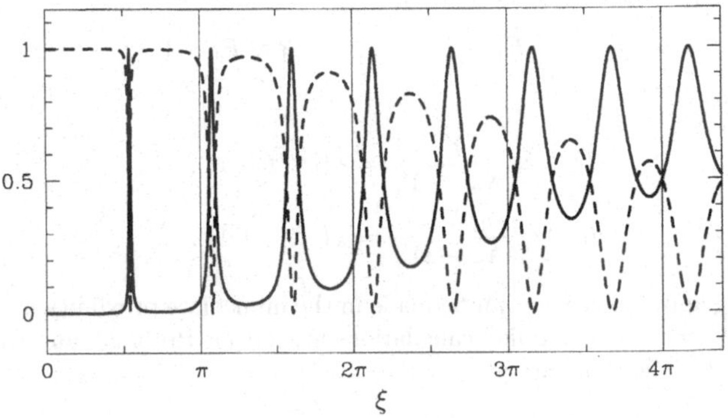

FIG. 4.7. Transmission and reflection when $\lambda = 1/4\pi$; the dashed curve is $|R(\xi)|^2$, the solid curve $|T(\xi)|^2$, and $\xi = ka$.

It is noteworthy that according to the approximate form (306), $R \to 1$ when ξ is distant from ξ_n^+. This agrees with the fact that $R = -e^{-2ika}$ when the walls are impenetrable, because for the values of ξ where (306) holds $2ka \simeq n\pi$ with n odd.

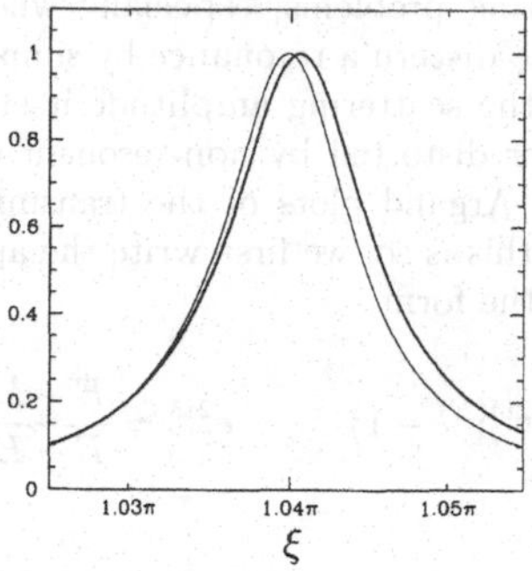

FIG. 4.8. The Lorentzian or Breit-Wigner approximation to the transmission resonance at $\xi \simeq \pi$, as given by Eq. 307 and shown as a light line, compared with the exact result shown as a heavy line, for $\lambda = 1/8\pi$.

It is often convenient to express the resonance forms in terms of energies. Let

$$E_n = \frac{\hbar^2 (\xi_n^+)^2}{2ma^2} \simeq E_n^0 (1 + 2\lambda) \tag{309}$$

be the energy at resonance. Bearing in mind that $|\xi - \xi_n^+| \ll \xi_n^+$ throughout, and multiplying numerator and denominator by $(\hbar^2/2ma^2)(\xi + \xi_n^+) \simeq (\hbar^2 \xi_n^+ / ma^2)$, gives

$$T(E) = -i \frac{\frac{1}{2}\Gamma_n}{E - E_n + \frac{1}{2} i \Gamma_n} , \tag{310}$$

$$|T(E)|^2 = \frac{\frac{1}{4}\Gamma_n^2}{(E - E_n)^2 + \frac{1}{4}\Gamma_n^2} , \tag{311}$$

where

$$\Gamma_n = \frac{2\hbar^2}{ma^2} \xi_n^+ \eta_n \simeq 2 E_n n \pi \lambda^2 \tag{312}$$

is called the *width* of the resonance, a name that stems from the fact that when $|T|^2 = \frac{1}{2}$ the width of the resonance curve is Γ_n (which goes with the acronym FWHM for full width at half maximum). In nuclear physics the expressions (310), (311) and (313) are known as Breit-Wigner formulas for elastic scattering.

A comment about the reflection amplitude is appropriate here. According to (300), it can be written as

$$R(E) = 1 - i \frac{\frac{1}{2}\Gamma_n}{E - E_n + \frac{1}{2} i \Gamma_n} , \tag{313}$$

that is, as the coherent superposition of an amplitude that varies slowly — in this case just the constant 1 in the neighborhood of the resonance, and the rapidly varying Breit-Wigner resonance amplitude $T(E)$. The corresponding rate is what might be called an anti-resonance, as already remarked in connection with (308). The general point illustrated by $R(E)$ is that the appearance of a resonance in a probability (or cross section in scattering theory) can be quite different from a symmetric peak if the resonant *amplitude* interferes with a slowly varying "background" amplitude.

In more complex scattering problems, especially when inelastic processes are present, it can be difficult to discern a resonance by simply looking at probabilities or cross sections, whereas the scattering amplitude itself can often reveal a weak resonance that is masked or distorted by non-resonant effects. This is illustrated with this simple model by Argand plots of the transmission or reflection amplitudes. To understand why this is so, we first write the approximation (310) to the transmission amplitude in the form

$$T(E) = \tfrac{1}{2}\left(e^{2i\delta(E)} - 1\right) , \qquad e^{2i\delta} = \frac{E - E_n - \frac{1}{2}i\Gamma_n}{E - E_n + \frac{1}{2}i\Gamma_n} , \tag{314}$$

and therefore

$$\tan\delta(E) = \frac{\frac{1}{2}\Gamma_n}{E_n - E} . \tag{315}$$

This establishes two facts about the Breit-Wigner form of this amplitude: (i) when plotted as an Argand diagram, it is on a circle of radius $\frac{1}{2}$ centered at $\mathrm{Re}\, T = -\frac{1}{2}$; (ii) the phase $\delta(E)$ rises rapidly through $\pi/2$ (mod 2π) as the energy rises through the resonance, and the location of $T(E)$ therefore traverses the circle in the anticlockwise sense with a "speed" $d\delta/dE$ that is maximum at resonances, i.e., at $T = -1$.

Equation (310) is only an approximation to $T(E)$, and the former will only have the ideal behavior just described in the limit $\lambda \to 0$. This is brought out by Fig. 4.9, which shows the Argand plots for the $\xi \simeq \pi$ resonance near the weak coupling limit of $\lambda = 1/32\pi$, and then the same resonance for $\lambda = 1/8\pi$, where the trajectory departs significantly from the ideal case.

The discussion just given is not just a curiosity about this one-dimensional model, but illustrates a generic feature of scattering amplitudes. In three-dimensional purely elastic scattering, the amplitude can be expressed in terms of phases $\delta_l(E)$ for each separate angular momentum state (see § 8.1) , and if there are narrow resonances, these "phase shifts" have the same behavior as just described. If, moreover, inelastic scattering is possible, the accompanying elastic amplitude will still have such phase shifts, though they will then be complex (see §9.3).

Another important and generic property of scattering amplitudes should be mentioned here: *The amplitudes $T(E)$ and $R(E)$ have a mathematical property that turns out to be of broad physical significance: considered as a function of the complex variable E, they have simple poles at $E_n - i\Gamma_n/2 \equiv W_n$* . As $\Gamma_n \to 0$, (i.e., as $\lambda \to 0$), these poles move onto the real axis at the points E_n^0 where there would be stationary states trapped permanently in the region $|x| \leq a$ (recall Eq. 277). It is therefore natural to suppose that these states have complex energy eigenvalues W_n, and to guess that they would have the time dependence

$$e^{-iW_n t/\hbar} = e^{-iE_n t/\hbar} e^{-\Gamma_n t/2\hbar} , \tag{316}$$

that is, *probabilities that decays exponentially like* $e^{-\Gamma_n t/\hbar}$. On second thought, however, there can be no complex energy eigenvalue because our Hamiltonian is Hermitian! So if this guess (316) is correct, it must mean something more subtle than the existence of an energy eigenstate with a complex energy. The guess is in essence correct, but what it really means remains to be understood.

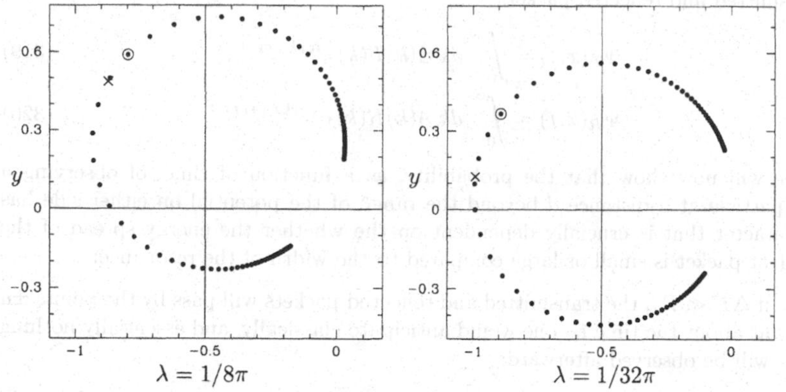

FIG. 4.9. The transmission amplitude $T(E)$ in the resonant region $\xi \simeq \pi$ shown as Argand plots, i.e., as $T = x + iy$, with points at constant intervals of $\Delta ka = 0.002\pi$ and running anticlockwise with increasing E. The true maximum, where $|T| = 1$, is shown as $\times$, and as expected this is where the phase $\delta(E)$, as defined in Eq. 315, increases most rapidly; the location of the resonance in the Breit-Wigner approximation is at $\odot$. If the Breit-Wigner form were exact, these Argand plots would be circles of radius $\frac{1}{2}$ centered at $x = -\frac{1}{2}$. This approximate shape is reasonably close to the truth for the weak coupling case $\lambda = 1/32\pi$, but much less so for $\lambda = 1/8\pi$. Note from Fig. 4.8 that the departure from the Breit-Wigner form is much less visible in $|T(E)|^2$.

(d) The Exponential Decay Law

As we will now learn, the subtle feature is that the exponential decay law is a property of a state having an uncertainty in energy ΔE that is *at least* of order the width Γ_n of the resonance, and *not* of any stationary state.

To establish this important fact, it is necessary to extend the theory to states that provide a realistic description of how a collision develops in time, which, by definition, the stationary states of the preceding treatment cannot do. In outline, we consider a free particle wave packet incident from the left,

$$\Phi(x,t) = \int_0^\infty dk\, A(k)\, e^{i(kx-\epsilon_k t)}, \tag{317}$$

where $\epsilon_k = \hbar k^2/2m$. The momentum space wave function $A(k)$ has a distribution whose mean momentum $\hbar k_n$ correspond to a mean energy at resonance, $\hbar^2 k_n^2/2m = E_n$, and a spread σ in k corresponding to an energy uncertainty ΔE. Because the stationary states $\{\psi_{k,L}\}$ form a complete set of states incident from the left, we know that

$$\Psi(x,t) = \int_0^\infty dk\, A(k)\psi_{k,L}(x)\, e^{-i\epsilon_k t} \tag{318}$$

is an exact solution of the Schrödinger equation for all times — before, during, and after the collision, which reduces to (317) initially. After some time, there will be transmitted and reflected packets:

$$\Psi_T(x,t) = \int_0^\infty dk\, A(k)T(k)\, e^{i(xk-\epsilon_k t)} , \tag{319}$$

$$\Psi_R(x,t) = \int_0^\infty dk\, A(k)R(k)\, e^{-i(kx+\epsilon_k t)} . \tag{320}$$

We will now show that the probability, as a function of time, of observing a free particle at a distance d beyond the range of the potential on either side has a character that is crucially dependent on the whether the energy spread of the incident packet is small or large compared to the width of the resonance:

1. if $\Delta E \ll \Gamma_n$, the transmitted and reflected packets will pass by the points $\pm d$ at about the time t_C one would anticipate classically, and essentially nothing will be observed afterward;

2. on the other hand, if $\Delta E \gg \Gamma_n$, then at about t_C transients packets will quickly pass by $\pm d$, but this prosaic phenomenon will be followed by an exponentially falling probability $\propto \exp(-\Gamma_n t/\hbar)$, which can be made to persist for as long as one wishes by choosing potential parameters that make the width correspondingly small.

The demonstration of these claims is greatly simplified by taking advantage of the fact that we are only concerned with states that have a spread σ in k that is small compared to the resonant value k_n even when $\Delta E \gg \Gamma_n$, which means that the integrals on k can be extended to $-\infty$ so that Cauchy's residue theorem can be used. To not leave doubt that the result is insensitive to the choice of initial state, we assume the following quite general form for the wave packet:

$$A(k) = \frac{1}{2\pi i} \frac{f(k)}{(k-k_n+i\sigma)^{p+1}}\, e^{i\phi} , \qquad \phi = kx_0 + \epsilon_k t_0 , \tag{321}$$

where $f(k)$ is a polynomial, p is an integer that is large compared to the degree of f so as to make the packet sharply peaked about k_n, and the phase ϕ places the packet at $x \simeq -x_0$ at $t \simeq t_0$. In view of this peaking in k, we approximate the energy in the phase $\exp(-i\epsilon_k t)$:

$$\hbar\epsilon_k \simeq \frac{\hbar^2}{2m}[k_n^2 + 2k_n(k-k_n)] = -E_n + \hbar v k , \tag{322}$$

where $v = \hbar k_n/m$ is the velocity at the resonant energy. The time evolution of the incident packet is then

$$\Phi(x,t) = e^{iE_n t/\hbar} \int_{-\infty}^{\infty} \frac{dk}{2\pi i} \frac{e^{-iku} f(k)}{(k-k_n+i\sigma)^{p+1}} , \qquad u \equiv v(t-t_0) - (x+x_0) . \tag{323}$$

Cauchy's theorem, and the fact that the only singularity is the $(p+1)^{\text{th}}$ order pole in the lower half-plane, means that this integral vanishes for $u < 0$, and for $u > 0$ has the value

$$\Phi(x,t) = \theta(u) e^{iE_n t/\hbar} \frac{1}{p!} \left(\frac{\partial}{\partial k}\right)^p [f(k)e^{-iku}]_{k=k_n-i\sigma} \tag{324}$$

$$= \theta(u) e^{i\alpha}\, Q_p(u)\, e^{-ik_n u}\, e^{-\sigma u} , \tag{325}$$

where $Q_p(u)$ is a polynomial of degree p and α an irrelevant real phase. In words, this is a packet with a sharp wavefront moving without distortion at velocity v which, after passing any fixed point, has a probability that falls off like $(t^p e^{-\sigma v t})^2$ at that point as $t \to \infty$.

The transmitted packet is computed from (319) by using (306) for the transmission amplitude $T(k)$:

$$\Psi_T(x,t) = e^{iE_n t/\hbar} \int_{-\infty}^{\infty} \frac{dk}{2\pi i} \frac{e^{-iku} f(k)}{(k - k_n + i\sigma)^p} \frac{-i\nu_n}{k - k_n + i\nu_n} , \tag{326}$$

where $\nu_n = \eta_n / a$. Now there are two singularities in the lower half-plane: the one in the initial wave function $A(k)$ at $k = k_n - i\sigma$ which led to (325), and the other in the transmission amplitude at $k = k_n - i\nu_n$, giving a result with two terms:

$$\Psi_T(x,t) = \theta(u) e^{i(\alpha - k_n u)} [\psi_R(u) + \psi_A(u)] , \tag{327}$$

$$\psi_R(u) = \frac{f(k_n - i\nu_n)}{(i\sigma - i\nu_n)^{p+1}} e^{-\nu_n u} , \qquad \psi_A(u) = \tilde{Q}_p(u) e^{-\sigma u} , \tag{328}$$

where ψ_R is due to the resonance pole in the transmission amplitude, and ψ_A due to the pole in $A(k)$, with $\tilde{Q}_p$ a somewhat different polynomial of degree p than the one in (325).

These two wave functions have a very different behavior in time. The term due to the resonance, ψ_R, is a pure exponential,

$$|\psi_R|^2 = Ce^{-t/\tau} , \tag{329}$$

where $\tau = 1/2v\nu_n$ is called *the lifetime of the resonance*, the relation between lifetime and width being

$$\tau = \frac{\hbar}{\Gamma_n} \tag{330}$$

Note that aside from an overall constant, *this function of time does not depend on the initial state and is completely determined by the resonance itself.* The other term gives a probability that depends in detail on the incident state,

$$|\psi_A|^2 \propto t^{2p} e^{-t/\Delta t} , \qquad \Delta t = 1/2v\sigma \sim \hbar/\Delta E , \tag{331}$$

where Δt is the time the incident packet requires to pass a point.[1]

Thus (327) confirms our claim: *if the incident state has an energy uncertainty large compared to the width of the resonance, $\Delta E \gg \Gamma_n$, the probability $P_T(t)$ for observing the particle on the far side of the potential vanishes until the time when*

---

[1] To maintain focus on the central issue, the preceding discussion has purposely glossed over two facts: (i) the free particle Schrödinger equation cannot produce sharp wavefronts because the dispersion law $E \propto k^2$ implies velocities that have no upper bound; (ii) wave forms that do not change their shape in time can only be produced by a restricted set of potentials (see "coherent states" in the index). The seemingly timid approximation (322) does not respect these truths. However, in §2.7(e) (see also Prob. 20) we learned that the unavoidable smearing of the wavefront takes a time of order $m\hbar/(\Delta p)^2 \simeq m/\hbar\sigma^2$, and an appropriate combination of the mass and spread σ can make that a large enough interval for the packet to travel from x_0 to the observation points while maintaining roughly its initial shape, apart of course from the dramatic effect due to the resonance.

the packet would be expected to arrive classically; there is then a complicated rise and fall-off in P_T *that lasts for a time* Δt *and whose detailed structure depends on the initial wave function; and a pure exponential* $e^{-\Gamma_n t/\hbar}$ *that persists long after the first transient structure has subsided. On the other hand, if the incident energy uncertainty is small compared to the width, the entire packet will pass any point in a time of order* $\hbar/\Delta E$ *and not display the lifetime.*

These conclusions provide another illustration of the *energy-time uncertainty principle:* the characteristic time τ can only be measure in a state that has an uncertainty in energy that is at least of order $\hbar/\tau$ (recall §2.4(d)).

The reflected amplitude Ψ_R is evaluated in the same way from (320), as sketched in Prob. 24. If $\Delta E \gg \Gamma_n$, there is again a transient that quickly subsides, followed by the same long-lived pure exponential as in the transmitted amplitude Ψ_T. But Ψ_R has an additional and important feature: *if the resonance has even parity, the long-lived portion of the full amplitude* $\Psi_R + \Psi_T$ *is symmetric about* $x = 0$*, whereas it is antisymmetric if the parity is odd.* This illustrates the important fact that after the transients associated with the state that produces the resonance has passed by, the amplitude associated with the resonance displays the symmetries that characterize the resonance, making it legitimate to think of it as a quasi-stationary state that exists for a time of order the lifetime.

Our simple example of resonant scattering serves as a schematic model of long-lived states, such as nuclei that undergo α-decay. Imagine bombarding the stable nucleus Pb^{206} with an α-particle beam of energy 5.4 MeV and an incredibly narrow spread in energy of ~ 1 eV, corresponding to a passage time $\Delta t \sim 10^{-16}$ sec. After the cyclotron is turned off, the transient will disappear almost instantly, and then be followed by an exponentially falling number of α-particles emanating from the target with a lifetime of 138 days. The target can be shipped off to a distant point where chemical analysis will reveal that it contains some amount of the element $Z = 84$, and mass spectroscopy will identify the isotope Po^{210}. It would be possible, indeed perfectly correct, to say that this is an $\alpha-Pb^{206}$ resonance of width 10^{-18} eV, but it is far more sensible to think of the phenomenon as lead nuclei having "captured" an α-particle to form nuclei of polonium, because for times of order 20 weeks the latter is to all intents and purposes a stationary state.

4.5 The Semiclassical Approximation

The semiclassical approximation is especially powerful in problems that can be treated as if just one dimension were in play. There are realistic situations where this does not entail any approximation, the most common being two-body scattering due to a spherically symmetric interaction. In virtue of the symmetry, this problem separates into dynamically independent problems for each angular momentum state (or partial wave), described by one-dimensional Schrödinger equations in the radial coordinate. The same holds whenever a symmetry allows such a separation of variables. There are also many phenomena in which a one-dimensional model can serve as a reasonable or even accurate surrogate for the actual problem. In short, the one-dimensional problem can provide invaluable insights for attacking far more difficult multi-dimensional cases.

(a) The WKB Approximation

The time-independent Schrödinger equation of interest is

$$\frac{d^2\Psi}{dx^2} + \frac{p^2(x)}{\hbar^2}\Psi = 0 , \tag{332}$$

where

$$p(x) = \Big(2m[E - V(x)]\Big)^{\frac{1}{2}} , \tag{333}$$

is the local momentum and E is the energy eigenvalue. The one-dimensional semi-classical approximation is also called the WKB approximation after its inventors Brillouin, Kramers and Wentzel. As we learned in §2.8(b) (see Eq. 468), this approximation requires the local momentum $p(x)$ to satisfy

$$\hbar|dp(x)/dx| \ll |p(x)|^2 , \tag{334}$$

which is equivalent to the requirement that the local de Broglie wavelength varies slowly enough to satisfy $|d\lambda/dx| \ll 1$. An equivalent inequality is

$$\frac{1}{2}\left|\lambda(x)\frac{dV}{dx}\right| \ll \frac{p^2(x)}{2m}, \tag{335}$$

which states that the local kinetic energy must be large compared to the change of potential energy when the position changes by a de Broglie wavelength.

For those who have not yet read §2.8, a derivation of the results just stated runs as follows. Assume that the wave function has the form

$$\Psi(x) = \exp\left\{\frac{i}{\hbar}\left(W(x) + \frac{\hbar}{i}W_1(x) + O(\hbar^2)\right)\right\} . \tag{336}$$

Substituting into (332) gives

$$(W')^2 - i\hbar(W'' + 2W_1'W') + O(\hbar^2) = p^2 . \tag{337}$$

If the $O(\hbar)$ term is dropped,[1]

$$|W'(x)| = |p(x)| , \tag{338}$$

and the approximation is valid provided (334) is satisfied.

Equation (338) can be integrated:

$$W(x) = \pm\int^x dx'\, p(x') , \tag{339}$$

where, for the moment, it is assumed that the integral is in a classically allowed region, so that $E \geq V(x)$. The coefficient of the $O(\hbar)$ term in (337) must also vanish, which means that

$$W_1'(x) = \frac{d}{dx}\ln(W')^{-\frac{1}{2}} , \tag{340}$$

[1] Those who have read §2.8(a) should note that when the action S, in the one-dimensional Hamilton-Jacobi equation, is replaced by $W - Et$, it reduces to (338) when $p(x)$ is given by (333).

and therefore

$$e^{W_1(x)} = 1/\sqrt{p(x)} \,. \tag{341}$$

To summarize, in classically allowed regions, the linearly independent WKB solutions to (332) are

$$\psi_\pm(x) = \frac{1}{|p(x)|^{\frac{1}{2}}} \exp\left(\pm\frac{i}{\hbar}\int^x dx'\, p(x')\right) \equiv \frac{1}{|p(x)|^{\frac{1}{2}}}\, e^{\pm iw(x)} \,, \tag{342}$$

where $w(x)$ is real. In classically forbidden regions, $w(x)$ is pure imaginary and the functions $\psi_\pm$ fall or rise exponentially. The general solution is

$$\psi = A\psi_+ + B\psi_- \,. \tag{343}$$

Note that $|\psi_\pm(x)|^2 = 1/p(x)$ was to be expected: the probability of being in the segment $(x, x+dx)$ is proportional to the time required to traverse it, and therefore inversely proportional to the velocity.

(b) Connection Formulas

The approximations $\psi_\pm$ suffer from two diseases. The first is obvious: $\psi_\pm$ diverge when $p(x) = 0$, i.e., at classical turning points, so in the neighborhood of such points the basic condition (334) is egregiously violated. As turning points occur in the majority of problems, the approximate solutions will only be useful if it is possible to *connect* solutions that are valid on either side of a turning point and sufficiently far away from it so that (334) is satisfied.

The second disease only becomes apparent when the differential equation (332) is extended from real values of x to the complex z-plane. This may sound arcane and irrelevant, but that is not so: it must be understood if the first disease is to be cured, because the cure involves a detour into the complex plane around the turning point. This maneuver will produce the so-called *WKB connection formulas*, which must be used in any problem involving turning points.

A theorem about differential equations of the type (332) states that if $V(z)$ is an entire function, then the exact solution $\Psi(z)$ is also entire, where "entire" means a function that is analytic everywhere except for singularities at infinity. Assume that $V(z)$ is entire; it will quickly be evident that this restriction is of little consequence here. The approximations $\psi_\pm(z)$ are not single-valued functions, and violate the theorem. If there is a turning point at x_0, the pre-factor $[E - V(z)]^{-1/4}$ is of the form $(z - x_0)^{-1/4}$ near that point, which introduces a branch cut at $z = x_0$. As we shall soon see, the functions $\exp[\pm iw(z)/\hbar]$ also have cuts emanating from turning points. All such cuts are artifacts of the approximations that led to (342). The connection formulas are precisely those relationships between approximate solutions in various regions of the z-plane which insure that the global approximate solution, applicable in both the classically allowed and forbidden regions, is single-valued.[1]

That the assumed analyticity of V has no bearing on our problem is now clear. If $V(z)$ is not entire, that will be reflected in Ψ. However, such singularities exist

[1]There is a connection between this issue and the problem of stitching together semiclassical propagators when there are crossings between classical paths, which was mentioned in §2.8; see the bibliography at the end of chapter 2.

whether or not there are turning points, whereas the approximate WKB expressions $\psi_\pm$ always have singularities at turning points.

Consider one isolated turning point which, without loss of generality, is at $x = 0$, and the case where $x > 0$ is a classically allowed region. Assume also that $dV(x)/dx$ does not vanish by accident at the turning point, so that $p^2(z) = a^2 z + O(z^2)$ near $z = 0$, where a is real. Call $\bar{p}^2(z) = a^2 z$, and *define* the function

$$\bar{w}(z) = \frac{1}{\hbar}\int_0^z \bar{p}(z')\,dz' = \frac{2}{3\hbar}|a|r^{3/2}e^{3i\phi/2} \equiv b(r)\,e^{3i\phi/2}\,, \tag{344}$$

where the integration runs from the turning point to $z = re^{i\phi}$; do not forget that $b(r)$ is proportional to $1/\hbar$. This function is not $w(z)$ in Eq. 342 because the higher-order terms in the Taylor expansion of $p^2(z)$ have been dropped. But it has the same analytic structure as the true function, which is all that matters to the issue at hand, and therefore we ignore the distinction between $\bar{w}(z)$ and $w(z)$ for now. In this sense, the linearly independent WKB solutions are then

$$\psi_\pm(z) = z^{-\frac{1}{4}}\,\varphi_\pm(z)\,, \tag{345}$$

$$\varphi_+(z) = e^{ib(r)\cos\frac{3}{2}\phi}\,e^{-b(r)\sin\frac{3}{2}\phi}, \qquad \varphi_-(z) = e^{-ib(r)\cos\frac{3}{2}\phi}\,e^{b(r)\sin\frac{3}{2}\phi}\,. \tag{346}$$

The key to the problem is the dependence of these functions on $\arg z = \phi$: they are not single-valued. Take the linear combination

$$\psi_I(z) = A_I\psi_+(z) + B_I\psi_-(z)\,. \tag{347}$$

When $z \to ze^{2\pi i}$, it does not return to itself:

$$\psi_I(z) = r^{-\frac{1}{4}}\left(A_I e^{ib(r)} + B_I e^{-ib(r)}\right) \qquad (\phi = 0)\,, \tag{348}$$

$$= r^{-\frac{1}{4}}\,e^{-i\pi/2}\left(A_I e^{-ib(r)} + B_I e^{ib(r)}\right) \qquad (\phi = 2\pi)\,. \tag{349}$$

As already stated, this multivalued property has two sources: from the obvious branch point in the pre-factor $z^{-1/4}$, and from $\varphi_\pm(z)$, that is, from the the action $w(z)$.

The latter is the subtle matter. According to (346), $w(z)$ is real and $|\varphi_+| = |\varphi_-|$ at $\phi = (0, 2\pi/3, 4\pi/3)$, but $w(z)$ is complex between these values of ϕ, so that in the sectors S_I, S_{II}, S_{III}, one of the functions falls, the other grows exponentially as $\left(ar^{3/2}/\hbar\right) \to \infty$. The exponentially growing function is called dominant, the other sub-dominant; Fig. 4.10 shows which one dominates in each sector. Here 'dominance' is not a vague figure of speech, but has a specific meaning. The WKB approximants $\psi_\pm(z)$ are the leading terms in the asymptotic expansions of the linearly independent solutions of Eq. 332 as $z \to \infty$, i.e., as $\hbar \to 0$. The general theory of differential equations shows that the corrections to the asymptotic forms (345) are smaller by inverse powers of z,

$$\psi_\pm(z) = z^{-\frac{1}{4}}\,\varphi_\pm(z)\Big(1 + C_\pm(z) + \ldots\Big)\,, \qquad C_\pm(z) \sim z^{-k_\pm} \text{ as } z \to \infty\,. \tag{350}$$

Calling, for example, $\psi_+(z)$ dominant in S_{II} is shorthand for the statement that $|\varphi_-(z)| \ll |\varphi_+(z)C_+(z)|$ as $z \to \infty$ for $\frac{2}{3}\pi + \Delta < \phi < \frac{4}{3}\pi - \Delta$. Therefore in any

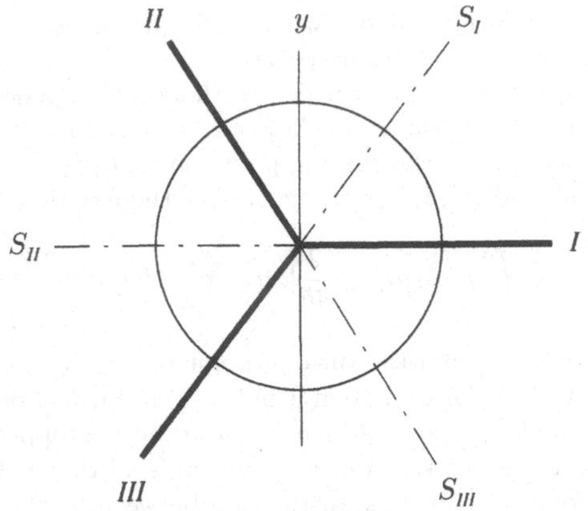

FIG. 4.10. The Stokes phenomenon, which is involved in the formulas that connect WKB solutions $\psi_\pm(x)$ across a classical turning point, which is here at $x = 0$. As functions of complex z, these solutions have different asymptotic behavior, as $\hbar \to 0$, in the three sectors with boundaries at arg $z = 0, 2\pi/3, 4\pi/3$. To connect a solution for $x > 0$ to another for $x < 0$ requires a traversal of the boundary between sectors.

sector and not too near its edges, the sub-dominant WKB function is overshadowed by the *unknown error* in the dominant function! Furthermore, in moving from S_I to S_{II}, and from S_{II} to S_{III}, the dominant and sub-dominant roles of ψ_+ and ψ_- interchange. This remarkable character of asymptotic expansion of the solutions of differential equations of type (332) is called the Stokes phenomenon.

A valid approximate solution of (332) must be single-valued, and thus be stitched together from solutions in the three sectors so that the resulting quilt is single-valued. Let $\psi_K(z)$ be the solution in S_K:

$$\psi_K(z) = A_K\psi_+ + B_K\psi_- \ . \tag{351}$$

The coefficients A_K and B_K will be chosen to produce the single-valued global solution. The first step is to relate (A_{II}, B_{II}) to (A_I, B_I) by analytically continuing ψ_I from boundary I to boundary II of S_I. Because (332) is a second-order linear equation, any pair of linearly independent solutions can be expressed as a linear combination of any other pair, and therefore the two sets of coefficients are linear combinations of each other. But $\psi_-(z)$ is dominant in S_I, and therefore continues unambiguously from one boundary to the other, so $B_{II} = B_I$. This is not so for $\psi_+(z)$; deep in S_I it is submerged by the unknown error in $\psi_-(z)$, so when it reappears with a magnitude comparable to $\psi_-(z)$ as II is approached it has become an unknown combination of the two independent functions originally defined on I, and therefore $A_{II} = cA_I + \alpha B_{II}$. There is a special, and as we shall see very important case, however, namely, $B_{II} = 0$. When that is so the dominant function in S_I has been eliminated, and there is no ambiguity in continuing the sub-dominant $\psi_+(z)$, which thus requires $A_I = A_{II}$ when $B_{II} = 0$, or $c = 1$. The connection between ψ_I and ψ_{II} is therefore

$$A_{II} = A_I + \alpha B_I \ , \qquad B_{II} = B_I \ , \tag{352}$$

In S_{II} the dominant function is $\psi_+(z)$, while in S_{III} it is $\psi_-(z)$, and the same argument therefore gives

$$A_{III} = A_{II} \ , \quad B_{III} = B_{II} + \beta A_{II} \ ; \quad A_{I'} = A_{III} + \gamma B_{III} \ , \quad B_{I'} = B_{III} \ , \tag{353}$$

where $(A_{I'}, B_{I'})$ are the coefficients on I following the continuation in ϕ from 0 to 2π. From (349), however,

$$\psi_{I'}(z) = -i\left(A_{I'}e^{-ib(r)} + B_{I'}e^{ib(r)}\right) , \qquad (\phi = 2\pi) , \tag{354}$$

and comparing with (348) shows that $(A_{I'} = iB_I, B_{I'} = iA_I)$ will produce the desired single-valued result. A simple exercise now yields the complete answer:

$$\alpha = \beta = \gamma = i \ . \tag{355}$$

We can now connect solutions in the classically allowed region, which is the boundary I, to those in the forbidden region, the negative real axis, which lies in S_{II}. From (352) and (355), the solution for $x < 0$ is

$$\psi_{II}(x) = z^{-\frac{1}{4}}\left((A_I + iB_I)\psi_+(z) + B_I\psi_-(z)\right) , \qquad z = re^{i\pi} . \tag{356}$$

When the energy is positive, i.e., in scattering problems, both rising and falling real exponentials are needed when the solution enters into a forbidden region, as illustrated by Fig. 4.11. The solution that decreases to the left of the turning point

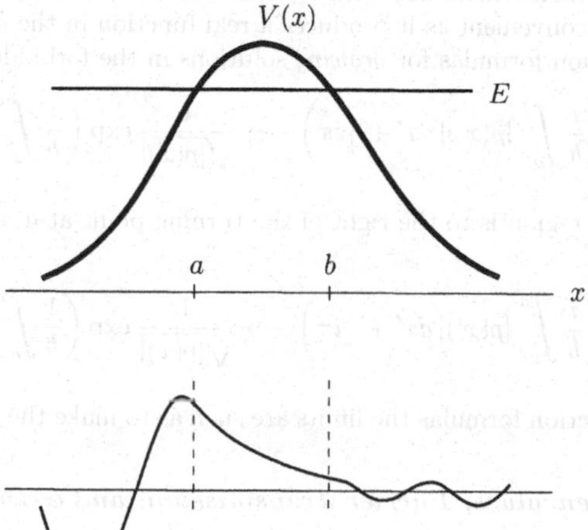

FIG. 4.11. The WKB wave function for a particle incident from the left on a barrier with energy E insufficient to overcome it classically.

is sub-dominant in S_{II}, so $A_I + iB_I = 0$ is required. Setting $A_I = e^{-i\pi/4}$ gives

$$\psi_I(x) = 2B_I[p(x)]^{-1/2}\cos[w(x) - \tfrac{1}{4}\pi] , \qquad \psi_{II}(x) = B_I[p(x)]^{-1/2}\exp[-|w(x)|] . \tag{357}$$

The relationship between these two forms is expressed by the *connection formula*

$$\frac{1}{\sqrt{|p(x)|}}\exp\left(-\frac{1}{\hbar}\int_x^a |p(x')|dx'\right) \longrightarrow \frac{2}{\sqrt{|p(x)|}}\cos\left(\frac{1}{\hbar}\int_a^x |p(x')|\,dx' - \tfrac{1}{4}\pi\right). \tag{358}$$

Here the turning point is at an arbitrary place $x = a$, the allowed region is $x > a$ to the right of the turning point, and the actual momentum $p(z)$ has replaced the approximation $\propto z^{1/2}$ valid only near the turning point. The corresponding connection formula for the case where the allowed region is to the left of a turning point is

$$\frac{1}{\sqrt{|p(x)|}}\exp\left(-\frac{1}{\hbar}\int_a^x |p(x')|dx'\right) \longrightarrow \frac{2}{\sqrt{|p(x)|}}\cos\left(\frac{1}{\hbar}\int_x^a |p(x')|\,dx' - \tfrac{1}{4}\pi\right). \tag{359}$$

The arrow in these connection formulas means that when the left-hand side is known in the forbidden region, it continues to the right-hand side in the allowed region, but that the converse is false because even a minute change in the phase on the right side will produce a dominant, exponentially growing component on the left which would eclipse the desired sub-dominant function. It should also not be forgotten that in these relationships x refers to *two different* values of the x-coordinate on either side of a turning point and at a finite distance from it.

The formula for a solution which grows exponentially in a forbidden region as the distance to the turning point increases is found by selecting the solution in S_I that is purely sub-dominant there, i.e., with $B_I = 0$. The choice of A_I is now arbitrary, but $A_I = e^{i\pi/4}$ is convenient as it produces a real function in the forbidden region. Then the connection formulas for *growing* solutions in the forbidden region are

$$\frac{1}{\sqrt{|p(x)|}}\exp\left(\frac{i}{\hbar}\int_a^x |p(x')|\,dx' + \tfrac{1}{4}i\pi\right) \longrightarrow \frac{1}{\sqrt{|p(x)|}}\exp\left(\frac{1}{\hbar}\int_x^a |p(x')|dx'\right), \tag{360}$$

when the allowed region is to the right of the turning point at a, and when it is to the left, it reads

$$\frac{1}{\sqrt{|p(x)|}}\exp\left(\frac{i}{\hbar}\int_x^a |p(x')|\,dx' + \tfrac{1}{4}i\pi\right) \longrightarrow \frac{1}{\sqrt{|p(x)|}}\exp\left(\frac{1}{\hbar}\int_a^x |p(x')|dx'\right). \tag{361}$$

In all these connection formulas the limits are such as to make the integral positive.

(c) Energy Eigenvalues, Barrier Transmission, and α-Decay

In the case of a bound state, we have two expressions for the wave function in the region between the turning points. To the right of $x = a$ it is given by the right-hand side of (358), while to the left of $x = b$ by the right-hand side of (359). These must be two forms of the same function, or

$$C\cos\left(\frac{1}{\hbar}\int_a^x p\,dx' - \tfrac{1}{4}\pi\right) - C'\cos\left(\frac{1}{\hbar}\int_x^b p\,dx' - \tfrac{1}{4}\pi\right) = 0\,, \tag{362}$$

where clearly $|C| = |C'|$. In one dimension the wave functions can always be chosen to be real, so $C' = \pm C$, and therefore (362) is of the form $\cos A \pm \cos B$, which vanishes when $A+B=n\pi$. Hence whenever

$$\frac{1}{\hbar}\int_a^b p(x)\,dx=(n+\tfrac{1}{2})\pi\,, \qquad n=0,1,2,\dots\,, \tag{363}$$

there is a bound state. Another way of expressing this is

$$\frac{1}{2\pi\hbar}\oint p(x)\,dx=n+\tfrac{1}{2}\,, \tag{364}$$

where the integral runs over a complete orbit in phase space, i.e., $a\to b\to a$. For this reason it is called the Bohr-Sommerfeld quantization rule, though the original rule by that name in the Old Quantum Theory had n on the right, a difference that is of questionable significance because the correspondence principle warns us that the semiclassical approximation is not reliable for low-lying states. In this connection, note that n is the number of zeros, or nodes, in the wave function. As is usually the case for the oscillator, the semiclassical result (363) is exact.

The amplitude for transmission through a barrier also follows from the connection formulas *provided* the two turning points in Fig. 4.11 are not too close together, as is intuitively clear from the considerations related to Fig. 4.10. When two or more turning points are in proximity, their cuts run into each other, so to say, and a more sophisticated analysis is required. In view of this, the simple WKB approximation as developed here cannot handle the problem when the energy is close to the top of the barrier, let alone account for imperfect transmission when E is above the barrier.

Let the transmitted wave run to the right, with a phase such that its wave function is the left-hand side of (360):

$$\psi_{\rm tr}(x)=\frac{T}{\sqrt{|p(x)|}}\exp\left(\frac{i}{\hbar}\int_b^x|p(x')|\,dx'+\tfrac{1}{4}\pi\right), \tag{365}$$

where T is the transmission amplitude, which will be determined by requiring unit incident amplitude. In the forbidden region $|\psi(x)|$ increases as x moves to the left of the turning point, and (360) is therefore the appropriate connection formula:

$$\begin{aligned}\psi_{\rm tr}&\longrightarrow\frac{T}{\sqrt{|p(x)|}}\exp\left(\frac{1}{\hbar}\int_x^b|p(x')|\,dx'\right)\\&=\frac{T}{\sqrt{|p(x)|}}\exp\left(\frac{1}{\hbar}\int_a^b|p(x')|\,dx'\right)\exp\left(-\frac{1}{\hbar}\int_a^x|p(x')|\,dx'\right).\end{aligned} \tag{366}$$

This then connects to the wave function $\psi_L(x)$ on the left of the barrier via Eq. 359:

$$\psi_L(x)=T\exp\left(\frac{1}{\hbar}\int_a^b|p(x')|\,dx'\right)\frac{2}{\sqrt{|p(x)|}}\cos\left(\frac{1}{\hbar}\int_a^x p(x')\,dx'+\tfrac{1}{4}\pi\right). \tag{367}$$

This is a superposition of waves of equal amplitude incident on and reflected by the barrier. Requiring unit incident amplitude determines T,

$$T=\exp\left(-\frac{1}{\hbar}\int_a^b|p(x')|\,dx'\right). \tag{368}$$

When $V(x) \to 0$ as $|x| \to \infty$, the velocity of the incident and transmitted beams are equal, and the ratio of transmitted to incident flux is

$$\frac{i_{\rm tr}}{i_{\rm inc}} = |T|^2 . \tag{369}$$

The energy dependence of T is implicit, but usually decreases exponentially because as E drops the distance between the turning points increases as does the local momentum $|p(x)|$.

Why does our result (367) claim that the reflected and incident fluxes are equal? This is obviously incorrect if $T \neq 0$. The error is a consequence of the WKB approximation, which cannot keep track of the sub-dominant amplitudes: they are of higher order in the approximation. As we know from §4.4, probability conservation requires the sum of the reflected and transmitted fluxes to equal the incident flux. Hence the exact wave function far to the left of the barrier is

$$\psi_L(x) = e^{i[(px/\hbar)+\frac{1}{4}\pi]} + (1-|T|^2)^{\frac{1}{2}}\, e^{i\gamma}\, e^{-i[(px/\hbar)+\frac{1}{4}\pi]} , \quad x \to -\infty , \tag{370}$$

where $E = p^2/2m$, and γ cannot be determined from this argument. Thus our result is only valid if the barrier is almost perfectly reflective, and therefore allows little transmission, or

$$1 \gg \exp\left(-\frac{2}{\hbar}\int_a^b |p(x')|\, dx'\right) . \tag{371}$$

This is a much less stringent requirement on the exponent, of course. When (371) is satisfied, the reflected amplitude is $(1-\frac{1}{2}|T|^2)$, i.e., the departure from perfect reflection is of higher order than the transmitted amplitude. Equation (371) is also consistent with the requirement that the turning points be widely separated. When E is near the top of the barrier, the turning points are close, transmission is large, and this solution fails. It also fails when E is not far above the top, because it does not give any reflected wave when there are no turning points.

For energies far above the barrier's top, however, the WKB approximation does give a shift in phase of the transmitted wave, namely $\psi \sim e^{i\Phi}e^{ikx}$ as $x \to \infty$, where

$$\Phi = k\int_{-\infty}^{\infty}\left(\left[1-\frac{V(x)}{E}\right]^{\frac{1}{2}} - 1\right) dx \simeq -\frac{\hbar}{v}\int_{-\infty}^{\infty} V(x)\, dx . \tag{372}$$

The absence of a reflected wave is consistent with the equality of incident and transmitted flux in this approximation.

The approximation Eq. 368 for the transmission amplitude was originally used by Gamow to explain that α-decay is due to the tunneling of the decay product through the nuclear Coulomb barrier. Transmission through this barrier is also the crucial factor in thermonuclear reactions. In a decay process, there is only a Coulomb interaction between daughter and parent when they have a separation larger then the nuclear radius R,

$$V(r) = \frac{zZe^2}{4\pi r} , \tag{373}$$

where z is the charge of the decay product, i.e., $z = 2$ in α-decay. For separations smaller than R the interaction involves many degrees of freedom. Nevertheless, it is possible to get a crude but instructive estimate of the decay rate $1/\tau$ by expressing it as the product of the probability of transmission through the Coulomb barrier, as given by (367), times the frequency ω with which the imprisoned α-particle strikes the barrier:

$$\frac{1}{\tau} = \omega \exp\left\{-2\int_R^b dr \sqrt{\frac{2\mu}{\hbar^2}[V(r)-E]}\right\}, \tag{374}$$

where μ is the reduced mass of the parent and daughter, E is the energy of the decay product and b the exterior classical turning point (see Fig. 4.12). The frequency

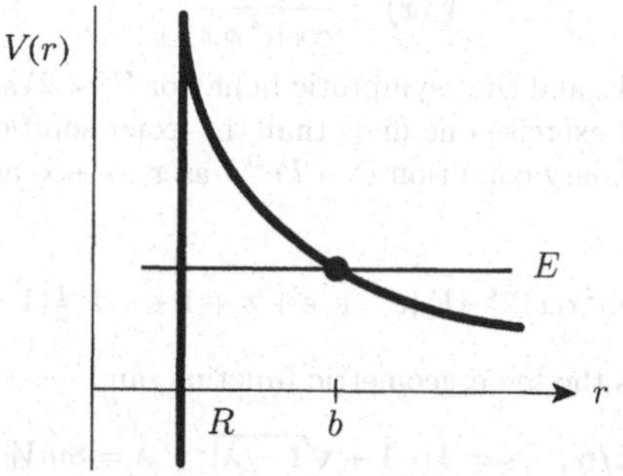

FIG. 4.12. The kinetic and potential energies in α-decay. Here and in Eq. 374 we assume S-wave decay; for $L \neq 0$ the centrifugal barrier must be added to $V(r)$.

ω can be expressed as the ratio $\bar{v}/2R$, where $\bar{v}$ is a characteristic velocity inside the nucleus. The virtue of the expression (374) is that the barrier transmission probability, which is respectably estimated by the WKB approximation, captures the very rapid variation of the lifetime as a function of the decay energy,[1] while the ill-defined frequency ω is seen to be relatively insensitive and can be treated as a phenomenological parameter. If the potential in Fig. 4.12 is assumed to hold as its stands down to $r = R$, the transmission amplitude can be evaluated analytically, which then yields a simple formula for the very rapid variation of the lifetime with energy and nuclear radius (see Prob. 25).

(d) Exactly Solvable Examples

Given the intricacy of the WKB approximation, and its somewhat disquieting answer for the wave reflected from a barrier, it is fortunate that there are very instructive exactly solvable problems.[2]

One such problem is reflection from a parabolic barrier, $V(x) = -\frac{1}{2}kx^2$, an upside-down harmonic potential. It is solvable because the Hamiltonian is a quadratic form, though not positive definite. The transmission coefficient in this case is

$$|T|^2 = \frac{1}{1+\exp[-2\pi E(m/k)^{\frac{1}{2}}/\hbar]}, \tag{375}$$

[1]See J.M. Blatt and V.F. Weisskopf, *Theoretical Nuclear Physics*, Wiley (1952), p. 575.
[2]The following material draws on *LLQM*, pp. 73–74, 80–81.

which holds for both positive and negative E, and therefore gives a reflected wave even when the energy is well above the top of the barrier. When the energy equals the top of the barrier $|T|^2 = \frac{1}{2}$, which is independent of $\hbar$ and therefore completely beyond the WKB approximation. For E large and negative, (375) agrees with the WKB result. The solution to the parabolic barrier can also be used as the basis for an extension of the WKB approach to two nearby turning points (see the bibliography).

A more realistic but still solvable example is provided by the repulsive Morse potential

$$V(x) = \frac{V_0}{\cosh^2 \alpha x} , \tag{376}$$

which has the height V_0 and the asymptotic behavior $V \sim 2V_0 \exp(-2\alpha|x|)$. After a bracing mathematical exercise one finds that the exact solution of the Schrödinger equation, for the boundary condition $\psi \sim Te^{ikx}$ as $x \to +\infty$ and $\psi \sim e^{ikx} + Re^{-ikx}$ as $x \to -\infty$, is

$$\psi(x) = C\,(1 - \tanh^2 \alpha x)^{\frac{1}{2}\epsilon}\; {}_2F_1(\epsilon - s, \epsilon + s + 1; \epsilon + 1; \tfrac{1}{2}(1 - \tanh \alpha x)) , \tag{377}$$

where ${}_2F_1(a, b; c; z)$ is the hypergeometric function, and

$$\epsilon = -ik/\alpha , \quad s = \tfrac{1}{2}[-1 + \sqrt{1 - \lambda}] , \quad \lambda = 8mV_0/\hbar^2\alpha^2 . \tag{378}$$

Note that s is complex if $\lambda > 1$, and that the classical limit $\hbar \to 0$ amounts to $\lambda \gg 1$. The exact transmission and reflection amplitudes are

$$T = \frac{\Gamma(\epsilon - s)\Gamma(\epsilon + s + 1)}{\Gamma(-\epsilon)\Gamma(\epsilon + 1)} , \tag{379}$$

$$R = \frac{\Gamma(\epsilon)\Gamma(\epsilon - s)\Gamma(\epsilon + s + 1)}{\Gamma(-\epsilon)\Gamma(s + 1)\Gamma(-s)} ; \tag{380}$$

these expressions are valid for all values of λ. By using properties of the gamma function it is possible to express the transmitted flux in terms of elementary functions:

$$|T|^2 = \frac{\sinh^2(\pi k/\alpha)}{\sinh^2(\pi k/\alpha) + \cos^2(\frac{1}{2}\pi\sqrt{1 - \lambda})} , \qquad \lambda < 1 , \tag{381}$$

$$= \frac{\sinh^2(\pi k/\alpha)}{\sinh^2(\pi k/\alpha) + \cosh^2(\frac{1}{2}\pi\sqrt{\lambda - 1})} , \qquad \lambda > 1 . \tag{382}$$

It is rather surprising that T does not have any special behavior as the energy passes through the top of the barrier V_{max} in either of these exact solutions. One might have expected something unusual given that there is no reflection when $E > V_{\mathrm{max}}$ in the WKB approximation.

The WKB approximation (368) for transmission through the barrier (376) gives

$$|T_0|^2 = \exp\Big(-\pi\lambda^{\frac{1}{2}}\,[1 - (E/V_0)^{\frac{1}{2}}]\Big) = \exp\left[-\pi\left(\lambda^{\frac{1}{2}} - 2\frac{k}{\alpha}\right)\right] , \tag{383}$$

which is only valid if the wavelength is short ($k \gg \alpha$) and the exponent is appreciably larger than 1 because $|T_0|^2 \ll 1$ is required. The exact result (382) agrees with

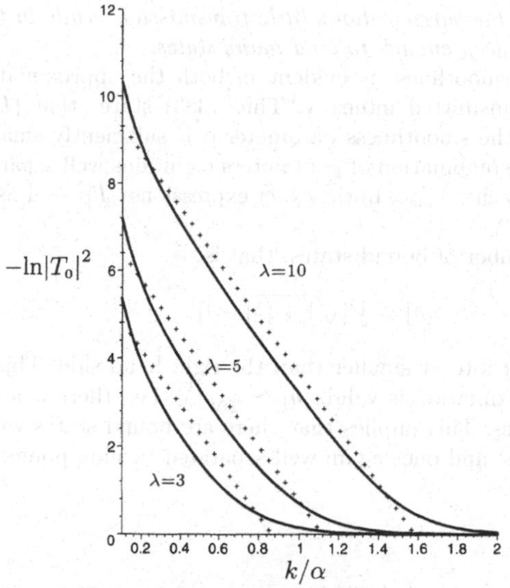

FIG. 4.13. Comparison between the exact transmission probability, Eq. 382, shown as solid lines, and the WKB approximation, Eq. 383, shown as crosses, for the Morse potential, Eq. 376.

this WKB answer in the limit $\lambda \gg 1$, i.e., $\hbar \to 0$, $\pi k \gg \alpha$, provided $\lambda^{\frac{1}{2}} \gg 2k\alpha$. Comparisons between the exact and approximate results are shown in Fig. 4.13. It is quite remarkable that the WKB approximation gives such a good account of $\ln|T|^2$ (though of course much less so of $|T|^2$ itself) for parameter values that do not satisfy the inequalities just given.

Given that the Schrödinger equation is solvable for the barrier (376), it must also be solvable when $V_0 \to -V_0$, in which case it can have bound states and permit another comparison with the WKB approximation. The exact bound-state energy eigenvalues are

$$E_n = -\frac{\hbar^2\alpha^2}{2m}\left[\tfrac{1}{2}\sqrt{1+|\lambda|} - (n+\tfrac{1}{2})\right]^2 , \tag{384}$$

and $n < s$ is required. The WKB result is

$$E_n^0 = -\frac{\hbar^2\alpha^2}{2m}\left[\tfrac{1}{2}\sqrt{|\lambda|} - (n+\tfrac{1}{2})\right]^2 , \tag{385}$$

which again agrees with the former when $|\lambda| \gg 1$.

It is important to recognize that the condition $|\lambda| \gg 1$, i.e.,

$$\frac{8m|V_0|}{\alpha^2\hbar^2} \gg 1 , \tag{386}$$

does not require the unphysical limit $\hbar \to 0$ in either the bound-state spectrum or the transmission amplitude.

The physically meaningful statements are that the WKB approximation is valid if the potential varies little in a characteristic wavelength; furthermore, it is valid in the repulsive case if the barrier allows little transmission, while in the attractive case if the potential is deep enough to bind many states.

The crucial role of smoothness is evident in both the approximate and exact expressions for the transmitted intensity. Thus (383) states that $|T_0|^2 \ll 1$ for any $E < V_0$ provided the smoothness parameter α is sufficiently small to make λ sufficiently large. This combination of parameters maintains well separated turning points even as E approaches V_{max}. In the exact expressions, $|T|^2 \to 1$ as $(k/\alpha) \to \infty$ for any fixed values of λ.

As for the exact number of bound states, that is

$$[n] = \tfrac{1}{2}\left[\sqrt{1+|\lambda|} - 1\right], \tag{387}$$

where $[n]$ is the largest integer smaller than the right-hand side. This means that when the WKB approximation is valid, $[n] \simeq \frac{1}{2}\sqrt{|\lambda|}$, i.e., there is an appreciable number of bound states. This implies that there are bound states with an appreciable number of nodes, and once again well-separated turning points.

4.6 Problems

1. The resonance spectroscopy problem, as solved in the rotating wave approximation in §4.1, can also be solved without resorting to the fancy unitary transformation technique, and it is instructive to redo the calculations in this less sophisticated way. (The technique of §4.1 is important for higher spin, for which it goes through as is, whereas the simple-minded approach becomes increasingly cumbersome.)

(a) Confirm that the transition probability $P_{ab}(t)$ (Eq. 41) can be found by writing the state as the spinor

$$|t\rangle = \begin{pmatrix} u(t) \\ v(t) \end{pmatrix},$$

setting up the Schrödinger equation for the Hamiltonian

$$H = \tfrac{1}{2}\hbar\Omega\sigma_3 + \tfrac{1}{2}\hbar\lambda(\sigma_1 \cos\omega t + \sigma_2 \sin\omega t),$$

and solving the differential equations for $u(t)$ and $v(t)$.

(b) Find the probability that the higher state is occupied when the system is in a temperature bath (Eq. 54) without the trace and density matrix techniques, by simply using this result for $P_{ba}(t)$ and the Boltzmann weights p_a and p_b.

2. Confirm Eq. 52 for the rotation of the Pauli vector $\boldsymbol{\sigma}$. (Hint: In $U^\dagger \boldsymbol{\sigma} U$, commute $\boldsymbol{\sigma}$ to the right.) Provide the argument as to why this holds for any vector operator $\boldsymbol{V}$ provided the appropriate angular momentum is used to generate the rotation, i.e.,

$$e^{-i\theta \boldsymbol{J}\cdot\boldsymbol{n}}\, \boldsymbol{V}\, e^{i\theta \boldsymbol{J}\cdot\boldsymbol{n}} = \boldsymbol{V}\cos\theta + \boldsymbol{n}\,(\boldsymbol{V}\cdot\boldsymbol{n})\,(1-\cos\theta) + (\boldsymbol{n}\times\boldsymbol{V})\sin\theta\,.$$

This problem demonstrates that rotations in 3-space are most efficiently handled with spin $\frac{1}{2}$, as of course Hamilton discovered long, long ago. This observation also holds for Lorentz transformations, as shown in §13.1.

3. The unitary transformation that solves the forced harmonic oscillator problem must not just produce the solution of Heisenberg's equation of motion, Eq. 101, but also solve the Schrödinger equation; Eq. 109 does not! Show that the missing phase factor is given by (110). (Hint: Consider $\langle 0|i\partial U/\partial t|0\rangle$.)

4. Verify (128) directly from (123).

5. Consider the displaced excited state $\exp(-iH_0t)D(z)|n\rangle$. Show that the coordinate and momentum dispersions in these states are the same for all time as in the stationary state $|n\rangle$, i.e., $\Delta q = \Delta p = (n+\frac{1}{2})^{\frac{1}{2}}$.

6. The displacements used to define coherent states do not form an Abelian group. Why not? Show that their composition law is mighty close to Abelian however, namely,

$$D(z_1)D(z_2) = D(z_1+z_2)\exp[i\,\mathrm{Im}(z_1 z_2^*)] .$$

This is an Abelian composition law for certain subsets of the parameters—which and why?

7. Prove Eq. 164.

8. Show that the propagator for the forced oscillator in the coherent state basis is

$$K_F(z,z';t) = e^Z\,\theta(t) ,$$

$$Z = -\tfrac{1}{2}|z|^2 - \tfrac{1}{2}|s(t)+z'|^2 + z^*[s(t)+z']e^{-it} + i\,\mathrm{Im}\,[s(t)z'^*] .$$

9. Confirm the expression (172) for the classical action of the driven oscillator. The following may be useful intermediate steps. Let $S = C(Q_a+Q_b)+I$; show that

$$2C = \Phi(1-\cos T) + (\partial\Phi/\partial t)\sin T , \quad 2I = \Phi^2 - \sin T\int_{t_a}^{t_b} dt\;\Phi(t,t_a)f(t) ,$$

$$\Phi(t_b,t_a) \equiv \int_{t_a}^{t_b} dt\; f(t)\sin(t_b-t) ;$$

and that the two terms in I combine to give the double integral in Eq. 172.

10. As in the case of the free particle, the factor F in $K = F\exp(iS_{\mathrm{cl}}/\hbar)$ can be derived form unitarity. Show that this leads to

$$\frac{F(t_b-t_a)}{F(t_b-t)F(t-t_a)} = \sqrt{2\pi i\,\frac{\sin(t_b-t)\,\sin(t-t_a)}{\sin(t_b-t_a)}} ,$$

and therefore to (173).

11. Show that the energy eigenvalues and eigenfunctions for the harmonic oscillator can be found from the propagator by use of the generating function for the Hermite polynomials[1]

$$\exp\left(\frac{2xyz-(x^2+y^2)z^2}{1-z^2}\right) = \sqrt{1-z^2}\sum_{n=0}^{\infty}\frac{z^n}{2^n n!}H_n(x)\,H_n(y) .$$

12. Let $|\psi\rangle = \sum_n c_n|n\rangle$ be any linear combination of stationary states of the harmonic oscillator. Show that it can be represented in terms of coherent states as follows:

$$|\psi\rangle = \int\frac{dz}{\pi}\,\chi(z^*)|z\rangle ,$$

where

$$\chi(z) = \sum_n c_n\, e^{-\frac{1}{2}|z|^2}\frac{(z^*)^n}{\sqrt{n!}} .$$

---

[1]Bateman Manuscript Project, *Higher Transcendental Functions, Vol. II*, McGraw-Hill (1953), p. 194.

13. Derive the Wigner distributions given in §4.2(e).

14. Find the Wigner distribution of the thermal equilibrium state of the harmonic oscillator (Eq. 159) by using the generating function for the Hermite polynomials, given in Prob. 11.

15. As mentioned in connection with motion in a magnetic field, one may wonder why operators such as $v_x + iv_y$ raise the eigenvalue of L_z because v_x and v_y are linear combinations of canonical variables along two different directions. This is related to the fact that the definition of angular momentum is gauge-dependent. Show, first, that if

$$\boldsymbol{A} \to \boldsymbol{A} + \boldsymbol{\nabla}\chi, \quad \text{then } \boldsymbol{L} \to \boldsymbol{L} + (e/c)\boldsymbol{r}\times\boldsymbol{\nabla}\chi.$$

Furthermore, show that in an arbitrary gauge

$$[v_x, L_z] = \frac{\hbar}{im}p_y + \frac{e\hbar}{mc}\mathcal{L}_z A_x\,, \quad [v_y, L_z] = -\frac{\hbar}{im}p_x + \frac{e\hbar}{mc}\mathcal{L}_z A_y\,,$$

where

$$\mathcal{L}_z = \frac{1}{i}\left(x\frac{\partial}{\partial y} - y\frac{\partial}{\partial x}\right).$$

How does this resolve the matter?

16. Show that the ground-state eigenfunctions with angular momentum μ for motion in a magnetic field are

$$\Psi_{0\mu}(\xi,\eta) = \frac{1}{\sqrt{2\pi\,\mu!}}\left(\frac{\xi + i\eta}{\sqrt{2}}\right)^{\mu} e^{-\rho^2/4}\,,$$

where $\xi = x/l_B, \eta = y/l_B, \rho^2 = \xi^2 + \eta^2$.

17. Motion in a magnetic field can also be treated by using the gauge $A_x = -By, A_y = A_z = 0$, which will be called L-gauge in honor of Landau. The gauge used in the text will be called the symmetric, or S-gauge.

(a) Show in this gauge the convenient pair of constants of motion are p_x and the Hamiltonian for motion in the $x - y$ plane,

$$H_L = \frac{1}{2m}p_y^2 + \frac{1}{2}m\omega_c^2(y - y_0)^2\,,$$

where $y_0 = -cp_x/eB$.

(b) Show that the spectrum of this Hamiltonian is

$$E_n = \hbar\omega_c(n + \tfrac{1}{2}), \qquad n = 0, 1, \ldots,$$

where $\hbar k$ is the eigenvalue of p_x, which ranges continuously over $(-\infty, \infty)$, but which, please note, does not affect the energy! Show that the eigenfunctions are

$$\Phi_{n\kappa}(x,y) = (2\pi)^{-\frac{1}{2}}e^{i\kappa\xi}\phi_n(\eta + \kappa)\,,$$

where $\kappa = kl_B$, and ϕ_n is the nth eigenfunction of the one-dimensional oscillator.

NB: (i) Although the eigenvalue k does not appear in the energy, it does specify the point about which the probability distribution in the y-direction is symmetric; (ii) whereas in the S-gauge the degenerate states belonging to one energy level are distinguished by a discrete quantum number (μ), here that is done by the continuous quantum number k; (iii) it is astonishing that the subspace of a given energy is spanned by a set of wave functions whose absolute values all have the same extent in one coordinate and centered about any point on the corresponding axis; and (iv) equally astonishing that the same subspace can be spanned by a set of wave functions whose absolute values are cylindrically symmetric about a single point, even though the physical situation has no preferred origin.

(c) It is now clear that the relationship between the eigenstates in the two gauges is non-trivial. The reason is that these are degenerate sets, and an energy eigenstate in one gauge will, in general, be a linear combination of states in another — though of states with the same energy, of course. (Why?) The task is to determine this relationship for the set of lowest energy. Let $\tilde{\Psi}_{n\mu}$ be the eigenfunction of definite energy and angular momentum after transformation to the L-gauge. The problem is then to find the expansion coefficients in

$$\Phi_{0\kappa} = \sum_{\mu} c_\mu(\kappa) \tilde{\Psi}_{0\mu} .$$

Show that these are given by the following Taylor expansion

$$e^{(\frac{1}{4}z^2 + i\kappa z)} = \pi^{\frac{1}{4}} e^{\kappa^2/4} \sum_{\mu} c_\mu(\kappa) \frac{z^\mu}{\sqrt{\mu!}} .$$

18. A charged particle is trapped between two infinitely long concentric cylinders; the inner cylinder encloses the magnetic flux Φ_B. Find the dependence on Φ_B of the lowest-energy eigenvalue.

19. Prove Eq. 253 by imposing the continuity equation on $a\psi_{k,L} + b\psi_{k,R}$.

20. Let $\psi(x,t)$ be an arbitrary one-dimensional wave packet with mean momentum $\hbar k_0 = mv$. Show that

$$\psi(x,t) = e^{i\hbar k_0^2/2m} \left[1 + \sum_{n=1}^{\infty} \frac{1}{n!} \left(\frac{t}{2mi\hbar} \right)^n \left(\frac{\hbar}{i} \frac{\partial}{\partial x} - \hbar k_0 \right)^{2n} \right] \psi(x - vt, 0) .$$

Use this to show that the packet will maintain its original shape for a time short compared to $m\hbar/(\Delta p)^2$.

21. This problem concerns the potential $U(x) = -\gamma\delta(x)$ of §4(b). Let $u_k(q)$ be the momentum space eigenfunction

$$u_q(k) = \frac{1}{2\pi} \int_{-\infty}^{\infty} dx\, e^{-iqx}\, \psi_k(x) .$$

(i) After converting Eq. 242 to an integral equation in momentum space, show that the solution that satisfies outgoing wave boundary conditions for both signs of the incident momentum is

$$u_k(q) = \delta(k-q) - \frac{\gamma}{2\pi} \frac{1}{D(k)} \frac{1}{k^2 - q^2 + i\epsilon}$$

$$D(k) = 1 + \frac{\gamma}{2\pi} \int_{-\infty}^{\infty} \frac{dq}{k^2 - q^2 + i\epsilon} .$$

(ii) Confirm the orthogonality relation

$$\int_{-\infty}^{\infty} dq\, u_k^*(q) u_{k'}(q) = \delta(k - k') .$$

(iii) Define

$$\Delta(q, q') = \int_{-\infty}^{\infty} dk\, u_k(q) u_k^*(q') .$$

By direct calculation in the case of repulsion ($\gamma < 0$), show that the scattering states form a complete set, i.e., $\Delta(q,q') = \delta(q - q')$, whereas in the case of attraction they do not, because

$$\Delta(q, q') = \delta(q - q') - \frac{\gamma^2}{4\pi} \frac{1}{(q^2 + \frac{1}{4}\gamma^2)(q'^2 + \frac{1}{4}\gamma^2)} .$$

Observe that you have thereby derived the bound-state wave function.

22. Show that the exact expressions for the transmission and reflection amplitudes in the case of two delta-function potentials, as given by Eq. 294, agree with the Born approximation in the appropriate limit.

23. The existence of simple poles in these transmission and reflection amplitudes is not an artifact of the approximations that led to Eq. 306. To confirm this, let $\zeta = \xi + i\eta$ be the complex variable whose real part is ka, and define the functions

$$F_+(\zeta) = i\cos^2\zeta / D_+(\zeta)\,, \qquad F_-(\zeta) = i\sin^2\zeta / D_-(\zeta)\,.$$

Show (i) that the only singularities of $F_\pm(\zeta)$ in the finite ζ-plane are due to the zeroes of $D_\pm(\zeta)$; (ii) that the real parts of these zeroes are the roots of

$$D_\pm \;:\quad e^{-2\xi\cot 2\xi} = \mp\frac{1}{\lambda}\,e^{-1/\lambda}\left(\frac{\sin 2\xi}{2\xi}\right)\;;$$

(iii) that as a consequence of this relationship there are an infinity of such roots, and that they give rise to simple (or first-order) poles in $F_\pm(\zeta)$; (iv) that if ξ_n is the real part of a zero, then the imaginary part is given by

$$D_\pm \;:\quad 2\lambda\eta_n = 1 \pm e^{-2\eta_n}\cos 2\xi_n\;;$$

(v) that all the poles of $F_\pm(\zeta)$ lie in the lower half-plane; (vi) that if $\lambda \lll 1$, and $\frac{1}{2}n\pi\lambda \ll 1$,

$$\zeta_n = \tfrac{1}{2}n\pi(1+\lambda) - i(\tfrac{1}{2}n\pi\lambda)^2\,,$$

where n is odd for poles of F_+ and even for poles of F_-, which agrees with the approximate forms in Eq. 306.

Those who enjoy complex variable theory might also prove that

$$T(\zeta) = 1 + \sum_{n=1}^{\infty}\frac{\rho_n}{\zeta - \zeta_n}$$

where the ζ_n are the exact poles of $(D_\pm)^{-1}$ and ρ_n the residues, which involves showing that in addition to the displayed expression there is not also an entire function of ζ.

24. Compute the amplitude $\Psi_R(x,t)$ from Eq. 320, using the technique that led to Eq. 327. Show that it is a function of $\bar{u} = v(t-t_0) + (x - x_0)$, and that the claims made in the text, in particular concerning parity, are correct.

25. Write the transmission probability involved in α-decay as e^{-2I}, and show that

$$I = \frac{zZe^2}{4\pi}\sqrt{\frac{2\mu}{\hbar^2 E}}\left\{\cos^{-1}\sqrt{x} - \sqrt{x - x^2}\right\}, \qquad x = E/V(R)\,.$$

26. Evaluate the transmission amplitude for the parabolic potential in the WKB approximation and compare the result with Eq. 375.

27. Confirm the WKB results for the potentials $\pm|V_0|/\cosh^2\alpha x$ as given in Eq. 383 and Eq. 385.

28. Find an exact expression for $|T|^2$ when E is close to V_0 from (382), compare this with Eq. 375, and give an intuitive explanation of the circumstances when the two results agree and disagree.

29. Derive the expression (372) for the phase shift of the transmitted wave. Compare this result with the exact answer (379) for the potential (376) when $E \gg V_0$.

Endnotes

The harmonic oscillator propagator is discussed in B.R. Holstein, *Am. J. Phys.* **66**, 583 (1998).

Our treatment of motion in a magnetic field follows A. Friedman and A.H. Kahn, *Phys. Rev. B* **1**, 4584 (1970). Many interesting elaborations of the Aharonov-Bohm effect are given in M.P. Silverman, *More than One Mystery: Explorations in Quantum Interference,* Springer (1995).

The discussion of connection formulas follows W.H. Furry, *Phys. Rev.* **71**, 360 (1947). Many interesting and sophisticated semiclassical problems are treated in *LLQM*, chapter VII, and in M.V. Berry and K.E. Mount, *Rep. Prog. Phys.* **35**, 315 (1972). For an instructive comparison of different semiclassical techniques as applied to one-dimensional problems, see A. Garg, *Am. J. Phys.* **68**, 430 (2000). Connection formulas for two nearby turning points are given in J. Heading, *An Introduction to Phase Integral Methods*, Methuen (1962). For a discussion of the Stokes phenomenon, see P.M. Morse and H. Feshbach, *Methods of Theoretical Physics*, McGraw-Hill (1953), pp. 609-611; and G.F. Carrier, M. Krook and C.E. Pearson, *Functions of a Complex Variable*, Hod Books, Ithaca NY (1983), pp. 286–299, which also treats connection formulas.

5

Hydrogenic Atoms

5.1 Qualitative Overview

Hydrogenic atoms have played a central role in the development of physics ever since Bohr recognized that Rutherford's model of the nuclear atom could account for atomic dimensions if Planck's constant is brought into the picture. Today we have a much larger cast of characters than in 1913: muonium, the bound states of an electron e and the positively charged muon μ^+; positronium, the $e^+ - e^-$ system; muonic atoms, in which μ^- plays the role of the electron; and mesic atoms, bound states of π^- and K^- in nuclear Coulomb fields. The discussion will, however, be restricted to atoms in which the satellite only interacts electromagnetically with the nucleus.

First, a few familiar facts will be derived *ab initio*. Recall from §3.3 that in nonrelativistic quantum mechanics, as in classical mechanics, the central-force two-body problem reduces to an effective one-body problem, that of a particle with the reduced mass

$$\mu = \frac{mM}{m + M} \tag{1}$$

moving about the center of mass. For a satellite of charge $-e$ and a nucleus treated as a *point* of charge $+Ze$, this one-body problem has the Hamiltonian[1]

$$H = \frac{\boldsymbol{p}^2}{2\mu} - \frac{Ze^2}{4\pi r} , \tag{2}$$

where $\boldsymbol{p}$ is the relative momentum and $\boldsymbol{r}$ the separation between the particles. *Note the use of rationalized units,* which avoids clouds of 4π's in Maxwell's equations and radiation theory. In these units, *the fine structure constant* is

$$\alpha = \frac{e^2}{4\pi\hbar c} \simeq \frac{1}{137.04} . \tag{3}$$

The characteristic energy and radius of the ground state can be estimated by using the uncertainty principle to caricaturize the Hamiltonian as

$$H \sim \frac{\hbar^2}{2\mu(\Delta r)^2} - \frac{Ze^2}{4\pi\Delta r} . \tag{4}$$

Minimizing with respect to Δr, and assuming $\Delta r \sim r$, gives the following estimate for the ground-state radius:

$$r \sim \frac{4\pi\hbar^2}{Ze^2\mu} \equiv \frac{1}{Z}\frac{m_e}{\mu} a_0 , \tag{5}$$

---

[1] If the satellite has the charge $-ze$, then $Z \to Zz$ in all formulae.

where m_e is the electron mass and a_0 the Bohr radius

$$a_0 = \frac{4\pi\hbar^2}{m_e e^2} = 0.52918 \cdot 10^{-8} \text{ cm} . \tag{6}$$

The estimate for the ground-state energy is then

$$E_0 \sim -\frac{1}{2}\frac{\mu}{\hbar^2}\left(\frac{Ze^2}{4\pi}\right)^2 \equiv -Z^2\frac{\mu}{m_e}\text{Ry} , \tag{7}$$

where Ry is the Rydberg unit of energy,[1]

$$\text{Ry} = \frac{1}{2}\frac{e^2}{4\pi a_0} = 13.606 \text{ eV} . \tag{8}$$

This estimate is on target for the ground-state energy, because the correct spectrum is given by

$$E_n = -Z^2\frac{\mu}{m_e}\frac{\text{Ry}}{n^2} , \qquad n = 1, 2, \ldots , \tag{9}$$

where n is the principal quantum number. The expectation values of r in the ground state is 1.5 times larger than the estimate (5).

The binding energies and Bohr radius have simple and important relationships to other fundamental energies and lengths related to the electron. These all involve the speed of light, whereas the definitions given thus far do not as they have a nonrelativistic origin. In terms of the electron's rest mass, the Rydberg is

$$\text{Ry} = \tfrac{1}{2}\alpha^2 m_e c^2 , \tag{10}$$

and the binding energy of the ground state is

$$|E_0| = \tfrac{1}{2}Z^2\alpha^2\mu c^2 . \tag{11}$$

The electron's Compton wavelength $\lambda\!\!\!^{-}_C$, and the classical electron radius r_e, are

$$\lambda\!\!\!^{-}_C = \hbar/m_e c \simeq 3.8616 \cdot 10^{-11} \text{ cm} , \quad r_e = e^2/4\pi m_e c^2 \simeq 2.8179 \cdot 10^{-13} \text{ cm} ; \tag{12}$$

thus they are related to the Bohr radius by powers of the fine structure constant:

$$a_0 = \frac{\lambda\!\!\!^{-}_C}{\alpha} = \frac{r_e}{\alpha^2} . \tag{13}$$

Hence, as Bohr realized, no theory that only involves the fundamental constants e, m_e, and c of classical physics can produce a length of order atomic dimensions, i.e., of order Ångstroms, unless a non-electromagnetic electron-proton interaction with a range of order a_0 is used in the atomic model.

The nucleus is not a point charge, and the true energy levels therefore depart from those of Eq. 9 (see Probs. 1 and 2). This is a very small effect for electrons unless Z is very large, but not in atoms in which a particle much heavier than the electron is bound to a nucleus, the lightest example being the muon with mass $m_\mu \simeq 207\, m_e$.

[1]The term "Rydberg" is also used for the wavenumber of the radiation required to ionize the ground state of hydrogen, and then stated as an inverse length. We always use the definition (8).

The tightly bound states of such a particle lie well inside the electronic charge distribution and therefore see a nuclear charge that is almost unscreened. By the same token, they are far more sensitive to the detailed shape of the nuclear charge distribution than are electronic states. The mean radius r_m of the lowest energy bound state for a heavy satellite of mass m is related to the nuclear radius R_N by the rough formula

$$\frac{r_m}{R_N} \sim 0.7\, Z^{-1} A^{-\frac{1}{3}} (m_e/m) \cdot 10^5 \,, \tag{14}$$

where A is the mass number, and R_N was approximated by $1.2\ A^{\frac{1}{3}}$ fm, where fm $= 10^{-13}$ cm is the Fermi unit of length.

Thus far we have sketched the so-called *gross structure* of the hydrogenic spectrum, which ignores the effects due to relativity and spin. Relativistic corrections are of order $(v/c)^2$. The expectation value of this quantity follows directly from the virial theorem (see Prob. 3), which for the Hamiltonian (1) gives

$$\langle p^2/2\mu \rangle_n = -E_n \,, \tag{15}$$

and therefore

$$\langle v^2/c^2 \rangle_n = Z^2\alpha^2/n^2 \,. \tag{16}$$

In short, velocities decrease like $1/n$, and are small compared to c for all nuclei except those with very large Z. Spin effects are of the same magnitude, as we will learn in §5.3.

Relativistic kinematics and spin-dependent effects produce the *fine structure* of the hydrogenic spectrum, small splittings of states that are degenerate in the non-relativistic limit. The fine structure energy is of order the relativistic correction to the kinetic energy, i.e., the p^4 term in $\sqrt{p^2\mu^2 + \mu^2 c^4}$, and therefore of order $(p^4/8\mu^3c^2) \sim \alpha^2 Z^2 E_n/n^2$. This crude estimate of the n-dependence is not correct, however, because there are cancellations between the various contributions, and the correct if crude order of magnitude relationship is

$$E_{\text{fs}} \sim \frac{(\alpha Z)^2}{n} E_n \,. \tag{17}$$

Hence the fine structure produces level shifts that are very small compared to the gross structure splittings as long as $(Z/137) \ll 1$.

The next level of refinement is *hyperfine structure*, due to interactions between the satellite and the nucleus that depend on the latter's spin. Of greatest importance is the interaction between the magnetic moments of the satellite and the nucleus. Magnetic moments have a magnitude of order $e\hbar/m$, and are best expressed in terms of the Bohr magneton

$$\mu_B \equiv \frac{e\hbar}{2m_e c} = 5.7884 \cdot 10^{-5}\ \text{eV/Tesla} \,, \tag{18}$$

which is specific to the electron. Nuclear magnetic moments are far smaller by a factor of $m_e/m_N \sim 1/1,800$, where m_N is the mass of *one* nucleon, not of the nucleus as a whole, because all but a few nucleons in a complex nucleus have their angular momenta coupled to zero. The interaction energy between two magnetic moments is of order

$$E_{\text{hfs}} \sim \frac{\mu_1\mu_2}{4\pi r^3} \sim \alpha^2 E_0 Z \, \frac{m_e}{m_N} \,, \tag{19}$$

where the latter estimate refers to a $1s$ electron state. The ratio of hyperfine to fine structure energies is thus roughly

$$\frac{E_{\rm hfs}}{E_{\rm fs}} \sim \frac{m_e}{m_N Z} , \tag{20}$$

which does not yet account for the important n dependence. In positronium, $m_N \to m_e$ and $Z = 1$, and therefore fine and hyperfine effects have the same magnitudes.

Fine and hyperfine interactions are not the only causes of departures from the naive spectrum. The discussion thus far has ignored the fact that the electromagnetic field is a dynamical system with its own degrees of freedom. These produce so-called radiative corrections, which manifest themselves as shifts of levels that are small compared to the fine structure unless $Z\alpha$ is not small. This topic is briefly discussed in §10.3(b). Still more subtle effects stem from the weak interaction, which are too small to be seen in the spectrum itself, but reveal themselves in certain radiative transitions that do not obey the parity selection rules required by quantum electrodynamics.

The lengths and energies that set the scale in hydrogenic atoms make it both natural and efficient to introduce *atomic units*, in which lengths, momenta and energies are stated in terms of the following quantities:

$$\text{length: } 4\pi\hbar^2/m_e e^2 \equiv a_0 \ ; \quad \text{momentum: } \hbar/a_0 \ ; \quad \text{energy: } e^2/4\pi a_0 \equiv 2\,\text{Ry} \ . \tag{21}$$

Note that the unit of energy is *twice* the Rydberg, i.e., twice the ionization energy of the $1s$ state of an electron bound to a infinitely heavy unit charge. These are the units most appropriate to hydrogen itself, and to many electron atoms (see Chap. 6). In analyzing hydrogenic atoms with $Z \neq 1$ and arbitrary reduced mass, these parameters can be absorbed into the units by use of (5) and (9).

5.2 The Kepler Problem

The wave mechanical solution of the Kepler problem offers no obvious explanation for the remarkable degeneracies in the spectrum, which appear as an infinite set of "accidents." In physics such accidents must have a logical explanation. In this case it is that the Coulomb field has constants of motion beyond those that follow merely from rotational invariance, and it is easy to see what they are from classical physics: In a $1/r^2$ attractive field of force, all particles move along orbits that stay fixed in space for all time — there is no precession. Any deviation from the $1/r^2$ law no matter how small, such as the famous example from General Relativity, produces a precession or some other time-dependent distortion of the orbit. In particular, the axes of the Kepler ellipse do not move if Newton's law of gravitation is exact, and for that reason there are further conserved quantities, an observation that dates back to Laplace. The symmetry that permits these extra constants depends on the specific form of the force, and is therefore called a *dynamical symmetry.*

(a) The Lenz Vector

The classical problem is a good introduction to that of quantum mechanics. Let $(\boldsymbol{P}_c, \boldsymbol{Q}_c)$ be the classical relative momentum and coordinate, and H_c the Hamiltonian in the center-of-mass frame, expressed in atomic units modified to eliminate

Z and the reduced mass μ:

$$\text{length}: \ (m_e/Z\mu)a_0 \ ; \quad \text{momentum}: \ \hbar/\text{length} \ ; \quad \text{energy}: \ 2(Z^2\mu/m_e)\,\text{Ry} \ . \tag{22}$$

The Hamiltonian is then

$$H_c = \frac{1}{2}P_c^2 - \frac{1}{Q_c} \, , \tag{23}$$

and the classical equations of motion are

$$\dot{\boldsymbol{Q}}_c = \boldsymbol{P}_c \, , \qquad \dot{\boldsymbol{P}}_c = -\boldsymbol{Q}_c/Q_c^3 \, . \tag{24}$$

The angular momentum, $\boldsymbol{L}_c = \boldsymbol{Q}_c \times \boldsymbol{P}_c$ is, of course, a constant of motion. There are an infinity of coplanar ellipses that have the same energy and angular momentum. The sought-for dynamical constant selects a particular member of this set, so it must lie in the orbital plane, and therefore be perpendicular to $\boldsymbol{L}_c$. As a first guess try $\boldsymbol{P}_c \times \boldsymbol{L}_c$. Equation (24) gives

$$\frac{d}{dt}(\boldsymbol{P}_c \times \boldsymbol{L}_c) = \frac{d}{dt}\frac{\boldsymbol{Q}_c}{Q_c} \, , \tag{25}$$

showing that this guess is not quite correct, but that

$$\boldsymbol{M}_c = (\boldsymbol{P}_c \times \boldsymbol{L}_c) - \hat{\boldsymbol{Q}}_c \tag{26}$$

is a constant of motion. Nowadays $\boldsymbol{M}_c$ is usually called the the *Lenz vector*, and with somewhat less historical inaccuracy the Runge-Lenz vector, but it was actually discovered long ago by Laplace.

For a bound orbit, i.e., when $H_c < 0$, the Lenz vector points along the major axis of the Kepler ellipse, and $|\boldsymbol{M}_c|$ is the eccentricity. The magnitude M_c can also be stated in a form with dynamical content, which has an important quantum mechanical counterpart:

$$M_c^2 = 1 + 2H_cL_c^2 \, . \tag{27}$$

Consequently $H_c = (M_c^2 - 1)/2L_c^2$, and therefore $M_c^2 < 1$ for bound orbits.

Historically, the first successful quantum mechanical calculation of the energy spectrum in a Coulomb field, due to Pauli, exploited the existence of the Lenz vector. It was completed in December 1925, just before Schrödinger discovered his equation and the solution given in §3.6(c), and convinced many that Heisenberg's five-month-old baby had a healthy future. Among other things, Pauli showed that the degeneracies are due to this non-geometric constant of motion. His solution is not only of historical interest, however. It is phrased entirely in terms of operators, and in that sense has a much more characteristic quantum mechanical flavor than does Schrödinger's solution. Furthermore, it is the prototype for solutions of other problems that have a dynamical symmetry. As we will see in §5.4, it also provides a powerful approach to the problem of a hydrogenic atom in a static external electromagnetic field (the Stark and Zeeman effects). On the other hand, Pauli's original solution did not produce wave functions, and many other detailed and important properties of the hydrogenic system.

Quantization is accomplished in the usual fashion — the Hamiltonian is still (23), and the canonical variables satisfy the dimensionless commutation rule

$$[Q_i, P_j] = i\delta_{ij} \, . \tag{28}$$

The quantum mechanical counterpart of the classical Lenz vector $\boldsymbol{M}_c$ is presumably the Hermitian operator $\boldsymbol{M}$ constructed as the obvious generalization of (26):

$$\boldsymbol{M} = \boldsymbol{P} \circledast \boldsymbol{L} - \hat{\boldsymbol{Q}} \,, \tag{29}$$

where for any two Hermitian vector operators the Hermitian cross-product is defined as

$$\boldsymbol{A} \circledast \boldsymbol{B} \equiv \tfrac{1}{2}(\boldsymbol{A}\times\boldsymbol{B} - \boldsymbol{B}\times\boldsymbol{A}) \,. \tag{30}$$

(b) The Energy Spectrum

The qualification "presumably" in the preceding sentence is a reminder that it is necessary to show that the operator defined by Eq. 29 commutes with H, which entails a rather complicated calculation. The proofs of the other algebraic properties of $\boldsymbol{M}$ required for the diagonalization of H are also lengthy. To keep the eye on the essentials, we first state these basic relationships, and leave proofs for later.

The conservation law that uniquely characterizes the Kepler problem is

$$[\boldsymbol{M}, H] = 0 \,, \tag{31}$$

the counterpart of (25). Because $\boldsymbol{M}$ is a vector operator, its commutator with the angular momentum is known:

$$[L_i, M_j] = i\epsilon_{ijk} M_k \,. \tag{32}$$

As M^2 is a scalar,

$$[\boldsymbol{L}, M^2] = 0 \,. \tag{33}$$

The different components of $\boldsymbol{M}$ do not commute with each other. However, if Eq. 31 is correct, we have identified the following compatible constants of motion:

$$L^2, L_3, M^2, M_3 \,. \tag{34}$$

L^2 and M^2 should be expected to determine the energy spectrum if there is a quantum counterpart to Eq. 27. The eigenvalues of L_3 and M_3 cannot appear in the formula for the energy eigenvalues because the very definition of these operators is contingent on a choice of coordinate frame, which is arbitrary in view of the rotational symmetry; these eigenvalues do, however, distinguish among states of equal energy.

Three relations involving $\boldsymbol{M}$ beyond $[\boldsymbol{M}, H] = 0$ will be needed:

1. the quantum analogue of (27):

$$M^2 = 1 + 2H(L^2 + 1) \,, \tag{35}$$

   which reduces to the classical expression when $l(l+1) \gg 1$;

2. the Lenz and angular momentum vectors are orthogonal, as in classical theory:

$$\boldsymbol{L} \cdot \boldsymbol{M} = 0 \,; \tag{36}$$

3. the commutation rule between different components of $\boldsymbol{M}$,

$$[M_i, M_j] = -2i\epsilon_{ijk} H L_k \,. \tag{37}$$

Equation (37) is the key that unlocks the safe, because it establishes that the operators $(H, \boldsymbol{M}, \boldsymbol{L})$ form a closed algebra.

Because $\boldsymbol{L}$ and $\boldsymbol{M}$ commute with H, it is possible to separate the analysis into subspaces $\mathfrak{H}_E$ in which H has the (still unknown!) bound-state eigenvalue $-E$. In $\mathfrak{H}_E$ the commutator (37) becomes

$$[M_i, M_j] = 2i\epsilon_{ijk} E L_k \,. \tag{38}$$

By introducing a differently normalized Lenz vector

$$\boldsymbol{N} = \boldsymbol{M}/\sqrt{2E}\,, \tag{39}$$

the basic commutator becomes

$$[N_i, N_j] = i\epsilon_{ijk} L_k \,. \tag{40}$$

According to (35) and (39), the energy eigenvalue in terms of N^2 and L^2 is

$$E = \frac{1}{2(L^2 + N^2 + 1)} \,. \tag{41}$$

This already has the appearance of the Bohr formula if L^2 and N^2 are understood to stand for their eigenvalues.

Only the eigenvalues of N^2 remain to be determined. They are found by a clever but simple device used repeatedly in quite different contexts (the most important being the theory of Lorentz transformations). To whit, define two new vector operators

$$\boldsymbol{F}^{(\pm)} = \tfrac{1}{2}(\boldsymbol{L} \pm \boldsymbol{N}) \,. \tag{42}$$

A short calculation shows that they satisfy the following commutation rules:

$$[F_i^{(\pm)}, F_j^{(\pm)}] = i\epsilon_{ijk} F_k^{(\pm)} \,, \tag{43}$$

$$[F_i^{(+)}, F_j^{(-)}] = 0 \,. \tag{44}$$

That is, $\boldsymbol{F}^{(+)}$ *and* $\boldsymbol{F}^{(-)}$ *have the algebra of two commuting angular momenta.* The true angular momentum is their sum:

$$\boldsymbol{L} = \boldsymbol{F}^{(+)} + \boldsymbol{F}^{(-)} \,. \tag{45}$$

The expression that appears in the energy eigenvalue is

$$L^2 + N^2 = 2\left([\boldsymbol{F}^{(+)}]^2 + [\boldsymbol{F}^{(-)}]^2\right) \,. \tag{46}$$

This reduces the energy eigenvalue problem to that of of angular momentum theory. Specifically, introduce products of pseudo angular momentum eigenkets

$$|f^+\mu^+\rangle \otimes |f^-\mu^-\rangle \equiv |f^+\mu^+ f^-\mu^-\rangle \equiv |\phi\rangle \,, \tag{47}$$

satisfying

$$[\boldsymbol{F}^{(\pm)}]^2|\phi\rangle = f^\pm(f^\pm + 1)|\phi\rangle \,, \qquad F_3^{(\pm)}|\phi\rangle = \mu^\pm|\phi\rangle \,, \tag{48}$$

where

$$f^{\pm} = 0, \tfrac{1}{2}, 1, \ldots, \qquad \mu^{\pm} = f^{\pm}, f^{\pm} - 1, \ldots, -f^{\pm} . \tag{49}$$

There is a further important simplification, however:

$$[\boldsymbol{F}^{(+)}]^2 - [\boldsymbol{F}^{(-)}]^2 = \tfrac{1}{2}\{\boldsymbol{L}\cdot\boldsymbol{N} + \boldsymbol{N}\cdot\boldsymbol{L}\} = 0 , \tag{50}$$

thanks to (36), and therefore the pseudo angular momenta have equal magnitudes,

$$f^+ = f^- = f . \tag{51}$$

Hence $L^2 + N^2 = 4f(f+1)$ from (46), and (41) then gives the final formula for the bound-state energy eigenvalues:

$$E_f = \frac{1}{2(2f+1)^2} , \qquad f = 0, \tfrac{1}{2}, 1 \ldots , \tag{52}$$

or in the traditional Bohr form,

$$E_n = \frac{1}{2n^2} , \qquad n = 2f + 1 = 1, 2, \ldots . \tag{53}$$

Which angular momenta are allowed for a given E_n? Because of Eq. 45, the answer is given by the theory of angular momentum addition for the special case of two angular momenta of equal magnitude f. Therefore

$$l = 0, 1, \ldots, 2f , \quad \text{i.e.,} \quad l = 0, 1, \ldots, n-1 . \tag{54}$$

Equation (48) tells us that the dimension of $\mathfrak{H}_E$ is $(2f+1)^2 = n^2$, which agrees with

$$\sum_{l=0}^{2f} (2l+1) = n^2 . \tag{55}$$

The kets $|f\mu^+ f\mu^-\rangle$ are eigenstates of energy but not of the angular momentum $\boldsymbol{L}$. In view of (45), however, simultaneous eigenstates of energy (i.e., f) and L^2, L_3 can be constructed with Clebsch-Gordan coefficients:

$$|nlm\rangle = \sum_{\mu^+\mu^-} |f\mu^+ f\mu^-\rangle\langle f\mu^+ f\mu^- | lm\rangle , \tag{56}$$

where $m = \mu^+ + \mu^-$.

(c) The Conservation of M

The derivation of the basic conservation law $[\boldsymbol{M}, H] = 0$ will now be given; that of the other identities that have been used involve similar manipulations. Everything follows from the canonical commutation rule Eq. 28, and its consequences

$$[Q_i, f(P)] = i\partial f/\partial P_i , \qquad [P_i, g(Q)] = -i\partial g/\partial Q_i . \tag{57}$$

The identities

$$\begin{aligned} [\boldsymbol{A}\times(\boldsymbol{B}\times\boldsymbol{C})]_i &= A_j B_i C_j - (\boldsymbol{A}\cdot\boldsymbol{B})C_i \\ [(\boldsymbol{A}\times\boldsymbol{B})\times\boldsymbol{C}]_i &= A_j B_i C_j - A_i(\boldsymbol{B}\cdot\boldsymbol{C}) \end{aligned} \tag{58}$$

will be useful; they hold for any three vectors, whether they commute or not. Another simplifying point: $[A, B]$ is anti-Hermitian if A and B are Hermitian, and in computing commutators it therefore pays to construct expressions that are anti-Hermitian, because any Hermitian term must cancel out.

To compute $[\boldsymbol{M}, H]$, start with

$$[\boldsymbol{P}\times\boldsymbol{L}, H] = [\boldsymbol{P}, H]\times\boldsymbol{L} = -[\boldsymbol{P}, Q^{-1}]\times\boldsymbol{L} = -i\frac{1}{Q^3}(\boldsymbol{Q}\times\boldsymbol{L}), \tag{59}$$

where rotational invariance and (57) were used. Furthermore

$$[\boldsymbol{P}\times\boldsymbol{L}, H]^\dagger = [\boldsymbol{L}\times\boldsymbol{P}, H] = -i\frac{1}{Q^3}(\boldsymbol{L}\times\boldsymbol{Q}). \tag{60}$$

Therefore

$$[\boldsymbol{P}\circledast\boldsymbol{L}, H] = -i\frac{1}{Q^3}\boldsymbol{Q}\circledast\boldsymbol{L}. \tag{61}$$

This last expression must be anti-Hermitian. Using this fact, and (58),

$$\begin{aligned}\frac{i}{Q^3}(\boldsymbol{Q}\circledast\boldsymbol{L})_i &= \frac{i}{2Q^3}\{(\boldsymbol{Q}\times(\boldsymbol{Q}\times\boldsymbol{P}))_i - ((\boldsymbol{Q}\times\boldsymbol{P})\times\boldsymbol{Q})_i\} \\ &= \frac{i}{2}\left(\frac{Q_i}{Q^3}\boldsymbol{Q}\cdot\boldsymbol{P} + \boldsymbol{P}\cdot\boldsymbol{Q}\frac{Q_i}{Q^3} - \frac{1}{Q}P_i - P_i\frac{1}{Q}\right) \\ &\quad + i\,([Q_i, \boldsymbol{P}\cdot\boldsymbol{Q}] - [Q_j, P_i]Q_j)\,\frac{1}{2Q^3}.\end{aligned} \tag{62}$$

The term $\boldsymbol{P}\circledast\boldsymbol{L}$ in the Lenz vector has now been dealt with. The other term gives

$$[\hat{Q}_i, H] = \tfrac{1}{2}[\hat{Q}_i, P^2] = \frac{1}{2}\left([\hat{Q}_i, P_j]P_j + P_j[\hat{Q}_i, P_j]\right). \tag{63}$$

But

$$[Q_i/Q, P_j] = \frac{i\delta_{ij}}{Q} + iQ_i\frac{\partial}{\partial Q_j}\frac{1}{Q} = \frac{i\delta_{ij}}{Q} - \frac{iQ_iQ_j}{Q^3}, \tag{64}$$

and therefore

$$[\hat{Q}_i, H] = \frac{i}{2}\left(\frac{1}{Q}P_i + P_i\frac{1}{Q} - \boldsymbol{P}\cdot\boldsymbol{Q}\frac{Q_i}{Q^3} - \frac{Q_i}{Q^3}\boldsymbol{Q}\cdot\boldsymbol{P}\right). \tag{65}$$

To evaluate $[\boldsymbol{M}, H]$, (65) and (62) must be combined; in the latter the last term vanishes, and the first term cancels (65), which confirms that $\boldsymbol{M}$ is conserved.

(d) Wave Functions

The wave functions can also be constructed by use of the Lenz vector by essentially the same technique as that used for the harmonic oscillator in §4.2(b).

First, consider the ground state, for which $n = 1$, and call it $|1\rangle$. For this state $f = 0$, and therefore $\boldsymbol{F}^{(\pm)}|1\rangle = 0$; hence

$$\boldsymbol{M}|1\rangle = (\boldsymbol{P}\circledast\boldsymbol{L} - \hat{\boldsymbol{Q}})|1\rangle = 0. \tag{66}$$

Here (and sometimes elsewhere) it is more convenient to use the expression

$$\boldsymbol{M} = \boldsymbol{P}\times\boldsymbol{L} - i\boldsymbol{P} - \hat{\boldsymbol{Q}} . \tag{67}$$

As $\boldsymbol{L}|1\rangle = 0$, (66) reduces to

$$(i\boldsymbol{P} + \hat{\boldsymbol{Q}})|1\rangle = 0 . \tag{68}$$

Calling $\boldsymbol{r}$ the eigenvalue of $\boldsymbol{Q}$, and $\psi_{10}(\boldsymbol{r}) = \langle \boldsymbol{r}|1\rangle$ the ground-state wave function, Eq. 68 becomes the differential equation

$$(\boldsymbol{\nabla} + \hat{\boldsymbol{r}})\psi_{10}(\boldsymbol{r}) = 0 . \tag{69}$$

As ψ_{10} is spherically symmetric, $\boldsymbol{\nabla}$ can be replaced by $\hat{\boldsymbol{r}}d/dr$, and therefore

$$\left(\frac{d}{dr} + 1\right)\psi_{10} = 0 . \tag{70}$$

Thus the normalized ground-state wave function is

$$\psi_{10}(\boldsymbol{r}) = \frac{1}{\sqrt{\pi}}e^{-r} , \tag{71}$$

in agreement with Eq. 3.246.

Wave functions for excited states are constructed by first selecting the state which, for a given n, has maximal $l = n - 1$ and $m = l$. It is unique, also satisfies a first-order equation, and wave functions for the other states in the multiplet can then be constructed from it by differentiation. Call this state $|n;+\rangle$. It has the eigenvalues $\mu^{\pm} = f = \frac{1}{2}(n-1)$, and therefore satisfies

$$(F_1^{(\pm)} + iF_2^{(\pm)})|n;+\rangle = 0, \quad (L_1 + iL_2)|n;+\rangle = 0 , \tag{72}$$

whence

$$(M_1 + iM_2)|n;+\rangle = 0 . \tag{73}$$

Using (67),

$$M_1 + iM_2 = -i(P_1 + iP_2)(L_3 + 1) + iP_3(L_1 + iL_2) - (\hat{Q}_1 + i\hat{Q}_2) . \tag{74}$$

Acting on $|n;+\rangle$, the second term gives zero, and $L_3 + 1 = n$, so (73) becomes the differential equation

$$\left[\frac{x+iy}{r} + n\left(\frac{\partial}{\partial x} + i\frac{\partial}{\partial y}\right)\right]\psi_{n;+}(\boldsymbol{r}) = 0 . \tag{75}$$

In the notation of Eq. 3.207, the radial wave function is $R_{n,n-1}$. As sketched in Prob. 5, (75) leads to the result

$$R_{n,n-1}(r) = Cr^{n-1}e^{-r/n} , \tag{76}$$

where C is the normalization constant. The general result, Eq. 3.245, gives this answer much more easily, however.

5.3 Fine and Hyperfine Structure

The exact degeneracies in the spectrum of the unadorned Kepler problem imply that any perturbation that has matrix elements between the degenerate states will lift degeneracies not mandated by symmetry requirements. And no matter how small these matrix elements may be, because of the exact initial degeneracy they will, in general, produce eigenstates that are, crudely put, 50-50 mixtures of previously degenerate state vectors.

The qualitative discussion of §5.1 explained that the largest such correction, which leads to what is called the *fine structure (fs)* of the spectrum, is due to relativistic effects in the orbital motion of the satellite, with the nucleus treated as if it were a point charge. Nuclei are not point charges, and those that have spin have electromagnetic moments. If R_N is the characteristic radius of the charge distribution, the potential deviates from $1/r$ when $r \lesssim R_N$, which produces what are called *isotope shifts* in the spectrum. This is a rather straightforward problem, and its essential features can be learned by solving Prob. 2. Nuclear electromagnetic moments cause *hyperfine structure — hfs*, a splitting of levels that would be degenerate if the nucleus had no such moments. Although hyperfine structure is minute compared to fine structure, the former determines important nuclear properties and is measured to very high precision. We will, therefore, discuss hfs in some detail.

The analysis of both fine and hyperfine structure is greatly simplified because, in the absence of external fields, an atom is in a rotationally invariant environment, and its energy eigenstates are eigenstates of total angular momentum. As both fs and hfs are due, almost entirely, to the electromagnetic interaction, which is reflection invariant, the corrected eigenstates are eigenstates of angular momentum and parity. The proviso "almost entirely" alludes to the violation of reflection invariance by the weak interaction, which produces minute though detectable effects that we will not consider.

Ever since Sommerfeld derived the fine structure formula from the Old Quantum Theory in 1916, the study of both fine and hyperfine structure in hydrogenic atoms has received an enormous amount of experimental and theoretical attention because it offers a stringent testing ground for quantum mechanics, quantum electrodynamics, and more recently the theory of the electroweak interaction. The discussion that follows only deals with the leading-order corrections; refinements beyond this level fall well outside the scope of this volume.

(a) Fine Structure

A sound derivation of the leading relativistic corrections, of order v^2/c^2, to the Schrödinger description of hydrogenic atoms requires the Dirac equation, and will be given in §13.5. Here these corrections are simply taken as prescribed inputs.

The correction to the kinetic energy is

$$H_{\text{kin}} = -\frac{1}{8c^2}\left(\frac{p_1^4}{m^3} + \frac{p_2^4}{M^3}\right) = -\frac{1}{4}\frac{p^2}{2m}\left(\frac{p}{mc}\right)^2\left(1 + \frac{m^3}{M^3}\right), \tag{77}$$

where the last expression holds in the frame where the total momentum vanishes, $\boldsymbol{p}$ is the relative momentum, m the mass of the satellite and M that of the nucleus. Henceforth we restrict ourselves to electronic atoms, take $M \to \infty$, and therefore use the following units (recall Eq. 22):

$$\text{length}: a_0/Z\,; \quad \text{momentum}: \hbar Z/a_0\,; \quad \text{energy}: 2Z^2\,\text{Ry}\,. \tag{78}$$

Writing $\bar{Q}$ for any quantity Q in these units, (77) reads

$$\bar{H}_{\text{kin}} = \frac{H_{\text{kin}}}{(Z\alpha)^2 m_e c^2} = -\frac{1}{8}(Z\alpha)^2\,\bar{p}^{\,4}\,. \tag{79}$$

There are two other order $(Z\alpha)^2$ corrections to the Hamiltonian responsible for fine structure. The first is the spin-orbit interaction,[1]

$$H_{\text{LS}} = \frac{1}{2}\left(\frac{\hbar}{m_e c}\right)^2 \frac{dV}{r\,dr}\boldsymbol{L}\cdot\boldsymbol{s} = \frac{1}{2}\lambda_C^2\,\frac{Ze^2}{4\pi r^3}\boldsymbol{L}\cdot\boldsymbol{s}\,, \tag{80}$$

where V is the electrostatic potential due to the nucleus. H_{LS} describes the interaction between the electron's magnetic moment with the magnetic field that it experiences due to the motion of the nuclear charge as seen in the electron's rest frame; in dimensionless form it is

$$\bar{H}_{\text{LS}} = \tfrac{1}{2}(Z\alpha)^2\frac{1}{\bar{r}^3}\,\boldsymbol{L}\cdot\boldsymbol{s}\,. \tag{81}$$

In (80) and (81), the electron's magnetic moment has been written as

$$\boldsymbol{\mu}_e = \frac{e\hbar}{2m_e c}\,g_e\boldsymbol{s}\,, \tag{82}$$

with the g-factor 2, as given by the Dirac equation. This is not quite correct, however, because vacuum fluctuations of the electromagnetic field produce an anomalous magnetic moment, expressed as

$$g_e = 2\left(1 + \frac{\alpha}{2\pi} + \ldots\right)\,, \tag{83}$$

where ... alludes to higher-order terms in α. In what follows, we shall usually approximate g_e by the naive value of 2.

The last leading-order correction is called the Darwin term, which has no classical counterpart:

$$H_{\text{D}} = -\frac{1}{8}e\lambda_C^2\,\boldsymbol{\nabla}\cdot\boldsymbol{E} = -\frac{1}{8}Ze^2\lambda_C^2\,\delta^3(\boldsymbol{r})\,; \tag{84}$$

here $\boldsymbol{E}$ is the electric field due to the nucleus, assumed to be a point charge in the second expression. In dimensionless form

$$\bar{H}_{\text{D}} = -\tfrac{1}{2}\pi(Z\alpha)^2\,\delta^3(\bar{\boldsymbol{r}})\,. \tag{85}$$

The overbar notation will now be dropped, it being understood that all quantities are in the units of Eq. 78.

The fine structure is thus due to the perturbation of the gross structure spectrum by the Hamiltonian

$$H_{\text{fs}} = H_{\text{kin}} + H_{\text{LS}} + H_{\text{D}}\,. \tag{86}$$

---

[1] This interaction can be derived from classical electrodynamics; see Jackson, §11.8.

H_{fs} must be diagonalized in the space $\mathfrak{H}$ formed by the eigenstates of the nonrelativistic Hamiltonian. This space can be divided into subspaces $\mathfrak{H}_n$ having the bound-state eigenvalues $-1/2n^2$, and a continuum of positive energies. H_{fs} has matrix elements *within* the subspaces $\mathfrak{H}_n$ and between such subspaces. Because H_{fs} as a whole carries the factor $(Z\alpha)^2$, its matrix *within* any degenerate subspace $\mathfrak{H}_n$ will produce splittings proportional to $(Z\alpha)^2$. Corrections due to matrix elements to other subspaces $\mathfrak{H}_m$ and the continuum of unbound states will be of the form

$$\sum_{n \neq m} \frac{(\langle n|H_{\text{fs}}|m\rangle)^2}{E_n - E_m} \sim (Z\alpha)^4 , \tag{87}$$

and are therefore of higher order.[1]

The problem is thus reduced to diagonalizing H_{fs} within the degenerate subspaces $\mathfrak{H}_n$, which have dimension $2n^2$, the factor 2 being due to spin. This would be a quite formidable task if it were to be attempted by brute force, but is actually quite simple because symmetry points directly to a representation in which H_{fs} is diagonal.

H_{kin} and H_{D} are both invariant under rotation of $\boldsymbol{r}$, but H_{LS} is not, nor is the latter invariant under rotation of $\boldsymbol{s}$. Hence H_{LS} is not diagonal in the representation used to solve for the gross structure, in which some one component of $\boldsymbol{L}$ is diagonal. On the other hand, the operator $\boldsymbol{L}\cdot\boldsymbol{s}$ is a scalar product between two angular momenta, and therefore invariant under rotations in the combined space of orbital and spin degrees of freedom, in which rotations are generated by their sum

$$\boldsymbol{j} = \boldsymbol{L} + \boldsymbol{s} ; \tag{88}$$

that is, $[H_{\text{fs}}, \boldsymbol{j}] = 0$. To take advantage of this, introduce simultaneous eigenstates of j^2, L^2, j_3:

$$|ljm\rangle = \sum_{m_l m_s} |lm_l s m_s\rangle\langle lm_l s m_s|jm\rangle . \tag{89}$$

Here $s = \frac{1}{2}$, and as it is an immutable property of the electron it is suppressed, as on the left-hand side of (89), unless clarity calls for showing it. The explicit CG-coefficients will not be needed to find the fine structure splittings, but they illustrate the opening remark that due to the degeneracy the new eigenstates are "50-50" mixtures of the old.

The standard spectroscopic notation for one-electron (in general, one-fermion) states of this type is nl_j, where n and j have their obvious meaning, but for reasons buried in the distant past, the orbital angular momentum quantum number l is stated as $s, p, d, f, g, h \ldots$ for $l = 0, 1, 2 \ldots$. Thus $3d_{\frac{5}{2}}$ stands for the sextet of states with $n = 3, l = 2, j = \frac{5}{2}$.

The spin-orbit interaction can be put into diagonal form by use of

$$\boldsymbol{L}\cdot\boldsymbol{s} = \tfrac{1}{2}[j^2 - L^2 - s^2] = \tfrac{1}{2}[j(j+1) - l(l+1) - \tfrac{3}{4}] . \tag{90}$$

Hence

$$\langle H_{\text{LS}}\rangle_{nlj} = \tfrac{1}{4}\,(Z\alpha)^2\langle r^{-3}\rangle_{nl} \times \begin{cases} l & j = l + \frac{1}{2} \\ -(l+1) & j = l - \frac{1}{2} \end{cases} \tag{91}$$

[1] It should be mentioned here that *consistent* higher-order calculations involve physics significantly more sophisticated than higher-order corrections to the kinetic energy and other higher-order terms from the Dirac equation because effects due to the quantum nature of the electromagnetic field arise at that point.

if $l > 0$, and zero otherwise. The radial expectation value is

$$\langle r^{-3}\rangle_{nl} = \frac{1}{n^3(l+1)(l+\frac{1}{2})l} \,. \tag{92}$$

The kinetic energy correction is evaluated with the following trick:

$$\langle H_{\rm kin}\rangle_{nl} = -\tfrac{1}{2}(Z\alpha)^2\langle(\tfrac{1}{2}p^2)^2\rangle_{nl} = -\tfrac{1}{2}(Z\alpha)^2\langle\left(E_n - \frac{1}{r}\right)^2\rangle_{nl} \,, \tag{93}$$

and

$$\langle r^{-1}\rangle_{nl} = \frac{1}{n^2}\,, \qquad \langle r^{-2}\rangle_{nl} = \frac{1}{n^3(l+\frac{1}{2})} \,. \tag{94}$$

The Darwin term can only contribute to s states because all states with $l \neq 0$ have wave functions that vanish at the origin, whereas for s states

$$|\psi_{n0}(0)|^2 = \frac{1}{4\pi}|R_{n0}(0)|^2 = \frac{1}{\pi n^3} \,. \tag{95}$$

The spin-orbit correction has the peculiar property of appearing to be finite for $l = 0$, and if taken at face value it gives the same correction as the Darwin term. This is a hoax, of course, for there is no spin-orbit interaction when $l = 0$, yet it provides a fortuitous "derivation" of the fine structure based on the classically expected corrections, $H_{\rm kin}$ and $H_{\rm LS}$.

The three terms in Eq. 86 then combine into the following formula for the spectrum, including fine structure to order $(Z\alpha)^2$:

$$E_{\rm fs}(nlj) = E_n\left[1 + \frac{(Z\alpha)^2}{n^2}\left(\frac{n}{j+\frac{1}{2}} - \frac{3}{4}\right)\right] \,. \tag{96}$$

This formula, first derived by Sommerfeld from the "old" quantum theory, motivated him to call α the fine structure constant. It is remarkable that although the spin-orbit and relativistic corrections both depend on l, their l-dependence has canceled exactly in this final result, leaving a spectrum that depends only on the principal quantum number and the total angular momentum. This means, of course, that the degeneracy in the gross structure has not been completely lifted. For example, in the $n = 2$ multiplet the states $s_{\frac{1}{2}}$ and $p_{\frac{1}{2}}$ are degenerate. The Dirac equation, which produces a fine structure formula to all orders in $Z\alpha$ for an infinitely heavy nucleus, retains this same degeneracy (see §13.5).

The observed spectrum, however, does not only depend on n and j: the s states lie above the $p_{1/2}$ states, and quite substantially so in the $n = 2$ multiplet (see Fig. 5.1). This effect is called the *Lamb shift* after its discoverer, and is completely accounted for by quantum electrodynamics as being due to vacuum fluctuations of the electromagnetic field, and other smaller effects due to virtual e^+e^- pairs. In §10.3 a heuristic derivation of the leading-order Lamb shift will be given, which suffices to reveal its general form and to show that it is dominantly an s-state effect. The complete leading-order Lamb shift in s states is

$$\Delta E_{\rm L}(ns) = \frac{8}{3\pi}\,\frac{(\alpha Z)^3 Z}{n^3}\left[\ln\left(\frac{m_e c^2}{\bar{W}_n}\right) + \frac{19}{30}\right]\,{\rm Ry} \,. \tag{97}$$

There is also a much smaller shift in p states, as shown in Fig. 5.1. The Lamb shift is of higher-order in αZ than fine structure, but the energies $\bar{W}_n$ are surprisingly large ($\bar{W}_2 \simeq 33Z^2$ Ry), and therefore the effect is substantially more important than an order of magnitude estimate would indicate.

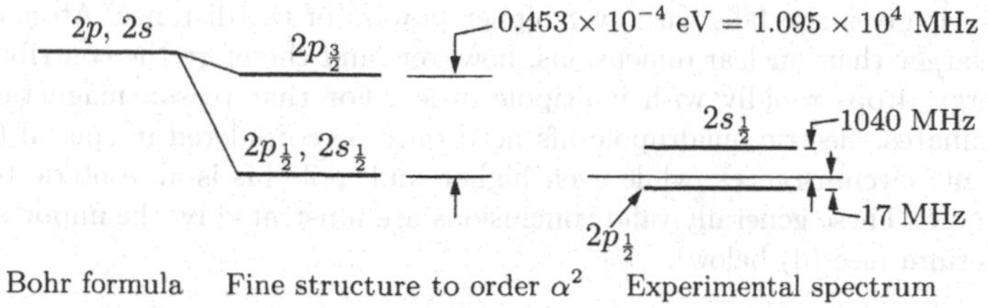

FIG. 5.1. The $n = 2$ multiplet in hydrogen.

(b) Hyperfine Structure — General Features

The spectrum displays hyperfine structure when the electron experiences deviations from a central field caused by nuclear electric and/or magnetic multipole moments. The electron's total angular momentum $\boldsymbol{j}$ is then no longer a constant of motion. In addition to its intrinsic importance, hfs therefore offers an instructive exercise in angular momentum theory, which is missed if one first learns unnecessarily sophisticated techniques (Wigner-Eckart and all that) and extracts the final results from opaque tables of Racah coefficients.

The only nuclear degree of freedom that can impact hyperfine structure is the nuclear spin $\boldsymbol{I}$, because all other nuclear degrees of freedom are frozen unless energies comparable to nuclear excitation energies are available, and these are enormous in comparison to atomic energies. In §3.3 we learned that states of angular momentum $I = \frac{1}{2}$ can only have dipole moments; states of angular momentum $I = 1$ can have both dipole and quadrupole moments, but not higher moments; and so forth. In short, a nucleus must have $I > 0$ if anything beyond an electric monopole moment is to have a nonzero expectation value, and by that token anything other than a central field is to exist. Therefore hyperfine structure is contingent on I not vanishing.

Space reflection invariance imposes further restrictions if nuclear states are parity eigenstates, i.e., if the strong interaction is reflection invariant, as it is known to be to rather high accuracy. The non-central energy of interaction between a charge and current-carrying object and the electromagnetic field is of the form $\boldsymbol{d} \cdot \boldsymbol{E} + \boldsymbol{\mu} \cdot \boldsymbol{B} + \ldots$, where $\boldsymbol{d}$ and $\boldsymbol{\mu}$ are the electric and magnetic dipole moments, and $+ \ldots$ refers to higher-order multipoles. Now $\boldsymbol{E}$ is a polar vector, and $\boldsymbol{B}$ an axial vector, and therefore $\boldsymbol{d}$ and $\boldsymbol{\mu}$ are, respectively, polar and axial vectors, i.e., odd and even under reflection, respectively. Let A be any operator, and $|\psi\rangle$ any state of definite parity, whether even or odd. Then $\langle\psi|A|\psi\rangle$ must vanish if A is odd under reflection. As a consequence, $\boldsymbol{d}$ cannot have an expectation value in any state of definite parity, but $\boldsymbol{\mu}$ can (provided $I \geq \frac{1}{2}$, of course). This story continues as follows; the electric quadrupole moment is a quadratic form in the coordinates, and therefore does not change sign under reflections, so states of definite parity can have an electric quadrupole moment provided $I \geq 1$. Magnetic quadrupole moments change sign under reflection and therefore have vanishing expectation values in parity eigenstates.

To summarize, only odd-order magnetic moments and even-order electric moments of the nucleus can produce hyperfine structure. From classical electrodynamics we know that as the order of the multipole moment increases, the field to which it gives rise falls with ever higher powers of the distance. Atomic states are far larger than nuclear dimensions, however, and therefore the contribution to the energy drops rapidly with multipole order. For that reason magnetic dipole hfs dominates, electric quadrupole hfs need only be considered in special (though important) circumstances, while even higher multipole hfs is an esoteric topic we shall ignore. These generally valid conclusions are illustrated by the important case of deuterium (see (d) below).

(c) Magnetic Dipole Hfs

The nuclear magnetic dipole moment $\boldsymbol{\mu}$ is a vector operator, and therefore proportional to $\boldsymbol{I}$, the nuclear spin. The conventional definition is

$$\boldsymbol{\mu} = g_N \mu_N \, \boldsymbol{I} \,, \tag{98}$$

where g_N is called the gyromagnetic ratio, or *g-factor*, of the nucleus in question; μ_N is the nuclear magneton,

$$\mu_N = \frac{e\hbar}{2m_p c} = \frac{\mu_B}{1836.52} \,; \tag{99}$$

m_p is the mass of the proton (*not* the nucleus); and μ_B is the Bohr magneton. By definition, the value of the nuclear magnetic moment is the expectation value $\langle \boldsymbol{\mu} \rangle$ in the state with $I_3 = I$, i.e., $g_N \mu_N I$.

The dependence of hfs on the angular momentum quantum numbers I and j is completely determined by rotational symmetry. This is most apparent in magnetic dipole hfs, because in this case $\boldsymbol{I}$ (or equivalently $\boldsymbol{\mu}$) is the only nuclear observable relevant to the problem. As this is a vector operator, rotational invariance requires the interaction between the electron and the nuclear magnetic moment to have the form

$$H_{M1} = \boldsymbol{\mu} \cdot \boldsymbol{\Lambda} \,, \tag{100}$$

where $\boldsymbol{\Lambda}$ is a vector operator in the electron's Hilbert space. With this interaction neither the nuclear nor the electronic angular momenta are constants of motion, and only the atom's total angular momentum,

$$\boldsymbol{F} = \boldsymbol{I} + \boldsymbol{j} \,, \tag{101}$$

is conserved.

In general $\boldsymbol{\Lambda}$ is not proportional to the electron's spin operator $\boldsymbol{s}$, because the orbital motion produces a current that also interacts with the magnetic field produced by the nucleus. For now, it is not necessary to know *anything* about $\boldsymbol{\Lambda}$ other than that it is a vector operator, and that the energies involved in hfs are minute compared to fine structure, so that only matrix elements of H_{M1} *within* a single fine structure level are of concern, i.e., matrix elements between states of the same j and l.

Recall now the general theorems in §3.3 regarding vector operators within subspaces of specified total angular momentum; in particular,

$$\boldsymbol{\Lambda} = \lambda(r) \boldsymbol{j} \,, \qquad \lambda(r) = \frac{\langle \boldsymbol{j} \cdot \boldsymbol{\Lambda} \rangle_{lj}}{j(j+1)} \,, \tag{102}$$

where the expectation value is only over the angular and spin variables. Hence λ is, in general, a function of the electron's radial coordinate $|\boldsymbol{r}|$ because rotational invariance, which is all that has been brought to bear thus far, cannot determine any dependence on rotationally invariant quantities. Such "details" as the r-dependence of λ need not concern us as yet.

In sum, the term in the atom's Hamiltonian responsible for the magnetic dipole contribution to hfs is

$$H_{M1} = g_N \mu_N \lambda(r) \boldsymbol{I} \cdot \boldsymbol{j} . \tag{103}$$

The familiar expression for the scalar product of two vector operators then gives *the F dependence of the magnetic dipole hyperfine structure*:

$$\langle H_{M1} \rangle_{FInlj} = \tfrac{1}{2} g_N \mu_N \langle \lambda_m(r) \rangle_{nlj} [F(F+1) - I(I+1) - j(j+1)] , \tag{104}$$

where $\langle \ldots \rangle_{nlj}$ is the expectation value in the indicated electronic state. Experiments measure hyperfine splittings, not absolute energies, and in any given hyperfine multiplet, j is fixed, as of course is the nuclear spin I for any one isotope, so these splittings are proportional to $F(F+1)$.

A bit of reflection shows that Eq. 104 also applies to a complex atom once j is replaced by the total angular momentum J of the electrons in the level of interest. Of course, the calculation of the coefficient that takes the place of $\lambda(r)$ is then far more difficult than in a hydrogenic atom.

The explicit form of the operator $\boldsymbol{\Lambda}$ remains to be found. This is the hardest part of the calculation, because magnetic interactions are complicated. There are two parts to the problem: determining the interaction due to the nuclear dipole moment; and computing the expectation value of that operator. The first is really a problem in classical electromagnetism. The second part is truly quantum mechanical.

The vector potential and magnetic field at $\boldsymbol{r}$ due to a nuclear magnetic dipole moment $\boldsymbol{\mu}$ at the origin are[1]

$$\boldsymbol{A} = \frac{1}{4\pi r^3}(\boldsymbol{\mu} \times \boldsymbol{r}) , \qquad \boldsymbol{B} = \frac{1}{4\pi r^3}[3\boldsymbol{n}(\boldsymbol{n} \cdot \boldsymbol{\mu}) - \boldsymbol{\mu}] + \frac{2}{3}\boldsymbol{\mu}\, \delta(\boldsymbol{r}) , \tag{105}$$

where $\boldsymbol{n} = \boldsymbol{r}/|\boldsymbol{r}|$. The delta function in $\boldsymbol{B}$ stems from the singularity in $\boldsymbol{A}$ at the origin, and as we will shortly see it is crucial. The energy of interaction between the electron and this magnetic field is therefore

$$H_{M1} = \frac{|e|}{mc}\boldsymbol{p} \cdot \boldsymbol{A} - \boldsymbol{\mu}_e \cdot \boldsymbol{B} = \frac{\mu_B}{2\pi r^3}\boldsymbol{\mu} \cdot \boldsymbol{L} - \boldsymbol{\mu}_e \cdot \boldsymbol{B} , \tag{106}$$

where the negligible higher-order term quadratic in $\boldsymbol{A}$ has been dropped, and $\boldsymbol{\mu}_e = -g_e \mu_B \boldsymbol{s}$ is the electron's magnetic moment. Hence

$$H_{M1} = \frac{2}{3} g_N g_e \mu_N \mu_B (\boldsymbol{s} \cdot \boldsymbol{I})\, \delta(\boldsymbol{r}) + \frac{g_N \mu_N \mu_B}{4\pi r^3}\Big\{ 2\boldsymbol{L} \cdot \boldsymbol{I} + g_e [3(\boldsymbol{n} \cdot \boldsymbol{s})(\boldsymbol{n} \cdot \boldsymbol{I}) - \boldsymbol{s} \cdot \boldsymbol{I}] \Big\} \tag{107}$$

$$\simeq \frac{4}{3} g_N \mu_N \mu_B (\boldsymbol{s} \cdot \boldsymbol{I})\, \delta(\boldsymbol{r}) + \frac{g_N \mu_N \mu_B}{2\pi r^3}\Big\{ \boldsymbol{L} \cdot \boldsymbol{I} + [3(\boldsymbol{n} \cdot \boldsymbol{s})(\boldsymbol{n} \cdot \boldsymbol{I}) - \boldsymbol{s} \cdot \boldsymbol{I}] \Big\} , \tag{108}$$

[1] Jackson, §5.7.

where $g_e = 2$ instead of 2.0023 was used in the second expression.

The complicated expression inside $\{\ldots\}$ is proportional to the scalar product between the second-rank tensors

$$N_{ij} = 3n_i n_j - \delta_{ij} , \qquad \Sigma_{ij} = \tfrac{3}{2}(s_i I_j + s_j I_i) - \delta_{ij} \boldsymbol{s} \cdot \boldsymbol{I} . \tag{109}$$

The two terms in H_{M1} thus have very distinct properties: the first term only contributes to $l = 0$ states because only such wave functions have a nonzero value at $r = 0$, whereas $\boldsymbol{L}$ and the tensor N_{ij} only have nonzero expectation values in states with $l \geq 1$. In consequence, the hyperfine shifts in s-states, and in particular in the all-important ground state, are given by Fermi's formula

$$\Delta E_{ns} = \frac{4}{3} \frac{\alpha^2 Z^3}{n^3} \frac{g_N m_e}{m_p} \left[F(F+1) - I(I+1) - \tfrac{3}{4}\right] \text{Ry} , \tag{110}$$

where $|\psi_{ns}(0)|^2 = Z^3/\pi n^3 a_0^3$ was used.

The $l \geq 1$ hyperfine interaction is

$$H^l_{M1} = \frac{g_N \mu_N \mu_B}{2\pi} \frac{\boldsymbol{\Lambda} \cdot \boldsymbol{I}}{r^3} , \qquad \boldsymbol{\Lambda} = \boldsymbol{L} + 3\boldsymbol{n}(\boldsymbol{n} \cdot \boldsymbol{s}) - \boldsymbol{s} . \tag{111}$$

The calculation is greatly simplified by noting that the only component of any vector operator $\boldsymbol{V}$ that contributes to its matrix elements in a subspace of definite total angular momentum j is the component parallel to $\boldsymbol{j}$; this fact is expressed by the identity

$$j(j+1)\langle jm|\boldsymbol{V}|jm'\rangle = \langle jm|(\boldsymbol{V} \cdot \boldsymbol{j})\, \boldsymbol{j}|jm'\rangle , \tag{112}$$

whose proof is Prob. 8. In the subspace of interest, therefore,

$$\boldsymbol{\Lambda} \cdot \boldsymbol{I} = \frac{(\boldsymbol{j} \cdot \boldsymbol{\Lambda})(\boldsymbol{I} \cdot \boldsymbol{j})}{j(j+1)} = \frac{F(F+1) - I(I+1) - j(j+1)}{2j(j+1)} \left[\boldsymbol{j} \cdot (\boldsymbol{L} - \boldsymbol{s}) + 3(\boldsymbol{j} \cdot \boldsymbol{n})(\boldsymbol{n} \cdot \boldsymbol{s})\right] \tag{113}$$

$$= \frac{l(l+1)}{2j(j+1)} \left[F(F+1) - I(I+1) - j(j+1)\right] , \tag{114}$$

where $\boldsymbol{L} \cdot \boldsymbol{n} = 0$ and $(\boldsymbol{s} \cdot \boldsymbol{n})^2 = \frac{1}{4}$ were used. The expectation value of $1/r^3$ (Eq. 92) then gives the complete expression for the hfs shifts in hydrogenic states with $l \geq 1$:

$$\Delta E_{nlj} = \frac{\alpha^2 Z^3}{n^3} \frac{g_N m_e}{m_p} \frac{F(F+1) - I(I+1) - j(j+1)}{j(j+1)(2l+1)} \text{ Ry} . \tag{115}$$

By what appears to be a sheer accident, this formula reduces to Eq. 110 for s states.

(d) Electric Quadrupole Hfs

The electric quadrupole ($E2$) contribution to hfs, when it exists, has a very different F-dependence from that of the $M1$ contribution. It can, therefore, be identified in the spectrum without knowledge of its overall coefficient, which, as in the $M1$ case, depends on the radial structure of the electronic state in question.

The energy of interaction between the nuclear charge distribution $\rho_N(\boldsymbol{R})$ and the charge distribution of the electron $\rho_e(\boldsymbol{r})$ is

$$H_E = \int d\boldsymbol{r} \int d\boldsymbol{R}\, \frac{\rho_N(\boldsymbol{R})\rho_e(\boldsymbol{r})}{4\pi|\boldsymbol{R} - \boldsymbol{r}|} . \tag{116}$$

As $r \gg R$, the appropriate form of the multipole expansion is

$$\frac{1}{|\boldsymbol{R}-\boldsymbol{r}|}=\sum_{lm}\frac{4\pi}{2l+1}\frac{R^l}{r^{l+1}}Y_{lm}(\hat{\boldsymbol{R}})Y^*_{lm}(\hat{\boldsymbol{r}})\,. \tag{117}$$

The $l=0$ term is just the central field responsible for gross structure, so only the $l>0$ terms are relevant to hfs. In evaluating their contribution, the expectation value is taken in the zeroth-order states, which are products of a nuclear and an electronic state. Under space reflection, $Y_{lm}(\hat{\boldsymbol{r}})\to(-1)^lY_{lm}(\hat{\boldsymbol{r}})$, so in a nuclear state of definite parity only the even l survive, as already stated at the outset. Hence the leading electrostatic contribution to hfs is produced by the energy

$$H_{E2}=\frac{1}{4\pi}\int d\boldsymbol{r}\int d\boldsymbol{R}\;\rho_N(\boldsymbol{R})\rho_e(\boldsymbol{r})\,\frac{R^2}{r^3}\,P_2(\hat{\boldsymbol{R}}\cdot\hat{\boldsymbol{r}})\,. \tag{118}$$

This expression can be written as the scalar product of the nuclear and electronic quadrupole tensors by recalling $P_2(x)=\frac{1}{2}(3x^2-1)$, and then noting that

$$P_2(\hat{\boldsymbol{R}}\cdot\hat{\boldsymbol{r}})=\frac{1}{6}\sum_{k,l=1}^{3}(3\hat{R}_k\hat{R}_l-\delta_{kl})(3\hat{r}_k\hat{r}_l-\delta_{kl})\,. \tag{119}$$

The nuclear quadrupole tensor is defined as

$$Q^{(N)}_{kl}=\int(3\hat{R}_k\hat{R}_l-\delta_{kl})\,R^2\,\rho_N(\boldsymbol{R})\,d\boldsymbol{R}\,. \tag{120}$$

This second-rank tensor is traceless (i.e., $Q^{(N)}_{kk}=0$), and according to the general result Eq. 3.99, its matrix elements in the $(2I+1)$-dimensional space of the nuclear ground state is

$$Q^{(N)}_{kl}=\lambda_N[I_kI_l+I_lI_k-\tfrac{2}{3}\delta_{kl}I(I+1)]\,. \tag{121}$$

By convention, what is called and tabulated as "the" quadrupole moment Q_N is $\langle Q_{33}\rangle$ in the state with $I_3=I$, and therefore

$$\lambda_N=\frac{3Q_N}{2I(2I-1)}\,. \tag{122}$$

Note that Q_N has the dimension (length)^2, where the length is at most of order the nuclear radius if the charge distribution is highly aspherical, but is usually far smaller than that.

For the electron it is convenient to use a dimensionless definition of the quadrupole moment:

$$Q^e_{kl}=-e\langle 3\hat{r}_k\hat{r}_l-\delta_{kl}\rangle\,, \tag{123}$$

where e is positive throughout. By the argument that led to (121) and (122),

$$Q^{(e)}_{kl}=-\frac{3Q_e}{2j(2j-1)}[j_kj_l+j_lj_k-\tfrac{2}{3}\delta_{kl}j(j+1)]\,. \tag{124}$$

Here Q_e is again the expectation value of $Q^{(e)}_{33}$ in the state with $j_3=j$,

$$Q_e=\langle 3\cos^2\theta-1\rangle_{jj}\,. \tag{125}$$

The level displacement due to the quadrupole interaction is therefore

$$\Delta_{E2}(Ij;F) = \langle H_{E2}\rangle = -\frac{1}{6}\cdot\frac{1}{4\pi}\langle r^{-3}\rangle_{nl}\,\frac{3Q_N}{2I(2I-1)}\frac{3Q_e}{2j(2j-1)}\Lambda(Ij;F)\,, \tag{126}$$

where

$$\begin{aligned}\Lambda(Ij;F) &= \langle[I_kI_l + I_lI_k - \tfrac{2}{3}\delta_{kl}I(I+1)][j_kj_l + j_lj_k - \tfrac{2}{3}\delta_{kl}j(j+1)]\rangle \\ &= A_F(A_F+1) - \tfrac{4}{3}I(I+1)j(j+1)\,, \qquad (127)\\ A_F &\equiv 2\boldsymbol{I}\cdot\boldsymbol{j} = F(F+1) - I(I+1) - j(j+1)\,. \qquad (128)\end{aligned}$$

As we see, the quadrupole interaction leads to a quite different F-dependence from the dipole hfs interaction, which produces level spacings proportional to A_F.

5.4 The Zeeman and Stark Effects

The Zeeman and Stark effects are the splittings and shifts seen in the spectrum when the atom (or molecule) is in a uniform applied magnetic or electric field, respectively. For a hydrogenic atom, the Hamiltonian that incorporates these effects is

$$H = \frac{1}{2m}\left(\boldsymbol{p} + \frac{e}{c}\boldsymbol{A}\right)^2 - \frac{Ze^2}{4\pi r} + e\boldsymbol{r}\cdot\boldsymbol{E} + g_e\mu_B\boldsymbol{s}\cdot\boldsymbol{B} + H_{\text{fs}}\,. \tag{129}$$

Here H_{fs} is the energy that produces the fine structure, the nuclear mass is treated as if it were infinite, m is to be the electron mass throughout, and $e > 0$, i.e., the electron's charge is $-|e|$. Although the effect of electromagnetic fields on hyperfine multiplets is an important topic, we will confine ourselves to fields that produce effects large compared to hyperfine splittings. Once this regime is understood, it is straightforward to analyze the effects produced in hyperfine spectra.

As in §4.3, we use the gauge $\boldsymbol{A} = \frac{1}{2}(\boldsymbol{B}\times\boldsymbol{r})$. Equation (129) then has the form

$$H = H_0 + H_{\text{fs}} + H_{\text{fld}}\,, \tag{130}$$

$$H_{\text{fld}} = H_{B1} + H_{B2} + H_{B,s} + H_E\,. \tag{131}$$

Here H_{Bk} are the terms of order B^k due to the interaction of the magnetic field with the electron's convection current, $H_{B,s}$ the interaction with the electron's magnetic moment, and the notation makes the meaning of the other terms obvious. The explicit expressions for H_{Bn} are

$$H_{B1} = \mu_B\boldsymbol{B}\cdot\boldsymbol{L} = 4.254\cdot 10^{-6}\,\boldsymbol{B}_{\text{T}}\cdot\boldsymbol{L}\ \text{Ry} \tag{132}$$

$$H_{B2} = \frac{e^2}{8mc^2}\left[B^2r^2 - (\boldsymbol{B}\cdot\boldsymbol{r})^2\right] = 4.524\cdot 10^{-12}\left[B_T^2r_\circ^2 - (\boldsymbol{B}\cdot\boldsymbol{r}_\circ)^2\right]\text{Ry}\,, \tag{133}$$

where $\boldsymbol{B}_{\text{T}}$ is the magnetic field strength in Tesla, $\boldsymbol{L}$ is the electron's orbital angular momentum, and $r_\circ = r/a_0$.

(a) Order of Magnitude Estimates

With a Hamiltonian having as many non-commuting terms as does Eq. 130, an understanding of the relative importance of the various terms is indispensable. We therefore begin by estimating their orders of magnitude.

When the applied fields are absent, the features of the spectrum are characterized by two energies: the distance between gross structure multiplets Δ_n, and the spread of fine structure within one such multiplet, ϵ_n. According to Eq. 9 and Eq. 96, these are

$$\Delta_n \simeq \frac{\partial |E_n|}{\partial n} = \frac{2Z^2}{n^3} \text{Ry} , \tag{134}$$

$$\epsilon_n = |E_{\text{fs}}(nj_{\text{max}}) - E_{\text{fs}}(nj_{\text{min}})| = Z^2 (Z\alpha)^2 \frac{n-1}{n^4} \text{Ry} . \tag{135}$$

We will assume that Z is modest so that $\epsilon_n \lll \Delta_n$.

The major divide is set by the field strengths required to make the perturbation H_{fld} comparable to the gross structure splittings Δ_n. If the fields are weak compared to this, it suffices to diagonalize H_{fld} in each separate gross structure multiplet. If, however, the fields are stronger than that, the unperturbed spectrum and states only provide a useful starting point for attacking the problem numerically. This difficult situation arises for any strength of the applied fields in a hydrogenic atom, of course, because the splittings Δ_n shrink like $1/n^3$ as the ionization threshold is approached.

If the magnetic field is sufficiently strong to make H_{Bk} large compared to the ground-state ionization energy, another tractable situation arises, for then motion in a uniform magnetic field in empty space provides a natural starting point, with the Coulomb attraction being a relatively small perturbation, even though this "small" central attraction has the crucial effect of producing bound states. Such enormous magnetic fields occur on the surface of neutron stars, and for that reason this problem has received much attention.

The ratio of the first-order magnetic energy to the gross structure splittings is

$$\frac{H_{B1}}{\Delta_n} \sim 2 \cdot 10^{-6} \frac{n^4}{Z^2} B_{\text{T}} \tag{136}$$

because $l_{\text{max}} = n - 1$. The ratio of the second to first-order energies is

$$\frac{H_{B2}}{H_{B1}} = B \frac{e^2}{8mc^2} \frac{\langle r^2 \sin^2 \theta \rangle}{\mu_B l} . \tag{137}$$

Now

$$\langle r^2 \rangle_{nl} = (a_0/Z)^2 \tfrac{1}{2} n^2 \left[5n^2 + 1 - 3l(l+1) \right] = (a_0/Z)^2 \, 2n^4 \, C_{nl} , \tag{138}$$

where $C_{nl} \sim 1$, while the variation of $\langle \sin^2 \theta \rangle$ with n is undramatic, so we use

$$\frac{H_{B2}}{H_{B1}} \sim 10^{-6} \frac{n^4}{Z^2 l} B_{\text{T}} \tag{139}$$

as a rough estimate. According to (136), as long as the applied field satisfies the condition

$$B \ll \frac{Z^2}{2n^4} \, 10^6 \text{ T} \tag{140}$$

the Zeeman effect does not mix gross structure multiplets. As expected, this crucial condition is inexorably violated by states with sufficiently large n. When (140) is satisfied, H_{B2} is negligible compared to H_{B1} because of Eq. 139 unless $l \lll n$. Consequently we will ignore the term quadratic in B (though it must not be forgotten as there are cases in which H_{B1} has a vanishing effects, whereas H_{B2} does not).

If the field satisfies condition (140), and is yet strong enough to make H_{B1} large compared to the fine structure splitting, a further simplification occurs, because then the perturbation due to the fields can be handled as if there is no fine structure splitting. The ratio that determines whether this circumstance holds is

$$\frac{H_{B1}}{\epsilon_n} \gtrsim \frac{4 \cdot 10^{-6}}{Z^2(Z\alpha)^2} n^4 \, B_{\mathrm{T}} \,, \tag{141}$$

where l has been replaced by $l_{\max}$. As we will see, the spectrum in the presence of both electric *and* magnetic fields in arbitrary directions can be determined exactly when their perturbation is large compared to the fine structure splitting (while still small compared to gross structure splittings).

Finally, a far simpler situation prevails when the field is weak enough to make H_{B1} small compared to the fine structure splittings, because then the mixing between different fine structure levels is small, and simple perturbation theory will do.

Turning to the Stark effect, it helps to use

$$ea_{\circ}E = 3.889 \cdot 10^{-12} \, E_{\mathrm{V}} \, \mathrm{Ry} \,, \tag{142}$$

where E_{V} is the electric field strengths in Volt/m. Consequently, the magnitude of the electric perturbation is

$$H_E \sim \frac{4n^2}{Z} \, 10^{-12} \, E_{\mathrm{V}} \, \mathrm{Ry} \,, \tag{143}$$

where the exact formula

$$\langle r \rangle_{nl} = (a_{\circ}/2Z)[3n^2 - l(l+1)] \tag{144}$$

was crudely approximated by $a_{\circ}n^2/Z$. On referring to (132) we see that the electric field does not mix gross structure multiplets as long as it satisfies the condition

$$E \ll \frac{Z^3}{2n^5} \, 10^{12} \, \mathrm{V/m} \,. \tag{145}$$

The ratio of the electric to magnetic perturbations is of order

$$\frac{H_E}{H_{B1}} \sim \frac{n^2}{Zl} \, 10^{-6} \, \frac{E_{\mathrm{V}}}{B_{\mathrm{T}}} \,. \tag{146}$$

Consequently, conditions under which the Stark and Zeeman effects are of comparable magnitude are easily attained.

Condition (145) is deceptive, however, because the electric field changes the character of the potential dramatically no matter how small E may be. Let $\boldsymbol{E}$ point along z; then the potential experienced by the electron is

$$V_E = -\frac{Ze^2}{4\pi\sqrt{x^2+y^2+z^2}} + eEz \,. \tag{147}$$

Hence, the applied field has turned the purely attractive Coulomb potential into one with a barrier through which the bound states can leak to $z \to -\infty$. This situation is in essence identical to that produced by adding a cubic term to the harmonic oscillator. That is, the potential V_E only has a continuous spectrum, and the field will eventually ionize the atom! In practice, however, the modification of the spectrum is not dramatic for states lying well below the top of the barrier, because then the ionization probability is exponentially small.[1] The position of these energy levels is correctly given by treating H_E with conventional perturbation theory, ignoring the miniscule leakage through the barrier.

(b) The $n = 2$ Multiplet

A detailed calculation of the Zeeman and Stark effects in the $n = 2$ multiplet of hydrogen ($Z = 1$) is of interest in its own right, and will also illustrate general features of such perturbation calculations. In order to maintain the flow of the presentation, and to give the reader some hands-on experience, quite a few details are not spelled out in the text and are left for Probs. 10 and 12.

First, *the Zeeman effect.* The Hamiltonian for a magnetic field in the z-direction is

$$H_Z = H_0 + H_{\text{fs}} + \mathcal{B}(L_z + 2s_z) , \qquad \mathcal{B} = \mu_B |\boldsymbol{B}| . \tag{148}$$

This form of the Hamiltonian holds for *any* atom, not just hydrogen, if $\boldsymbol{L}$ and $\boldsymbol{s}$ are replaced by the total orbital and spin ($\boldsymbol{S}$) angular momenta of all the electrons. Hence the following statements apply to the Zeeman energy in general:

1. H_Z is invariant under rotations about $\boldsymbol{B}$, and therefore only has matrix elements between states with the same magnetic quantum number M_J, where M_J is the eigenvalue of $\boldsymbol{J} \cdot \boldsymbol{B}/B$.

2. Both $\boldsymbol{L}$ and $\boldsymbol{S}$ are axial vector operators, and therefore H_Z only has matrix elements between states of the same parity. Furthermore, $L_z + 2S_z = J_z + S_z$, so only the matrix elements of $\mathcal{B}S_z$ are needed, and they satisfy the selection rule $\Delta J = 0, 1$ if $\boldsymbol{S}$ is not a constant of motion.

3. If $\boldsymbol{L}$ and $\boldsymbol{S}$ are separate constants of motion (called Russel-Saunders coupling in complex atoms — see §6.3), the selection rule is $\Delta L = \Delta S = 0$.

Now on to hydrogen. If the field is too weak to mix different fine structure states, the expectation value of the perturbation in the states $|nljm\rangle$ gives the shift:

$$\Delta E_{nljm} = \mathcal{B}\langle j_z + s_z \rangle_{nljm} . \tag{149}$$

But $\langle j_z \rangle = m$, and by transforming from the $(lsjm)$ to the $(lm_l s m_s)$ basis one finds that

$$\langle s_z \rangle_{nljm} = \pm \frac{m}{2l+1} \qquad (j = l \pm \tfrac{1}{2}) . \tag{150}$$

Hence the low-field Zeeman effect for all hydrogen levels is

$$\Delta E_{nljm} = \mathcal{B}\, m \, \frac{2l + 1 \pm 1}{2l+1} \qquad (j = l \pm \tfrac{1}{2}) . \tag{151}$$

---

[1] See *LLQM*, §77.

If the field is strong enough to mix fine structure levels, a separate calculation must be done for each multiplet. The selection rules are now $\Delta j = 0, 1$, and $\Delta l = 0$ because $\Delta l = 1$ is forbidden by parity.

Consider the $n = 2$ multiplet, with unperturbed levels $2p_{\frac{1}{2}}, 2s_{\frac{1}{2}}, 2p_{\frac{3}{2}}$ in order of increasing energy. Because of the parity selection rule the s state cannot mix with the p states, and the angular momentum selection rule forbids mixing of $p_{\frac{3}{2}}$ states with $m = \pm\frac{3}{2}$ with the $p_{\frac{1}{2}}$ states. But the p states with $m = \pm\frac{1}{2}$ do mix:

$$\langle p_{\frac{1}{2}}, m | s_z | p_{\frac{3}{2}}, m \rangle = -\tfrac{1}{3}\sqrt{\tfrac{9}{4} - m^2}\,. \tag{152}$$

That is, (151) applies as it stands to the s states, and the p states with $m = \pm\frac{3}{2}$, but the energies of the p states with $m = \pm\frac{1}{2}$ must be found by diagonalizing the following matrix,

$$\begin{pmatrix} \Delta + \frac{4}{3}\mathcal{B}m & -\frac{1}{3}\mathcal{B}\sqrt{2} \\ -\frac{1}{3}\mathcal{B}\sqrt{2} & \frac{2}{3}\mathcal{B}m \end{pmatrix}, \tag{153}$$

where $\Delta = E(p_{\frac{3}{2}}) - E(p_{\frac{1}{2}})$ when there is no field. The energies of the p states with $m = \pm\frac{1}{2}$ are therefore

$$E_m^{\pm} = \tfrac{1}{2}\Delta + \mathcal{B}m \pm \tfrac{1}{2}\sqrt{\Delta^2 + \tfrac{4}{3}\mathcal{B}m\Delta + (\mathcal{B})^2}\,. \tag{154}$$

This spectrum is plotted in Fig. 5.2.

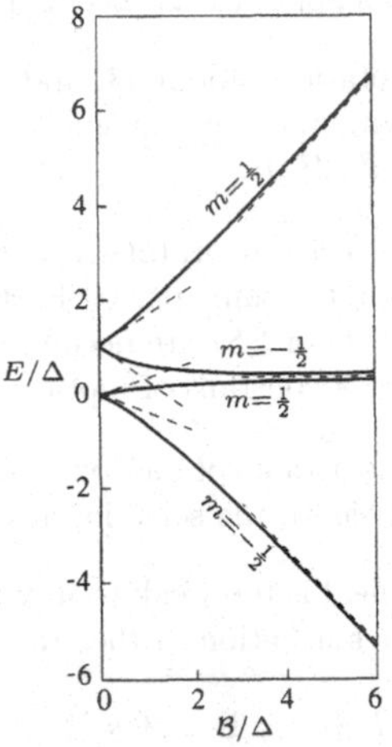

FIG. 5.2. The Zeeman effect in the $n = 2$ multiplet. Only the $m = \pm\frac{1}{2}$ levels are shown because the $m = \pm\frac{3}{2}$ levels are given by Eq. 151 for all values of $\mathcal{B} = \mu_B|\boldsymbol{B}|$. The solid curves are the exact energies, while the dashed lines are the low and high field limits.

The limit when $\mathcal{B}$ is small compared to the fine structure splitting Δ is instructive. Calling $\lambda = \mathcal{B}/\Delta$,

$$E_m^{+} = \Delta\left\{1 + \tfrac{4}{3}\lambda m + \lambda^2(\tfrac{1}{4} - \tfrac{1}{9}m^2) + \ldots\right\} \tag{155}$$

$$E_m^{-} = \tfrac{2}{3}\mathcal{B}m - \mathcal{B}\lambda(\tfrac{1}{4} - \tfrac{1}{9}m^2) + \ldots\,. \tag{156}$$

The terms linear in $\mathcal{B}$ agree, as they must, with Eq. 151, while those quadratic in $\mathcal{B}$ are what is found by using the next order in perturbation theory using the fs spectrum as the starting point so that Δ is the energy denominator. Diagonalizing (153) amounts to summing the perturbation series involving only the two p states to all orders.

The other limit, where $\mathcal{B}$ is large compared to the fine structure splitting is of considerable interest. The energies are

$$E_m^{\pm} = \mathcal{B}\left\{(m \pm \tfrac{1}{2}) + \lambda^{-1}(\tfrac{1}{2} \pm \tfrac{1}{3}m) \pm \lambda^{-2}(\tfrac{1}{4} - \tfrac{1}{9}m^2) + \ldots\right\}; \tag{157}$$

keeping only the leading terms gives what is known as the Paschen-Back effect:

$$E_{\frac{1}{2}}^{+} = \mathcal{B} + \tfrac{2}{3}\Delta\,, \quad E_{-\frac{1}{2}}^{-} = -\mathcal{B} + \tfrac{2}{3}\Delta\,, \quad E_{-\frac{1}{2}}^{+} = E_{\frac{1}{2}}^{-} = \tfrac{1}{3}\Delta\,. \tag{158}$$

In this limit the Zeeman effect is trivial in the $(lm_l sm_s)$ basis, where it is diagonal and gives the energies $\mathcal{B}(m_l + 2m_s)$, while the fine structure is accounted for by the expectation value of $H_{\rm fs}$ in this basis.

The Hamiltonian for *the Stark effect* in hydrogen is

$$H_S = H_0 + H_{\rm fs} + eEz\,. \tag{159}$$

In a complex atom, z is replaced by $D_z = \sum_i z_i$, the component of the electric dipole moment $\boldsymbol{D}$ along $\boldsymbol{E}$. The perturbation is invariant under rotations about $\boldsymbol{E}$, so as in the Zeeman effect $\Delta M_J = 0$. On the other hand, $\boldsymbol{D}$ is a polar vector and therefore has vanishing matrix elements between states of the same parity; as a vector it imposes the selection rule $\Delta J = 0, 1$ on matrix elements of H_S. Hence the expectation value of H_S vanishes in any state of definite parity, and the shift of any nondegenerate state of definite parity only appears in second order perturbation theory. As a consequence, *in the Stark effect the level shifts are quadratic in E as $E \to 0$, unless there are "accidental" degeneracies between states of opposite parity.* If $\boldsymbol{L}$ and $\boldsymbol{S}$ are separate constants of motion of the atom, there are the further selection rules $\Delta L = 1, \Delta S = 0$, unless two states of the same L and opposite parity exist.

There is another consequence of symmetry: states with opposite values of M_J have the same energy in the Stark effect.This follows from time-reversal invariance, which will only be treated in §7.2, so a heuristic argument will have to do here. The electric dipole moment does not change if the direction of motion of the of the electrons in the atom are reversed, and for that reason H_S is invariant under time reversal. Under time reversal, furthermore, magnetic quantum numbers, such as M_J, change sign, because the sense of rotation is reversed, and as this reversal is a symmetry the energies of states with opposite M_J must be the same. This argument fails for the Zeeman effect, because the angular momenta $\boldsymbol{L}$ and $\boldsymbol{S}$ change sign under time reversal. For that reason there is no degeneracy between states with opposite magnetic quantum numbers in the Zeeman effect, as was evident in the preceding calculations.

Once more we consider the $n = 2$ multiplet in hydrogen. According to the selection rules just given, the perturbing operator z has matrix elements between the s state and the $p_{\frac{1}{2}}$ and $p_{\frac{3}{2}}$ states, for $m = |\frac{1}{2}|$, and as a consequence the energies of these states are found by diagonalizing a three-dimensional matrix. On the other hand, because of the parity selection rule the $m = |\frac{3}{2}|$ states are oblivious to the electric field. Prob. 12 sketches the calculation, and the final result is shown in Fig. 5.3. As in the Zeeman effect, the spectrum is simple in the limit where the fine

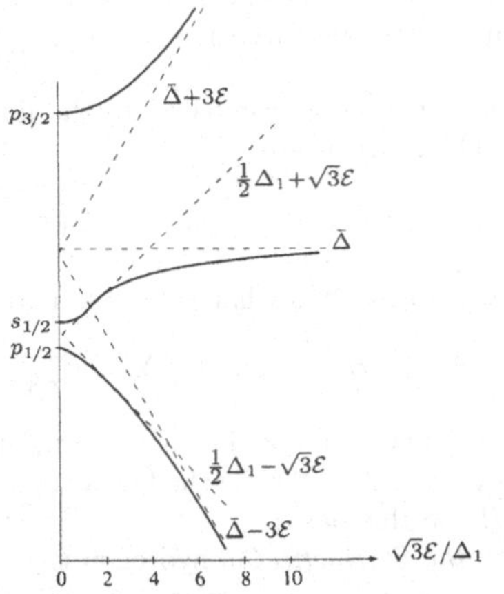

FIG. 5.3. The Stark effect in the $n = 2$ multiplet. The solid curves are the eigenvalues of the $|m| = \frac{1}{2}$ states as a function of field strength; $\mathcal{E} = ea_0|\boldsymbol{E}|$, Δ_2, and Δ_1 are the zero field energies of $p_{3/2}$ and $s_{1/2}$ relative to $p_{1/2}$, respectively, and $\bar{\Delta} = \frac{1}{3}(\Delta_1 + \Delta_2)$. The dashed lines are the low and high field limits. These easily computable limits suffice to guide the eye to the complete solution, which can be useful in more complicated spectra, such as the one shown in Fig. 6.2.

structure splittings are small compared to the energy ea_0E. In that limit the gross structure spectrum has "accidental" degeneracies between states of opposite parity, and as a consequence the Stark shifts become linear in E for large E. For weak fields, on the other hand, the shifts are quadratic in E, as is true quite generally for complex atoms as degeneracies between states of opposite parity are a special feature of hydrogenic atoms.

(c) Strong Fields

In his paper which gave the first solution of the Kepler problem, Pauli also calculated the spectrum in the presence of *electric and magnetic fields with arbitrary orientations* in the strong field case, that is, where the Zeeman and Stark splittings are large compared to fine structure but small compared to gross structure. Pauli's *tour de force* exploited the Lenz vector technique described in §5.2.

The Hamiltonian, in terms of the units and notation used in §5.2, is

$$H = \tfrac{1}{2}P^2 - (1/Q) + \gamma_B\,\boldsymbol{B}_{\mathrm{T}}\cdot\boldsymbol{L} + \gamma_E\,\boldsymbol{E}_{\mathrm{V}}\cdot\boldsymbol{Q} + \gamma_B\,\boldsymbol{B}_{\mathrm{T}}\cdot\boldsymbol{\sigma} + H_{\mathrm{fs}}\,. \tag{160}$$

The coefficients are found from (132) and (142), bearing in mind that the units of energy and length are $Z^2e^2/4\pi a_0$ and a_0/Z,

$$\gamma_B = 2.127\cdot 10^{-6}/Z^2\,, \qquad \gamma_E = 1.945\cdot 10^{-12}/Z^3\,. \tag{161}$$

The subscripts will be dropped from the fields from now on, it being understood that they are in Tesla and Volt/meter.

$H_{\rm fs}$ is assumed to be negligible, and will be dropped now though it will be taken into account approximately at the end. Once $H_{\rm fs}$ is gone, the spin term in (160) commutes with the rest of the Hamiltonian, and turns every level into a doublet of splitting $2\gamma_B B$. Hence it too will be ignored for now. The perturbation that we must deal with is thus

$$H_1 = \gamma_B \boldsymbol{B}\cdot\boldsymbol{L} + \gamma_E \boldsymbol{E}\cdot\boldsymbol{Q}\,. \tag{162}$$

Let $\mathfrak{H}_E$ be the subspace spanned by the degenerate states with principal quantum number n and energy $-|E_n|$. The objective is to diagonalize H_1 in each $\mathfrak{H}_E$. The essential tool is the following lemma, whose one-line proof is left to the reader:

> $\mathbb{A}$. *Matrix elements between states of equal energy of any operator $\dot{A}$ vanish, where $\dot{A}$ is the time derivative of A in the Heisenberg picture.*

By using this fact, $\boldsymbol{Q}$ can be related to the Lenz vector $\boldsymbol{M}$, whose spectrum is known.

To exploit $\mathbb{A}$, it is essential to bear in mind that H_1 is a perturbation, and that the operators appearing in H_1 can therefore be taken to satisfy equations of motion governed by the unperturbed Hamiltonian H_0, and not by $H_0 + H_1$. Hence it is legitimate to write

$$\boldsymbol{M} = \boldsymbol{P} \circledast \boldsymbol{L} - \hat{\boldsymbol{Q}} = \frac{d}{dt}(\boldsymbol{Q}\times\boldsymbol{L}) - \hat{\boldsymbol{Q}}\,, \tag{163}$$

though this is only correct if $\dot{\boldsymbol{L}} = 0$ and $\dot{\boldsymbol{P}} = \boldsymbol{Q}$, neither of which hold when the Hamiltonian is $H_0 + H_1$. Now let P_E be the projection operator onto $\mathfrak{H}_E$, and for any operator A, define

$$\langle A\rangle_E \equiv P_E A P_E\,. \tag{164}$$

Then (163) and $\mathbb{A}$ give

$$\langle \boldsymbol{M}\rangle_E = -\langle\hat{\boldsymbol{Q}}\rangle_E\,. \tag{165}$$

This is not yet the desired relation between $\boldsymbol{M}$ and $\boldsymbol{Q}$, which is

$$\langle \boldsymbol{Q}\rangle_E = -\frac{3}{4E}\,\langle \boldsymbol{M}\rangle_E\,, \tag{166}$$

where $E = 1/2n^2$ is the energy in atomic units; the proof of (166) is the subject of Prob. 14. From §5.2(b), recall $\boldsymbol{M} = \sqrt{2E}\,(\boldsymbol{F}^{(+)} - \boldsymbol{F}^{(-)})$, $\boldsymbol{L} = \boldsymbol{F}^{(+)} + \boldsymbol{F}^{(-)}$, where the operators $\boldsymbol{F}^{(\pm)}$ have the commutation rules of two commuting angular momenta. Hence

$$\begin{aligned} H_1 &= \gamma_B\,\boldsymbol{B}\cdot(\boldsymbol{F}^{(+)} + \boldsymbol{F}^{(-)}) - \tfrac{3}{4}\,(2/E)^{\frac{1}{2}}\,\gamma_E\,\boldsymbol{E}\cdot(\boldsymbol{F}^{(+)} - \boldsymbol{F}^{(-)}) \\ &= \boldsymbol{U}_n^{(+)}\cdot\boldsymbol{F}^{(+)} + \boldsymbol{U}_n^{(-)}\cdot\boldsymbol{F}^{(-)}\,, \end{aligned} \tag{167}$$

$$\boldsymbol{U}_n^{(\pm)} = \gamma_B\,\boldsymbol{B} \mp \tfrac{3}{2}n\gamma_E\,\boldsymbol{E}\,. \tag{168}$$

The fact that $\boldsymbol{F}^{(+)}$ commutes with $\boldsymbol{F}^{(-)}$ now comes to the fore, because their vector coefficients in (168) point in different n-dependent directions unless $\boldsymbol{B}$ and $\boldsymbol{E}$ are parallel. We may, nevertheless, diagonalize the component of $\boldsymbol{F}^{(+)}$ along $\boldsymbol{U}_n^{(+)}$, with eigenvalues μ^+, and similarly for $\boldsymbol{F}^{(-)}$ with eigenvalues μ^-. As a result, the eigenvalues of the complete Zeeman-Stark energy H_1 are

$$\bar{\Delta}_n(\mu^+,\mu^-) = \mu^+ U_n^{(+)} + \mu^- U_n^{(-)}\,, \quad \mu^\pm = f, f-1, \ldots -f, \quad f = \tfrac{1}{2}(n-1)\,. \tag{169}$$

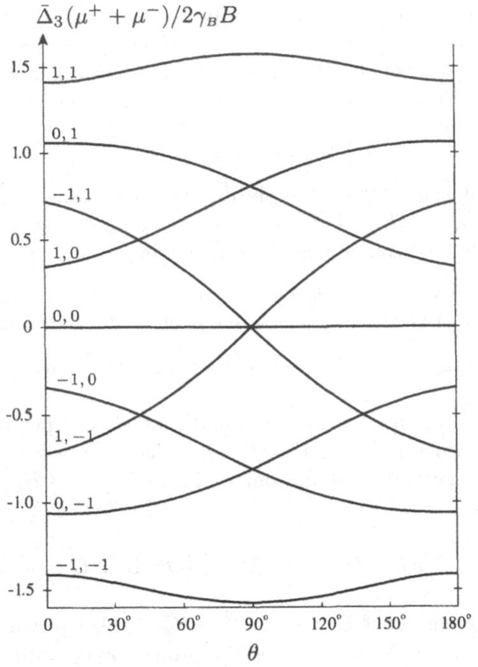

FIG. 5.4. The hydrogenic $n = 3$ levels in the presence of homogeneous electric and magnetic fields when the perturbation is large compared to the fine structure, as a function of the angle $\theta\angle(\boldsymbol{B}, \boldsymbol{E})$, for the case $3n\gamma_E E/2\gamma_B B = 0.5$. The energy eigenvalues $\bar{\Delta}$ shown are given by Eq. 169, and do not include the effects included in Eq. 171; the eigenvalues μ^+, μ^- are shown on each level. Note the symmetries $\mu^+ \leftrightarrow \mu^-$ and $\theta \rightarrow \pi - \theta$, and the degeneracy when the fields are "crossed," or $\boldsymbol{B} \cdot \boldsymbol{E} = 0$; in the latter regard see Prob. 15.

These energies are shown in Fig. 5.4 for the $n = 3$ multiplet. It is important to note that the eigenvalues $\mu^\pm$ refer to different directions, so that $\mu^+ + \mu^-$ is not an eigenvalue of any component of $\boldsymbol{L}$ unless the fields are parallel. The case where $\boldsymbol{E}$ and $\boldsymbol{B}$ are perpendicular ("crossed") is special, however, as can be seen from degeneracies at $\theta = \pi/2$ in Fig. 5.4. In this case the meaning of $\mu^+ + \mu^-$ is taken up in Prob. 15.

Although (169) gives the eigenvalues of H_1, it has a zero of energy that does not include the fine structure. The flaw can be seen in Figs. 5.2 and 5.3, where the asymptotes for large B and E would not be correct if the fine structure splittings were simply neglected outright. The reason is the following. The trace of H_1 vanishes for all values of the field strengths, but $\mathrm{Tr}\, H_{\mathrm{fs}} \neq 0$. Diagonalization of $H_{\mathrm{fs}} + H_1$ leaves its trace invariant, and therefore the correct zero of energy will be maintained by adding $d_n^{-1}\mathrm{Tr}\, H_{\mathrm{fs}}$ to (169), where d_n is the dimension of $\mathfrak{H}_E$. Ignoring the Lamb shift, this shift is

$$K_n = -\frac{(Z\alpha)^2}{8n^5}(5n-4)\,. \tag{170}$$

The final result for the energies in this strong field situation is therefore

$$\Delta_n(\mu^+, \mu^-, m_s) = K_n + \mu^+ U_n^{(+)} + \mu^- U_n^{(-)} + 2\gamma_B \, B \, m_s \,, \tag{171}$$

where $m_s = \pm\frac{1}{2}$ is the spin projection along $\boldsymbol{B}$.

5.5 Problems

1. Verify Eq. 14 using the fact that nuclear radii are roughly $1.2\,A^{\frac{1}{3}}$ fm. For heavy nuclei, $Z \sim 0.4\,A$; use this to show that in muonic atoms the lowest bound state has a radius of order the nuclear radius when $Z \sim 60$.

2. This problem deals with the *isotope shift*, the shift in energy eigenvalues in a Coulomb field due to the finite size of the nuclear charge distribution. This is a very small effect in electronic atoms of moderate Z, but not when the satellite is heavy, such as a muon.

(a) Assume that the bound state has a radius a that is much larger than the nuclear radius R. Show that only s states have a significant shift and that the effect *decreases* the binding energy. Show that the shift for a satellite of mass m is roughly

$$\Delta E \sim \left(\frac{R}{a}\right)^3 \frac{Ze^2}{4\pi R} \sim \left(\frac{R}{a_0}\right)^2 \left(\frac{m}{m_e}\right)^3 Z^4 \text{ Ry} .$$

(b) Show that the following formula is exact in the limit $a \gg R$:

$$\Delta E_n = |\psi_{ns}(0)|^2 \int \delta V(\boldsymbol{r}) \, d\boldsymbol{r} \,,$$

where δV is the change in electrostatic energy due to a distributed charge density $e\rho_N$:

$$\delta V(\boldsymbol{r}) = \frac{Ze^2}{4\pi r} - e^2 \int d\boldsymbol{r}' \, \frac{\rho_N(\boldsymbol{r}')}{4\pi|\boldsymbol{r} - \boldsymbol{r}'|} \,.$$

Note that ρ_N is effectively spherically symmetric no matter what shape or angular momentum the nucleus may have. Exploit this to show that

$$\Delta E_n = \frac{1}{6}\, Ze^2 \langle r^2 \rangle_N \, |\psi_{ns}(0)|^2 = \frac{4}{3}\, Z^4 \, \frac{m^3}{m_e^3} \, \frac{\langle r^2 \rangle_N}{a_0^2} \, \frac{1}{n^3} \text{ Ry} \,,$$

where $\langle r^2 \rangle_N$ is the mean square radius of the nuclear charge distribution.

(c) Evaluate the $1s$ shift explicitly for electrons and muons in H, Fe, and Pb, and compare it to the ground-state hfs, which is of order $\lambda_C^2(m_e^2/mm_N)|\psi_{1s}(0)|^2$. Is the formula for the shift valid for all these nuclei in the case of an electron? of a muon?

3. *The virial theorem* holds in both classical and quantum mechanics. In the former, it relates the time-averaged kinetic and potential energies. In quantum mechanics it relates the corresponding expectation values in any stationary state $|\Psi\rangle$, and is a consequence of $\langle [A, H] \rangle_\Psi = 0$ for any observable A. Consider a system with a Hamiltonian of the form $T + V(q_i)$, where T is the conventional kinetic energy. Taking $A = \sum q_i p_i$ show that

$$2\langle T \rangle = \sum_i \langle q_i \, \partial V / \partial q_i \rangle \,.$$

4. After studying the proof that $\boldsymbol{M}$ is a constant of motion, confirm the relationship between the energy, angular momentum, and Lenz vector as given by Eq. 35.

5. This problem fills in the steps leading to (76). Show, first, that

$$\frac{\partial}{\partial x} + i\frac{\partial}{\partial y} = e^{i\phi}\sin\theta\frac{\partial}{\partial r} + \frac{1}{r\cos\theta}\left(L_+ - e^{i\phi}\sin^2\theta\frac{\partial}{\partial\theta}\right) ;$$

that (75) therefore reduces to

$$nR' + \left(1 - \frac{n(n-1)}{r}\right) R = 0 ;$$

and that this equation has the stated solution $R_{n,n-1}(r)$.

6. Ignoring the Lamb shift, show that the trace of the fine structure Hamiltonian is

$$\mathrm{Tr}\, H_{\mathrm{fs}} = -\frac{Z^2(Z\alpha)^2}{4n^3}(5n-3) .$$

7. Show that the anomalous magnetic moment, which is ignored in Eq. 96, cannot contribute to any $s_{\frac{1}{2}} - p_{\frac{1}{2}}$ splittings.

8. Using the fact that the matrix elements of a vector operator $\boldsymbol{V}$ in a subspace of definite total angular momentum are proportional to those of $\boldsymbol{J}$, show that

$$j(j+1)\,\langle jm|\boldsymbol{V}|jm'\rangle = \langle jm|(\boldsymbol{V}\cdot\boldsymbol{J})\,\boldsymbol{J}|jm'\rangle .$$

(Hint: Don't look in the literature, which tends to give surprisingly complicated derivations.)

9. The deuteron has spin 1, magnetic moment $0.857\,\mu_N$, and electric quadrupole moment 0.28 fm^2. Evaluate and plot the hyperfine splittings of the $n = 2$ levels in deuterium.

10. This problems concerns the detailed calculations involved in the *Zeeman effect* in hydrogen.

(a) By using explicit expressions for CG-coefficients, confirm Eq. 150 and Eq. 152; diagonalize the matrix (153); and plot the spectrum as in Fig. 5.2.

(b) Show that the eigenstates of the matrix (153), when $\mathcal{B} \gg \Delta$, are linear combination of the states $|lsjm\rangle$ with coefficients that are precisely the CG-coefficients for the transformation to the states $|lm_l sm_s\rangle$.

11. This problem addresses *fine and hyperfine structure in positronium* (Ps). Because this atom consists of two particles of equal mass, there is actually no distinction between fs and hfs, but it is customary to call the ground-state splitting an hfs effect.

(a) In Ps the two spins are on an equal footing, in contrast to other atoms where the spin of the satellite and the nucleus play very different dynamical roles. Hence Ps states are called $n\,{}^1L_J$ and $n\,{}^3L_J$ for spin singlets and triplets, where n is the principle quantum number, J the total angular momentum, and L the orbital angular momentum about the center of mass (with $L = 0, 1, 2$ designated by S, P, D, etc.). List the states in this notation for $n \leq 3$.

(b) The leading correction to the ground state, called the hyperfine correction even though it is far bigger than in hydrogen, is given by the Hamiltonian[1]

$$H_{\mathrm{hfs}} = 4\pi\mu_B^2\left[\tfrac{3}{2} + \tfrac{7}{8}\boldsymbol{\sigma}^+\cdot\boldsymbol{\sigma}^-\right]\delta^3(\boldsymbol{r}) ,$$

where $\boldsymbol{\sigma}^{\mp}$ are the spins of $e^{\mp}$. Show that the ground-state splitting is $\Delta_{\mathrm{hfs}} = \frac{7}{6}\alpha^2$ Ry .

---

[1] In addition to the effects that contribute to hfs in ordinary atoms, this includes a contribution from virtual pair annihilation.

(c) Consider the Zeeman effect in the $n = 1$ multiplet of Ps. Take advantage of the symmetry of the Zeeman interaction to show that only two of the four states participate in the effect. Show that their energies in the presence of the magnetic field are

$$E_{\pm} = \tfrac{1}{2}\Delta_{\mathrm{hfs}} \left(1 \pm \sqrt{1 + (4\mu_B B/\Delta_{\mathrm{hfs}})^2}\,\right).$$

12. This problem concerns *the Stark effect* in the $n = 2$ multiplet of hydrogen, the objective being Fig. 5.3. Because the Lamb shift is small compared to the fs splitting, the spectrum as a function of field strength has several interesting features.

(a) Calculate the matrix elements of eEz in the basis that diagonalizes H_{fs}. Show that the $|m| = \frac{3}{2}$ states are unaffected, whereas the perturbed energies and eigenstates in the space of the $|m| = \frac{1}{2}$ states can found by diagonalizing

$$\mathcal{H} = \begin{pmatrix} 0 & \sqrt{3}\mathcal{E} & 0 \\ \sqrt{3}\mathcal{E} & \Delta_1 & -\sqrt{6}\mathcal{E} \\ 0 & -\sqrt{6}\mathcal{E} & \Delta_2 \end{pmatrix},$$

where $\mathcal{E} \equiv eEa_0$, $\Delta_1 = E(s_{1/2}) - E(p_{1/2})$, $\Delta_2 = E(p_{3/2}) - E(p_{1/2})$.

(b) $\mathcal{H}$ can of course be diagonalized as it stands, but that is not an illuminating exercise. It is more instructive to find the energies in two limits: when the electric perturbation is small or large compared to the fs separation. Consider the first case now, and show that the level shifts are quadratic in $\mathcal{E}$ when $\mathcal{E} \ll \Delta_1$, but linear in $\mathcal{E}$ for $\mathcal{E} \gg \Delta_1$ (though of course $\mathcal{E} \ll \Delta_2$ throughout). This behavior illustrates the remarks following Eq. 159, and can be seen in Fig. 5.3. Show that the quadratic shift of the $p_{\frac{3}{2}}$ state can also be found easily with perturbation theory.

(c) Finally, consider the strong field limit, $\mathcal{E} \gg \Delta_2$. Care must be taken here, however, to keep the correct zero of energy. To that end, note that $\mathrm{Tr}\,\mathcal{H}$ is the same for all values of $\mathcal{E}$, including $\mathcal{E} = 0$. Bearing this in mind, and writing the high field eigenvalues as $\frac{1}{3}(\Delta_1 + \Delta_2) + \delta E$, show that $\delta E = 0, \pm 3\mathcal{E}$. These limits are also shown in Fig. 5.3.

13. Consider an arbitrary atom with a non-degenerate $J = 0$ ground state $|0\rangle$ which has been placed in a uniform electric field $\boldsymbol{E}$. The *induced electric dipole moment* D can be found in two ways: (i) from the ground-state energy as a function of E (ii) by a direct calculation of the dipole moment in the disturbed ground state. Show that both methods give the same expression for D:

$$D = 2e^2\, E \sum_{n \neq 0} \frac{|\langle 0|X|n\rangle|^2}{E_n - E_0},$$

where $X = \sum_i x_i$, and E_n and $|n\rangle$ are the unperturbed energies and stationary states.

14. Derive Eq. 166 by considering $d[(\boldsymbol{Q}\cdot\boldsymbol{P})\boldsymbol{Q}]/dt$ and $d(Q^2\boldsymbol{P})/dt$, and showing that

$$\langle \tfrac{1}{2}P^2\boldsymbol{Q} - \hat{\boldsymbol{Q}} + E\boldsymbol{Q}\rangle_E = 0.$$

15. When the electric and magnetic fields are perpendicular (or "crossed"), the spectrum shown in Fig. 5.4 has the same degeneracies as in the Zeeman effect, though the spacing depends on both B and E. This is "arithmetically" obvious from Eq. 169 because $\bar{\Delta}$ is proportional to $\mu^+ + \mu^-$ in this case. Give an explanation that makes the result physically obvious and reveals the meaning of the quantum number $\mu^+ + \mu^-$.

Endnotes

The authoritative treatment of hydrogenic spectra can be found in Bethe and Salpeter. There is, of course, a huge literature on the theory of atoms. A useful introduction is provided by B.H. Bransden and C.J. Joachain, *Quantum Mechanics* (2nd. ed.), Prentice-Hall (2000).

An extensive group theoretical treatment of bound and scattering states in the Coulomb field is provided by M. Bander and C. Itzykson, *Rev. Mod. Phys.* **38**, 330–358 (1966). Green's function for the Schrödinger equation in the case of a Coulomb field is known; see J. Schwinger, *J. Math. Phys.* **5**, 1606 (1965); this also gives an elegant derivation of the wave functions in momentum space. Coherent states in the Coulomb field are described by M. Nauenberg in *The Physics and Chemistry of Wave Packets,* J.A. Yeazell and T. Uzer, eds., J. Wiley (2000), pp. 1–30.

6

Two-Electron Atoms

It is not possible to find exact eigenstates for the helium atom by analytical techniques. This is to be expected, because the classical counterpart, two satellites moving in each other's gravitational field and that of a massive star, is already a problem that cannot be solved analytically. But very good and rather simple approximations give a good rendition of the helium spectrum, and reveal a profound consequence of the identity of the two electrons.

In classical mechanics the concept of indistinguishability, or identity, can be defined by requiring the Hamiltonian of a two-particle system to be invariant under the interchange of the coordinates of the pair. This requires the masses to be the same, and any internal or externally applied force to be the same for each member. But little of consequence ensues if one of the masses is changed slightly, or if the force on one particle differs somewhat from that on the other; the motions will undergo some mundane modification that tends to zero smoothly as the alterations are removed. And for this reason "identity" receives hardly any attention in classical mechanics.

The consequences of identity are dramatic and far-reaching in quantum mechanics, however. In this chapter we only deal with the simplest example, that of two identical, or indistinguishable, particles, and in particular, with two-electron atoms.[1] As we shall see, indistinguishability requires particular superpositions of two-electron states, and the ensuing interference effects then produce energy level patterns which, to a pre-quantum physicist, would appear to be due to enormous interactions between magnetic moments, but which are actually of electrostatic origin.

6.1 Two Identical Particles

(a) Spin and Statistics

By definition, the Hamiltonian of a system of two indistinguishable particles is invariant under interchange of the particles, $H(1,2) = H(2,1)$, where 1 stands for the coordinate and spin operators of the first particle, etc. This symmetry of H holds whether or not the system is isolated. As the solutions of the time-independent Schrödinger equation

$$H(1,2)\Psi_E(1,2) = E\Psi_E(1,2) \tag{1}$$

form a complete set, it suffices to consider the case where H is time-independent.

Let P_{12} be the operator that interchanges (1, 2). By hypothesis, P_{12} commutes with H, and is therefore a constant of motion. Furthermore, $(P_{12})^2 = 1$, and there-

[1] Chapter 11 deals with systems of many identical particles.

fore the eigenvalues of P_{12} are $\varpi = \pm 1$. This eigenvalue can be specified together with the energy, so that the states will written as $\Psi_{E\varpi}$. It is intuitively evident, and will be demonstrated in the next section, that the energy is not, in general, independent of ϖ. To understand why, consider particles of spin 0. Then for $\varpi = 1$ the wave functions are symmetric under interchange of the spatial coordinates, while for $\varpi = -1$ antisymmetric. The latter therefore vanish at zero separation, while the former do not, and for almost any interaction the two types of states will surely lead to distinct energy eigenvalues.

Now let $A(1,2)$ be any observable associated with the system. It too must be symmetric under $1 \leftrightarrow 2$, for if it were not it would serve to distinguish between the two supposedly identical particles. As a consequence,

$$\langle \Psi_{E\varpi} | A | \Psi_{E'\varpi'} \rangle = 0 \qquad \text{if } \varpi \neq \varpi' \,. \tag{2}$$

Thus the stationary states of such a system would separate the Hilbert space into two rigorously disconnected subspaces, one for the symmetric states, the other for the antisymmetric states, because there would be no observables whatever with matrix elements between these subspaces. If a non-stationary state is initially in one subspace, it will stay forever in that subspace. In principle, both subspaces could exist in nature, even though there would be no communication between them. But nature is not so weird:

> *The states of a pair of indistinguishable integer spin particles are symmetric under interchange, while the states of a pair of indistinguishable half-integer spin particles are antisymmetric under interchange.*

This statement is called *the connection between spin and statistics.* Nonrelativistic quantum mechanics is incapable of deriving this empirical observation, but it is a theorem in quantum field theory, and one of the latter's great accomplishments. The word "statistics" appears here because the statistical mechanics of systems with symmetric as compared to antisymmetric states are strikingly different, and different also from classical (Boltzmann) statistics. That of particles with integer spin is known as *Bose-Einstein statistics*, and such particles are called *bosons*, while particles with half-integer spin are called *fermions* and obey *Fermi-Dirac statistics.* Knowledge of these statistics is not needed in the considerations that now follow, and will therefore be postponed until chapter 11.

It is now clear why there is such an enormous difference between the classical and quantum mechanical concepts of identity. To repeat, nothing earth-shaking happens in classical mechanics if a teeny term that is not invariant under $1 \leftrightarrow 2$ is added to the Hamiltonian. But in quantum mechanics such a term would lay waste to the argument just given: energy eigenstates would not have a definite signature under P_{12}, and the symmetric and antisymmetric subspaces would mingle. Should a state be either symmetric or antisymmetric initially, with time it would develop a component with the other symmetry.

A word about the meaning of "spin" and "particle" in the preceding statements. As always in quantum physics, when speaking of a "particle" there is the implicit understanding that the energy regime in question is such that any internal degrees of the "particle" that may exist cannot be excited, so that its only degrees of freedom are its position and rest frame total angular momentum (or "spin").

The point is best made with a concrete example: the neutral helium atoms with the isotopes He^3 and He^4 as nuclei. Both have two electrons, and two protons in their nuclei, but the former has one and the latter two neutrons. Chemically they are identical and therefore experience the same inter-atomic forces; at first glance, therefore, they only differ in mass. But their nuclei have different spin, because the first contains three spin $\frac{1}{2}$ particles and therefore has half-integer spin, whereas the latter has an additional neutron and therefore integer spin. Hence He^3 is a fermion and He^4 a boson, and in many-particle systems the first has antisymmetric and the second symmetric wave functions. Although the interaction between atoms of the first and second type are identical, the energetics will be very different, and it is indeed the case that fluids composed of one atom or the other undergo phase transitions at very different temperatures, and below these critical temperatures display entirely different properties. It is most remarkable — among the most remarkable consequences of quantum mechanics, that these deep differences are wholly due to one additional neutron, which does not play any role in the interaction between the atoms, and "only" produces a different nuclear spin.

(b) The Exclusion Principle

Consider first two identical particles in the absence of interaction. Let H_1 be the Hamiltonian for particle 1, and $\{\varphi_n(1;t)\}$ be a set of orthonormal states satisfying the one-particle Schrödinger equation. Then

$$\Phi(1,2;t) = \varphi_{n_1}(1;t)\varphi_{n_2}(2;t) \tag{3}$$

is a solution of the two-body equation with Hamiltonian $H_1 + H_2$ (recall §1.2). This state is neither symmetric nor antisymmetric, however, unless $n_1 = n_2$, in which case it is symmetric, and would be an allowed state only if the particles are bosons.

If the particles are fermions, legal antisymmetric states can easily be constructed:

$$\Psi_A(1,2;t) = \frac{1}{\sqrt{2}}[\varphi_{n_1}(1;t)\varphi_{n_2}(2;t) - \varphi_{n_1}(2;t)\varphi_{n_2}(1;t)] \,. \tag{4}$$

But this state vanishes identically — does not exist, if $n_1 = n_2$. This is *Pauli's exclusion principle: two fermions cannot be in the same state.* Furthermore, although $\Psi_A(1,2;t)$ does exist if $n_1 \neq n_2$, it *vanishes* when $1 = 2$, i.e., *the probability of finding two identical fermions in the same spin state and at the same position is zero.*

The probability distribution just alluded merits a closer look:

$$|\Psi_A|^2 = \tfrac{1}{2}[|\varphi_{n_1}(1)|^2|\varphi_{n_2}(2)|^2 + (1 \leftrightarrow 2)] - \tfrac{1}{2}[\varphi^*_{n_1}(1)\varphi^*_{n_2}(2)\varphi_{n_1}(2)\varphi_{n_2}(1) + (1 \leftrightarrow 2)] \,. \tag{5}$$

The first term is what classical statistics would give for the joint probability distribution for two non-interacting indistinguishable particles — the symmetrized product of the distributions for each particle. The second term is deeply mysterious from that perspective, and is due to interference between the two terms in the antisymmetric wave function (4). This interference term (and its generalizations in more complicated situations) underlies many of the surprising consequences of indistinguishability. It should be noted that the interference term vanishes if the two one-particle wave functions do not overlap, for then $\phi_{n_1}(1)\phi_{n_2}(1) \equiv 0$. This is intuitively "obvious": when there is no overlap there is no ambiguity as to which particle is where, so to speak.

Thus far the exclusion principle has been put in terms of one-particle wave functions, but it can be stated without reference to any representation or Hamiltonian. For *any* antisymmetric state $\Psi(1,2;t)$, whether or not it is of the form (4) appropriate to non-interacting particles, the probability for the two fermions to have the same position and spin state (i.e., $1 = 2$) is identically zero. Furthermore, the labels 1 and 2 need not be spatial coordinates and spin projection — they can be the eigenvalues of any four compatible observables. As an example that will shortly be examined in detail, take the two electrons in helium. The four quantum numbers can be $(nljm)$ as used in the hydrogen spectrum, defined by a set of one-particle states $\varphi(nljm;1)$ appropriate to the $Z=2$ nucleus, even though $(nljm)$ are not the eigenvalues of constants of motion when the interaction between the electrons is taken into account. If $\Psi_E(1,2)$ is any exact energy eigenstate of the helium atom, it can be expressed as a linear superposition of amplitudes

$$\Psi_E(n_1l_1j_1m_1;n_2l_2j_2m_2) \equiv \int d1\,d2\,\varphi^*(n_1l_1j_1m_1;1)\,\varphi^*(n_2l_2j_2m_2;2)\Psi_E(1,2)\,, \tag{6}$$

which, by antisymmetry, will all vanish if all four quantum numbers are identical. This then provides the general statement of the exclusion principle:

> *The probability for finding a pair of identical fermions with the same values of any four compatible quantum numbers is identically zero, whatever the nature of the Hamiltonian or state.*

(c) Symmetric and Antisymmetric States

We now turn to a more detailed description of stationary states for isolated but interacting pairs of identical particles of spin $0, \frac{1}{2}, 1$. By working in the center of mass frame, we need only concern ourselves with the relative coordinate $\boldsymbol{r}$ and the two-particle spin states. The states will be of the form

$$\Psi = R_{EL}(r)\,Y_{LM_L}(\boldsymbol{r})\,\chi_{SM_S}\,, \tag{7}$$

where (LM_L) are the quantum numbers of the orbital angular momentum about the center of mass and its projection along some axis, while (SM_S) are the total spin and its projection. In writing Eq. 7, we assumed that the orbital and spin angular momenta are separate constants of motion, and from the fine structure of hydrogen we already know that there can be interactions that would contradict this hypothesis. Nevertheless, we will ignore this possible complication as it is readily dealt with once the argument that follows has been understood. Dynamics will only enter into the determination of the radial function R_{EL}, which is a "technical detail" of no concern to us here.

When the particles are interchanged, $\boldsymbol{r} \to -\boldsymbol{r}$, and therefore

$$P_{12}Y_{LM_L(\boldsymbol{r})} = Y_{LM_L}(-\boldsymbol{r}) = (-1)^L\,Y_{LM_L}(\boldsymbol{r})\,. \tag{8}$$

But pairs of identical bosons can only exist in symmetric states, and if they have no spin, $\boldsymbol{r}$ is the only variable. Therefore *pairs of identical spin zero bosons only have even total angular momentum.*

This theorem has many important application. As an example, take a diatomic molecule with identical spin zero nuclei (e.g., O_2, with O^{16} nuclei). The rotational levels of this molecule will therefore differ dramatically from those where one nucleus is a different isotope of oxygen, in that the former will only have even angular momenta whereas the latter will have odd ones as well.

Consider next a pair of identical spin $\frac{1}{2}$ fermions. Then the total spin is 0 or 1; these states χ_{SM_S} were constructed explicitly in §3.6(a), where it was shown that $S = 1$ is symmetric under interchange while $S = 0$ is antisymmetric (crudely put, parallel and antiparallel spins, respectively). The orbital wave function Y_{LM_L} must then have the opposite symmetry to make an overall antisymmetric state. As a consequence, *two identical spin $\frac{1}{2}$ fermions only exist in spin singlets (S=0) with even orbital angular momentum or in spin triplets (S=1) with odd orbital angular momenta.* Again a large fraction of all the angular momentum possibilities do not exist.

In view of the preceding statement, in $S = 1$ states the spatial wave function vanishes at zero separation, whereas it does not in $S = 0$ states. As a consequence, the Pauli principle will produce effects that appear to be due to a spin-dependent interaction when the interaction has no spin dependence whatsoever! For example, if the interaction is a attractive potential, the spatially symmetric states with $S = 0$ will take better advantage of the interaction and have lower energies than the antisymmetric $S = 1$ states, and vice versa if the interaction is repulsive. This is but one example of how the Pauli principle can produce effects that, to anyone who does not know quantum mechanics, would be due to fictitious forces.

Finally a pair of identical spin 1 bosons. Then $S = 0, 1, 2$. In §3.6(b) we learned that the $S = 0$ & 2 states are symmetric while $S = 1$ is antisymmetric. Hence a pair of unit spin bosons only exists in spin triplets with odd orbital angular momenta and in spin singlets or quintets with even orbital angular momenta.

The parity of the states just described is $(-1)^L$ throughout. Space reflection produces $\boldsymbol{r} \to -\boldsymbol{r}$, and the reflection of any internal coordinates in the individual particles. The latter produced a factor $(-1)^\pi$, where π is called the *intrinsics parity* of the particle, but as there are two identical particles this factor is squared, leaving just $(-1)^L$ as the parity. It must be recognized that space reflection and interchange of identical particles is only related in the center of mass frame of a two-particle state; for the two electrons in helium, space reflection in the frame of the nucleus does not interchange the electrons.

On the other hand, in the center of mass frame, parity and the exclusion principle do combine to restrict the possible states greatly. To see how, it is best to introduce the standard notation for such states: ${}^{2S+1}L_J$, where $J = |L - S|, \ldots, L + S$, and L is specified in the traditional spectroscopic hieroglyphics: $S, P, D, F, \ldots$ for $L = 0, 1, \ldots$, with the capital letters indicating a many-particle angular momentum. The states allowed by the Pauli principle are then

$$ {}^1S_0\ , \quad {}^3P_{0,1,2}\ , \quad {}^1D_2\ , \quad {}^3F_{2,3,4}\ , \ldots\ . \tag{9} $$

Thus there are unique states of parity and total angular momentum 0 and 1 even when the orbital angular momentum is not a constant of motion. The first case where there are two options is $J = 2$ and odd parity, i.e., ${}^3P_2$ and ${}^3F_2$, and the energy eigenstates will then be some linear combination of these if the forces do not conserve $\boldsymbol{L}$. This is the full story for two protons. For a neutron-proton system (e.g., the deuteron), all the other states are allowed, and the lowest case where states of different L have the same parity and total angular momentum is ${}^3S_1$ and ${}^3D_1$. The deuteron is actually in a combination of these states.

The list (9) can also be used to explain the meaning of intrinsic parity. Imagine an "elementary particle" of spin 0 which is actually a composite of two identical fermions. If they are in the state $^1S_0$ the "particle" has even intrinsic parity, but if the state is $^3P_0$ the intrinsic parity is odd.

6.2 The Spectrum of Helium

The helium spectrum has a simplifying feature that can be understood by making the naive assumption that the electron-electron interaction plays a secondary role in excitation energies. If that is so, states can be specified by assigning the electrons to hydrogenic levels. Then the ground state has both electrons in the $1s$ level, and an excited state would, say, have one in $2p$ and another in $3d$. Such assignments are called "configurations" in spectroscopy, and these two would be designated by $1s^2$ and $2p\,3d$, respectively.

The simplifying feature just mentioned is that *all* bound states have one electron in the $1s$ level. The empirical binding energy of He is -5.8 Ry. If one electron is removed, the ground-state energy of the ion is that of an electron in the $1s$ state when $Z = 2$, which is -4Ry. Hence 1.8 Ry is the ionization energy. This is to be compared with the energy required to lift both electrons from the $n = 1$ to $n = 2$ level, which is $2 \times 4(\frac{1}{1} - \frac{1}{4}) = 6$ Ry, ignoring any change in the electron-electron interaction energy. But 6 Ry is so large compared to the ionization energy that it does not take much courage to draw the (correct) conclusion that all the discrete excited states are well described by assigning one electron to the $1s$ level. The bound states can, therefore, be designated by the quantum numbers of the excited electron, e.g., 4^3D for the $S = 1$ state with the configuration $1s4d$. This is an approximate description, of course, because it is not strictly correct that the bound states only contain an amplitude for these simple configuration: the exact states will have small components where both electrons are in excited levels, including states in the continuum. But what we are confining ourselves to here is a fine approximation, and more than adequate for our purpose.

The Hamiltonian of helium is

$$H = \frac{1}{2}\left(p_1^2 + p_2^2\right) - 2\left(\frac{1}{r_1} + \frac{1}{r_2}\right) + \frac{1}{r_{12}} \ ; \tag{10}$$

we are using atomic units for infinite nuclear mass and $Z = 1$, i.e., lengths in units of the Bohr radius and energies in units of 2 Ry.

If the assignment of hydrogenic quantum numbers to the levels is a good approximation, one might suppose that the electron-electron interaction $1/r_{12}$ could be treated straightforwardly with perturbation theory. The unperturbed wave functions from which such a calculation would start would then be the Coulomb wave functions for $Z = 2$. But a moment's thought indicates that this is unreasonable, for in a state with one electron in the $1s$ level and another in a higher level, the latter will see a central charge that is greatly reduced due to screening by the inner electron. This is an effect that perturbation theory cannot handle. But the variational method can do so easily, by using the effective central charge as a variational parameter. Of course, the variational principle can only address a very limited set of states: the ground state, of course, but as will be recalled from §2.7(e), also the lowest state of a given exact symmetry, i.e., $2P, 3D, 4F$, etc. That suffices to establish a very important fact stemming from the exclusion principle: large spin-dependent level splittings are due to the nature of two-fermion states, and not to any spin-dependent forces.

In light of the preceding remarks, as trial functions we choose to use hydrogenic eigenfunctions with the nuclear charge a parameter to be determined by minimizing the energy. We shall restrict ourselves to the $1s$ and $2p$ levels, and to the normalized functions (see Eq. 3.246)

$$1s: \qquad u(r) = (4\pi)^{-1/2}\, 2\alpha^{3/2}\, e^{-\alpha r}\,, \tag{11}$$

$$2p: \qquad v_m(\mathbf{r}) = (2\sqrt{6})^{-1}\beta^{5/2} r e^{-\frac{1}{2}\beta r}\, Y_{1m}(\hat{\mathbf{r}})\,. \tag{12}$$

Such one-electron wave functions are called *orbitals* in spectroscopy.

For the ground state the trial function is

$$\Psi_0(\mathbf{r}_1\mathbf{r}_2) = u(r_1)u(r_2)\chi_0\,. \tag{13}$$

The calculation of the expectation value and its minimization are left as an exercise. The results are

$$\alpha_0 = 2 - \tfrac{5}{16}\,, \qquad E_0 = -5.69 \text{ Ry}\,. \tag{14}$$

This is just 2% above the observed binding energy, a remarkably good result given the crudity of the trial function. The parameter α_0 is the effective nuclear charge; as was to be expected, it is substantially smaller than 2 because each electron partially screens the nucleus as seen by its partner.

Turning to the $2P$ states, we would anticipate $\alpha \simeq 2$ and $\beta \simeq 1$, because an electron in the $2p$ orbital is well outside the charge distribution of the $1s$ orbital. The outer orbital should thus see a nuclear charge reduced to roughly $Z = 1$, and the inner essentially the full nuclear charge.

The orbitals u and v_m are used to form antisymmetric trial functions:

$$\Psi(2^1P; m) = 2^{-\frac{1}{2}}[u(r_1)v_m(\mathbf{r}_2) + (\mathbf{r}_1 \leftrightarrow \mathbf{r}_2)]\chi_0\,, \tag{15}$$

$$\Psi(2^3P; mM_S) = 2^{-\frac{1}{2}}[u(r_1)v_m(\mathbf{r}_2) - (\mathbf{r}_1 \leftrightarrow \mathbf{r}_2)]\chi_{1M_S}\,. \tag{16}$$

Writing the Hamiltonian as

$$H = H_1 + H_2 + \frac{1}{r_{12}}\,, \tag{17}$$

the expectation values of H in these two states are

$$\langle H\rangle_\pm = \tfrac{1}{2}\int d^3r_1\, d^3r_2 [u^*(r_1)v^*(\mathbf{r}_2) \pm (\mathbf{r}_1 \leftrightarrow \mathbf{r}_2)]H[u(r_1)v(\mathbf{r}_2) \pm (\mathbf{r}_1 \leftrightarrow \mathbf{r}_2)] \tag{18}$$

$$= \langle H_1 + H_2\rangle + I \pm J\,. \tag{19}$$

The first term is due to the kinetic energy and the attraction by the nuclear charge; I, called the direct integral, stems from the electrostatic repulsion between the electrons,

$$I = \int d^3r_1\, d^3r_2\, \frac{|u(1)v(2)|^2}{r_{12}}\,, \tag{20}$$

and J is called the *exchange integral:*

$$J = \int d^3r_1\, d^3r_2 \,\frac{u^*(1)v^*(2)u(2)v(1)}{r_{12}} \,. \tag{21}$$

It has this name because it stems from the requirement of antisymmetry under exchange of identical fermions.

The origin of J resides in the two-electron probability distribution, which was already discussed in general terms in §6.1(b). In this instance, the probability that the two electrons are at $\boldsymbol{r}_1$ and $\boldsymbol{r}_2$ when they have total spin S is

$$P_S(\boldsymbol{r}_1\boldsymbol{r}_2) = \tfrac{1}{2}\Big[|u(1)|^2|v(2)|^2 + (1 \leftrightarrow 2)\Big] + \tfrac{1}{2}(-1)^S\Big[u^*(1)v^*(2)u(2)v(1) + \text{c.c.}\Big] \,. \tag{22}$$

The second term, due to interference between the two terms in the antisymmetric wave functions (15) and (16), describes correlations in position between the electrons, which influence the electrostatic energy, an effect that is accounted for by J. In the $S = 1$ states, the joint probability vanishes when the two electrons are at the same point and thereby suppresses the effect of the repulsion between the electrons, whereas the opposite effect occurs in the spatially symmetric $S = 0$ state. These correlations in the joint probability distribution produce spin-dependent effects that are of the same order of magnitude as the spin-independent "classical" part of the electrostatic energy accounted for by I.

The angular integrations in I and J are done by using

$$\frac{1}{r_{12}} = \sum_{lm} \frac{4\pi}{2l+1}\,\frac{r_<^l}{r_>^{l+1}} Y_{lm}^*(\hat{\boldsymbol{r}}_1)Y_{lm}(\hat{\boldsymbol{r}}_2) \,, \tag{23}$$

where $r_>$ ($r_<$) is the larger (smaller) of (r_1, r_2). Then

$$I = 4\pi \int_0^\infty r_1^2 dr_1 \int_0^\infty r_2^2 dr_2 \,\frac{1}{r_>}\, |u(1)R(2)|^2 \,, \tag{24}$$

$$J = \frac{4\pi}{3} \int_0^\infty r_1^2 dr_1 \int_0^\infty r_2^2 dr_2 \,\frac{r_<}{r_>^2}\, u^*(1)R^*(2)R(1)u(2) \,, \tag{25}$$

where $R(r)$ is the $2p$ radial function. In this case all the wave functions have no nodes, so J is positive and the spin triplet is more bound than the singlet, as one would have surmised from the vanishing probability at $\boldsymbol{r}_1 = \boldsymbol{r}_2$ in the former. This holds in most cases and is captured in *Hund's rule: states of highest multiplicity lie lowest.* The ferromagnet is the extreme illustration of Hund's rule.

Evaluation of the integrals and minimization with respect to α and β is straightforward. The theoretical results for the ionization energies and the comparison with data are shown in the following table.[1]

	α	β	Theory	Experiment
2^3P	1.99	1.09	0.262 Ry	0.266 Ry
2^1P	2.003	0.965	0.245 Ry	0.248 Ry

[1] C. Eckart, *Phys. Rev.* **36**, 878 (1930).

The values of α and β are astonishingly close to the intuitive expectations, and the agreement with experiment is at the 1% level. The $2^3P - 2^1P$ energy difference is 0.017 Ry, which is small as such things go, but still enormous compared to energies due to the interaction between the electrons' magnetic moments, which are of order $(v/c)^2$ Ry, or $\sim (1/137)^2$ Ry. The reason for this small value of J is that there is little overlap between the $1s$ and $2p$ orbitals, and as already emphasized in §6.1(b), interference due to symmetry or antisymmetry can only take place in regions where wave functions overlap. That $\alpha \simeq 2$ and $\beta \simeq 1$ is also a consequence of the small overlap between the two orbitals.

6.3 Atoms with Two Valence Electrons

(a) The Shell Model and Coupling Schemes

The atomic shell model assumes that any one electron can be approximately described by a single-particle wave function that conforms to the average electrostatic field produced by all the others. It then assigns electrons to levels in this field, filled in accordance with the Pauli principle. Completely filled levels are called closed shells, and are effectively not in play when excitation energies are low. Electrons in unfilled shells are called valence electrons because they suffice to describe the chemical phenomena in which the atom participates. To a good approximation, the low-lying excited states of the atom also only involve excitations of the valence electrons.[1]

Because the shell model is known to give a reliable if rough understanding of chemical and spectroscopic properties, the theory of atomic structure and spectroscopy can be divided into two rather distinct problems: that of determining the average field and its one-particle states, and that of determining the low-lying excitations. The first problem will be discussed in chapter 11. Here we concentrate on the second problem, which at least in atoms with few valence electrons, is soluble to a large extent with symmetry arguments and simple perturbation theory provided various integrals over the wave functions in the mean field are treated as parameters to be fitted to data. This is not an empty exercise, for as we will see, the theory makes a large enough number of predictions that yield meaningful comparisons with experiment.

The problem thus defined is therefore to find eigenstates of an effective Hamiltonian H_v that only refers to the v valence electrons. H_v is to describe the interactions between the valence electrons themselves, and between them and the remainder of the atom. The former has two contributions: the electrostatic potential $V(\boldsymbol{r})$ between the valence and closed-shell electrons, and the spin-orbit coupling H_{LS} experienced by the valence electrons due to their motion in this potential. The closed shells have no angular momentum, so that V is a central field if the polarization of

[1]The shell model also works quite well for many nuclei, but the situation is more complex. First, there is no heavy central object to which a potential like V can be anchored. Second, the interaction between the valence nucleons is much stronger than between electrons, so that configuration mixing is more important. Third, and of greater significance, nuclei well removed from closed shells are ellipsoidal, not spherical like atoms, and their spectroscopy is more akin to that of molecules than of atoms. See A. Bohr and B.R. Mottelson, *Nuclear Structure*, W.A. Benjamin, Vol. I (1969), Vol. II (1975).

the closed shells by the valence electrons is ignored. In short, in atomic units

$$H_v = \sum_{i=1}^{v} \left(\tfrac{1}{2} p_i^2 + V(r_i) \right) + H_{ee} + H_{LS} \;, \tag{26}$$

where

$$H_{ee} = \sum_{i<j=1}^{v} \frac{1}{r_{ij}} \;, \qquad H_{LS} = \sum_{i=1}^{v} \xi(r_i)\, \boldsymbol{L}_i \cdot \boldsymbol{s}_i \;, \quad \xi(r) = \frac{1}{2} \left(\frac{\hbar}{mc} \right)^2 \frac{1}{r} \frac{dV}{dr} \;. \tag{27}$$

As already indicated, the actual form of $V(r)$ and that of $\xi(r)$ will be buried in parameters to be defined shortly.

If the electron-electron and spin-orbit interactions in H_v are set aside for a moment, the first term of H_v defines a set of pseudo-hydrogenic one-electron states or "orbitals," and the assignment of the electrons to these defines a configuration, with all states in any one configuration being degenerate. The interactions just set aside produce splittings of these states, and these are usually small compared to the energy difference between the lowest and next higher configurations. As a consequence, *the problem of describing the atom's low-lying states is reduced to diagonalizing the truncated valence Hamiltonian H_v in the subspace $\mathfrak{H}_v$ spanned by just one configuration.*

With $\mathfrak{H}_v$ being defined by one configuration, the first term of (26) contributes the same energy to all levels. The task then comes down to diagonalizing $H_{ee} + H_{LS}$ in $\mathfrak{H}_v$. These two operators have quite different symmetries, and if one or the other dominates it is important to choose the basis in $\mathfrak{H}_v$ that diagonalizes the constants of motion of the dominant term. H_{ee} is invariant under rotation of the electrons' spatial coordinates and spins *separately;* hence the total orbital and spin angular momenta,

$$\boldsymbol{L} = \sum_i \boldsymbol{L}_i \;, \qquad \boldsymbol{S} = \sum_i \boldsymbol{s}_i \;, \tag{28}$$

are both constants of motion with respect to H_{ee}. Thus when H_{ee} dominates, the simultaneous eigenstates of $(\boldsymbol{L}^2, L_z, \boldsymbol{S}^2, S_z)$ form the appropriate basis. This is called *LS or Russel-Saunders coupling*, and it is the scheme suited to lighter atoms. The spin-orbit interaction grows rapidly with Z, and for the heaviest atoms H_{LS} is somewhat larger than H_{ee}. If H_{LS} were dominant, the appropriate basis is the one where the orbitals are eigenstates of H_{LS}, i.e., where each valence electron has a definite eigenvalue of its total angular momentum $\boldsymbol{j}_i$, and these are then added to form $\boldsymbol{J}$. This scheme is called *jj coupling.*

(b) The Configuration p^2

As an illustration of the approach just sketched, consider the configurations $(np)^2$, which describe the low-lying states of C, Si, Ge, Sn, Pb when $n = 2, \ldots, 6$. At first sight, this configuration appears to define a 36-dimensional space $\mathfrak{H}_v$, but this ignores the Pauli principle. The symmetry under interchange of the eigenstates of $(\boldsymbol{L}^2, L_z)$ formed by adding two angular momenta of $l = 1$ was shown in §3.6(b) to be even for $L = 0$ and $L = 2$, and odd for $L = 1$. Hence the former must be combined with the spin singlet (S=0) and the latter with the triplet ($S = 1$) to produce fully antisymmetric states. In short, in the LS coupling scheme the states allowed by the Pauli principle are

$$^1S_0 \quad ^3P_{0,1,2} \quad ^1D_2 \,, \tag{29}$$

and the dimension d_v of $\mathfrak{H}_v$ is "only" 15. Nevertheless, we will see that in both the LS and jj limits, symmetry alone suffices to produce diagonalization; in any event, as there are only two $J = 0$ and two $J = 2$ states, at worst only two 2-dimensional matrices have to be diagonalized.

In the basis $|LM_LSM_S\rangle$, the expectation value of H_{ee} only depends on L:

$$\Delta_{ee}(L) \equiv \langle LM_LSM_S|H_{ee}|LM_LSM_S\rangle \tag{30}$$

$$= \sum_{k=0}^{2l} \langle llLM_L| \frac{r_<^k}{r_>^{k+1}} P_k(\hat{\boldsymbol{r}}_1 \cdot \hat{\boldsymbol{r}}_2)|llLM_L\rangle \,, \tag{31}$$

where $l = 1$ in our example. Reflection invariance requires k to be even, so only $k = 0$ and $k = 2$ contribute. The $k = 0$ term represents the interaction of one electron with the spherical average of the other's charge distribution, and produces an L-independent shift, whereas the $k = 2$ term is the interaction between the quadrupole moments of the two electronic charge distributions. It can be handled by the technique developed for quadrupole hyperfine structure in §5.3(d), and therefore the details are left to Prob. 4. The result is

$$\Delta_{ee}(L) = F_0 + \frac{1}{6}F_2\left(\frac{3Q}{2l(2l-1)}\right)^2 [A(A+1) - \tfrac{4}{3}l^2(l+1)^2] \,, \tag{32}$$

where

$$A = L(L+1) - 2l(l+1) \,, \tag{33}$$

$$F_k = \int_0^\infty r_1^2 dr_1 \int_0^\infty r_2^2 dr_2 \, \frac{r_<^k}{r_>^{k+1}} \, |R_{np}(r_1)R_{np}(r_2)|^2 \,; \tag{34}$$

R_{nl} is the radial wave function for the configuration in question; and in our $l = 1$ case

$$Q = \langle 3\cos^2\theta - 1\rangle_{l=m_l=1} = -2/5 \,. \tag{35}$$

The integrals F_k are treated as free parameters in comparing to the data. The final result in LS coupling for the electrostatic energy in the configuration p^2 is then

$$\Delta_{ee}(L) = F_0 + \tfrac{1}{25}F_2\lambda_L, \qquad \lambda_0 = 10, \ \lambda_1 = -5, \ \lambda_2 = 1 \,. \tag{36}$$

If the spin-orbit coupling is assumed to be a small perturbation, it is only necessary to evaluate the expectation values of H_{LS} in the states listed in Eq. 29. Clearly, only the ^3P states can be effected by the LS interaction. Their shifts can be calculated by the following sequence of steps, valid in any multiplet of v-electron states with quantum numbers $LSJM$:

$$\Delta_{LS}(J) \equiv \langle H_{LS}\rangle_{LSJM} = v\langle \xi(r_1)\boldsymbol{L}_1 \cdot \boldsymbol{s}_1\rangle_{LSJM} = \zeta_{LS}\langle \boldsymbol{L}\cdot\boldsymbol{S}\rangle_{LSJM} \tag{37}$$

$$= \tfrac{1}{2}\zeta_{LS}[J(J+1) - L(L+1) - S(S+1)] \,. \tag{38}$$

The first step exploits indistinguishability, and the second that $\boldsymbol{L}_1 \cdot \boldsymbol{s}_1$ is the scalar product of vector operators. ζ_{LS} is treated as a parameter like the integrals F_k; it involves the radial wave functions and the function $\xi(r)$ defined by (27), but does *not* depend on J. In our $l = 1$ case,

$$\Delta_{LS}(J) = \zeta, \quad -\zeta, \quad -2\zeta, \tag{39}$$

for $J = 2, 1, 0$ respectively, and

$$\zeta = \tfrac{1}{2} \int_0^\infty r^2 dr \; \xi(r) |R_{np}(r)|^2 \,. \tag{40}$$

The final result for the configuration p^2 in the LS-coupling limit is the level scheme shown in Fig. 6.1a.

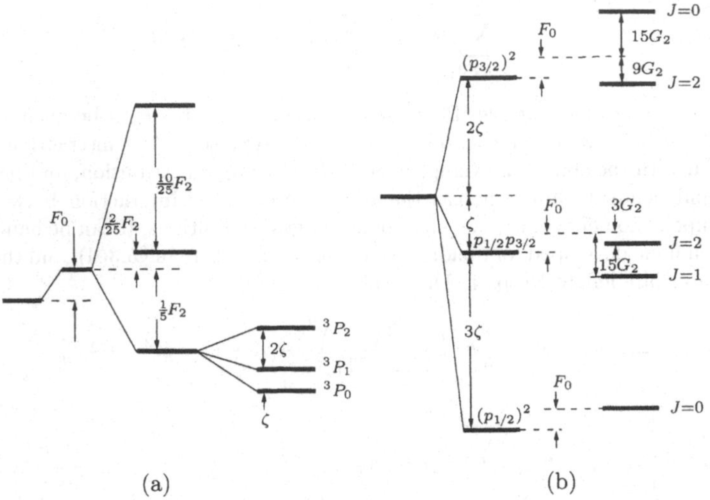

FIG. 6.1. The influence of the electron-electron interaction H_{ee} and spin-orbit coupling H_{LS} on the configuration p^2. In (a) the results are for LS or Russel-Saunders coupling; in (b) for jj coupling. The transition between the two schemes must smoothly connect the states of given J, as shown in Fig. 6.2.

The corresponding calculation in the jj-coupling scheme requires antisymmetric states $|j_1 j_2 JM\rangle$. The identity

$$\langle j_1 m_1 j_2 m_2 | JM \rangle = (-1)^{j_1 + j_2 - J} \langle j_2 m_2 j_1 m_1 | JM \rangle \tag{41}$$

identifies the allowed states, which are listed below. The evaluation of $\langle H_{LS} \rangle$ in these states is immediate, but that of $\langle H_{ee} \rangle$ involves some labor, and is left for Prob. 4. The results are

$$\Delta_{jj}(\tfrac{3}{2}\tfrac{3}{2};2) = F_0 + 2\zeta - 9G_2 \tag{42}$$
$$\Delta_{jj}(\tfrac{3}{2}\tfrac{3}{2};0) = F_0 + 2\zeta + 15G_2 \tag{43}$$
$$\Delta_{jj}(\tfrac{1}{2}\tfrac{3}{2};2) = F_0 - \zeta - 3G_2 \tag{44}$$
$$\Delta_{jj}(\tfrac{1}{2}\tfrac{3}{2};1) = F_0 - \zeta - 15G_2 \tag{45}$$
$$\Delta_{jj}(\tfrac{1}{2}\tfrac{1}{2};0) = F_0 - 4\zeta \,, \tag{46}$$

where $G_2 = F_2/75$. The level scheme in the jj-coupling limit is shown in Fig. 6.1b. These theoretical results are compared in Fig. 6.2 to the data on three atoms whose low-lying states are in the configuration p^2. The fit is remarkably good given

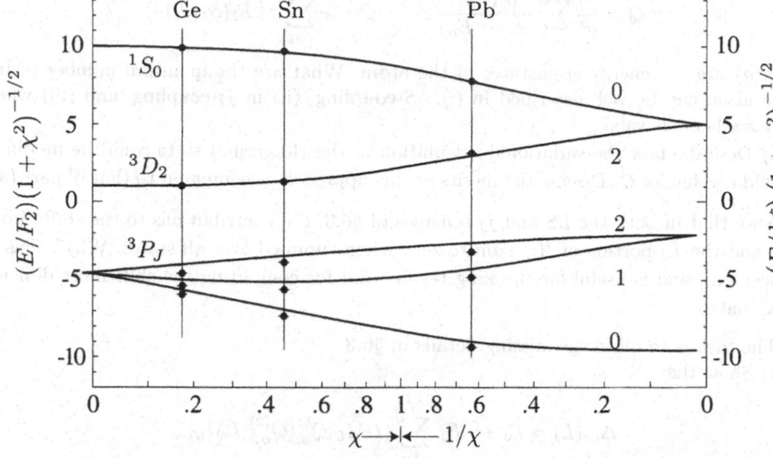

FIG. 6.2. Comparison between theory and experiment for the p^2 configuration in Ge, Sn, and Pb, taken from E.U. Condon and G.H. Shortley, *The Theory of Atomic Spectra,* Cambridge University Press (1935); p.275. The solid curves result from diagonalizing $H_{ee} + H_{LS}$ exactly by the method of Prob. 6. Their tangents on the left is the spectrum in the LS-limit, as shown in Fig. 6.1a, and the tangents on the right is the jj-limit of Fig. 6.1b. The transition from LS to jj coupling is governed by $\chi = \frac{1}{5}\zeta/F_2$. The parameters F_0, ζ and F_2 are chosen so that $^3P_1$ and the means of the $J = 0$ and $J = 2$ states fit the data exactly; the theory then predicts the separations between the pairs of $J = 0$ and $J = 2$ levels.

that mixing with other configurations is ignored completely. The lighter atoms Ge and Sn are seen to be well described by LS coupling, but Pb is not, nor does it have a sufficiently high Z to be at the jj-coupling limit. It is, however, quite straightforward to find the spectrum in intermediate coupling, i.e., when only J is a good quantum number. In the case of p^2, this requires the diagonalization of $H_{ee}+H_{LS}$ in the two 2-dimensional subspaces for $J = 0$ and $J = 2$. The figure shows the results of this calculation, which requires the evaluation of the transformation from the LS to the jj scheme; see Prob. 6(b).

6.4 Problems

1. Consider the influence on the He ground state of a uniform magnetic field. Recall the Hamiltonian $H_{\rm fld}$ derived in §5.4, and show that the energy shift of this state is quadratic in B. The magnetic susceptibility χ of a system with energy E is defined by $\partial E/\partial B = -\chi B$. Show that for this state, and in atomic units, $\chi = \frac{1}{6}\alpha^2\langle r_1^2 + r_2^2\rangle_0$.

2. A Z-electron atom with a $J = 0$ ground state has its nucleus fixed at the origin, and is in the electrostatic field of two identical point charges q on the z-axis at $z = \pm b$.

(a) Show that as $b \to \infty$, the leading effect on the atom is that it acquires the induced quadrupole moment

$$Q = \frac{4q}{b^3} \sum_n \frac{|\langle 0|\mathcal{Q}|n\rangle|^2}{E_0 - E_n} , \qquad \mathcal{Q} = e \sum_{i=1}^{Z} r_i^2 \, P_2(\cos\theta_i) .$$

Here $|n\rangle$ are the energy eigenstates of the atom. What are the quantum number of $|n\rangle$ if the atom can be well described in (i) LS-coupling, (ii) in jj-coupling, and (iii) when neither scheme is valid?

(b) Describe how the variational calculation of the He ground state could be modified to yield a value for Q. Discuss the merits of this approach as compared to that of part (a).

3. Note that in both the LS and jj schemes of §6.3, the contributions to the shifts from H_{LS} and the P_2 portion of H_{ee} sum to zero when summed over all states. Why? This is no accident, and is useful for checking results and for computing the shift most difficult to evaluate.

4. The task is to fill in the missing details in §6.3.

(a) Show that

$$\Delta_{ee}(L) = F_0 + \tfrac{1}{6} F_2 \sum_{\alpha\beta=1}^{3} \langle LM_L|Q^{(1)}_{\alpha\beta} Q^{(2)}_{\alpha\beta}|LM_L\rangle ,$$

where $Q^{(i)}_{\alpha\beta} = 3\hat{\boldsymbol{r}}_{i\alpha}\hat{\boldsymbol{r}}_{j\beta} - \delta_{\alpha\beta}$; confirm (32), (35) and (39).

(b) Construct the states in the jj-coupling scheme and confirm the results stated in the text for the energy shifts.

5. The low-lying states of many nuclei are well described by an extension of the atomic shell model, with most nucleons being in closed and effectively inert shells of zero total angular momentum, and a handful of valence nucleons. This problem considers a toy version of such a spectrum — toy, because the nucleons are assumed to have *no spin!* The model has two valence nucleons, a neutron and a proton, occupying two orbitals, an s and p level separated by the energy $E_s - E_p = \Delta$, with all higher orbitals ignored. The interaction between these nucleons is of zero range:

$$V(\boldsymbol{r}_n, \boldsymbol{r}_p) = -g\delta^3(\boldsymbol{r}_n - \boldsymbol{r}_p) .$$

(a) Let J be the total angular momentum of the valence nucleons, and write their eigenstates as $|l_1 l_2 J^{\pm}\rangle$, where, for example, $3^{\pm}$ means $J = 3$ and parity $\pm$, and l_i will be designated by s or p. Show that J and parity are constants of motion when V is included. When V is ignored, show that the spectrum and states are as follows:

$$E^0 = 2\Delta,\ \ 0^+ ; \qquad E^0 = \Delta,\ \ 1^-, 1^- ; \qquad E^0 = 0,\ \ 0^+, 1^+, 2^+ .$$

(b) Now include V. You will see that only one Clebsch-Gordon coefficient is needed. Fist show that in the 1^- subspace V is the matrix

$$-C \begin{pmatrix} 1 & 1 \\ 1 & 1 \end{pmatrix} , \qquad C = \frac{g}{4\pi} \int_0^\infty [R_a(r) R_p(r)]^2 \, r^2 \, dr ,$$

and that the energy shifts are therefore $0, -2C$.

(c) *Using only symmetry* (i.e., no calculation), show that the 1^+ sate suffers no shift, whereas that of 2^+ is

$$\delta(2^+) = -\frac{6}{5} B , \qquad B = \frac{g}{4\pi} \int_0^\infty [R_p(r)]^4 \, r^2 \, dr .$$

(d) Show that the 0^+ energies are the eigenvalues of

$$\begin{pmatrix} 2\Delta - A & \sqrt{3}C \\ \sqrt{3}C & -3B \end{pmatrix}, \qquad A = \frac{g}{4\pi}\int_0^\infty [R_s(r)]^4 \, r^2 \, dr \, .$$

6. A more challenging and instructive example than that of Prob. 5 is provided by a neutron and proton with spin, and as in real nuclei exposed to strong spin-orbit coupling. Once again, the *np* interaction is taken to have zero range, but is now spin-dependent.

The configuration to be analyzed is *sd*, and thus $\mathfrak{H}_v$ is 20-dimensional. The perturbation is $H_{LS} + H_{np}$, where

$$H_{LS} = \tfrac{1}{2}[\xi(r_n)\, \boldsymbol{L}_n \cdot \boldsymbol{\sigma}_n + \xi(r_p)\, \boldsymbol{L}_p \cdot \boldsymbol{\sigma}_p] \, , \qquad H_{np} = -g\boldsymbol{\sigma}_n \cdot \boldsymbol{\sigma}_p \, \delta^3(\boldsymbol{r}_n - \boldsymbol{r}_p) \, .$$

H_{LS} is diagonal in *jj*-coupling, but H_{np} is not.

(a) Working in *jj*-coupling, the states may be labeled by the j of the nucleon in the d orbital and written as $|jJM\rangle$. They are linear combinations of the states in the *LS* scheme:

$$|jJM\rangle = \sum_S C^S_{jJ} |lSJM\rangle \, ,$$

where $S = (0, 1)$, and $l = 2$ throughout. Show first that the states with $J = 1$ and 3 are simply

$$|\tfrac{5}{2}\, 3\rangle \equiv |{}^3D_3\rangle \, , \qquad |\tfrac{3}{2}\, 1\rangle \equiv |{}^3D_1\rangle \, ,$$

and that their energy shifts E_J are $E_1 = -\frac{3}{2}\zeta - G$ and $E_3 = \zeta - G$, where

$$\zeta = \int_0^\infty \xi(r)\, r^2 dr \, |R_d(r)|^2 \, , \qquad G = g\int_0^\infty r^2 dr \, |R_s(r) R_d(r)|^2 \, .$$

(b) Prove that the coefficients[1] C^S_{jJ} cannot depend on M. Then, by repeated angular momentum addition, show that

$$C^S_{jJ} = \sum \langle l m_l S M_S | JM \rangle \langle S M_S | s m_s s m_s' \rangle \langle l m_l s m_s' | j m_j \rangle \langle s m_s j m_j | JM \rangle \, ,$$

where $s = \frac{1}{2}$, and in this problem $l = 2$. Evaluate the C^S_{jJ} and show that

$$|\tfrac{3}{2}\, 2\rangle = \tfrac{3}{\sqrt{15}}\, |{}^3D_2\rangle + \sqrt{\tfrac{2}{5}}\, |{}^1D_2\rangle \, , \qquad |\tfrac{5}{2}\, 2\rangle = \sqrt{\tfrac{2}{5}}\, |{}^3D_2\rangle - \tfrac{3}{\sqrt{15}} |{}^1D_2\rangle \, ,$$

with the common eigenvalue M suppressed. (Hint: As C^S_{jJ} is independent of M, some choices of M are smarter than others.)

(c) Finally, show that the $J = 2$ energy shifts are the eigenvalues of

$$\begin{pmatrix} -\frac{3}{2}\zeta + \frac{3}{5}G & \frac{4}{5}\sqrt{6}\,G \\ \frac{4}{5}\sqrt{6}\,G & \zeta + \frac{7}{5}G \, . \end{pmatrix}$$

Endnotes

For the spectrum and properties of He, see the authoritative review by Bethe and Salpeter. For complex spectra, see the classic and still very useful treatise by E.U. Condon and G.H. Shortley, *The Theory of Atomic Spectra*, Cambridge University Press (1935).

---

[1] They are actually Racah coefficients; see §7.6(c). For this problem it is more efficient to just use the CG-coefficients, rather than learning how to use Racah coefficients.

7

Symmetries

7.1 Equivalent Descriptions and Wigner's Theorem

It is customary to talk as if Hilbert space were the arena in which the states of quantum mechanics live, and on the whole we have used this terminology. But it is unjustified:

> If $|\alpha\rangle$ and $|\beta\rangle$ are vectors in a system's Hilbert space, then only the absolute value of their scalar product $\langle\beta|\alpha\rangle$ has physical significance, and therefore all vectors $\lambda|\alpha\rangle$, etc., where $|\lambda| = 1$, represent the same physical state; furthermore, all these vectors have the same expectation value for any Hermitian operator.

The set of vectors that differ only by a phase is called a ray, and therefore *the state space of quantum mechanics is a ray space — not a Hilbert space.*

Once this is recognized, it is no longer a foregone conclusion that physical actions, such as evolution in time or rotations in space, are to be represented by unitary transformations, because this conclusion stems from the assumption that scalar products between states are to be invariants. This is now seen to be naive, and dire consequences might be expected. However, an important theorem due to Wigner shows that with but rare exceptions, only unitary transformations need to be considered.

A more scrupulous discussion of symmetry transformations than presented thus far is now in order. To be concrete, consider first a system that is described in two coordinate frames F and F' related to each other by a rotation. These descriptions must be equivalent. For example, if the rays $\lambda_\alpha|\alpha\rangle$ and $\lambda_\beta|\beta\rangle$ describe any two states of the system in F, where λ is a generic symbol for an arbitrary phase factor, and $\lambda'_\alpha|\alpha'\rangle$ and $\lambda'_\beta|\beta'\rangle$ describe the same two states of the system in F', then it must be that

$$|\langle\beta'|\alpha'\rangle| = |\langle\beta|\alpha\rangle| \,. \tag{1}$$

This is *not* the requirement that rotational symmetry holds *dynamically;* it is a requirement imposed solely by the equivalence of the two descriptions. The condition (1) must, therefore, be imposed on any pair of equivalent descriptions, whether related by a rotation, translation, reflection, interchange of names given to indistinguishable particles, and the like.

Wigner's theorem. Let $\mathcal{T}$ stand for the transformation that relates a specific pair of equivalent descriptions, to which there corresponds the mapping of the rays $\lambda_\alpha|\alpha\rangle$ into $\lambda'_\alpha|\alpha'\rangle$, etc. The arbitrary phases can be chosen so that under $\mathcal{T}$ *either*

$$\mathbf{I}: \qquad c_\alpha|\alpha\rangle + c_\beta|\beta\rangle \to c_\alpha|\alpha'\rangle + c_\beta|\beta'\rangle \,, \tag{2}$$

$$\langle\beta'|\alpha'\rangle = \langle\beta|\alpha\rangle \,, \tag{3}$$

or

$$\textbf{II}: \qquad c_\alpha|\alpha\rangle + c_\beta|\beta\rangle \to c_\alpha^*|\alpha'\rangle + c_\beta^*|\beta'\rangle \,, \tag{4}$$

$$\langle\beta'|\alpha'\rangle = \langle\beta|\alpha\rangle^* \,, \tag{5}$$

where the c's are complex numbers. Case I is the familiar unitary transformation. Case II is called *antiunitary*; it is not a linear transformation, and would be ruled out if the state space were a Hilbert space.

Case II is rare. It cannot be used for any $\mathcal{T}$ that is a member of a continuous group, because applying $\mathcal{T}$ twice in case II results in a unitary transformation. However, if $\mathcal{T}$ belongs to a continuous group, such as rotations, there exists a transformation $\mathcal{T}_{1/2}$ such that $[\mathcal{T}_{1/2}]^2 = \mathcal{T}$, and self-consistency therefore requires $\mathcal{T}$ to be represented by unitary transformations. By the same argument, candidates for an antiunitary transformation must be such that $\mathcal{T}^2$ reproduces the original description. The most familiar examples of this type are space reflection and time reversal. Others are charge conjugation, which is the interchange between particles and antiparticles, and the interchange of identical particles. It turns out, as we will see in the next section, that time reversal is represented by an antiunitary transformation. The others just mentioned are all represented by unitary transformation.

The proof of Wigner's theorem for an arbitrary space of states is quite complicated and not sufficiently instructive to warrant its reproduction here. We shall, instead, give an illuminating proof due to Wick for the simplest case, that of a two-dimensional complex vector space $\mathfrak{C}_2$, and refer in the endnotes to his extension of this proof to $\mathfrak{C}_n$ as well as to Wigner's proof.

We know from §3.3 that the theory of spin $\frac{1}{2}$ provides a complete description of $\mathfrak{C}_2$. As a basis in this space we choose the eigenkets $|\pm\rangle$ of σ_3 with eigenvalues ± 1. An arbitrary ket is constructed from $|+\rangle$ by the rotation

$$|\zeta\rangle = \lambda\, e^{-\frac{1}{2}i\sigma_3\varphi}\, e^{-\frac{1}{2}\,i\sigma_2\vartheta}\, |+\rangle \tag{6}$$

$$= \lambda \left(e^{-\frac{1}{2}i\varphi} \cos\tfrac{1}{2}\vartheta\, |+\rangle + e^{\frac{1}{2}i\varphi}\, \sin\tfrac{1}{2}\vartheta\, |-\rangle \right) , \tag{7}$$

where $0 \leq \vartheta \leq \pi$ and $-\pi \leq \varphi \leq \pi$. It is intuitively clear that $|\zeta\rangle$ is eigenket of $\boldsymbol{n}\cdot\boldsymbol{\sigma}$ with eigenvalue $+1$, where $\boldsymbol{n}$ is a unit vector with polar and azimuthal angles (ϑ, φ). Of course, the rotation (6) and the vector $\boldsymbol{n}$ are in an abstract Euclidean 3-space, not in any physical space. If $\boldsymbol{n}$ has the meaning claimed, it must be that

$$\langle\zeta|\sigma_i|\zeta\rangle \equiv n_i = (\sin\vartheta\cos\varphi, \sin\vartheta\sin\varphi, \cos\vartheta) \,. \tag{8}$$

To confirm this, write (6) as $\lambda D(\boldsymbol{n})|+\rangle$; then

$$\langle\zeta|\sigma_i|\zeta\rangle = \mathrm{Tr}\; D^\dagger(\boldsymbol{n})\sigma_i D(\boldsymbol{n}) P_+ \,, \tag{9}$$

where $P_+ = \frac{1}{2}(1+\sigma_3)$ is the projection operator onto $|+\rangle$. The algebra of the Pauli matrices and the fact that they are traceless then leads directly to (8).

Define another ket $|\zeta'\rangle = \lambda' D(\boldsymbol{n}')|+\rangle$. Its scalar product with $|\zeta\rangle$ is

$$\langle\zeta'|\zeta\rangle = \lambda'^*\lambda\; \mathrm{Tr}\; D^\dagger(\boldsymbol{n}')D(\boldsymbol{n})\, P_+ \tag{10}$$

$$= \lambda'^*\lambda \left[\cos\tfrac{1}{2}(\varphi-\varphi')\cos\tfrac{1}{2}(\vartheta-\vartheta') - i\sin\tfrac{1}{2}(\varphi-\varphi')\cos\tfrac{1}{2}(\vartheta+\vartheta')\right] . \tag{11}$$

A simple exercise in trigonometry then yields

$$|\langle\zeta'|\zeta\rangle| = \sqrt{\tfrac{1}{2}(1+\boldsymbol{n}\cdot\boldsymbol{n}')}\ . \tag{12}$$

Equation (12) is central to the argument. It shows that in $\mathfrak{C}_2$ the absolute value of the scalar product is invariant under (i) rotations of both unit vectors, and (ii) under a reflection of both. An arbitrary unitary transformation in $\mathfrak{C}_2$, i.e., $|\zeta\rangle \to U|\zeta\rangle$ and $|\zeta'\rangle \to U|\zeta'\rangle$, produces a rotation of both $\boldsymbol{n}$ and $\boldsymbol{n}'$ through some angle about some axis, and therefore leaves (12) unchanged. *The reflection $\boldsymbol{n} \to -\boldsymbol{n}$, $\boldsymbol{n}' \to -\boldsymbol{n}'$, cannot be accomplished by any such unitary transformation, however.* If there were such a U, it would have to satisfy

$$U^\dagger \sigma_i U = -\sigma_i\ . \tag{13}$$

Any matrix in $\mathfrak{C}_2$ is a linear combination of the Pauli matrices and 1, and therefore there is no matrix in $\mathfrak{C}_2$ that anticommutes with all three Pauli matrices! This simple algebraic statement has remarkably deep consequences. Here it plays a crucial role in establishing Wigner's theorem, but it also is the reason that Dirac's relativistic theory of spin $\frac{1}{2}$ particles requires 4-component wave functions, as we shall see in §13.2.

Is it possible to exploit the reflection symmetry of (12)? It is, because $|\langle\zeta'|\zeta\rangle|$ is not changed if the coefficients of the kets are complex-conjugated, for then $\varphi \to -\varphi$ in (6). That is, complex-conjugation produces

$$(n_1, n_2, n_3) \to (n_1, -n_2, n_3)\ , \tag{14}$$

and similarly for $\boldsymbol{n}'$, which is a reflection in the 13 plane. As for the basis $|\pm\rangle$, it is left unchanged under complex conjugation because σ_3 is real.[1]

To summarize, the absolute value of all scalar products in a two-dimensional Hilbert space are left invariant by

- any unitary transformation applied to all kets;
- by the antiunitary transformation which replaces the expansion coefficients of all kets by their complex conjugates.

Wigner's theorem is the extension of this statement to an arbitrary Hilbert space. In the general case, given an orthonormal basis $\{|n\rangle\}$ that transform to $\{|n'\rangle\}$ under an antiunitary $\mathcal{T}$, the transformation law for an arbitrary ket is

$$|\alpha\rangle = \sum_n |n\rangle\,\langle n|\alpha\rangle \to |\alpha'\rangle = \sum_n |n'\rangle\,\langle n|\alpha\rangle^*\ , \tag{15}$$

and therefore

$$\langle\beta|\alpha\rangle \to \langle\beta'|\alpha'\rangle = \sum_n \langle\alpha|n\rangle\langle n|\beta\rangle = \langle\alpha|\beta\rangle\ , \tag{16}$$

as stated in (5).

Whether a particular transformation $\mathcal{T}$ (satisfying the requirement that $\mathcal{T}^2$ reproduces the original description) is represented by a unitary or antiunitary transformation can only be determined by further considerations that draw on the physical or geometrical meaning of $\mathcal{T}$. This issue will be addressed in the next section.

---

[1] Change of basis is taken up in §7.2(a).

7.2 Time Reversal

There are no mystical notions involved in time reversal. One need not be equipped with clocks that suddenly run backward to comprehend it. Consider a particle obeying the laws of classical mechanics and subject to some unknown, static forces. Let $\boldsymbol{r}(t_0), \boldsymbol{v}(t_0)$ be its position and velocity at time t_0, and allow it to proceed undisturbed for a time t to the position and velocity $\boldsymbol{r}(t_0+t), \boldsymbol{v}(t_0+t)$. At that time, start another identical particle off with the initial conditions $\boldsymbol{r}(t_0+t), -\boldsymbol{v}(t_0+t)$. If after another interval of duration t, i.e, at the time $t_0 + 2t$, this second particle has the position and velocity $\boldsymbol{r}(t_0), -\boldsymbol{v}(t_0)$, the basic laws governing motion in these circumstances are said to be invariant under time reversal, and if those are not the position and velocity at that time, the laws do not possess this invariance. Clearly, if the forces are the gradient of a potential, the set of all orbits will satisfy this invariance condition. If, on the other hand, the particle is charged and subjected to a magnetic field, the motions will not display any symmetry of this kind (see Fig. 7.1). Of course, the symmetry would be restored if the motion of the charges that produce the magnetic field were also reversed, but that is not relevant if the object of concern is a system in given external conditions.

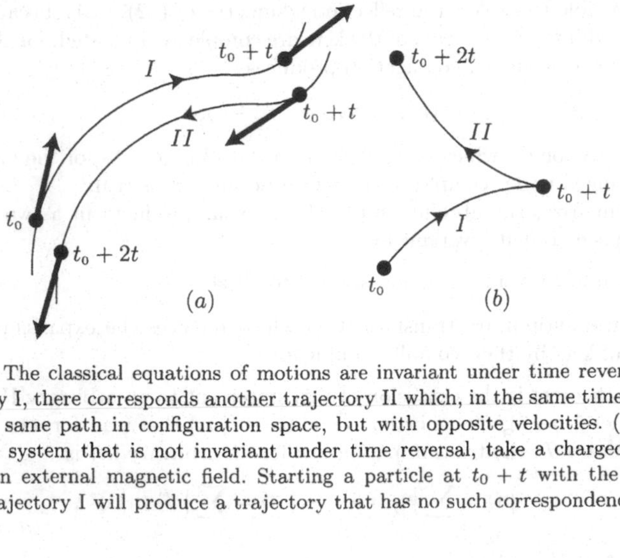

FIG. 7.1. (a) The classical equations of motions are invariant under time reversal if, to any trajectory I, there corresponds another trajectory II which, in the same time interval, traverses the same path in configuration space, but with opposite velocities. (b) As an example of a system that is not invariant under time reversal, take a charged particle exposed to an external magnetic field. Starting a particle at $t_0 + t$ with the opposite velocity of trajectory I will produce a trajectory that has no such correspondence to I.

To summarize, the classical equations of motion are said to be invariant under time reversal if the manifold of all possible orbits can be divided into two subsets related by a one-to-one correspondence effected by a simultaneous reversal of all the momenta. This correspondence can be stated formally by designating the canonical variables in one subset by $(q_I(t), p_I(t))$, and in the other by $(q_{II}(t), p_{II}(t))$. Then the condition of that defines time reversal symmetry is $q_I(t_0) = q_{II}(t_0 + 2t)$, $p_I(t_0) = -p_{II}(t_0+2t)$. As t_0 is arbitrary, it is convenient to put $t_0 = -t$, so that the condition reads

$$q_I(-t) = q_{II}(t) \,, \qquad p_I(-t) = -p_{II}(t) \,. \tag{17}$$

(a) The Time Reversal Operator

Now to quantum mechanics. Let $|\psi\rangle$ be any state, and $|\psi^R\rangle$ be its motion-reversed counterpart, both at, say, $t = 0$. For example, if the system in question is a set of N spin 0 particles with definite momenta, then $|\psi\rangle = |\boldsymbol{p}'_1, \ldots, \boldsymbol{p}'_N\rangle$ and $|\psi^R\rangle = |-\boldsymbol{p}'_1, \ldots, -\boldsymbol{p}'_N\rangle$. Thus far no symmetry requirement has been imposed because at any instant such a transformation can always be made. Now define the *time reversal operator* I_t by $I_t|\psi\rangle = |\psi^R\rangle$. Clearly, $(I_t)^2$ acting on any ket must reproduce that ket to within at most a phase, that is, $|\psi\rangle$ and $I_t^2|\psi\rangle$ must belong to the same ray. Hence by Wigner's theorem, I_t can be unitary or antiunitary.

If $\boldsymbol{p}_i$ is a momentum operator, then $\boldsymbol{p}_i|\psi^R\rangle = -\boldsymbol{p}'_i|\psi^R\rangle$, and I_t applied to this equation gives

$$I_t\boldsymbol{p}_iI_t^{-1}I_t|\psi^R\rangle = -\boldsymbol{p}'_iI_t|\psi^R\rangle = -\boldsymbol{p}_iI_t|\psi^R\rangle\,, \tag{18}$$

or

$$I_t\boldsymbol{p}_iI_t^{-1} = -\boldsymbol{p}_i\,. \tag{19}$$

Thus far, this argument would apply just as well to space reflection. But now comes the crucial point, which is the stipulation that as the system evolves, any state $|\psi\rangle$ and its motion-reversed counterpart $I_t|\psi\rangle$ satisfy the correspondence analogous to that defined for classical motions in Eq. 17:

$$I_t\left(e^{iHt/\hbar}|\psi\rangle\right) = e^{-iHt/\hbar}\left(I_t|\psi\rangle\right)\,. \tag{20}$$

As $|\psi\rangle$ is any state, this requires

$$I_te^{iHt/\hbar}I_t^{-1} = e^{-iHt/\hbar}\,. \tag{21}$$

Now if A is antiunitary and B Hermitian, $Ae^{-iB}A^{-1} = \exp(iABA^{-1})$, which leaves two possibilities:

$$I_t \quad \text{unitary}: \qquad I_tH + HI_t = 0\,, \tag{22}$$

$$I_t \quad \text{antiunitary}: \qquad I_tH - HI_t = 0\,. \tag{23}$$

But for free particles H is quadratic in the momenta and commutes with I_t, and therefore *the time reversal operator I_t is antiunitary.*

Space reflection could, in principle, also be antiunitary because I_s^2 produces no change of description. However, if I_s were antiunitary, and given the requirement $I_s\boldsymbol{P}I_s^{-1} = -\boldsymbol{P}$, then I_s would not change the spatial translation operator,

$$I_se^{i\boldsymbol{a}\cdot\boldsymbol{P}/\hbar}I_s^{-1} = e^{i\boldsymbol{a}\cdot\boldsymbol{P}/\hbar}\,, \tag{24}$$

which is false; clearly the right-hand side must be a translation in the opposite sense, as it is when I_s is unitary.

The transformation of angular momenta under time reversal is of great importance. Orbital angular momenta $\boldsymbol{x}\times\boldsymbol{p}$ are odd, but what about spins? The question is answered by requiring time reversal to not have any effect on geometrical transformations. This condition is met by spatial translations, $\mathcal{T} = \exp(i\boldsymbol{a}\cdot\boldsymbol{P})$, which satisfies $I_t\mathcal{T}I_t^{-1} = \mathcal{T}$ since I_t reverses both i and the total momentum $\boldsymbol{P}$. The same consideration for the unitary rotation operator shows that its generator, the total angular momentum $\boldsymbol{J}$, must be odd under time reversal,

$$I_t\boldsymbol{J}I_t^{-1} = -\boldsymbol{J}\,, \tag{25}$$

which requires the spin operators to also be odd.

Requiring time reversal invariance will, in general, restrict the types of terms that can appear in a system's Hamiltonian. For example, a one-body Hamiltonian cannot contain the term $\boldsymbol{x} \cdot \boldsymbol{\sigma}$, and this cannot be saved by a factor i because that would make it non-Hermitian. True enough, this operator is also not space reflection invariant. In the two-body case there is, in principle, an interaction that is reflection invariant but odd under time reversal, namely,

$$U(r)(\boldsymbol{\sigma}_1 \times \boldsymbol{\sigma}_2) \cdot \boldsymbol{L} \,, \tag{26}$$

where $\boldsymbol{L}$ is the orbital angular momentum in the center-of-mass frame. Scattering experiments setting upper limits on violations of time reversal in the nucleon-nucleon system have been done.

Because time reversal is an antiunitary transformation, the formulas involving it do not flow effortlessly in the Dirac notation. Transformation theory cannot be used with eyes shut when antiunitary operators are on the road.

Let $\{|n\rangle\}$ be a fixed orthonormal basis, and $\{|n^R\rangle\}$ its motion-reversed counterpart. Two such bases are related by a unitary transformation:

$$U = \sum_n |n^R\rangle\langle n| \,. \tag{27}$$

An example would be the basis formed by the states of a spin-zero particle in a central field, $\{|nlm\rangle\}$; their motion-reversed counterparts are obtained by $m \to -m$, as is intuitively clear and will be proven in the next subsection.

Let $|\psi\rangle$ be an arbitrary state. Because I_t is antiunitary, we know from case II of §7.1 that

$$I_t|\psi\rangle \equiv |\psi^R\rangle = I_t \left(\sum_n |n\rangle\langle n|\psi\rangle \right) = \sum_n |n^R\rangle\langle n|\psi\rangle^* = \sum_n U|n\rangle\langle n|\psi\rangle^* \,. \tag{28}$$

Furthermore, if $|\phi\rangle$ is an another arbitrary ket, then as a consequence of (28)

$$\langle\psi|\phi\rangle = \langle\phi^R|\psi^R\rangle \,. \tag{29}$$

Equation (28) implies that I_t has the form

$$I_t = UK \,, \qquad I_t^{-1} = KU^\dagger \,, \tag{30}$$

where K is the operator that takes the complex conjugate of all the coefficients in the basis on which U acts. The operator U is basis-dependent, as is evident from the fact that $U = 1$ if the basis kets are the coordinate eigenkets, whereas if the basis kets are the momentum eigenkets U must reverse their momenta. In general, if the basis is changed by the unitary transformation W, the new unitary operator $\tilde{U}$ in $I_t = \tilde{U}K$ is related to U by

$$\tilde{U} = W^T U W \,, \tag{31}$$

which is *not* a unitary transformation unless W is orthogonal. The proof of this statement is left for Prob. 1.

The matrix elements of operators between time-reversed states are often needed, for example, in collision theory, where time reversal invariance, when applicable, places constraints on scattering amplitudes by relating the process $i \to f$ to its motion-reversed counterpart $f \to i$. Let A be *any* operator, and $|\psi\rangle$ be any state. Then from (28)

$$I_t A|\psi\rangle = I_t \left(A I_t^{-1}|\psi^R\rangle\right) = \sum_n |n^R\rangle\langle n|A I_t^{-1}|\psi^R\rangle^* = \sum_n |n^R\rangle\langle\psi|A^\dagger|n\rangle \,. \qquad (32)$$

Let $|\phi\rangle$ be any other state; then

$$\langle\phi^R|I_t A I_t^{-1}|\psi^R\rangle = \sum_n \langle\psi|A^\dagger|n\rangle\langle n|\phi\rangle \,, \qquad (33)$$

and therefore

$$\langle\phi^R|I_t A I_t^{-1}|\psi^R\rangle = \langle\psi|A^\dagger|\phi\rangle \,. \qquad (34)$$

Hence if A is Hermitian and invariant under time reversal

$$\langle\phi^R|I_t A I_t^{-1}|\psi^R\rangle = \langle\phi^R|A|\psi^R\rangle = \langle\psi|A|\phi\rangle \,. \qquad (35)$$

If we think of $|\phi\rangle$ and $|\psi\rangle$ as initial and final states in some process produced by the operator A, we see that time reversal does indeed relate the process to its inverse.

(b) Spin 0

The form of the time reversal operator depends on spin, and spin 0 is of course the simplest case. Let $\psi(\boldsymbol{r}, t)$ be the Schrödinger wave function of a spin 0 particle in the coordinate representation, $\psi^R(\boldsymbol{r}, t)$ the motion-reversed counterpart, and H a time reversal invariant Hamiltonian. By definition

$$\psi(\boldsymbol{r}, t) = \langle\boldsymbol{r}|e^{-iHt/\hbar}|\psi\rangle \,, \quad \psi^R(\boldsymbol{r}, t) = \langle\boldsymbol{r}|e^{-iHt/\hbar}|\psi^R\rangle \,. \qquad (36)$$

Because $|\boldsymbol{r}\rangle$ is unchanged under time reversal, (34) implies that

$$\psi^R(\boldsymbol{r}, t) = \langle\psi|I_t e^{iHt/\hbar} I_t^{-1}|\boldsymbol{r}\rangle = \langle\psi|e^{-iHt/\hbar}|\boldsymbol{r}\rangle \,, \qquad (37)$$

and therefore

$$\psi^R(\boldsymbol{r}, t) = \psi^*(\boldsymbol{r}, -t) \,. \qquad (38)$$

The generalization to more than one particle is straightforward. In consequence, *for spin 0 and in the coordinate representation, time reversal is accomplished by complex conjugation and $t \to -t$.*

Momentum eigenstates are not invariant under time reversal, $I_t|\boldsymbol{p}\rangle = |-\boldsymbol{p}\rangle$. Therefore the momentum space wave function

$$\chi(\boldsymbol{p}, t) \equiv \langle\boldsymbol{p}|e^{-iHt/\hbar}|\psi\rangle \qquad (39)$$

has the following behavior under time reversal if the Hamiltonian is time reversal invariant:

$$\chi^R(\boldsymbol{p}, t) = \chi^*(-\boldsymbol{p}, -t) \,. \qquad (40)$$

Consider some implications of Eq. 38. If the wave function is the spherical harmonic $Y_{lm}(\theta\phi)$, then time reversal produces

$$I_t : \;\; Y_{lm}(\theta\phi) \to Y^*_{lm}(\theta\phi) = (-1)^m Y_{l,-m}(\theta\phi) \,. \qquad (41)$$

In the time-reversed situation the particle travels in the opposite sense, and therefore its angular momentum projected onto the quantization axis is reversed. As the whole vector operator $\boldsymbol{L}$ is odd under time reversal, it follows that $\boldsymbol{n}\cdot\boldsymbol{L}$ must reverse along any direction $\boldsymbol{n}$. Incidentally, the phase $(-1)^m$ is representation dependent; a different phase convention for the spherical harmonics could remove this factor in (41), but $m \to -m$ of course holds in any phase convention.

A more interesting illustration of time reversal is provided by states that describe inherently time-dependent processes, such as scattering. As a simple example, consider scattering in one dimension, §4.5, and in particular Eq. 243:

$$\psi_k(x) = e^{ikx} + \int dx'\, G_0(x,x')U(x')\psi_k(x') \,, \tag{42}$$

which represents an incident plane wave propagating toward the right which is then scattered into *outgoing* waves by U. The complex conjugate of this equation has a plane wave propagating toward the left and the Green's function

$$G_0^*(x,x') = -\frac{1}{2ik} e^{-ik|x-x'|} \,, \tag{43}$$

which produces waves *incoming toward* U. In a fully time-dependent description, as given in §4.5(d), this second situation is the one where for early times there are waves converging on the scattering potential from different directions which subsequently combine coherently to form a wave packet moving away from U to the left. Of course, such states cannot be prepared in the laboratory, but they play an important role in scattering theory, as we will see in §8.2 and §9.2.

That complex conjugation is equivalent to time reversal in the case of zero spin has an interesting implication for non-degenerate stationary states ψ_E. For if ψ_E is a stationary state, and $I_t H I_t^{-1} = H$, then ψ_E^* is also a stationary state of energy E. There are then two possibilities: the level is degenerate and ψ_E^* does not belong to the same ray as ψ_E, as is the case with the spherical harmonics for $l \neq 0$; or there is no degeneracy and $\psi_E^* \propto \psi_E$. Hence *the wave function of a nondegenerate stationary state of a spin 0 particle is real apart from an overall phase.*

(c) Spin $\frac{1}{2}$

Consider first a single spin $\frac{1}{2}$ particle. According to the argument leading to (25), the operator I_t must reverse the spin operators, or in the notation of (30)

$$UK\boldsymbol{\sigma}KU^\dagger = U\boldsymbol{\sigma}^*U^\dagger = -\boldsymbol{\sigma} \,. \tag{44}$$

Complex conjugation — the antiunitary character of time reversal, plays a crucial role here, because without it the *sine qua non* requirement $\boldsymbol{\sigma} \to -\boldsymbol{\sigma}$ could not be satisfied, for there is no two-dimensional matrix U that anticommutes with all three Pauli matrices, a fact that already played a central role in our demonstration of Wigner's theorem in the preceding section.

In the standard representation, σ_x and σ_z are real, while σ_y is pure imaginary, so U must produce

$$U\sigma_x U^\dagger = -\sigma_x \,,\; U\sigma_y U^\dagger = \sigma_y \,,\; U\sigma_z U^\dagger = -\sigma_z \,. \tag{45}$$

This is satisfied by

$$U = e^{i\delta}\,\sigma_y\,, \tag{46}$$

which is a rotation through π about the y-axis, and accords with intuition in that it makes a system running clockwise about the z-axis run counter-clockwise. Acting on the spin basis $|\pm\frac{1}{2}\rangle$, and with the standard phase conventions,

$$I_t|\tfrac{1}{2}\rangle = ie^{i\delta}|-\tfrac{1}{2}\rangle\,, \qquad I_t|-\tfrac{1}{2}\rangle = -ie^{i\delta}|\tfrac{1}{2}\rangle\,. \tag{47}$$

Let $|\mathbf{p}m_s\rangle$ and $|\mathbf{r}m_s\rangle$, with $m_s = \pm\frac{1}{2}$, be momentum and coordinate representation eigenkets. Then[1]

$$I_t|\mathbf{p}m_s\rangle = e^{i\delta}\,i^{2m_s}|-\mathbf{p}-m_s\rangle\,, \tag{48}$$

$$I_t|\mathbf{r}m_s\rangle = e^{i\delta}\,i^{2m_s}|\mathbf{r}-m_s\rangle\,. \tag{49}$$

In the coordinate representation, therefore,

$$I_t = e^{i\delta}\,\sigma_y K\,. \tag{50}$$

The square of the time reversal operator is of surprising importance. That $I_t^2 = \pm 1$ is shown as follows. We already know that $I_t^2 = c$, where $|c| = 1$. But $[I_t^2, I_t] = 0$, so $(c - c^*)I_t = 0$, and therefore $c = c^* = \pm 1$. In the case of spin 0, where $I_t = K$, clearly $I_t^2 = 1$. In the case of spin $\frac{1}{2}$,

$$I_t^2 = e^{i\delta}\,\sigma_y K e^{i\delta}\,\sigma_y K = e^{i\delta}\sigma_y e^{-i\delta}\sigma_y^* K^2 = -\sigma_y^2 = -1\,. \tag{51}$$

The phase factor $e^{i\delta}$ was kept until now to show that it disappears from I_t^2; indeed, it can be shown that the value of I_t^2 has no dependence whatever on the representation (Prob. 2). In short,

$$I_t^2 = +1 \qquad \text{for one spin 0 particle}\,, \tag{52}$$

$$I_t^2 = -1 \qquad \text{for one spin } \tfrac{1}{2} \text{ particle}\,. \tag{53}$$

Consider a system of N particles. If they have spin zero, then $I_t^2 = 1$ for any N, which is of no consequence. But that is not so for for a system of spin $\frac{1}{2}$ particles. I_t must reverse all the spins $\boldsymbol{\sigma}^{(k)}$, and therefore

$$I_t = \prod_{k=1}^{N} \sigma_y^{(k)} \times K\,, \tag{54}$$

aside from an arbitrary phase. Therefore

$$I_t^2 = (-1)^N\,. \tag{55}$$

This fact has an interesting consequence. Let $\{|E,N\rangle\}$ be the stationary states of a system of N spin $\frac{1}{2}$ particles, which need not be identical, and assume that

[1]These and other relations involving I_t can be "cleaned up" by adopting a different phase convention, e.g., by departing from the convention $J_\pm|jm\rangle = \sqrt{j(j+1) - m(m\pm 1)}\,|jm\pm 1\rangle$. For example, with an appropriate choice of phases, $Y_{lm}^* = Y_{l,-m}$. Naturally, physically meaningful results do not change.

the Hamiltonian is invariant under time reversal. Then $I_t|E,N\rangle$ must also be an eigenstate of H with energy E. The question now arises as to whether this state is linearly independent of $|E,N\rangle$ or not, for if it is, there is a degeneracy. If $I_t|E,N\rangle$ is not degenerate, $I_t|E,N\rangle = \lambda|E,N\rangle$; from this it follows that

$$I_t^2|E,N\rangle = \lambda^*\lambda|E,N\rangle = (-1)^N|E,N\rangle \,, \tag{56}$$

which is false if N is odd. This is *Kramer's theorem: if a system composed of an odd number of spin $\frac{1}{2}$ particles has a Hamiltonian that is invariant under time reversal, then all its stationary states are degenerate.* It is also possible to show that the degree of degeneracy is even. This theorem is useful in complicated systems if there are no symmetries other than time reversal. The theorem also implies that when the theorem's assumptions are violated by exposing the system to a magnetic field, the levels will split into multiplets with an even number of members.

7.3 Galileo Transformations

The theory developed thus far stands on the Newtonian conception of space and time, and it should, therefore, be invariant under Galileo transformations. This section is devoted to these transformations which, as we will see, have some remarkable features not shared by the other continuous transformations of space and time.

(a) Transformation of States: Galileo Invariance

Let $\mathcal{S}(t)$ be the state of a classical system composed of point masses m_n which have the coordinates and momenta $\boldsymbol{r}_n$ and $\boldsymbol{k}_n$ as seen in an inertial frame F. Consider a second state $\mathcal{S}'(t)$ of this same system in the same frame, having the coordinates $\boldsymbol{r}_n + \boldsymbol{v}t$ and momenta $\boldsymbol{k}_n + m_n\boldsymbol{v}$. We call $\mathcal{S}'(t)$ a boosted replica of $\mathcal{S}(t)$. If one prefers, $\mathcal{S}'(t)$ is the first state $\mathcal{S}(t)$ as seen in a frame F' which at $t = 0$ coincides with F and is moving with velocity $-\boldsymbol{v}$ as seen from F.

The Galileo transformation is then defined as the boost

$$\boldsymbol{k}_n \to \boldsymbol{k}'_n = \boldsymbol{k}_n + m_n\boldsymbol{v} \,, \quad \boldsymbol{r}_n \to \boldsymbol{r}'_n = \boldsymbol{r}_n + \boldsymbol{v}t \,, \quad t' = t \,. \tag{57}$$

If $\mathcal{S}''(t)$ is a state boosted with velocity $\boldsymbol{v}'$ with respect to $\mathcal{S}'(t)$, the boost from the original state $\mathcal{S}(t)$ to this doubly boosted state $\mathcal{S}''(t)$ is effected by a single boost with velocity $\boldsymbol{v}+\boldsymbol{v}'$. Because these velocities simply add, successive Galileo transformations (57) form an Abelian group.

A spatial translation through $\boldsymbol{a}$ does not alter the momenta, but replaces the coordinates by $\boldsymbol{r}_n+\boldsymbol{a}$; hence spatial translations commute with Galileo transformations. On the other hand, displacements in time do not commute with Galileo transformations, nor do rotations. In short, the complete Galileo group is not Abelian (see Prob. 3).

Turning to quantum mechanics, this same system has coordinate and momentum operators $\boldsymbol{x}_n$ and $\boldsymbol{p}_n$, with eigenvalues $\boldsymbol{r}_n$ and $\boldsymbol{k}_n$, which are taken for now to be in the Schrödinger picture. Because the Galileo transformations form a continuous group, they are carried out by unitary operators $G(\boldsymbol{v})$ which are to produce the counterpart of (57),

$$G^\dagger(\boldsymbol{v})\,\boldsymbol{x}_n\,G(\boldsymbol{v}) = \boldsymbol{x}_n + \boldsymbol{v}t \,, \qquad G^\dagger(\boldsymbol{v})\,\boldsymbol{p}_n\,G(\boldsymbol{v}) = \boldsymbol{p}_n + m_n\boldsymbol{v} \,. \tag{58}$$

As with other continuous transformations, we write G in the form[1]

$$G(\boldsymbol{v}) = e^{-i\boldsymbol{v}\cdot\boldsymbol{N}/\hbar} . \tag{59}$$

The Hermitian generator $\boldsymbol{N}$ will be called the boost. It is determined by requiring the infinitesimal transformation $(1 - i\delta\boldsymbol{v}\cdot\boldsymbol{N}/\hbar)$ to produce (58):

$$\frac{i}{\hbar}[N_i, x_{nj}] = \frac{\partial N_i}{\partial p_{nj}} = \delta_{ij} t \,, \qquad \frac{i}{\hbar}[N_i, p_{nj}] = -\frac{\partial N_i}{\partial x_{nj}} = \delta_{ij} m_n \,. \tag{60}$$

Therefore

$$\boldsymbol{N} = \sum_n (\boldsymbol{p}_n t - m_n \boldsymbol{x}_n) = \boldsymbol{P}t - M\boldsymbol{X} \,, \tag{61}$$

where M is the system's total mass, $\boldsymbol{P}$ its total momentum and $\boldsymbol{X}$ its center of mass coordinate.

Because $\boldsymbol{P}$ and $\boldsymbol{X}$ do not commute, care must be used in working with $G(\boldsymbol{v})$. Recall the BCH theorem for the case of two operators A and B whose commutator is a c-number,

$$e^A\, e^B = e^{A+B}\, e^{\frac{1}{2}[A,B]} \,. \tag{62}$$

But $[X_i, P_j] = i\hbar\delta_{ij}$, and therefore

$$G(\boldsymbol{v}) = e^{iM\boldsymbol{v}\cdot\boldsymbol{X}/\hbar}\, e^{-it\boldsymbol{v}\cdot\boldsymbol{P}/\hbar}\, e^{-iwt/\hbar} = e^{-it\boldsymbol{v}\cdot\boldsymbol{P}/\hbar}\, e^{iM\boldsymbol{v}\cdot\boldsymbol{X}/\hbar}\, e^{iwt/\hbar} \,, \tag{63}$$

where $w = \frac{1}{2}Mv^2$.

The transformation law for states now follows directly. Consider an arbitrary one particle state $|\Psi(t)\rangle$, and $|\boldsymbol{r}\rangle$ a position eigenstate; call its boosted counterpart $|\Psi'(t)\rangle$. (The relationship between these states is the analogue of the one between the classical states $\mathcal{S}(t)$ and $\mathcal{S}'(t)$ introduced at the outset.) Because the Galileo transformation of the coordinates has the appearance of a displacement by $\boldsymbol{v}t$, one might guess that the relationship between the wavefunctions has the same form as in a spatial translations, namely,

$$\langle \boldsymbol{r}|\Psi'(t)\rangle = \langle \boldsymbol{r} - \boldsymbol{v}t|\Psi(t)\rangle \,, \quad \text{or} \quad \psi'(\boldsymbol{r}, t) = \psi(\boldsymbol{r} - \boldsymbol{v}t, t) \,. \tag{64}$$

Not so, however. According to (63), the wave function of the boosted state is

$$\langle \boldsymbol{r}|\Psi'(t)\rangle = \langle \boldsymbol{r}|G(\boldsymbol{v})|\Psi(t)\rangle = \langle \boldsymbol{r}|e^{im\boldsymbol{v}\cdot\boldsymbol{x}/\hbar}\, e^{-it\boldsymbol{v}\cdot\boldsymbol{p}/\hbar}|\Psi(t)\rangle\, e^{-\frac{1}{2}imv^2t/\hbar} \,. \tag{65}$$

But $\boldsymbol{x}|\boldsymbol{r}\rangle = \boldsymbol{r}|\boldsymbol{r}\rangle$ and $e^{i\boldsymbol{p}\cdot\boldsymbol{a}/\hbar}|\boldsymbol{r}\rangle = |\boldsymbol{r} - \boldsymbol{a}\rangle$, and therefore the correct result is not (64) but

$$\psi'(\boldsymbol{r}, t) = e^{im(\boldsymbol{v}\cdot\boldsymbol{r} - \frac{1}{2}v^2 t)/\hbar}\, \psi(\boldsymbol{r} - \boldsymbol{v}t,\, t) \,. \tag{66}$$

Eq. 66 is important. The phase factor drops out of the probability density, of course,

$$|\psi'(\boldsymbol{r}, t)|^2 = |\psi(\boldsymbol{r} - \boldsymbol{v}t,\, t)|^2, \tag{67}$$

which conforms with the intuition that provoked the incorrect guess (64).

---

[1]In this section only we reintroduce $\hbar$ to make it clear where quantum mechanics enters the story.

The transformation law (66) was derived without any reference to the Schrödinger equation or a Hamiltonian. As it stands, therefore, it says *nothing* about invariance under Galileo transformation, but is merely a consequence of the definition of the transformation. Of course, when applied to a one particle momentum eigenstate it had better give the obvious result, and it does. Thus if $\psi(\boldsymbol{r},t) = \exp[i(\boldsymbol{k}\cdot\boldsymbol{r} - k^2t/2m)/\hbar]$, then according to (66) the boosted wave function is

$$\psi'(\boldsymbol{r},t) = e^{im(\boldsymbol{v}\cdot\boldsymbol{r}-\frac{1}{2}v^2t)/\hbar}\, e^{i[\boldsymbol{k}\cdot(\boldsymbol{r}-\boldsymbol{v}t)-k^2t/2m]/\hbar} \tag{68}$$

$$= \exp\left\{\frac{i}{\hbar}\left[(\boldsymbol{k}+m\boldsymbol{v})\cdot\boldsymbol{r} - \frac{(\boldsymbol{k}+m\boldsymbol{v})^2\,t}{2m}\right]\right\}, \tag{69}$$

which is the eigenfunction for momentum $\boldsymbol{k}+m\boldsymbol{v}$. In retrospect, one could have inferred a need for the phase factor in (66) from the little calculation that led to (69), but that would not have sufficed to show that it holds for any state.

For an arbitrary many body state the transformation law follows in the same way because the generator $\boldsymbol{N}$ is a sum of commuting operators for the separate particles. Hence

$$G(\boldsymbol{v}) = \prod_n e^{-i\boldsymbol{v}\cdot(\boldsymbol{p}_n t - m_n\boldsymbol{x}_n)/\hbar}, \tag{70}$$

and therefore

$$\psi'(\boldsymbol{r}_1,\ldots,\boldsymbol{r}_N;t) = e^{iM(\boldsymbol{v}\cdot\boldsymbol{R}-\frac{1}{2}v^2t)/\hbar}\,\psi(\boldsymbol{r}_1-\boldsymbol{v}t,\ldots,\boldsymbol{r}_N-\boldsymbol{v}t;\,t), \tag{71}$$

where $\boldsymbol{R}$ is the eigenvalue of $\boldsymbol{X}$.

Now we address the question of what condition the Hamiltonian must satisfy to produce invariance under Galileo transformations. This requirement can be stated in several equivalent ways. The most direct is to demand that if the system is isolated from external forces, i.e., if the total momentum $\boldsymbol{P}$ is a constant of motion and the Hamiltonian H does not depend on time, then its center of mass must move with the uniform velocity $\boldsymbol{P}/M$. In the Heisenberg picture, this is the requirement

$$\frac{d}{dt}\boldsymbol{X}(t) = -\frac{i}{\hbar}[\boldsymbol{X}(t),H] = \frac{\boldsymbol{P}}{M}. \tag{72}$$

If H has the conventional form $T(\boldsymbol{p}_1,\ldots)+V(\boldsymbol{x}_1,\ldots)$, then (72) becomes

$$\sum_n\left(m_n\frac{\partial T}{\partial\boldsymbol{p}_n} - \boldsymbol{p}_n\right) = 0, \tag{73}$$

which results in the familiar form

$$T = \sum_n \frac{p_n^2}{2m_n}. \tag{74}$$

When H satisfies this condition, all the states $G(\boldsymbol{v})|\Psi(t)\rangle$ are solutions of the Schrödinger equation if $|\Psi(t)\rangle$ is a solution (see Prob. 4). But H itself, being the energy, is of course not invariant:

$$G^\dagger(\boldsymbol{v})\,H\,G(\boldsymbol{v}) = H + \boldsymbol{P}\cdot\boldsymbol{v} + \tfrac{1}{2}Mv^2. \tag{75}$$

In terms of coordinates relative to the center of mass, and their canonically conjugate momenta $\boldsymbol{\pi}_n$, the Hamiltonian takes the form

$$H = \sum_n \frac{\boldsymbol{\pi}_n^2}{2m_n} + V + \frac{\boldsymbol{P}^2}{2M} \equiv H_{\text{int}} + \frac{\boldsymbol{P}^2}{2M} , \tag{76}$$

where H_{int} is the energy in the rest frame, which is invariant by definition.

(b) Mass Differences

In the foregoing we glossed over what appears to be a fatal inconsistency: *from a purely kinematical point of view, Galileo transformations and spatial translations commute, whereas the corresponding generators* $\boldsymbol{N}$ *and* $\boldsymbol{P}$ *do not!* According to (61)

$$[P_i, N_j] = i\hbar M \delta_{ij} . \tag{77}$$

This "paradox" is related to what would seem to be a rather arcane matter we have also glossed over. If (g_1, g_2) are group elements of some group $\mathfrak{G}$, such as the rotation group, and $U(g_i)$ are the corresponding unitary operators, then these operators must satisfy

$$U(g_1)U(g_2) = e^{i\lambda}\, U(g_1 g_2) \tag{78}$$

so as to properly represent $\mathfrak{G}$ in ray space. We have, up to now, ignored the phase factor in (78), because it can, almost always, be removed by replacing the Hilbert space vectors belonging to one ray by other vectors in the same ray, that is, by a suitable phase convention. Galileo transformations are the exception; the phase cannot be removed. This property of the Galileo group is remarkable because it is not shared by the Lorentz group.

That the commutation rule (77) produces an irremovable phase follows from the unitary transformation

$$W = T^\dagger(\boldsymbol{a})G^\dagger(\boldsymbol{v})T(\boldsymbol{a})G(\boldsymbol{v}) \tag{79}$$

where $T(\boldsymbol{a}) = e^{-i\boldsymbol{P}\cdot\boldsymbol{a}/\hbar}$ is the translation through $\boldsymbol{a}$. According to their *kinematical and geometrical* definitions, Galileo transformations and translations commute, and therefore this sequence of transformations maps any space-time point $(\boldsymbol{r}, t)$ into itself. On the other hand, use of (63) shows that the unitary operator is

$$W = e^{-iM\boldsymbol{a}\cdot\boldsymbol{v}/\hbar} . \tag{80}$$

For any state $|\Psi_M\rangle$ of definite mass M, therefore,

$$W|\Psi_M\rangle = e^{-iM\boldsymbol{a}\cdot\boldsymbol{v}/\hbar}\, |\Psi_M\rangle . \tag{81}$$

While the phase is just an overall factor for a mass eigenstate, it is not harmless for a superposition of such states, for then M is in effect an operator that assumes different values in subspaces of various masses. For such a superposition, after transformation by W,

$$W\Big(\alpha_1|\Psi_{M_1}\rangle + \alpha_2|\Psi_{M_2}\rangle\Big) = e^{-iM_1\boldsymbol{a}\cdot\boldsymbol{v}/\hbar}\Big(\alpha_1|\Psi_{M_1}\rangle + \alpha_2 e^{-i\Delta M(\boldsymbol{a}\cdot\boldsymbol{v})/\hbar}|\Psi_{M_2}\rangle\Big) , \tag{82}$$

where $\Delta M = M_2 - M_1$. As W maps any spatial point $\boldsymbol{r}$ into itself, and does not alter the time, the superposition before and after application of W should represent the same physical state, i.e., the same ray. But that is not true of Eq. 82: the relative phase between the states of differing mass produced by the transformation W can lead to detectable interference terms, for example in alpha decay, which is described by a superposition of nuclear states having different masses.

It was long believed that this inconsistency must be removed by imposing a superselection rule on mass[1] — by demanding nonrelativistic quantum mechanics to not admit *any* observables that have matrix elements between states of different mass. While this mass superselection rule would be consistent with Newtonian mechanics, which does not admit changes of mass, to require no changes of mass in nonrelativistic quantum mechanics would be a stifling restriction. In nuclear physics (where the rule has been largely ignored), the momenta of the reaction products would be wrong unless the change in rest mass energy is accounted for in energy conservation even when the velocities of the reaction products are small compared to c. This illustrated by the model in §9.3 of a nonrelativistic inelastic reaction involving particles of different mass, which requires that mass changes be permitted, and which uses superpositions of states of various masses.

Greenberger has, however, discovered that the relative phase in (82) does not imply that nonrelativistic quantum mechanics outlaws interference between states of differing mass.[2] On the contrary, the relative phase is precisely what is required by the mass difference when the Galileo transformation in W is replaced by the corresponding Lorentz transformation *to first order in* v/c, for then the resulting unitary operator W_L does *not* map $(\boldsymbol{r}, t)$ into itself, but produces a change in time in the amount $\boldsymbol{a} \cdot \boldsymbol{v}/c^2$. Therefore W_L should *not* map a superposition of states of differing masses into itself. It is remarkable that nonrelativistic quantum mechanics, as it stands, accounts correctly for this lowest-order relativistic effect. The widespread use of superpositions of various mass eigenstates, especially in nuclear physics, is therefore legitimate.

Because time does not change in a Galileo transformation, how the time dependence of wave functions is expressed is immaterial. In carrying out Lorentz transformations, on the other hand, space and time must be on an equal footing. The time dependence of a wave function can be expressed either in the Schrödinger picture by a moving system state vector $|\Psi(t)\rangle$, or in the Heisenberg picture by moving basis kets:

$$\psi(\boldsymbol{r}, t) = \langle \boldsymbol{r} | \Psi(t) \rangle = \langle \boldsymbol{r} | e^{-iHt/\hbar} | \Psi_H \rangle = \langle \boldsymbol{r}, t | \Psi_H \rangle , \tag{83}$$

where $\boldsymbol{r}$ stands for all coordinates. Thus in the Heisenberg picture, space and time are treated symmetrically in terms of time-dependent coordinate eigenkets.[3] The time dependence of this basis is

$$|\boldsymbol{r}, t + \tau\rangle = e^{iH\tau/\hbar} |\boldsymbol{r}, t\rangle , \tag{84}$$

i.e., opposite to that of the system state $|\Psi(t)\rangle$ in the Schrödinger picture.

---

[1] V. Bargmann, *Ann. Math.* **59**, 1 (1954).

[2] D.M. Greenberger, *Phys. Rev. Lett.* **87**, 100405 (2001).

[3] This would be apparent in the second quantization formalism, for then the time dependence would be expressed in terms of the creation operator $\psi^\dagger(\boldsymbol{r}, t)$ in the Heisenberg picture.

The Lorentz transformation of $|\boldsymbol{r}, t\rangle$ is

$$L(\boldsymbol{v})|\boldsymbol{r}, t\rangle = |\boldsymbol{r} + \boldsymbol{v}t, t + \boldsymbol{v}\cdot\boldsymbol{r}/c^2\rangle + O(v^2/c^2) , \tag{85}$$

where the sign of $\boldsymbol{v}$ is set to agree with the convention (66) for Galileo transformations bearing in mind (84). The phase factors can be ignored because (80) already tells what the final phase will be. Thus

$$W_L^\dagger |\boldsymbol{r}, t\rangle = L^\dagger(\boldsymbol{v})T^\dagger(\boldsymbol{a})L(\boldsymbol{v})T(\boldsymbol{a})\,|\boldsymbol{r}, t\rangle \tag{86}$$

$$= L^\dagger(\boldsymbol{v})T^\dagger(\boldsymbol{a})\,|\boldsymbol{r} + \boldsymbol{a} + \boldsymbol{v}t, t + \boldsymbol{v}\cdot(\boldsymbol{r} + \boldsymbol{a})/c^2\rangle \tag{87}$$

$$= |\boldsymbol{r}, t + \boldsymbol{v}\cdot\boldsymbol{a}/c^2\rangle = e^{i\boldsymbol{v}\cdot\boldsymbol{a}H/c^2\hbar}\,|\boldsymbol{r}, t\rangle \tag{88}$$

in view of (84). But $H = Mc^2 + O(v^2)$, and therefore (80) is

$$\langle \boldsymbol{r}, t|W_L|\Psi_M\rangle = \langle \boldsymbol{r}, t|e^{-iMc^2\tau/\hbar}\,|\Psi_M\rangle = \psi_M(\boldsymbol{r}, t + \tau) , \qquad \tau = \boldsymbol{a}\cdot\boldsymbol{v}/c^2 . \tag{89}$$

Therefore the relative phase between states of different mass produced by the sequence of transformations W_L is exactly what is required by the change in time induced by these transformations, and for that reason (82) should be written as

$$\langle \boldsymbol{r}, t|W_L\Big(\alpha_1|\Psi_{M_1}\rangle + \alpha_2|\Psi_{M_2}\rangle\Big) = e^{i\lambda}\Big\{\alpha_1\,\psi_{M_1}(\boldsymbol{r}, t) + \alpha_2 e^{-i\tau\Delta E/\hbar}\,\psi_{M_2}(\boldsymbol{r}, t)\Big\}, \tag{90}$$

where $\Delta E = (M_2 - M_1)c^2$ is the energy difference between the two states when they are at rest.

7.4 The Rotation Group

The consequences of rotational symmetry play so large a role in quantum mechanics that the considerable attention that has already been devoted to it in earlier chapters still leaves a great deal of important material untouched. Beyond rotational symmetry *per se*, there are other more complicated symmetries, and in particular Lorentz invariance, whose analysis is a more or less direct extension of the theory of rotations. A more elaborate knowledge of rotations is therefore required.

(a) The Group SO(3)

As our earlier discussion of rotations has appeared in several places, a précis is in order. From §2.5(d), recall that the set of all proper rotations R of a vector $\boldsymbol{K}$ in Euclidean 3-space $\mathfrak{E}_3$ form a non-Abelian Lie group, and that the terms in this sentence have the following meaning: *proper* excludes reflections, i.e., includes only rotations that can be achieved by a continuous motion of $\boldsymbol{K}$; *group* means that a sequence of two rotations R_2R_1 is again a rotation, and that to every rotation R there is an inverse R^{-1} such that $RR^{-1} = R^{-1}R \equiv I$ produces no rotation; *non-Abelian* means that in general $R_2R_1 \neq R_1R_2$; and a *Lie group* is one whose elements can be specified by a continuous set of parameters. There are several common choices of these parameters for R, the one used thus far having been an axis of rotation specified by a unit vector $\boldsymbol{n}$ and an angle θ of rotation about $\boldsymbol{n}$. Two other important parametrizations will be introduced in this section.

The mathematicians' name for this group is $SO(3)$, the special three-dimensional orthogonal group, because if $\boldsymbol{K}$ is written as a column vector with three entries, the rotations are produced by 3×3 orthogonal matrices, while "special" has the same meaning as "proper," i.e., unit determinant. Equivalently, this is the group of 3×3 orthogonal matrices with unit determinant, whereas a matrix that produces a reflection (and in general also a rotation) has determinant -1.

Thus far the discussion has been purely geometrical. Quantum mechanics enters by the argument set out in §2.5 and §7.1: to every element R there corresponds a unitary transformation $D(R)$ on the states in the Hilbert space, which satisfy the composition law

$$D(R_2)D(R_1) = D(R_2R_1) , \tag{91}$$

and the inverse relationship

$$D(R^{-1}) = D^{\dagger}(R) . \tag{92}$$

In mathematics, one calls *any* set of matrices that satisfies these laws a *representation of the group $SO(3)$*. One such representation is formed by the 3×3 matrices that transform the 3-vector $\boldsymbol{K}$. As we shall learn, this is but one example of matrices of every possible dimensions that satisfy the same composition law, and are therefore representations of $SO(3)$.

In terms of the parameters $(\boldsymbol{n}, \theta)$, the unitary operators $D(R)$ is

$$D(R) = e^{-i\boldsymbol{\omega} \cdot \boldsymbol{J}} , \tag{93}$$

where

$$\boldsymbol{\omega} = \theta \boldsymbol{n} . \tag{94}$$

The orientation of $\boldsymbol{n}$ is given by the polar and azimuthal angles (ϑ, φ). (See §3.3(a).)

The three Hermitian operators J_i are called the *infinitesimal generators* of the group, and the geometrical composition law for rotations in $\mathfrak{E}_3$ requires them to satisfy the angular momentum commutation rules,

$$[J_i, J_j] = i\epsilon_{ijk}J_k . \tag{95}$$

This set of commutation rules is called the *Lie algebra* of the group (in this instance, of $SO(3)$).

The three-dimensional orthogonal matrices do not form the lowest-dimensional nontrivial representation. As we learned in §3.3(a), this is formed by the two-dimensional unitary matrices

$$D^{\frac{1}{2}}(R) = e^{-\frac{1}{2}i\boldsymbol{\omega} \cdot \boldsymbol{\sigma}} ; \tag{96}$$

the 2×2 matrices $\frac{1}{2}\sigma_i$ also obey (95), of course. Because the Pauli matrices and the 2×2 unit matrix form a complete set of two-dimensional matrices,

$$D^{\frac{1}{2}}(R) = \cos \tfrac{1}{2}\theta - i\boldsymbol{n} \cdot \boldsymbol{\sigma} \sin \tfrac{1}{2}\theta . \tag{97}$$

This simplification only holds in this two-dimensional case.

Although $D^{\frac{1}{2}}(R)$ satisfies the basic composition law (91), it has a very significant peculiarity in that it is a double-valued representation of $SO(3)$. This is so because for the rotations through $\theta = 2\pi$ about any axis, which returns any object in $\mathfrak{E}_3$ to its original orientation, $D^{\frac{1}{2}}$ is not 1 as one might have supposed, but rather -1. What lies behind this remarkable fact will become clearer in what follows.

(b) SO(3) and SU(2)

In 1843, Hamilton discovered that rotations in $\mathfrak{E}_3$ can be represented by a set of four non-commuting roots of -1, which he called quaternions; in modern notation they are $(i, i\boldsymbol{\sigma})$. This way of representing rotations does not only elucidate the origin of the double-valued representations, of which $D^{\frac{1}{2}}$ is the simplest example, but it also is a method that generalizes directly to the analysis of Lorentz transformations. Furthermore, it is much easier to compose rotations with this method than with the more familiar three-dimensional matrices.

The basic idea is to represent any real vector in $\mathfrak{E}_3$, say, the position vector $\boldsymbol{r}$, by the two-dimensional matrix

$$\mathcal{R} = \boldsymbol{r} \cdot \boldsymbol{\sigma} = \begin{pmatrix} z & x - iy \\ x + iy & -z \end{pmatrix} . \tag{98}$$

$\mathcal{R}$ has the following properties:

$$\det \mathcal{R} = -(x^2 + y^2 + z^2) , \quad \mathrm{Tr}\ \mathcal{R} = 0 , \quad \mathcal{R} = \mathcal{R}^\dagger . \tag{99}$$

Hence any continuous linear transformation of $\mathcal{R}$ that leaves its determinant and trace unchanged, and keeps it Hermitian, corresponds to a rotation.

Consider

$$\mathcal{R}' = \Omega^\dagger \mathcal{R} \Omega ; \tag{100}$$

then $\mathcal{R}'$ is Hermitian, Tr $\mathcal{R}'$ = Tr $\Omega\Omega^\dagger \mathcal{R}$, and

$$\det \mathcal{R}' = \det \Omega^\dagger \cdot \det \mathcal{R} \cdot \det \Omega = |\det \Omega|^2 \det \mathcal{R} . \tag{101}$$

So if $\Omega\Omega^\dagger = 1$, i.e., if Ω is unitary, it will satisfy these conditions. Hence *any two-dimensional unitary matrix will perform a rotation of a 3-vector via Eq. 100.* To be totally explicit: in view of the definition (98), the components of the rotated vector will be the coefficients of σ_i on the right-hand side of (100). An illustration of what has just been said is the rotation of $\boldsymbol{\sigma}$ itself as carried out in §3.3(a).

As is true of any two-dimensional matrix, Ω can be written as a linear combination of 1 and $\boldsymbol{\sigma}$:

$$\Omega = Q_0 + i\boldsymbol{\sigma} \cdot \boldsymbol{Q} . \tag{102}$$

If Ω is to be unitary, the quantities $(Q_0, \boldsymbol{Q})$ must satisfy

$$\Omega\Omega^\dagger = |Q_o|^2 + |\boldsymbol{Q}|^2 + i\boldsymbol{\sigma} \cdot (\boldsymbol{Q}Q_0^* - \mathrm{c.c}) + i\boldsymbol{\sigma} \cdot (\boldsymbol{Q} \times \boldsymbol{Q}^*) = 1 , \tag{103}$$

where $(\boldsymbol{\sigma} \cdot \boldsymbol{a})(\boldsymbol{\sigma} \cdot \boldsymbol{b}) = \boldsymbol{a} \cdot \boldsymbol{b} + i\boldsymbol{\sigma} \cdot (\boldsymbol{a} \times \boldsymbol{b})$ was used. The coefficient of $\boldsymbol{\sigma}$ vanishes if Q_0 and $\boldsymbol{Q}$ are real, apart from a common phase,

$$\Omega = e^{i\alpha}(Q_0 + i\boldsymbol{\sigma} \cdot \boldsymbol{Q}) , \tag{104}$$

and Ω will be unitary if

$$Q_0^2 + |\boldsymbol{Q}|^2 = 1 . \tag{105}$$

The overall phase factor in (104) will be dropped because it does not survive in (100).

The condition (105) reduces the four real numbers $(Q_0, \boldsymbol{Q})$ to three. There is any infinity of choices for these. Ours is identified by letting $\boldsymbol{r} = (x, y, 0)$, and requiring the rotation to be about the z-axis through θ in the anticlockwise sense when looking toward the origin. A short calculation using the algebra of the Pauli matrices will confirm that this is accomplished with $Q_0 = \cos\frac{1}{2}\theta, \boldsymbol{Q} = -\hat{\boldsymbol{z}}\sin\frac{1}{2}\theta$. For the same rotation about any other other axis,

$$Q_0 = \cos\tfrac{1}{2}\theta\,, \qquad \boldsymbol{Q} = -\boldsymbol{n}\sin\tfrac{1}{2}\theta\,, \tag{106}$$

and so we have shown that

$$\Omega(R) = D^{\frac{1}{2}}(R)\,. \tag{107}$$

The composition law for rotations can be found much more readily from this two-dimensional representation than from the more familiar three-dimensional orthogonal matrices. The most straightforward way is to write R_1 and R_2 in terms of two axes and two angles, and to use the algebra of the Pauli matrices to compute the axis and angle for the composite rotation. As it happens, this is not well suited to our purposes. It is more convenient to write (104) out explicitly:

$$D^{\frac{1}{2}}(R) = \begin{pmatrix} Q_0 + iQ_3 & iQ_1 + Q_2 \\ iQ_1 - Q_2 & Q_0 - iQ_3 \end{pmatrix} \equiv \begin{pmatrix} a & b \\ -b^* & a^* \end{pmatrix}, \tag{108}$$

with (105) requiring

$$|a|^2 + |b|^2 = 1\,. \tag{109}$$

The complex numbers (a, b), as constrained by (109), are known as the Cayley-Klein parameters. Their relation to $(\boldsymbol{n}, \theta)$ is

$$a = \cos\tfrac{1}{2}\theta - i\sin\tfrac{1}{2}\theta\cos\vartheta\,, \quad b = -ie^{-i\varphi}\sin\tfrac{1}{2}\theta\sin\vartheta\,. \tag{110}$$

In terms of these parameters, the inverse is

$$D^{\frac{1}{2}}(R^{-1}) = \begin{pmatrix} a^* & -b \\ b^* & a \end{pmatrix}. \tag{111}$$

Let (a_1, b_1) and (a_2, b_2) be the parameters for the rotations $R_{1,2}$. Then

$$D^{\frac{1}{2}}(R_2)D^{\frac{1}{2}}(R_1) = D^{\frac{1}{2}}(R_2R_1) = \begin{pmatrix} a_2a_1 - b_2b_1^* & a_2b_1 + b_2a_1^* \\ -a_2^*b_1^* - b_2^*a_1 & a_2^*a_1^* - b_2^*b_1 \end{pmatrix}, \tag{112}$$

which retains the form of (108), as it must. The composition law is therefore

$$(a_2, b_2) \times (a_1, b_1) = (a_2a_1 - b_2b_1^*\,, a_2b_1 + b_2a_1^*)\,. \tag{113}$$

The right-hand side is not symmetric in 1 and 2, confirming that the group is non-Abelian. The inverse element is

$$(a, b)^{-1} = (a^*, -b)\,. \tag{114}$$

It is tempting to suppose that (113) is the composition law for $SO(3)$, but that is false! The reason has already been given: $\theta = 0$ and $\theta = 2\pi$ both correspond to the identity element of $SO(3)$, but to two distinct parameters in (108), namely, $a = \pm 1, b = 0$. Quite generally, the *pair* of matrices $D^{\frac{1}{2}}(\pm a, \pm b)$ give the *same* rotation of a 3-vector, because the transformation $\mathcal{R} \to \mathcal{R}'$ is bilinear in $D^{\frac{1}{2}}$, and therefore *the matrices* $D^{\frac{1}{2}}(R)$ *are a double-valued representation of the rotation group* $SO(3)$.

These matrices do, however, form a single-valued representation of a different group. This group is defined by the linear transformations of complex 2-vectors, with components (u_1, u_2), that span the two-dimensional complex vector space $\mathfrak{C}_2$,

$$D^{\frac{1}{2}}(a,b)\begin{pmatrix} u_1 \\ u_2 \end{pmatrix} \equiv \begin{pmatrix} u_1' \\ u_2' \end{pmatrix} = \begin{pmatrix} au_1 + bu_2 \\ -b^*u_1 + a^*u_2 \end{pmatrix}, \tag{115}$$

and which, because Ω is unitary, leave invariant the Hermitian form

$$|u_1|^2 + |u_2|^2 . \tag{116}$$

This group is called $SU(2)$, the special two-dimensional unitary group,with "special" referring to the fact that the phase factor $e^{i\alpha}$ has been dropped from (104), so that det $D^{\frac{1}{2}} = 1$. When the phase is included, the group is called $U(2)$, the two-dimensional unitary group. The representation is single-valued in $\mathfrak{C}_2$ because $D^{\frac{1}{2}}(\pm a, \pm b)$ produce distinct complex 2-vectors $(\pm u_1', \pm u_2')$ while producing the same real 3-vector when used as Ω in Hamilton's relation $\Omega^\dagger \mathcal{R} \Omega$. A more detailed discussion of the relationship between $SO(3)$ and $SU(2)$ will be given in §7.4(f).

(c) Irreducible Representations of SU(2)

For any group $\mathfrak{G}$, with elements $(g_1, g_2, \ldots)$, a set of K-dimensional matrices $\{M_K(g)\}$ that satisfy the composition law

$$M_K(g_2)M_K(g_1) = M_K(g_2 g_1) , \tag{117}$$

form a representation of $\mathfrak{G}$. If there exists a unitary transformation W that does *not* depend on g, and which for *all* g, reduces M_K to block-diagonal form,

$$W^\dagger M_K W = \begin{pmatrix} M_m & 0 \\ 0 & M_n \end{pmatrix}, \tag{118}$$

the representation is called reducible, and if there exists no such W, it is *irreducible.* In view of (118), if all the irreducible representations are known, any transformation of $\mathfrak{G}$ can be constructed from them.

We therefore address the task of determining all the irreducible representations of $SU(2)$, which will include those of $SO(3)$. This information will enable us to rotate the states of any system. The task is greatly simplified by the theory of angular momentum, and by our knowledge of $D^{\frac{1}{2}}(a,b) \equiv \Omega(a,b)$, for this is the two-dimensional irreducible representation of $SU(2)$. We know it is irreducible because $\exp(-\frac{1}{2}i\theta \boldsymbol{n} \cdot \boldsymbol{\sigma})$ can only be brought to diagonal form by a unitary transformation that depends on $\boldsymbol{n}$. As we shall see, all the higher dimensional irreducible representations can be constructed from $D^{\frac{1}{2}}$ by what is essentially angular momentum addition.

Consider an arbitrary system possessing an angular momentum operator $\boldsymbol{J}$, and a Hilbert space spanned by the states $|\alpha jm\rangle$, where α designates the eigenvalues of all other operators which, together with $\boldsymbol{J}$, form a compatible set of observables, and where (j, m) are standard angular momentum quantum numbers, $j = 0, \frac{1}{2}, 1 \ldots$,

$m = j, j-1, \ldots, -j$. No dynamical knowledge is involved here — no Hamiltonian is specified, and whether $\boldsymbol{J}$ is a constant of motion or not is irrelevant.

Under the active rotation R,

$$|\alpha jm\rangle \to |\alpha jm; R\rangle = D(R)|\alpha jm\rangle \ , \tag{119}$$

and therefore

$$D(R)|\alpha jm\rangle = \sum_{\alpha'} \sum_{j'm'} |\alpha' j' m'\rangle \langle \alpha' j' m'|D(R)|\alpha jm\rangle \ . \tag{120}$$

But $D(R) = \exp(-i\theta \boldsymbol{n} \cdot \boldsymbol{J})$ only contains the operator $\boldsymbol{J}$, which does not change j. Furthermore, all matrix elements of $\boldsymbol{J}$ are completely determined by $\boldsymbol{J}$ itself because of the nonlinear angular momentum commutation rules. Hence the matrix elements of $D(R)$ in (120) are diagonal in j, and are not only diagonal in α but totally independent of these other quantum numbers. They are written as

$$D^j_{mm'}(R) = \langle jm|D(R)|jm'\rangle \ , \tag{121}$$

so that (120) reads

$$|jm; R\rangle \equiv D(R)|jm\rangle = \sum_{m'} |jm'\rangle \, D^j_{m'm}(R) \ . \tag{122}$$

The other quantum numbers α will be suppressed as they play no role in this whole discussion.

The quantities $D^j_{mm'}(R)$ form the $(2j+1)$-dimensional rotation matrix $D^j(R)$. These matrices are universal: they depend only on the eigenvalues of the angular momentum operator that generates the rotation R, and are the same for any and all systems that possess states with the angular momentum in question.

The composition law (112) for the $j = \frac{1}{2}$ representation holds for all j, of course:

$$D^j_{mm'}(R_2 R_1) = \sum_{m''} D^j_{mm''}(R_2) \, D^j_{m''m'}(R_1) \ . \tag{123}$$

If $R_1 = R_2^{-1}$, the left-hand side becomes

$$D^j_{mm'}(I) = \delta_{mm'} \ , \tag{124}$$

and therefore

$$\sum_{m''} D^j_{mm''}(R) \, D^j_{m''m'}(R^{-1}) = \delta_{mm'} \ . \tag{125}$$

But D^j is a unitary matrix, i.e.,

$$\sum_{m''} D^j_{mm''}(R) \, [D^j_{m'm''}(R)]^* = \delta_{mm'} \ , \tag{126}$$

whence

$$D^j_{mm'}(R^{-1}) = [D^j_{m'm}(R)]^* \ . \tag{127}$$

We know already that $D^{\frac{1}{2}}$ is irreducible, but so are all the D^j. This follows from an example, $D = \exp(-i\theta \hat{\boldsymbol{x}} \cdot \boldsymbol{J})$. When expanded in powers of J_x, it terminates with $(J_x)^{2j}$, and therefore has just enough raising and lowering operators $J_\pm$ so that in acting on any state $|jm\rangle$ it will produce a linear combination of all $(2j+1)$ states, and thus populate all elements of D^j. *QED*.

The computation of D^j from $D^{\frac{1}{2}}$ involves two steps:

1. construction of polynomials $u_1^p u_2^q$ which, with proper choice of the integers p and q, transform like the states $|jm\rangle$ in virtue of the fact that the quantities (u_1, u_2) introduced in (115) transform under rotations like the $(j = \frac{1}{2}, m = \pm\frac{1}{2})$ states;
2. evaluation of the transformation of these polynomials in terms of the transformation of (u_1, u_2), which then determines D^j.

The first step exploits that (u_1, u_2) transforms in the manner just stated by its definition and that of $\Omega = D^{\frac{1}{2}}$. But if u_1 transforms under rotations like $|j = m = \frac{1}{2}\rangle$, it follows that u_1^{2j} transforms like a product of $2j$ spin-$\frac{1}{2}$ states, each with projection $m = \frac{1}{2}$, so u_i^{2j} transform like $|jj\rangle$. Under an arbitrary rotation, $|jj\rangle$ will become a linear combination of all the $|jm\rangle$, while according to (115), $u_1^{2j} \to (au_1 + bu_2)^{2j}$ under rotation, and this is a linear combination of all polynomials $u_1^p u_2^q$ with $p+q = 2j$. Furthermore, as u_2 transforms like $|j = \frac{1}{2}, m = -\frac{1}{2}\rangle$, it follows that $p - q = 2m$. Hence

$$\Phi_{jm} = c_{jm} u_1^{j+m} u_2^{j-m} \tag{128}$$

must transform like $|jm\rangle$ provided the constants c_{jm} are chosen to satisfy the familiar requirement

$$J_\pm \Phi_{jm} = \sqrt{j(j+1) - m(m \pm 1)}\, \Phi_{jm\pm 1} \,. \tag{129}$$

To make use of this condition, the angular momentum operators must be expressed in terms of the variables (u_1, u_2). This is left for Prob. 5, the results being

$$J_+ = u_1 \frac{\partial}{\partial u_2}\,, \quad J_- = u_2 \frac{\partial}{\partial u_1}\,, \quad J_z = \frac{1}{2}\left(u_1 \frac{\partial}{\partial u_1} - u_2 \frac{\partial}{\partial u_2} \right) . \tag{130}$$

Clearly, $J_z \Phi_{jm} = m\Phi_{jm}$, and $J_\pm$ change the powers of u_1 and u_2 as required. To determine c_{jm} it helps to notice that $j(j+1) - m(m \pm 1) = (j \mp m)(j \pm m + 1)$. Apart from an arbitrary and irrelevant constant the final result is

$$\Phi_{jm}(u_1, u_2) = \frac{u_1^{j+m} u_2^{j-m}}{\sqrt{(j+m)!(j-m)!}} \,. \tag{131}$$

Now to the second step, the transformation of Φ_{jm}. Here care must be taken to use precisely the same transformation law for (u_1, u_2) as in the defining relation (122). Equation (115) does not have this form, however; rather, the form required is

$$\begin{pmatrix} u_1' \\ u_2' \end{pmatrix} = \begin{pmatrix} u_1 D_{11} + u_2 D_{21} \\ u_1 D_{12} + u_2 D_{22} \end{pmatrix} = \begin{pmatrix} u_1 a - u_2 b^* \\ u_1 b + u_2 a^* \end{pmatrix} . \tag{132}$$

Transforming Φ_{jm} now determines D^j:

$$\Phi_{jm}(u_1', u_2') = \sum_{m'} \Phi_{jm'}(u_1, u_2)\, D^j_{m'm}(a, b) \tag{133}$$

$$= \frac{(u_1 a - u_2 b^*)^{j+m}\, (u_1 b + u_2 a^*)^{j-m}}{\sqrt{(j+m)!(j-m)!}} \,. \tag{134}$$

The binomial theorem gives

$$\Phi'_{jm} = c_{jm} \sum_{\mu\nu} \binom{j+m}{\mu} \binom{j-m}{\nu} (u_1 a)^{j+m-\mu} (-u_2 b^*)^\mu (u_1 b)^{j-m-\nu} (u_2 a^*)^\nu \,. \tag{135}$$

This contains the factor $(u_1)^{2j-\mu-\nu}(u_2)^{\mu+\nu}$, which should be proportional to $\Phi_{jm'}$, so set $2j-\mu-\nu = j+m'$ and $\mu+\nu = j-m'$. Then, on comparing with (133), and interchanging m and m', one finds the sought-for result

$$D^j_{mm'}(a,b) = \sqrt{\frac{(j+m)!(j-m)!}{(j+m')!(j-m')!}}\ \Sigma \tag{136}$$

$$\Sigma = \sum_\mu \binom{j+m'}{\mu}\binom{j-m'}{j-m-\mu}(a)^{j+m'-\mu}(a^*)^{j-m-\mu}(-b^*)^\mu(b)^{m-m'+\mu}\ . \tag{137}$$

(d) D(R) in Terms of Euler Angles

In essence, Eq. 136 is the complete answer to the problem of finding the irreducible representations. However, the parameters (a,b) do not have a simple and direct geometrical meaning in $\mathfrak{E}_3$, and it is desirable to have D^j in terms of parameters that do.

An especially convenient parametrization is provided by the Euler angles (α,β,γ), defined in Fig. 7.2. By the definition of these angles, the rotation R of the frame

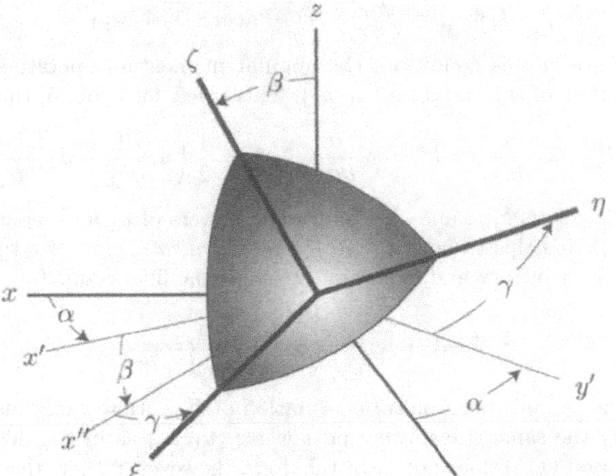

FIG. 7.2. The Euler angles.

$K=(x,y,z)$ to $K'=(\xi,\eta,\zeta)$ is performed by three successive rotations:

$$D(R) = D_\zeta(\gamma)D_{y'}(\beta)D_z(\alpha)\ , \tag{138}$$

where $D_z(\alpha) = \exp(-i\alpha J_z)$, etc. Any orientation of K' with respect to K in $\mathfrak{E}_3$ can be attained by angles in the ranges

$$0 \le \alpha,\gamma \le 2\pi\ , \quad 0 \le \beta \le \pi\ , \tag{139}$$

but this manifold has to be extended for double-valued representations of $SO(3)$, as we shall see in §7.4(f).

The rotations in (138) are with respect to three *distinct* coordinate frames, whereas the desired matrix elements $\langle jm|D(R)|jm'\rangle$ are between states which are *both* referred to the original frame K. As it stands, (138) is therefore in a form that is difficult to evaluate. However, the rotation $K \to K'$ can also be performed by three rotations all of which are with respect to the *original* frame K, namely,

$$D(R) = D_z(\alpha)D_y(\beta)D_z(\gamma) = e^{-i\alpha J_z}\, e^{-i\beta J_y}\, e^{-i\gamma J_z}\,. \tag{140}$$

To not interrupt the discussion, the proof that the two expressions (138) and (140) are identical will be postponed; note now though that the order of the rotations in (140) is opposite to that of (138).

Thanks to (140), the rotation matrix is

$$D^j_{mm'}(R) = \langle jm|e^{-i\alpha J_z}\, e^{-i\beta J_y}\, e^{-i\gamma J_z}|jm'\rangle\,. \tag{141}$$

Now J_z acts on its eigenstates, which demonstrates that the dependence on the angles α and γ is trivial:

$$D^j_{mm'}(R) = e^{-i\alpha m}\, d^j_{mm'}(\beta)\, e^{-i\gamma m'}\,. \tag{142}$$

This confirms what we already know: if j is an integer, the representation is single-valued for $SO(3)$, but if j is a half-integer, it is double-valued, because it changes sign if either α or γ is increased by 2π.

The β-dependence of D^j is entirely in the functions

$$d^j_{mm'}(\beta) = \langle jm|e^{-i\beta J_y}|jm'\rangle\,. \tag{143}$$

This "gut" part of the rotation matrix is known from (136), however. From (110), the Cayley-Klein parameters for a rotation about the y-axis through β are $a = \cos\frac{1}{2}\beta$, $b = -\sin\frac{1}{2}\beta$, and therefore

$$d^j_{mm'}(\beta) = (-1)^{m-m'}\sqrt{\frac{(j+m)!(j-m)!}{(j+m')!(j-m')!}}(\cos\tfrac{1}{2}\beta)^{2j}\Sigma \tag{144}$$

$$\Sigma = \sum_\mu (-1)^\mu \binom{j+m'}{\mu}\binom{j-m'}{j-m-\mu}(\tan\tfrac{1}{2}\beta)^{m-m'+2\mu} \tag{145}$$

Note that this is a real function. This completes the evaluation of all the irreducible representations of $SU(2)$ in terms of the Euler angles.

The inverse rotation matrices in terms of the Euler angles follow from the latters' geometrical definition, or equivalently from Eq. 140:

$$D^{-1}(\alpha,\beta,\gamma) = D(-\gamma,-\beta,-\alpha)\,. \tag{146}$$

Because d^j is orthogonal, and $d^j(\beta)\,d^j(-\beta) = 1$,

$$d^j_{mm'}(\beta) = d^j_{m'm}(-\beta)\,. \tag{147}$$

The matrix $D^{\frac{1}{2}}$, in terms of the Euler angles, is

$$D^{\frac{1}{2}}(R) = \begin{pmatrix} e^{-\frac{1}{2}i(\alpha+\gamma)}\cos\frac{1}{2}\beta & -e^{-\frac{1}{2}i(\alpha-\gamma)}\sin\frac{1}{2}\beta \\ e^{\frac{1}{2}i(\alpha-\gamma)}\sin\frac{1}{2}\beta & e^{\frac{1}{2}i(\alpha+\gamma)}\cos\frac{1}{2}\beta \end{pmatrix}\,. \tag{148}$$

The relationship between the Euler angles and the Cayley-Klein parameters is therefore

$$a = e^{-\frac{1}{2}i(\alpha+\gamma)} \cos \tfrac{1}{2}\beta \,, \qquad b = -e^{-\frac{1}{2}i(\alpha-\gamma)} \sin \tfrac{1}{2}\beta \,. \tag{149}$$

Equation (144) actually shows that the polynomial $d^j_{mm'}$ is a hypergeometric function, and that when j is the integer l, and $m' = 0$, it is proportional to the associated Legendre polynomial $P^l_m(\cos\beta)$. In this case there is then the following relation to the spherical harmonics:

$$D^l_{m0}(\alpha\beta\gamma) = \sqrt{\frac{4\pi}{2l+1}}\, Y^*_{lm}(\beta\alpha) = \sqrt{\frac{4\pi}{2l+1}}\, Y^*_{lm}(\boldsymbol{n}) \,, \tag{150}$$

where $\boldsymbol{n}$ is the unit vector in the original frame K with polar angle β and azimuth α. This relation will be derived in a less obscure way in §7.5(a).

Formulas for higher values of j are given in the Appendix, as well as important identities satisfied by $d^j(\beta)$ and $D^j(R)$.

It still remains to show that the two expressions (138) and (140) for $D(R)$ are identical. Let R_n be the rotation that takes the frame K_{n-1} into K_n, and let $D(R_i; K_j)$ be the unitary operator for a rotation R_i with respect to the frame K_j. Then

$$D(R_2; K_1) = D(R_1; K_0) D(R_2; K_0) D^\dagger(R_1; K_0) \,; \tag{151}$$

therefore

$$D(R_2; K_1) D(R_1; K_0) = D(R_1; K_0) D(R_2; K_0) \,, \tag{152}$$

and by the same token

$$D(R_3; K_2) D(R_2; K_1) D(R_1; K_0) = D(R_1; K_0) D(R_2; K_0) D(R_3; K_0) \,, \tag{153}$$

QED.

(e) The Kronecker Product

We have just exploited the fact that products of $j = \frac{1}{2}$ representations produce higher-dimensional representations. This is a special case of a broader truth, one which is intimately related to that of angular momentum addition (see especially §3.5(c)). Let the angular momentum $\boldsymbol{J}$ that appears in (93) and many subsequent formulas be the sum of two independent angular momenta:

$$\boldsymbol{J} = \boldsymbol{J}_1 + \boldsymbol{J}_2 \,. \tag{154}$$

Because $[\boldsymbol{J}_1, \boldsymbol{J}_2] = 0$,

$$e^{-i\boldsymbol{\omega}\cdot\boldsymbol{J}} = e^{-i\boldsymbol{\omega}\cdot\boldsymbol{J}_1}\, e^{-i\boldsymbol{\omega}\cdot\boldsymbol{J}_2} \,, \tag{155}$$

or in an obvious notation,

$$D(R) = D_1(R) D_2(R) \,. \tag{156}$$

We now take matrix elements of this relationship in the basis sets $\{|j_1 j_2 jm\rangle\}$ and $\{|j_1 m_1 j_2 m_2\rangle\}$, which are related by the Clebsch-Gordan coefficients $\langle jm|j_1 m_1 j_2 m_2\rangle$. By taking matrix elements in the first representation, the left-hand side of (156) gives

$$\langle j_1 j_2 jm|D(R)|j_1 j_2 jm'\rangle = D^j_{mm'}(R) \,. \tag{157}$$

This cannot depend on (j_1, j_2) by the argument related to Eq. 121, because the matrix elements of $D(R)$ can depend only on the eigenvalues of the angular momentum that generates the rotation R. The right-hand side of (156), after transforming to the product basis, gives

$$\sum_{m_1 m_2 m_1' m_2'} \langle jm|j_1m_1j_2m_2\rangle\langle j_1m_1|D_1(R)|j_1m_1'\rangle\langle j_2m_2|D_2(R)|j_2m_2'\rangle\langle j_1m_1'j_2m_2'|jm'\rangle . \tag{158}$$

Thus

$$D^j_{mm'}(R) = \sum_{m_1\ldots m_2'} \langle jm|j_1m_1j_2m_2\rangle\, D^{j_1}_{m_1m_1'}(R)\, D^{j_2}_{m_2m_2'}(R)\, \langle j_1m_1'j_2m_2'|jm'\rangle . \tag{159}$$

It is noteworthy that in (159) the quantum numbers (j_1, j_2) appear only on the right-hand side. The matrix elements of (156) in the product representation give

$$D^{j_1}_{m_1m_1'}(R)\, D^{j_2}_{m_2m_2'}(R) = \sum_{jmm'} \langle j_1m_1j_2m_2|jm\rangle\, D^j_{mm'}(R)\, \langle jm'|j_1m_1'j_2m_2'\rangle . \tag{160}$$

In both (159) and (160), j_1 and j_2 must satisfy the triangular inequality of angular momentum addition, $(j_1+j_2) \geq j \geq |j_1-j_2|$, and of course $m_1+m_2 = m,\ m_1'+m_2' = m'$.

Equation (160) is called the Kronecker (or direct) product of representations, and is usually written in terms of the matrices, and not their elements, in a notation that is essentially the one introduced in §3.5(c), namely,

$$D^{j_1} \otimes D^{j_2} = D^{j_1+j_2} \oplus D^{j_1+j_2-1} \ldots \oplus D^{|j_1-j_2|} . \tag{161}$$

Put equivalently, the direct product of the matrices on the left-hand side of (161), by a particular change of basis that does not depend on R, namely, the one produced by the Clebsch-Gordan coefficients, is recast into block-diagonal form, with the various blocks being irreducible representations of $SU(2)$:

$$D^{j_1} \otimes D^{j_2} = \begin{pmatrix} D^{j_1+j_2} & 0 & \cdots & 0 \\ 0 & D^{j_1+j_2-1} & \cdots & 0 \\ \vdots & & & \vdots \\ 0 & \cdots & & D^{|j_1-j_2|} \end{pmatrix} . \tag{162}$$

(f) Integration over Rotations

The rotation matrices arise in many applications, such as the wave functions of molecules and deformed nuclei, scattering amplitudes for particles with spin, and in the angular distributions of final states produced by radiative transitions and nuclear decays. Integrals over products of rotation matrices are then needed, and these can be evaluated by exploiting the group properties.

Integrations over all rotations R involve the relationship between $SO(3)$ and $SU(2)$ at a deeper level. Readers who prefer to accept the final result for the integral over the group as a tested and tasty recipe can proceed directly to Eq. 174.

To carry out an integration over the elements of a Lie group, it is necessary to understand the manifold defined by the parameters that specify the elements. Let

$\mathfrak{M}_O$ and $\mathfrak{M}_U$ be these manifolds for $SO(3)$ and $SU(2)$, respectively. From what has already been said, it will come as no surprise that these manifolds are not identical.

The Cayley-Klein parameters ($a = a_1 + ia_2$, $b = b_1 + ib_2$) define $\mathfrak{M}_U$ through the condition (109):

$$a_1^2 + a_2^2 + b_1^2 + b_2^2 = 1 \; ; \tag{163}$$

i.e., *the parameter manifold* $\mathfrak{M}_U$ *of* $SU(2)$ *is the surface of the unit sphere in four dimensions.* The parametrization of $SO(3)$ is actually a more subtle matter. It is best to start with $\boldsymbol{\omega} = \theta \boldsymbol{n}$, where, as before, the axis of rotation $\boldsymbol{n}$ is specified by the polar and azimuthal angles (ϑ, φ), and the ranges are

$$0 \leq \theta \leq \pi \, , \quad 0 \leq \vartheta \leq \pi \, , \quad 0 \leq \varphi \leq 2\pi \, . \tag{164}$$

One might think that $0 \leq \theta \leq 2\pi$ is required to encompass all rotations, but that is not so because, for a given $\boldsymbol{n}$, the points that cannot be reached in $0 \leq \theta \leq \pi$ can be reached by using $-\boldsymbol{n}$ as the axis. At first sight, the manifold defined by (164) is just a sphere of radius π, with a point in the sphere having the polar coordinates $(\theta, \vartheta, \varphi)$, with θ acting as the radial coordinate. But this is no everyday sphere, because the points $(\pi, \boldsymbol{n})$ and $(\pi, -\boldsymbol{n})$ give the same rotation! Hence *the manifold* $\mathfrak{M}_O$ *of* $SO(3)$ *is a 3-sphere of radius* π *with diametrically opposed points on its surface identified.*

Because of this identification of diametrically opposite points, the manifold $\mathfrak{M}_O$ is not simply connected. In contrast, the surface of a sphere in any number of dimensions, and in particular $\mathfrak{M}_U$, is simply connected. As we shall now see, the relation between $\mathfrak{M}_U$ and $\mathfrak{M}_O$ is analogous to that of the double-sheeted Riemann surface for a function of a complex variable with a square-root branch cut, with $\mathfrak{M}_U$ playing the role of the whole surface and $\mathfrak{M}_O$ that of one sheet.

The objective is to determine how many representation matrices $D^{(j)}(R)_n$ ($n =$ 1, 2, ...) can be associated with a single group element R of $SO(3)$. The basic notion used in the argument is that the continuous nature of the group requires that any representation $D^{(j)}(R)_n$ changes continuously if R is varied in a continuous fashion. The continuity requirement is that when R is in the neighborhood of R_0, each and every matrix element $D^{(j')}_{mm'}(R)_n$ remains close in *value* to $D^{(j')}_{mm'}(R_0)_n$. Consider now a matrix $D^{(j)}(R_0)_n$, and follow its value as we move along some path in the space of group elements. $D^{(j)}(R)_n$ will, as we have stated, vary continuously along such a path, but when the path returns to the starting point R_0, $D^{(j)}$ could assume any one of the values $D^{(j)}(R)_{n'}$, i.e., with n' not necessarily equal to n. The possibility $n \neq n'$ can be ruled out on the basis of continuity if the path in question can be continuously deformed until it shrinks to a point. If the path cannot be shrunk down, however, the continuity argument does not apply because we cannot stay in the neighborhood of the starting point on such a path; hence on paths of this type $D^{(j)}$ may (it need not) end up having a different value when we return to the beginning of the circuit. This argument indicates that a group whose parameter manifold is simply connected and of finite extent (i.e., compact) only has single valued representations; $SU(2)$ is an example of such a group. On the other hand, a compact group with a multiply connected parameter manifold can be expected to have multivalued representations, and the maximum degree of multiplicity is clearly given by the number of topologically distinct path classes. Fig. 7.3 shows that there are two such distinct classes in the manifold $\mathfrak{M}_O$ for $SO(3)$, and that its representations can be at most double-valued.

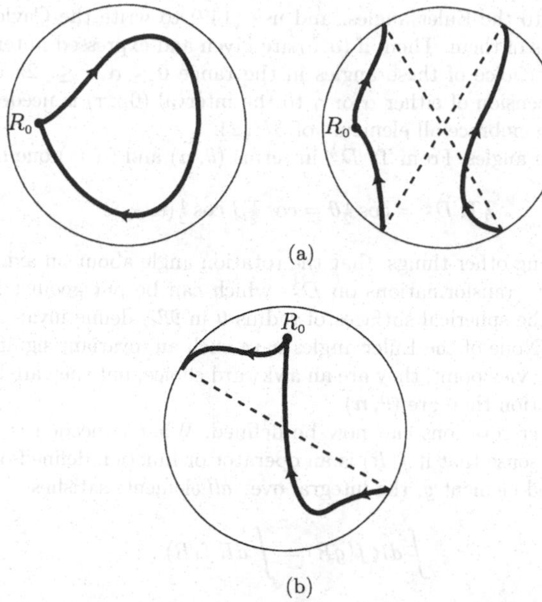

FIG. 7.3. Closed paths in the parameter manifold $\mathfrak{M}_O$ of $SO(3)$. Because the diametrically opposite points on the surface of the sphere are identified, a path that touches the surface continues at the opposite point. Consequently, there are two types of paths. Those shown in (a) can be shrunk to a point; they never touch the surface, or do so twice (or any even number of times). In the second example the path can be shrunk to a point after rotating the dashed lines until they coincide. In (b), on the other hand, the path cannot be shrunk to a point.

The detailed relation between the manifolds is clarified further by writing the Cayley-Klein parameters in terms of polar coordinates on the unit 4-sphere:[1]

$$a_1 = \cos\chi \,, \tag{165}$$

$$a_2 = \sin\chi\cos\vartheta \,, \tag{166}$$

$$b_1 = \sin\chi\sin\vartheta\sin\varphi \,, \tag{167}$$

$$b_2 = \sin\chi\sin\vartheta\cos\varphi \,, \tag{168}$$

$$0 \le \chi \le \pi \,, \quad 0 \le \vartheta \le \pi \,, \quad 0 \le \varphi \le 2\pi \,. \tag{169}$$

Now compare this with (110). After changing the conventional definition of θ to $-\theta$, with $0 \le \theta \le \pi$, and setting $\chi = \frac{1}{2}\theta$, we see that the $SO(3)$ manifold $\mathfrak{M}_O$ only contains positive a_1, and not the whole surface of the 4-sphere. Therefore the manifold $\mathfrak{M}_U$ of $SU(2)$ can only be encompassed by *doubling* the range of the rotation angle θ to the interval $(0, 2\pi)$. Hence $\mathfrak{M}_U$ covers $\mathfrak{M}_O$ twice, and the simply connected group $SU(2)$ is called the *covering group* of the doubly connected group $SO(3)$.

---

[1] A. Sommerfeld, *Partial Differential Equations in Physics,* Academic Press (1964), p. 227.

We now turn to the Euler angles, and use (149) to write the Cayley-Klein parameters in terms of them. Then, if (a, b) are given and expressed in terms of Euler angles, no other choice of these angles in the range $0 \leq \alpha, \gamma \leq 2\pi$ can produce $(-a, -b)$. An extension of either α or γ to the interval $(0, 4\pi)$ is needed to accomplish this, i.e., to embrace all elements of $SU(2)$.

A last word on angles. From Tr $D^{\frac{1}{2}}$ in terms $(\theta, \boldsymbol{n})$ and $(\alpha\beta\gamma)$ one has

$$\tfrac{1}{2}\mathrm{Tr}\, D^{\frac{1}{2}} = \cos \tfrac{1}{2}\theta = \cos \tfrac{1}{2}\beta \, \cos \tfrac{1}{2}(\alpha + \gamma) \, . \tag{170}$$

This shows, among other things, that the rotation angle about an axis is invariant under all unitary transformations on $D^{\frac{1}{2}}$, which can be put geometrically as the statement that the spherical surfaces of radius θ in $\mathfrak{M}_O$ define invariant classes of group elements. None of the Euler angles have such an invariant significance; from a group-theoretic viewpoint, they are an awkward choice, but they are better suited to describing motion than are $(\theta, \boldsymbol{n})$.

Integration over rotations can now be defined. What is needed is an invariant definition in the sense that if $f(R)$ is an operator or function defined on the group, then for any fixed element g, the integral over *all* elements satisfies

$$\int dR\, f(gR) = \int dR\, f(R) \, . \tag{171}$$

The measure must therefore be an invariant. The obvious invariant definition is in terms of the Cayley-Klein parameters:

$$\int dR\, f(R) = \int_{-\infty}^{\infty} da_1 \ldots \int_{-\infty}^{\infty} db_2 \; \delta(|a|^2 + |b|^2 - 1)\, f(a, b) \, . \tag{172}$$

Hence the invariant measure is the element of surface area of the unit 4-sphere, which is proportional to $\sin^2\chi \equiv \sin^2 \frac{1}{2}\theta$. The integral over $SU(2)$, expressed in terms of the parametrization $\boldsymbol{\omega} = \theta \boldsymbol{n}$ and normalized to one, is therefore

$$\int dR = \int_0^{\pi} \frac{d\vartheta}{\pi} \int_0^{2\pi} \frac{d\varphi}{2\pi} \int_0^{2\pi} \frac{d\theta}{\pi} \, \sin^2 \tfrac{1}{2}\theta \, . \tag{173}$$

In the corresponding integral over Euler angles it is only necessary to double the range of either α or γ, but the choice where both are doubled is more convenient:

$$\int dR = \int_0^{4\pi} \frac{d\alpha}{4\pi} \int_0^{4\pi} \frac{d\gamma}{4\pi} \int_0^{\pi} \frac{d\beta}{2} \, \sin\beta \, . \tag{174}$$

That $\sin\beta$ enters here is no surprise, but given that the Euler angles have a somewhat peculiar character from the perspective of group theory it is wise to confirm that this is correct by evaluating the Jacobian relating (172) to (174).

The integrals of interest to us all devolve from a simple statement: *Let $|a\rangle$ be an arbitrary state, and $|a; R\rangle$ its rotated counterpart, then when $|a; R\rangle$ is averaged over all orientations R, the only component in $|a\rangle$ that survives is $\langle 0|a\rangle$, its projection onto the spherically symmetric $j = 0$ state $|0\rangle$:*

$$\int dR\; D(R)|a\rangle = |0\rangle\langle 0|a\rangle \, . \tag{175}$$

As a consequence

$$\int dR\, D^j_{mm'}(R) = \delta_{j0}\,\delta_{m0}\,\delta_{m'0}\,. \tag{176}$$

Because of the half-angles in D^j when j is half-integral, (176) would be false if the integration were only over the parameter manifold of $SO(3)$. Lastly, there is no other factor in (175) because the measure has been chosen to make $\int dR = 1$.

Integrating (160) over R and using (176) then gives

$$\int dR\, D^{j_1}_{m_1 m_1'}(R)\, D^{j_2}_{m_2 m_2'}(R) = \langle j_1 m_1 j_2 m_2|00\rangle\langle 00|j_1 m_1' j_2 m_2'\rangle\,. \tag{177}$$

Clearly $\langle j_1 m_1 j_2 m_2|00\rangle \propto \delta_{j_1 j_2}\delta_{m_1,-m_2}$; the Clebsch-Gordan recursion relations of §3.6(a) determine the coefficient:

$$\langle j_1 m_1 j_2 m_2|00\rangle = (-1)^{j_1-m_1}(2j_1+1)^{-\frac{1}{2}}\,\delta_{j_1 j_2}\,\delta_{m_1,-m_2}\,. \tag{178}$$

From $D(R^{-1}) = D^\dagger(R)$ and Eq. 144 it can be shown that

$$D^j_{mm'}(R)^* = (-1)^{m-m'}\,D^j_{-m,-m'}(R)\,. \tag{179}$$

The last two identities show that (177) is the orthogonality relationship for the rotation matrices:

$$\int dR\, D^{j_1}_{m_1 m_1'}(R)\, D^{j_2}_{m_2 m_2'}(R)^* = \frac{\delta_{j_1 j_2}\,\delta_{m_1 m_2}\,\delta_{m_1' m_2'}}{2j_1+1}\,. \tag{180}$$

Returning to (160), multiplying by $D^J_{MM'}(R)^*$, and integrating, gives

$$\int dR\, D^J_{MM'}(R)^*\, D^{j_1}_{m_1 m_1'}(R)\, D^{j_2}_{m_2 m_2'}(R) = \frac{\langle j_1 m_1 j_2 m_2|JM\rangle\langle j_1 m_1' j_2 m_2'|JM\rangle}{2J+1}\,. \tag{181}$$

In view of (150), the integral over three spherical harmonics is a special case:

$$\int d\Omega\, Y^*_{LM}(\Omega) Y_{l_1 m_1}(\Omega) Y_{l_2 m_2}(\Omega) = \sqrt{\frac{(2l_1+1)(2l_2+1)}{4\pi(2L+1)}}\langle l_1 m_1 l_2 m_2|LM\rangle\langle l_1 0 l_2 0|L0\rangle\,; \tag{182}$$

here, as usual, $d\Omega = d\alpha \sin\beta\, d\beta$, and the integration is over the unit sphere. The last Clebsch-Gordan coefficient vanishes unless (l_1+l_2+L) is even, which is a parity selection rule, and an explicit formula for it is given in the Appendix.

7.5 Some Consequences of Rotation and Reflection Invariance

Now that we know how states transform under rotation and reflection, many opportunities for applying this knowledge have opened. Most will arise in the context of particular applications in later chapters. Here the object is to illustrate the power of symmetry arguments with three applications: (i) to the rotation of spherical harmonics; (ii) to states that arise in the scattering and decay of particles with spin; and (iii) to the rotational spectra of molecules.

(a) Rotation of Spherical Harmonics

The behavior of spherical harmonics under rotation is not, *per se*, an application to a problem of direct physical interest, but knowledge of this transformation is basic to much of what follows.

Let $|lm\rangle$ be the angular momentum eigenstate of a single particle of spin 0, with m the projection of the orbital angular momentum $\boldsymbol{l}$ on the z-axis of the frame K in the preceding section.[1] Under the rotation R, the transformation of such a state is given by Eq. 120 with all eigenvalues being integers:

$$|lm; R\rangle = \sum_{m'} |lm'\rangle \, D^l_{m'm}(R) \, . \tag{183}$$

Let $r\boldsymbol{n}$ be the position eigenvalue, with the unit vector $\boldsymbol{n}$ having the orientation (ϑ, φ) in K. From §3.4, we know that the scalar product $\langle \boldsymbol{n}|lm\rangle$ is the quantum mechanical definition of a spherical harmonic:

$$\langle \boldsymbol{n}|lm\rangle = Y_{lm}(\vartheta, \varphi) \equiv Y_{lm}(\boldsymbol{n}) \, . \tag{184}$$

If the orientation of $\boldsymbol{n}$ in the rotated frame K' is (Θ, Φ), then

$$\langle \boldsymbol{n}|lm; R\rangle = Y_{lm}(\Theta, \Phi) \, . \tag{185}$$

Thus under the rotation $R: \; K \to K'$, the spherical harmonics transform as follows:

$$Y_{lm}(\Theta, \Phi) = \sum_{m'} Y_{lm'}(\vartheta, \varphi) \, D^l_{m'm}(R) \, . \tag{186}$$

Bear in mind that here m is an eigenvalue of $\boldsymbol{n}_\zeta \cdot \boldsymbol{l}$, where $\boldsymbol{n}_\zeta$ is the unit vector along the ζ axis in K' (see Fig. 7.4), whereas m' is an eigenvalue of l_z.

It is useful to have (186) restated so that it only refers to the frame K. This can be done by exploiting

$$\langle \boldsymbol{n}|lm; R\rangle = \langle \boldsymbol{n}|D(R)|lm\rangle = \langle R^{-1}\boldsymbol{n}|lm\rangle \, , \tag{187}$$

where $R^{-1}\boldsymbol{n}$ is the vector which, when rotated through R becomes $\boldsymbol{n}$. Thus (186) is also

$$Y_{lm}(R^{-1}\boldsymbol{n}) = \sum_{m'} Y_{lm'}(\boldsymbol{n}) \, D^l_{m'm}(R) \, , \tag{188}$$

or using (127),

$$Y_{lm}(R\boldsymbol{n}) = \sum_{m'} D^l_{mm'}(R)^* \, Y_{lm'}(\boldsymbol{n}) \, . \tag{189}$$

In this equation, m and m' are both eigenvalues of l_z, and the spherical harmonics are functions of the orientation of *two* vectors $\boldsymbol{n}$ and $R\boldsymbol{n}$ in a *single* frame K, whereas in (186) they were functions of the orientation of a *single* vector $\boldsymbol{n}$ in *two* frames K and K', the latter being obtained from the former by the rotation R.

---

[1] In this section we use lowercase symbols for the angular momentum operators of a single particle.

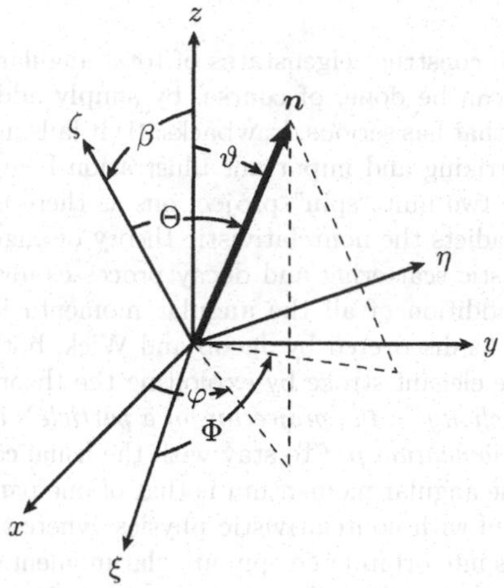

FIG. 7.4. The orientation of $\boldsymbol{n}$ in two frames that are related by a rotation.

If $\boldsymbol{n}$ in (189) is along the z-axis in K, the relation Eq. 150 between spherical harmonics and rotation matrices follows immediately. When $\vartheta = 0$, as it is for this orientation of $\boldsymbol{n}$,

$$Y_{lm}(0,\varphi) = \sqrt{(2l+1)/4\pi}\, P_l^m(1) = \delta_{m,0}\sqrt{(2l+1)/4\pi}\,; \tag{190}$$

hence $m' = 0$ in (189), so γ does not enter and can be set to zero. The orientation of $R\boldsymbol{n}$ is then given by the polar and azimuthal angles (β, α), respectively, which results in the sought after relationship

$$Y_{lm}(\vartheta,\varphi) = \sqrt{\frac{2l+1}{4\pi}}\, D^l_{m0}(\varphi,\vartheta,0)^*\,. \tag{191}$$

The special case $m = 0$ of (186) is of particular interest because it is the familiar *addition theorem for spherical harmonics*. When $m = 0$, the spherical harmonic on the left-hand side does not depend on Φ and is proportional to a Legendre polynomial, while on the right D^l_{m0} is given by (191). As a consequence

$$Y_{l0}(\Theta) = \sqrt{\frac{4\pi}{2l+1}} \sum_m Y_{lm}(\vartheta,\varphi)\, Y^*_{lm}(\beta,\alpha) \tag{192}$$

But $Y_{l0}(\theta) = \sqrt{(2l+1)/4\pi}\, P_l(\cos\theta)$, whence

$$P_l(\boldsymbol{n}\cdot\boldsymbol{n}_\zeta) = \frac{4\pi}{2l+1} \sum_m Y_{lm}(\vartheta,\varphi)\, Y^*_{lm}(\beta,\alpha)\,, \tag{193}$$

which is the addition theorem. In the notation of (184), which will be used frequently, this reads as follows for any two unit vectors:

$$P_l(\boldsymbol{n}_1\cdot\boldsymbol{n}_2) = \frac{4\pi}{2l+1} \sum_m Y_{lm}(\boldsymbol{n}_1)\, Y^*_{lm}(\boldsymbol{n}_2)\,. \tag{194}$$

(b) Helicity States

Here the objective is to construct eigenstates of total angular momentum for particles with spin. This can be done, of course, by simply adding orbital and spin angular momenta, but that has serious drawbacks: (i) it fails in relativistic quantum mechanics, the most striking and important illustration being that of one-photon states which only have two unit "spin" projections as there is no longitudinal polarization, which contradicts the nonrelativistic theory of angular momentum; and (ii) even in nonrelativistic scattering and decay processes involving particles with spin, the brute-force addition of all the angular momenta leads to disgracefully complicated formulas. As discovered by Jacob and Wick, both of these drawbacks can be overcome in one elegant stroke by exploiting the theory of rotations.

The key concept is helicity — the projection of a particle's total angular momentum $\boldsymbol{j}$ onto its linear momentum $\boldsymbol{p}$. (To stay with the usual conventions, lowercase is used because here the angular momentum is that of one particle.) Until we come to photons, we only deal with nonrelativistic physics, where there is an unambiguous separation $\boldsymbol{j} = \boldsymbol{l}+\boldsymbol{s}$ into orbital and spin angular momenta, so that $\boldsymbol{j}\cdot\boldsymbol{p} = \boldsymbol{s}\cdot\boldsymbol{p}$. The helicity is defined as the eigenvalue of the operator $\mathcal{H} = \boldsymbol{j}\cdot\boldsymbol{p}/p = \boldsymbol{s}\cdot\boldsymbol{p}/p$, the eigenvalues being

$$\lambda = s, s-1, \ldots, -s\,. \tag{195}$$

The crucial property of $\mathcal{H}$ is that it is a pseudoscalar, because $\boldsymbol{p}$ is a polar and $\boldsymbol{s}$ an axial vector. Hence $\mathcal{H}$ commutes with $\boldsymbol{j}$, though it does not commute with either $\boldsymbol{l}$ or $\boldsymbol{s}$. Hence a compatible set of one-particle observables is the energy or magnitude of the momentum, $\mathcal{H}, \boldsymbol{s}^2, \boldsymbol{j}^2, j_z$, with eigenvalues $p, \lambda, j(j+1), m$. This is to be compared with the more obvious compatible set $\boldsymbol{p}, \boldsymbol{l}^2, \boldsymbol{j}^2, j_z$.

The momentum eigenvalue of a single particle will be designated by $p\boldsymbol{n}$, with the unit vector $\boldsymbol{n}$ having the polar and azimuthal angles (ϑ, φ), and $\boldsymbol{n}\cdot\boldsymbol{n}_z = \cos\vartheta$. Let $|p\boldsymbol{n}_z, \lambda\rangle$ be a state with helicity λ and linear momentum p along the z-axis. The spin value s is not shown explicitly because it is an intrinsic and unchanging property of the particle, like its mass. The state $|p\boldsymbol{n}, \lambda\rangle$ with momentum $p\boldsymbol{n}$ in any other direction and the *same* helicity is obtained by rotating $|p\boldsymbol{n}_z, \lambda\rangle$ by the rotation $\boldsymbol{n}_z \to \boldsymbol{n}$ because λ is unaffected by rotations:

$$|p\boldsymbol{n}, \lambda\rangle = e^{-i\varphi j_z}\, e^{-i\vartheta j_y} |p\boldsymbol{n}_z, \lambda\rangle = D(\varphi, \vartheta, 0)|p\boldsymbol{n}_z, \lambda\rangle\,. \tag{196}$$

The choice $(\alpha = \varphi, \beta = \vartheta)$ for the Euler angles is required by the definition of these angles in Eq. 138, but γ is arbitrary. Many authors use $\gamma = -\varphi$, which only amounts to a different phase convention. Because $D(\varphi, \vartheta, 0)$ appears so frequently, we introduce the abbreviations

$$D(\boldsymbol{n}) \equiv D(\varphi, \vartheta, 0)\,, \qquad D^j_{mm'}(\varphi, \vartheta, 0) \equiv D^j_{mm'}(\boldsymbol{n}) \tag{197}$$

already used in §7.1, there for the special case of $j = \frac{1}{2}$. Beware of the following: the $D(\boldsymbol{n})$ do not form a group; $D(\boldsymbol{n}_2)D(\boldsymbol{n}_1)$ cannot, in general, be written in the form $D(\boldsymbol{n})$ because all three Euler angles are needed.

As it stands, $|p\boldsymbol{n}_z, \lambda\rangle$ is an eigenstate of j_z with eigenvalue λ. It can, therefore, be decomposed into eigenstates $|pjm\rangle$ with $m = \lambda$:

$$|p\boldsymbol{n}_z, \lambda\rangle = \sum_{j=|\lambda|}^{\infty} |pj\lambda\rangle\langle pj\lambda|p\boldsymbol{n}_z, \lambda\rangle\,. \tag{198}$$

The sum starts with $j = |\lambda|$ because $|m| \leq j$, and runs over integers or half-integers according to whether s is integer or half-integer. The state with momentum in an arbitrary direction $\boldsymbol{n}$ is obtained from (198) by the rotation (196), and use of (122):

$$|p\boldsymbol{n}, \lambda\rangle = \sum_{j=|\lambda|}^{\infty} \sum_{m=-j}^{j} |pjm\lambda\rangle\, D^j_{m\lambda}(\boldsymbol{n})\, \langle pj\lambda|p\boldsymbol{n}_z, \lambda\rangle\,. \tag{199}$$

Here the states are, with good reason, written as $|pjm\lambda\rangle$ as they are helicity eigenstates with eigenvalue λ because the helicity is unchanged by the rotation.

Equation (199) is actually the generalization to particles with spin of the expansion of a plane wave in terms of spherical waves (recall §3.4):

$$e^{i\boldsymbol{k}\cdot\boldsymbol{r}} = 4\pi \sum_{lm} i^l j_l(kr) Y^*_{lm}(\hat{\boldsymbol{k}}) Y_{lm}(\hat{\boldsymbol{r}})\,. \tag{200}$$

This can be seen in full detail by taking the scalar product of (199) with a position eigenket in the $s = 0$ case.

The helicity state $|pjm\lambda\rangle$ can be extracted from (199) by using the orthogonality relation (180) for the rotation matrices. The latter can be simplified as $\gamma = 0$ throughout:

$$\int d\boldsymbol{n}\, D^j_{m\lambda}(\boldsymbol{n})\, D^{j'}_{m'\lambda}(\boldsymbol{n})^* = \frac{4\pi}{2j+1}\delta_{jj'}\delta_{mm'}\,, \tag{201}$$

where $d\boldsymbol{n} = \sin\vartheta d\vartheta d\varphi$; the integration need only run once over the unit sphere because both j and j' are either integer or half-integer. Thus

$$|pjm\lambda\rangle = N \int d\boldsymbol{n}\, D^j_{m\lambda}(\boldsymbol{n})^*\, |p\boldsymbol{n}, \lambda\rangle\,, \tag{202}$$

where N is to be determined by the normalization convention. We demand that

$$\langle pjm\lambda|p'j'm'\lambda'\rangle = \frac{\delta(p-p')}{pp'}\, \delta_{jj'}\delta_{mm'}\,, \tag{203}$$

which has the associated completeness relation

$$1 = \sum_{jm\lambda} \int_0^{\infty} p^2 dp\, |pjm\lambda\rangle\langle pjm\lambda|\,. \tag{204}$$

If the momentum eigenstates are normalized as follows,

$$\langle p\boldsymbol{n}, \lambda|p'\boldsymbol{n}', \lambda'\rangle = \delta^3(p\boldsymbol{n} - p'\boldsymbol{n}')\, \delta_{\lambda\lambda'}\,. \tag{205}$$

these definitions lead to $N = \sqrt{(2j+1)/4\pi}$.

To summarize, with the normalization conventions as just defined, the simultaneous free-particle eigenstate of energy, angular momentum and helicity is

$$|pjm\lambda\rangle = \sqrt{\frac{2j+1}{4\pi}} \int d\boldsymbol{n}\, D^j_{m\lambda}(\boldsymbol{n})^*\, |p\boldsymbol{n}, \lambda\rangle\,, \tag{206}$$

and the transformation from the linear to the angular momentum representations is

$$\langle pjm\lambda|p\boldsymbol{n}, \lambda'\rangle = \frac{\delta(p-p')}{pp'}\, \delta_{\lambda\lambda'} \sqrt{\frac{2j+1}{4\pi}}\, D^j_{m\lambda}(\boldsymbol{n})\,. \tag{207}$$

(c) Decay Angular Distributions

This entire discussion extends readily to states of two particles a and b in the frame where they have zero total momentum. Let $p\boldsymbol{n}$ now be the relative momentum in that frame, and $\boldsymbol{J}$ the total angular momentum, i.e., orbital plus both spins. The eigenvalue of $\boldsymbol{J}\cdot\boldsymbol{n}$ is

$$\Lambda = \lambda_a - \lambda_b , \tag{208}$$

where (λ_a, λ_b) are the individual helicities, which by the definition of helicity contribute to this projection of $\boldsymbol{J}$ with opposite signs. The individual helicities are, of course, again invariant under rotation, and therefore the two-body state is specified by both. In short, (206) and (207) generalize to

$$|pJM\lambda_a\lambda_b\rangle = \sqrt{\frac{2J+1}{4\pi}} \int d\boldsymbol{n}\, D^J_{M\Lambda}(\boldsymbol{n})^* \, |p\boldsymbol{n}, \lambda_a\lambda_b\rangle , \tag{209}$$

$$\langle pJM\lambda_a\lambda_b|p\boldsymbol{n}, \lambda'_a\lambda'_b\rangle = \frac{\delta(p-p')}{pp'}\, \delta_{\lambda_a\lambda'_a}\delta_{\lambda_b\lambda'_b} \sqrt{\frac{2J+1}{4\pi}}\, D^J_{M\Lambda}(\boldsymbol{n}) , \tag{210}$$

where (J, M) are the eigenvalues of the total angular momentum in the frame where the total momentum vanishes.

The angular distribution in the process $a \to b+c$ for particles of arbitrary spins is completely determined by symmetry if the interaction responsible for the process is rotationally invariant. Let s_a be the spin of the parent; then this is the total angular momentum of the decay products in the rest frame of the parent. In that frame the daughters b and c have momenta in the directions $\pm\boldsymbol{n}$ and helicities (λ_b, λ_c). A detector will thus select a final state $|f\rangle$, which in the notation just defined is $|p\boldsymbol{n}, \lambda_b\lambda_c\rangle$. But by angular momentum conservation, only the projection of $|f\rangle$ with $J = s_a$ and some one value of M can participate. Hence the amplitude for the process is

$$A(M \to f) \propto \langle s_a M|p\boldsymbol{n}, \lambda_b\lambda_c\rangle \propto D^{s_a}_{M\Lambda}(\boldsymbol{n}) , \quad \Lambda = \lambda_b - \lambda_c , \tag{211}$$

which confirms the claim that the angular distribution is completely determined by symmetry for specified values of the quantum numbers M, λ_b, λ_c.

Symmetry also restricts the unspecified coefficients in (211). Rotational invariance requires them to not depend on M, because M is undefined without the choice of a coordinate system, which must be irrelevant if the interaction is rotationally invariant. A formal way of seeing this is to define an operator X whose matrix element $\langle s_a M|X|f\rangle$ equals the amplitude $A(M \to f)$; if perturbation theory is valid X is simply the term H_I in the Hamiltonian responsible for the decay, but if it is not it is some unknown but rotationally invariant function of H_I, and that suffices to prove that the coefficient in (211) is independent of M. An equivalent proof is one corollary of Prob. 11.

The coefficients in (211) can, however, depend on the helicities because they are rotationally invariant. Call them $C(\lambda_b\lambda_c)$. Nothing further can be said if only rotational invariance is assumed. However, if the interaction is reflection invariant, then there are consequence because helicities, being the eigenvalues of a pseudoscalar, change sign under reflection. At first sight this requires $C(\lambda_b\lambda_c) = C(-\lambda_b\,, -\lambda_c)$. But this is only correct if the issue of intrinsic parity is overlooked. If these parities are π_a etc., then the correct relationship is

$$C(\lambda_b\lambda_c) = \pi_a\pi_b\pi_c\, C(-\lambda_b\,, -\lambda_c) , \tag{212}$$

which keeps track of what happens to the internal states of the particles under reflection.

In most experiments, the initial state is not pure in that it does not have a completely known value of M, but is a mixture with the density matrix $\langle M|\rho|M'\rangle$. Nevertheless, symmetry still produces considerable restrictions on the angular distribution, which is is the subject of Prob. 12.

(d) Rigid-Body Motion

The concept of a perfectly rigid body is incompatible with the fundamental principles of quantum mechanics. Nevertheless, the idealization it embodies is often very useful, with the most important example being that of molecular dynamics. Because nuclei are so heavy compared to electrons, they can, with good accuracy, be treated as if they had fixed positions with respect to each other so that their only possible motion is a rigid rotation, while the far more rapidly moving electrons contribute to the mean electrostatic field that rotates together with the nuclear framework. The preceding sentence is, in essence, the Born-Oppenheimer approximation. Of course, the nuclei are not really in fixed relative positions; they oscillate, approximately harmonically, and the "fixed" positions are actually the expectation values of their coordinates. Given this contextual background, it is evident that the analysis of molecular dynamics does require an understanding of rigid body motion. As we shall now see, symmetry considerations largely determine the stationary states and spectrum of this motion.

A rigid body's position in space is specified by the mutually orthogonal unit vectors $(\boldsymbol{a}, \boldsymbol{b}, \boldsymbol{c})$ along its principal axes, which are taken to be the (ξ, η, ζ) axes in the definition of the Euler angles (Fig. 7.2). *These vectors are dynamical variables* — three commuting vector operators, which can be envisaged as being formed from the coordinate operators of a set of immovable mass points in the body-fixed frame. The Hamiltonian of the rigid body is taken over from classical mechanics:

$$H = \frac{J_a^2}{2I_a} + \frac{J_b^2}{2I_b} + \frac{J_c^2}{2I_c} , \tag{213}$$

where $I_{a,b,c}$ are the principal moments of inertia, and $J_{a,b,c}$ the projection of the angular momentum onto the principal axes:

$$J_a = \boldsymbol{a} \cdot \boldsymbol{J} , \quad \text{etc.} \tag{214}$$

It is essential to realize that J_a is *not* a component of a vector operator, but a *pseudoscalar* operator! Hence

$$[\boldsymbol{J}, J_a] = 0 , \quad \text{etc.} \tag{215}$$

On the other hand, as the relative positions of the body's constituents are assumed to be fixed, the moments of inertia are numbers, not operators.

The angular momentum can be expressed in two forms: as a superposition of the projections onto the body-fixed frame, and onto space fixed unit vectors $\boldsymbol{e}_i$, which are *not* operators:

$$\boldsymbol{J} = J_a\boldsymbol{a} + J_b\boldsymbol{b} + J_c\boldsymbol{c} = \sum_{i=x,y,z} J_i\boldsymbol{e}_i . \tag{216}$$

Now

$$[J_a, J_b] = [J_i a_i, J_j b_j] = J_i[a_i, J_j]b_j + [J_i, J_j]a_i b_j + J_j[J_i, b_j]a_i \tag{217}$$

because $[a_i, b_j] = 0$. As $\boldsymbol{a}$ is a vector operator, $[J_i, a_j] = i\epsilon_{ijk}a_k$; furthermore, $\boldsymbol{a}\times\boldsymbol{b} = \boldsymbol{c}$, and therefore

$$[J_a, J_b] = -iJ_c\ , \qquad \text{(cyclic)}\ . \tag{218}$$

The minus sign on the right is not an error!

Consider first the case of a symmetric rigid body, with $\boldsymbol{c}$ as the symmetry axis and $I_a = I_b$. The Hamiltonian is then

$$H = W_a(J^2 - J_c^2) + W_c J_c^2\ , \qquad W_a = 1/2I_a\ ,\ \text{etc.} \tag{219}$$

Because $[J^2, J_c] = 0$, these operators can be diagonalized simultaneously. Call K the eigenvalue of J_c, the projection of $\boldsymbol{J}$ along the symmetry axis. Then the spectrum is

$$E_{JK} = W_a[J(J+1) - K^2] + W_c K^2\ , \tag{220}$$

where of course $K = J, J-1, \ldots, -K$.

The degeneracy in $\pm K$ is imposed by symmetry; K is the eigenvalue of the pseudoscalar J_c, and therefore $K \to -K$ under reflection. But the Hamiltonian is invariant under reflection, which thus requires states of opposite K to have the same energy.

The wave functions of the symmetric rigid body are also determined completely by symmetry. In the frame in which the body is instantaneously at rest, the states are $|JK; B\rangle$, where B stands for body-fixed. The wave function Ψ is the amplitude for finding $|JK; B\rangle$ in the space-fixed state $|JM\rangle$, where M is the eigenvalue of the projection of $\boldsymbol{J}$ onto the space fixed axis $\boldsymbol{e}_z$:

$$\Psi_{JMK} = \langle JK; B|JM\rangle\ . \tag{221}$$

The body and space-fixed states are, however, just related by the rotation $R(\alpha\beta\gamma)$ as parametrized by the Euler angles:

$$|JK; B\rangle = D(\alpha\beta\gamma)|JK\rangle\ . \tag{222}$$

Therefore $\Psi = \langle JM|D(R)|JK\rangle^*$, and so the wave functions are

$$\Psi_{JMK}(\alpha\beta\gamma) = \sqrt{2J+1}\ D^J_{MK}(\alpha\beta\gamma)^*\ , \tag{223}$$

which, because of (180), are normalized as follows:

$$\int dR\ \Psi_{JMK}(\alpha\beta\gamma)\ \Psi^*_{J'M'K'}(\alpha\beta\gamma) = \delta_{JJ'}\delta_{MM'}\delta_{KK'}\ . \tag{224}$$

The question of whether half-integer values of J are permitted is actually a subtle matter, and hinges on whether the electrons' spins are "locked" to the nuclear framework by the electrons' spin-orbit interaction. At this point we set this issue aside and assume that only integer values of J exist. The hyperfine interaction between the nuclear spins and the electrons is so weak that these spins are decoupled from the rotation of the molecule, and are of no consequence for this question.

The wave function Ψ_{JMK} can also be derived by setting up the Schrödinger equation for the symmetric rotor. This is a quite involved exercise (see Prob. 13), and produces a differential equation in the Euler angles that, after separation of variables, reduces to the following equation for the function of β:

$$\left\{\frac{d^2}{d\beta^2} + \cot\beta\frac{d}{d\beta} - \frac{M^2 + K^2 - 2MK\cos\beta}{\sin^2\beta} + J(J+1)\right\} u_{JMK}(\beta) = 0 \,. \tag{225}$$

The solution is the trigonometric polynomial $d^J_{MK}(\beta)$ as given in (144).

The diatomic molecule composed of spin zero nuclei is the special case of the symmetric body shrunk down to an infinitely thin rod along the axis $\boldsymbol{c}$. The transverse moments of inertia vanish so only $K = 0$ is allowed, whence the wave function is simply $Y_{JM}(\beta, \alpha)$, and the energy spectrum is $W_a J(J+1)$. For nuclei with spins s_1 and s_2, the complete nuclear wave function is

$$\Psi_{JM,m_1m_2}(\beta, \alpha) = Y_{JM}(\beta, \alpha)\chi^{(1)}_{m_1}\chi^{(2)}_{m_2}, \tag{226}$$

where (m_1, m_2) are the spin projections onto space-fixed axes. If the nuclei are not identical, nothing further is to be said. For identical nuclei, however, the whole wave function must be symmetric or antisymmetric under nuclear interchange according to whether the nuclei are bosons or fermions. The implications are then identical to those spelled out in §6.1(c). In particular, if the nuclei have zero spin, only states even under reflection through the origin are allowed, i.e., only $J = 0, 2, \ldots$, and the rotational spectrum of the molecule has half its states missing.

The parities of the stationary states of the symmetric rigid body have interesting consequences for the Stark effect. If the body has a charge distribution of the same symmetry, it will have an electric dipole moment $\boldsymbol{d}$ that, by symmetry, must point along $\boldsymbol{c}$. It will then display a linear Stark effect in an applied electric field $\boldsymbol{E}$ due to the interaction $H_1 = -\boldsymbol{E}\cdot\boldsymbol{d}$. A system with a reflection invariant Hamiltonian and no "accidental" degeneracies between states of opposite parity cannot have a linear Stark effect, however, as we learned for hydrogen when the degeneracies due to the fine structure interaction are lifted (recall §5.4). Here, on the other hand, there is a degeneracy between states of opposite parity forced by symmetry, i.e., that of K and $-K$, and a linear Stark effect is to be expected. Indeed, the result for the level shifts in the weak field limit is (see Prob. 14)

$$\Delta_{JMK} = -dE\,\frac{MK}{J(J+1)} \,. \tag{227}$$

Note that for the $K = 0$ state, which has the definite parity $(-1)^J$, there is no linear Stark effect.

Symmetry does not suffice to determine the spectrum or eigenfunctions when the body is asymmetric. The wave functions Ψ_{JMK} for various values of K are, however, a complete set for given values of J and M, but the coefficients in the linear combination must be found by diagonalizing the asymmetric term in the Hamiltonian. Reflection through the origin is still a symmetry, so the energy eigenstates will have a definite parity unless there are still degeneracies between states of opposite parity. The case of the $J = 1$ states illustrates the points just made; see Prob. 15.

7.6 Tensor Operators

Tensors arise first in both classical and quantum physics as polynomials composed of vectors. The example encountered thus far is that of the quadrupole tensors due to the nuclear and electronic charge distributions in hyperfine structure (§5.4). In that context, advantage was taken of the fact that the matrix elements of these electrostatic tensor are proportional to those of a tensor of the same rank composed of angular momentum operators. This is not an accidental relationship; it holds because both tensors transform in the same manner under rotations, and all objects that transform in the same way under rotations have the same matrix elements between angular momentum eigenstates, apart from overall constants. This last statement is the Wigner-Eckart theorem, which will be proven in this section.

(a) Definition of Tensor Operators

Although tensor operators will be defined strictly in terms of their transformation property, and not by how they may perhaps be constructed out of other operators (e.g., coordinates, momenta, spins, electromagnetic field operators, etc.), it is important to understand the motivation for the definition, which does stem from the transformation law for vector operators.

Given any observable A, its rotated counterpart A^R is defined by the unitary transformation

$$A \to A^R = D^\dagger(R)\, A\, D(R)\,. \tag{228}$$

This is an active rotation of the observable, as shown by the following relation for matrix elements between arbitrary states,[1]

$$\langle b|A^R|c\rangle = \langle b; R|A|c; R\rangle\,. \tag{229}$$

For a vector operator $\boldsymbol{V}$, this transformation is

$$V_i \to V_i^R = \sum_j a_{ij}(R)\, V_j\,, \tag{230}$$

where $a_{ij}(R)$ are the elements of the 3×3 orthogonal matrix that gives the rotated vector in terms of its Cartesian coordinates. As before, $R\boldsymbol{V}$ will designate the rotated vector with components given by the right-hand side of (230). In the case of an infinitesimal rotation $\delta\boldsymbol{\omega} \equiv \boldsymbol{n}\delta\theta$, the unitary rotation operator is

$$D(R) \simeq 1 - i\delta\boldsymbol{\omega}\cdot\boldsymbol{J}\,, \tag{231}$$

and

$$\boldsymbol{V}^R \simeq \boldsymbol{V} + i[\delta\boldsymbol{\omega}\cdot\boldsymbol{J}, \boldsymbol{V}] = \boldsymbol{V} + \delta\boldsymbol{\omega}\times\boldsymbol{V}\,. \tag{232}$$

Equation (230) does not generalize neatly to higher-rank tensors because it produces objects that do not transform with irreducible representations of $SO(3)$. The way to avoid this is suggested by recalling that the Cartesian components of the position vector $\boldsymbol{r}$ have the following linear relation to the spherical harmonics of order 1,

$$Y_{10} = N\,\frac{z}{r}\,, \qquad Y_{1,\pm1} = \mp\frac{N}{\sqrt{2}}\,\frac{x \pm iy}{r}\,, \tag{233}$$

---

[1]Note that many authors use the definition $A \to \bar{A}^R = DAD^\dagger$, or $\langle b; R|\bar{A}^R|c; R\rangle = \langle b|A|c\rangle$.

where $N = \sqrt{3/4\pi}$. The rotation $\boldsymbol{r} \to R\boldsymbol{r}$, or for *any* vector, as expressed in terms of Cartesian components, is therefore equivalent to the $l = 1$ case of the transformation law for spherical harmonics (Eq. 189), where "equivalent" means that the transformation matrix $a(R)$ in (230) and the rotation matrix $D^1(R)$ differ only by a similarity transformation with constant coefficients (recall §3.2). In short, for any vector operator $\boldsymbol{V}$, we can define the three operators

$$T_0^{(1)} = V_z \,, \qquad T_{\pm 1}^{(1)} = \mp \frac{1}{\sqrt{2}}(V_x \pm iV_y) \,; \tag{234}$$

then under the finite rotation R this operator transforms as follows:

$$D^\dagger(R)\, T_\kappa^{(1)}\, D(R) = \sum_{\kappa=0,\pm 1} D^1_{\kappa\kappa'}(R)^*\, T_{\kappa'}^{(1)} \,. \tag{235}$$

This last relation generalizes readily to tensors of higher rank. Consider first the operator $Y_{lm}(\boldsymbol{V})$ defined as the same polynomial in the components of the operator $\boldsymbol{V}$ as is $Y_{lm}(\boldsymbol{n})$ in terms of those of an ordinary numerical 3-vector $\boldsymbol{r} = r\boldsymbol{n}$. Under the rotation R, it will therefore transform in precisely the same way as the numerical function $Y_{lm}(\boldsymbol{n})$, namely,

$$D^\dagger(R)\, Y_{lm}(\boldsymbol{V})\, D(R) = Y_{lm}(R\boldsymbol{V}) = \sum_{m'} D^l_{mm'}(R)^*\, Y_{lm'}(\boldsymbol{V}) \,. \tag{236}$$

A tensor operator $\mathsf{T}^{(k)}$ *of rank* k*, with components* $\kappa = k, k-1, \ldots, -k$*, is then defined as a set of* $2k+1$ *objects* $T_\kappa^{(k)}$ *that transform under rotation in the exactly same manner as* $Y_{k\kappa}(\boldsymbol{V})$*, i.e.,*

$$D^\dagger(R)\, T_\kappa^{(k)}\, D(R) = \sum_{\kappa'} D^k_{\kappa\kappa'}(R)^*\, T_{\kappa'}^{(k)} \,. \tag{237}$$

Once again: any set $\mathsf{T}^{(k)}$ of $2k+1$ objects that transform in accordance with this law form a tensor operator of rank k; whether or how they may be constructed out of other operators is irrelevant for the discussion to follow.

Tensor operators defined in this way are "irreducible"; $\mathsf{T}^{(k)}$ does not contain pieces that transform with any representation other than D^k. In the case of second-rank tensors this will be familiar from §3.3(b): the second-rank Cartesian tensor $T_{ij} = u_i v_j + u_j v_i$ is reducible because it contains the scalar $\boldsymbol{u} \cdot \boldsymbol{v}$, whereas the traceless tensor $T_{ij} - \frac{2}{3}\delta_{ij}\boldsymbol{u} \cdot \boldsymbol{v}$ is irreducible because it contains no scalar, and transforms under $SO(3)$ with the 5-dimensional representation D^2.

Tensor operators of half-integer rank can also be defined as sets of objects that transform in accordance with (237), but they cannot be observables because such objects are double-valued. Such operators do appear in the theory of many-fermion systems, and field theories such as quantum electrodynamics involving fermions, but the observables (such as current and energy densities) are even-order polynomials in the double-valued fermion operators.

Commutation rules between tensor operators and the angular momentum can be obtained from (237) by taking the limit of an infinitesimal rotation:

$$\delta T_\kappa^{(k)} = i[\delta\boldsymbol{\omega} \cdot \boldsymbol{J}, T_\kappa^{(k)}] = \sum_{\kappa'} \langle k\kappa| - i\delta\boldsymbol{\omega} \cdot \boldsymbol{J} |k\kappa'\rangle^*\, T_{\kappa'}^{(k)} \,. \tag{238}$$

The matrix elements of J_z and $J_\pm$ then give

$$[J_z, T_\kappa^{(k)}] = \kappa\, T_\kappa^{(k)} , \tag{239}$$

$$[J_\pm, T_\kappa^{(k)}] = \sqrt{k(k+1) - \kappa(\kappa \pm 1)}\, T_{\kappa\pm 1}^{(k)} . \tag{240}$$

For a vector operator, (239) and (240) reduce to the familiar commutation rule

$$[V_i, J_j] = i\epsilon_{ijk} V_k \tag{241}$$

for the Cartesian components.

Products of tensor operators can be combined to form tensor operators of both higher an lower rank. This can be done by means of angular momentum addition for the following reason:

> $T_\kappa^{(k)}|0\rangle$, *where* $|0\rangle$ *is any* $j = 0$ *state, transforms under rotations like the angular momentum eigenstate* $|k\kappa\rangle$;

this can be verified by carrying out a rotation. Thus $\mathsf{T}^{(k)}$ is an operator that "manufactures" k units of angular momentum, and as a consequence, if two such operators, $\mathsf{U}^{(k_1)}$ and $\mathsf{V}^{(k_2)}$, act in succession, they will produce changes in angular momentum that range in unit steps from $k_1 + k_2$ down to $|k_1 - k_2|$. This argument implies that

$$U_{\kappa_1}^{(k_1)} V_{\kappa_2}^{(k_2)} = \sum_{k\kappa} \langle k_1\kappa_1 k_2\kappa_2 | k\kappa\rangle\, T_\kappa^{(k)} , \tag{242}$$

$$T_\kappa^{(k)} = \sum_{\kappa_1\kappa_2} \langle k\kappa | k_1\kappa_1 k_2\kappa_2\rangle\, U_{\kappa_1}^{(k_1)} V_{\kappa_2}^{(k_2)} . \tag{243}$$

A formal proof involves the defining transformation law (237) and the direct product of representations. Whether $\mathsf{U}^{(k_1)}$ and $\mathsf{V}^{(k_2)}$ commute with each other does not matter.

An especially important example of (242) is the *scalar product of two tensor operators*. This is the invariant ($k = 0$) case of (242):

$$U^{(k)} \cdot V^{(k)} = (-1)^k \sqrt{2k+1} \sum_{\kappa_1\kappa_2} \langle 00 | k\kappa_1 k\kappa_2\rangle\, U_{\kappa_1}^{(k)} V_{\kappa_2}^{(k)} , \tag{244}$$

where the normalization is chosen so that for vector operators

$$U^{(1)} \cdot V^{(1)} = \boldsymbol{U} \cdot \boldsymbol{V} . \tag{245}$$

For higher ranks, the explicit form of (244) is

$$U^{(k)} \cdot V^{(k)} = \sum_\kappa (-1)^\kappa\, U_\kappa^{(k)}\, V_{-\kappa}^{(k)} . \tag{246}$$

(b) The Wigner-Eckart Theorem

Matrix elements of tensor operators between angular momentum eigenstates arise in many practical problems, such as radiative transitions, nuclear β-decay and studies of atomic spectra. To a large extent, such matrix elements are determined by the properties of the rotation group. This again follows from the fact that $\mathsf{T}^{(k)}$ adds

k units of angular momentum to any state on which it acts, so that $T_\kappa^{(k)}|j'm'\rangle$ transforms under rotations in the same way as does the product state $|k\kappa\rangle \otimes |j'm'\rangle$, which then implies that $\langle jm|T_\kappa^{(k)}|j'm'\rangle$ must be proportional to the Clebsch-Gordan coefficient $\langle jm|k\kappa j'm'\rangle$. That is to say, the dependence of the matrix element on the quantum numbers (m, m', κ), whose meaning is contingent on the orientation of the coordinate system, is completely determined by geometry, i.e., group theory. This statement is the *Wigner-Eckart theorem.*

To confirm the theorem, rewrite the basic transformation law (237) in the form

$$T_\kappa^{(k)} = \sum_{\kappa'} D_{\kappa\kappa'}^k(R)^* \, D(R) \, T_{\kappa'}^{(k)} \, D^\dagger(R) \, . \tag{247}$$

The matrix element of this identity is

$$\langle \alpha jm|T_\kappa^{(k)}|\alpha' j'm'\rangle = \sum_{\kappa'\mu\mu'} D_{\kappa\kappa'}^k(R)^* D_{m\mu}^j(R) D_{m'\mu'}^{j'}(R)^* \langle \alpha j\mu|T_{\kappa'}^{(k)}|\alpha' j'\mu'\rangle \, , \tag{248}$$

where (α, α') again stand for the eigenvalues of all compatible observables other than angular momentum. Integrating this over R, and recalling Eq. 181, gives

$$\langle \alpha jm|T_\kappa^{(k)}|\alpha' j'm'\rangle = \frac{\langle jm|j'm'k\kappa\rangle}{2j+1} \sum_{\kappa'\mu\mu'} \langle \alpha j\mu|T_{\kappa'}^{(k)}|\alpha' j'\mu'\rangle \langle j\mu|j'\mu' k\kappa'\rangle \, . \tag{249}$$

This states that the entire dependence of the matrix element on the quantum numbers (m, m', κ) is carried by the first Clebsch-Gordan coefficient. *QED.*

As was already implicit in the argument preceding Eq. 247, important *selection rules* follow from the Wigner-Eckart theorem via the laws of angular momentum addition. This property is embodied in the Clebsch-Gordon coefficient in (249), which states that *the matrix element* $\langle \alpha jm|T_\kappa^{(k)}|\alpha' j'm'\rangle$ *vanishes unless*

$$|j - j'| \leq k \leq (j + j') \, , \qquad m - m' = \kappa \, . \tag{250}$$

The selection rule already derived in §5.4(d) for vector operators is the special case $k = 1$.

It is conventional to write the Wigner-Eckart theorem in the following form:

$$\langle \alpha jm|T_\kappa^{(k)}|\alpha' j'm'\rangle = \frac{1}{\sqrt{2j+1}} \langle \alpha j \,\|T^{(k)}\|\, \alpha' j'\rangle \, \langle jm|j'm'k\kappa\rangle \tag{251}$$

$$= (-1)^{2k+j+m} \langle \alpha j \,\|T^{(k)}\|\, \alpha' j'\rangle \begin{pmatrix} j & k & j' \\ -m & \kappa & m' \end{pmatrix} . \tag{252}$$

The object $\langle \alpha j \,\|\, T^{(k)} \,\|\, \alpha' j'\rangle$ is called *the reduced matrix element;* it is evaluated by computing the complete matrix element for some convenient choice of (m, m', κ) and dividing by the appropriate Clebsch-Gordan coefficient.

The specific and frame-independent properties of the states, as expressed by the rotationally invariant operators with eigenvalues (α, j, α', j'), only enter into the reduced matrix elements. That the dependence on the orientation-dependent quantum numbers is fully captured by the Clebsch-Gordan coefficient leads to many simplifications in practical calculations. Perhaps the most important example is that of transition rates in a system that is not spin-polarized. As we shall learn in §10.4, in that case the observed rate Γ due to a perturbation involving an operator $\mathsf{T}^{(k)}$ is proportional to the square of the matrix element summed over the final eigenvalues of J_z and averaged over the initial ones,

$$\Gamma = \frac{1}{2j_i+1} \sum_{m_i m_f} |\langle \alpha_f j_f m_f | T_\kappa^{(k)} | \alpha_i j_i m_i \rangle|^2 = |\langle \alpha_f j_f || T^{(k)} || \alpha_i j_i \rangle|^2 \,, \tag{253}$$

where the identity

$$\langle j_1 m_1 j_2 m_2 | jm \rangle = (-1)^{j_1+j_2-j} \sqrt{\frac{2j+1}{2j_2+1}} \langle j_1 - m_1 jm | j_2 m_2 \rangle \tag{254}$$

was used.

As a quite important example, consider the spherical harmonic operator $Y_{k\kappa}(\boldsymbol{x})$, where $\boldsymbol{x}$ is the position operator of one particle, and the matrix element of this operator between one-particle orbital angular momentum states. Then from Eq. 182 it follows that

$$\langle l || \mathsf{Y}^{(k)} || l' \rangle = \sqrt{\frac{(2l'+1)(2k+1)}{4\pi}} \langle l0 | k0l'0 \rangle \,. \tag{255}$$

Because observables (as compared to operators) that transform like tensor operators of half-integer rank cannot exist, transitions between states of integer and half-integer angular momentum are not possible in principle, and the Hilbert spaces spanned by states of integer and half-integer are rigorously decoupled. This is an example of a *superselection rule.* A superselection rule differs profoundly from a conventional selection rule, such as Eq. 250, because the former cannot be broken by any system under any circumstances. In particular, the relative phases of states belonging to spaces separated by superselection rule cannot be determined by any measurement and are therefore arbitrary.

(c) Racah Coefficients and 6-j Symbols

The Wigner-Eckart theorem can be extended to matrix elements of tensor operators involving more than one angular momentum. The first example to be discussed is that of a scalar product between two tensor operators pertaining to two different subsystems, e.g., operators that pertain separately to one of the two electrons in helium.

Consider then a system that has two angular momenta, $\boldsymbol{J}_1$ and $\boldsymbol{J}_2$, with $\boldsymbol{J} = \boldsymbol{J}_1 + \boldsymbol{J}_2$. Let $\mathsf{U}^{(k)}$ be a tensor operator that is rotated in the standard way by $\boldsymbol{J}_1$ but is invariant under rotations generated by $\boldsymbol{J}_2$, while $\mathsf{V}^{(k)}$ is another tensor operator that commutes with $\boldsymbol{J}_1$ and is rotated by $\boldsymbol{J}_2$. The object of interest is the matrix element of $S \equiv U^{(k)} \cdot V^{(k)}$ in the basis $|\alpha j_1 j_2 jm\rangle$, where the eigenvalues have their usual meaning. As S is an invariant, its matrix elements vanish unless $j = j'$ and $m = m'$, and cannot depend on m. On the other hand, S is not invariant under rotations generated separately by $\boldsymbol{J}_1$ and $\boldsymbol{J}_2$, and is therefore not diagonal in the separate angular momenta.

The first step in the evaluation of the matrix element uses the theory of angular momentum addition:

$$\langle j_1j_2jm|S|j_1'j_2'jm\rangle = \sum_{\kappa m_1\dots m_2'} (-1)^\kappa \langle jm|j_1m_1j_2m_2\rangle\langle j_1m_1|U_\kappa^{(k)}|j_1'm_1'\rangle$$
$$\times \langle j_2m_2|V_{-\kappa}^{(k)}|j_2'm_2'\rangle\langle j_1'm_1'j_2'm_2'|jm\rangle\,, \tag{256}$$

where α and α' have been suppressed. The Wigner-Eckart theorem then produces

$$\langle j_1j_2j|U^{(k)}\cdot V^{(k)}|j_1'j_2'j\rangle = \frac{\langle j_1||U^{(k)}||j_1'\rangle\langle j_2||V^{(k)}||j_2'\rangle}{\sqrt{(2j_1+1)(2j_2+1)}}\,X\,, \tag{257}$$

where

$$X = \sum_{\kappa m_1\dots m_2'} (-1)^\kappa \langle jm|j_1m_1j_2m_2\rangle\langle j_1'm_1'k\kappa|j_1m_1\rangle\langle j_2'm_2'k-\kappa|j_2m_2\rangle\langle j_1'm_1'j_2'm_2'|jm\rangle\,. \tag{258}$$

As X cannot depend on m, it is only a function of k and the angular momenta; apart from an overall factor, X is a *Racah coefficient*. It is most convenient to express the result in terms of *Wigner's 6-j symbol*:

$$\langle j_1j_2j|U^{(k)}\cdot V^{(k)}|j_1'j_2'j\rangle = (-1)^{j_1'+j_2+j}\begin{Bmatrix} j_1 & j_2 & j \\ j_2' & j_1' & k \end{Bmatrix}\langle j_1||U^{(k)}||j_1'\rangle\langle j_2||V^{(k)}||j_2'\rangle\,, \tag{259}$$

where the 6-j symbol is

$$\begin{Bmatrix} j_1 & j_2 & j_3 \\ j_1' & j_2' & j_3' \end{Bmatrix} = \sum_{m_1\dots m_3'} (-1)^s \begin{pmatrix} j_1 & j_2 & j_3 \\ m_1 & m_2' & m_3' \end{pmatrix}\begin{pmatrix} j_1 & j_2' & j_3' \\ -m_1 & m_2' & -m_3' \end{pmatrix}$$
$$\times \begin{pmatrix} j_1' & j_2 & j_3' \\ -m_1' & -m_2 & m_3' \end{pmatrix}\begin{pmatrix} j_1' & j_2' & j_3 \\ m_1' & -m_2' & -m_3 \end{pmatrix}, \tag{260}$$

and $s = \sum_i (j_i + j_i' + m_i')$. Clearly, the 6-$j$ symbol vanishes unless the triads

$$(j_1j_2j_3)\,,(j_1j_2'j_3')\,,(j_1'j_2j_3')\,,(j_1'j_2'j_3)\,, \tag{261}$$

all simultaneously satisfy the triangular inequalities $|j_1-j_2| \leq j_3 \leq j_1+j_2$, etc. The relation between the 6-j symbol and Racah's original W-coefficient is

$$W(j_1j_2j_2'j_1';j_3j_3') = (-1)^{j_1+j_2+j_1'+j_2'}\begin{Bmatrix} j_1 & j_2 & j_3 \\ j_1' & j_2' & j_3' \end{Bmatrix}\,. \tag{262}$$

Another important class of matrix elements that can be simplified substantially by this technique is that of a tensor operator $\mathsf{T}^{(k)}$ that pertains to a subsystem carrying the angular momentum $\boldsymbol{J}_1$, but is unaffected by rotations generated by the angular momentum $\boldsymbol{J}_2$ of the remainder of the system. In this case the reduced matrix element in question is

$$\langle j_1j_2j \parallel T^{(k)} \parallel j_1'j_2j'\rangle = (-1)^{j_1+j_2+j'+k}\sqrt{(2j+1)(2j'+1)}$$
$$\times\begin{Bmatrix} j_1 & j_1' & k \\ j' & j & j_2 \end{Bmatrix}\langle j_1 \parallel T^{(k)} \parallel j_1'\rangle\,. \tag{263}$$

Note that in this formula, and also in (259), the dependence on the eigenvalues of $\boldsymbol{J}^2$ is completely contained in the 6-j symbol, i.e., is fully determined by symmetry.

The 6-j symbols also play a central role in the addition of three angular momenta, $\boldsymbol{J} = \sum_i \boldsymbol{J}_i$. This addition can be carried out in several distinct ways. Thus $\boldsymbol{J}_{12} \equiv \boldsymbol{J}_1 + \boldsymbol{J}_2$ can first be formed, and $\boldsymbol{J}_3$ added subsequently:

$$|j_1 j_2 (j_{12}) j_3 m\rangle = \sum |j_1 m_1 j_2 m_2 j_3 m_3\rangle \langle j_1 m_1 j_2 m_2 | j_{12} m_{12}\rangle \langle j_{12} m_{12} j_3 m_3 | jm\rangle \,. \tag{264}$$

On the other hand, $\boldsymbol{J}_{13}$ could be formed first:

$$|j_1 j_3 (j_{13}) j_2 j m\rangle = \sum |j_1 m_1 j_2 m_2 j_3 m_3\rangle \langle j_1 m_1 j_3 m_3 | j_{13} m_{13}\rangle \langle j_{13} m_{13} j_2 m_2 | jm\rangle \,. \tag{265}$$

These states are not identical. Their scalar product cannot depend on m, and involves the product of four 3-j symbols, so it is not astonishing that it is proportional to a 6-j symbol:

$$\langle (j_{13}) j_2 j | (j_{12}) j_3 j\rangle = (-1)^{j_2+j_3+j_{12}+j_{13}} \sqrt{(2j_{12}+1)(2j_{13}+1)} \left\{ \begin{array}{ccc} j_1 & j_2 & j_{12} \\ j & j_3 & j_{13} \end{array} \right\} . \tag{266}$$

Transformations between different angular momentum coupling schemes often arise in spectroscopy, and can thus be calculated with this technique and its extension to larger numbers of angular momenta.

Because (266) demonstrates that the 6-j symbol is, apart from normalization, a transformation from one basis to another, it satisfies orthogonality relations such as

$$\sum_{\{j\}} (2j+1) \left\{ \begin{array}{ccc} j_1 & j_2 & j \\ j_3 & j_4 & J \end{array} \right\} \left\{ \begin{array}{ccc} j_1 & j_2 & j \\ j_3 & j_4 & J' \end{array} \right\} = \frac{\delta_{JJ'}}{2J+1} \,. \tag{267}$$

7.7 Geometric Phases

Geometry plays a basic role in the quantum mechanical treatment of rotations and translations, one that is largely familiar from classical physics. Geometrical concepts also arise in a quite different context, as we will see here: in adiabatic processes. In §4.1(a) we already had a very cursory look at such a process, but did not even mention the topic that we take up here.

The simplest example — the only one we will consider in any detail, is a particle with spin s in an external magnetic field $\boldsymbol{B}(t)$. Imagine that *the field varies adiabatically* — that is, with a characteristic frequency $\nu \sim \dot{B}/B$ that is *very small* compared to the characteristic excitation frequency $\bar{\omega}$. If, at $t = 0$, the system is in the energy eigenstate $|m\rangle$, where $m = s, s-1, \ldots, -s$, transitions to states with a different spin projection are highly improbable — as we shall see, of order $\nu/\bar{\omega}$ in amplitude. As a consequence, at later times the system will with high probability be in the state $|m(t)\rangle$, the eigenstate of the instantaneous Hamiltonian $H(t)$ at that time. We now ask whether the system will be in the original state $|m\rangle$ if, at some later time T, the field $\boldsymbol{B}(T)$ returns to its original value $\boldsymbol{B}(0)$. The answer, in general, is No! The state will acquire a phase $\exp i\gamma(C)$ which depends only on the closed path C traversed by the moving point $\boldsymbol{B}(t)$ in the space S_B whose coordinates are (B_x, B_y, B_z). This *Berry phase* is a purely geometric property of the path C, and can be observed by interference between a state that has traversed the adiabatic circuit and another that has not.

The preceding example is a case where an adiabatic change in the environment of a system induces an observable phase change that is geometric in nature. Another class of phenomena in which geometric phases appear is where a system has both slowly and swiftly moving degrees of freedom. The most familiar example is provided by molecules, for the heavy nuclei move very slowly as compared to the light electrons. This motivates the Born-Oppenheimer approximation, in which the electronic states are found by holding the nuclei in fixed positions, while averages over these electronic states determining the configuration and motion of the nuclei. The moving nuclear configuration can return to itself, and thereby lead to geometric phases and to observable consequences, but this is a much more complicated problem and will not be treated here.

(a) Spin in Magnetic Field

The Hamiltonian of this system is

$$H(\boldsymbol{B}) = -\mu \boldsymbol{s} \cdot \boldsymbol{B} \,, \tag{268}$$

and for any *fixed value* of the magnetic field $\boldsymbol{B}$ the energy eigenvalues and eigenstates are are

$$E_m(\boldsymbol{B}) = -\mu B m \,, \qquad [H(\boldsymbol{B}) - E_m(\boldsymbol{B})]\, |m(\boldsymbol{B})\rangle \,, \quad m = s, s-1, -s \,. \tag{269}$$

When the field varies, we seek a solution of the Schrödinger equation

$$\Big(i\frac{\partial}{\partial t} - H(\boldsymbol{B}(t))\Big)|\psi(t)\rangle = 0 \,, \qquad |\psi(0)\rangle = |m(0)\rangle \,, \tag{270}$$

where $|m(0)\rangle$ is an eigenstate of $\boldsymbol{s} \cdot \boldsymbol{B}(0)$.

Recall from the WKB approximation that when the local momentum $p(x)$ varies slowly, the spatial phase of the wave function is given by $\int^x dx' p(x')$. Here it is the energy $E_m(t)$ that varies slowly, so it is natural to suppose that this variation produces an accumulated phase that is the analogous integral over time:

$$|\psi_0(t)\rangle = e^{-i\int_0^t dt' E_m(t')}\, |m(t)\rangle \,. \tag{271}$$

This does not quite work, however; an additional factor $a_m(t)$ is needed to produce a solution of the Schrödinger equation under the basic assumption that mixing with states $m' \neq m$ is negligible when the perturbation is adiabatic:

$$|\psi(t)\rangle = a_m(t)\, e^{-i\int_0^t dt' E_m(t')}\, |m(t)\rangle \,. \tag{272}$$

Substituting (272) into (270) shows that $a_m(t)$ satisfies

$$\frac{d}{dt}\, a_m(t) + a_m(t)\, \langle m(t)|\dot{m}(t)\rangle = 0 \,, \qquad |\dot{m}(t)\rangle \equiv \frac{d}{dt}\, |m(t)\rangle \,, \tag{273}$$

which has the solution

$$a_m(t) = a(0)\, e^{-\int_0^t dt' \langle m(t')|\dot{m}(t')\rangle} \,. \tag{274}$$

From the normalization condition $\langle m(t)|m(t)\rangle = \text{const.}$

$$\langle \dot{m}(t)|m(t)\rangle + \langle m(t)|\dot{m}(t)\rangle = 0\,, \tag{275}$$

and therefore the exponent in (274) is pure imaginary. Hence

$$a_m(t) = a_m(0)\,e^{i\gamma_m(t)}\,, \qquad \gamma_m(t) = i\int_0^t dt'\,\langle m(t')|\dot{m}(t')\rangle\,, \tag{276}$$

where $\gamma_m(t)$ is real, and is called the Berry phase when it *cannot* be removed by a change of phase convention.

One might think that such a removal could be accomplished by redefining the phases of the states $|m(t)\rangle$ as follows:

$$|m'(t)\rangle = e^{i\chi_m(t)}\,|m(t)\rangle\,, \qquad \chi_m(t) = i\int^t dt'\,\langle m(t')|\dot{m}(t')\rangle\,, \tag{277}$$

because then

$$i\langle m'(t)|\dot{m}'(t)\rangle = i\langle m(t)|\dot{m}(t)\rangle - \dot{\chi}_m(t) = 0\,. \tag{278}$$

This is not always possible, however. In particular, if after an interval T the magnetic field returns to its original value, then requiring states to be single-valued, i.e., $|m'(T)\rangle = |m'(0)\rangle$, implies that

$$e^{i\chi_m(T)}|m(T)\rangle = e^{i\chi_m(0)}|m(0)\rangle = e^{i\chi_m(0)}|m(T)\rangle\,, \tag{279}$$

or $\chi_m(T) - \chi_m(0) = 2n\pi$, where n is an integer, and therefore no choice of $\chi_m(t)$ can remove the Berry phase under this circumstance.

The geometrical character of the Berry phase emerges when the variation of the instantaneous energy eigenstates with time is restated as their variation with field:

$$\frac{d}{dt}\,|m(t)\rangle = \frac{d\boldsymbol{B}(t)}{dt}\cdot\frac{\partial}{\partial\boldsymbol{B}}|m(\boldsymbol{B})\rangle \equiv \dot{\boldsymbol{B}}\cdot|\boldsymbol{\nabla}_B m(\boldsymbol{B})\rangle\,. \tag{280}$$

This then expresses the phase as an integral over field values:

$$\gamma_m(C) = i\int_C d\boldsymbol{B}\cdot\langle m(\boldsymbol{B})|\boldsymbol{\nabla}_B m(\boldsymbol{B})\rangle\,, \tag{281}$$

where C is a curve in the space $\mathcal{S}_B$ defined at the outset. If C is a closed path, Stokes's theorem turns this into an integral over the surface $S(C)$ enclosed by C,

$$\gamma_m(C) = i\int_{S(C)} d\boldsymbol{S}\cdot\langle\boldsymbol{\nabla}_B m(\boldsymbol{B})|\times|\boldsymbol{\nabla}_B m(\boldsymbol{B})\rangle \equiv -\int_{S(C)} d\boldsymbol{S}\cdot\boldsymbol{V}_m(\boldsymbol{B})\,, \tag{282}$$

where $d\boldsymbol{S}$ is normal to $S(C)$, $|d\boldsymbol{S}|$ is an element of area on $S(C)$, and $\boldsymbol{V}_m$ is real.

To evaluate $\boldsymbol{V}_m$, we first insert the complete set of states:

$$\boldsymbol{V}_m(\boldsymbol{B}) = \mathrm{Im}\sum_{n\neq m}\langle\boldsymbol{\nabla}_B m(\boldsymbol{B})|n(\boldsymbol{B})\rangle\times\langle n(\boldsymbol{B})|\boldsymbol{\nabla}_B m(\boldsymbol{B})\rangle\,; \tag{283}$$

the term $n = m$ does not contribute because $\langle m|\boldsymbol{\nabla}_B m\rangle$ is pure imaginary by the argument of (275). From differentiating (269),

$$(\boldsymbol{\nabla}_B H(\boldsymbol{B}))|m\rangle + H(\boldsymbol{B})|\boldsymbol{\nabla}_B m\rangle = (\boldsymbol{\nabla}_B E_m(\boldsymbol{B}))|m\rangle + E_m(\boldsymbol{B})|\boldsymbol{\nabla}_B m\rangle\,; \tag{284}$$

but the $\{|n(\boldsymbol{B})\rangle\}$ are an orthogonal set, so for $n \neq m$

$$\langle n(\boldsymbol{B})|\boldsymbol{\nabla}_B m(\boldsymbol{B})\rangle = \frac{\langle n(\boldsymbol{B})|\boldsymbol{\nabla}_B H(\boldsymbol{B})|m(\boldsymbol{B})\rangle}{E_m(\boldsymbol{B}) - E_n(\boldsymbol{B})} \, . \tag{285}$$

Hence[1]

$$\boldsymbol{V}_m(\boldsymbol{B}) = \mathrm{Im} \sum_{n \neq m} \frac{\langle m(\boldsymbol{B})|\boldsymbol{\nabla}_B H(\boldsymbol{B})|n(\boldsymbol{B})\rangle \times \langle n(\boldsymbol{B})|\boldsymbol{\nabla}_B H(\boldsymbol{B})|m(\boldsymbol{B})\rangle}{[E_m(\boldsymbol{B}) - E_n(\boldsymbol{B})]^2} \, , \tag{286}$$

and on recalling (268) and (269),

$$\boldsymbol{V}_m(\boldsymbol{B}) = \frac{1}{B^2} \, \mathrm{Im} \sum_{n \neq m} \frac{\langle m(\boldsymbol{B})|\boldsymbol{s}|n(\boldsymbol{B})\rangle \times \langle n(\boldsymbol{B})|\boldsymbol{s}|m(\boldsymbol{B})\rangle}{(m-n)^2} \, , \tag{287}$$

where m and n are eigenvalues of $\boldsymbol{s} \cdot \boldsymbol{B}/B$. The vector operator $\boldsymbol{s}$ only has off-diagonal matrix elements when $n = m \pm 1$, so $(m-n)^2 = 1$; furthermore, the $n = m$ term can be added to (287) because $\langle m|\boldsymbol{s}|m\rangle \times \langle m|\boldsymbol{s}|m\rangle = 0$. In short,

$$\boldsymbol{V}_m(\boldsymbol{B}) = \frac{1}{B^2} \, \mathrm{Im} \, \langle m(\boldsymbol{B})|\boldsymbol{s} \times \boldsymbol{s}|m(\boldsymbol{B})\rangle = \frac{1}{B^2} \langle m(\boldsymbol{B})|\boldsymbol{s}|m(\boldsymbol{B})\rangle, \tag{288}$$

where the commutation rule $\boldsymbol{s} \times \boldsymbol{s} = i\boldsymbol{s}$ was used. Finally, therefore,

$$\boldsymbol{V}_m(\boldsymbol{B}) = \frac{m}{B^2} \, \boldsymbol{u}_B \, , \tag{289}$$

where $\boldsymbol{u}_B = \boldsymbol{B}/B$.

Returning to (282),

$$\gamma_m(C) = -m \int_{S(C)} \frac{\boldsymbol{u}_B \cdot d\boldsymbol{S}}{B^2} = -m \, \Omega(C) \, , \tag{290}$$

where $\Omega(C)$ is the solid angle subtended by the closed curve C as seen from the point $\boldsymbol{B} = 0$. This demonstrates that the Berry phase is a purely geometrical quantity; in this case it does not depend on the magnitude of the spin $\boldsymbol{s}$ nor of the magnetic field $\boldsymbol{B}$, only on the projection eigenvalue m of the state. By an appropriate choice of C, any desired phase change $e^{i\gamma_m(C)}$ can be produced, and made observable by superposing a beam which has passed through an adiabatically varying magnetic field with another that has not. This can produce the interference pattern $I_m = \frac{1}{2}(1 + \cos\gamma_m)$; in particular, if C is the circuit around a cone of opening angle ϑ,

$$I_m(\vartheta) = \cos^2[m\pi(1 - \cos\tfrac{1}{2}\vartheta)] \, . \tag{291}$$

This effect has been observed (cf. endnotes).

(b) Correction to the Adiabatic Approximation

The preceding discussion assumed the validity of the adiabatic approximations, and we now turn to estimating the error the approximation entails. What follows holds for any slowly varying Hamiltonian $H(t)$, not just the one-particle spin problem.

---

[1] This shows again that $\gamma(C)$ does not depend on the phases of the basis.

We generalize the Ansatz (272) to

$$|\psi(t)\rangle = \sum_n a_n(t)\, e^{-i\int_0^t dt'\, E_n(t')}|n(t)\rangle \,, \tag{292}$$

where the instantaneous kets are the orthonormal solutions of

$$H(t)|n(t)\rangle = E_n(t)|n(t)\rangle \,. \tag{293}$$

Substituting (292) into the time-dependent Schrödinger equation gives

$$\sum_n e^{-i\int_0^t dt'\, E_n(t')} \left\{ \dot{a}_n(t)|n(t)\rangle + a_n(t)|\dot{n}(t)\rangle \right\} = 0 \,, \tag{294}$$

and the scalar product of (294) with $|k(t)\rangle$ is

$$\dot{a}_k(t) = -\sum_n a_n(t)\, \langle k(t)|\dot{n}(t)\rangle\, e^{i\int_0^t dt'\, \omega_{kn}(t')} \,, \qquad \omega_{kn}(t) = E_k(t) - E_n(t) \,. \tag{295}$$

From the time derivative of (293) we have

$$\langle k|\partial H/\partial t|n\rangle + (E_k - E_n)\langle k|\dot{n}\rangle = \delta_{kn}\dot{E} \,, \tag{296}$$

and therefore

$$\dot{a}_k(t) + a_k(t)\langle k(t)|\dot{k}(t)\rangle = \sum_{n\neq k} a_n(t) \frac{\langle k(t)|\partial H/\partial t|n(t)\rangle}{\omega_{kn}(t)}\, e^{i\Phi_{kn}(t)} \,, \tag{297}$$

where

$$\Phi_{kn}(t) = \int_0^t dt'\, \omega_{kn}(t') \,. \tag{298}$$

Observe now that

$$\begin{aligned} \dot{a}_k + a_k\langle k|\dot{k}\rangle &= e^{-\int_0^t dt'\langle k(t')|\dot{k}(t')\rangle} \frac{\partial}{\partial t}\left(a_k(t)\, e^{\int_0^t dt'\langle k(t')|\dot{k}(t')\rangle} \right) \\ &= e^{i\gamma_k(t)} \frac{\partial}{\partial t}\left(a_k(t) e^{-i\gamma_k(t)} \right) , \end{aligned} \tag{299}$$

where $\gamma_k(t)$ is the Berry phase (recall Eq. 276). Therefore

$$\frac{\partial}{\partial t}\left(a_k(t)\, e^{-i\gamma_k(t)} \right) = \sum_{n\neq k} a_n(t)\, \frac{\langle k(t)|\partial H/\partial t|n(t)\rangle}{\omega_{kn}(t)}\, e^{-i\gamma_k(t)}\, e^{i\Phi_{kn}(t)} \,. \tag{300}$$

No approximation has yet been made. We now assume that the state whose evolution is being followed is $|m(0)\rangle$ at the outset, that it is in the discrete spectrum of $H(t)$ and is nondegenerate. If the adiabatic approximation is a valid first approximation, the sum in (300) is dominated by the term $n = m$, for which the amplitude is known from (276). Consequently,

$$\frac{\partial}{\partial t}\left(a_k(t)\, e^{-i\gamma_k(t)} \right) \simeq a_m(0)\, e^{i\gamma_m(t)}\, \frac{\langle k(t)|\partial H/\partial t|m(t)\rangle}{\omega_{km}(t)}\, e^{-i\gamma_k(t)}\, e^{i\Phi_{km}(t)} \,. \tag{301}$$

The assumption that we are dealing with a slowly varying $H(t)$ is expressed by setting

$$t = sT\,, \qquad T \to \infty\,, \qquad 0 \le s \le 1\,. \tag{302}$$

We assume that as $T \to \infty$, the energy difference $\omega_{km}(t)$ remains nonzero and is also bounded. Let $H(s)$ be the Hamiltonian at the indicated value of s, and $\omega_{km}(s)$ be the energy difference at that point. Then (301) becomes

$$\frac{\partial}{\partial s}\left(a_k(s)\, e^{-i\gamma_k(s)}\right) \simeq a_m(0)\, \frac{\langle k(s)|\partial H/\partial s|m(s)\rangle}{\omega_{km}(s)}\, e^{-i[\gamma_m(s)-\gamma_k(s)]}\, e^{iT\int_0^s ds'\,\omega_{km}(s')}\,. \tag{303}$$

Thus

$$a_k(s)\, e^{-i\gamma_k(s)} = \int_0^s ds'\, F(s')\, e^{iT\int_0^{s'} ds''\,\omega_{km}(s'')} \tag{304}$$

$$F(s) \equiv a_m(0)\, \frac{\langle k(s)|\partial H(s)/\partial s|m(s)\rangle}{\omega_{km}(s)}\, e^{-i[\gamma_m(t)-\gamma_k(s)]}\,. \tag{305}$$

Because of the rapidly oscillating exponential in (304), the right-hand side tends to zero like $1/T$ as $T \to \infty$. This is demonstrated by integrating by parts:

$$F(s')e^{iT\Phi_{km}(s')} = \frac{1}{iT}\left\{\frac{d}{ds'}\left(\frac{F(s')}{\omega_{km}(s')}\, e^{iT\Phi_{km}(s')}\right) - e^{iT\Phi_{km}(s')}\,\frac{d}{ds'}\left(\frac{F(s')}{\omega_{km}(s')}\right)\right\}, \tag{306}$$

and therefore

$$a_k(s)e^{-i\gamma_k(s)} = \frac{i}{T}\left\{\frac{F(0)}{\omega_{km}(0)} - \frac{F(s)}{\omega_{km}(s)}\, e^{iT\Phi_{km}(s)} - \int_0^s ds'\, e^{iT\Phi_{km}(s')}\,\frac{d}{ds'}\,\frac{F(s')}{\omega_{km}(s')}\right\} \tag{307}$$

$$\to \frac{i}{T}\,\frac{F(0)}{\omega_{km}(0)} + O(1/T^2) \qquad (T \to \infty)\,. \tag{308}$$

To summarize, when a state undergoes an adiabatic transition, then for long times t it remains in the instantaneous eigenstate of the Hamiltonian $H(t)$ apart from admixtures with other states that are of order $1/t$.

7.8 Problems

1. Confirm Eq. 31 by defining a new basis $|\nu\rangle \equiv W|n\rangle$, where $WW^\dagger = 1$.

2. Show that I_t^2 does not depend on the representation by using Eq. 31.

3. The pure Galileo transformations (57) considered in the text are a subgroup of the transformations

$$\boldsymbol{r}' = R\boldsymbol{r} + \boldsymbol{v}t + \boldsymbol{a}\,, \quad t' = t + s\,,$$

where R is a spatial rotation. Designate this transformation's parameters by $(R, s, \boldsymbol{v}, \boldsymbol{a})$. Show that the composition law is

$$(R', s', \boldsymbol{v}', \boldsymbol{a}') \cdot (R, s, \boldsymbol{v}, \boldsymbol{a}) = (R'R, s + s', R\boldsymbol{v} + \boldsymbol{v}', R'\boldsymbol{a} + \boldsymbol{a}' + s\boldsymbol{v}')\,.$$

Under what restrictions is this an Abelian group?

4. Assume that the Hamiltonian of a system is of the form $H = T + V$, where $T(\boldsymbol{p}_1, \ldots)$ satisfies (73) and $V(\boldsymbol{x}_1, \ldots)$ is translation invariant. Supposing that $|\Psi(t)\rangle$ satisfies the Schrödinger equation, show that the boosted state $G(v)|\Psi(t)\rangle$ does also. In carrying this out, note that $G(\boldsymbol{v})$ is a function of time.

5. (a) By using Eq. 132 for an infinitesimal rotation of $\delta\varphi$ about the z-axis, and writing the transformation as $(1 - i\delta\varphi J_z)$, show that

$$J_z = \frac{1}{2}\left(u_1 \frac{\partial}{\partial u_1} - u_2 \frac{\partial}{\partial u_2}\right) .$$

Confirm the other expressions in Eq. 130 in the same way.

(b) In the text above Eq. 128, it was argued that in the expression $u_1^p u_2^q$, the powers (p, q) are related to the angular momentum eigenvalues by $j = \frac{1}{2}(p+q)$ and $m = \frac{1}{2}(p-q)$. Use the differential operators derived in (a) to verify this.

6. Use the differential operators of Eq. 130 to derive

$$\langle jm|(J_+)^{m-m'}|jm'\rangle = \sqrt{\frac{(j+m)!(j-m')!}{(j+m')!(j-m)!}} .$$

7. Show that the Euler angles in the range defined by Eq. 139 do not encompass the parameter manifold of $SU(2)$, and that either α or γ must be extended $(0, 4\pi)$.

8. Evaluate the Jacobian that is required to express $\int dR$ in terms of the Euler angles.

9. By taking the scalar product of Eq. 199 with a position eigenket in the $s = 0$ case, derive (200) by showing that $\langle \boldsymbol{r}|plm\rangle \propto j_l(kr)Y_{lm}(\hat{\boldsymbol{r}})$.

10. Verify that the helicity states are normalized in accordance with Eq. 203 if $N^2 = (2j+1)/4\pi$.

11. Show that if the initial state in the process $a \to b+c$ is a random mixture of states with all values of M, the angular distribution is isotropic *provided* the unspecified coefficient in Eq. 211 is independent of M. In contrast, show that if M is specified, and the helicities of b and c are *not* measured, the angular distribution is, in general, not isotropic, whether or not reflection invariance is valid.

12. Consider the decay $a \to b+c$ due to an interaction that is invariant under rotations and reflection, and an initial state that is described by an arbitrary density matrix $\langle M|\rho|M'\rangle$. Show that the angular distribution has the property

$$W(\theta, \varphi) = W(\pi - \theta, \pi + \varphi).$$

(*Hint:* This involves identities satisfied by the rotation matrices listed in the Appendix.)

13. To arrive at the Schrödinger equation for the rigid symmetric rotor, derive, in succession, the following relationships:

$$\partial D(R)/\partial\alpha = -iJ_z D(R) , \quad \partial D(R)/\partial\beta = -iJ_{y'} D(R) , \quad \partial D(R)/\partial\gamma = -iJ_\zeta D(R) ;$$

$$J_\zeta = J_z \cos\beta + \tfrac{1}{2}(e^{-i\alpha} J_+ + e^{i\alpha} J_-) \sin\beta ;$$

$$J_{y'} = \tfrac{1}{2} i(e^{i\alpha} J_- - e^{-i\alpha} J_+) , \quad [J_\zeta, J_z] = -iJ_{y'} \sin\beta ,$$

where the notation is defined in (138). Finally, show that

$$\boldsymbol{J}^2 = \operatorname{cosec}^2\beta\, (J_z^2 + J_\zeta^2 - 2\cos\beta\, J_\zeta J_z) + J_{y'}^2 - iJ_{y'} \cot\beta ,$$

and use this to derive the differential equation (225).

14. The interaction between an applied electric field and the electric dipole moment d of a symmetric rigid body is $H_1 = -d\boldsymbol{E}\cdot\boldsymbol{c}$, where $\boldsymbol{c}$ is the symmetry axis. Show that the matrix elements of H_1 are

$$\langle JMK|H_1|J'M'K'\rangle = -dE\sqrt{\frac{2J+1}{2J'+1}}\langle JM10|J'M'\rangle\langle JK10|J'K'\rangle .$$

This has the unsurprising selection rules $M = M'$, $|J-J'| = 0, 1$, but also requires $K = K'$; what is the reason for the latter? Use this result to verify Eq. 227.

15. (a) This problem concerns the asymmetric rigid body. Write the Hamiltonian in the form

$$H = \tfrac{1}{2}(W_a + W_b)(J_a^2 + J_b^2) + W_c J_c^2 + H' .$$

By using the commutation rule (218), show that the only off-diagonal matrix elements of H' in the representation defined by (J^2, J_c) are

$$\langle JK \pm 2|H'|JK\rangle = \tfrac{1}{4}(W_a - W_b) \times \sqrt{[J(J+1)-(K\pm 1)(K\pm 2)][J(J+1)-K(K\pm 1)]} .$$

Show that the matrix $\langle K|H|K'\rangle$ separates into pieces for even and odd K, and that these can be reduced further by using a basis in which states are either even or odd under $K \to -K$.

(b) Consider now the states with $J = 1$. Show that the energy eigenvalues are

$$E_a = W_b + W_c\,, \quad E_b = W_c + W_a\,, \quad E_c = W_a + W_b\,,$$

and that the eigenstates, in terms of the basis $|K\rangle$, are

$$|E_a\rangle = 2^{-\frac{1}{2}}[|1\rangle - |-1\rangle]\,, \tag{309}$$

$$|E_b\rangle = 2^{-\frac{1}{2}}[|1\rangle + |-1\rangle]\,, \tag{310}$$

$$|E_c\rangle = |0\rangle\,. \tag{311}$$

Note that these are all states of definite parity, so that there is no linear Stark effects if the body is asymmetric.

(c) Extend this analysis to $J = 2$ states. Show that the energies are

$$E(2^-) = 4W_c + W_a + W_b\,, \qquad E(1^+) = 4W_a + W + b + W_c\,,$$

$$E(1^-) = 4W_b + W_a + W_c\,,$$

$$E_\pm = 2(W_a + W_b + W_c) \pm 2\sqrt{(W_a + W_b + W_c)^2 - 3(W_aW_b + W_bW_c + W_aW_c)}\,,$$

where the notation $E(K^\pm)$ indicates the value of $|K|$ and whether the state is even or odd under $K \to -K$, while $E_\pm$ are the energies of states that are superpositions of $K = 0$ and 2^+.

16. Prove that $T_\kappa^{(k)}|0\rangle$, where $|0\rangle$ is a state of zero angular momentum, transforms under rotations like the angular momentum eigenstate $|k\kappa\rangle$.

17. Prove the tensor operator composition laws (242) and (243).

18. Show that Eq. 263 is correct, or at least, that this reduced matrix element is proportional to the indicated 6-j symbol.

19. This problem concerns rotations in Euclidean 3-space, $\mathfrak{E}_3$. Write the coordinate vector $\boldsymbol{r}$ as a column vector with components (x, y, z), and define the matrix

$$I_x = \begin{pmatrix} 0 & 0 & 0 \\ 0 & 0 & -1 \\ 0 & 1 & 0 \end{pmatrix} .$$

(a) Show that $\exp(\phi_x I_x)$ rotates $\boldsymbol{r}$ through the angle ϕ_x about the x-axis. Define (I_y, I_z) in the same way for the y and z axes, and show that they satisfy the commutation rules

$$[I_x, I_y] = I_z ,$$

and cyclic permutations.

(b) Define

$$\boldsymbol{r}_1 = \exp(\phi_y I_y) \exp(\phi_x I_x) \boldsymbol{r} , \quad \boldsymbol{r}_2 = \exp(\phi_x I_x) \exp(\phi_y I_y) \boldsymbol{r} .$$

Show that

$$\lim_{\phi_x, \phi_y \to 0} (\boldsymbol{r}_2 - \boldsymbol{r}_1) = \phi_x \phi_y I_z \boldsymbol{r} + O(\phi^3) .$$

(c) Given that there must be a one-one correspondence between rotations in $\mathfrak{E}_3$ and unitary transformations in the Hilbert space, show that this requires

$$[e^{-i\phi_x J_x} - 1, e^{-i\phi_y J_y} - 1] = e^{-i\phi_x \phi_y J_z} - 1$$

in the limit $\phi_x \to 0$, $\phi_y \to 0$, and that this imposes the angular momentum commutation rules on the components of $\boldsymbol{J}$.

Endnotes

G.C. Wick's proof of Wigner's theorem, and the extension to spaces of arbitrary dimension, is in *Preludes in Theoretical Physics,* A. de-Shalit, H. Feshbach and L. Van Hove (eds.), North-Holland (1966), pp. 231–239. For the original proof, see E.P. Wigner, *Group Theory,* Academic Press (1959), §26.

For further discussion of Galileo transformations see M. Hammermesh, *Group Theory,* Addison-Wesley (1962), Chapter 12; T.F. Jordan, *Linear Operators for Quantum Mechanics,* Wiley (1969); and J.-M. Lévy-Leblond, *Rivista Nuovo Cimento* **4**, 99 (1974).

Many of the original papers on angular momentum theory are reprinted in *Quantum Theory of Angular Momentum,* L.C. Biedenharn and H. Van Dam (eds.), Academic Press (1965). See especially the otherwise unpublished and important papers by J. Schwinger (pp. 229–280) and E.P. Wigner (pp. 89–133). For an excellent introduction to group theory in general, and detailed information about the rotation group in particular, see A. Messiah, *Quantum Mechanics,*Vol.II, Wiley (1961), Appendices C and D.

For an authoritative discussion of rotational motion in various contexts, see A. Bohr and B.R. Mottelson, *Nuclear Structure,* Vol. II, W.A. Benjamin (1975), Chapter 7.

For the original papers on geometric phases, and subsequent developments, see A. Shapere and F. Wilczek (eds.), *Geometric Phases in Physics,* World Scientific (1989). Our treatment closely follows M.V. Berry's original paper *Proc. Roy. Soc. A* **392**, 45 (1984). For an extensive treatment of the adiabatic approximation, see Messiah, pp. 744–761, *ibid.*

8

Elastic Scattering

Ever since Rutherford inferred the existence of the the atomic nucleus from data on the scattering of α particles by metallic foils, collision experiments have played a central role in physics. By one of the fortunate accidents associated with motion in the Coulomb field of a point charge, Rutherford's classical calculation gives the same answer for the cross section as does quantum mechanics, which did not yet exist in 1911. But this is an isolated accident, and quantum mechanics is indispensable.

The quantum theory of collisions is an enormous and complex edifice. Even a bird's eye sketch would take many pages, and serve no real purpose at this point. Chapter 4 provided a kindergarten introduction to collision theory because it was restricted to one-dimensional problems. This chapter addresses a more realistic situation, elastic scattering of nonrelativistic particles due to rotationally invariant interactions. Later chapters take up inelastic processes, and relativistic collisions to a limited degree. A full-fledged theory of relativistic collisions is inherently a topic in quantum field theory.

8.1 Consequences of Probability and Angular Momentum Conservation

A surprising amount can be learned about scattering by postponing the problem of actually computing the scattering amplitude, and just exploring the consequences of probability and angular momentum conservation. Of course, these laws are only applicable if the system has the properties from which these laws stem: a Hamiltonian H that is Hermitian and invariant under rotations. Beyond this, only one detailed property of H is needed — that the interaction falls off faster than a Coulomb field as the separation between the colliding systems tends to infinity. There is also the issue of whether the projectile, target or both have spin. If so, the consequences of rotational invariance are more complicated than if both have no spin, but this is really just a technical problem, and to avoid indigestion this will also be postponed.

(a) Partial Waves

As we know from §3.4, when H is invariant under translations the two-body problem reduces to an interacting one-body problem. It is easier, however, to visualize the situation where the the target is infinitely massive in comparison to the projectile, and fixed at the origin of the coordinate system. We will therefore talk as if this were the case, though the analysis holds with straightforward changes when the masses are arbitrary and the scattering is elastic. The relations between the kinematics describing the true two-body problem and the one-body problem are given in §8.2(d).

Consider first a state of the projectile in which it is free and has momentum k along the z-direction.[1] The corresponding wave function, expressed as a coherent superposition of angular momentum eigenstates, was derived in §3.2:

$$e^{ikz} = \sum_{l=0}^{\infty}(2l+1)i^l j_l(kr)P_l(\cos\theta) . \tag{1}$$

The forms of the radial function as $r \to \infty$ and $r \to 0$ will play an essential role:

$$j_l(kr) \sim \frac{1}{2ikr}\left(e^{i(kr-\frac{1}{2}l\pi)} - e^{-i(kr-\frac{1}{2}l\pi)}\right) \qquad (r \to \infty) , \tag{2}$$

$$\longrightarrow \frac{(kr)^l}{(2l+1)!!} \qquad (r \to 0) . \tag{3}$$

The asymptotic form of the free particle wave function is so important in collision theory because, by hypothesis, the interaction vanishes as $r \to \infty$, which means that the free particle Schrödinger equation must provide the *complete* solution asymptotically. Of course, there is the question of how rapidly the interaction must fade away for it to be ignorable asymptotically; the answer, as we will see in §8.4, is faster than a Coulomb field, i.e., more rapidly than $1/r$.

The momentum eigenstate (1) is a superposition of incoming and outgoing spherical waves, which are linearly independent solutions of the free particle Schrödinger equation as $r \to \infty$. Causality requires the incoming wave to be unaltered by the interaction, so that part of the plane wave must be unaltered in the asymptotic wave function when the interaction is at work.

What about the outgoing wave? Its two essential properties follow from the conservation laws:

- If the interaction is rotationally invariant, no mixing of angular momentum states is possible; that is, if one imagines an incident state with just one angular momentum component, then only that same angular momentum state can be present in the outgoing state. Furthermore, as the incident state is invariant under rotations about the incident direction, the scattered state must be a superposition of angular momentum states having projection zero on that direction (because we ignore spin for now).
- If probability is to be conserved, the amplitude of the outgoing wave must be equal to or smaller than that of the incoming wave.

These conclusions determine the asymptotic form of the wave function. For all distances, angular momentum conservation requires it to have the same form as (1), but with the Bessel function replaced by some unknown function of r:

$$\psi_{\rm el}(\boldsymbol{r}) = \sum_{l=0}^{\infty}(2l+1)i^l R_l(k;r)P_l(\cos\theta) . \tag{4}$$

The separate terms in this sum are called *partial waves*. It is common to use spectroscopic notation for partial waves of low l: s-wave, p-wave, etc. Although $R_l(k;r)$

[1] In this chapter $\hbar = 1$, unless otherwise noted.

is unknown in the region where the interaction exists, its asymptotic form is, by the preceding argument, known apart from the amplitude η_l of its outgoing wave:

$$R_l(k;r) \sim \frac{1}{2ikr}\left(\eta_l(k)\, e^{i(kr-\frac{1}{2}l\pi)} - e^{-i(kr-\frac{1}{2}l\pi)}\right) \qquad (r \to \infty)\,. \tag{5}$$

Probability conservation requires

$$|\eta_l(k)| \leq 1\,. \tag{6}$$

It is important to note that we have not assumed that elastic scattering is the only possible process resulting from the interaction of the projectile with the target. Whether inelastic processes occur at the energy in question depends on the structure of the target and projectile, and the interaction between them, but this information is not needed to establish (5) and (6), and as we will now see, these two equations suffice to express the cross section for elastic scattering in terms of the parameters η_l. To actually calculate these parameters, it is necessary to solve the Schrödinger equation in the region where the interaction is at work, and that can run the gamut from requiring the solution of a simple ordinary differential equation to a very difficult or even intractable problem.

If, for whatever reason, only elastic scattering is possible, $|\eta_l(k)| = 1$, because the probability of emerging with the incident energy from the collision must equal that of having been incident. In this case, it is convenient to introduce the *partial wave phase shift*:

$$\eta_l(k) = e^{2i\delta_l(k)}\,. \tag{7}$$

The motivation for this terminology is that the asymptotic radial function is

$$R_l(k;r) \sim \frac{1}{kr}\, e^{i\delta_l} \sin(kr - \tfrac{1}{2}l\pi + \delta_l) \qquad (r \to \infty) \tag{8}$$

when the scattering process is purely elastic.

There is thus an elastic wave function $\psi_{\text{el}}(r)$ whether or not inelastic scattering is taking place. If there is only elastic scattering, then ψ_{el} is the only wave function needed to describe the collision process. Should there be inelastic scattering, the wave function is more complicated, but outside the interaction region the elastic process itself is fully described by ψ_{el}. A familiar example, to which we often turn, will elucidate this point. Consider a beam of visible monochromatic light impinging on a coin. This will result in a diffraction pattern formed by light of the incident frequency, i.e., elastically scattered light. In addition the coin will absorb light, heat up, and emit in the infrared. Nevertheless, the diffraction pattern is completely described by the elastically scattered wave. The inelastic processes determine the opacity of the coin, which in turn determines the diffraction pattern. In this analogy, the opacity is related to the parameters $|\eta_l|$, as we shall see in Eq. 56.

The elastically scattered wave, ψ_{sc}, is defined by[1]

$$\psi_{\text{el}}(\mathbf{r}) = e^{ikz} + \psi_{\text{sc}}(\mathbf{r})\,. \tag{9}$$

[1] In this section the equations are neater if the factor $(2\pi)^{-3/2}$ is not included in the normalization.

Its asymptotic form follows from comparing (10) with (2):

$$\psi_{sc}(\boldsymbol{r}) \sim \frac{1}{2ik}\frac{e^{ikr}}{r}\sum_{l=0}^{\infty}(2l+1)(\eta_l - 1)P_l(\cos\theta)\ . \tag{10}$$

Thus the asymptotic form of the whole elastic wave function is

$$\psi_{el}(\boldsymbol{r}) \sim e^{ikz} + \frac{e^{ikr}}{r} f_{el}(k;\theta)\ , \tag{11}$$

where f_{el}, *the elastic scattering amplitude*, is

$$f_{el}(k;\theta) = \frac{1}{2ik}\sum_{l=0}^{\infty}(2l+1)(\eta_l - 1)P_l(\cos\theta) \equiv \sum_{l=0}^{\infty}(2l+1)f_l(k)P_l(\cos\theta)\ ; \tag{12}$$

$f_l(k)$ is called the partial wave amplitude. It is important to remember that the scattering amplitude has the dimension of length.

To determine the cross section from (11), a bit of trickery is called for because this wave function is not normalizable. The flux of the incident state is v, the incident velocity, and does not have the dimension $L^{-2}T^{-1}$, as it should. To repair this, multiply the wave function by $V^{-\frac{1}{2}}$, where $V^{1/3}$ is a length vastly larger than any length in the problem, and renormalize the wave function to unity. The incident flux is then $F_{inc} = v/V$, which has the correct dimension. The flux of elastically scattered particles is outwards radially, and of magnitude

$$F_{sc} = \frac{v}{V}\frac{|f_{el}|^2}{r^2}\ . \tag{13}$$

Let $d\dot{N}_{sc}$ be the number of particles scattered per unit time into an element of area dA subtending the solid angle $d\Omega = \sin\theta d\theta d\phi$ at the distance r from the target:

$$d\dot{N}_{sc} = F_{sc}dA = \frac{v}{V}|f_{el}|^2 d\Omega\ . \tag{14}$$

The differential cross section for elastic scattering is defined as

$$\frac{d\sigma_{el}}{d\Omega} = \frac{d\dot{N}_{sc}/d\Omega}{F_{inc}}\ , \tag{15}$$

and therefore

$$\frac{d\sigma_{el}}{d\Omega} = |f_{el}(k;\theta)|^2\ . \tag{16}$$

This derivation of (16) suffers not only from the rather minor defect of a somewhat questionable normalization, but has also ignored the interference between the incident and scattered waves.

The various partial waves interfere in the differential cross section. But the Legendre polynomials are orthogonal functions,

$$\int d\Omega\, P_l(\cos\theta)P_{l'}(\cos\theta) = \frac{4\pi}{2l+1}\,\delta_{ll'}\ . \tag{17}$$

so there is no such interference in the total elastic cross section:

$$\sigma_{el} = \frac{\pi}{k^2}\sum_{l=0}^{\infty}(2l+1)|\eta_l - 1|^2\ . \tag{18}$$

The maximum contribution of any partial wave is $|\eta_l - 1| = 2$, and therefore the partial wave cross sections satisfy the following bound:

$$\sigma_{\rm el}(l) \leq 4\pi\lambda\!\!\!^{-2}(2l+1) , \tag{19}$$

where $\lambda\!\!\!^- = 1/k$ is the reduced wave length (i.e., $\lambda/2\pi$). This bound is set by probability conservation, and to reach the upper limit $\eta_l = -1$ there must be no inelastic scattering in that partial wave.

If there is only elastic scattering, the expressions in terms of phase shifts are often more useful than those in terms of η_l:

$$f_{\rm el}(k;\theta) = \lambda\!\!\!^- \sum_l (2l+1)\, e^{i\delta_l} \, \sin\delta_l \, P_l(\cos\theta) , \tag{20}$$

$$\sigma_{\rm el} = 4\pi\lambda\!\!\!^{-2} \sum_l (2l+1) \sin^2\delta_l . \tag{21}$$

Intuitively, one would expect a qualitative relation between the quantum mechanical concept of partial wave and the classical concept of impact parameter, which is the distance between the incident trajectory and the parallel line through the target's center. The orbital angular momentum of a classical particle at impact parameter b is $bp_{\rm inc}$, or $kb \simeq l$. If the range of the interaction is a, one would therefore expect the partial waves with $l \lesssim ka$ to be scattered substantially, in contrast to those with larger values of l. The behavior (3) of the free particle state as $r \to 0$ strengthens this expectation. Sterling's formula for the asymptotic form of the gamma function gives

$$(2l+1)!! \equiv \frac{\Gamma(2l+2)}{2^l\Gamma(l+1)} \sim 2^{l+\frac{3}{2}}\,(l/e)^{l+1}\left(1+\frac{1}{l}\right)^{l+1} \quad (l \gg 1) , \tag{22}$$

and therefore

$$j_l(kr) \longrightarrow \frac{\sqrt{2}}{4}\, e\, \frac{1}{l}\left(1+\frac{1}{l}\right)^{-l-1}\left(\frac{ekr}{2l}\right)^l \quad (r \to 0,\ l \gg 1) . \tag{23}$$

Hence if $ka \equiv l_0$, then for $l \gg l_0$ the incident wave hardly reaches the interaction, and we would anticipate that $\eta_l \to 1$ for such values of l.

The partial wave expansion is especially useful at low energies, therefore, because only a few partial waves contribute to the cross section. For example, if the energy and interaction range are such as to only produce substantial s-wave scattering,

$$\frac{d\sigma_{\rm el}}{d\Omega} = \frac{1}{4k^2}\Big|(\eta_0 - 1) + 3(\eta_1 - 1)\cos\theta\Big|^2 \simeq \frac{1}{4k^2}\left(|\eta_0 - 1|^2 + \epsilon\cos\theta\right) , \tag{24}$$

where $\epsilon = 6\,{\rm Re}\,[(\eta_0 - 1)(\eta_1^* - 1)] \ll 1$. Hence we have learned a good deal with but little work. Low-energy elastic scattering experiments thus amount to the determination of a few complex parameters η_l from the angular distribution, which reduces to finding the real phase shifts δ_l if only elastic scattering is possible. This problem is not as straightforward as might appear, however, because more than one set of parameters will usually fit a given angular distribution, but it would take us too far afield to more than mention this fact.

(b) Hard Sphere Scattering

A simple example will elucidate the general theory. Consider purely elastic scattering by a repulsive potential that vanishes for $r > a$, and does not allow the projectile to reach values of r smaller than a. Assume, furthermore, that $ka \ll 1$, so that scattering is only appreciable in the s-wave. For $r > a$, the radial wave function is a linear combination of free particle functions:

$$rR_0(r) = \alpha e^{ikr} + \beta e^{-ikr} \,. \tag{25}$$

When there is no scattering, this function is proportional to $\sin kr$, so we set $\beta = -1/2i$, and in accordance with (5) and (7), $\alpha = e^{2i\delta_0}/2i$:

$$rR_0(r) = e^{i\delta_0} \sin(kr + \delta_0) \,. \tag{26}$$

In short, the $l = 0$ asymptotic form (8) holds for all distances $r \geq a$ in this case. The interaction requires $R_0(a) = 0$, and therefore

$$\delta_0 = -ka \,. \tag{27}$$

Thus the phase shift is just the shift that pushes the particle out of harm's way.

For $l \neq 0$, the phase shifts involve zeroes of Bessel functions, but two limiting cases are simple:

$$\delta_l \simeq -ka + \tfrac{1}{2} l\pi \qquad (ka \gg l) \tag{28}$$

$$\simeq -\frac{(ka)^{2l+1}}{(2l+1)!!(2l-1)!!} \qquad (ka \ll l) \,. \tag{29}$$

This confirms the intuitive argument based on (23).

(c) Time-Dependent Description and the Optical Theorem

Scattering is an inherently time-dependent phenomenon, and the preceding treatment in terms of stationary states does not do justice to this fact. For that reason, one could ask whether the cross section (Eq. 16) has been inferred correctly from (11). Furthermore, was it legitimate to ignore the interference between the incident and scattered waves in (11) ? The answer is No! The interference term leads to the optical theorem, a very important relationship between the scattering amplitude and the total cross section.

We therefore turn to a more realistic time-dependent description of the collision process. Consider, first, a free-particle wave packet that is localized along the z-direction, but is of infinite extent in the perpendicular plane,

$$\varphi(z,t) = N \int_{-\infty}^{\infty} dk \; e^{i(kz - E_k t)} \, e^{-(k-k_0)^2/\lambda^2} \,, \tag{30}$$

where $E_k = k^2/2m$, λ is the momentum spread about the mean k_0, and N the normalization factor. It is very common in formulations of time-dependent scattering theory to use wave packets that are also localized in the transverse direction, but this is no more realistic and also leads to more complicated formulas that call for approximations we shall avoid.

Equation (30) is evaluated by using

$$\int_{-\infty}^{\infty} dx\, e^{-(ax^2+bx)} = \sqrt{\pi/a}\, e^{b^2/4a} \qquad (\mathrm{Re}\, a > 0)\,, \tag{31}$$

which gives

$$\varphi(z,t) = e^{i(k_0 z - E_0 t)} \frac{N'}{\sqrt{1+i\epsilon t}} \exp\left\{ -\frac{\lambda^2}{4} \frac{(z-vt)^2}{1+i\epsilon t} \right\} \tag{32}$$

$$\equiv e^{i(k_0 z - E_0 t)} \Phi(z,t)\,, \tag{33}$$

where $v = k_0/m$ and $E_0 = k_0^2/2m$ are the velocity and energy at the mean momentum, and $\epsilon = \lambda^2/2m$.

The derivations that follow use the Gaussian weight of (30) in the belief that readers will find a completely explicit wave function most convincing. It is possible to derive the final results for any wave packet with a narrow momentum distribution, but that requires more elaborate but less visualizable mathematics.

The function $|\Phi(z,t)|^2$ vanishes rapidly for values of z outside the interval

$$|z - vt| \lesssim \frac{1}{\lambda}\sqrt{1+\epsilon^2 t^2}\,. \tag{34}$$

At $t = 0$ the packet has a spread of order $1/\lambda$, which grows linearly when $|t| \gg \epsilon^{-1}$, and it does so both at times before and after $t = 0$. This unrealistic behavior could be removed by a slightly different definition from (30), so that the spread grows throughout the collision process, but that would not change anything significant and only complicate the formulas. All that counts is that throughout the collision the spread be negligible compared to distance from the origin to the initial and final locations of the particles. For such large values of $|t|$, (34) is

$$|z - vt| \lesssim \tfrac{1}{2}(\lambda/k_0)v|t| \sim (\lambda/k_0)|z|\,, \tag{35}$$

and therefore the requirement just stated will be satisfied if $\lambda \lll k_0$.

The free wave packet (30) can be decomposed into angular momentum eigenstates by expanding the plane wave in the integrand into partial waves:

$$\varphi(z,t) \sim \sum_l (2l+1)\, i^l\, P_l(\cos\theta)[\varphi_l^+(r,t) + \varphi_l^-(r,t)]\,, \tag{36}$$

$$\varphi_l^{\pm}(r,t) = \pm\frac{N}{2ir}\int_{-\infty}^{\infty} \frac{dk}{k}\, e^{\pm i(kr - \frac{1}{2}l\pi)}\, e^{-iE_k t}\, e^{-(k-k_0)^2/\lambda^2}\,, \tag{37}$$

where $\sim$ always indicates the asymptotic form as $r \to \infty$. While the phase factors in (37) cannot be tampered with, $1/k$ varies slowly over the integration range and can be approximated by $1/k_0$, so that the integral is proportional to (30) with $z \to \pm r$:

$$\varphi_l^{\pm}(r,t) = \pm\frac{(\mp i)^l}{2ik_0 r}\, e^{\pm ik_0 r}\, e^{-iE_0 t}\, \Phi(\pm r,t)\,. \tag{38}$$

According to (32),

$$|\Phi(\pm r,t)| \propto \exp\left\{ -\frac{\lambda^2}{4} \frac{(\pm r - vt)^2}{1+\epsilon^2 t^2} \right\}\,. \tag{39}$$

Here the crucial point is that r is positive by definition, and therefore *only* φ_l^- is present when t is large and negative, while *only* φ_l^+ is present when t is large and positive. That is, *the incoming and outgoing spherical waves separately superpose to form the incident plane wave packet well before and well after the projectile reaches the target.*

This justifies fully the argument that culminated in (5), because the effect of the interaction is to multiply each outgoing Fourier component in $\varphi_l^+(r,t)$ by $\eta_l(k)$. Therefore, the asymptotic form of the wave function that reduces initially to the incident packet $\varphi(z,t)$ is

$$\Psi(\boldsymbol{r},t) \sim \sum_l (2l+1)i^l P_l(\cos\theta)[\eta_l(k_0)\,\varphi_l^+(r,t) + \varphi_l^-(r,t)] , \tag{40}$$

provided $\eta_l(k)$ varies negligibly over the interval $|k-k_0| \lesssim \lambda$, so that it can be taken outside the integral in (37). In principle, it is always possible to construct a packet that is narrow enough to obey this stricture, but in practice that could be impossible, especially if there is an exceedingly sharp resonance. In that case, a separate discussion along the lines of §4.4(c) is needed, and as shown there, if the incident packet is broad in energy compared to the width of the resonance the final state will obey the exponential decay law.

For large positive time, the functions φ_l^- in (40) vanish, so at such time this wave function is

$$\Psi(\boldsymbol{r},t) = \Psi_{\text{inc}}(z,t) + \Psi_{\text{sc}}(\boldsymbol{r},t) , \tag{41}$$

where, because of (32) and (38),

$$\Psi_{\text{inc}}(z,t) = e^{ik_0 z}\,\Phi(z,t)\,e^{-iE_0 t} , \tag{42}$$

$$\Psi_{\text{sc}}(\boldsymbol{r},t) \sim \frac{e^{ik_0 r}}{r}\,f_{\text{el}}(k_0,\theta)\,\Phi(r,t)\,e^{-iE_0 t} . \tag{43}$$

The evaluation of the flux, $\boldsymbol{i} = (1/m)\,\text{Im}\,(\Psi^*\boldsymbol{\nabla}\Psi)$, is now unambiguous. That due to the incident packet involves

$$\frac{1}{m}\boldsymbol{\nabla}\Psi_{\text{inc}}(z,t) = \left(iv\Psi_{\text{inc}} + e^{i(k_0 z - E_0 t)}\frac{1}{m}\frac{\partial}{\partial z}\Phi(z,t) \right)\hat{\boldsymbol{z}} . \tag{44}$$

From (30) it follows that $\partial\Phi/\partial z$ is of order $\lambda\Phi$, and as $\lambda \ll k_0$, the second term is negligible, so that

$$\boldsymbol{i}_{\text{inc}} = v|\Phi(z,t)|^2\hat{\boldsymbol{z}} . \tag{45}$$

The angular derivatives in the gradient, when applied to Ψ_{sc}, produce $1/r^2$, which is negligible as $r \to \infty$ compared to the radial derivative whose leading contribution to $\boldsymbol{\nabla}\Psi_{\text{sc}}$ varies like $1/r$. Thus

$$\frac{1}{m}\boldsymbol{\nabla}\Psi_{\text{sc}} \sim iv\Psi_{\text{sc}}\,\hat{\boldsymbol{r}} , \tag{46}$$

and therefore the flux carried by this part of the wave function is

$$\boldsymbol{i}_{\text{sc}} \sim \frac{v}{r^2}\,|f_{\text{el}}(k_0,\theta)|^2\,|\Phi(r,t)|^2\,\hat{\boldsymbol{r}} . \tag{47}$$

The total number of particles incident per unit area is

$$\frac{dN_{\rm inc}}{dA} = \int_{-\infty}^{\infty} i_{\rm inc}(z,t)\,dt = v\int_{-\infty}^{\infty} |\Phi(z,t)|^2\,dt\,. \tag{48}$$

The total number of particles scattered into the area that subtends the solid angle $d\Omega$ at the distance r is[1]

$$dN_{\rm sc} = r^2 d\Omega \int_{-\infty}^{\infty} (\boldsymbol{i}_{\rm sc}\cdot\hat{\boldsymbol{r}})\,dt = v\,d\Omega\,|f_{\rm el}(k_0,\theta)|^2 \int_{-\infty}^{\infty} |\Phi(r,t)|^2\,dt\,. \tag{49}$$

The differential cross section is defined as

$$\frac{d\sigma_{\rm el}}{d\Omega} = \frac{dN_{\rm sc}/d\Omega}{dN_{\rm inc}/dA}\,. \tag{50}$$

The integrals in (48) and (49) are equal, and (50) therefore agrees with our earlier result, Eq. 16.

We now take up the previously ignored interference between $\Psi_{\rm inc}$ and $\Psi_{\rm sc}$. Its contribution to the probability flux is

$$\boldsymbol{i}_{\rm int} = \frac{1}{m}\,{\rm Im}\left(\Psi_{\rm sc}^*\boldsymbol{\nabla}\Psi_{\rm inc} + \Psi_{\rm inc}^*\boldsymbol{\nabla}\Psi_{\rm sc}\right). \tag{51}$$

The evaluation of this flux is rather intricate, so for clarity's sake we begin with a sketch of the central feature in the derivation, and a statement of the optical theorem to which it leads. The stationary state wave function (Eq. 11) will do for this purpose, and it gives

$$\boldsymbol{i}_{\rm int} \sim \frac{v}{r}\,{\rm Re}\left[\hat{\boldsymbol{z}}\,e^{ik(z-r)}f_{\rm el}^*(\theta) + \hat{\boldsymbol{r}}\,e^{-ik(z-r)}f_{\rm el}(\theta)\right]. \tag{52}$$

As r is tends to infinity, this expression oscillates with ever higher frequency unless $|z-r| = 2r\sin^2\frac{1}{2}\theta \to 0$, or to be more precise, unless $\theta^2 \lesssim (kr)^{-1}$. Therefore, when $\boldsymbol{i}_{\rm int}$ is integrated over a surface at large r, only the immediate vicinity of $\theta = 0$ can yield a finite result. We may therefore set $\theta = 0$ everywhere except in the phase factors,[2]

$$\boldsymbol{i}_{\rm int} \sim \hat{\boldsymbol{z}}\,\frac{v}{r}\,{\rm Re}\left[f_{\rm el}^*(0)\,e^{-\frac{1}{2}ikr\theta^2} + {\rm c.c.}\right]. \tag{53}$$

Thus the flux due to the interference term only has a component parallel to the incident direction as $r \to \infty$. After it is integrated over an infinitesimal area normal to the incident direction on the downstream side of the target, it subtracts from the incident flux, so that the total downstream flux is *smaller* than what is incident upstream. In short, $\boldsymbol{i}_{\rm int}$ accounts for the loss of particles from the incident beam due to collisions. As a consequence, the interference term leads to a relation between the *elastic* amplitude in the forward direction and the *total* cross section, elastic plus inelastic, namely,

$$\sigma_{\rm tot} = \frac{4\pi}{k}\,{\rm Im}\,f_{\rm el}(0)\,. \tag{54}$$

[1] Eq. 14 concerns the number of particles per unit time, $\dot{N}_{\rm sc}$, whereas here $N_{\rm sc}$ is the total number of particles.

[2] This equation, and many to follow, only make sense if the elastic amplitude is finite in the forward direction. This holds for interactions that fall off more rapidly than the Coulomb field.

This is the optical theorem, which is due to Bohr, Peierls and Placzek.

Two facts lie behind the theorem: (i) only the states scattered elastically in the forward direction have the same energy and momentum as the incident state, which is required for interference between them to survive in the limit $r \to \infty$; (ii) the resulting depletion of the forward flux must account for *all* the processes resulting from the collision, whether they be elastic or inelastic. The term "optical" theorem is motivated by the fact that in optics this interference between the incident and forward scattered waves is responsible for the shadow cast by an opaque object.

The optical theorem and the partial wave expansion lead to important expressions for the total and inelastic cross sections in terms of the amplitudes η_l. Combining (54) and (12) gives

$$\sigma_{\text{tot}} = \frac{2\pi}{k^2} \sum_l (2l+1)(1 - \text{Re}\,\eta_l) . \tag{55}$$

The inelastic cross section, $\sigma_{\text{inel}} = \sigma_{\text{tot}} - \sigma_{\text{el}}$, then follows from Eq. 18,

$$\sigma_{\text{inel}} = \frac{\pi}{k^2} \sum_l (2l+1)(1 - |\eta_l|^2) . \tag{56}$$

Here it is again evident that *inelastic scattering is always accompanied by elastic scattering,* for $|\eta_l - 1| \neq 0$ if $|\eta_l| < 1$, but the converse is not true because $|\eta_l| = 1$ does not require $\eta_l = 1$.

It should be recognized that whereas the parameters η_l determine the angular distribution of elastic scattering and the amount of inelastic scattering in each angular momentum state, they provide no information about the angular distribution of inelastic scattering. This is hardly surprising, because degrees of freedom beyond those of the projectile come into play when inelastic processes take place. Even if the process is relatively simple, as in inelastic electron scattering by an atom with zero angular momentum, the angular momentum of the electron in the final state, and therefore its angular distribution, depends on the angular momentum of the state excited in the atom.

We now return to the wave packet description and a proper derivation of the optical theorem. Substituting (44) and (46) into (51) gives

$$\boldsymbol{i}_{\text{int}} \sim \frac{v}{r} \,\text{Re}\left[\hat{\boldsymbol{z}}\, e^{ik_0(z-r)} f_{\text{el}}^*(\theta) \Phi^*(r,t)\Phi(z,t) + \hat{\boldsymbol{r}}\, e^{-ik_0(z-r)} f_{\text{el}}(\theta)\Phi(r,t)\Phi^*(z,t)\right] . \tag{57}$$

From (32)

$$\Phi^*(r,t)\Phi(z,t) = \xi^{-\frac{1}{2}}\, e^{-K(r,z,t)}\, e^{i(z-r)\chi(r,z,t)} , \tag{58}$$

where $\xi = 1 + \epsilon^2 t^2$ and

$$K = \frac{\lambda^2}{4\xi}[(z - vt)^2 + (r - vt)^2] , \tag{59}$$

$$\chi = \frac{\lambda^2 \epsilon t}{4\xi}[(r - vt) + (z - vt)] . \tag{60}$$

Hence,

$$\boldsymbol{i}_{\text{int}} \sim \frac{v}{r} \xi^{-\frac{1}{2}} e^{-K}\, \text{Re}\left[\hat{\boldsymbol{z}}\, e^{-2i(\chi+k_0)r\sin^2\frac{1}{2}\theta} f_{\text{el}}^*(\theta) + \hat{\boldsymbol{r}}\, e^{2i(\chi+k_0)r\sin^2\frac{1}{2}\theta} f_{\text{el}}(\theta)\right] . \tag{61}$$

As in (52), and for the same reason, only the immediate neighborhood of $\theta = 0$ survives as $r \to \infty$, and θ can be set to zero wherever deviation from $\theta = 0$ cannot produce rapid oscillations, i.e.,

$$\boldsymbol{i}_{\text{int}} \sim \hat{\boldsymbol{z}} \frac{v}{r} |\Phi(z,t)|^2 \operatorname{Re} \left\{ f_{\text{el}}^*(0) \exp\left(-\tfrac{1}{2} i[\bar{\chi}(z,t) + k_0] r\theta^2\right) + \text{c.c.} \right\} , \quad (62)$$

where $\bar{\chi}(z,t)$ is the function defined in (60) when $r = z$.

The number of particles passing through the cap C of a cone of opening angle 2Δ about $\theta = 0$ due to $\boldsymbol{i}_{\text{int}}$ is proportional to

$$2\pi r^2 \int_0^{\Delta} e^{-\frac{1}{2} i(\bar{\chi}+k_0) r\theta^2}\, \theta d\theta = \frac{2\pi r}{i(\bar{\chi}+k_0)} \left(1 - e^{-\frac{1}{2} i k_0 r \Delta^2} e^{-\frac{1}{2} i r \Delta^2 \bar{\chi}(z,t)}\right) . \quad (63)$$

As in (48) and (49), this must still be integrated over t with the weight $|\Phi(z,t)|^2$. For large values of t, $\bar{\chi} \simeq (mv/z)(z - vt)$, which assumes both positive and negative values in the integration interval., and as $r \to \infty$ the factor $\exp(-\frac{1}{2} i r \bar{\chi} \Delta^2)$ will therefore not survive the integration over t. Furthermore, $|\bar{\chi}| \sim \lambda$ according to (35), and is negligible in the denominator of (63). Hence the total number of particles passing through the cap C due to the interference term is

$$N_{\text{int}} = \int_C dA \int_{-\infty}^{\infty} dt\; \hat{\boldsymbol{z}} \cdot \boldsymbol{i}_{\text{int}}(\boldsymbol{r},t) = -\frac{4\pi v}{k_0} \operatorname{Im} f_{\text{el}}(0) \int_{-\infty}^{\infty} |\Phi(z,t)|^2\, dt . \quad (64)$$

From the partial wave expansion (12)

$$\operatorname{Im} f_{\text{el}}(0) = \frac{1}{2k} \sum_l (2l+1)(1 - \operatorname{Re} \eta_l) ; \quad (65)$$

but probability conservation requires $|\eta_l| \leq 1$, and therefore $\operatorname{Im} f_{\text{el}}(0) > 0$. Hence N_{int} is a negative number, or better put, *the flux $\boldsymbol{i}_{\text{int}}$ flows in the direction opposite to that of the incident beam.*

The total cross section σ_{tot} can be found by measuring the attenuation of the incident beam in traversing the target. If N_{lost} is the number of particles lost due to elastic and inelastic scattering,

$$N_{\text{lost}} = \sigma_{\text{tot}} \frac{dN_{\text{inc}}}{dA} . \quad (66)$$

Identifying N_{lost} with $|N_{\text{int}}|$ then establishes the optical theorem (Eq. 54).

8.2 General Properties of Elastic Amplitudes

The study of elastic scattering of a particle by a target that cannot be excited can have at least two rather distinct objectives: to learn how to set up scattering theory by first taking on a simple situation; or to learn how actually to calculate elastic scattering cross sections. This section attacks the first goal. The actual evaluation of scattering amplitudes is, on the whole, difficult, and almost always requires some combination of analytical approximation techniques and numerical computation. The next section will introduce several analytical approximation methods.

In scattering theory it is natural to incorporate the boundary conditions at the outset, and as we already learned in the one-dimensional case in §4.4(a), this can be done by replacing the Schrödinger equation by an integral equation. Our task is to extend this to three dimensions. For three reasons, we shall emphasize an abstract operator formalism: (i) it is actually simpler and far easier to remember than one where differential equations are explicitly translated into integral equations; (ii) it leads directly to important general properties of scattering amplitudes; and (iii) the formalism generalizes readily to problems that are far more complex than scattering by an inert target.

In this section we continue to assume that the projectile and target have no spin. Once more we speak as if the target were infinitely heavy and fixed at the origin, and the projectile had mass m. As in classical collision theory, the results extend directly to the case where the target has a finite mass and can recoil, as summarized below in subsection (d).

(a) Integral Equations and the Scattering Amplitude

Let H_0 be the kinetic energy operator for the projectile, whose energy is $E = k^2/2m$, where again $\hbar = 1$. The Schrödinger equation is

$$(E - H_0)|\boldsymbol{k}^{(+)}\rangle = V|\boldsymbol{k}^{(+)}\rangle \tag{67}$$

The notation means that this solution of energy E is to have an incident state of momentum $\boldsymbol{k}$ and outgoing scattered waves. Detailed properties of the interaction V, beyond it being a Hermitian operator, will often not be needed to develop the general theory.

The incident state, called $|\boldsymbol{k}\rangle$, satisfies

$$(E - H_0)|\boldsymbol{k}\rangle = 0\ . \tag{68}$$

We try to "solve" (67) by inverting the operator $(E - H_0)$:

$$|\boldsymbol{k}^{(+)}\rangle = \frac{1}{E - H_0} V|\boldsymbol{k}^{(+)}\rangle + |\boldsymbol{k}\rangle\ . \tag{69}$$

It was permissible to add $|\boldsymbol{k}\rangle$ here because (67) will be recovered thanks to (68) when (69) is multiplied by $(E - H_0)$, and as we know from §4.4(a), this will impose the requirement that $|\boldsymbol{k}\rangle$ be incident. But (69) makes no sense as it stands because $(E - H_0)^{-1}$ is not defined given that H_0 has a continuous spectrum within which E must lie. Recall, however, that in §2.6(b) this was overcome by replacing E by a complex variable z, and defining the resolvent for the operator H_0,

$$\mathcal{G}_0(z) = \frac{1}{z - H_0}\ . \tag{70}$$

This is well defined for Im $z \neq 0$, and as we learned there, as z approaches the real axis from above (below) it leads to outgoing (incoming) scattered waves. Thus the equation replacing (67) which incorporates the requirement that $|\boldsymbol{k}\rangle$ be incident is

$$|\boldsymbol{k}^{(+)}\rangle = |\boldsymbol{k}\rangle + \mathcal{G}_0(E + i\epsilon)V|\boldsymbol{k}^{(+)}\rangle \tag{71}$$

$$\equiv |\boldsymbol{k}\rangle + \frac{1}{E - H_0 + i\epsilon} V|\boldsymbol{k}^{(+)}\rangle\ . \tag{72}$$

This is the *Lippmann-Schwinger equation.* Clearly, the second term in these expressions is an abstract representation of the scattered wave.

As with any equation in Dirac notation, (71) can be written as an integral equation in any representation. The momentum representation is obtained by multiplying by a momentum eigenbra $\langle \boldsymbol{q}|$, and using

$$\langle \boldsymbol{q}|\mathcal{G}_0(E+i\epsilon)|\boldsymbol{q}'\rangle = \delta(\boldsymbol{q}-\boldsymbol{q}')\,\frac{2m}{k^2-q^2+i\epsilon}\,. \tag{73}$$

Hence

$$\Phi_{\boldsymbol{k}}(\boldsymbol{q}) = \delta(\boldsymbol{q}-\boldsymbol{k}) + \frac{1}{k^2-q^2+i\epsilon}\int d\boldsymbol{q}'\,\langle \boldsymbol{q}|U|\boldsymbol{q}'\rangle\,\Phi_{\boldsymbol{k}}(\boldsymbol{q}')\,, \tag{74}$$

where $\Phi_{\boldsymbol{k}}(\boldsymbol{q}) = \langle \boldsymbol{q}|\boldsymbol{k}^{(+)}\rangle$ is the wave function in momentum space, and $U = 2mV$. If V is a conventional potential, its coordinate matrix is diagonal,

$$\langle \boldsymbol{r}|V|\boldsymbol{r}'\rangle = \delta(\boldsymbol{r}-\boldsymbol{r}')V(\boldsymbol{r})\,, \tag{75}$$

in which case

$$\langle \boldsymbol{q}|U|\boldsymbol{q}'\rangle = \int \frac{d\boldsymbol{r}}{(2\pi)^3}\,e^{-i(\boldsymbol{q}-\boldsymbol{q}')\cdot\boldsymbol{r}}\,U(\boldsymbol{r})\,. \tag{76}$$

We always assume that $V(r)\to 0$ as $r\to\infty$.

The coordinate space equation follows by multiplying (71) by a coordinate eigenstate $\langle \boldsymbol{r}|$:

$$\psi_{\boldsymbol{k}}(\boldsymbol{r}) = \varphi_{\boldsymbol{k}}(\boldsymbol{r}) + \int d\boldsymbol{r}'\,\langle \boldsymbol{r}|\mathcal{G}_0(E+i\epsilon)|\boldsymbol{r}'\rangle\,V(\boldsymbol{r}')\,\psi_{\boldsymbol{k}}(\boldsymbol{r}')\,, \tag{77}$$

where the incident and complete wave functions are[1]

$$\varphi_{\boldsymbol{k}}(\boldsymbol{r}) = (2\pi)^{-3/2}\,e^{i\boldsymbol{k}\cdot\boldsymbol{r}}\,, \qquad \psi_{\boldsymbol{k}}(\boldsymbol{r}) = \langle \boldsymbol{r}|\boldsymbol{k}^{(+)}\rangle\,. \tag{78}$$

In §2.6(c) we derived

$$\langle \boldsymbol{r}|\mathcal{G}_0(E+i\epsilon)|\boldsymbol{r}'\rangle = -\frac{2m}{4\pi}\,\frac{e^{ikR}}{R}\,, \qquad R = |\boldsymbol{r}-\boldsymbol{r}'|\,. \tag{79}$$

Therefore the basic integral equation in the coordinate representation is

$$\psi_{\boldsymbol{k}}(\boldsymbol{r}) = \varphi_{\boldsymbol{k}}(\boldsymbol{r}) - \frac{1}{4\pi}\int d\boldsymbol{r}'\,\frac{e^{ik|\boldsymbol{r}-\boldsymbol{r}'|}}{|\boldsymbol{r}-\boldsymbol{r}'|}\,U(\boldsymbol{r}')\,\psi_{\boldsymbol{k}}(\boldsymbol{r}')\,. \tag{80}$$

Sometimes it is also useful to pose the bound-state problem in terms of an integral equation. In this situation there is no solution of the equation $(E-H_0)|\;\rangle = 0$ for negative energy. Setting $k\to i\alpha$, or $E=-\alpha^2/2m$, turns (80) into

$$\psi_b(\boldsymbol{r}) = -\frac{1}{4\pi}\int d\boldsymbol{r}'\,\frac{e^{-\alpha|\boldsymbol{r}-\boldsymbol{r}'|}}{|\boldsymbol{r}-\boldsymbol{r}'|}\,U(\boldsymbol{r}')\,\psi_b(\boldsymbol{r}')\,. \tag{81}$$

If the potential falls off sufficiently to insure that $(r'/r)\to 0$ as $r\to\infty$, the asymptotic form of bound-state wave functions is

$$\psi_b(\boldsymbol{r}) \sim \frac{e^{-\alpha r}}{r}\,C\,, \qquad E_b = -\frac{\alpha^2}{2m}\,. \tag{82}$$

---

[1] Note that in contrast to the preceding section our states are normalized to $\langle \boldsymbol{k}|\boldsymbol{k}'\rangle = \delta(\boldsymbol{k}-\boldsymbol{k}')$.

As was shown in §3.6(b), this is the universal relation between the binding energy and the tail of bound-state wave functions valid for all fields that fall off more rapidly than the Coulomb field. In other words, a potential that falls off like $1/r$ as $r \to \infty$ does *not* confine r' sufficiently to permit the extraction of $e^{-\alpha r}/r$ from the integral in (81).

Returning to the scattering problem, if $\boldsymbol{k}$ is in the z-direction, (80) has the form $e^{ikz} + \psi_{sc}$ of Eq. 9. As shown there, the scattering amplitude characterizes ψ_{sc} as $r \to \infty$, and so we take this limit in (80). We again assume that the potential falls off more rapidly than a Coulomb field, so the phase kR can then be approximated on the assumption that $(r'/r) \to 0$:

$$kR = kr|\hat{\boldsymbol{r}} - (\boldsymbol{r}'/r)| = kr\sqrt{1 - 2\frac{\hat{\boldsymbol{r}} \cdot \boldsymbol{r}'}{r} + \left(\frac{r'}{r}\right)^2} \sim kr - k\hat{\boldsymbol{r}} \cdot \boldsymbol{r}' + O(r^{-1}) . \tag{83}$$

Thus the asymptotic form of the integral equation (80) is

$$\psi_{\boldsymbol{k}} \sim \varphi_{\boldsymbol{k}}(\boldsymbol{r}) - \frac{e^{ikr}}{4\pi r} \int d\boldsymbol{r}' \, e^{-i\boldsymbol{k}' \cdot \boldsymbol{r}'} \, U(\boldsymbol{r}')\psi_{\boldsymbol{k}}(\boldsymbol{r}') , \tag{84}$$

where $\boldsymbol{k}' \equiv k\hat{\boldsymbol{r}}$, i.e., a vector of magnitude k pointing from the origin toward $\boldsymbol{r}$. The wave fronts tangent to the spherical wave in the direction $\hat{\boldsymbol{r}}$ form a momentum eigenfunction with eigenvalue $k\hat{\boldsymbol{r}}$ (Fig. 8.1). Therefore $\boldsymbol{k}'$ is the momentum

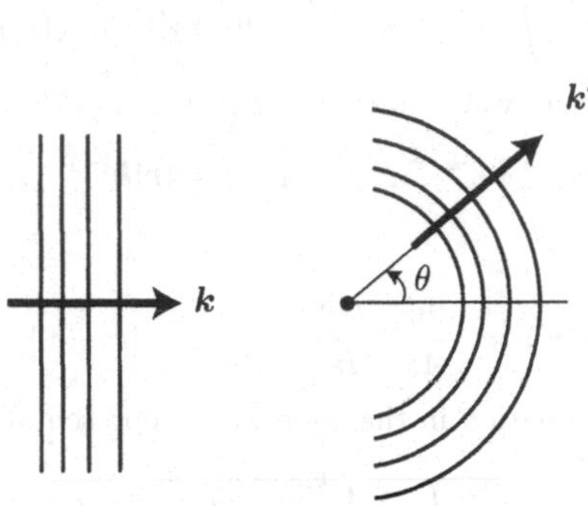

FIG. 8.1. Kinematics in scattering.

of particles scattered elastically in the direction of an observer situated at $\boldsymbol{r}$. The scattering angle θ is then related to the initial and final momenta by

$$\boldsymbol{k} \cdot \boldsymbol{k}' = k^2 \cos\theta . \tag{85}$$

To retain the definition of the scattering amplitude f used in §8.1, (84) is written in the form

$$\psi_{\boldsymbol{k}}(\boldsymbol{r}) \sim (2\pi)^{-3/2} \left\{ e^{i\boldsymbol{k} \cdot \boldsymbol{r}} + \frac{e^{ikr}}{r} f(\boldsymbol{k}'|\boldsymbol{k}) \right\} , \tag{86}$$

where

$$f(\boldsymbol{k}'|\boldsymbol{k}) = -4\pi^2 m \int d\boldsymbol{r} \, \varphi^*_{\boldsymbol{k}'}(\boldsymbol{r}) \, V(\boldsymbol{r}) \, \psi_{\boldsymbol{k}}(\boldsymbol{r}) = -4\pi^2 m \, \langle \boldsymbol{k}'|V|\boldsymbol{k}^{(+)}\rangle . \tag{87}$$

The last expression for the amplitude is notable because, as we shall see, scattering amplitudes have this form in a far wider range of situations than elastic scattering by a potential.[1]

Equation (87) for the scattering amplitude still contains the unknown solution to the integral equation. Nevertheless, it can be used to derive general and exact properties of such amplitudes, and it is also a convenient point of departure for various approximation methods.

The integral equation for the complete state $|\boldsymbol{k}^{(+)}\rangle$ can be used to derive integral equations for each partial wave, which then give useful formulas for the phase shifts. To accomplish this, the angular momentum decomposition of the three-dimensional Green's function is needed:

$$-\frac{e^{ikR}}{4\pi R} = \sum_{lm} G_k^{(l)}(r;r') Y_{lm}^*(\hat{\boldsymbol{r}}) Y_{lm}(\hat{\boldsymbol{r}}') , \tag{88}$$

where

$$G_k^{(l)}(r;r') = -ikj_l(kr_<)h_l(kr_>) , \tag{89}$$

h_l is the spherical Hankel function, and $r_<(r_>)$ means the smaller (larger) of r and r'. We postpone the derivation of (89) to sustain the flow of the argument. For now it suffices to note that as (88) is a function only of $|\boldsymbol{r} - \boldsymbol{r}'|$, it can depend only on (r, r', Θ), where $\boldsymbol{r} \cdot \boldsymbol{r}' = rr' \cos\Theta$, and can therefore be expanded in terms of Legendre polynomials which, via the addition theorem, then decompose into spherical harmonics whose arguments are the orientation of the two vectors.

As in §8.1(a), the wave function is expanded in spherical harmonics $Y_{l0}(\theta)$, where θ is the angle between $\boldsymbol{k}$ and $\boldsymbol{r}$:

$$\psi_{\boldsymbol{k}}(\boldsymbol{r}) = \sum_l \sqrt{\frac{2l+1}{2\pi^2}}\, i^l R_l(k;r) Y_{l0}(\theta) . \tag{90}$$

The corresponding expansion for the incident state $\varphi_{\boldsymbol{k}}(\boldsymbol{r})$ has j_l instead of the unknown radial function R_l. The normalization differs from Eq. 4 by the factor $(2\pi)^{-3/2}$ because in §8.1 the incident wave is $e^{i\boldsymbol{k}\cdot\boldsymbol{r}}$.

When (88) and the expansions for the wave functions are substituted into the basic equation (80), the following integral equations for the radial functions emerge:

$$R_l(k;r) = j_l(kr) + \int_0^\infty r'^2 dr'\, G_k^{(l)}(r;r')\, U(r)\, R_l(k;r') . \tag{91}$$

As always, the scattering amplitude is found by taking the limit $r \to \infty$. The asymptotic forms of the Bessel and Hankel functions were given in §3.4:

$$j_l(\rho) \sim \frac{\sin(\rho - \frac{1}{2}l\pi)}{\rho} , \qquad h_l(\rho) \sim \frac{e^{i\rho}}{i^{l+1}\rho} . \tag{92}$$

Hence the asymptotic form of the scattered wave in (91) is $-(ke^{ikr}/i^l kr)A_l(k)$, where

$$A_l(k) = \int_0^\infty r^2 dr\, j_l(kr)\, U(r)\, R_l(k;r) , \tag{93}$$

[1]Note that the transmission and reflection amplitudes also have this form; see §4.4(a).

and the asymptotic form of the radial function can then be put into the form

$$R_l(k;r) \sim -\frac{1}{2ikr}\left(e^{-i(kr-\frac{1}{2}l\pi)} - e^{i(kr-\frac{1}{2}l\pi)}\,e^{2i\delta_l(k)}\right) , \tag{94}$$

where

$$e^{2i\delta_l(k)} = 1 - 2ikA_l(k) . \tag{95}$$

An equivalent but sometimes more useful expression is the partial wave scattering amplitude

$$f_l \equiv \frac{1}{2ik}\left(e^{2i\delta_l} - 1\right) = \frac{1}{k}e^{i\delta_l}\,\sin\delta_l = -A_l(k) , \tag{96}$$

in terms of which the full scattering amplitude (Eq. 20) is

$$f(k,\theta) = \sum_l (2l+1) f_l P_l(\cos\theta) . \tag{97}$$

To actually find the phase shifts, the integral equation (91), or its counterpart ordinary differential equation, must be solved.

We must still derive Eq. 89. Starting from

$$-\frac{e^{ikR}}{4\pi R} = \int \frac{d\boldsymbol{q}}{(2\pi)^3}\frac{e^{i\boldsymbol{q}\cdot\boldsymbol{R}}}{k^2 - q^2 + i\epsilon} , \tag{98}$$

and decomposing the plane wave in the integrand into spherical harmonics, gives

$$G_k^{(l)}(r;r') = \frac{1}{2\pi}\int_0^\infty q^2 dq\,\frac{j_l(kr)j_l(kr')}{k^2 - q^2 + i\epsilon} . \tag{99}$$

The function $j_l(z)$ is entire, so the only singularities come from the denominator; also $j_l(-z) = (-1)^l j_l(z)$, so the integral can be extended to $(\infty, -\infty)$. Furthermore, $j_l = \frac{1}{2}[h_l + h_l^*]$, and $h_l \sim e^{iz}/z$ as $|z| \to \infty$; the integral can therefore be closed in the half-plane where the integrand decreases exponentially. The identity $h_l^*(z) = (-1)^l h_l(-z)$ then yields (89).

(b) A Solvable Example

Much of the development in this section is rather abstract, and an easily solvable example is therefore in order. As is to be expected, examples that are both easily solvable and realistic do not grow on trees, and our illustration is certainly unrealistic. It assumes an interaction that is maximally non-local, and is called the separable "potential":

$$\langle \boldsymbol{r}|V|\boldsymbol{r}'\rangle = -\frac{\lambda}{2m}u(r)u(r') . \tag{100}$$

While no real-life situation has such an interaction, beyond providing a nice sitting-up exercise it can sometimes be used to mimic a two-body interaction to simplify a many-body problem.

In this problem, the basic equation (80) is

$$\psi_{\boldsymbol{k}}(\boldsymbol{r}) = \varphi_{\boldsymbol{k}}(\boldsymbol{r}) + \frac{\lambda}{4\pi}\int d\boldsymbol{r}'\,\frac{e^{ik|\boldsymbol{r}-\boldsymbol{r}'|}}{|\boldsymbol{r}-\boldsymbol{r}'|}u(r')\,C(k) , \tag{101}$$

where

$$C(k) = \int d\boldsymbol{r}\, u(r)\, \psi_{\boldsymbol{k}}(\boldsymbol{r}) \,. \tag{102}$$

Thus the problem has been reduced to finding the constant $C(k)$. Furthermore, because the integration over $\boldsymbol{r}'$ has the spherically symmetric weight $u(r')$, the scattering term cannot depend on the direction of $\boldsymbol{r}$, so there is only s-wave scattering.

The asymptotic form of (101) is

$$\psi_{\boldsymbol{k}}(\boldsymbol{r}) \sim \varphi_{\boldsymbol{k}}(\boldsymbol{r}) + \lambda C(k) v(k)\, \frac{e^{ikr}}{r} \,, \tag{103}$$

where

$$v(k) = \int \frac{d\boldsymbol{r}}{4\pi} e^{-i\boldsymbol{k}'\cdot\boldsymbol{r}} u(r) = \int_0^\infty r^2 dr\, \frac{\sin kr}{kr}\, u(r) \,. \tag{104}$$

Equation (103) shows explicitly that the scattered wave is spherically symmetric, and that the scattering amplitude is

$$f(k) = (2\pi)^{3/2} \lambda C(k) v(k) \,. \tag{105}$$

To find $C(k)$, multiply (101) by $u(r)$ and integrate:

$$C(k) = 4\pi (2\pi)^{-3/2} v(k) + \lambda C(k) I(k) \,, \tag{106}$$

where

$$I(k) = \int d\boldsymbol{r} d\boldsymbol{r}'\, u(r) \frac{e^{ikR}}{R} u(r') = -\int \frac{d\boldsymbol{q}}{(2\pi)^3} \int d\boldsymbol{r} d\boldsymbol{r}'\, \frac{u(r) e^{i\boldsymbol{q}\cdot\boldsymbol{R}} u(r')}{k^2 - q^2 + i\epsilon} \tag{107}$$

$$= -8 \int_0^\infty q^2 dq\, \frac{|v(q)|^2}{k^2 - q^2 + i\epsilon} \,. \tag{108}$$

The final result for the scattering amplitude is therefore

$$f(k) = \frac{4\pi\lambda |v(k)|^2}{D(k^2 + i\epsilon)} \,, \tag{109}$$

where

$$D(z) = 1 + 8\lambda \int_0^\infty q^2 dq\, \frac{|v(q)|^2}{z - q^2} \,. \tag{110}$$

The bound-state equation (81) can be solved with equal ease because it reduces to (106) when the inhomogeneous term is dropped, and $k \to i\alpha$. The bound-state eigenvalues are the positive values of α^2 that are roots of

$$D(-\alpha^2) = 1 - 8\lambda \int_0^\infty q^2 dq\, \frac{|v(q)|^2}{\alpha^2 + q^2} = 0 \,. \tag{111}$$

The integral is a monotonically decreasing function for $\alpha^2 > 0$. Hence there is at most one bound state, and the interaction must be strong enough to satisfy

$$8\lambda \int_0^\infty dq\, |v(q)|^2 > 1 \tag{112}$$

for there to be such a state.

Although the scattering amplitude has now been found, it is not apparent that it corresponds to a real s-wave phase shift $\delta(k)$, i.e., that it has the required form

$$1 + 2ikf(k) = e^{2i\delta(k)} . \tag{113}$$

If this is correct,

$$\frac{8\pi i\lambda k \left|v(k)\right|^2 + D(k^2 + i\epsilon)}{D(k^2 + i\epsilon)} \tag{114}$$

must have unit modulus.

Clearly, it is time to learn something about functions of a complex variable z defined by integrals of the form

$$\Phi(z) = \int_a^b dx' \, \frac{f(x')}{z - x'} , \tag{115}$$

as they arise repeatedly in scattering theory and in many other topics. The integration is along the real axis, though this is not essential, and $f(x)$ is assumed to be continuous in the interval (a, b). The function $\Phi(z)$ is analytic except for a branch cut from $z = a$ to $z = b$, which can be taken as the real line between them. The discontinuity across this cut is

$$\lim_{\epsilon \to 0} [\Phi(x + i\epsilon) - \Phi(x - i\epsilon)] = -2\pi i f(x) . \tag{116}$$

This fact follows from

$$\Phi(x + i\epsilon) - \Phi(x - i\epsilon) = -2i \int_a^b dx' \, \frac{\epsilon}{(x - x')^2 + \epsilon^2} \, f(x') \tag{117}$$

$$\to -2if(x) \int_a^b dx' \, \frac{\epsilon}{(x - x')^2 + \epsilon^2} = -2\pi i f(x) , \tag{118}$$

where in the step to the second line f was taken outside because the integrand is $O(\epsilon)$ unless $|x - x'| \sim \epsilon$, and f is continuous in the neighborhood of x, while in the last step the integration limits can be extended to $\pm\infty$. Hence the discontinuity is purely imaginary if $f(x)$ is real.

To evaluate $\Phi(z)$ just above the cut, which is needed in (114), one evaluates

$$\tfrac{1}{2}[\Phi(x + i\epsilon) + \Phi(x - i\epsilon)] = \int_a^b \frac{(x - x')f(x')\, dx'}{(x - x')^2 + \epsilon^2} \tag{119}$$

$$\to \left(\int_a^{x-\delta} + \int_{x+\delta}^b \right) \frac{f(x')\, dx'}{x - x'} + \int_{x-\delta}^{x+\delta} \frac{(x - x')f(x')\, dx'}{(x - x')^2 + \epsilon^2} \tag{120}$$

where $\delta \ggg \epsilon$ as both δ and ϵ tend to zero. In the last term f can again be taken outside, and the integral then vanishes because the integrand is odd, while the first term is the definition of the Cauchy principle value of the integral, which is designated by $\mathrm{P}\!\int$.

To summarize,

$$\Phi(x \pm i\epsilon) = \mp\pi i f(x) + \mathrm{P} \int_a^b dx' \, \frac{f(x')}{x - x'} . \tag{121}$$

If $f(x)$ is real, $\Phi(z)$ can be written as

$$\Phi(z) = -\frac{1}{\pi}\int_a^b dx\, \frac{\operatorname{Im}\Phi(x+i\epsilon)}{z-x} . \tag{122}$$

In mathematics this is known as a Hilbert transform, and it appears often in physics in dispersion relations. Note that what has been learned can be put as the very useful mnemonic

$$\frac{1}{x-x'\pm i\epsilon} = \mp i\pi\,\delta(x-x') + \mathrm{P}\frac{1}{x-x'} , \tag{123}$$

with the understanding that this relationship is to be used for the purpose of integration. This can also be stated in operator form:

$$\frac{1}{E\pm i\epsilon - H} = \mp\pi i\delta(E-H) + \mathrm{P}\,\frac{1}{E-H} . \tag{124}$$

Applying this to (110) gives

$$D(k^2+i\epsilon) = 1 - 4\pi i\lambda k\,|v(k)|^2 + 8\lambda\,\mathrm{P}\int_0^\infty q^2 dq\,\frac{|v(q)|^2}{k^2-q^2} , \tag{125}$$

which shows that (114) has the form Z/Z^* and therefore a real phase shift does emerge.

An important point about a connection between scattering and bound states can now be seen by comparing the formula for the scattering amplitude (Eq. 109) with the formula that determines the bound-state eigenvalue (Eq. 111), assuming of course that (112) is satisfied and there is such a state. Namely, if we consider f to be a function of the complex variable z whose real part is $k^2 = 2mE_k$, then $f(z)$ has a simple pole at $z = -2mE_b$. Furthermore, in the light of what has been learned about the function $\Phi(z)$, the scattering amplitude has a branch cut on the positive real axis ($0 \le E < \infty$).

(c) Bound-State Poles

The appearance of a pole in scattering amplitude at a pure imaginary value of the momentum, i.e., at a negative energy, is not a bizarre peculiarity of the weird separable potential, but a general property of scattering amplitudes. To see why this is so, consider the integral equation for a bound-state radial function $F_l(\alpha; r)$. This equation follows from performing an angular momentum decomposition of (81), but there is no need to go through this exercise because the sought-for equation must be obtainable from (91) by setting $k \to i\alpha$ and deleting the incident wave:

$$F_l(\alpha;r) = \alpha\int_0^\infty r'^2\,dr'\; j_l(i\alpha r_<)h_l(i\alpha r_>)U(r')F_l(\alpha;r') . \tag{126}$$

Here we have adopted the convention $\alpha > 0$; if the convention $\alpha < 0$ is used, h_l^* must appear in the kernel.

To understand where the bound-state pole comes from, it is only necessary to think more carefully about the recipe that led to (126). The inhomogeneous integral equation for the radial functions in the continuous spectrum (Eq. 91) defines the function $R_l(z;r)$ of the complex variable z when k is replaced by z. However, $R_l(z;r)$ is not square integrable because $j_l(zr)$ diverges exponentially as $|z| \to \infty$ unless $\text{Im } z = 0$. The scattering term in (91), on the other hand, converges as $|zr| \to \infty$ if $\text{Im } z > 0$ because h_l falls off exponentially in the upper half plane. Consider now the scattering integral equation after multiplication by $(z - z_0)$:

$$(z-z_0)R_l(z;r) = (z-z_0)j_l(zr) - iz \int_0^\infty r'^2\, dr'\, j_l(zr_<)h_l(zr_>)U(r')(z-z_0)R_l(z;r') \,. \tag{127}$$

For arbitrary values of z_0, this equation just reads $0 = 0$ at $z = z_0$. However, if R_l has a simple pole at z_0, then in the neighborhood of this point it has the form

$$R_l(z;r) = \frac{F_l(z_0;r)}{z - z_0} + \text{reg.} \,, \tag{128}$$

where "reg." stands for a function that is regular at z_0. Given that $j_l(z)$ is an entire function, the residue $F_l(z_0;r)$ at the assumed pole must satisfy the equation

$$F_l(z_0;r) = -iz_0 \int_0^\infty r'^2\, dr'\, j_l(z_0 r_<)h_l(z_0 r_>)U(r')F_l(z_0;r') \,. \tag{129}$$

This is (126) after the replacement $\alpha \to -iz_0$. Hence $F_l(z_0;r)$ is a square integrable solution of the Schrödinger equation with eigenvalue $z_0^2/2m$. But all square integrable solutions of the Schrödinger equation have real, negative eigenvalues, and therefore z_0 must lie on the positive imaginary axis.

In short, the radial continuum solutions $R_l(k;r)$ must have poles on the positive imaginary k-axis at the points $i\sqrt{2m|E_l|}$, where E_l is a bound-state eigenvalue of angular momentum l, with the residue being the radial bound-state wave function. In the argument we assumed that these poles are simple, and in fact they are, though this was not proven. On the other hand, the argument demonstrates that $R_l(k;r)$ cannot have singularities at any other points in the finite upper-half of the k-plane (but it says nothing about the lower half-plane).

Returning to the partial wave scattering amplitude, recall that in (96) $f_l(k)$ is expressed as a radial integral over the function $R_l(k;r)$; that then is how the bound-state poles appear as poles in the scattering amplitude. The eigenvalue continuum also appear — as a cut in the scattering amplitude. The amplitude can, however, also have singularities that have nothing to do with the eigenvalues of the Hamiltonian, but stem from the interaction itself. This is evident in in the case of the separable potential, because singularities in the function $v(k)$ produce singularities directly in $f(k)$, as shown by (109).

(d) Symmetry Properties of the Amplitude

Some general properties of $f(\mathbf{k}'|\mathbf{k})$ can be derived without any formula for the amplitude. If the interaction is rotationally invariant, then the amplitude can only depend on rotationally invariant quantities, i.e., $|\mathbf{k}|^2 = k^2 = |\mathbf{k}'|^2$ and $\mathbf{k} \cdot \mathbf{k}'$. In this important case, therefore, f is only a function of the momentum k and the scattering angle θ, or

$$f(\mathbf{k}'|\mathbf{k}) = f(k, \theta) \,, \tag{130}$$

as in §8.1 for the same reason. Note that this argument did not require V to be a rotationally invariant *potential*; what counts is that the non-local kernel $\langle \boldsymbol{r}|V|\boldsymbol{r}'\rangle$ be a function only of r, r' and $|\boldsymbol{r}-\boldsymbol{r}'|$. By an essentially identical argument, if the interaction is reflection invariant,

$$f(\boldsymbol{k}'|\boldsymbol{k}) = f(-\boldsymbol{k}'|-\boldsymbol{k}) \,. \tag{131}$$

Symmetry under time reversal is less straightforward and more interesting. Recall from §7.2 that the time reversal operator I_t is antiunitary, and that for any operator A and any two states $|a\rangle$ and $|b\rangle$

$$\langle a|A|b\rangle = \langle b^R|I_t A^\dagger I_t^{-1}|a^R\rangle \,, \tag{132}$$

where $|a^R\rangle$ is the motion-reversed counterpart of $|a\rangle$.

When I_t is applied to the integral equation (72) it produces

$$I_t\,|\boldsymbol{k}^{(+)}\rangle = |-\boldsymbol{k}\rangle + \frac{1}{E-H_0-i\epsilon} V I_t\,|\boldsymbol{k}^{(+)}\rangle \,. \tag{133}$$

Here it is assumed that V is invariant under time reversal, which only requires V to be Hermitian as the position operator is, by assumption, the only degree of freedom. The appearance of $-i\epsilon$ in the resolvent means that the scattered wave in (133) is *incoming*, not outgoing, as we learned in §2.6(c) and is clear from (78). This is what time reversal should do: if $|\boldsymbol{k}^{(+)}\rangle$ is a stationary-state caricature of a process that begins with a plane wave packet of mean momentum $\boldsymbol{k}$ running toward the target and which ends with spherical waves running away from it, then the time-reversed process begins with a spherical wave converging on a target and ends with a plane wave packet running away from it with momentum $-\boldsymbol{k}$. Admittedly, this second process is not one that accelerator laboratories can produce (at least as of today), but it turns out that such states play a very important role in collision theory.

In the light of this, we now define the state $|\boldsymbol{k}^{(-)}\rangle$ as the solution of

$$|\boldsymbol{k}^{(-)}\rangle = |\boldsymbol{k}\rangle + \frac{1}{E-H_0-i\epsilon} V|\boldsymbol{k}^{(-)}\rangle \,. \tag{134}$$

In words, this is a state with an incoming spherical wave that scatters into the momentum eigenstate $|\boldsymbol{k}\rangle$. According to (133), time reversal produces

$$I_t\,|\boldsymbol{k}^{(+)}\rangle = |-\boldsymbol{k}^{(-)}\rangle \,. \tag{135}$$

Time reversal must also interchange the incoming and outgoing states, and the formula we now have for the amplitude, $\langle \boldsymbol{p}|V|\boldsymbol{k}^{(+)}\rangle$, shows no hint of any symmetry between these states. This defect is overcome most efficiently by means of another integral equation for scattering states. To derive it, first rewrite $(E-H_0)|\boldsymbol{k}\rangle = 0$ as

$$(E-H)|\boldsymbol{k}\rangle = -V|\boldsymbol{k}\rangle \,. \tag{136}$$

Now inverting $(E-H)$ permits the addition of a solution of the "homogeneous" equation $(E-H)|\;\rangle = 0$, for which we choose the scattering states $|\boldsymbol{k}^{(\pm)}\rangle$. Then the maneuver that led from (67) to (72) yields

$$|\boldsymbol{k}^{(\pm)}\rangle = \left(1 + \frac{1}{E-H\pm i\epsilon} V\right)|\boldsymbol{k}\rangle \,. \tag{137}$$

This equation does not, of course, really give an explicit solution for the scattering state in terms of the incoming state, because it involves the resolvent (or Green's function) for the complete Hamiltonian, $(z - H)^{-1}$, which is much more difficult to find explicitly than any single scattering state. Equation (137) is very useful for deriving theorems about scattering states and for developing approximation methods, but only for explicitly solving scattering problems exactly in very simple models.

We now define the operator $T(E)$

$$T = V + V \frac{1}{E - H + i\epsilon} V \, . \tag{138}$$

This provides the desired symmetric formulation of the scattering amplitude because

$$\langle \boldsymbol{p}|V|\boldsymbol{k}^{(+)}\rangle = \langle \boldsymbol{p}|T(E)|\boldsymbol{k}\rangle \qquad \text{iff } E_k = E_p = E \, , \tag{139}$$

$$= \langle \boldsymbol{p}^{(-)}|V|\boldsymbol{k}\rangle \qquad \text{iff } E_p = E_k \, . \tag{140}$$

Note the provisos: the identities *only* hold if two states have the *same* energy, $E_k = E_p$, or to use the standard jargon, provided the matrix element is on *the energy shell*. This restriction is underscored here because it is often convenient, or even necessary, to extend scattering amplitudes and related quantities to states of unequal energy. On the energy shell, the scattering amplitude is

$$f(\boldsymbol{p}|\boldsymbol{k}) = -4\pi^2 m \langle \boldsymbol{p}|T|\boldsymbol{k}\rangle) \, , \tag{141}$$

where the argument of $T(E)$ is usually suppressed.

From (132), the result of the time reversal operation is

$$\langle \boldsymbol{p}|T|\boldsymbol{k}\rangle = \langle -\boldsymbol{k}|I_t T^\dagger I_t^{-1}| - \boldsymbol{p}\rangle \tag{142}$$

because $|\boldsymbol{k}^R\rangle = | - \boldsymbol{k}\rangle$. If, however, V is invariant under time reversal

$$I_t T^\dagger I_t^{-1} = T \, , \tag{143}$$

and therefore the consequence of time reversal invariance for the elastic scattering amplitude is

$$f(\boldsymbol{p}|\boldsymbol{k}) = f(-\boldsymbol{k}| - \boldsymbol{p}) \, . \tag{144}$$

Once again, this is the result expected by intuition. An easy application may be noted. In one-dimensional scattering by an arbitrary potential $V(x)$ that is *not* reflection invariant, it may not be intuitively obvious that the transmission and reflection amplitudes do not depend on whether the incident state comes from the left or right; Eq. 144 shows that this is so.

(e) Relations Between Laboratory and Center-of-Mass Quantities

The treatment given in this and the preceding section extends to the elastic scattering and bound states of two particles of arbitrary masses m_1 and m_2 if, in the dynamical equations, the mass m is replaced by the reduced mass $m_1 m_2/(m_1+m_2)$, $\boldsymbol{r}$ is taken to be the separation between the particles, and $\boldsymbol{k}$, $\boldsymbol{q}$, etc., are relative momenta. The scattering angle θ is the angle of deflection of either particle in the center-of-mass frame, $\mathcal{F}_{\text{com}}$. The relation between this angle, and those in the "laboratory" frame $\mathcal{F}_{\text{lab}}$ where the particle of mass m_1 is incident and the other is initially at rest, are identical to those in classical collision theory.[1] Let ϑ_1 and ϑ_2 be the angles between the final momenta of particles 1 and 2 with respect to the incident direction in $\mathcal{F}_{\text{lab}}$; then

$$\tan\vartheta_1 = \frac{m_2 \sin\theta}{m_2 + m_1\cos\theta}\,, \qquad \vartheta_2 = \tfrac{1}{2}(\pi - \theta)\,. \tag{145}$$

The total energy in $\mathcal{F}_{\text{com}}$ is what appears as the eigenvalue in the Schrödinger equation, and it is related to k in the foregoing formulas by $E = k^2/2\mu$.

The differential cross section in $\mathcal{F}_{\text{com}}$ is as before:

$$\frac{d\sigma_{\text{com}}}{d\Omega_{\text{com}}} = |f(\theta)|^2\,. \tag{146}$$

The differential cross section in $\mathcal{F}_{\text{lab}}$ is defined as in Eq. 15, though with all quantities in that frame, i.e.,

$$\frac{d\sigma_{\text{lab}}(\vartheta_1)}{d\Omega_{1,\text{lab}}} = \frac{d\dot{N}_{\text{sc,lab}}/d\Omega_{\text{lab}}}{F_{\text{inc,lab}}}\,, \tag{147}$$

where $d\Omega_{1,\text{lab}} = \sin\vartheta_1 d\vartheta_1 d\varphi_1$. Then

$$\frac{d\sigma_{\text{lab}}(\vartheta_1)}{d\Omega_{1,\text{lab}}} = |f(\theta)|^2\,\frac{(1 + 2\rho\cos\theta + \rho^2)^{3/2}}{1 + \rho\cos\theta}\,, \tag{148}$$

where $\rho = m_1/m_2$.

In the important case of equal masses, $\vartheta_1 = \frac{1}{2}\theta$, so there is no scattering in $\mathcal{F}_{\text{lab}}$ beyond 90°. The cross sections are related through

$$\frac{d\sigma_{\text{lab}}(\vartheta_1)}{d\Omega_{1,\text{lab}}} = 4\cos\vartheta_1 |f(2\vartheta_1)|^2\,. \tag{149}$$

8.3 Approximations to Elastic Amplitudes

Scattering by an unscreened Coulomb field due to a point charge is the only three-dimensional scattering problem describing a real-life situation that can be solved exactly by analytical means. Analytical and numerical approximation methods are therefore indispensable in scattering theory, and as the latter are often based on the former, this introduction to the topic is devoted to analytical approximations.

The most obvious approximation is to assume (or hope) that the force responsible for scattering is sufficiently weak to alter the incident wave but little. This is the Born approximation. It was used by him in the paper that proposed the statistical interpretation of the Schrödinger wave function, which arose rather naturally in this

[1] Cf., e.g., H. Goldstein, *Classical Mechanics*, 2nd. ed., (1980), §3–11. The kinematic relations and cross sections are actually most easy to derive in a relativistic formulation for both elastic and inelastic collisions, including the case where more than two particles are in the final state.

very first paper on scattering in quantum mechanics. The Born approximation has been used in an enormous range of problems, but like virtually all approximations, it introduces important conflicts with properties dictated by general laws — in this case the optical theorem. This defect can be repaired by calculating phase shifts approximately and then using these in the partial wave expansion, but it is often unclear whether this improvement is numerically reliable. Furthermore, if many partial waves are important, this is a cumbersome approach unless it can be done efficiently by numerical means. This last hurdle can be overcome quite elegantly if the incident momentum is sufficiently high, for that allows the use of a technique borrowed from physical optics which produces a complete scattering amplitude that satisfies the optical theorem. In potential scattering itself there are powerful mathematical methods that can often spell out conditions under which one or another approximation method is valid, but there are important interactions and kinematic regimes which are beyond the power of this approach.

These remarks should make it clear that we now enter uncertain territory where intuition and experience are indispensable guides — qualities that can only be acquired by getting one's hands dirty.

(a) The Born Approximation

The elastic scattering amplitude, as derived in §8.2(a), is

$$f(\boldsymbol{k}'|\boldsymbol{k}) = -4\pi^2 m \, \langle \boldsymbol{k}'|V|\boldsymbol{k}^{(+)}\rangle \,, \tag{150}$$

where $|\boldsymbol{k}^{(+)}\rangle$ is the exact solution of the Schrödinger equation for incident momentum $\boldsymbol{k}$ and outgoing scattered waves, and $|\boldsymbol{k}'\rangle$ is the final momentum eigenstate.

Equation (150) makes it tempting to replace the exact but unknown scattering state by the incident state $|\boldsymbol{k}\rangle$ in the hope that this is permissible if the potential is sufficiently weak or the energy sufficiently high, and to worry afterwards about the conditions under which this may be valid. This is the Born approximation:

$$f_B(\boldsymbol{k}'|\boldsymbol{k}) = -4m\pi^2 \langle \boldsymbol{k}'|V|\boldsymbol{k}\rangle = -\frac{m}{2\pi}\int d\boldsymbol{r}\, e^{i\boldsymbol{q}\cdot\boldsymbol{r}}\, V(\boldsymbol{r}) \equiv -\frac{m}{2\pi}\,\tilde{V}(\boldsymbol{q}) \,, \tag{151}$$

where $\boldsymbol{q}$ is called the momentum transfer, i.e., the change in momentum of the projectile and thus the momentum given to the target:

$$\boldsymbol{q} = \boldsymbol{k} - \boldsymbol{k}' \,, \qquad q^2 = 4k^2 \sin^2 \tfrac{1}{2}\theta \,. \tag{152}$$

In words, the scattering amplitude in the Born approximation is proportional to the Fourier transform of the potential, $\tilde{V}(\boldsymbol{q})$, evaluated at the momentum transfer $\boldsymbol{q}$ in the collision. This statement is, of course, tied to potential scattering, but the first expression in (151) will, once again, turn out to be of far greater generality if it is written as $\langle f|H_{\text{int}}|i\rangle$ in terms of the initial and final states and the interaction responsible for the process.

For an arbitrary potential, the exact amplitude (150) will depend on both the incident and final momenta. In contrast, the Born approximation to the amplitude is only a function of the single vector $\boldsymbol{q}$. If the potential is spherically symmetric, or $V(\boldsymbol{r}) = V(r)$, the Born amplitude is only a function of the single variable $|\boldsymbol{q}|^2$, and not a function of the momentum and angle separately. This is evident from symmetry, and it is apparent after the integration over the orientations of $\boldsymbol{q}$:

$$f_B(q^2) = -2m \int_0^\infty r^2 dr \, \frac{\sin qr}{qr} \, V(r) \, . \tag{153}$$

As a concrete and important example, consider the Yukawa potential,

$$V(r) = V_0 \, \frac{e^{-\alpha r}}{\alpha r} \, , \tag{154}$$

whose Fourier transform is

$$\tilde{V}(q) = \frac{4\pi V_0}{\alpha^3} \, \frac{\alpha^2}{q^2 + \alpha^2} \, . \tag{155}$$

The differential cross section is therefore

$$\frac{d\sigma}{d\Omega} = \left(\frac{2V_0 m}{\alpha^3} \right)^2 \left| \frac{\alpha^2}{4k^2 \sin^2 \frac{1}{2}\theta + \alpha^2} \right|^2 . \tag{156}$$

Here, for once, $\hbar$ has been restored to show how this can be done by a dimensional argument, if desired. The prefactor must be an area as this is a cross section; α is an inverse length, and therefore $V_0 m/\alpha^2$ must be dimensionless, which is uniquely achieved by $\alpha \to \hbar\alpha$ as this converts this inverse length or wave number into a momentum.

This differential cross section has features that recur in many situations. For wavelengths $1/k$ long compared to the range $1/\alpha$, it is approximately isotropic. This was to be expected from the qualitative discussion in §8.1(a), which concluded that for such momenta only the s-wave scatters appreciably, giving a nearly isotropic scattering amplitude (cf. Eq. 24). When $k \to 0$, the cross becomes energy independent, which cannot be inferred from the partial wave expansion alone. As k grows, the cross section becomes progressively more peaked toward the forward direction, and behaves like $1/q^4$, or $1/\theta^4$ at small angles, down to $q \simeq \alpha$, where it tops off to the energy independent value $(2V_0 m/\alpha^3)^2$. The angular distribution (156) is shown in Fig. 8.2.

The total cross section is also interesting. Because $q^2 = 2k^2(1 - \cos\theta)$,

$$\int d\Omega = \frac{\pi}{k^2} \int_0^{4k^2} dq^2 \, , \tag{157}$$

and therefore

$$\sigma(k) = \left(\frac{2V_0 m}{\alpha^3} \right)^2 \frac{\alpha^2 \pi}{k^2} \, \frac{4k^2}{4k^2 + \alpha^2} \, . \tag{158}$$

This total cross section tends to zero as $k \to \infty$, which is a general feature in potential scattering because the interaction becomes irrelevant as the kinetic energy goes to infinity. This statement is false in relativistic quantum mechanics for a variety of reasons.

The Yukawa potential turns into the Coulomb field in the limit where the range $1/\alpha$ goes to infinity, or $\alpha \to 0$, provided the replacement $(V_0/\alpha) \to Z_1 Z_2 e^2/4\pi$ is

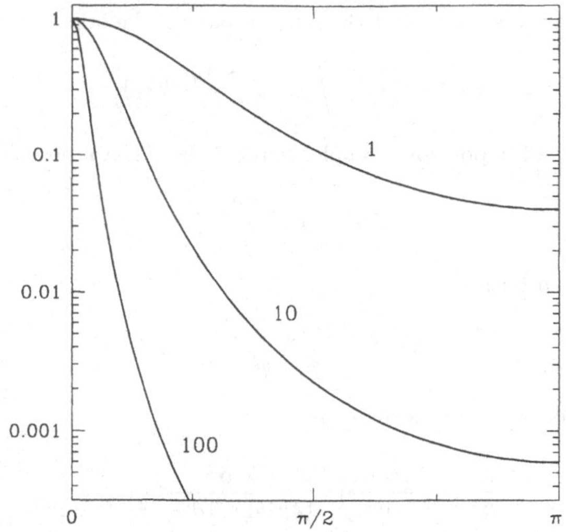

FIG. 8.2. The angular distribution for elastic scattering from a Yukawa potential in the Born approximation. Write Eq. 156 as $d\sigma/d\Omega = (2V_0 m/\alpha^3)^2 F(\theta)$; then the plot shows $F(\theta)$ for $(k/\alpha)^2 = 1$, 10, 100.

also made (we always use rationalized units in which $e^2/4\pi\hbar c \simeq 1/137$). Hence the Born approximation for scattering from a Coulomb field is

$$f_B = -m\frac{Z_1 Z_2 e^2}{4\pi}\frac{1}{2p^2 \sin^2 \frac{1}{2}\theta}\,, \tag{159}$$

where p is the incident momentum. Eq. 159 diverges as $\theta \to 0$, and produces an infinite total cross section at all energies. This is due to the infinite range of the Coulomb field.

The Coulomb cross section formula is a classical expression because it does not contain $\hbar$ provided one understands p to really mean momentum, i.e., $p = \hbar k$. Indeed, it is *the Rutherford cross section,* as first found by him from classical collision theory. In §8.4 it will be shown that this cross section formula is *exact*, because the exact scattering amplitude is the square root of the Rutherford cross section multiplied by a quantum mechanical phase factor which of course disappears in the cross section. This is another example of remarkable "accidental" agreements between inconsistent theoretical considerations related to the Coulomb field, which is due at least in part to the fact that this field contains no parameter having the dimension of length, indeed no dimensionful parameter whatever in units where $\hbar = c = 1$.

Rutherford's formula, as written in (159), is somewhat opaque for atomic or nuclear scattering. A more useful expression is

$$\frac{d\sigma}{d\Omega} = \frac{1}{4}\,Z_1^2 Z_2^2\, a_m^2\, \frac{W_m^2}{E^2 \sin^4 \frac{1}{2}\theta}\,, \tag{160}$$

where $a_m = 4\pi/me^2$ and $W_m = e^2/8\pi a_m$ are the Bohr radius and Rydberg for mass m, and E is the incident energy.

The Born approximation can also be applied to the partial waves. In the exact expression (96) the exact radial wave function $R_l(k:r)$ is replaced by the incident wave $j_l(kr)$, and the phase shift is assumed to be small, or $e^{i\delta_l}\sin\delta_l \simeq \delta_l$, which results in

$$\delta_l(k) = -k\int_0^\infty r^2 dr\,[j_l(kr)]^2\,U(r)\,. \tag{161}$$

Were this to be done for all partial waves, and only the term linear in δ_l kept in the partial wave amplitudes, the sum over l would merely be a perverse computation of the Fourier transform (151). Perhaps the most sensible use of this version of Born approximation is to use it only for partial waves that have small phase shifts, and to combine it with a better approximation or an accurate numerical computation of the phase shifts that are large. Another approach is to take the phase shift from (161), and to then use this value in the correct expression $e^{i\delta}\sin\delta$; this leaves unsettled whether this really produces a reliable estimate of higher order terms in the potential.

(b) Validity of the Born Approximation

Having seen that the Born approximation is easy to use, we now address the much harder question of finding the conditions under which it is valid. The following discussion only offers rather superficial answers.

It is evident that the Born approximation disagrees with the optical theorem, and therefore fails to conserve probability! From (151) it follows that the Born amplitude is real if $V(\boldsymbol{r})$ is invariant under reflection, as exemplified by the amplitude for the Yukawa potential (Eq. 155). Strictly speaking, a real amplitude makes no sense because the optical theorem would then say that the total cross section vanishes. The reason for this "paradox" is that the optical theorem relates the amplitude itself to a quadratic expression in the amplitude, whereas the Born approximation is of first order in the interaction. Clearly, the optical theorem implies that the exact amplitude is an expression that, at a minimum, must contain all powers of the interaction.

With this remark as a hint, it becomes easy to see that the Born approximation is the first term in an expansion of the scattering amplitude as a power series in V. The basic integral equation (71) can be expanded in powers of V by iteration:

$$|\boldsymbol{k}^{(+)}\rangle = (1 + \mathcal{G}_0 V + \mathcal{G}_0 V \mathcal{G}_0 V + \ldots)\,|\boldsymbol{k}\rangle\,. \tag{162}$$

Substituting into the exact expression (150) for the amplitude, this gives

$$f(\boldsymbol{k}'|\boldsymbol{k}) = -4m\pi^2\left(\langle\boldsymbol{k}'|V|\boldsymbol{k}\rangle + \langle\boldsymbol{k}'|V\mathcal{G}_0 V|\boldsymbol{k}\rangle + O(V^3)\right)\,. \tag{163}$$

Here the first term is what we have thus far called the Born approximation, but when confusion could arise it will be called the first Born approximation. The explicit expression for the matrix element appearing in the second Born approximation is

$$\langle\boldsymbol{k}'|V\mathcal{G}_0 V|\boldsymbol{k}\rangle = 2m\int d\boldsymbol{p}\,\frac{\langle\boldsymbol{k}'|V|\boldsymbol{p}\rangle\langle\boldsymbol{p}|V|\boldsymbol{k}\rangle}{k^2 - p^2 + i\epsilon}\,. \tag{164}$$

The integral (164) is complex even when the matrix elements of V are real, as they are for a reflection-invariant interaction. As we will now show, the imaginary part of this *second*-order Born approximation to the amplitude agrees with the total cross section as given by the *first*-order Born approximation. This is not the end of the story, of course: if the Born series is terminated at any finite order in V, it will violate the optical theorem in the next order. The underlying reason for this behavior, as we will eventually learn, is that the scattering amplitude is the matrix element of an operator whose schematic form is $(e^{iV}-1)$, which the Born approximation replaces by a power series in V.

To compute the imaginary part of (164), recall the rule (123). For the optical theorem only the forward amplitude is needed:

$$\mathrm{Im}\,\langle \boldsymbol{k}|V\mathcal{G}_0 V|\boldsymbol{k}\rangle = -2\pi m \int d\boldsymbol{k}'\;\delta(k^2-k'^2)\,|\langle \boldsymbol{k}|V|\boldsymbol{k}'\rangle|^2\,. \tag{165}$$

Now $d\boldsymbol{k}' = \frac{1}{2}k'dk'^2 d\Omega'$, and therefore

$$\mathrm{Im}\,f(\boldsymbol{k}|\boldsymbol{k}) = 4\pi^3 m^2 k \int d\Omega'\;|\langle \boldsymbol{k}|V|\boldsymbol{k}\rangle|^2 = \frac{k}{4\pi}\sigma_B\,, \tag{166}$$

where σ_B is the total cross section in the first-order Born approximation, i.e., for the amplitude (151). As promised, this is consistent with the optical theorem if third-order terms in V are ignored.

The optical theorem implies that the Born series for the scattering amplitude must contain all powers in V, and this raises the following issue. If V is replaced by gV, the Born series becomes a power series in g, and in developing it we have, whether we like it or not, tacitly assumed that the amplitude is an analytic function of g that converges in the unit circle in the complex g plane! In most problems of interest, it is at best difficult to prove that the amplitude is such a function, although in potential scattering a great deal is known about this matter. By and large, intuitive considerations, experience with simple and if possible solvable problems, and/or quasi-mathematical arguments must be used to decide whether the Born series makes sense. In the great majority of applications the issue is rather moot because it is not possible to compute more than a few terms in the series. The most famous example is quantum electrodynamics, where the series gives superb agreement with high-accuracy measurements; nevertheless, there are quite convincing arguments indicating that the amplitudes do not have the analytic property as function of the fine structure constant that is required for a convergent series. One can, so to say, believe that it is an asymptotic series, and that if one could compute a large enough number of terms it would start to disagree with experiment!

A quantitative insight into the validity of the Born approximation can be obtained by computing the distortion of the wave function on the assumption that it must be small if the approximation is to be reliable. One way of characterizing the distortion is to compute the difference between the incoming and scattered wave to first order in V. The distortion should be maximum well inside the potential, and the origin is the most convenient place to do this calculation. We therefore define

$$C(k) \equiv [\varphi_{\boldsymbol{k}}(0) - \psi_{\boldsymbol{k}}(0)]/\varphi_{\boldsymbol{k}}(0) \simeq (\pi/2)^{\frac{1}{2}} \int d\boldsymbol{r}\;\frac{e^{ikr}}{r}\,U(r)\varphi_{\boldsymbol{k}}(\boldsymbol{r}) \tag{167}$$

$$= \frac{1}{k}\int_0^\infty dr\, e^{ikr}\sin kr\, U(r)\,. \tag{168}$$

If the Born approximation is to be valid, it is necessary, but not sufficient, that $|C(k)| \ll 1$.

As the energy increases, $|C(k)|$ decreases because of the explicit factor of $1/k$ and even more so because of the increasing oscillations of the integrand, *provided* of course that $U(r)$ is not so singular at short distances as to make the integral divergent. In fact, the Riemann-Lebesgue lemma states that the integral in (168) tends to zero as $k \to \infty$ provided the function $U(r)\sin kr$ is square integrable. A broad class of potentials, such as the Yukawa potential, can therefore be shown to have a convergent Born series if the energy is high enough. On the other hand, there are important potentials, such as those that describe collisions between atoms, which are so violently repulsive at short distances (e.g., $\sim r^{-12}$) as to not even allow an incorrect calculation with the Born approximation because their Fourier transform does not exist.

The most stringent condition is at $k = 0$, where the Born approximation requires

$$\left|\int_0^\infty U(r) r dr\right| \ll 1 . \tag{169}$$

Therefore scattering by the Coulomb field cannot be treated with the Born approximation at low energies. For a potential characterized by a depth V_0 and range a, (169) can be rendered as the rough inequality

$$ma^2|V_0| \ll 1 \tag{170}$$

This has a simple and important meaning. If an attractive potential of depth V_0 and range a is to bind one state, it will have no nodes, have a momentum of order $1/a$ and kinetic energy of order $1/ma^2$, and therefore (170) is really the statement that the potential must be far too weak to bind a state if the Born approximation is work at zero energy. Observe that the absolute value of V_0 appears in this argument, and therefore a repulsive potential will not submit to the Born approximation at low energy if its attractive counterpart can bind a state. This, of course, is just an illustration of our earlier conclusion that the Born series tacitly assumes convergence in the complex plane of a variable such as V_0, and therefore does not distinguish between attraction and repulsion.

It is instructive to examine how $C(k)$ behaves as the energy rises in a concrete case. The Yukawa potential offers such an example:

$$C(k) = \frac{2mV_0}{\alpha^2}\left(\frac{\alpha}{k}\right)\left[\frac{1}{2}\tan^{-1}\frac{2k}{\alpha} + \frac{i}{4}\ln\left(1 + \frac{4k^2}{\alpha^2}\right)\right] . \tag{171}$$

The high- and low-energy limits of this function are

$$C(k) \simeq \frac{2mV_0}{\alpha^2}\left(\frac{\alpha}{k}\right)\left(\frac{\pi}{4} + \frac{i}{2}\ln\frac{2k}{\alpha}\right) \qquad k \gg \alpha , \tag{172}$$

$$\simeq \frac{2mV_0}{\alpha^2}\left(1 + \frac{ik}{\alpha} - \frac{4k^2}{3\alpha^2} + \ldots\right) \qquad k \ll \alpha . \tag{173}$$

As expected from the discussion related to Eq. 170, $m|V_0|/\alpha^2 \ll 1$ is required to make the approximation valid at all energies.

The requirement that the incident kinetic energy be large compared to the some sort of mean of the potential energy does not suffice to make the Born approximation valid. For example, if the potential is weak but extends over a large region $\mathcal{R}$, the wave function will still be close to a plane wave in $\mathcal{R}$ but accumulate a large shift in phase, which is beyond the power of the Born approximation. This is familiar from optics, where refraction cannot be accounted for by expanding the wave function in powers of $n-1$, where n is the index of refraction.

The partial wave Born approximation (161) can be used to calculate the small phase shifts that are expected for angular momenta $l \gg ka$, but in so doing it is essential to not forget the value of ka. If $ka \ll 1$, the limiting form of $j_l(x)$ as $x \to 0$ can be used, which gives

$$\delta_l \simeq -\frac{2mk^{2l+1}}{[(2l+1)!!]^2} \int_0^\infty r^{2l+2}\, V(r)\, dr \, . \tag{174}$$

Clearly, this can only be used for potentials that fall off faster than any power. If, however, ka is itself $O(1)$ or larger, which is the more interesting situation as higher partial waves are then more important, the problem is not straightforward. In the case of the Yukawa potential (154), the Born approximation phase shifts, as computed from (161), are

$$\delta_l(k) = -mV_0 a^2 \left(\frac{1}{ka}\right) Q_l\left(1 + \frac{1}{2k^2a^2}\right) , \tag{175}$$

where $Q_l(z)$ is Legendre's function of the second kind, and $a = 1/\alpha$. The Q_l are not polynomials, and have very different behaviors along different directions in the (l, ka) plane for large values of these parameters (see Fig. 8.3).

(c) Short-Wavelength Approximations

When the incident wavelength is short compared to the lengths that characterize the interaction, the concepts and techniques of geometrical optics can be used to compute the wave function in the interaction region. The ensuing propagation of the wave to infinity is, in essence, then identical to how a diffraction pattern is described in physical optics.

The idea just expressed is captured by the *Ansatz*

$$\psi_{\boldsymbol{k}}(\boldsymbol{r}) = e^{ikz}\, \Phi(\boldsymbol{r})/(2\pi)^{3/2} \, , \tag{176}$$

the assumption being that the potential only produces a slowly varying modulation of the incident wave, so that Φ varies slowly over distances of order $1/k$. In terms of (176), the exact Schrödinger equation is

$$\left(\frac{d}{dz} + \frac{i}{v} V(\boldsymbol{r})\right) \Phi(\boldsymbol{r}) = -\frac{1}{2ik} \nabla^2 \Phi(\boldsymbol{r}) \, , \tag{177}$$

where $v = k/m$ is the incident velocity. If this assumption about Φ is valid, for sufficiently large k the right-hand side is small compared to the first derivative on the left, or

$$\left(\frac{d}{dz} + \frac{i}{v} V(\boldsymbol{r})\right) \Phi(\boldsymbol{r}) = 0 \, . \tag{178}$$

That is, the three-dimensional second-order Schrödinger equation has been reduced to a first-order ordinary differential equation for every value of the transverse coordinates (x, y), which we will specify with the vector $\boldsymbol{b}$ perpendicular to the incident

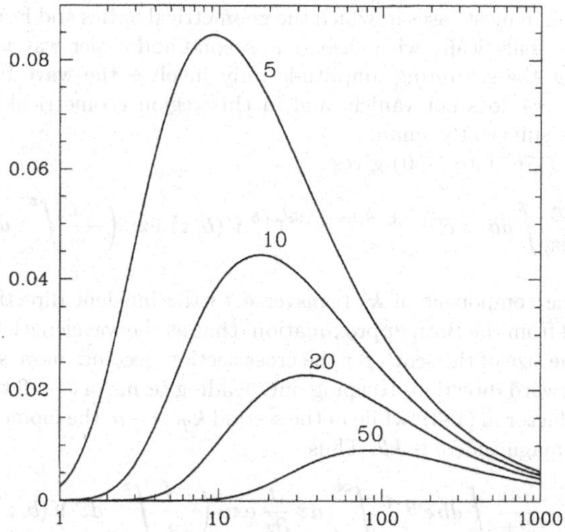

FIG. 8.3. The Born approximation phase shifts for a Yukawa potential as a function of l and k (Eq. 175). The quantity plotted is $\frac{1}{ka}Q_l\left(1+\frac{1}{2k^2a^2}\right)$ as a function of ka for $l = 5$, 10, 20, 50.

direction. This mathematical statement says that in this approximation the wave function has been replaced by a set of rays parallel to the incident direction, with various impact parameters $\boldsymbol{b}$, which propagate without refraction or reflection, so that each ray only acquires a phase shift. This phase is given by the solution of (178) that satisfies the requirement that ψ reduce to the incident wave as $z \to -\infty$, i.e.,

$$\Phi(\boldsymbol{b}, z) = \exp\left(-\frac{i}{v}\int_{-\infty}^{z} dz'\, V(\boldsymbol{b}, z')\right). \tag{179}$$

The approximations embodied in (179) are too drastic to describe scattering; an observer far downstream would just see the untouched incident wave at distances off the z-axis larger than the force range a. This is a well-known issue in optics. As the wave propagates beyond an aperture of characteristic size a, there are three regimes in which it has quite distinct behaviors:

$$\frac{z}{a} \ll ka \qquad \text{geometrical optics} \tag{180}$$

$$\simeq ka \qquad \text{Fresnel diffraction} \tag{181}$$

$$\gg ka \qquad \text{Fraunhofer diffraction}\,. \tag{182}$$

The conventional scattering amplitude describes the wave as $z \to \infty$, and therefore belongs to the Fraunhofer regime, whereas the approximation (179) only holds in the geometrical optics regime. Given the latter, there are two ways of finding the Fraunhofer diffraction pattern. The first is to calculate the evolution of the wave as a function of z; this is a difficult exercise in the Fresnel region, and cannot be done analytically even in most cases in which the geometrical optics and Fraunhofer limits can be handled analytically with ease. The second and easier way is to remember that Eq. 150 for the scattering amplitude only involves the wave function in the region where $V(\boldsymbol{r})$ does not vanish, and in this region geometrical optics is valid provided $1/k$ is sufficiently small.

Substituting (176) into (150) gives

$$f(\boldsymbol{k}'|\boldsymbol{k}) = -\frac{m}{2\pi}\int d\boldsymbol{b}\, dz\, e^{i(1-\cos\theta)kz}\, e^{-i\boldsymbol{k}'_T\cdot\boldsymbol{b}}\, V(\boldsymbol{b},z)\, \exp\left(-\frac{i}{v}\int_{-\infty}^{z} dz'\, V(\boldsymbol{b},z')\right), \tag{183}$$

where $\boldsymbol{k}'_T$ is the component of $\boldsymbol{k}'$ transverse to the incident direction. It is clear intuitively, and from the Born approximation, that as the wavelength becomes much shorter than the size of the scatterer the cross section becomes more sharply peaked towards the forward direction. Keeping only leading terms in $\theta \to 0$ then eliminates the first phase factor in (183), while in the second $\boldsymbol{k}'_T \simeq -\boldsymbol{q}$, the momentum transfer, which has the magnitude $q \simeq k\theta$. Thus

$$f(\boldsymbol{k}'|\boldsymbol{k}) \simeq -\frac{imv}{2\pi}\int d\boldsymbol{b}\, e^{i\boldsymbol{q}\cdot\boldsymbol{b}}\int_{-\infty}^{\infty} dz\, \frac{\partial}{\partial z}\exp\left(-\frac{i}{v}\int_{-\infty}^{z} dz'\, V(\boldsymbol{b},z')\right) ; \tag{184}$$

now the integration over z is trivial,

$$f(\boldsymbol{k}'|\boldsymbol{k}) = \frac{k}{2\pi i}\int d\boldsymbol{b}\, e^{i\boldsymbol{q}\cdot\boldsymbol{b}}\left(e^{2i\Delta(\boldsymbol{b})} - 1\right), \tag{185}$$

where the phase shift is

$$\Delta(\boldsymbol{b}) = -\frac{1}{2v}\int_{-\infty}^{\infty} V(\boldsymbol{b},z)\, dz\,. \tag{186}$$

Equation (185), which is due to Glauber, is often called the eikonal approximation to the scattering amplitude, a term borrowed from geometrical optics.

Should the potential be spherically symmetric, Δ will only depend on $|\boldsymbol{b}|$, and the integration over the angle $\chi = \angle(\boldsymbol{b},\boldsymbol{q})$ can be done by use of

$$J_0(x) = \frac{1}{2\pi}\int_0^{2\pi} d\chi\, e^{-ix\cos\chi}\,, \tag{187}$$

where $J_0(x)$ is a Bessel function. Hence in the rotationally invariant case

$$f(k,\theta) = -ik\int_0^{\infty} bdb\, J_0(qb)\left(e^{2i\Delta(b)} - 1\right). \tag{188}$$

The total cross section, which in this instance is purely elastic, when computed under the assumption that the amplitude is sharply forward peaked, requires the evaluation of

$$\int_0^{\infty} bdb\int_0^{\infty} b'db'\left(e^{2i\Delta(b)} - 1\right)\left(e^{-2i\Delta(b')} - 1\right)\int_0^{\infty}\theta d\theta\, J_0(qb)J_0(qb')\,. \tag{189}$$

The identity

$$\int_0^{\infty}\theta d\theta\, J_0(x\theta)J_0(x'\theta) = \frac{1}{x}\delta(x-x') \tag{190}$$

then leads to the result

$$\sigma = 8\pi \int_0^\infty bdb \sin^2 \Delta(b) . \tag{191}$$

These expressions for the amplitude and cross section have the same form as the exact partial wave expansion, with the integral over impact parameters playing the role of the sum over angular momenta. As a consequence they satisfy the optical theorem — indeed, the theorem is also obeyed when the potential is not spherically symmetric, as it should because probability conservation is certainly not contingent on angular momentum conservation.

The approximate relation between impact parameter and angular momentum just mentioned permits an extension of the eikonal approximation to situations where there is both elastic and inelastic scattering. In this case the partial wave expansion for the elastic amplitude was derived in §8.1(a):

$$f_{\rm el}(k,\theta) = \frac{1}{2ik} \sum_l (2l+1)[\eta_l(k) - 1]\, P_l(\cos\theta) . \tag{192}$$

The parameters η_l are only phases when there is no inelastic scattering, as in potential scattering. The idea now is to set $l \simeq kb$, to assume that many partial waves contribute and that η_l varies smoothly with l, so that η_l can be replaced by an interpolating function $\eta(k,b)$, which would allow the sum on l to be approximated by an integral over b. To relate this to the preceding discussion, in the case of potential scattering $\eta(k,b) = \exp[i\Delta(b)]$.

What is still needed is an approximation for P_l when the integers l are replaced by the continuous variable b. This is provided by the asymptotic approximation

$$P_l(\cos\theta) \simeq J_0(2l \sin \tfrac{1}{2}\theta) \simeq J_0(qb), \qquad (l \gg 1\,,\ \sin^2 \tfrac{1}{2}\theta \ll 1) . \tag{193}$$

The conversion of the sum on l to an integral over $dl = kdb$ converts (192) into

$$f_{\rm el}(k,\theta) = -ik \int_0^\infty bdb\, J_0(qb)\, [\eta(k,b) - 1] . \tag{194}$$

The total elastic cross section can again be found by use of (190), and the total cross section from the optical theorem. The results are

$$\sigma_{\rm el} = 2\pi \int_0^\infty bdb\, |1 - \eta(b)|^2 , \tag{195}$$

$$\sigma_{\rm tot} = 4\pi \int_0^\infty bdb\, [1 - \mathrm{Re}\, \eta(b)] , \tag{196}$$

$$\sigma_{\rm inel} = 2\pi \int_0^\infty bdb\, [1 - |\eta(b)|^2] . \tag{197}$$

As an instructive example, consider a spherical target that absorbs the incident wave perfectly for all $b \le R$. This actually is a useful if crude approximation for elastic scattering of energetic hadrons (π, K, nucleons) by heavy nuclei, because if such a projectile strikes the target it will excite the target with a probability close to unity. That is, we assume

$$\eta(b) = 0, \quad (b < R)\,; \qquad \eta(b) = 1, \quad (b > R) . \tag{198}$$

Then the elastic amplitude is

$$f_{\rm el}(k,\theta) = ik \int_0^R J_0(qb)\, bdb = ikR^2\, \frac{J_1(qR)}{qR}\,. \tag{199}$$

This is the Fraunhofer diffraction pattern due to a black disc, as one would expect from what has already been said. Note that the elastic amplitude is pure imaginary, as it is whenever the scatterer is purely absorptive, i.e., when $\eta(b)$ is real, which corresponds in optics to an index of refraction that is pure imaginary.

The total cross section follows immediately from (196), or if one prefers, from $J_1(x) \to \frac{1}{2}x$ as $x \to 0$:

$$\sigma_{\rm tot} = 2\pi R^2\,. \tag{200}$$

This is *twice* the area presented to the beam by the target! The "obvious" area is equal to the *inelastic* cross section, $\sigma_{\rm inel}$, which on second thought is what it should be. The other πR^2 needed to give $\sigma_{\rm tot}$ is the total *elastic* cross section, which, in the Fresnel regime, is the shadow cast by the black disc.

8.4 Scattering in a Coulomb Field

Scattering in a Coulomb field is an important phenomenon in many branches of physics. In most situations, the interaction is only that of a Coulomb field over a more or less restricted range of distances. For example, in scattering of a point charge, such as an electron, by a neutral atom, the field is not even approximately Coulombic for distances comparable to or larger than the radius R_K of the electronic K-shell, nor for distances comparable to or smaller than the nuclear radius R_N. Nevertheless, the scattering amplitude in a pure Coulomb field accounts accurately for scattering through momentum transfers q in the interval between $qR_K \gg 1$ and $qR_N \ll 1$, which is a large kinematic region because $R_K \ggg R_N$. Knowledge of the Coulomb amplitude is therefore of importance in its own right, and as the point of departure for various techniques that incorporate scattering due to a non-Coulombic electrostatic fields, and due to interactions that are not electromagnetic at all. The latter is a common situation, and will be illustrated here by scattering due to the combination of a short-range nuclear interaction and a Coulomb field, as in collisions between a proton and a nucleus.

It is fortunate that scattering in a pure Coulomb field can be solved exactly, as we shall now see. Readers should, however, be warned that this solution relies on a more sophisticated command of complex variable theory than all other parts of this volume. The final result for the scattering amplitude in a pure Coulomb field is given by Eq. 225, and when there is also a short-range interaction by Eq. 250.

(a) The Coulomb Scattering Amplitude

The two-body Schrödinger equation to be solved can be written as

$$\left(\nabla^2 + k^2 + \frac{2\gamma k}{r}\right)\psi(\boldsymbol{r}) = 0\,. \tag{201}$$

Here

$$\gamma = \frac{Z_1 Z_2 e^2}{4\pi\hbar v} \simeq \frac{Z_1 Z_2}{137}\,\frac{c}{v}\,, \tag{202}$$

where v is the incident relative velocity, $\hbar k = \mu v$, μ is the reduced mass and $\boldsymbol{r}$ the separation. To make the relations to classical theory more transparent, in this section we do not use atomic units and leave $\hbar$ explicit. The field is *attractive* when $\gamma > 0$, and loosely speaking weak when $|\gamma| \ll 1$, where the qualification alludes to the fact that a field that varies like $1/r$ as $r \to \infty$ will produce a divergent scattering amplitude in the forward direction no matter how small its strength may be.

The Coulomb field has a symmetry that is more extensive than rotational invariance, which was taken advantage of in treating the bound state problem in §5.2. As expected, this symmetry also has an impact on the scattering problem. Because of this symmetry, (201) is reduced to an ordinary differential equation not only in spherical coordinates but also in parabolic coordinates, and the latter are better adapted to the scattering problem. This can be understood from an observation due to Gordon, who first solved this problem. At distances large compared to the incident wavelength, i.e., $kr \gg 1$, the field varies slowly and only produces a slow modulation of the incident wave. The asymptotic wavefronts are then expected to be the family of surfaces normal to the set of classical hyperbolic orbits having the incident momentum $\hbar\boldsymbol{k}$. This is the family of surfaces

$$\boldsymbol{k}\cdot\boldsymbol{r} + \gamma \ln(kr - \boldsymbol{k}\cdot\boldsymbol{r}) = \text{const.}\,, \tag{203}$$

which suggests the *Ansatz*

$$\psi_{\boldsymbol{k}}(\boldsymbol{r}) = e^{i\boldsymbol{k}\cdot\boldsymbol{r}}\,\chi(u)\,, \tag{204}$$

where u is a parabolic coordinate:

$$u = i(kr - \boldsymbol{k}\cdot\boldsymbol{r}) = ik(r-z) = 2ikr\sin^2\tfrac{1}{2}\theta\,. \tag{205}$$

Substitution of (204) into (201) yields the ordinary differential equation

$$\left(\frac{d^2}{du^2} + \frac{1-u}{u}\frac{d}{du} - \frac{i\gamma}{u}\right)\chi(u) = 0\,. \tag{206}$$

The solutions of such second-order linear equations have properties that are determined by the singularities of the functions that multiply χ and its derivatives. The coefficient of χ'' is a constant, while the others are analytic everywhere except for simple poles at $u = 0$. Consequently, χ must be analytic throughout the finite u plane except for a pole or branch point at $u = 0$. Only solutions that are regular at $u = 0$ are physically admissible, and therefore such solutions are either polynomials in u or entire functions of u. Which of the latter two possibilities obtains is determined by the behavior of the solution as $u \to \infty$. This behavior is revealed by introducing the variable $w = 1/u$, which gives

$$\left\{\frac{d^2}{dw^2} + \frac{1}{w}\left(1 + \frac{1}{w}\right)\frac{d}{dw} - \frac{i\gamma}{w^3}\right\}\chi(w) = 0. \tag{207}$$

Now the coefficients of χ and $d\chi/dw$ have higher-order poles at $w = 0$, which implies that $\chi(w)$ has an essential singularity at $w = 0$, i.e., at $u = \infty$, and $\chi(u)$ is therefore an entire function of u.

The Laplace transform,

$$\chi(u) = \int_{t_1}^{t_2} dt\; e^{ut}\, f(t)\,, \tag{208}$$

reduces equations of the type (206) to first order. The integration path in the t-plane remains to be specified. When (208) is substituted into (206), the latter turns into

$$\int_{t_1}^{t_2} [ut^2 + (1-u)t - i\gamma]e^{ut} f(t)\, dt = \int_{t_1}^{t_2} f(t) \left[(t - i\gamma) + t(t-1)\frac{d}{dt}\right] e^{ut}\, dt\,, \quad (209)$$

which, after integrating by parts, becomes

$$t(t-1)f(t)e^{ut}\Big|_{t_1}^{t_2} + \int_{t_1}^{t_2} e^{ut} \left[(t-i\gamma)f(t) - \frac{d}{dt}[t(t-1)f(t)]\right] . \quad (210)$$

Thus a solution of the original equation is achieved if $f(t)$ satisfies

$$(t - i\gamma)f(t) - \frac{d}{dt}[t(t-1)f(t)] = 0\,, \quad (211)$$

and the integration path makes the first term of (210) vanish.

The solution of (211) is

$$f(t) = At^{i\gamma-1}(1-t)^{-i\gamma}\,. \quad (212)$$

The sought-for function $\chi(u)$ is therefore given by

$$\chi(u) = A\int_C e^{ut}\, t^{i\gamma-1}(1-t)^{-i\gamma}\, dt\,, \quad (213)$$

where the contour C is either closed or such as to give the function $e^{ut}\, t^{i\gamma}(1-t)^{1-i\gamma}$ the same value at both ends of C.

As we are dealing with a second-order equation, there is not just one solution; different choices of the contour C satisfy different boundary conditions. The integrand of (213) contains a function of the form $t^x(1-t)^y$, and thus has branch points at $t = 1$ and $t = 0$. The cuts can be taken to run from the branch points to infinity, but as $x + y$ is an integer, they can also be merged into one cut from 0 to 1. Any C that is open, and on which $e^{ut} \to 0$ at its end points will satisfy the stated requirements, as will a closed curve that runs around the merged cuts. Equation (205) implies that u is on the positive imaginary axis for physical values of the kinematical variables, which suggest that C be chosen to run from $i\infty$ around one or the other cut, or both, and then back to $i\infty$. The choice that satisfies the boundary condition appropriate to scattering will be shown to be the one in which C encloses both cuts, as shown in Fig. 8.4a.

The scattering boundary condition emerges from the asymptotic form of χ, though it should be noted that $u \to \infty$ only implies $r \to \infty$ if the scattering angle is kept away from $\theta = 0$. The asymptotic behavior becomes more apparent after the substitution $s = ut$:

$$\chi(u) = A\int_C e^s\, s^{i\gamma-1}(u-s)^{-i\gamma}\, ds\,. \quad (214)$$

As $u \to \infty$, s/u is small on the lower leg of C, so for that portion of the integral expansion in powers of $1/u$ is permissible, but this is not true of the upper leg because the branch point at $s = u$ recedes to $i\infty$ with u. For that reason C is

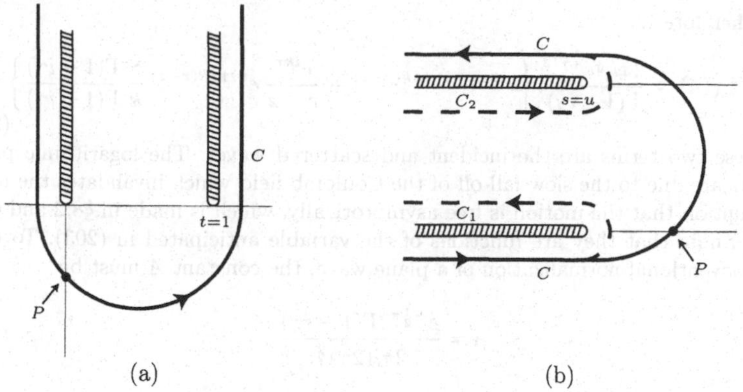

FIG. 8.4. The contour in the t-plane for Eq. 213 is on the left, those in the s-plane are on the right.

deformed into C_1 and C_2, as shown in Fig. 8.4b, and in the integral along C_2 the substitution $s \to s+u$ is made. This translates C_2 into C_1, giving

$$\chi(u) = A \int_{C_1} [e^s\, s^{i\gamma-1}(u-s)^{-i\gamma} + e^{s+u}(s+u)^{i\gamma-1}(-s)^{-i\gamma}]\, ds\,. \tag{215}$$

Now the expansion in powers of s/u can be done. The coefficients are evaluated by use of Hankel's formula[1]

$$\frac{2\pi i}{\Gamma(z)} = \int_{C_1} e^s\, s^{-z} ds\,, \tag{216}$$

which holds for all non-integer z. To leading order as $u \to \infty$, the result is

$$\chi(u) \sim 2\pi i A \left\{ \frac{u^{-i\gamma}}{\Gamma(1-i\gamma)} + \frac{e^u}{u}\, \frac{u^{i\gamma}(-1)^{-i\gamma}}{\Gamma(i\gamma)} + \cdots \right\} \tag{217}$$

$$\sim 2\pi i A \left\{ \frac{u^{-i\gamma}}{\Gamma(1-i\gamma)} + \frac{e^u}{u}\, \frac{(u^*)^{i\gamma}}{\Gamma(i\gamma)} \right\} \tag{218}$$

$$\sim \frac{2\pi i A u^{-i\gamma}}{\Gamma(1-i\gamma)} \left\{ 1 + \frac{e^u}{u}\, (uu^*)^{i\gamma}\, i\gamma\, \frac{\Gamma(1-i\gamma)}{\Gamma(1+i\gamma)} \right\}, \tag{219}$$

where we have used $\Gamma(z+1) = z\Gamma(z)$, and $u = -u^*$ because in the scattering context $u = iR$, with R real and positive.[2] In view of this, set $u = e^{\ln|u|}\, e^{i\pi/2}$, $|u| = k(r-z)$. Then

$$(uu^*)^{i\gamma} = e^{2i\gamma \ln k(r-z)}\,, \qquad u^{-i\gamma} = e^{\pi\gamma/2}\, e^{-i\gamma \ln k(r-z)}\,, \tag{220}$$

[1] E.T. Whittacker and G.N. Watson, *Modern Analysis*, Cambridge University Press (1950), p. 245.

[2] The function $\chi(u)$ is multivalued and displays the Stokes phenomenon (see §4.5(b)), which is typical for asymptotic expansions about an essential singularity. Eq. 219 gives a correct rendition of $\chi(u)$ as $|u| \to \infty$ in the sector $0 \le \arg u \le \pi$, as required because $\arg u = \frac{1}{2}\pi$ for real θ.

and therefore

$$e^{ikz}\,\chi(u) \sim \frac{2\pi i A e^{\pi\gamma/2}}{\Gamma(1-i\gamma)} \left\{ e^{ikz}\,e^{-i\gamma\ln k(r-z)} + \frac{e^{ikr}}{r-z}\,e^{i\gamma\ln k(r-z)}\,\frac{\gamma}{k}\,\frac{\Gamma(1-i\gamma)}{\Gamma(1+i\gamma)} \right\} . \tag{221}$$

These two terms are the incident and scattered waves. The logarithmic phase factors are due to the slow fall-off of the Coulomb field which invalidates the naive assumption that the motion is free asymptotically, which is made in §8.2 and elsewhere; note that they are functions of the variable anticipated in (203). To have our conventional normalization of a plane wave, the constant A must be

$$A = \frac{e^{-\frac{1}{2}\pi\gamma}\Gamma(1-i\gamma)}{2\pi i(2\pi)^{\frac{3}{2}}} . \tag{222}$$

Finally, therefore, the asymptotic form of the wave function for scattering by an attractive Coulomb field is

$$\psi_{\boldsymbol{k}}(\boldsymbol{r}) \sim \frac{1}{(2\pi)^{\frac{3}{2}}} \left\{ e^{ikz}e^{-i\gamma\ln k(r-z)} + \frac{e^{ikr}}{r}\,e^{i\gamma\ln k(r-z)}\,\frac{\gamma\Gamma(1-i\gamma)}{k(1-\cos\theta)\Gamma(1+i\gamma)} \right\} . \tag{223}$$

The amplitude $f_C(k,\theta)$ for scattering in a pure Coulomb field is defined by writing (223) in the form

$$\psi_{\boldsymbol{k}}(\boldsymbol{r}) \sim \frac{1}{(2\pi)^{\frac{3}{2}}} \left\{ e^{ikz}e^{-i\gamma\ln k(r-z)} + f_C(k,\theta)\,\frac{e^{i(kr+\gamma\ln 2kr)}}{r} \right\} , \tag{224}$$

so that

$$f_C(k,\theta) = \gamma\,\frac{\Gamma(1-i\gamma)}{\Gamma(1+i\gamma)}\,\frac{e^{2i\gamma\ln\sin\frac{1}{2}\theta}}{2k\sin^2\frac{1}{2}\theta} . \tag{225}$$

The ratio of gamma functions is a phase. The amplitude therefore has the form

$$f_C(k,\theta) = f_C^B(k,\theta)\,e^{i\Phi(k,\theta)} , \tag{226}$$

where f_C^B is the Born approximation amplitude, and the phase Φ is

$$\Phi(k,\theta) = 2\arg\Gamma[(1-i\gamma)] + 2\gamma\ln\sin\tfrac{1}{2}\theta . \tag{227}$$

$\Phi \to 0$ as $\gamma \to 0$, so to leading order in γ the whole wave function reduces to that given by the Born approximation.

There are important problems, especially in nuclear physics, involving the probability that two charged particles will, in essence, come into contact and thereby induce a process. This is given by the ratio of $|\psi_{\boldsymbol{k}}(0)|$ to the incident amplitude, $|\psi_{\text{inc}}| = 1/(2\pi)^{3/2}$. From (214)

$$\chi(0) = (-1)^{-i\gamma}A\int_C e^s s^{-1}\,ds = (-1)^{-i\gamma}2\pi i A = 2\pi i e^{\pi\gamma}A . \tag{228}$$

Hence

$$\frac{|\psi_{\boldsymbol{k}}(0)|^2}{|\psi_{\text{inc}}|^2} = e^{\pi\gamma}\Gamma(1-i\gamma)\Gamma(1+i\gamma) = \frac{2\pi\gamma}{1-e^{-2\pi\gamma}} , \tag{229}$$

where $\Gamma(z)\Gamma(1-z) = \pi/\sin \pi z$ was used. This expression holds for both signs of γ, but the important application is in nuclear reactions — the repulsive case, where γ is negative. In this case it is better to write the formula as

$$\frac{|\psi_k(0)|^2}{|\psi_{\text{inc}}|^2} = \frac{2\pi|\gamma|}{e^{2\pi|\gamma|} - 1} \to 2\pi\gamma\, e^{-2\pi|\gamma|} , \tag{230}$$

where the latter holds when $|Z_1 Z_2 c/137v| \gg 1$, showing that for collisions at sufficiently low velocities or between sufficiently large charges of the same sign, the probability that the particles will "meet" is exponentially small.

To evaluate the cross section the incident and scattered currents must be found. The interference between the two only persists at $\theta \to 0$ as $r \to \infty$, as in other situations. But in this case f_C diverges as $\theta \to 0$, reflecting the infinite cross section in fully unscreened Coulomb scattering, and therefore the very forward direction is sensitive to the screening mechanism, which we do not treat. The current due to the incident wave is the same as for a free particle, plus a correction term from the logarithmic factor proportional to $1/r$ as $r \to \infty$, which is negligible asymptotically. The scattering term also gives a current equal to that from the conventional and familiar factor e^{ikr}/r, plus a negligible correction that varies like $1/r^3$. In consequence, the differential cross section in the center of mass frame is $|f_C(k,\theta)|^2$, i.e.,

$$\frac{d\sigma_R}{d\Omega} = \frac{\gamma^2}{4k^2 \sin^4 \frac{1}{2}\theta} = \left(\frac{Z_1 Z_2 e^2}{16\pi E \sin^2 \frac{1}{2}\theta} \right)^2 . \tag{231}$$

This is the classical Rutherford cross section. The quantum mechanical phase Φ drops out in the cross section, but it survives in many situations more complicated than scattering from a single point charge. An important example is scattering of identical particles, because the amplitude is then a *coherent* superposition of two terms in which the role of the projectile and target are interchanged (see §8.7). Another is scattering by two spatially separated charges, as in a diatomic molecule, in which case the cross section will have an angular variation that depends on Φ and the dimensionless variable $\hbar ka$, where a is the separation. Yet another is scattering from the combined effects of a Coulomb field and a short range force, and we now turn to this important problem.

(b) The Influence of a Short-Range Interaction

We consider scattering due to the combined effect of the Coulomb interaction and a short-range interaction, and assume the energy is low enough to make the latter insignificant in all but some finite set of low partial waves. The appropriate way of handling this problem is to expand the wave function into partial waves, and to modify only those changed by the short-range interaction. As one might expect from the argument that led to the partial wave phase shifts in §8.1(a), the final result is merely a change of phase.

The starting point is (204) and (213):

$$\psi_{\mathbf{k}}(\mathbf{r}) = A \int_C e^{i\mathbf{k}\cdot\mathbf{r}(1-t)}\, e^{ikrt}\, t^{i\gamma-1}(1-t)^{-i\gamma}\, dt . \tag{232}$$

The first exponential is then expanded into spherical harmonics,

$$\psi_{\boldsymbol{k}}(\boldsymbol{r}) = \sum_{l=0}^{\infty} \sqrt{\frac{2l+1}{2\pi^2}}\, i^l C_l(k;r)\, Y_{l0}(\theta) , \tag{233}$$

where the radial wave function is defined as in §8.2. In the present case it is

$$C_l(k;r) = A(2\pi)^{\frac{3}{2}} \int_C e^{ikrt} j_l(kr(1-t))\, t^{i\gamma-1}(1-t)^{-i\gamma}\, dt . \tag{234}$$

If the entire function $e^{ikrt}j_l(kr(1-t)$ were to be expressed as a power series, each term could be integrated easily. For this purpose set $j_l(z) = e^{iz}z^l f_l(z)$ in Bessel's equation, which then gives the following equation for f_l:

$$zf_l'' + 2(iz + l + 1)f_l' + 2i(l+1)f_l = 0 . \tag{235}$$

This is solved as a power series to give

$$j_l(z) = 2^l z^l e^{iz} \sum_{n=0}^{\infty} \frac{(l+n)!}{(2l+1+n)!} \frac{(-2iz)^n}{n!} . \tag{236}$$

When substituted into (234), the exponentials conveniently cancel, leaving integrals of the type

$$I = \int_C t^{a-1}(1-t)^{b-1}\, dt . \tag{237}$$

The branch cuts from $t = 0$ and $t = 1$ can be merged because $a + b$ is an integer, and the contour C can run around the cut as shown in Fig. 8.5. In evaluating I,

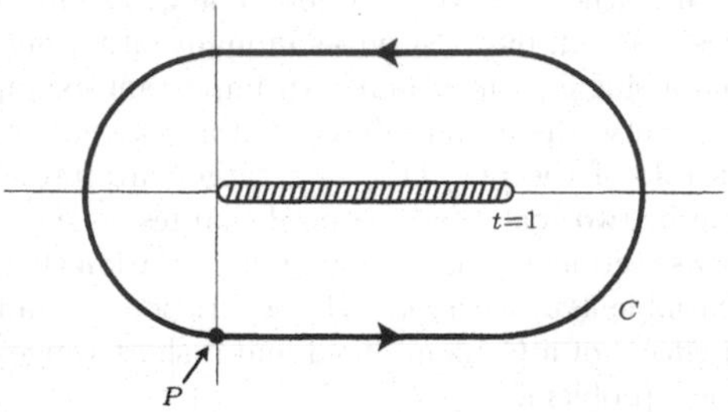

FIG. 8.5. Contour in the t-plane for the integral in Eq. 237.

care must be taken to choose the phase of the multivalued integrand consistently. In (214) the phase of s is taken to be 0 at the point P in Fig. 8.4b, which corresponds to P in the t-plane as shown in Figs. 8.4a and 8.5. Hence

$$I = (1 - e^{2\pi ib}) \int_0^1 t^{a-1}(1-t)^{b-1}\, dt = (1 - e^{2\pi ib}) \frac{\Gamma(a)\Gamma(b)}{\Gamma(a+b)} , \tag{238}$$

and

$$C_l(k;r) = A(2\pi)^{\frac{3}{2}} (2kr)^l e^{ikr} (1 - e^{2\pi\gamma})\Gamma(i\gamma) \sum_n \frac{\Gamma(l+n+1-i\gamma)}{(2l+1+n)!} \frac{(-2ikr)^n}{n!} . \tag{239}$$

The confluent hypergeometric function is defined by the power series

$$ {}_1F_1(a;c;z) = 1 + \frac{a}{c}\frac{z}{1} + \frac{a(a+1)}{c(c+1)}\frac{z^2}{2!} + \ldots ; \tag{240} $$

it too is an entire function of z. On comparing with (239) we see that the radial wave function is

$$ C_l(k;r) = \frac{(2kr)^l e^{ikr} e^{\frac{1}{2}\pi\gamma}}{(2l+1)!} \Gamma(l+1-i\gamma)\, {}_1F_1(l+1-i\gamma; 2l+2; -2ikr) . \tag{241} $$

The asymptotic form of this function is needed to determine the Coulomb phase shifts. This form can be found by the technique that led from (214) to (218), and the direct analogue of the former equation is

$$ {}_1F_1(a;c;z) = \frac{\Gamma(c)}{2\pi i} \int_C e^s (s-z)^{-a} s^{a-c}\, ds . \tag{242} $$

By expanding in powers of z and using (216) one recovers (240). The argument that led to (219) applies here again, and gives

$$ \begin{aligned} {}_1F_1(a;c;z) &= \frac{\Gamma(c)}{2\pi i} \int_{C_1} [(s-z)^{-a} s^{a-c} + e^z (s+z)^{a-c} s^{-a}] e^s\, ds \\ &\sim \frac{\Gamma(c)}{\Gamma(c-a)} (-z)^{-a} + \frac{\Gamma(c)}{\Gamma(a)} e^z z^{a-c} . \end{aligned} \tag{243} $$

When used in (241), this gives the asymptotic form for the radial wave function:

$$ C_l(k;r) \sim -\frac{1}{2ikr} \left\{ e^{-i(kr - \frac{1}{2}l\pi + \gamma \ln 2kr)} - e^{2i\mu_l(k)} e^{+i(kr - \frac{1}{2}l\pi + \gamma \ln 2kr)} \right\} , \tag{244} $$

where the Coulomb phase shifts μ_l are defined by

$$ e^{2i\mu_l(k)} = \frac{\Gamma(l+1-i\gamma)}{\Gamma(l+1+i\gamma)} . \tag{245} $$

Apart from the logarithm in the phases, the asymptotic form (244) of the partial waves is the same as that for interactions that fall off faster than $1/r$, as given by Eq. 94:

$$ \pm\frac{1}{2ikr} e^{\pm i(kr - \frac{1}{2}l\pi)} \to \pm\frac{1}{2ikr} e^{\pm i(kr - \frac{1}{2}l\pi + \gamma \ln 2kr)} \equiv u_l^{(\pm)}(kr) . \tag{246} $$

The effect on the asymptotic wave function produced by an additional short-range interaction is now evident. As $r \to \infty$, only the Coulomb field survives and the asymptotic wave function must be a linear combination of $u_l^{(-)}(kr)$ and $u_l^{(+)}(kr)$, where the former — the incident wave, cannot be influenced by the interaction, while the latter — the outgoing wave, must have an amplitude that cannot be larger than that of the incident wave. Call $\bar{C}_l(k;r)$ the radial wave function when there is a short-range interaction in addition to a Coulomb field. Then the asymptotic form of this function has the general form

$$ \bar{C}_l(k;r) \sim u_l^{(-)}(kr) + e^{2i\Delta_l(k)} u_l^{(+)}(kr) . \tag{247} $$

In those partial waves where the short-range interaction has a negligible effect, $\Delta_l = \mu_l$, but Δ_l will of course differ from the Coulomb phase shift in lower partial waves. There Δ_l will be real if elastic scattering is the only process at the energy in question, but will have a positive imaginary part if there is any inelastic scattering.

A phase shift δ'_l due to the short-range interaction can be defined by

$$\Delta_l(k) = \mu_l(k) + \delta'_l(k) . \tag{248}$$

The function $\delta'_l(k)$ is *not* the phase shift $\delta_l(k)$ produced by the short-range interaction when the projectiles are uncharged, however, because the two interactions produce scattered waves that interfere. Nevertheless, if the energy is such as to make δ_l small, δ'_l will also be small.

The scattering amplitude can now be found in the familiar way by replacing C_l in (233) by $\bar{C}_l$, and taking the asymptotic limit. If $\Psi_{\boldsymbol{k}}$ is the wave function when both interactions are taken into account,

$$\Psi_{\boldsymbol{k}}(\boldsymbol{r}) \sim \psi_{\boldsymbol{k}}(\boldsymbol{r}) + \frac{1}{2ikr} e^{i(kr+\gamma \ln 2kr)} \sum_l \sqrt{\frac{2l+1}{2\pi^2}} e^{2i\mu_l}(e^{2i\delta'_l} - 1) Y_{l0}(\theta) \tag{249}$$

$$= \frac{1}{(2\pi)^{\frac{3}{2}}} \left\{ e^{ikz}\, e^{-i\gamma \ln k(r-z)} + \frac{e^{i(kr+\gamma \ln 2kr)}}{r} [f_C(\theta) + f_{\rm sr}(\theta)] \right\} \tag{250}$$

where

$$f_{\rm sr}(\theta) = \frac{1}{2ik} \sum_l \sqrt{4\pi(2l+1)} e^{2i\mu_l}(e^{2i\delta'_l} - 1)\, Y_{l0}(\theta) . \tag{251}$$

As a consequence the differential cross section is

$$\frac{d\sigma}{d\Omega} = \frac{d\sigma_R}{d\Omega} + |f_{\rm sr}(\theta)|^2 + 2{\rm Re}\,[f_C(\theta) f^*_{\rm sr}(\theta)] . \tag{252}$$

The interference between scattering due to the Coulomb and short range interactions provides information about the latter that is unavailable when it acts alone; in particular, $|f_{\rm sr}|$ does not change if all the phase shifts are reversed. Yet the sign of the phase shift is meaningful, because the interference term depends on the sign of δ'_l, as shown in Prob. 8. This fact has often been used in nuclear and particle physics to gain information about unknown interactions.

8.5 Scattering of Particles with Spin

Scattering often involves particles with spin, and an extension of the formalism developed thus far is needed to handle such collisions. The generalization follows directly from §8.2, and the formulas needed for this purpose will now be developed with a straightforward elaboration of notation.

Once again the target is assumed to be infinitely heavy in comparison to the projectile, the extension to arbitrary masses being the same as in §8.2(d). The incident state will be called $|\boldsymbol{k}_i \nu_i\rangle \equiv |i\rangle$, where ν_i specifies the spin state of the projectile *and* the target. To be specific, if the projectile and target have spins s_1 and s_2 with projections m_1 and m_2, then ν_i is a pair of such eigenvalues specifying the initial state. Experiments select a similarly defined final state $|\boldsymbol{k}_f \nu_f\rangle \equiv |f\rangle$, and the goal is the set of amplitudes $i \to f$. For every incident and final direction, this set forms a matrix of dimension

$$N_S = (2s_1 + 1)(2s_2 + 1)\,. \tag{253}$$

The Hamiltonian is again $H = H_0 + V$, but where V is now an operator that involves the projectile's and target's spin operators $\boldsymbol{s}_1$ and $\boldsymbol{s}_2$. The formalism of §8.2 that culminated in Eq. 138 did not hinge on the assumption that the particles have no spin, and generalizes immediately by enlarging the Hilbert space to include the spin degrees of freedom. The scattering amplitudes are now $f(\boldsymbol{k}_f\nu_f|\boldsymbol{k}_i\nu_i)$, and are again proportional to matrix elements between states of equal energy of the operator

$$T = V + V\frac{1}{E - H + i\epsilon}V\,. \tag{254}$$

The relationship between f and T is that of (141),

$$f(\boldsymbol{k}_f\nu_f|\boldsymbol{k}_i\nu_i) = -4\pi^2 m\langle\boldsymbol{k}_f\nu_f|T|\boldsymbol{k}_i\nu_i\rangle\,. \tag{255}$$

The normalization of the amplitude f is such as to make the asymptotic form of the scattering state

$$\Psi \sim \frac{1}{(2\pi)^{\frac{3}{2}}}\left(e^{i\boldsymbol{k}_i\cdot\boldsymbol{r}}|\nu_i\rangle + \frac{e^{ikr}}{r}\sum_{\nu_f}|\nu_f\rangle\, f(\boldsymbol{k}_f\nu_f;\boldsymbol{k}_i\nu_i)\right)\,, \tag{256}$$

so that the differential cross section for scattering from $|i\rangle$ to $|f\rangle$ is

$$\frac{d\sigma_{i\to f}}{d\Omega} = |f(\boldsymbol{k}_f\nu_f|\boldsymbol{k}_i\nu_i)|^2\,. \tag{257}$$

(a) Symmetry Properties

The symmetry properties of f are more complicated than when there is no spin, and the form of these symmetry statements depend on the definition of the spin projections. We define these projections as being along some fixed axis $\boldsymbol{n}$ in space; the operators $\boldsymbol{s}_i\cdot\boldsymbol{n}$ are then even under space reflection and odd under time reversal. Hence if $\{m_i\}$ designates the eigenvalues of these operators, space reflection provides relationships between scattering amplitudes with the same values of the m_i, while time reversal does so for opposite values of the m_i. (When the helicity basis is used, the rules change in detail because the helicity operators are odd under space reflection and even under time reversal.)

When V is invariant under space reflection, so is T. Furthermore, $I_s|\boldsymbol{k}\nu\rangle = \lambda|-\boldsymbol{k}\nu\rangle$, where λ is the product of the intrinsic parities of the two particles. Therefore reflection invariance requires

$$f(-\boldsymbol{k}_f\nu_f|-\boldsymbol{k}_i\nu_i) = f(\boldsymbol{k}_f\nu_f|\boldsymbol{k}_i\nu_i) \qquad \text{(reflection invariance)}\,. \tag{258}$$

It is much more convenient to express this and other symmetry properties of the scattering amplitudes in terms of a scattering operator M that acts only in the spin space, and which is defined by

$$\langle\nu_f|M(\boldsymbol{k}_f,\boldsymbol{k}_i)|\nu_i\rangle \equiv -4\pi^2 m\langle\boldsymbol{k}_f\nu_f|T|\boldsymbol{k}_i\nu_i\rangle\,. \tag{259}$$

M is a function of the initial and final momenta, and the two spin operators $\boldsymbol{s}_1$ and $\boldsymbol{s}_2$; in terms of M, the asymptotic state (256) is

$$\Psi \sim (2\pi)^{-\frac{3}{2}} \left(e^{i\boldsymbol{k}_i \cdot \boldsymbol{r}} + \frac{e^{ikr}}{r} M(\boldsymbol{k}_f, \boldsymbol{k}_i) \right) |\nu_i\rangle \,. \tag{260}$$

Rotational invariance requires M to be a scalar or pseudoscalar formed from the four vectors $(\boldsymbol{k}_i, \boldsymbol{k}_f, \boldsymbol{s}_1, \boldsymbol{s}_2)$. Because the momenta are polar vectors while the spin operators are axial vectors, space reflection invariance requires

$$M(\boldsymbol{k}_f, \boldsymbol{k}_i; \boldsymbol{s}_1, \boldsymbol{s}_2) = M(-\boldsymbol{k}_f, -\boldsymbol{k}_i; \boldsymbol{s}_1, \boldsymbol{s}_2) \qquad \text{(reflection invariance)} \,. \tag{261}$$

Time reversal invariance is, as always, more complicated, until the intuitively "obvious" final result is reached. The identity relating matrix elements under time reversal, whether or not it is a symmetry, is (132),

$$\langle f|T|i\rangle = \langle i^R|I_t T^\dagger I_t^{-1}|f^R\rangle \,, \tag{262}$$

where $|i^R\rangle$ and $|f^R\rangle$ are the motion-reversed states, e.g., $|i^R\rangle = |-\boldsymbol{k}_i - \nu_i\rangle$. If V is invariant under time reversal, $I_t T^\dagger I_t^{-1} = T$, and consequently time reversal invariance requires the intuitively expected relation

$$\langle f|T|i\rangle = \langle i^R|T|f^R\rangle \,. \tag{263}$$

In terms of M, this is

$$\langle \nu_f|M(\boldsymbol{k}_f, \boldsymbol{k}_i; \boldsymbol{s}_1, \boldsymbol{s}_2)|\nu_i\rangle = \langle \nu_i^R|M(-\boldsymbol{k}_i, -\boldsymbol{k}_f; \boldsymbol{s}_1, \boldsymbol{s}_2)|\nu_f^R\rangle \,. \tag{264}$$

To obtain a relationship for M itself, the matrix element on the right side must be between the same spin states as on the left; to bring this about, we introduce the time reversal operator in the combined spin space, $\mathcal{T}$, in terms of which

$$\langle \nu_i^R|M(-\boldsymbol{k}_f, -\boldsymbol{k}_i; \boldsymbol{s}_1, \boldsymbol{s}_2)|\nu_f^R\rangle = \langle \nu_f|\mathcal{T} M^\dagger(-\boldsymbol{k}_f, -\boldsymbol{k}_i; \boldsymbol{s}_1, \boldsymbol{s}_2)\mathcal{T}^{-1}|\nu_i\rangle \,. \tag{265}$$

But $\boldsymbol{s}_i$ is Hermitian, and $\mathcal{T}\boldsymbol{s}_i\mathcal{T}^{-1} = -\boldsymbol{s}_i$. The Hermitian conjugation required by $M^\dagger$ produces a complex conjugation of numerical functions of the momenta in M, which is then undone by the antiunitary operator $\mathcal{T}$. Hence the final result is again intuitively "obvious":

$$M(\boldsymbol{k}_f, \boldsymbol{k}_i; \boldsymbol{s}_1, \boldsymbol{s}_2) = M(-\boldsymbol{k}_i, -\boldsymbol{k}_f; -\boldsymbol{s}_1, -\boldsymbol{s}_2) \qquad \text{(time reversal)} \,. \tag{266}$$

(b) Cross Section and Spin Polarization

Scattering due to a spin-dependent interaction can produce a spin-polarized final state even when the incident state is unpolarized. This sentence already indicates that the density matrix in the spin space will be needed to describe such processes.

Let $\{\nu_n\}$ be an orthonormal basis in the two-particle spin space, and $p_{i,n}$ the probability that the incident state is in $|\nu_n\rangle$. The incident state is then described fully by the density matrix

$$\rho_i = \sum_n |\nu_n\rangle p_{i,n} \langle \nu_n| \,. \tag{267}$$

If this is a totally unpolarized state,

$$\rho_i = \frac{1}{N_S} . \tag{268}$$

Because of the linearity of the Schrödinger equation, the density matrix ρ_f of the final state follows directly from ρ_i once the scattering amplitude M is known. That is, when the incident spin state is $|\nu_n\rangle$, the scattered spin state in the final direction $\boldsymbol{k}_f$ will be $M(\boldsymbol{k}_f, \boldsymbol{k}_i)|\nu_n\rangle$. Hence the superposition principle implies that the density matrix of the final state is

$$\rho_f = \frac{M\,\rho_i\,M^\dagger}{\mathrm{Tr}\ \rho_i M^\dagger M} , \tag{269}$$

where the denominator guarantees Tr $\rho_f = 1$. Of course, (269) is just shorthand for

$$\rho(\boldsymbol{k}_f) = \frac{M(\boldsymbol{k}_f, \boldsymbol{k}_i)\,\rho(\boldsymbol{k}_i)\,M^\dagger(\boldsymbol{k}_f, \boldsymbol{k}_i)}{\mathrm{Tr}\ \rho_i M^\dagger M} . \tag{270}$$

Compact and convenient expressions now follow for the cross section and the expectation values of the spins after scattering. The cross section for scattering from $|\boldsymbol{k}_i\nu_n\rangle$ to $|\boldsymbol{k}_f\nu_m\rangle$ is $|\langle\nu_m|M|\nu_n\rangle|^2$; as the probability of being in this initial state is $p_{i,n}$, the cross section when no spin orientations are selected after scattering is

$$\frac{d\sigma}{d\Omega} = \sum_{n,m} |\langle\nu_m|M(\boldsymbol{k}_f, \boldsymbol{k}_i)|\nu_n\rangle|^2\, p_{i,n} , \tag{271}$$

which is

$$\frac{d\sigma}{d\Omega} = \mathrm{Tr}\ \rho_i\, M^\dagger M . \tag{272}$$

The expectation value of the projectile's spin after scattering is

$$\langle \boldsymbol{s}_1 \rangle = \mathrm{Tr}\ \boldsymbol{s}_1\, \rho_f = \frac{\mathrm{Tr}\ \boldsymbol{s}_1\, M \rho_i\, M^\dagger}{d\sigma/d\Omega} \tag{273}$$

according to (269), with a similar expression for the spin of the target should it also have spin. When both particles have spin, there will, in general, be correlations between the two spins following scattering even if there is none to begin with. Examples of this will appear in §8.6(a) and §8.7(b).

(c) Scattering of a Spin $\frac{1}{2}$ Particle by a Spin 0 Target

This is the simplest example of scattering involving spin, and it has many applications aside from being instructive.

The matrix M is now two-dimensional, and can be expressed most conveniently in terms of the Pauli matrix $\boldsymbol{\sigma}$ belonging to the projectile. It is at most linear in $\boldsymbol{\sigma}$, and a scalar or pseudoscalar under simultaneous rotation of $\boldsymbol{\sigma}, \boldsymbol{k}_i$ and $\boldsymbol{k}_f$. Hence it has the form

$$M = g_1 + \boldsymbol{\sigma}\cdot(\boldsymbol{k}_i \times \boldsymbol{k}_f)\, g_2 + \boldsymbol{\sigma}\cdot(\boldsymbol{k}_i + \boldsymbol{k}_f)\, g_3 + \boldsymbol{\sigma}\cdot(\boldsymbol{k}_i - \boldsymbol{k}_f)\, g_4 , \tag{274}$$

where the g_i are functions of the rotationally invariant quantities $k = |\boldsymbol{k}_i| = \boldsymbol{k}_f|$ and $\theta = \angle(\boldsymbol{k}_f, \boldsymbol{k}_i)$.

Invariance under space reflection and time reversal eliminates some of the terms in (274). As the momenta are polar vectors and the spin is an axial vector, reflection invariance requires $g_3 = g_4 = 0$. Under time reversal, $\boldsymbol{\sigma} \to -\boldsymbol{\sigma}$, and $\boldsymbol{k}_i \to -\boldsymbol{k}_f, \boldsymbol{k}_f \to -\boldsymbol{k}_i$, so this invariance requires only $g_4 = 0$. In this example, therefore, space reflection invariance is more restrictive, in that it also eliminates the term ruled out by time reversal. For that reason we will examine the former, i.e., restrict ourselves to discussing the scattering matrix

$$M = g(k,\theta) + \boldsymbol{\sigma} \cdot \boldsymbol{n}\, h(k,\theta) \,, \tag{275}$$

where

$$\boldsymbol{n} = \frac{\boldsymbol{k}_i \times \boldsymbol{k}_f}{|\boldsymbol{k}_i \times \boldsymbol{k}_f|} \tag{276}$$

is a unit vector normal to the scattering plane.

The functions g and h can only be found by solving the Schrödinger equation for a given interaction V. The form of this interaction is, of course, also severely restricted in that it is a linear form in $\boldsymbol{\sigma}$ which is invariant under space reflection. As $\boldsymbol{L}$, the orbital angular momentum of the projectile, is the only axial vector available in the Hilbert space of the spatial coordinate, the interaction that leads to (275) is the familiar spin-orbit coupling,

$$V = V_0(r) + \boldsymbol{\sigma} \cdot \boldsymbol{L}\, V_1(r) \,. \tag{277}$$

When the incident beam is unpolarized, $\rho_i = \frac{1}{2}$. The cross section (272) is then

$$\frac{d\sigma}{d\Omega} = \tfrac{1}{2}\mathrm{Tr}\; (g^* + \boldsymbol{\sigma} \cdot \boldsymbol{n}\, h^*)(g + \boldsymbol{\sigma} \cdot \boldsymbol{n}\, h) \,. \tag{278}$$

But $\mathrm{Tr}\; \boldsymbol{\sigma} = 0$ and $(\boldsymbol{\sigma} \cdot \boldsymbol{n})^2 = 1$, and therefore

$$\frac{d\sigma}{d\Omega} = |g(k,\theta)|^2 + |h(k,\theta)|^2 \,. \tag{279}$$

Hence for an unpolarized incident beam, the cross section does not depend on the azimuthal angle, as is clear from cylindrical symmetry. This result holds much more generally for arbitrary spins (see Prob. 9).

We define the polarization of the scattered beam as

$$\boldsymbol{P}_f = \mathrm{Tr}\; \boldsymbol{\sigma} \rho_f \,. \tag{280}$$

Equation (273) can be used to find $\boldsymbol{P}_f$. If the incident beam is unpolarized, the numerator of (273) is

$$\tfrac{1}{2}\mathrm{Tr}\; \boldsymbol{\sigma}\, (g + \boldsymbol{\sigma} \cdot \boldsymbol{n}\, h)(g^* + \boldsymbol{\sigma} \cdot \boldsymbol{n}\, h^*) = \boldsymbol{n}\, (gh^* + g^* h) \,, \tag{281}$$

and therefore

$$\boldsymbol{P}_f = \boldsymbol{n}\, \frac{2\,\mathrm{Re}\, gh^*}{|g|^2 + |h|^2} \,. \tag{282}$$

Thus the polarization after scattering, $\boldsymbol{P}_f$, is normal to the scattering plane. A short calculation will show that if terms that are not reflection invariant are kept in M, then $\boldsymbol{P}_f$ will have a component in the scattering plane.

That $\boldsymbol{P}_f$ is normal to the scattering plane when the interaction is reflection invariant can be understood from a simple geometrical consideration. Arguments of this type are important in that they can often lead directly to results of great generality without resorting to a cloud of formalism. Consider the scattering experiment depicted in Fig. 8.6. The z-axis is toward the reader, unmarked arrows are the

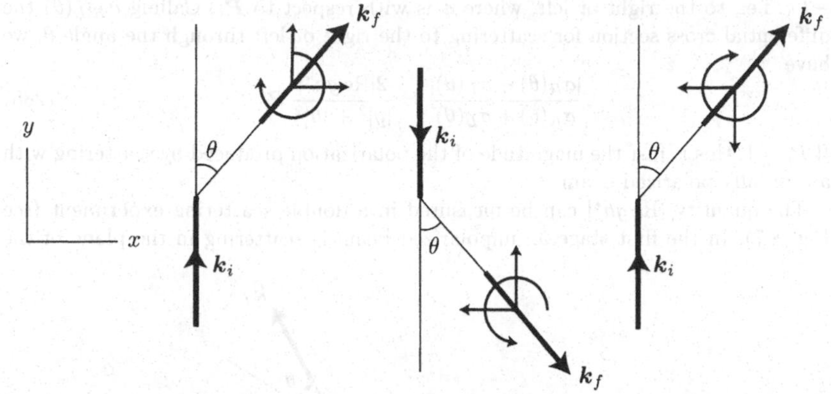

FIG. 8.6. Implications of rotation and reflection invariance for scattering of a particle with spin from a spin 0 target: starting from the configuration on the left, the reflection $y \to -y$, followed by a rotation through π about x-axis. In the text, (b) refers to the central configuration.

components of $\boldsymbol{P}_f$ in the scattering plane, and the circular arrows give the sense of $\boldsymbol{P}_f$ along the z-axis, that is, the sign of $\boldsymbol{P}_f \cdot \boldsymbol{n}$. That $\boldsymbol{P}_f$ is an axial vector will be essential. Now perform the reflection $y \to -y$ followed by the rotation π about the x-axis. If rotation and reflection invariance holds, then the probabilities for the collision processes shown in these figures must *all be equal.* Observe, however, that the components of $\boldsymbol{P}_f$ in the scattering plane have both changed sign after these operations, and must therefore vanish, whereas the component normal to the scattering plane has not. By rotating (b) through π about the z-axis, we also see that the probability for scattering through some angle θ to the right and having $\boldsymbol{P}_f$ up (down) must equal that for scattering to left through the same angle and finding $\boldsymbol{P}_f$ down (up). Equation (282) incorporates all these facts, as it must.

This geometric argument establishes a much more general result: *When an unpolarized beam of arbitrary spin is scattered from an unpolarized target of arbitrary spin, the projectiles' final spin polarizations are normal to the scattering plane if the interaction is reflection invariant.*

Finally we address the question of measuring the polarization induced by scattering. Various techniques are available, but we restrict ourselves to using scattering for this purpose. The principle on which this technique is based is that the scattering cross section of a polarized beam has an azimuthal dependence even when no spins are actually measured. This can be seen by simply evaluating (272) when $\boldsymbol{P}_i \neq 0$:

$$\frac{d\sigma}{d\Omega} = \tfrac{1}{2}\mathrm{Tr}\ (1 + \boldsymbol{\sigma} \cdot \boldsymbol{P}_i)(g^* + \boldsymbol{\sigma} \cdot \boldsymbol{n}\, h^*)(g + \boldsymbol{\sigma} \cdot \boldsymbol{n}\, h)\,. \tag{283}$$

The identities satisfied by the Pauli matrices reduce this expression to

$$\frac{d\sigma}{d\Omega} = |g|^2 + |h|^2 + 2(\boldsymbol{P}_i \cdot \boldsymbol{n})\mathrm{Re}\, gh^* . \tag{284}$$

When the scattering plane is perpendicular to $\boldsymbol{P}_i$, the cross section for a fixed value of θ differs depending on whether the scattering is through the azimuth $\phi = \frac{1}{2}\pi$ or $-\frac{1}{2}\pi$, i.e., to the right or left, where ϕ is with respect to $\boldsymbol{P}_i$. Calling $\sigma_{R,L}(\theta)$ the differential cross section for scattering to the right or left through the angle θ, we have

$$\frac{|\sigma_R(\theta) - \sigma_L(\theta)|}{\sigma_R(\theta) + \sigma_L(\theta)} = \frac{2|\mathrm{Re}\, gh^*|}{|g|^2 + |h|^2} P_i . \tag{285}$$

If $P_i = 1$, this is just the magnitude of the polarization produced by scattering with an initially polarized beam.

The quantity $|\mathrm{Re}\, gh^*|$ can be measured in a double scattering experiment (see Fig. 8.7). In the first stage an unpolarized beam is scattering in the plane of the

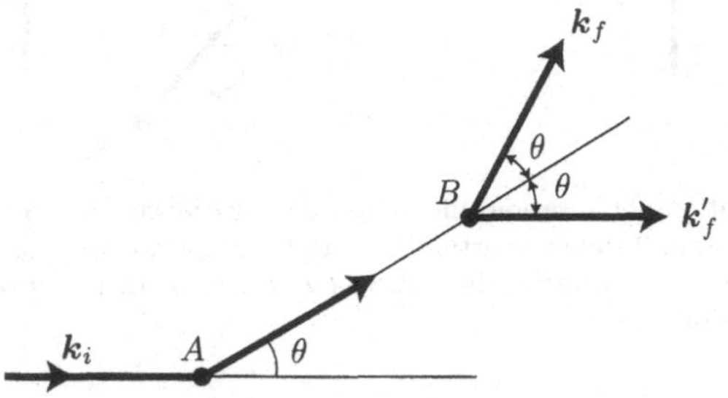

FIG. 8.7. Double scattering experiment for measuring polarization; A and B are the first and second target, which are both unpolarized.

paper, with the interaction playing the role of a polarizer. The scattered beam is then scattered again through the same angle to the right and left from an identical target, which now acts as the analyzer. The polarization produced in the first stage is (282), and therefore the left-right asymmetry in the second is

$$\left(\frac{2\mathrm{Re}\, gh^*}{|g|^2 + |h|^2} \right)^2 . \tag{286}$$

The denominator $(|g|^2 + |h|^2)$ is given by the differential cross section in the first stage, which then yields the promised result. The phase of $\mathrm{Re}\, gh^*$ cannot be found in this way, however; some further interference effect is needed.

8.6 Neutron-Proton Scattering and the Deuteron

The deuteron is, so to say, the hydrogen atom of nuclear physics, and for that reason it was studied extensively a long time ago. Nevertheless, it continues to provide a very instructive illustration of how scattering and bound state data can be used to gain information about an unknown interaction.

The basic facts about the deuteron (d) are that it is the only bound state of the np system. It has binding energy 2.23 MeV, spin 1, mean square charge radius $\bar{r} \simeq 2$ fm, magnetic moment 0.857 μ_N (where μ_N is the nucleon Bohr magneton), and quadrupole moment 0.29 fm^2 e.

To the trained eye, this information is very revealing. The characteristic binding energy of nuclei is about 8 MeV per nucleon, so d is loosely bound by nuclear standards. As there is no $J = 0$ bound state, the interaction must be spin-dependent. The sum of the neutron and proton magnetic moments is $0.879\mu_N$, which leaves but little room for any contribution to the magnetic moment from orbital motion of the proton. And the nonzero quadrupole moment is especially intriguing.

The most general np state with $J = 1$ has the form

$$\Psi = a|^3S_1\rangle + b|^3P_1\rangle + c|^1P_1\rangle + d|^3D_1\rangle \,. \tag{287}$$

Of these only $^3S_1$ has the magnetic moment $\mu_p + \mu_n$; the other have contributions of order μ_N from orbital motion of the proton, so the first term in Ψ must dominate. Furthermore, if the interaction is reflection invariant, either $b = c = 0$ or $a = d = 0$. In view of the preceding sentence, the only credible choice is $b = c = 0$, and $|d| \ll |a|$, so that $|a| \simeq 1$. The state $^3S_1$ has a spherically symmetric charge distribution and cannot have a quadrupole moment, so $d \neq 0$. The expectation value of the quadrupole moment will then be

$$\bar{Q} = d\langle ^3S_1|Q|^3D_1\rangle + \text{c.c.} + O(d^2) \,, \tag{288}$$

and $|d| \ll 1$ is consistent with the fact that $\bar{Q} \ll \bar{r}^2 e$.

The necessity of having both an S and D component in Ψ means that the np interaction does not conserve the orbital angular momentum $\boldsymbol{L}$, only the total angular momentum $\boldsymbol{J}$. But the small value of the quadrupole moment implies that to a first approximation the violation of $\boldsymbol{L}$ can be ignored, and for that reason we first ask what can be learned about the nuclear interaction from np scattering on the assumption that $\boldsymbol{L}$ is a constant of motion.

(a) Low-Energy Neutron-Proton Scattering

We work in the center-of-mass system. The reduced mass is[1]

$$\frac{m_n m_p}{m_n + m_p} \equiv \tfrac{1}{2}m \,, \tag{289}$$

where $m \simeq 939$ MeV. The following are convenient kinematic relationships:

$$E \equiv E_{\text{com}} = \tfrac{1}{2}E_{\text{lab}} = k^2/m \,, \tag{290}$$

and

$$k(\text{fm}^{-1}) = 0.1098\,\sqrt{E_{\text{lab}}(\text{MeV})} \,, \qquad \lambda\!\!\!^{-} \equiv k^{-1} = 0.567\,\sqrt{300°\text{K}/kT}\ \text{Å} \,, \tag{291}$$

where the latter is useful when slow neutrons are being used. Note that $k\bar{r} \ll 1$ if $E_{\text{lab}} \ll 100$ MeV, so s-wave scattering dominates for such low energies.

---

[1] $m_p = 938.3$ MeV and $m_n - m_p = 1.29$ MeV.

Given that ^3S is bound while ^1S is not, we consider first the following interaction which distinguishes between spin singlets and triplets, indicated by s and t:

$$V = V_s(r)P_s + V_t(r)P_t \, , \tag{292}$$

where the singlet and triplet projection operators are defined already in §3.5(b):

$$P_s = \tfrac{1}{4}(1 - \boldsymbol{\sigma}_n \cdot \boldsymbol{\sigma}_p) \, , \qquad P_t = \tfrac{1}{4}(3 + \boldsymbol{\sigma}_n \cdot \boldsymbol{\sigma}_p) \, . \tag{293}$$

With this interaction the scattering problem separates into two uncoupled "channels," one for spin singlets and the other for spin triplets. For some arbitrary initial np spin state $|\nu_i\rangle$, the asymptotic state will have the form of Eq. 260, with the scattering matrix being

$$M(\boldsymbol{k}_f, \boldsymbol{k}_i) = f_s(k,\theta)P_s + f_t(k,\theta)P_t \, . \tag{294}$$

At low energies it is appropriate to express the amplitudes $f_{s,t}$ as partial wave expansions,

$$f_a(k,\theta) = \frac{1}{k}\sum_l (2l+1)e^{i\delta_l^a} \sin\delta_l^a \, P_l(\cos\theta) \, , \tag{295}$$

where the δ_l^a are the phase shifts produced separately by the potentials V_a, with a being s or t. (Here θ is the scattering angle in the center-of-mass, and $\theta = 2\theta_{\text{lab}}$.)

For an unpolarized beam and target, the initial spin-space density matrix is $\rho_i = \frac{1}{4}$. According to Eq. 272, the differential cross section is then

$$\frac{d\sigma}{d\Omega} = \tfrac{1}{4}\text{Tr}\ \left(P_s|f_s|^2 + P_t|f_t|^2\right) \, . \tag{296}$$

But Tr $P_s = 1$, Tr $P_t = 3$, and therefore

$$\frac{d\sigma}{d\Omega} = \tfrac{1}{4}|f_s(k,\theta)|^2 + \tfrac{3}{4}|f_t(k,\theta)|^2 \, , \tag{297}$$

where the relative weights (1:3) where to be expected.

The density matrix following scattering is instructive. In all generality, the spin-space density matrix of the np system has the form

$$\rho = \frac{1}{4}\left(1 + \boldsymbol{\sigma}_n \cdot \boldsymbol{P}_n + \boldsymbol{\sigma}_p \cdot \boldsymbol{P}_p + \sum_{i,j=1}^{3} \sigma_{n,i}\sigma_{p,j}C_{ij}\right) \, , \tag{298}$$

where $\boldsymbol{P}_{n,p}$ are the separate spin polarizations and C_{ij} is a measure of the correlations between the spins. In conventional scattering experiments, one or both polarizations may exist in the initial state, but it will not have any correlations. It is interesting, therefore, that even when the initial spin state is totally random, the interaction (292) will produce such correlations following scattering. To see this, we compute the density matrix ρ_f following scattering from Eq. 269:

$$\rho_f = \tfrac{1}{4}\left(d\sigma/\Omega\right)^{-1}\{|f_s|^2P_t + |f_t|^2P_t\} \tag{299}$$

$$= \frac{(d\sigma/d\Omega) + \frac{1}{4}\boldsymbol{\sigma}_n \cdot \boldsymbol{\sigma}_p(|f_t|^2 - |f_s|^2)}{4\, d\sigma/d\Omega} \, . \tag{300}$$

After scattering, therefore,

$$C_{ij} = \delta_{ij} \frac{|f_t|^2 - |f_s|^2}{4(d\sigma/d\Omega)} , \qquad \boldsymbol{P}_n = \boldsymbol{P}_p = 0 . \tag{301}$$

The nonzero spin correlation, when the interactions in the singlet and triplet states differ, is to be expected because the interaction distinguishes between the "parallel" and "antiparallel" spin configurations, and therefore scatters one or the other more strongly, as is then reflected in the sign of C_{ij}. At first sight it is surprising, however, that this interaction produces no polarization of the target or the projectile spins. On further thought this result is seen to be a consequence of the symmetries of the interaction: it is invariant if *both* spins are rotated simultaneously, while $\boldsymbol{r}$ is fixed, or vice versa, and therefore there is no spatial direction with respect to which a polarization of the target or projectile spin could even be defined.

(b) The Deuteron and Low-Energy np *Scattering*

We consider the $L = 0$ radial equation for the function $u(r) = rR_0(r)$:

$$\left(\frac{d^2}{dr^2} - mV(r) + k^2 \right) u(r) = 0 , \tag{302}$$

where V is either V_s or V_t. An s-wave bound state, with energy $E = -\alpha^2/m$, is a solution of this equation provided k is replaced by $i\alpha$, with $\alpha \geq 0$.

For orientation, consider a square well of radius R and depth V_0. Continuity at $r = R$ leads to the eigenvalue condition

$$K \cot KR = -\alpha , \qquad K = \sqrt{mV_0 - \alpha^2} , \qquad \alpha^2 < mV_0 . \tag{303}$$

As the deuteron is weakly bound, we are interested in the case $\alpha^2 \ll mV_0$, so that $K \simeq \sqrt{mV_0} \equiv K_0$, and $\cot K_0 R \simeq 0$, or $K_0 R \simeq \frac{1}{2}\pi$. Expanding $K \cot KR$ about this value then gives the solution

$$\alpha R \simeq \tfrac{1}{2}\pi (K_0 R - \tfrac{1}{2}\pi) . \tag{304}$$

Scattering states have the asymptotic form $\sin(kr + \delta)$, where δ is the s-wave phase shift, with the subscript indicating $L = 0$ suppressed as for now we will only be dealing with this partial wave. In the case of the square well, for $r < R$ the solution is $\sin Kr$, where

$$K = \sqrt{mV_0 + k^2} , \tag{305}$$

while for $r > a$ it is $A \sin(kr + \delta)$. Matching the interior and exterior forms at $r = R$ then gives

$$\tan(\delta + kR) = \frac{k}{K} \tan KR . \tag{306}$$

As the energy tends to zero, the right-hand side of (306) tends to zero, and therefore the s-wave phase shift tends to zero linearly with k:

$$\delta = -ka \qquad (k \to 0) , \tag{307}$$

where a is the *scattering length*. The s-wave amplitude is $e^{i\delta}\sin\delta/k$, and its low-energy limit is therefore

$$f(k) = -a\,, \qquad (k \to 0)\,. \tag{308}$$

When only s-waves scatter, the total cross section is $(4\pi/k^2)\sin^2\delta$, and therefore the zero-energy limit of the total cross section is compactly expressed in terms of the scattering length:

$$\sigma_{\text{tot}} = 4\pi a^2 \qquad (k \to 0)\,. \tag{309}$$

Equations (308) and (309) hold in general for any interaction that falls off faster than $1/r$, though the scattering length does of course depend on the interaction. According to (306), for the square well it is

$$a = R\left(1 - \frac{1}{K_0 R \cot K_0 R}\right)\,. \tag{310}$$

This expression for the scattering length is specific to the square well potential, but it reveals a property of the zero-energy scattering cross section that is universal: *As a function of the characteristic strength u of the interaction, the scattering length has a singularity as u passes through the value u_0 needed to bind an s state, and therefore the zero-energy cross section tends to infinity as $u \to u_0$.* In the case of the square well, $u = K_0R$, and $u_0 = \frac{1}{2}\pi$, so in the vicinity of $K_0R = \frac{1}{2}\pi$, Eq. 310 becomes

$$a = \frac{2R}{\pi(K_0R - \frac{1}{2}\pi)}\,, \qquad K_0R \simeq \tfrac{1}{2}\pi\,. \tag{311}$$

Equation (304) states that there is a (weakly) bound state if K_0R is somewhat larger than $\frac{1}{2}\pi$, and therefore the scattering length tends to $+\infty$ as the binding energy tends to zero, while $a \to -\infty$ as the strength parameter climbs up to the critical value $\frac{1}{2}\pi$ (see Fig. 8.8). Put more succinctly, if K_0R is replaced by the complex

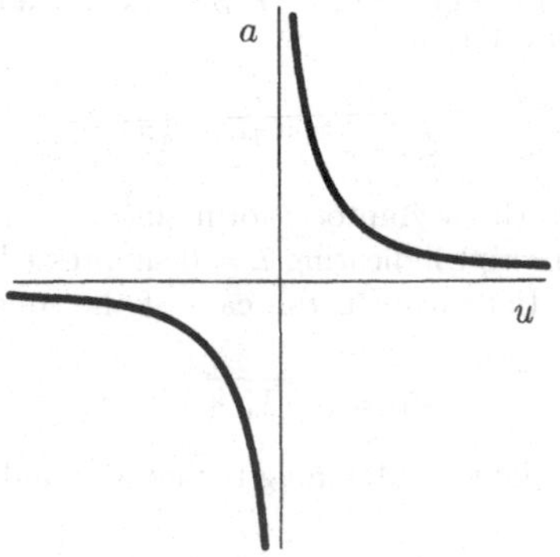

FIG. 8.8. The scattering length a as a function of the interaction strength u, when the latter is close to the value required to bind a state; the bound state exists above the singularity.

variable u, the scattering length, considered as a function of u, has a simple pole at $u = \frac{1}{2}\pi$, the value required to just bind a state. This behavior of the scattering length is not a property unique to the square well.[1] For example, the separable potential of §8.2(b) and Prob. 5 has the scattering length

$$a = \frac{2\xi b}{\xi - 1}, \tag{312}$$

where b is the range and $\xi = 2\pi\lambda b$, with λ the strength parameter; in contrast to (311), Eq. 312 holds for all values of the parameters. When $\xi \geq 1$ there is a bound state, and therefore the scattering length for the separable potential has the same singular behavior as for the square well. Another example arises in Prob. 6.

These examples show that the zero-energy cross section (Eq. 309) can be arbitrarily large in comparison to the square of the interaction range, the area one would expect classically. This is not a uniquely quantum mechanical feature, of course; it has analogues in all wave phenomena.

The fact that low-energy scattering by a finite range interaction is parameterized by just one number, the scattering length a, means that problems involving many scatterers can be greatly simplified by replacing the true interaction by an effective interaction, which is then handled with the Born approximation. This is the Fermi pseudopotential,

$$V_F(\mathbf{r}_1 - \mathbf{r}_2) = \frac{2\pi}{m_{\text{red}}}\, a\, \delta^3(\mathbf{r}_1 - \mathbf{r}_2). \tag{313}$$

When the Born approximation is used, this will immediately produce the correct s-wave scattering amplitude, as the reader should verify. To treat scattering, say, of neutrons, by a set of nuclei in a molecule, crystal or fluid, etc., with various neutron-nucleus scattering lengths, the actual interactions are replaced by such pseudo-potentials. The pseudo-potential must only be used with the first Born approximation, however, for that is how it is defined; indeed, the second Born approximation using V_F is a divergent integral!

Equation (307) is the first term in what is called *the effective range expansion*, namely,

$$k \cot \delta = -\frac{1}{a} + \tfrac{1}{2} r_0 k^2 + O(k^4). \tag{314}$$

The merit of this expansion is that the effective range, r_0, in contrast to the scattering length, is a parameter that is of order the range of the interaction and insensitive to whether or not there is a weakly bound state. The expansion therefore demonstrates that low-energy scattering data gives only limited information about the interaction, which is hardly surprising. The expression for r_0 is rather complicated for the square well potential, but is simply related to the scattering length when $|a| \gg R$:

$$r_0 \simeq R + \frac{1}{a}\frac{1}{mV_0} + O(a^{-2}). \tag{315}$$

In the case of the separable potential in Prob. 5, the exact formula for the effective range is $b(1+2\xi^{-1})$, which, like (315), shows that r_0 is indeed of order the interaction range and insensitive to its strength.

[1] See K. Gottfried, *Quantum Mechanics*, W. A. Benjamin (1966), §49.3.

Interesting consequences emerge when the low-energy s-wave scattering amplitude is expressed in terms of the scattering length and effective range:

$$f_0(k) \equiv \frac{1}{k} e^{i\delta} \sin\delta = \frac{1}{k\cot\delta - ik} \simeq \frac{1}{-a^{-1} + \frac{1}{2} r_0 k^2 - ik} \,. \tag{316}$$

In §8.2(b) we learned from the separable potential that the s-wave scattering amplitude, considered as a function of the complex variable z, with Re $z = k$, has a simple pole at $z = i\alpha$ if there is a bound s-state of energy $-\alpha^2/2\mu$, where the reduced mass μ in the present example is $\frac{1}{2}m$. This is not a quirk of this particular interaction, as will be demonstrated in a moment. If this claim is accepted for now, then (316) provides a relationship between the scattering length, the effective range and the binding energy of a weakly bound state. For if the binding is weak, the size $1/\alpha$ of the bound state is large compared to the range of the interaction, i.e., compared to r_0, and the small k expansion is valid in a circle about the origin in the z-plane that contains the bound state pole at $i\alpha$. This pole is at the zero of the denominator of Eq. 316, i.e.,

$$\frac{1}{a} = \alpha - \tfrac{1}{2} r_0 \alpha^2 \,. \tag{317}$$

The deuteron offers a test of this relationship. In the spin triplet channel the scattering length and effective range are 5.42 fm and 1.75 fm, respectively, while the binding energy of 2.23 MeV implies that $\alpha = 0.232$ fm^{-1}, and with these values both sides of Eq. 317 have the value 0.185 fm^{-1} !

The np singlet channel is also interesting because in it the scattering length is vary large and negative, -23.7 fm, which means that the spin-singlet interaction is nearly strong enough to also bind a state.

In §8.2(c) we already saw that a bound state appears as a singularity in the scattering amplitude, and we return to this from a slightly different perspective. From §8.2(a), recall that for any potential that falls off faster than $1/r$ the asymptotic form of an s-wave scattering state is

$$u(k;r) \sim e^{-ikr} - e^{2i\delta(k)} e^{ikr} \,, \tag{318}$$

apart from a normalization-dependent factor, while the asymptotic form of a bound state is $e^{-\alpha r}$. The two differential equations that determine these solutions are combined into one by extending the definition of k to the complex plane. Consider, then, the replacement $k \to i\alpha$ in (318):

$$e^{\alpha r} - e^{2i\delta(i\alpha)} e^{-\alpha r} \,. \tag{319}$$

If this is to become $e^{-\alpha r}$, it must be that $\exp[2i\delta(i\alpha)] \to \infty$ as α approaches the value appropriate to a bound state. But $\cot\delta \to i$ as $e^{2i\delta} \to \infty$. Recalling (316), this means that *a bound state of energy* $-\alpha^2/m$ *produces a simple pole at* $k = i\alpha$ *in the scattering amplitude.* This argument evidently applies to all partial waves — to bound states and scattering amplitudes of all angular momenta.

(c) Neutron Scattering by the Hydrogen Molecule

The sign of the np scattering length cannot be determined from low-energy np scattering itself; interference in some form or other is required. A particularly interesting example is neutron scattering by the H_2 molecule, because a crucial role is played by the identity of the two protons.

The essential features can be understood by modeling the molecule as two protons separated by a constant distance R, and only free to rotate about their center-of-mass, which is taken as fixed. The spatial wave functions of this toy molecule are $Y_{LM}(\boldsymbol{n})$, where $\boldsymbol{n}$ is a unit vector along the line between the protons. The Pauli principle requires the protons to be in the spin singlet state if L is even and in the triple state if L is odd. The incident neutron energy is assumed to be low enough so that the scattering is fully described by the singlet and triplet scattering lengths a_s and a_t.

In this low-energy regime, the scattering amplitude can be evaluated by replacing the neutron-molecule interaction by the appropriate Fermi pseudo-potential:

$$V_F(\boldsymbol{r}) = \frac{2\pi}{m}[A_1\delta(\boldsymbol{r} - \tfrac{1}{2}\boldsymbol{R}) + A_2\delta(\boldsymbol{r} + \tfrac{1}{2}\boldsymbol{R})] , \tag{320}$$

where $\boldsymbol{r}$ is the neutron's coordinate, m the neutron's mass, and A_i is the spin-dependent scattering length appropriate to proton i. Because the molecule has been nailed down by fiat, it is the neutron's mass that appears in (320).

Scattering can produce the transition $(LM\mu SM_S) \to (L'M'\mu' SM'_S)$, where (LM) are the orbital angular momentum quantum numbers of the molecular rotation, (SM_S) the *total* spin quantum numbers of the protons, and μ the neutron's spin projection. The pseudo-potential must, as explained in §8.6(b), be used with the Born approximation, which for neutron momentum transfer $\boldsymbol{q}$ gives the scattering amplitude

$$F = -\frac{m}{2\pi}\int d^3r \int d\boldsymbol{n}\, e^{i\boldsymbol{q}\cdot\boldsymbol{r}}\, \langle L'M'\mu' S'M'_S|V_F|LM\mu SM_S\rangle . \tag{321}$$

One check of this formula comes from elastic scattering with low momentum transfer by assuming that the scattering lengths are spin-independent, in which case (321) gives $F = -2a$, as it should.

In elastic collisions, L (and therefore S) does not change:

$$\begin{aligned} F_{\text{el}} = &- \langle \mu' SM'_S|A_1 + A_2|\mu SM_S\rangle \int d\boldsymbol{n}\, Y^*_{LM'}(\boldsymbol{n})Y_{LM}(\boldsymbol{n}) \cos(\tfrac{1}{2}\boldsymbol{q}\cdot\boldsymbol{R}) \\ &- i\langle \mu' SM'_S|A_1 - A_2|\mu SM_S\rangle \int d\boldsymbol{n}\, Y^*_{LM'}(\boldsymbol{n})Y_{LM}(\boldsymbol{n}) \sin(\tfrac{1}{2}\boldsymbol{q}\cdot\boldsymbol{R}) . \end{aligned} \tag{322}$$

Because the spherical harmonics have the same parity, the second term vanishes.

Let $P^i_{s,t}$ be the singlet and triplet projection operators for the spins of the pairs np_i, i.e.,

$$P^i_s = \tfrac{1}{4}(1 - \boldsymbol{\sigma}_n\cdot\boldsymbol{\sigma}_i) , \qquad P^i_t = \tfrac{1}{4}(3 + \boldsymbol{\sigma}_n\cdot\boldsymbol{\sigma}_i) ; \tag{323}$$

then $A_i = a_s P^i_s + a_t P^i_t$, and

$$A_1 + A_2 = \tfrac{1}{2}(a_s + 3a_t) + \tfrac{1}{2}(a_t - a_s)\boldsymbol{\sigma}_n\cdot\boldsymbol{S} , \tag{324}$$

where $\boldsymbol{S}$ is the total pp spin operator.

In the ground state of the molecule, $L = 0$ and therefore $S = 0$, so the elastic amplitude in this case is

$$F^0_{\text{el}} = -\delta_{\mu\mu'}\, \tfrac{1}{2}(a_s + 3a_t) \int \frac{d\boldsymbol{n}}{4\pi} \cos(\tfrac{1}{2}\boldsymbol{q}\cdot\boldsymbol{R}) , \tag{325}$$

where the integral equals $2\sin\frac{1}{2}qR/qR$. If the incident neutron state is unpolarized and the spin of the scattered neutrons are not measured, the differential cross section is

$$\frac{d\sigma^0_{\text{el}}}{d\Omega} = \tfrac{1}{4}(a_s + 3a_t)^2 \left(\frac{2\sin\frac{1}{2}qR}{qR}\right)^2 . \tag{326}$$

This is a remarkable result. Even though no spin polarizations are imposed by the experiment, the requirement that the protons in the molecule obey the Pauli principle imposes a spin correlation on them which leads to *coherence* between the triplet and singlet *np* scattering amplitudes, in contrast to the situation in scattering from free protons (Eq. 297) which gives $\frac{1}{4}(a_s^2 + 3a_t^2)$ as the low-energy cross section — an *incoherent* sum of scattering in the two spin states.

(d) The Tensor Force

In the introduction to this section we concluded that the existence of the deuteron's quadrupole moment requires the *np* interaction to contain a term that mixes the states ${}^3S_1$ and ${}^3D_1$, and therefore does not commute with the orbital angular momentum $\boldsymbol{L}$. Insofar as rotations generated by $\boldsymbol{L}$ are concerned, this interaction must be a second-rank tensor operator. It must, however, conserve the total angular momentum $\boldsymbol{J} = \boldsymbol{L} + \boldsymbol{S}$, and therefore be an invariant formed by contracting the aforementioned tensor with a second-rank tensor formed from the spins.

An obvious candidate for this interaction is the operator

$$(\boldsymbol{\sigma}_n \cdot \hat{\boldsymbol{r}})(\boldsymbol{\sigma}_p \cdot \hat{\boldsymbol{r}}) . \tag{327}$$

This is a second-rank tensor in coordinate space, and invariant under combined rotations in coordinates and spins. But it is not the scalar product of two irreducible tensors. This is annoying, as it would give a contribution to the energy of an S-state; i.e., in effect (327) is in part a central interaction. To find the irreducible counterpart it is only necessary to find the angular average of (327):

$$\int \frac{d\Omega}{4\pi} (\boldsymbol{\sigma}_n \cdot \hat{\boldsymbol{r}})(\boldsymbol{\sigma}_p \cdot \hat{\boldsymbol{r}}) = \tfrac{1}{3}\boldsymbol{\sigma}_n \cdot \boldsymbol{\sigma}_p . \tag{328}$$

The *tensor force* is then defined as

$$\mathcal{T}\, V_T(r) \equiv \Big(3(\boldsymbol{\sigma}_n \cdot \hat{\boldsymbol{r}})(\boldsymbol{\sigma}_p \cdot \hat{\boldsymbol{r}}) - \boldsymbol{\sigma}_n \cdot \boldsymbol{\sigma}_p\Big)\, V_T(r) , \tag{329}$$

where $V_T(r)$ is a spherically symmetric "potential."

By construction, the tensor force vanishes in S-states, so it cannot produce any difference between low-energy scattering in spin singlet and spin triplet states. In short, an interaction that accounts for all the low-energy *np* scattering and bound state data must be some combination of (292) and (329):

$$V = V_s(r)\, P_s + V_t(r)\, P_t + V_T(r)\, \mathcal{T} . \tag{330}$$

This statement does not mean that other types of interactions are logically untenable — for example, velocity-dependent interactions, — but that the interaction must be at least as complicated as (330). Velocity-dependent interactions, in particular, are by their nature unimportant at low energies.

We now confine our attention to the ground state of the deuteron. As it is a spin triplet, in this state the *np* interaction is

$$V_d = V_t(r) + V_T(r)\,\mathcal{T}\,. \tag{331}$$

The small size of the quadrupole moment implies that the tensor force plays only a subsidiary role in V_d. It is, therefore, tempting to treat it with perturbation theory, but a bit of scratch paper soon establishes that such a calculation is daunting as it requires sums over the scattering states due to V_t. A more efficient, and instructive, approach is to use the variational principle.

The first task is to express the ground state wave function in a convenient form. This is done by using the tensor operator $\mathcal{T}$, and *avoiding* the explicit but clumsy angular momentum addition machinery. The point is that $\mathcal{T}$ is a $(\Delta L = 2, \Delta S = 0, \Delta J = 0)$ operator, and therefore

$$\mathcal{T}|^3S_1; M\rangle \propto |^3D_1; M\rangle\,, \tag{332}$$

where M is the eigenvalue of J_z. As a consequence, the deuteron's state can be written as

$$\Psi_M = \frac{1}{r}\Big[u(r) + \mathcal{T}\,v(r)\Big]\chi_M \equiv \Theta_M + \mathcal{T}\Phi_M\,, \tag{333}$$

where $u(r)$ and $v(r)$ are spherically symmetric, and χ_M is the triplet spin function with eigenvalue M of S_z.

To use the variational method, the norm of Ψ_M and the expectation values of the various terms in the Hamiltonian must be evaluated. These do not depend on M, and $M = 0$ is to be understood from now until Eq. 341.

Let $A(r)$ be any spherically symmetric function. Its expectation value cannot have an $S - D$ interference term. The D-state contribution is

$$\langle\Phi|A|\Phi\rangle = \int_0^\infty dr\,|v(r)|^2 A(r)\int d\Omega\Big(\chi_0^*,\,\mathcal{T}^2\chi_0\Big)\,. \tag{334}$$

As shown in Prob. 13, in triplet states

$$\mathcal{T}^2 = 8 - 2\mathcal{T}\,. \tag{335}$$

The term linear in $\mathcal{T}$ does not contribute to (334) because its angular average vanishes, and therefore

$$\langle\Psi|A|\Psi\rangle = 4\pi\int_0^\infty dr\,A(r)\Big(|u(r)|^2 + 8|v(r)|^2\Big)\,. \tag{336}$$

This has reduced the norm and the expectation value of $V_t(r)$ to a one-dimensional integration over the trial functions.

The expectation value of the tensor interaction vanishes in the ^3S term of Ψ, so

$$\langle\Psi|V_T\mathcal{T}|\Psi\rangle = \langle\Theta|V_T\mathcal{T}^2|\Phi\rangle + \text{c.c.} + \langle\Phi|V_T\mathcal{T}^3|\Phi\rangle\,. \tag{337}$$

Using (335) again, and noting that only a scalar under rotations of $\mathbf{r}$ can contribute to these matrix elements, then results in

$$\langle\Psi|V_T\mathcal{T}|\Psi\rangle = 64\pi\int_0^\infty dr\,V_T(r)\Big(\mathrm{Re}\,u^*(r)v(r) - |v(r)|^2\Big)\,. \tag{338}$$

The final ingredient is the expectation value of the kinetic energy, $K = p^2/m$. A straightforward calculation leads to

$$\langle\Psi|K|\Psi\rangle = \frac{4\pi}{m}\int_0^\infty dr\left(|u'(r)|^2 + 8|v'(r)|^2 + \frac{48}{r^2}|v(r)|^2\right). \tag{339}$$

To carry the variational calculation through, it remains to choose specific forms for the "potentials" $V_t(r)$ and $V_T(r)$, and for the trial functions $u(r)$ and $v(r)$. This is straightforward in principle and will not be done here. Once done, the quadrupole moment $\bar{Q}$ can be calculated, which will then require a simultaneous tuning of the parameters in the interaction until the correct experimental values are obtained for the independent low-energy quantities, i.e., the ^3S scattering length, the binding energy and $\bar{Q}$.

Bearing in mind that only the proton contributes to the quadrupole moment, the operator that gives $\bar{Q}$ is

$$Q = e(\tfrac{1}{2}r)^2(3\cos^2\vartheta - 1) = e\,\tfrac{1}{2}r^2 P_2(\cos\vartheta). \tag{340}$$

By definition, $\bar{Q}$ is the expectation value in the $M = 1$ state. The evaluation is left for Prob. 13. On the assumption that the $S-D$ interference term dominates (recall Eq. 288), the result is

$$\bar{Q} = e\,\frac{8\pi}{5}\int_0^\infty r^2 dr\,\mathrm{Re}\left[u(r)v^*(r)\right]. \tag{341}$$

8.7 Scattering of Identical Particles

When the projectile and target are identical, the theory appropriate to distinguishable particles must be modified. In the case of spin 0, Bose statistics requires the scattering amplitude to be the coherent sum of the now familiar function $f(\boldsymbol{k}_f|\boldsymbol{k}_i)$ and a second term wherein the role of target and projectile are interchanged. For particles of spin $\frac{1}{2}$, the requirement of antisymmetry under interchange of coordinates and spins leads to a somewhat more complicated result.

(a) Boson-Boson Scattering

Consider elastic scattering of two identical spin 0 bosons. The relative and center-of-mass coordinates are $\boldsymbol{r} = \boldsymbol{r}_1 - \boldsymbol{r}_2$ and $\boldsymbol{R} = \frac{1}{2}(\boldsymbol{r}_1 + \boldsymbol{r}_2)$; under interchange, $\boldsymbol{r} \to -\boldsymbol{r}$, $\boldsymbol{R} \to \boldsymbol{R}$. If $\Psi(\boldsymbol{r}, \boldsymbol{R}, t)$ is a wave function in which the particles are described as if they were distinguishable, the corresponding wave function that obeys the requirement of symmetry under interchange is

$$\bar{\Psi}(\boldsymbol{r}, \boldsymbol{R}, t) = \tfrac{1}{\sqrt{2}}[\Psi(\boldsymbol{r}, \boldsymbol{R}, t) + \Psi(-\boldsymbol{r}, \boldsymbol{R}, t)]. \tag{342}$$

The definition of the cross section now requires some care because the detector of scattered particles cannot distinguish between those that were originally in the beam and target. An unambiguous definition of the counting rate at such a detector is therefore required. In the one-body collision problems handled thus far in this chapter, this rate was given by the Schrödinger current. In the present problem, however, the Schrödinger current is no longer in 3-dimensional coordinate space but in 6-dimensional configuration space, and hence not immediately related to what is counted by any detector of individual particles. To get a current that has a directly measurable meaning, we assume that the bosons of interest carry the electric charge Ze, and that the counter measures the charge deposited on it. (If the particles are not charged, another observable that triggers detection must be used; the final result for the cross section will be the same.)

The electric current carried by the state (342) is given by Eq. 2.233:

$$\boldsymbol{j}(\boldsymbol{r}_0,t) = \frac{Ze}{m}\int d^3r_1 d^3r_2 \sum_{i=1}^{2} \delta(\boldsymbol{r}_0 - \boldsymbol{r}_i)\, \mathrm{Im}\,[\bar{\Psi}^* \boldsymbol{\nabla}_i \bar{\Psi}] \tag{343}$$

$$= \frac{2Ze}{m}\int d^3r\, d^3R\, \delta(\boldsymbol{r}_0 - \boldsymbol{R} - \tfrac{1}{2}\boldsymbol{r})\, \mathrm{Im}\,[\bar{\Psi}^*(\boldsymbol{\nabla}_r + \tfrac{1}{2}\boldsymbol{\nabla}_R)\bar{\Psi}]\,, \tag{344}$$

where $\boldsymbol{\nabla}_i = \partial/\partial \boldsymbol{r}_i$, etc., and the last step used the symmetry of $\bar{\Psi}$.

The symmetric two-body scattering state can be constructed by slightly modifying the one-particle wave packets introduced in §8.1(c). First, consider a superposition of two-particle states of zero total momentum whose relative momenta are along the z axis:

$$N\int_{-\infty}^{\infty} d\bar{k}\, \exp\{i[\bar{k}(z_1 - z_2) - 2(\bar{k}^2/2m)t]\}\, e^{-(k-\bar{k})^2/\lambda^2} = e^{i(k-2E_k t)}\,\Phi(z,2t)\,, \tag{345}$$

where Φ is the function defined in Eq. 33, and $E_k = k^2/2m$. This packet is the incident state in the center-of-mass frame, but it is not yet symmetrized. As t increases the interaction comes into play, and (345) evolves into $\psi(\boldsymbol{r},t)$. The corresponding symmetrized state is then

$$\bar{\Psi}(\boldsymbol{r},\boldsymbol{R},t) = \tfrac{1}{\sqrt{2}}[\psi(\boldsymbol{r},t) + \psi(-\boldsymbol{r},t)]u(\boldsymbol{R},t)\,. \tag{346}$$

The wave function u is taken to have vanishing mean values of $\boldsymbol{R}$ and total momentum, with dispersions that are negligible in comparison to $|\boldsymbol{r}_0|$ and the mean incident relative momentum k, respectively. In (344), therefore, $\boldsymbol{\nabla}_R$ gives a negligible contribution, and $\boldsymbol{R} \simeq 0$ in the δ-function, so that u disappears from the calculation. (Even though the two terms in (346) overlap when $t \sim 0$, it is properly normalized because as $t \to -\infty$ there is no overlap and the norm does not change with time for any solution of the complete Schrödinger equation.)

The current at the detector is due to the scattered wave $\bar{\Psi}_{\rm sc}$. The scattered state, with the particles identified, follows from (345) by the argument that led to Eq. 43, viz.,

$$\psi_{\rm sc}(\boldsymbol{r},t) \sim \frac{e^{i(kr-2E_k t)}}{r} f(k,\theta)\Phi(r,2t)\,. \tag{347}$$

The symmetric two-particle scattered packet is therefore

$$\bar{\Psi}_{\rm sc}(\boldsymbol{r},t) \sim \frac{e^{i(kr-2E_k t)}}{\sqrt{2}\,r}[f(k,\theta) + f(k,\pi-\theta)]\Phi(r,2t)\,, \tag{348}$$

because $-\boldsymbol{r} = (r, \pi - \theta, \phi + \pi)$. As in §8.1(c), only the radial component of the gradient survives as $r \to \infty$, and therefore (344) gives the following result for the electric current at the detector:

$$\boldsymbol{j}_{\rm sc}(\boldsymbol{r}_0,t) = \frac{2Ze}{m}\frac{8}{(\sqrt{2})^2}\frac{k}{(2r_0)^2}|f(k,\theta)+f(k,\pi-\theta)|^2\,|\Phi(2r_0,2t)|^2\,\hat{\boldsymbol{r}}_0\,. \tag{349}$$

The total charge scattered into the area that subtends the solid angle $d\Omega$ at $\boldsymbol{r}_0$ is therefore

$$dQ_{\rm sc} = d\Omega\,\frac{k}{m}\,Ze\,|f(k,\theta)+f(k,\pi-\theta)|^2\int_{-\infty}^{\infty}|\Phi(2r_0,2t)|^2\,2dt\,. \tag{350}$$

Initially, there is no overlap between projectile and target, and no ambiguity regarding the incident flux. If, say, the right-moving particles in the initial state are taken to be those that are incident, the total incident charge per unit area is Eq. 48 multiplied by the charge per particle:

$$\frac{dQ_{\rm inc}}{dA} = \frac{k}{m}\,Ze\int_{-\infty}^{\infty}|\Phi(z,t)|^2\,dt\,. \tag{351}$$

As in the derivation of Eq. 50, the integrals in (350) and (351) are equal. Defining the cross section as

$$\frac{d\sigma}{d\Omega} \equiv \frac{dQ_{\rm sc}/d\Omega}{dQ_{\rm inc}/dA}\,, \tag{352}$$

then gives the important result

$$\frac{d\sigma}{d\Omega} = |f(k,\theta)+f(k,\pi-\theta)|^2 \tag{353}$$

$$\equiv |f(\boldsymbol{k}_f|\boldsymbol{k}_i)+f(-\boldsymbol{k}_f|\boldsymbol{k}_i)|^2\,. \tag{354}$$

It is instructive to express the interference term in the cross section explicitly:

$$\frac{d\sigma}{d\Omega} = |f(k,\theta)|^2+|f(k,\pi-\theta)|^2+2\,{\rm Re}\,f(k,\theta)f^*(k,\pi-\theta)\,. \tag{355}$$

The first two terms are what was to be expected from classical collision theory for identical particles:

$$\frac{d\sigma^{\rm id}(\theta)}{d\Omega} = \frac{d\sigma^{\rm dist}(\theta)}{d\Omega}+\frac{d\sigma^{\rm dist}(\pi-\theta)}{d\Omega}\,, \tag{356}$$

where $\sigma^{\rm id}$ and $\sigma^{\rm dist}$ are the naive (or classical) cross sections for identical and distinguishable particles. The interference term in (355) is a typical quantum mechanical effect.

An important illustration of the general theory is offered by Coulomb scattering. The Coulomb scattering amplitude is given by Eq. 226, and so in this case (353) is

$$\frac{d\sigma}{d\Omega} = \left(\frac{Z^2e^2}{4\pi}\right)^2\frac{1}{16E^2}\left|\frac{e^{2i\gamma\ln\sin\frac{1}{2}\theta}}{\sin^2\frac{1}{2}\theta}+\frac{e^{2i\gamma\ln\cos\frac{1}{2}\theta}}{\cos^2\frac{1}{2}\theta}\right|^2 \tag{357}$$

$$= \left(\frac{Z^2e^2}{4\pi}\right)^2\frac{1}{16E^2}\left\{\frac{1}{\sin^2\frac{1}{2}\theta}+\frac{1}{\cos^4\frac{1}{2}\theta}+8\csc^2\theta\,\cos(\gamma\ln\tan^2\tfrac{1}{2}\theta)\right\}\,. \tag{358}$$

where E is the total energy in the c.o.m. frame. The first two terms form the Rutherford cross section for scattering of identical particles in classical mechanics, whereas the interference term is manifestly quantum mechanical because $\gamma = Z^2e^2/4\pi\hbar v$, when $\hbar$ is reintroduced. For scattering angles around $\pi/2$, where indistinguishability is most flagrant, the interference term produces dramatic departures from the classical result, as discussed in Prob. 14.

In the scattering of two identical particles, whether bosons or fermions, the total cross section is *not* the integral of the differential cross section over all solid angles, but rather

$$\sigma_{\text{tot}} = \frac{1}{2}\int d\Omega \frac{d\sigma}{d\Omega}. \tag{359}$$

This is so because the event in which one of the particles is detected in the direction $\boldsymbol{n}$ in the center-of-mass frame, is indistinguishable from the event in which $\boldsymbol{n}$ is replaced by $-\boldsymbol{n}$. Hence all events are counted by integrating over half the unit sphere.

(b) Fermion-Fermion Scattering

We now turn to scattering of identical spin $\frac{1}{2}$ particles. Consider, first, the partial waves allowed by the Pauli principle:

$$J = 0: \quad {}^1S,\ {}^3P \tag{360}$$

$$J = 1: \quad {}^3P \tag{361}$$

$$J = 2: \quad {}^3P,\ {}^1D,\ {}^3F. \tag{362}$$

This list says that if the interaction is reflection invariant, there is no mixing of triplets and singlets; we will assume that this is the case.

The list also emphasizes the following remarkable fact: *Even when the interaction is spin-independent and the spins are totally random in the initial state, scattering will, in general, produce correlations between the spins.* To see this, it suffices to consider low energy scattering, in which case only the S-wave scatters, which means that the spins in the scattered wave are in the singlet state, or crudely put antiparallel. This is yet another example of how the Pauli principle produces spin-dependent (e.g., magnetic) effects which, from a pre-quantum perspective, would appear to be due to spin-dependent interactions.

The scattering matrix of §8.5 appropriate to this problem is four-dimensional, and by the preceding consideration it separates into singlet and triplet parts. If the particles were distinguishable, this matrix would then have the form

$$M(\boldsymbol{k}_f, \boldsymbol{k}_i) = M_s(\boldsymbol{k}_f, \boldsymbol{k}_i) + M_t(\boldsymbol{k}_f, \boldsymbol{k}_i). \tag{363}$$

As the singlet and triplet states are antisymmetric and symmetric in the spins, respectively, when the particles are identical, these amplitudes must be modified to have the complementary symmetry. Thus the scattering matrix for identical fermions is

$$\mathcal{M}(\boldsymbol{k}_f, \boldsymbol{k}_i) = [M_s(\boldsymbol{k}_f, \boldsymbol{k}_i) + M_s(-\boldsymbol{k}_f, \boldsymbol{k}_i)] + [M_t(\boldsymbol{k}_f, \boldsymbol{k}_i) - M_t(-\boldsymbol{k}_f, \boldsymbol{k}_i)]. \tag{364}$$

The general theory for scattering of particles with spin developed in §8.5(b) applies to this problem. If the initial state is described by the spin-space density matrix ρ_i, the differential cross section when no spins are determined subsequent to scattering is given by Eq. 272:

$$\frac{d\sigma}{d\Omega} = \text{Tr}\ \rho_i \mathcal{M}^\dagger \mathcal{M} \ , \tag{365}$$

and the density matrix of the scattered state is given by Eq. 269:

$$\rho_f = \frac{\mathcal{M}\rho_i\mathcal{M}^\dagger}{d\sigma/d\Omega} \ . \tag{366}$$

The most general form of the density matrix is that of the *np* problem, i.e., Eq. 298:

$$\rho = \frac{1}{4}\left(1 + \boldsymbol{\sigma}_1 \cdot \boldsymbol{P}_1 + \boldsymbol{\sigma}_2 \cdot \boldsymbol{P}_2 + \sum_{i,j} \sigma_{1,i}\sigma_{2,j} C_{ij}\right) \ . \tag{367}$$

If the initial state is random ($\rho_i = \frac{1}{4}$), the correlation coefficients after scattering are

$$C_{ij} = \frac{\text{Tr}\ \sigma_{1,i}\sigma_{2,j}\mathcal{M}^\dagger\mathcal{M}}{\text{Tr}\ \mathcal{M}^\dagger\mathcal{M}} \ . \tag{368}$$

To be specific, consider the spin-spin interaction already studied in connection with *np* scattering in §8.6(a):

$$V = V_0(r) + \boldsymbol{\sigma}_1 \cdot \boldsymbol{\sigma}_2\, V_1(r) \ . \tag{369}$$

The corresponding scattering matrix is then that of Eq. 293, and the cross section for distinguishable particles is therefore (297) when the initial state is completely unpolarized. For identical particles the rule given in (364) then gives the following cross section

$$\frac{d\sigma}{d\Omega} = \tfrac{1}{4}|f_s(k,\theta) + f_s(k,\pi-\theta)|^2 + \tfrac{3}{4}|f_t(k,\theta) - f_t(k,\pi-\theta)|^2 \ . \tag{370}$$

If the forces are spin-independent, $f_s = f_t$, and (370) simplifies to

$$\frac{d\sigma}{d\Omega} = |f(k,\theta)|^2 + |f(k,\pi-\theta)|^2 - \text{Re}\, f^*(k,\theta)f(k,\pi-\theta) \ . \tag{371}$$

Note the difference with respect to the boson-boson cross section, Eq. 355: *in fermion-fermion scattering the cross section is smaller than the incoherent sum at $\theta = \pi/2$, as one would expect from the exclusion principle, whereas it is larger in boson-boson scattering.*

Problem 15 deals with spin correlations. When the interaction is spin-independent, it gives the result

$$C_{ij} = -\delta_{ij}\, \frac{\text{Re}\, f(k,\theta)f^*(k,\pi-\theta)}{d\sigma/d\Omega} \ . \tag{372}$$

As explained at the outset, the exclusion principle imposes spin correlations on the scattered state even when the interaction does not depend on spin. In the low-energy limit, f does not depend on angle, and $C_{ij} = -\delta_{ij}$, corresponding to the antiparallel spin configuration in the *s*-wave.

8.8 Problems

1. Derive the boundary of the shaded region in Fig. 8.9, which shows the possible values of the elastic cross section for given values of the inelastic cross section in a single partial wave. In the figure $x = \sigma^l_{\text{inel}}/\pi\lambda\!\!\!^{-2}(2l+1)$, $y = \sigma^l_{\text{el}}/\pi\lambda\!\!\!^{-2}(2l+1)$.

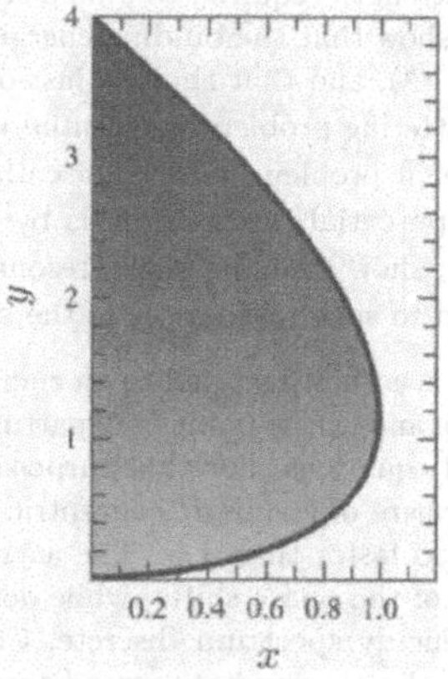

FIG. 8.9. The physically allowed (shaded) region in elastic and inelastic cross sections for any partial wave.

2. Apart from a small blemish, the optical theorem, though restricted to the case where only elastic scattering is occurring, can be derived from the stationary-state wave function (Eq. 9). Start from the conservation law

$$\int_S d^3r\, \nabla \cdot \boldsymbol{i} = 0 ,$$

where S is a large sphere of radius r centered on the target. Show that this imposes a relationship between the total flow crossing the surface of S due (i) to the scattered wave alone, and (ii) the interference between the scattered and incident waves. Show that the latter is confined to angles of order $(kr)^{-\frac{1}{2}}$; and that its total contribution to the flow is

$$-2\pi vr \,\text{Re}\left\{ f^*(0) \int_\epsilon^1 dx\, e^{ikr(x-1)} + \text{c.c.} \right\} \equiv X ,$$

where $kr\epsilon \gg 1$ but $\epsilon \ll 1$. This integral has an oscillating contribution which disappears in the wave packet treatment presented in the text, whereas here it must be eliminated by the handwaving argument that a wave packet treatment produces an average over k, which is somewhat oversimplified. Nevertheless, once this is accepted, show that it does lead to the optical theorem in the absence of inelastic scattering.

This restriction to elastic scattering stems from the stated conservation law; Why? Of course the argument in the text relating N_{int} to N_{lost} will from X also give the optical theorem in the general case.

3. As a three-dimensional analogue of the potentials studied in §4.4(c), consider

$$U(r) = -\lambda\delta(r-a) ,$$

where λ is a strength parameter having the dimension $(\text{length})^{-1}$.

(a) Show that the partial wave scattering amplitude is

$$\frac{1}{\xi}e^{i\delta_l(\xi)}\sin\delta_l(\xi) = \frac{g[j_l(\xi)]^2}{1 - i\xi g j_l(\xi)h_l(\xi)} ,$$

where $\xi = ka, g = a\lambda$.

(b) Show that the minimum strength required to bind a state of angular momentum l is $g = 2l + 1$. In the case of $l = 0$, show that the binding energies $-\eta_n^2/2ma^2$ are determined by the roots η_n of $2\eta = g(1 - e^{-2\eta})$, and that there is just one root if $g > 1$.

(c) Show that any s-wave scattering problem is essentially identical to that of the odd-parity states in a one-dimensional problem with a potential $V(|x|) \equiv V(r)$, where the latter is the three-dimensional potential. Illustrate this by comparing with the problem treated in §4.4(c). In particular, show that the sharp resonances found there in the one-dimensional problem correspond to such resonances in the s-wave scattering amplitude.

4. The partial wave phase shift is actually related to an energy shift, and this relationship is useful both as an intuitive tool and as the point of departure for treating scattering with the Fredholm theory of integral equations. For this purpose, imagine that the scattering particle is confined in a large sphere of radius R concentric with the potential $V(r)$, and assume that $V(r) \to 0$ as $r \to \infty$ faster than $1/r$. The asymptotic radial wave functions can then be expressed in terms of the phase shifts, while demanding that these functions vanish at $r = R$ renders the energy spectrum discrete. Consider a set of levels in the neighborhood of E, let $\epsilon_l(E)$ be the spacing between adjacent levels when $V = 0$, and for any given level, let $\Delta_l(E)$ be the difference in energy between its position in the presence and absence of V. Show that

$$\frac{\Delta_l(E)}{\epsilon_l(E)} = -\frac{1}{\pi}\delta_l(E).$$

This demonstrates that the phase shifts are all negative (positive) for a purely repulsive (attractive) potential.

The relationship can still be useful when the potential does not have one of these simple properties. As an important example that occurs in both atomic and nuclear physics, consider a potential that is strongly repulsive when $r \lesssim a$, but is attractive at longer distances, with the attraction sufficiently strong to bind an s-state. Then for $l \gtrsim ka$ the partial waves will be insensitive to the repulsion and have negative phase shifts. On the other hand, the low l states, and in particular $l = 0$, will at low energies ($ka \lesssim 1$) behave as if the interaction is overall attractive, but be dominated by the repulsion at much higher energies; such phase shifts will therefore start out positive at low energies and change sign at some high energy. This behavior is illustrated in Prob. 6 below.

5. Consider the separable potential with $u(r) = e^{-r/b}/r$.

(a) Show that

$$k\cot\delta(k) = [(k^2 + b^2)^2 + \xi(k^2b^2 - 1)]/2\xi b \equiv X ,$$

where $\xi = 2\pi\lambda b^3$.

(b) Determine the condition on ξ for the existence of a bound state, and confirm that there can only be one.

(c) For each value of k determine the largest value of ξ for which the Born series converges; in the light of part (ii), what is the significance of this value in the case of $k = 0$.

(d) Show that $k\cot\delta \to a^{-1} + O(k^2)$ as $k \to 0$, where a is called the scattering length. Express the zero energy limit of the cross section in terms of a. Evaluate a and discuss the behavior of the cross section as a function of ξ. How is the occurrence of the bound state reflected in a and the cross section?

6. The following is a solvable model of an interaction that can produce an s-wave phase shift of the type observed in nucleon-nucleon scattering in the ^1S state. This interaction has the coordinate space matrix

$$\langle \boldsymbol{r}|V|\boldsymbol{r}'\rangle = \delta(\boldsymbol{r}-\boldsymbol{r}')V_0(r) - \lambda u(r)u(r') .$$

The first term is a conventional, local potential, with V_0 an infinite repulsion at $r = r_c$ that vanishes for $r > r_c$.

(a) Consider the same $u(r)$ as in the preceding problem. Show that the s-wave phase shift is given by

$$k \cot(\delta_0 + kr_c) = X ,$$

but where now $\xi = 2\pi\lambda b^3 \exp(-2r_c/b)$.

(b) Evaluate the scattering length and effective range. Determine the value ξ_0 of ξ required to bind one state. Can there be more than one such state? Study the behavior of the phase shift at low energy as ξ passes through ξ_0.

(c) Show that the phase shift is positive at low and negative at high energies.

7. Prove directly from Eq. 185 that the eikonal approximation to the scattering amplitude satisfies the optical theorem whether or not the potential is spherically symmetric. [Hint: Scattering is confined to small angles.]

8. Consider scattering due to a short-range potential and a Coulomb field at energies sufficiently low to make the former unimportant in all but the s-wave, and yet high enough to produce small Coulomb phase shifts. These conditions can be met simultaneously, for example in proton-proton scattering below 10 MeV. Show that the differential cross section under these circumstances is

$$\frac{d\sigma}{d\Omega} = \frac{d\sigma_R}{d\Omega} + \frac{1}{k^2}\sin^2\delta_0 + \gamma\,\frac{\sin\delta_0\cos\delta_0}{k^2\sin^2\frac{1}{2}\theta} ,$$

where δ_0 is the s-wave phase shift due to the short range force alone.

9. Assume that the initial spin state in a collision process is completely random, and that no spin properties are measured in the final state. Show that the scattering cross section does not depend on the azimuthal angle, i.e., only on the angle between $\boldsymbol{k}_i$ and $\boldsymbol{k}_f$.

10. This problem develops the partial wave expansion for scattering due to the spin-orbit interaction (277).

(a) Show that the solution of the Schrödinger equation can be written as

$$\Psi = \sum_{l=0}^{\infty} \sqrt{4\pi(2l+1)}\, i^l [C_l^+ R_l^+(r)\Lambda_l^+ + C_l^- R_l^-(r)\Lambda_l^-] Y_{l0}(\theta)\,|\nu_i\rangle ,$$

where $\Lambda_l^\pm$ are the $j = l \pm \frac{1}{2}$ projection operators

$$\Lambda_l^+ = \frac{l+1+\boldsymbol{\sigma}\cdot\boldsymbol{L}}{2l+1} , \qquad \Lambda_l^- = \frac{l-\boldsymbol{\sigma}\cdot\boldsymbol{L}}{2l+1} ,$$

$C_l^\pm$ are constants to be fixed by the boundary conditions, the radial functions are the solutions of

$$\left\{\frac{1}{r^2}\frac{d}{dr}r^2\frac{d}{dr} - \frac{l(l+1)}{r^2} + k^2 - 2mV_l^\pm(r)\right\} R_l^\pm(r) = 0 ,$$

and

$$V_l^+(r) = V_0(r) + lV_1(r) , \qquad l = 0, 1, \dots ,$$
$$V_l^-(r) = V_0(r) - (l+1)V_1(r) , \qquad l = 1, 2, \dots .$$

(b) Let $\delta_l^\pm$ be the phase shifts in the states with total angular momentum $j = l \pm \frac{1}{2}$, i.e.,

$$R_l^\pm \sim \frac{\text{const.}}{kr} \sin(kr - \tfrac{1}{2}l\pi + \delta_l^\pm), \qquad (r \to \infty) .$$

Show that the functions g and h that appear in the scattering matrix M (Eq. 275) are

$$g(k,\theta) = \frac{1}{k}\sum_{l=0}^{\infty}\left(\frac{4\pi}{2l+1}\right)^{\frac{1}{2}} [(l+1)a_l^+ + la_l^-]Y_{l0}(\theta) ,$$

$$h(k,\theta) = \frac{1}{k}\sum_{l=1}^{\infty}\left(\frac{4\pi}{2l+1}\right)^{\frac{1}{2}} (a_l^+ - a_l^-)\, i\sin\theta \frac{d}{d(\cos\theta)} Y_{l0}(\theta) ,$$

where $a_l^\pm = e^{i\delta_l^\pm}\sin\delta_l^\pm$.

(c) Show that the total cross section is

$$\sigma = \frac{4\pi}{k^2}\sum_l [(l+1)\sin^2\delta_l^+ + l\sin^2\delta_l^-] .$$

11. Scattering at short wavelengths by the spin-orbit interaction can be treated with the eikonal approximation.

(a) For every impact parameter $\boldsymbol{b}$, introduce the projection operators

$$P_\pm = \tfrac{1}{2}(1 \pm \boldsymbol{u}\cdot\boldsymbol{\sigma}) , \qquad \boldsymbol{u} = (\boldsymbol{b}\times\boldsymbol{k}_i)/kb ,$$

and show that as $k \to \infty$, the Schrödinger equation separates into two decoupled first-order equations

$$\left(\frac{d}{dz} + \frac{i}{v}V_\pm(b,z)\right)\varphi_\pm(\boldsymbol{b},z) = 0 , \qquad V_\pm = V_0 \pm kbV_1 .$$

(b) Show that in this approximation the scattering matrix is

$$M = \frac{k}{2\pi i}\int d^2b\, e^{i\boldsymbol{q}\cdot\boldsymbol{b}} \{(e^{2i\Delta_+(b)} - 1)P_+ + (e^{2i\Delta_-(b)} - 1)P_-\} ,$$

$$\Delta_\pm(b) = -\frac{1}{2v}\int_{-\infty}^{\infty} dz\, V_\pm(b,z) .$$

(c) Finally, show that the functions g and h are now given by

$$g(k,\theta) = -ik\int_0^\infty b\,db\, J_0(kb\theta)\{e^{2i\Delta_0(b)}\cos[2kb\Delta_1(b)] - 1\} ,$$

$$h(k,\theta) = ik\int_0^\infty b\,db\, J_1(kb\theta)\, e^{2i\Delta_0(b)}\sin[2kb\Delta_1(b)] ,$$

$$\Delta_{0,1}(b) = -\frac{1}{2v}\int_{-\infty}^{\infty} V_{0,1}(\boldsymbol{b},z)\,dz .$$

Note that $h \to 0$ as $\theta \to 0$; show that this is not an artifact of this approximation but a general result.

12. (a) Consider elastic neutron scattering by the first excited rotational state of the toy hydrogen molecule introduced in §8.6(c). Show that in the low-momentum transfer limit, the total cross section is

$$\sigma_{\text{el}}^1 = \pi[(a_s + 3a_t)^2 + 2(a_s - a_t)^2] .$$

(b) Consider inelastic scattering from the ground to first excited state of the molecule. Show first that the incident energy threshold can be expressed as $kR > 2$. Then show that the scattering amplitude is

$$F_{\text{inel}} = -i\langle \mu' M_S | A_1 - A_2 | \mu S = 0\rangle \int \frac{dn}{\sqrt{4\pi}} Y^*_{1M}(n) \sin(\tfrac{1}{2}\boldsymbol{q}\cdot\boldsymbol{R}) .$$

As a step toward calculating the differential cross section, show that

$$\tfrac{1}{2} \sum_{\mu\mu' M_S} |\langle \mu' M_S | A_1 - A_2 | \mu \rangle|^2 = \tfrac{1}{32}(a_t - a_s)^2 \operatorname{Tr}\{P_t^{pp}(\boldsymbol{\sigma}_1 - \boldsymbol{\sigma}_2)^2\} ,$$

where P_t^{pp} is the projection operator onto the pp triplet state. Show that the final result for the total inelastic cross section (i.e., summed over M) is

$$\frac{d\sigma_{\text{inel}}}{d\Omega} = (a_t - a_s)^2 \left(\frac{\sin \frac{1}{2}qR - \frac{1}{2}qR \cos \frac{1}{2}qR}{qR} \right)^2 .$$

Note that this vanishes as $qR \to 0$, as it must; Why?

13. This problem fills in some of the steps missing from the discussion of the tensor force in §8.6(d).

(a) Show that when acting on spin singlets the operator $\mathcal{T}$ gives zero, and when acting on spin triplets it can be expressed in terms of the total np spin $\boldsymbol{S}$ as follows:

$$\mathcal{T} = 6(\boldsymbol{S}\cdot\hat{\boldsymbol{r}})(\boldsymbol{S}\cdot\hat{\boldsymbol{r}}) - 2\boldsymbol{S}^2 .$$

(b) Confirm Eq. 335.

(c) Derive Eq. 341, bearing in mind that $\bar{Q}$ is, by definition, the expectation value of (340) in the state with $M = 1$. [Hint: Show that in this state $\mathcal{T}$ can be replaced by $2P_2(\cos\vartheta)$.]

14. Fig. 8.10 shows actual data for the differential cross section in the c.o.m. frame for scattering between certain identical nuclei at an energy sufficiently low to make the Coulomb interaction dominant. Analyze these data. Do these nuclei have spin 0 or $\frac{1}{2}$? Determine Z, A, and the laboratory energy.

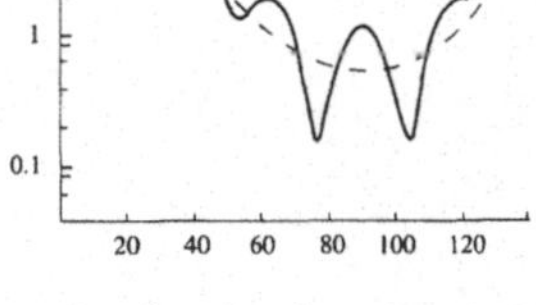

FIG. 8.10. Differential cross section for elastic scattering between identical nuclei.

15. Consider scattering of identical fermions due to the interaction of Eq. 369. If the initial state is unpolarized, show that the spin correlation coefficients are $C_{ij} = \delta_{ij}C$, where

$$C = \frac{|f_t(\theta) - f_t(\pi - \theta)|^2 - |f_s(\theta) + f_s(\pi - \theta)|^2}{3|f_t(\theta) - f_t(\pi - \theta)|^2 + |f_s(\theta) + f_s(\pi - \theta)|^2} .$$

Show also that $\frac{1}{3} \geq C \geq -1$, where these limits corresponds to scattering only in the triplet and singlet states. Provide an explanation of these limits.

16. An instructive derivation of the eikonal approximation (§8.3(c)) can be obtained from the exact integral equation for scattering, Eq. 77, by approximating Green's function.

(a) Show first that the eikonal approximation emerges from the *ad hoc* substitution

$$\langle \boldsymbol{r}|\mathcal{G}_0|0\rangle \equiv 2m \int \frac{d\boldsymbol{q}}{(2\pi)^3} \frac{e^{i\boldsymbol{q}\cdot\boldsymbol{r}}}{k^2 - q^2 + i\epsilon} \to \frac{i}{v}\theta(z)\,\delta^2(\boldsymbol{b})\,e^{ikz} ,$$

where $\boldsymbol{r} = (\boldsymbol{b}, z)$. Note how this makes the geometrical optics approximation evident. Next, show that this can be derived from this integral representation for $\mathcal{G}_0$ by assuming that at high enough energy the integration over $\boldsymbol{q}$ is dominated by the immediate neighborhood of the forward direction, i.e., by setting $\boldsymbol{q} = \boldsymbol{k} + \boldsymbol{\kappa}$ and keeping only the leading term in $\boldsymbol{\kappa}$.

(b) By retaining also the next to leading term in $\boldsymbol{\kappa}$ transverse to $\boldsymbol{k}$, show that Green's function becomes

$$\frac{i}{v}\,\theta(z)\,e^{ikz}\,\frac{k}{2\pi i z}e^{ikb^2/2z} .$$

Use this to show that Eqs. (180) – (182) give the correct conditions for the various regimes of optics.

9
Inelastic Collisions

A vast array of phenomena in virtually all branches of physics involve inelastic collisions. Our goal here is modest: to illustrate with concrete examples some of the generic features that arise, and to develop S matrix theory, which is a general approach to formulating collision problems

The chapter begins with a treatment of scattering of fast charged particles by atoms, where "fast" means swift enough to make first-order time-dependent perturbation theory applicable. The processes that result from the collision of various types of fast projectiles by atoms are of interest in their own right, and their understanding is also a basic input to many areas of physics, from particle detectors to astrophysics. In many situations, however, the restriction to leading-order perturbation theory is crippling, and this calls for a formulation of collision theory that is free of this limitation. This is provided by the S matrix, which is developed in the second section. The chapter then closes with a soluble model of resonant inelastic scattering.

9.1 Atomic Collision Processes

We will confine ourselves to collisions between fast (but nonrelativistic) particles and neutral atoms. By "fast" is meant energetic enough to make the Coulomb interaction energy small compared to the kinetic energy of the projectile so that the leading approximation to time-dependent perturbation theory is applicable. As we know from §8.3, this requires the particle's velocity to satisfy[1]

$$Z\alpha/v_i \ll 1 \,, \tag{1}$$

assuming the projectile has unit charge and v_i is its incident velocity. The restriction to fast particles is not just a pedagogic evasion; many important phenomena satisfy (1). Perturbation theory leads easily to simple results that provide much insight, and reveals features that persist in the relativistic regime and often appear at lower energies where the approximation is not reliable or even invalid. When (1) is not satisfied, complicated analytical and numerical calculations are needed.

Some comments on the preceding paragraph are called for. First, leading order time-dependent perturbation theory is often called the Born approximation; although this is historically unjustified, we often use this terminology. Second, the velocity in (1) stands for *both* the initial and final states, because perturbation theory, in this context, replaces the true incident and scattered states by momentum eigenstates, and this is not valid if a fast incident particle causes a highly inelastic

[1]In this whole chapter we use units in which $\hbar = c = 1$.

collision so that the scattered particle is too slow to satisfy (1). Third, the restriction to neutral atoms is only a convenient simplification; it is easily removed for when (1) holds so does $Z_{\text{ion}}\alpha/v_i \ll 1$, and it is this inequality that permits the replacement of the true wave functions by plane waves.

(a) Scattering Amplitudes and Cross Sections

The momentum transfer $\boldsymbol{q} = \boldsymbol{k}_i - \boldsymbol{k}_f$ is a crucial kinematic variable in this topic, where $\boldsymbol{k}_{i,f}$ are the projectile's momenta before and after the collision. Energy conservation relates them by

$$k_i^2 = k_f^2 + \gamma_\mu^2 , \qquad \gamma_\mu^2 = 2M\Delta_\mu , \tag{2}$$

where M is the mass of the projectile and Δ_μ the excitation energy of the atomic state χ_μ. The target atom is assumed to be sufficiently heavy to make its recoil ignorable. For scattering through the angle θ the square of the momentum transfer is

$$q^2 = 4k_i^2 \sin^2 \tfrac{1}{2}\theta - 2k_i[(k_i^2 - \gamma_\mu^2)^{\frac{1}{2}} - k_i] \cos\theta - \gamma_\mu^2 . \tag{3}$$

The collisions we deal with here are all due to the Coulomb interaction, which is a "soft" force that disfavors large excitation energies. This fact, in combination with the restriction to large incident momenta, means that $k_i^2 \gg \gamma_\mu^2$. A reliable expression for the momentum transfer is therefore given by dropping terms of relative order $(\gamma_\mu/k_i)^4$,

$$q^2 \simeq 4(k_i^2 - \tfrac{1}{2}\gamma_\mu^2) \sin^2 \tfrac{1}{2}\theta + q_{\text{min}}^2 \cos\theta , \tag{4}$$

where the minimum momentum transfer at $\theta = 0$ is

$$q_{\text{min}} = \tfrac{1}{2}\gamma_\mu^2/k_i = M\Delta_\mu/k_i . \tag{5}$$

As small-angle scattering dominates, for most purposes a further approximation is adequate:

$$q^2 \simeq 4k_i^2 \sin^2 \tfrac{1}{2}\theta + q_{\text{min}}^2 . \tag{6}$$

The interaction responsible for the collision is

$$V = -\sum_{s=1}^{Z} \frac{e^2}{4\pi|\boldsymbol{r} - \boldsymbol{r}_s|} + Z\int d\bar{\boldsymbol{r}}\, \frac{e^2 \rho_N(\bar{\boldsymbol{r}})}{4\pi|\boldsymbol{r} - \bar{\boldsymbol{r}}|} , \tag{7}$$

where ρ_N is the charge distribution of the nucleus, normalized to unity, and $\boldsymbol{r}$ is the coordinate of the projectile. (We assume that the incident energy is too low to excite the nucleus.)

In the Born approximation, the initial and final states are approximated by $\Phi_i = \varphi_{\boldsymbol{k}_i}\chi_0$ and $\Phi_f = \varphi_{\boldsymbol{k}_f}\chi_\mu$. The scattering amplitude T_{fi} will be written as $T_\mu(\boldsymbol{q})$, and is

$$T_\mu(\boldsymbol{q}) = \frac{1}{(2\pi)^3} \int d\boldsymbol{r}\, \langle \chi_\mu | e^{i\boldsymbol{q}\cdot\boldsymbol{r}}\, V | \chi_0 \rangle . \tag{8}$$

That the amplitude depends only on $\boldsymbol{q}$, and not separately on $\boldsymbol{k}_i$ and $\boldsymbol{k}_f$, is a feature special to the Born approximation (as it also is in potential scattering — recall §8.3).

The Fourier transform of the Coulomb field was already given in §8.3(a); it is

$$\int d\mathbf{r}\, e^{i\mathbf{q}\cdot\mathbf{r}} \frac{1}{4\pi|\mathbf{r}-\mathbf{x}|} = \frac{1}{q^2} e^{i\mathbf{q}\cdot\mathbf{x}} . \tag{9}$$

Hence T reduces to

$$T_\mu(\mathbf{q}) = -\frac{e^2}{q^2(2\pi)^3}\langle\chi_\mu|n_{\mathbf{q}}|\chi_0\rangle + \frac{Ze^2}{q^2(2\pi)^3}\delta_{\mu,0}\int d\bar{\mathbf{r}}\, e^{i\mathbf{q}\cdot\bar{\mathbf{r}}}\,\rho_N(\bar{\mathbf{r}}) , \tag{10}$$

where $n_{\mathbf{q}}$ is

$$n_{\mathbf{q}} = \sum_{s=1}^{Z} e^{i\mathbf{q}\cdot\mathbf{r}_s} . \tag{11}$$

The second term in (10) is the amplitude for scattering from the nucleus; it is purely elastic because the interaction responsible is not an operator on the electrons.

The matrix elements of $n_{\mathbf{q}}$, which are at the heart of all Born approximation scattering amplitudes, have a transparent physical meaning. This is so because $e^{i\mathbf{q}\cdot\mathbf{r}_s}$ is the translation operator in momentum space for particle s:

$$\langle\mathbf{p}_s|e^{i\mathbf{q}\cdot\mathbf{r}_s}|\mathbf{p}'_s\rangle = \delta(\mathbf{p}_s - \mathbf{q} - \mathbf{p}'_s) . \tag{12}$$

Therefore the matrix elements of $n_{\mathbf{q}}$ are

$$\langle\chi_\mu|n_{\mathbf{q}}|\chi_0\rangle = \sum_{s=1}^{Z}\int d\mathbf{p}_1 \ldots d\mathbf{p}_Z\, \phi_\mu^*(\mathbf{p}_1,\ldots,\mathbf{p}_s+\mathbf{q},\ldots,\mathbf{p}_Z)\phi_0(\mathbf{p}_1,\ldots,\mathbf{p}_s,\ldots,\mathbf{p}_Z) , \tag{13}$$

where $\phi_\mu(\mathbf{p}_1\cdots\mathbf{p}_Z)$ is the momentum space wave function of $|\chi_\mu\rangle$.

As we now see, the amplitude is a coherent superposition of Z terms, in each of which the collision imparts the full momentum transfer $\mathbf{q}$ to *one* target constituent; and each such term is the overlap of the amplitude for finding the constituent's momentum $\mathbf{p}_s+\mathbf{q}$ after this impact in the final state with the amplitude for finding the momentum $\mathbf{p}_s$ in the initial state. That the full momentum transfer is given to each constituent individually is only true in the first Born approximation; in second-order double scattering is included and $\mathbf{q}$ is shared by two constituents, and so forth in higher orders.

Another crucial feature of the first Born approximation is that it is impulsive: the collision is described as if the projectile and a target constituent were colliding in empty space, and not under the influence of the other constituents. The initial state of the target has the function of providing a cloud of momentarily free particles with momenta given by the amplitude $\phi_0(\mathbf{p}_1\ldots\mathbf{p}_Z)$; then the projectile impulsively alters the momentum of one or another constituents; and this altered cloud of still free particles must then "fit" into the momentum distribution of the final state.

This description of the Born amplitude leads naturally to the *impulse approximation.* It can be used in high-energy collisions when the basic interaction between the projectile and the target's constituents is too strong for the Born approximation, but the duration of the collision is short compared to the time between collisions of the target's constituents. The impulse approximation treats the projectile-constituent collision exactly, but also ignores the interaction with other constituents, so that the full scattering amplitude is then the same as in the Born approximation except for the replacement of the two-body Born amplitudes by exact two-body amplitudes. It is applicable to high-energy collisions between hadrons and nuclei, provided that the mean free path is larger than the dimensions of the nucleus.

The cross section is related to the Born amplitude by the *Golden Rule* , as derived §3.7(d), which gives the transition rate between states of perfectly defined energy:

$$\Gamma_{i\to f} = 2\pi |T_\mu(\boldsymbol{q})|^2 \delta(E_f - E_i) \,. \tag{14}$$

Here the symbol Γ for width has been used because the transition rate ($\dot{P}_{i\to f}$) and the width are identical when $\hbar = 1$. The rate for transitions into the interval $F = (E_f, E_f + dE_f)$ of final states is then

$$\Gamma_{i\to F} = 2\pi |T_\mu(\boldsymbol{q})|^2 \rho_f \,. \tag{15}$$

where ρ_f is the density in energy of the final states. By definition, the differential cross section is the transition rate divided by the incident flux F_{inc},

$$d\sigma = 2\pi |T_\mu(\boldsymbol{q})|^2 \frac{\rho_f}{F_{\text{inc}}} \,. \tag{16}$$

This expression has the dimensions of area, no matter how complex the final state may be.

In calculating cross sections from $\Gamma_{i\to f}$, care must be taken to use a consistent set of definitions for the initial and final states, and for the incident flux. In §8.2, we used states with a continuous momentum spectrum, normalized to δ-functions, and in (8) have continued to do so. With this definition the incident flux is $v_i/(2\pi)^3$, where v_i is the projectile's incident velocity (or the relative initial velocity if recoil of the target must be taken into account). True enough, this expression for F_{inc} does not have the dimension of a flux because the probability per unit volume for finding the projectile at a point in space is $(2\pi)^{-3}$, which does not have the dimension of 1/volume. This is an innocent flaw, however, because it cancels out in observable quantities such as cross sections. If the collision produces N free particles escaping to infinity in the final state, then in continuum normalization

$$\rho_f = \frac{1}{dE_f} \prod_{s=1}^{N} d^3p_s \,, \qquad E_f = \sum_s \epsilon_{p_s} \,, \tag{17}$$

where ϵ_{p_s} is the kinetic energy of particle s. The reader should confirm that this always gives $d\sigma$ the dimensions of area.

Doubts raised by a flux with the wrong dimensions can be put to rest in several ways. One is to recall §3.7(d), where the Golden Rule was written in terms of states satisfying periodic boundary conditions on the surface of a large cube of volume L^3, and normalized to unity. In this convention $F_{\text{inc}} = v_i/L^3$, which has correct dimensions, and each factor in the density of states is $(L/2\pi)^3 d^3p_s$. Naturally, L must disappear from all observable quantities, which therefore offers a (weak) clue for detecting errors. Another equivalent derivation uses stationary-state scattering theory; indeed, this was used to derive elastic cross sections in Chapter 8, and by applying (16) to potential scattering one readily confirms that it gives the same result as stationary-state perturbation theory.

Finally, therefore, the cross section when the atom remains in a bound state is

$$d\sigma_\mu = \frac{(2\pi)^3}{v_i} 2\pi |T_\mu(\boldsymbol{q})|^2 \frac{d^3k_f}{dE_f} , \tag{18}$$

and therefore

$$\frac{d\sigma_\mu}{d\Omega_f} = (2\pi)^4 M^2 \frac{k_f}{k_i} |T_\mu(\boldsymbol{q})|^2 . \tag{19}$$

This reduces, as it must, to the familiar result for elastic scattering in which case $k_f = k_i$. In collisions in which the atom is ionized, the differential cross section must also specify the momenta of the ejected electrons. This process is discussed below.

(b) Elastic Scattering

Elastic scattering amplitudes are usually expressed in terms of *elastic form factors*. There are such form factors for the nucleus and the electrons:

$$F_N(q) = \int d\boldsymbol{r}\, e^{i\boldsymbol{q}\cdot\boldsymbol{r}} \rho_N(r) , \tag{20}$$

$$F_0(q) = \frac{1}{Z} \langle \chi_0 | n_{\boldsymbol{q}} | \chi_0 \rangle . \tag{21}$$

The amplitude for elastic scattering off the atom is thus

$$T_{\text{el}}(q) = -\frac{Ze^2}{q^2\,(2\pi)^3} \left[F_0(q) - F_N(q) \right] . \tag{22}$$

Here it has been assumed that both charge distributions are spherically symmetric, so that the form factors depend only on $|\boldsymbol{q}|$. In practice, this is almost always valid; even when the atom and/or nucleus has nonzero angular momentum, which is required for an anisotropic charge distribution, one is usually concerned with scattering from a target that has random spin directions, and therefore an effectively isotropic distribution.

By definition, both form factors equal 1 when $q = 0$, and the factor $[F_0 - F_N]$ therefore cancels the divergent Rutherford factor $1/q^2$ in the forward direction. This, of course, is a consequence of our assumption that the atom is electrically neutral, and would not hold for elastic scattering by an ion. The behavior of the form factors when $q \to 0$ is therefore needed. Expanding the exponential and taking advantage of spherical symmetry yields

$$F_N(q) \simeq 1 - \tfrac{1}{6} q^2 \bar{R}_N^2 , \qquad (q \to 0) , \tag{23}$$

where $\bar{R}_N$ is the rms radius of the nuclear charge distribution,

$$\bar{R}_N^2 = \int d\boldsymbol{r}\, r^2 \rho_N(r) . \tag{24}$$

The electronic elastic form factor has the same behavior as $q \to 0$, of course:

$$F_0(q) \simeq 1 - \tfrac{1}{6} q^2 \bar{R}_e^2 , \tag{25}$$

$$\bar{R}_e^2 = Z^{-1} \sum_s \langle \chi_0 | r_s^2 | \chi_0 \rangle . \tag{26}$$

But the atom is far larger than the nucleus, so $\bar{R}_e^2 \ggg \bar{R}_N^2$. Consequently the forward elastic amplitude is

$$T_{\rm el}(0) = \frac{Ze^2\,\bar{R}_e^2}{48\pi^3}\ . \tag{27}$$

Note that this amplitude is proportional to Z, and that the elastic cross section in the forward direction is therefore proportional to Z^2. An amplitude that is proportional to the number of scatterers is said to be *coherent*, because the amplitudes from the individual scatterers must add coherently to produce this result. Coherence is not maintained for all momentum transfers, however: when the momentum transfer grows beyond the inverse dimensions of the target system, the amplitudes from different positions in the target are no longer in phase, the amplitude for the whole target drops off rapidly, and the overall factor of Z no longer characterizes the magnitude of the target's amplitude.

The preceding statement can be phrased in another instructive manner. As is evident from the preceding formulas, the inverse momentum transfer $1/q$ plays the role of an effective wavelength $\lambda_{\rm eff}$ probing the system. When $\lambda_{\rm eff}$ is large compared to the system, it acts like one object, and in our example produces a scattering amplitude proportional to its total charge. When $\lambda_{\rm eff}$ is of order the size of the region occupied by a single constituent of the target, the probe can in principle measure its coordinate, as in Heisenberg's γ-ray microscope, which "destroys" coherence with other similar regions. In the latter case, the cross section is an *incoherent* sum of the cross sections (*not* the amplitudes) due to the various regions. Quite generally, one refers to processes in which the cross section is proportional to the number of constituents as *incoherent scattering.*[1]

Because the nucleus is so small compared to the atom, the atom's elastic amplitude for momentum transfers small compared to $1/\bar{R}_N$ is given by

$$T_{\rm el}(q) = \frac{Ze^2}{q^2\,(2\pi)^3}\,[1 - F_0(q)]\ . \tag{28}$$

The elastic cross section is calculated in the manner that led to Eq. 19, which results in

$$\frac{d\sigma_{\rm el}}{d\Omega} = Z^2\,\frac{d\sigma_R}{d\Omega}\,|1 - F_0(q)|^2\ , \tag{29}$$

where $d\sigma_R/d\Omega$ is the Rutherford cross section for unit charges.

As is evident from (13), the elastic form factor drops off rapidly for momentum transfers large compared to the characteristic momenta k_C in the ground state.[2] For momentum transfers in the regime

$$(\bar{R}_N)^{-2} \gg q^2 \gg k_C^2\ , \tag{30}$$

---

[1]When absorption is important, as it is, for example, in the scattering of energetic hadrons by nuclei, the hallmark of incoherent scattering is a cross section that is proportional to the area presented to the beam by the nucleus, i.e., $\propto A^2$, where A is the mass number, because nuclear radii are proportional to $A^{1/3}$. For the same reason the cross section for coherent scattering by such a target is proportional to the square of this area.

[2]The value of k_C depends on the internal dynamics of the system, and need not be $\sim (1/D)$, where D is the the system's gross size. In a complex atom, the various shells have quite different k_C, and in (30) that for the K shell should be used, i.e., Z/a_0.

the electrons' form factor is therefore very small compared to one, while the nuclear form factor is approximately 1. Hence the cross section is entirely due to the nucleus and is given by the Rutherford formula for point charges. For larger momentum transfers, $q^2(\bar{R}_N)^2 \gg 1$, the cross section drops off rapidly in comparison to the Rutherford formula, and thereby reveals the nuclear charge distribution.

(c) Inelastic scattering

Inelastic scattering, in contrast to elastic scattering, is not described adequately by simple, explicit and widely applicable expressions for the cross section. If the excited state χ_μ is bound, the final state has just the scattered projectile in a continuum state, and the cross section is given by (19):

$$\frac{d\sigma_\mu}{d\Omega_f} = 4M^2 \frac{k_f}{k_i} \left(\frac{e^2}{4\pi q^2}\right)^2 |\langle\chi_\mu|n_{\boldsymbol{q}}|\chi_0\rangle|^2 \tag{31}$$

$$= \frac{d\sigma_R}{d\Omega_f} \frac{k_f}{k_i} |\langle\chi_\mu|n_{\boldsymbol{q}}|\chi_0\rangle|^2 \,, \tag{32}$$

where $d\sigma_R/d\Omega_f$ is again the Rutherford point cross section, but evaluated at the momentum transfer appropriate to the inelastic collision as given by (4) or (6). It is customary to express the inelastic cross section in terms of an *inelastic form factor*,

$$F_\mu(\boldsymbol{q}) = \frac{1}{Z} \langle\chi_\mu|n_{\boldsymbol{q}}|\chi_0\rangle \,. \tag{33}$$

The simplest example of practical interest is, once again, provided by hydrogen. In this case χ_0 is the $1s$ state, and if the excited state χ_μ is bound it has the quantum numbers (nlm); spin is irrelevant as it plays no role in nonrelativistic Coulomb scattering. By expanding $e^{i\boldsymbol{q}\cdot\boldsymbol{r}}$ into spherical waves one finds that

$$F_{nlm}(\boldsymbol{q}) = \sqrt{4\pi}\, i^l Y^*_{lm}(\hat{\boldsymbol{q}}) \int_0^\infty j_l(qr) u_{nl}(r) u_{10}(r)\, dr \equiv \sqrt{4\pi}\, i^l Y^*_{lm}(\hat{\boldsymbol{q}})\, I_{nl}(q) \,, \tag{34}$$

where u_{nl} is the radial wave function as defined in §3.6(b). In most experiments the angular momentum projection m is not determined, and the cross section is then proportional to

$$4\pi \sum_m |Y_{lm}(\hat{\boldsymbol{q}})|^2 = 2l+1 \,, \tag{35}$$

which follows from the addition theorem for spherical harmonics. The complete result for the differential cross section is therefore

$$\frac{d\sigma_{nl}}{d\Omega_f} = (2l+1) \frac{d\sigma_R}{d\Omega_f} \frac{k_f}{k_i} |I_{nl}(q)|^2 \,. \tag{36}$$

No general statement can be made about this angular distribution for all values of q and μ. Nevertheless, because $j_l(x) \sim x^l$ for $x \to 0$, the angular distribution does have a form that depends only on the angular momentum of the excited state for small q:

$$\frac{d\sigma_{nl}}{d\Omega_f} \propto q^{2l-4} \qquad q \to q_{\min} \,. \tag{37}$$

Clearly, this behavior is not just a peculiarity of scattering by hydrogen. It serves as a useful diagnostic tool in electron scattering by systems whose spectrum is not well understood, such as complex nuclei.

When the momentum transfer is large compared to the characteristic momenta of bound states, the overlap (13) between amplitudes is very small, and the most probable result of a collision will be the ionization of the atom. In the case of hydrogen, which has just one electron, the general matrix element (13) reduces to

$$\langle \boldsymbol{p}_f | n_{\boldsymbol{q}} | \chi_0 \rangle = \int d\boldsymbol{p}\, \delta(\boldsymbol{p} + \boldsymbol{q} - \boldsymbol{p}_f)\, \phi_0(\boldsymbol{p}) = \phi_0(\boldsymbol{p}_f + \boldsymbol{k}_f - \boldsymbol{k}_i) \,, \tag{38}$$

where $\boldsymbol{p}_f$ is the momentum of the ejected electron, which is assumed to be large enough to make the influence of the Coulomb field on the ejected electron negligible (i.e., $\alpha/v_f \ll 1$). If the collision were in empty space, the projectile's momenta $\boldsymbol{k}_i$ and $\boldsymbol{k}_f$, and that of the ejected electron, would determine the latter's initial momentum $\boldsymbol{p}_i$ to be $\boldsymbol{p}_f + \boldsymbol{k}_f - \boldsymbol{k}_i$. Eq. 38 thus establishes that in this process the cross section, being proportional to $|\phi_0(\boldsymbol{p}_i)|^2$, is proportional to nothing other than *the probability of finding the momentum in the target state required by momentum conservation.* This provides the most striking illustration of the impulsive nature of the Born approximation.

The cross section when an originally bound particle is ejected is calculated as before, but now the density of final states has a factor for the production product, $\rho_f = d^3k_f\, d^3p_f / dE_f$. Consequently the differential cross section is

$$\frac{d\sigma}{d\Omega_f\, d^3p_f} = \frac{d\sigma_R}{d\Omega_f} \frac{k_f}{k_i} \,|\phi_0(\boldsymbol{p}_f + \boldsymbol{k}_f - \boldsymbol{k}_i)|^2 \,, \tag{39}$$

where the Rutherford factor is again evaluated at $\boldsymbol{q} = \boldsymbol{k}_i - \boldsymbol{k}_f$.

When the projectile is an electron, its indistinguishability from electrons in the target must be considered. Inelastic scattering by hydrogen will suffice to show what is at stake. The "direct" amplitude is the one that would describe the collision if the the projectile were distinguishable. It describes the process in which electron 1 is incident and scattered while electron 2 is remains bound:

$$T_{\text{dir}} = e^2 \int d\boldsymbol{r}_1 d\boldsymbol{r}_2 \frac{\varphi^*_{\boldsymbol{k}_f}(\boldsymbol{r}_1)\chi^*_\mu(\boldsymbol{r}_2)\chi_0(\boldsymbol{r}_2)\varphi_{\boldsymbol{k}_i}(\boldsymbol{r}_1)}{4\pi|\boldsymbol{r}_1 - \boldsymbol{r}_2|} \tag{40}$$

$$= \frac{e^2}{(2\pi)^3} \frac{1}{|\boldsymbol{k}_i - \boldsymbol{k}_f|^2} \int d\boldsymbol{p}\, \phi^*_\mu(\boldsymbol{p} + \boldsymbol{q})\phi_0(\boldsymbol{p}) \,, \tag{41}$$

which is just the first term of (10). The "exchange" amplitude describes the indistinguishable process where electron 1 is incident and is left afterwards in the excited bound state χ_μ, while electron 2 ends up as the scattered particle even though it was originally bound; this amplitude is

$$T_{\text{ex}} = e^2 \int d\boldsymbol{r}_1 d\boldsymbol{r}_2 \frac{\varphi^*_{\boldsymbol{k}_f}(\boldsymbol{r}_2)\chi^*_\mu(\boldsymbol{r}_1)\chi_0(\boldsymbol{r}_2)\varphi_{\boldsymbol{k}_i}(\boldsymbol{r}_1)}{4\pi|\boldsymbol{r}_1 - \boldsymbol{r}_2|} \tag{42}$$

$$= \frac{e^2}{(2\pi)^3} \int d\boldsymbol{p}\, \frac{1}{|\boldsymbol{k}_f - \boldsymbol{p}|^2}\, \phi^*_\mu(\boldsymbol{p} + \boldsymbol{q})\phi_0(\boldsymbol{p}) \,. \tag{43}$$

The forms of (41) and (43) are understandable. In both amplitudes electrons 1 and 2 have the initial momenta $(\boldsymbol{k}_i, \boldsymbol{p})$, respectively; in the direct amplitude their respective final momenta are $(\boldsymbol{k}_f, \boldsymbol{p}')$, whereas in the exchange amplitude they are, respectively, $(\boldsymbol{p}', \boldsymbol{k}_f)$, where $\boldsymbol{p}'$ is the argument of the excited state $\phi_\mu(\boldsymbol{p}')$. In the direct process the momentum transfer experienced by 1 is $\boldsymbol{k}_i - \boldsymbol{k}_f$, but in the exchange process it is $\boldsymbol{k}_i - \boldsymbol{p}'$. In both cases $\boldsymbol{k}_i + \boldsymbol{p} = \boldsymbol{k}_f + \boldsymbol{p}'$, so in the exchange process the momentum transfer is $\boldsymbol{k}_f - \boldsymbol{p}$.

It is now evident that when the initial and final momenta $\boldsymbol{k}_{i,f}$ are large compared to those in the target, the exchange amplitude is very small compared to the direct amplitude because the target's momentum space wave function requires $\boldsymbol{p}$ to be small compared to $\boldsymbol{k}_{i,f}$. Therefore in T_{ex} the Rutherford factor is of order $(1/k_f)^2 \sim (1/k_i)^2$, which is small at all momentum transfers, whereas it is $1/q^2$ in T_{dir} irrespective of the magnitudes of $\boldsymbol{k}_{i,f}$.

The lesson taught by this example then is that the projectile can, to good approximation, be treated as if it were distinguishable from the constituents of the target when its initial and final momenta are large compared to those typical of the target. By and large, this condition and that for the validity of the Born approximation are met simultaneously in the scattering of energetic electrons by atoms.

While there are no formulas that give transparent and explicit expressions for the cross section in arbitrary inelastic processes from arbitrary targets, there is a simple and powerful result for *the angular distribution of total inelastic scattering at high energies*. By definition, and according to (31), this cross section is

$$\frac{d\sigma_{\text{inel}}}{d\Omega} = \frac{4M^2(e^2/4\pi)^2}{k_i} \sum_{\mu \neq 0} \frac{k_f}{q^4} |\langle \chi_\mu | n_{\boldsymbol{q}} | \chi_0 \rangle|^2 \,. \tag{44}$$

For specified values of (k_i, θ), both q and k_f are functions of μ, and the sum over excited states has an upper kinematic limit. For a "soft" interaction like the Coulomb field, the probability of excitation drops quite rapidly with excitation energy Δ_μ. Therefore, if the incident momentum is high enough, $k_f \simeq k_i$, and the dependence of q on Δ_μ can be ignored provided θ is not exceptionally small $[\theta^2 \gtrsim (M\Delta_{\max})^2/k_i^4]$. The sum on μ in (44) can then be done because $\{\chi_\mu\}$ is a complete set:[1]

$$\frac{d\sigma_{\text{inel}}}{d\Omega} = \frac{d\sigma_R}{d\Omega} \left\{ \langle n_{\boldsymbol{q}}^\dagger n_{\boldsymbol{q}} \rangle_0 - |\langle n_{\boldsymbol{q}} \rangle_0|^2 \right\} \,. \tag{45}$$

Thus this cross section is proportional to the mean-square fluctuations in the target's charge distribution. Because the electrons are indistinguishable,

$$\langle n_{\boldsymbol{q}}^\dagger n_{\boldsymbol{q}} \rangle_0 - |\langle n_{\boldsymbol{q}} \rangle_0|^2 = Z + Z(Z-1)\langle e^{i\boldsymbol{q}\cdot(\boldsymbol{r}_1 - \boldsymbol{r}_2)} \rangle_0 - Z^2 |\langle e^{i\boldsymbol{q}\cdot\boldsymbol{r}_1} \rangle_0|^2 \,. \tag{46}$$

Let $P_2(\boldsymbol{r}_1\boldsymbol{r}_2)$ be the probability for finding, in the target's ground state, one electron at $\boldsymbol{r}_1$ and another at $\boldsymbol{r}_2$, and $P_1(\boldsymbol{r}_1)$ that for finding only one at $\boldsymbol{r}_1$:

$$P_2(\boldsymbol{r}_1\boldsymbol{r}_2) = \int d\boldsymbol{r}_3 \ldots d\boldsymbol{r}_Z \, |\chi_0(\boldsymbol{r}_1 \ldots \boldsymbol{r}_Z)|^2 \,, \quad P_1(\boldsymbol{r}_1) = \int d\boldsymbol{r}_2 \, P_2(\boldsymbol{r}_1\boldsymbol{r}_2) \,. \tag{47}$$

Then

$$\langle n_{\boldsymbol{q}}^\dagger n_{\boldsymbol{q}} \rangle_0 - |\langle n_{\boldsymbol{q}} \rangle_0|^2 = Z + \int d\boldsymbol{r}_1 d\boldsymbol{r}_2 e^{i\boldsymbol{q}\cdot(\boldsymbol{r}_1 - \boldsymbol{r}_2)} \left[Z(Z-1)P_2(\boldsymbol{r}_1\boldsymbol{r}_2) - Z^2 P_1(\boldsymbol{r}_1)P_1(\boldsymbol{r}_2) \right] . \tag{48}$$

[1]Formulas of this type for the angular distribution of inelastic scattering are widely applicable — for example, to neutron scattering from atoms, molecules and extended systems.

The most naive assumption is that there are no correlations,

$$P_2(\boldsymbol{r}_1\boldsymbol{r}_2) = P_1(\boldsymbol{r}_1)P_1(\boldsymbol{r}_2)\ , \tag{49}$$

which would hold if the many electron wave function were simply a product of one-electron wave functions. When (49) is assumed,

$$\langle n_{\boldsymbol{q}}^{\dagger} n_{\boldsymbol{q}} \rangle_0 - |\langle n_{\boldsymbol{q}} \rangle_0|^2 = Z[1 - |F_0(q)|^2]\ , \tag{50}$$

where $F_0(q) = \langle e^{i\boldsymbol{q}\cdot\boldsymbol{r}} \rangle_0$ is the elastic form factor. Although the assumption of no correlations is invalid, the result reveals an essential fact: *the total inelastic cross section is proportional to Z, and not to Z^2 — it is an incoherent cross section.* In reality, the Coulomb interaction between electrons and the Pauli principle impose correlations on the ground-state wave function, so that (49) only holds for $|\boldsymbol{r}_1 - \boldsymbol{r}_2|$ larger than the mean separation d between electrons. These effects show up in the cross section as the Fourier transform of a correlation function, but this does not change the Z-dependence because d does not grow with Z.

(d) Energy Loss

When an energetic particle much heavier than the electron runs through a medium, it follows a rather straight path as it slows down because, as we have just seen, the cross sections for inelastic collisions are sharply peaked toward the forward direction, which makes the mean free path between large-angle scatterings long. Hence the quantity of interest is *the energy loss per unit path length, dE/dx*, due to all inelastic collisions in which the particle is scattered through angles smaller than some small angle Θ. While Θ is set by experimental conditions, it is clear that a formula for dE/dx will only be useful if it is rather insensitive to this angle, which will be shown to be the case.

The energy loss, or stopping power, is given by[1]

$$\frac{dE}{dx} = N \sum_{\mu} (E_\mu - E_0) \int_0^{2\pi} d\phi \int_0^{\Theta} \sin\theta\, d\theta\, \frac{d\sigma_\mu}{d\Omega}\ , \tag{51}$$

where N is the number of atoms per unit volume. Here E_μ is the energy of the atomic state χ_μ, and the momentum and velocity of the particle passing through matter will be designated by (k, v) instead of (k_i, v_i).

Then, using $qdq = kk_f\, d(\cos\theta)$ and (31),

$$\frac{dE}{dx} = \frac{e^4}{2\pi} N z^2 \frac{1}{v^2} \sum_{\mu} (E_\mu - E_0) \int_{q_{\text{min}}}^{q_{\text{max}}} |\langle \chi_\mu | n_{\boldsymbol{q}} | 0 \rangle|^2 \frac{dq}{q^3}\ , \tag{52}$$

where ze is the charge of the particle in question, q_{max} is expression (3) evaluated at $\theta = \Theta$, and

$$q_{\text{min}} = (E_\mu - E_0)/v\ , \tag{53}$$

as given by (5). As long as Θ is not so small as to make q_{max} comparable to q_{min},

$$q_{\text{max}} \simeq Mv\Theta\ . \tag{54}$$

---

[1]See e.g., E. Segré, *Nuclei and Particles,* 2nd ed., W.A. Benjamin (1977), §1-2.

The basic tool for the evaluation of (52) is the identity

$$\sum_\mu (E_\mu - E_0)|\langle\chi_\mu|n_{\boldsymbol{q}}|0\rangle|^2 = \tfrac{1}{2}\langle[[n_{\boldsymbol{q}}^\dagger, H], n_{\boldsymbol{q}}]\rangle_0 \,, \tag{55}$$

where H is the Hamiltonian of the atom; the proof is left for Prob. 2. Because the potential energy commutes with $n_{\boldsymbol{q}}$,

$$[n_{\boldsymbol{q}}^\dagger, H] = \frac{i}{2m}\sum_s \left(\frac{\partial n_{\boldsymbol{q}}^\dagger}{\partial \boldsymbol{r}_s}\cdot \boldsymbol{p}_s + \boldsymbol{p}_s \cdot \frac{\partial n_{\boldsymbol{q}}^\dagger}{\partial \boldsymbol{r}_s}\right) \tag{56}$$

$$= \frac{1}{2m}\sum_s \left(e^{-i\boldsymbol{q}\cdot\boldsymbol{r}_s}\, \boldsymbol{q}\cdot\boldsymbol{p}_s + \boldsymbol{q}\cdot\boldsymbol{p}_s\, e^{-i\boldsymbol{q}\cdot\boldsymbol{r}_s}\right) , \tag{57}$$

and therefore

$$\tfrac{1}{2}\langle[[n_{\boldsymbol{q}}^\dagger, H], n_{\boldsymbol{q}}]\rangle_0 = \frac{q^2}{2m} Z \,. \tag{58}$$

Note here and henceforth that m is the electron mass, whereas M is the mass of the particle losing energy!

It is not a straightforward matter to exploit (55) and (58), however, because $q_{\rm min}$ is a function of the excitation energy, and therefore the sum on μ is not (55). At this point it is instructive to be sloppy for a moment, and to replace $q_{\rm min}$ by some sort of average, for example, in view of (53), by Δ/v, where Δ is an average excitation energy, so that the integral over q and the sum over states can be interchanged. If this is done, the answer is

$$\frac{dE}{dx} = \frac{e^4}{4\pi} N Z z^2 \frac{1}{m v^2} \ln\left(\frac{M v^2 \Theta}{\Delta}\right) . \tag{59}$$

Clearly, this guesswork does not tell us whether Δ is a function of v and M, and the crucial, practical question is to determine the dependence of the energy loss on the velocity and mass of the particle. An ingenious treatment of the troublesome small momentum transfers due to Bethe answers this question.

The first step is to break the integration over q into to two intervals, $(q_{\rm min}, q_0)$ and $(q_0, q_{\rm max})$, where q_0 is independent of μ and k, and sufficiently small so that $q_0 r \ll 1$ for all distances in the atom. (Note that as $k \to \infty$, $q_{\rm min}/q_0 \to 0$.) The calculation leading to (59) is valid in the large q interval, and gives the contribution

$$\left(\frac{dE}{dx}\right)_{\rm large} = \frac{e^4}{4\pi} N Z z^2 \frac{1}{m v^2} \ln\left(\frac{M v \Theta}{q_0}\right) . \tag{60}$$

In the small q interval,

$$\langle\chi_\mu|e^{i\boldsymbol{q}\cdot\boldsymbol{r}_s}|\chi_0\rangle \simeq i\,\langle\chi_\mu|\boldsymbol{q}\cdot\boldsymbol{r}_s|\chi_0\rangle \,, \qquad (\mu \neq 0) \,, \tag{61}$$

so the contribution of this interval is

$$\left(\frac{dE}{dx}\right)_{\rm small} = \frac{e^4}{2\pi} N z^2 \frac{1}{v^2} J \,, \tag{62}$$

where

$$J = \sum_\mu (E_\mu - E_0) \int_{q_{\rm min}}^{q_0} |\langle\chi_\mu|\boldsymbol{q}\cdot\boldsymbol{D}|0\rangle|^2 \frac{dq}{q^3} \,, \tag{63}$$

and $\boldsymbol{D}$ is the atom's dipole moment. Unless the material through which the particle is moving is spin-polarized, the atom's state is isotropic, and this last expression cannot depend on the orientation of $\boldsymbol{q}$. Therefore

$$J = \sum_{\mu} (E_\mu - E_0)|\langle\chi_\mu|D_z|\chi_0\rangle|^2 \, \ln\left(\frac{vq_0}{E_\mu - E_0}\right) . \tag{64}$$

As q_0 is a fictitious quantity without physical significance, it should, if possible, be eliminated. This is done by combining the logarithms in (60) and (64):

$$J = \tfrac{1}{2}\langle[[D_z, H], D_z]\rangle_0 \ln\left(\frac{vq_0}{I}\right) + \sum_{\mu}(E_\mu - E_0)|\langle\chi_\mu|D_z|\chi_0\rangle|^2 \, \ln\left(\frac{I}{E_\mu - E_0}\right) , \tag{65}$$

where I is, for the moment an arbitrary constant, which, in the first term, allows the sum to be replaced by the double commutator. The latter has the value $Z/2m$. The two parts of dE/dx can now be combined:

$$\frac{dE}{dx} = \frac{e^4}{4\pi} NZz^2 \frac{1}{mv^2}\left\{\ln\left(\frac{Mv^2\Theta}{I}\right) + \frac{2m}{Z}\sum_{\mu}(E_\mu - E_0)|\langle\chi_\mu|D_z|\chi_0\rangle|^2 \, \ln\left(\frac{I}{E_\mu - E_0}\right)\right\} , \tag{66}$$

The last clever move is to note that as I is at our disposal, it can be chosen to eliminate the last term! This choice is

$$\ln I = \frac{Z}{2m}\sum_{\mu}(E_\mu - E_0)|\langle\chi_\mu|D_z|\chi_0\rangle|^2 \, \ln\left(E_\mu - E_0\right) . \tag{67}$$

The final result is then Bethe's formula for the energy loss,

$$\frac{dE}{dx} = \frac{e^4}{4\pi} NZz^2 \frac{1}{mv^2} \ln\left(\frac{Mv^2\Theta}{I}\right) . \tag{68}$$

The great complexity of the excitation spectrum has thus been reduced to just the single parameter I, which characterizes *only* the atom in question, and does *not* depend on the mass or velocity of the projectile. I is usually taken from energy loss measurements. The formula gives the crucial dependence of the energy loss (or stopping power) on velocity and mass of the projectile. That the mass only appears in the logarithm means that very accurate measurements of energy loss are required to distinguish particles of the same charge but different mass.

9.2 The S Matrix

The general theory of collisions is very elegant and compact — so much so that it is easy to lose sight of the complexity of the phenomena that are described by it. For that reason we first present a concrete, albeit rather artificial, example of a system that has inelastic processes. This will then provide a natural point of departure for the general formalism.

(a) Scattering by a Bound Particle

The illustrative system consists of two spin 0 particles, 1 and 2, which interact with each other through a potential V, while 2 is also subjected to an external potential U in which it has a number of bound states; particle 1 does not interact with U, however. The complete Hamiltonian is then

$$H = K_1 + K_2 + U + V \,, \tag{69}$$

where K_i is the kinetic energy of particle i.

The energy eigenstates of 2 when 1 is not present are the solutions of

$$(K_2 + U)\chi_\mu(\boldsymbol{r}_2) = W_\mu \chi_\mu(\boldsymbol{r}_2) \,. \tag{70}$$

The label μ stands for all the quantum numbers required to specify any member of this complete set, which can contain both bound and scattering states.

Let $\Psi_{\boldsymbol{k}}^{(+)}(\boldsymbol{r}_1\boldsymbol{r}_2)$ describe a collision where particle 1 is incident with momentum $\boldsymbol{k}$, and particle 2 is initially in the lowest bound state χ_0. This wave function is a solution of

$$(E - H_0)\Psi_{\boldsymbol{k}}^{(+)} = V\Psi_{\boldsymbol{k}}^{(+)} \,, \tag{71}$$

where H_0 is the part of H that cannot cause a collision between 1 and 2,

$$H_0 = K_1 + K_2 + U \,, \tag{72}$$

$E = \epsilon_k + W_0$, and ϵ_k is the initial kinetic energy of 1. As in potential scattering (recall §8.2(a)), the boundary conditions just stated are incorporated by replacing the differential equation (71) by the following integral equation:

$$\Psi_{\boldsymbol{k}}^{(+)}(\boldsymbol{r}_1\boldsymbol{r}_2) = \varphi_{\boldsymbol{k}}(\boldsymbol{r}_1)\chi_0(\boldsymbol{r}_2) + \int d\boldsymbol{r}_1'\, d\boldsymbol{r}_2'\, \langle \boldsymbol{r}_1\boldsymbol{r}_2|\mathcal{G}_0(E+i\epsilon)|\boldsymbol{r}_1'\boldsymbol{r}_2'\rangle V(\boldsymbol{r}_1'\boldsymbol{r}_2')\Psi_{\boldsymbol{k}}^{(+)}(\boldsymbol{r}_1'\boldsymbol{r}_2') \,, \tag{73}$$

where $\mathcal{G}_0$ is the resolvent for the Hamiltonian H_0:

$$\mathcal{G}_0(E + i\epsilon) = \frac{1}{\epsilon_k + W_0 - K_1 - K_2 - U + i\epsilon} \,. \tag{74}$$

Clearly, Green's function in Eq. 73 is a very complicated beast, but as we shall see immediately, the elastic and inelastic scattering amplitudes can be identified without knowing its actual form.

The two-particle wave function can be expanded in terms of the complete set $\{\chi_\mu\}$:

$$\Psi_{\boldsymbol{k}}^{(+)}(\boldsymbol{r}_1\boldsymbol{r}_2) = \sum_\mu \psi_{\boldsymbol{k},\mu}^{(+)}(\boldsymbol{r}_1)\,\chi_\mu(\boldsymbol{r}_2) \,. \tag{75}$$

The functions $\psi_{\boldsymbol{k},\mu}^{(+)}(\boldsymbol{r}_1)$ are the probability amplitudes for finding particle 1 at $\boldsymbol{r}_1$ when particle 2 is in the state χ_μ; they are not orthogonal. Equations that determine these amplitudes follow from (73) when it is multiplied by χ_μ^* and integrated over $\boldsymbol{r}_2$. In this way one encounters

$$\langle \boldsymbol{r}_1\chi_\mu|\mathcal{G}_0(E + i\epsilon)|\boldsymbol{r}_1'\chi_\nu\rangle = \delta_{\mu\nu}\, \langle \boldsymbol{r}_1|\frac{1}{\epsilon_k + W_0 - W_\mu + i\epsilon}|\boldsymbol{r}_1'\rangle \,, \tag{76}$$

which holds because χ_ν is an eigenfunction of $K_2 + U$, and the latter commutes with K_1. Thus the right-hand side is just Green's function $G_0(\boldsymbol{r}_1\boldsymbol{r}_1'; k_\mu)$ for a free particle (see §2.6(c)) with the momentum k_μ defined by

$$\frac{k_\mu^2}{2m} = \frac{k^2}{2m} + W_0 - W_\mu \equiv \frac{k^2}{2m} - \Delta_\mu \,. \tag{77}$$

That is,

$$\langle \boldsymbol{r}_1\chi_\mu|\mathcal{G}_0(E+i\epsilon)|\boldsymbol{r}_1'\chi_\nu\rangle = -2m\,\frac{e^{ik_\mu R}}{4\pi R}\,\delta_{\mu\nu}\,, \qquad R = |\boldsymbol{r}_1 - \boldsymbol{r}_1'|\,. \tag{78}$$

Hence the single integral equation (73) for the complete wave function $\Psi^{(+)}$ produces the following coupled system of equations for the amplitudes $\psi_\mu^{(+)}$:

$$\psi_{\boldsymbol{k},\mu}^{(+)}(\boldsymbol{r}_1) = \varphi_{\boldsymbol{k}}(\boldsymbol{r}_1)\,\delta_{\mu,0} - \frac{m}{2\pi}\sum_\nu \int d\boldsymbol{r}_1'\,\frac{e^{ik_\mu R}}{R} V_{\mu\nu}(\boldsymbol{r}_1')\psi_{\boldsymbol{k},\nu}^{(+)}(\boldsymbol{r}_1')\,, \tag{79}$$

where

$$V_{\mu\nu}(\boldsymbol{r}_1) = \int d\boldsymbol{r}_2\,\chi_\mu^*(\boldsymbol{r}_2)V(\boldsymbol{r}_1\boldsymbol{r}_2)\chi_\nu(\boldsymbol{r}_2)\,. \tag{80}$$

Equation (79) brings us no closer to an actual solution of the problem. Nevertheless, it does shed light on the processes produced by the collision and provides explicit expressions for the elastic and inelastic scattering amplitudes. Consider the asymptotic form of (79):

$$\psi_{\boldsymbol{k},\mu}^{(+)}(\boldsymbol{r}_1) \sim \varphi_{\boldsymbol{k}}(\boldsymbol{r}_1)\,\delta_{\mu,0} - \frac{m}{2\pi}\,\frac{e^{ik_\mu r_1}}{r_1}\sum_\nu \int d\boldsymbol{r}_1'\,e^{-i\boldsymbol{k}_\mu'\cdot\boldsymbol{r}_1'}V_{\mu\nu}(\boldsymbol{r}_1')\psi_{\boldsymbol{k},\nu}^{(+)}(\boldsymbol{r}_1')\,. \tag{81}$$

Here $\boldsymbol{k}_\mu'$ is the vector, with magnitude given by (77), pointing from the target to the observation point $\boldsymbol{r}_1$. The second term gives the amplitudes for finding 1 at large distance when 2 is in one of the states χ_ν. Elastic scattering is described by $\psi_{\boldsymbol{k},0}^{(+)}$; it has precisely the same form as in potential scattering. Comparing with Eq. 8.87 identifies *the elastic amplitude*:

$$f_{\text{el}}(\boldsymbol{k}'|\boldsymbol{k}) = -4\pi^2 m\sum_\nu \int d\boldsymbol{r}_1\,\varphi_{\boldsymbol{k}'}^*(\boldsymbol{r}_1)V_{0\nu}(\boldsymbol{r}_1)\psi_{\boldsymbol{k},\nu}^{(+)}(\boldsymbol{r}_1) \tag{82}$$

$$= -4\pi^2 m\int d\boldsymbol{r}_1 d\boldsymbol{r}_2\,\varphi_{\boldsymbol{k}'}^*(\boldsymbol{r}_1)\chi_0^*(\boldsymbol{r}_2)V(\boldsymbol{r}_1\boldsymbol{r}_2)\Psi_{\boldsymbol{k}}^{(+)}(\boldsymbol{r}_1\boldsymbol{r}_2)\,. \tag{83}$$

Equation (82) reveals that the amplitudes $\psi_\nu^{(+)}$ for *all* the excited states of the target particle 2 contribute to elastic scattering at every energy — even when the incident momentum of the projectile 1 is too low for any inelastic scattering. And Eq. 83 shows that the form of the scattering amplitude familiar from potential scattering holds here also.

Inelastic scattering is described by (81) when $\mu \neq 0$ *provided* k_μ is real, i.e., provided the incident kinetic energy ϵ_k suffices to excite the state χ_μ. The latter has excitation energy Δ_μ, and therefore $k > \sqrt{2m\Delta_\mu}$ is required. Reactions that can excite a certain state are often called *open inelastic channels*. Those forbidden by energy conservation ($\Delta_\mu > \epsilon_k$) are said to be closed; as shown by (81), their amplitudes decay exponentially as $r_1 \to \infty$.

The *inelastic scattering amplitude* for an open channel μ, in which particle 2 is still bound, is defined in the same way as Eq. 83:

$$f_\mu(\boldsymbol{k}'_\mu|\boldsymbol{k}) = -4\pi^2 m \int d\boldsymbol{r}_1 d\boldsymbol{r}_2\, \varphi^*_{\boldsymbol{k}'_\mu}(\boldsymbol{r}_1)\chi^*_0(\boldsymbol{r}_2)V(\boldsymbol{r}_1\boldsymbol{r}_2)\Psi^{(+)}_{\boldsymbol{k}}(\boldsymbol{r}_1\boldsymbol{r}_2) \tag{84}$$

$$= -4\pi^2 m \sum_\nu \int d\boldsymbol{r}_1\, \varphi^*_{\boldsymbol{k}'_\mu}(\boldsymbol{r}_1)V_{\mu\nu}(\boldsymbol{r}_1)\psi^{(+)}_{\boldsymbol{k},\nu}(\boldsymbol{r}_1)\,, \tag{85}$$

where $|\boldsymbol{k}'_\mu| = (k^2 - 2m\Delta_\mu)^{\frac{1}{2}}$. As in elastic scattering, all channels, whether open or closed, actually contribute indirectly to the scattering amplitude for any open channel.

From §8.2 we know that the differential cross section for elastic scattering is $|f_{\rm el}|^2$, but $|f_\mu|^2$ is *not* the differential cross section for inelastic scattering. The correct relation will be derived below (see Eq. 132).

(b) The S Matrix

We are now ready to address quite general collision problems by bearing the foregoing example in mind. The word quite in the preceding sentence refers to certain limitations on the theory that will be spelled out at the end of this section.

The Hamiltonian is again of the form $H = H_0 + V$, where H_0 governs the dynamics of the projectile and target when they do not interact. The degrees of freedom of the projectile and target will simply be called $\boldsymbol{r}$, which now stands for all the coordinate and spin variables of their constituents, and integrals over $\boldsymbol{r}$ can also include sums over spin projections as needed. Both projectile and target may be composite systems. Let $\Phi_i(\boldsymbol{r},t)$, for $t \lll 0$, be a product state, one factor being a packet of reasonably well defined momentum for the projectile moving toward the target, the other factor some stationary state of the isolated target. As V is assumed to be of finite range, for such times Φ_i is a solution of the "free" Schrödinger equation

$$(i\partial_t - H_0)\Phi_i(t) = 0\,, \qquad t \lll 0\,. \tag{86}$$

When $t \simeq 0$, projectile and target are within range, and (86) fails to describe the system. The correct wave function for all times, which describes the processes caused by V, is

$$\Psi^{(+)}_i(\boldsymbol{r},t) = \int d\boldsymbol{r}'\, K(\boldsymbol{r}t,\boldsymbol{r}'t')\Phi_i(\boldsymbol{r}',t')\,, \qquad t > t',\ t' \lll 0\,, \tag{87}$$

where K is the propagator for the complete Hamiltonian (recall §2.6(a)). For $t \ggg 0$, this wave function will contain outgoing scattered waves. The propagator for the complete system is related to K_0, the propagator in the absence of the interaction V, by Eq. 2.394:

$$K(\boldsymbol{r}t,\boldsymbol{r}'t') = K_0(\boldsymbol{r}t,\boldsymbol{r}'t') - i\int d\boldsymbol{r}_1 \int_{t'}^{t} dt_1\, K_0(\boldsymbol{r}t,\boldsymbol{r}_1t_1)V(\boldsymbol{r}_1)K(\boldsymbol{r}_1t_1,\boldsymbol{r}'t')\,. \tag{88}$$

Therefore

$$\Psi^{(+)}_i(\boldsymbol{r},t) = \Phi_i(\boldsymbol{r},t) - i\int d\boldsymbol{r}_1 \int_{t'}^{t} dt_1\, K_0(\boldsymbol{r}t,\boldsymbol{r}_1t_1)V(\boldsymbol{r}_1)\Psi^{(+)}_i(\boldsymbol{r}_1,t_1)\,. \tag{89}$$

This is the extension of (73), which dealt with stationary states and a specific model, to a time-dependent description of a general collision process.

In a scattering experiment a specific final state is selected after the whole system has separated spatially into the reaction products. Let $\Phi_f(\boldsymbol{r},t)$, for $t \ggg 0$, describe such a final state. It too is a solution of the "free" Schrödinger equation (86), because the interaction ceases to influence the dynamics at such times. We could, therefore, define the scattering amplitude by $\langle\Phi_f(t)|\Psi_i^{(+)}(t)\rangle$. This definition has the disadvantage of being time-dependent. A definition that does not have this defect replaces $\Phi_f(t)$ by the solution $\Psi_f^{(-)}(t)$ of the complete Schrödinger equation which, in the *future*, becomes the finally selected state $\Phi_f(t)$. This state therefore has incoming (*not* outgoing) scattered waves. We now use this description of the final state to define the *S matrix*:

$$S_{fi} = \langle\Psi_f^{(-)}|\Psi_i^{(+)}\rangle\,, \tag{90}$$

where

$$\Psi_{f,i}^{(\pm)} \equiv \Psi_{f,i}^{(\pm)}(0)\,. \tag{91}$$

Here $t = 0$ was chosen merely for convenience; the scalar product that defines S does not depend on t because both states are solutions of the same Schrödinger equation, which is not true of the more obvious definition $\langle\Phi_f(t)|\Psi_i^{(+)}(t)\rangle$.

As in potential scattering, and for that matter in the discussion in the first part of this very section, we now idealize by letting $\Psi_{i,f}^{(\pm)}(t)$ be stationary states. Equation (89), at $t = 0$, then becomes

$$\Psi_i^{(+)}(\boldsymbol{r}) = \Phi_i(\boldsymbol{r}) - i\int d\boldsymbol{r}_1 \int_{t'\to-\infty}^{0} dt_1\, \langle\boldsymbol{r}|e^{iH_0t_1}|\boldsymbol{r}_1\rangle V(\boldsymbol{r}_1)e^{-iE_it_1}\Psi_i^{(+)}(\boldsymbol{r}_1) \tag{92}$$

$$= \Phi_i(\boldsymbol{r}) + \int d\boldsymbol{r}_1\, \langle\boldsymbol{r}|(E_i - H_0 + i\epsilon)^{-1}|\boldsymbol{r}_1\rangle V(\boldsymbol{r}_1)\Psi_i^{(+)}(\boldsymbol{r}_1)\,, \tag{93}$$

or, recalling (8.137),

$$|\Psi_i^{(+)}\rangle = |\Phi_i\rangle + \frac{1}{E_i - H_0 + i\epsilon}V|\Psi_i^{(+)}\rangle \tag{94}$$

$$= \left(1 + \frac{1}{E_i - H + i\epsilon}V\right)|\Phi_i\rangle\,. \tag{95}$$

These equations have the same form as those in §8.2(c) concerning potential scattering, but the prototypical example of scattering by a bound particle has taught us that they apply to a far broader set of circumstances than potential scattering.

The next objective is to demonstrate that the S matrix is unitary, and that this property implies a powerful generalization of the optical theorem. The first step is to show that the set of states $|\Psi_a^{(+)}\rangle$ have the same scalar products as do their incident counterparts $|\Phi_a\rangle$. (They are not complete, of course; even in potential scattering they are not if there are bound states.) This result can be anticipated by thinking again in terms of the time-dependent functions defined by (89). Before the time when the interaction V can come into play, the states $\Psi_{a,b}^{(+)}(\boldsymbol{r}t)$ *were* $\Phi_{a,b}(t)$, and the latters' scalar products will not change as t grows provided they are are evolved with the exact Schrödinger equation into $\Psi_{a,b}^{(+)}(t)$. The formal proof uses

(95):

$$\langle \Psi_b^{(+)} | \Psi_a^{(+)} \rangle = \langle \Phi_b | \left(1 + V \frac{1}{E_b - H - i\epsilon} \right) | \Psi_a^{(+)} \rangle . \tag{96}$$

Now H acts on $|\Psi_a^{(+)}\rangle$, so $H \to E_a$, which turns the resolvent into a number and allows V to be moved to the right; thereupon E_b can be replaced by H_0 acting on $\langle \Phi_b|$, which gives

$$\langle \Psi_b^{(+)} | \Psi_a^{(+)} \rangle = \langle \Phi_b | \left(1 - \frac{1}{E_a - H_0 + i\epsilon} V \right) | \Psi_a^{(+)} \rangle . \tag{97}$$

Using (94) then gives

$$\langle \Psi_b^{(+)} | \Psi_a^{(+)} \rangle = \langle \Phi_b | \Phi_a \rangle = \delta_{ba} , \tag{98}$$

where the last equation is a shorthand for whatever normalization convention has been chosen. The same argument applies to the states with incoming scattered waves; they satisfy

$$|\Psi_a^{(-)}\rangle = \left(1 + \frac{1}{E_a - H - i\epsilon} V \right) |\Phi_a\rangle , \tag{99}$$

and

$$\langle \Psi_b^{(-)} | \Psi_a^{(-)} \rangle = \delta_{ba} . \tag{100}$$

The relationship between the S matrix and the scattering amplitudes, as they have been defined for example in (84), will now be derived. First, use (95) and (99) to write

$$\langle \Psi_f^{(+)} | - \langle \Psi_f^{(-)} | = \langle \Phi_f | V \left(\frac{1}{E_f - H - i\epsilon} - \frac{1}{E_f - H + i\epsilon} \right) . \tag{101}$$

Next, take the scalar product with $|\Psi_i^{(+)}\rangle$, and use (98):

$$\delta_{fi} - S_{fi} = \left(\frac{1}{E_f - E_i - i\epsilon} - \frac{1}{E_f - E_i + i\epsilon} \right) \langle \Phi_f | V | \Psi_i^{(+)} \rangle . \tag{102}$$

Recalling

$$\frac{1}{x + i\epsilon} = \mathrm{P}\frac{1}{x} - i\pi\delta(x) \tag{103}$$

then gives

$$\delta_{fi} - S_{fi} = 2\pi i \, \delta(E_f - E_i) \, \langle \Phi_f | V | \Psi_i^{(+)} \rangle . \tag{104}$$

By starting from $|\Psi_i^{(+)}\rangle - |\Psi_i^{(-)}\rangle$ one also finds that

$$S_{fi} - \delta_{fi} = -2\pi i \, \delta(E_f - E_i) \, \langle \Psi_f^{(-)} | V | \Phi_i \rangle . \tag{105}$$

The S matrix therefore has the form

$$S_{fi} = \delta_{fi} - 2\pi i \, \delta(E_f - E_i) \, T_{fi} . \tag{106}$$

The term δ_{fi} accounts for the possibility that there is no scattering whatever, whereas the the matrix T has as its elements the various scattering amplitudes,

$$T_{fi} = \langle \Phi_f | V | \Psi_i^{(+)} \rangle = \langle \Psi_f^{(-)} | V | \Phi_i \rangle . \tag{107}$$

Equation (106) is an important relationship. It demonstrates that *the S matrix only exists between states of equal energy — that energy is conserved in scattering from one asymptotic state to another.* It also shows that the two expressions for the scattering amplitude T_{fi} in (107) only agree for states of equal energy — on the energy shell. This last statement was already encountered in §8.2(c), where T was introduced only for potential scattering.

Because S is unitary, it is in principle possible to find a representation in which it is diagonal, with elements that are simply phase factors. We are already familiar with an important illustration of this fact: *in elastic scattering of spin 0 particles, the S matrix is diagonal in the total orbital angular momentum representation, with elements $e^{2i\delta_L(E)}$, where $\delta_L(E)$ is the phase shift in the partial wave L.* This statement generalizes to elastic collisions in which symmetry completely specifies the partial waves, such as in scattering of spin $\frac{1}{2}$ particles due to spin-orbit coupling, in which case the eigenvalues (L, S, J) suffice to decouple all the partial waves, so that the cross section is determined by phase shifts $\delta_{LSJ}(E)$. This situation does not, however, hold in elastic scattering if there are tensor forces, which couple spin triplets of the same parity, such as ^3S and $^3D_1$, or $^3P_2$ and $^3F_2$. In this case the S matrix reduces to diagonal elements for spin singlets, and spin triplets that have no partner of the same parity, such as ^1S, and 2×2 blocks for those that do. Of course, these blocks can be easily diagonalized, but the transformation depends on the interaction and will be energy dependent. It is evident, of course, that when inelastic scattering is possible, diagonalization of the S matrix by means of symmetry cannot be possible in general unless there is a dynamical symmetry at work. Hence diagonal phases do not provide a useful description of inelastic amplitudes.

Another general property of the S matrix should be pointed out here. If the interaction V is invariant under time reversal,

$$T_{fi} = T_{i^R f^R} , \tag{108}$$

where i^R and f^R are the motion reversed states, i.e., have opposite momenta and the same helicities. The proof is identical to the one in §8.2(c).

Now to the proof that S is unitary. To start with,

$$\sum_a S_{fa} S_{ia}^* = \sum_a [\delta_{fa} - 2\pi i\, \delta(E_f - E_a)\, T_{fa}][\delta_{ia} + 2\pi i\, \delta(E_a - E_i)\, T_{ia}^*] \tag{109}$$

$$= \delta_{fi} - 2\pi i\, \delta(E_f - E_i)\, X_{fi} , \tag{110}$$

where

$$X_{fi} = \langle \Phi_f | V | \Psi_i^{(+)} \rangle - \langle \Psi_f^{(+)} | V | \Phi_i \rangle + 2\pi i \sum_a \delta(E_i - E_a)\, \langle \Phi_f | V | \Psi_a^{(+)} \rangle \langle \Psi_a^{(+)} | V | \Phi_i \rangle . \tag{111}$$

Using (95) and (99) once again,

$$\langle \Phi_f | V | \Psi_i^{(+)} \rangle - \langle \Psi_f^{(+)} | V | \Phi_i \rangle = \langle \Phi_f | V \frac{1}{E_i - H + i\epsilon} V | \Phi_i \rangle - \langle \Phi_f | V \frac{1}{E_f - H - i\epsilon} V | \Phi_i \rangle; \tag{112}$$

on inserting the eigenstates $|\Psi_a^{(+)}\rangle$ of H, this becomes

$$\langle \Phi_f | V | \Psi_i^{(+)} \rangle - \langle \Psi_f^{(+)} | V | \Phi_i \rangle = -2\pi i \sum_a \delta(E_i - E_a)\, \langle \Phi_f | V | \Psi_a^{(+)} \rangle \langle \Psi_a^{(+)} | V | \Phi_i \rangle , \tag{113}$$

which shows that $X_{fi} = 0$. This then establishes that S *is unitary*,

$$SS^\dagger = 1 \,. \tag{114}$$

The generalized optical theorem is nothing other than the statement $X_{fi} = 0$. Written in terms of T matrix elements, i.e., of scattering amplitudes, (113) reads

$$T_{fi} - T_{if}^* = -2\pi i \sum_a \delta(E_i - E_a)\, T_{fa} T_{ia}^* \,. \tag{115}$$

This relationship only holds on the energy shell, as is clear from its derivation. It is valid for any pair of states (f, i) of equal energy that can be connected by a collision. In the special case of forward elastic scattering, i.e., $f = i$, Eq. 115 becomes

$$\text{Im}\; T_{ii} = -\pi \sum_a \delta(E_i - E_a)\, |T_{ia}|^2 \,. \tag{116}$$

We must confirm that Eq. 116 is indeed the optical theorem in the form familiar from potential scattering, because it must remain valid in collisions between complex systems when the energy is too low to open any inelastic channels. In complete generality, the elastic scattering amplitude is (83), i.e.,

$$f_{\text{el}}(\boldsymbol{p}|\boldsymbol{k}) = -4\pi^2 m\, \langle \Phi_{\boldsymbol{p}}|V|\Psi_{\boldsymbol{k}}^{(+)}\rangle = -4\pi^2 m\, \langle \Phi_{\boldsymbol{p}}|T|\Phi_{\boldsymbol{k}}\rangle \,, \tag{117}$$

so when inelastic scattering is forbidden (116) becomes

$$\text{Im}\; f_{\text{el}}(\boldsymbol{k}|\boldsymbol{k}) = \frac{1}{4\pi m} \int d^3q\; 2m\, \delta(k^2 - q^2)\, |f_{\text{el}}(\boldsymbol{k}|\boldsymbol{q})|^2 = \frac{k}{4\pi}\, \sigma_{\text{el}} \,; \quad QED \,. \tag{118}$$

(c) Transition Rates and Cross Sections

The connection between the amplitudes T_{fi} and directly observable quantities such as cross sections remains to be established. The argument is a straightforward generalization of the one given in §3.7(c). We return to (89), written in Dirac notation:

$$|\Psi_i^{(+)}(t)\rangle = |\Phi_i(t)\rangle - i \int_{-\infty}^{t} dt'\, e^{-iH_0(t-t')}\, V|\Psi_i^{(+)}(t)\rangle \,. \tag{119}$$

In a scattering experiment (as compared to one that simply measures the depletion of the incident beam), the final state $|\Phi_f(t)\rangle$ has particles running in directions that are not in the beam direction, and therefore it has no overlap with $|\Phi_i(t)\rangle$. Consequently the amplitude of interest is

$$A_{i\to f}(t) = \langle \Phi_f(t)|\Psi_i^{(+)}(t)\rangle = -i \int_{-\infty}^{t} dt' \langle \Phi_f(t)|e^{-iH_0(t-t')}\, V|\Psi_i^{(+)}(t)\rangle \,. \tag{120}$$

Once again, we idealize by using stationary states:

$$A_{i\to f}(t) = -i \langle \Phi_f|V|\Psi_i^{(+)}\rangle \int_{-\infty}^{t} dt\, e^{i(E_f - E_i)t'} \,. \tag{121}$$

Thus the only change vis-à-vis the first-order perturbation result is the replacement

$$\langle \Phi_f | V | \Phi_f \rangle \to \langle \Phi_f | V | \Psi_i^{(+)} \rangle \equiv T_{fi} \,. \tag{122}$$

Therefore the transition rate between states of perfectly defined energy is

$$\Gamma_{i \to f} = 2\pi \, |T_{fi}|^2 \, \delta(E_f - E_i) \,. \tag{123}$$

The rate for transitions in the interval $(E_f, E_f + dE_f)$ is then

$$\Gamma_{i \to F} = 2\pi \, |T_{fi}|^2 \, \rho_f \,, \tag{124}$$

where ρ_f is the density in energy of the final states as given by (17).

Equation (124) is *the generalization of the Golden Rule* (Eq. 15), to all orders in V. Indeed, perturbation theory was not used to derive it, and it holds also when the Born series does not converge, or even exist. The calculation of cross sections from $\Gamma_{i \to F}$ is identical to that of §9.1(a). In all generality, therefore, the differential cross section is

$$d\sigma = 2\pi |T_{fi}|^2 \frac{\rho_f}{F_{\text{inc}}} \,, \tag{125}$$

where F_{inc} is the incident flux. If f is a two-body state, this becomes

$$\frac{d\sigma}{d\Omega} = (2\pi)^4 |T_{fi}|^2 \frac{k_f^2}{v_i v_f} \,. \tag{126}$$

For interactions that are invariant under time reversal, the symmetry relation (108) between the amplitudes for $i \to f$ and $f^R \to i^R$ holds. As a consequence, if both i and f are two-body states, the differential reaction cross sections are related by

$$\frac{d\sigma_{i \to f}}{d\Omega_f} \frac{1}{k_f^2} = \frac{d\sigma_{f^R \to i^R}}{d\Omega_i} \frac{1}{k_i^2} \,. \tag{127}$$

This is known as *the principle of detailed balance;* it applies to states with specified helicities if the reactions involve particles with spin.

That (116) is the generalization of the optical theorem when there is inelastic scattering is a consequence of (115), (123) and (125):

$$-\text{Im } T_{ii} = \tfrac{1}{2} F_{\text{inc}} \sum_a \sigma_{ia} = \frac{v_i}{16\pi^3} \sigma_{\text{tot}} \,. \tag{128}$$

Here the sum runs over *all* open channels, σ_{ia} is the total cross section for the channel $i \to a$, and σ_{tot} is the total cross section, elastic plus inelastic. The optical theorem in terms of the elastic scattering amplitude f_{el}, instead of T matrix elements, follows from (117):

$$\text{Im } f_{\text{el}}(\boldsymbol{k}|\boldsymbol{k}) = \frac{k}{4\pi} \sigma_{\text{tot}} \,. \tag{129}$$

The generalization to the case where there is inelastic scattering therefore only requires the substitution $\sigma_{\text{el}} \to \sigma_{\text{tot}}$ in Eq. 118, a result already derived in §8.1(c). The reason, it will be recalled, is that the optical theorem is the statement that the shadow cast in the forward direction results from the diminution of the incident beam due to all possible scattering processes.

The cross section for inelastic scattering off the bound particle is also found by applying the Golden Rule. Using states with continuum normalization, and a self-evident notation,

$$d\sigma_\mu = \frac{(2\pi)^3}{v_i} 2\pi\, |T_\mu(\boldsymbol{k}_f|\boldsymbol{k}_i)|^2 \frac{d^3k_f}{dE_f} , \tag{130}$$

and therefore

$$\frac{d\sigma_\mu}{d\Omega_f} = (2\pi)^4\, m^2 \frac{k_f}{k_i} |T_\mu(\boldsymbol{k}_f|\boldsymbol{k}_i)|^2 , \qquad k_f = \sqrt{k_i^2 - 2m\Delta_\mu} . \tag{131}$$

The relation between the T matrix element and f is found by comparing (84) with (122). Therefore the cross section for both elastic and inelastic scattering to states in which particle 2 remains bound is

$$\frac{d\sigma_\mu}{d\Omega_f} = \frac{k_f}{k_i} |f_\mu(\boldsymbol{k}_f|\boldsymbol{k}_i)|^2 , \tag{132}$$

which reduces to the familiar result for elastic scattering where $k_f = k_i$. If particle 2 is ejected into a continuum state, the differential cross section specifies not only the direction and momentum of the scattered projectile 1 but also those of particle 2 (recall Eq. 39).

In many circumstances an evaluation of T that does not involve perturbation theory is unnecessary, impractical or impossible. In such cases a perturbation expansion is very useful. As in §8.2(c), we define the operator

$$T(E) = V + V \frac{1}{E - H + i\epsilon} V \tag{133}$$

$$= V + V \frac{1}{E - H_0 + i\epsilon} T(E) , \tag{134}$$

where we are no longer restricting ourselves to potential scattering. One readily verifies that

$$\langle \Phi_f | T(E) | \Phi_i \rangle = T_{fi} , \qquad \text{iff } E = E_i = E_f . \tag{135}$$

Iteration of (134) then produces the Born series for T:

$$T = V + V \frac{1}{E - H_0 + i\epsilon} V + \cdots . \tag{136}$$

This is a widely used expansion.

The other commonly needed observable quantity is the rate for decay or disintegration of an almost stationary state. The most familiar examples are emission of light and beta decay. Both are due to interactions that are very weak in comparison to the forces that bind atoms and nuclei, respectively, and they can therefore be described by perturbation theory. In first-order processes, such as the emission of one photon or one electron-neutrino pair, first-order perturbation theory suffices, and the simple Golden Rule of §3.7(d) then gives the rate for the process. But there are occasions when the first-order process is forbidden by a selection rule, as in the case of a radiative transition between two states of zero angular momentum, which cannot go via emission of one photon because the photon has spin 1. There are similar, but rarer, situations in beta decay. In such circumstances V must be replaced by T, as expanded to the order in V which can first produce the process (e.g., to second order for two photon emission).

Finally, a brief discussion of the limitations of the theory just developed. Clearly, the velocities of all particles involved must be such that $v^2/c^2 \ll 1$. We have also assumed that V is inoperative at large distances, which is not true of the Coulomb interaction. This is only a technical limitation; it can be taken care of by using Coulomb scattering states instead of plane waves for the projectile and/or the reaction products, though this of course leads to much more complicated formulas. We have also ignored the possibility that the projectile, or its constituents, may be indistinguishable from the constituents of the target; in §8.7 we learned that this a relatively straightforward problem in elastic scattering between structureless objects, but it can lead to considerable complications in more complex situations. A deeper problem is posed by rearrangement collisions, such as d+d = He^3+p, or chemical reactions. In such processes the "free" part of the complete Hamiltonian and the interaction between widely separated subsystems is not the same in the initial and final states, and major changes in the theory are required to handle such phenomena. The nuclear reaction just cited also demonstrates that the issue of indistinguishability usually accompanies rearrangement collisions. In short, the theory just presented is not the end of the story. Nevertheless, the concepts on which it is based are fundamental to scattering theory in these more sophisticated contexts and the most important properties of the S matrix hold in great generality.

9.3 Inelastic Resonances

Inelastic collisions dominated by resonances play a prominent role in atomic, nuclear and particle physics. Once again, this is a complex topic. Our goal is to provide a transparent introduction that is sufficiently realistic to illustrate important and generic features that appear in such processes, and at the same time to exploit and elucidate what we have learned about the S matrix.

For this purpose, think of a nucleus $\mathcal{A}$ with neutron and atomic numbers (N, Z), and reactions produced by neutron capture to an excited state $\mathcal{A}^*$ of the nucleus $(N+1, Z)$. The latter can, by hypothesis, decay into a neutron plus the original nucleus, and thus produce the elastic process $n + \mathcal{A} \to \mathcal{A}^* \to n + \mathcal{A}$. Imagine, in addition, that $\mathcal{A}^*$ can also decay into a proton and the nucleus $(N+1, Z-1)$ called $\mathcal{A}'$, producing the inelastic process $n + \mathcal{A} \to p + \mathcal{A}'$. If the incident energy suffices to form the state $\mathcal{A}^*$, both the elastic and inelastic cross sections will display resonances, and it is this phenomenon that we wish to explore.

(a) A Solvable Model

We now construct a solvable model motivated by this example. Imagine a stable spin 0 object called $\mathcal{A}$, and two spin 0 particles a and b with masses $m_a < m_b$, and interactions such that a or b can combine with $\mathcal{A}$ to form the spin 0 object $\mathcal{A}^*$ whose mass exceeds that of $\mathcal{A}$ by E_0. When a is incident, the possible reactions are taken to be

$$a + \mathcal{A} \to \mathcal{A}^* \to a + \mathcal{A} \tag{137}$$

$$\to b + \mathcal{A}\,, \qquad \text{iff} \quad e_a > (m_b - m_a)\,, \tag{138}$$

where e_a is the kinetic energy of a. When b is incident, the possible reactions are

$$b+\mathcal{A}\rightarrow\ \mathcal{A}^*\rightarrow(a,b)\ +\mathcal{A}\,, \tag{139}$$

but as b is heavier than a there is no threshold condition, as there is in (138). In short, in this model there are two continuous spectra, for e_a and e_b, plus a discrete state $\mathcal{A}^*$ suspended in these continua at E_0. This is reminiscent of the one-dimensional elastic scattering problem treated in §4.4. The analogue of the perfectly enclosed states inside impermeable walls is here $\mathcal{A}^*$; the analogue of the continuum outside the walls are the spectra of e_a and e_b; and the analogue of transmission through the walls are the couplings $\mathcal{A}^*\leftrightarrow(a,b)+\mathcal{A}$. Of course, in the example of §4.4 there is no inelastic scattering.

Because all the reactions (137–139) go through the spin zero state $\mathcal{A}^*$, there is only elastic and inelastic s wave scattering in this model. By that token, if a and b both have spin 0 and $\mathcal{A}^*$ has integer spin $l\neq0$, all scattering would be confined to the partial wave with angular momentum l.

It is important to recognize that the energy need not suffice to excite the intermediate state $\mathcal{A}^*$ for either the elastic or the inelastic process to exist; energy conservation only requires the initial and final states to have the same energy, and is silent about intermediate states. Of course, if the energy is too low to excite $\mathcal{A}^*$, no resonance will appear in the cross sections.

Because particles a and b disappear into the intermediate state $\mathcal{A}^*$ and then reappear, it is necessary to have a formalism that can handle this.[1] Let $|0\rangle$ and $|1\rangle$ denote the states $\mathcal{A}$ and $\mathcal{A}^*$ and *no* particle, and define the operator $\xi^\dagger$ that transform the target $\mathcal{A}$ into the unstable system $\mathcal{A}^*$:

$$\xi^\dagger=|1\rangle\langle0|\,,\qquad \xi=|0\rangle\langle1|\,; \tag{140}$$

note that $\xi^\dagger\xi$ is the projection operator onto $|1\rangle$. The other states needed have particle a or b of momentum $\boldsymbol{k}$ and the unexcited target, i.e.,

$$|a,\boldsymbol{k}\rangle\otimes|0\rangle\,,\qquad |b,\boldsymbol{k}\rangle\otimes|0\rangle\,. \tag{141}$$

These one particle states can be constructed from the state $|0\rangle$ which has no particle by means of creation operators $A_{\boldsymbol{k}}^\dagger$ and $B_{\boldsymbol{k}}^\dagger$:

$$A_{\boldsymbol{k}}^\dagger|0\rangle\equiv|a,\boldsymbol{k}\rangle\otimes|0\rangle\,,\qquad B_{\boldsymbol{k}}^\dagger|0\rangle\equiv|b,\boldsymbol{k}\rangle\otimes|0\rangle\,. \tag{142}$$

These operators can be used to describe the emission transitions in Eqs. 137–139:

$$\mathcal{A}^*\rightarrow a+\mathcal{A}\ :\ A_{\boldsymbol{k}}^\dagger\xi|1\rangle\,,\qquad \mathcal{A}^*\rightarrow b+\mathcal{A}\ :\ B_{\boldsymbol{k}}^\dagger\xi|1\rangle\,. \tag{143}$$

Transitions in which a particle is absorbed are described by the destruction operators $A_{\boldsymbol{k}}$ and $B_{\boldsymbol{k}}$. In combination with $\xi^\dagger$ they describe the transitions

$$a+\mathcal{A}\rightarrow\mathcal{A}^*\ :\ A_{\boldsymbol{k}}\,\xi^\dagger\left(|a,\boldsymbol{q}\rangle\otimes|0\rangle\right)=\delta(\boldsymbol{k}-\boldsymbol{q})|1\rangle\,, \tag{144}$$

and similarly for $b+\mathcal{A}\rightarrow\mathcal{A}^*$.

---

[1]Readers already familiar with the technique of second quantization (developed in the next section — §10.1) can skip over the material between here and Eq. 150.

The algebra of these creation and destruction operators is a generalization of those of the multi-dimensional harmonic oscillator, namely,

$$[A_{\boldsymbol{k}}, A_{\boldsymbol{q}}^{\dagger}] = \delta(\boldsymbol{k} - \boldsymbol{q}) \,, \quad [A_{\boldsymbol{q}}, A_{\boldsymbol{k}}] = 0 \,, \tag{145}$$

with the same relations among the operators for b, and vanishing commutators between all operators for a and those for b. Furthermore, the destruction operators acting on states containing no particles annihilate the state:

$$A_{\boldsymbol{k}}|1\rangle = 0 \,, \qquad B_{\boldsymbol{k}}|1\rangle = 0 \,. \tag{146}$$

Consider now the operators

$$N_{\boldsymbol{k}}^{a} = A_{\boldsymbol{k}}^{\dagger} A_{\boldsymbol{k}} \,, \qquad N_{\boldsymbol{k}}^{b} = B_{\boldsymbol{k}}^{\dagger} B_{\boldsymbol{k}} \,. \tag{147}$$

As a consequence of (145),

$$[N_{\boldsymbol{k}}^{a}, A_{\boldsymbol{q}}^{\dagger}] = \delta(\boldsymbol{k} - \boldsymbol{q}) A_{\boldsymbol{k}}^{\dagger} \,, \tag{148}$$

and similarly for b. The operators $N_{\boldsymbol{k}}^{a}$ acting on states with one or no particles therefore give the following result:

$$N_{\boldsymbol{k}}^{a} A_{\boldsymbol{q}}^{\dagger}|0\rangle = \delta(\boldsymbol{k} - \boldsymbol{q}) A_{\boldsymbol{q}}^{\dagger}|0\rangle \,, \quad N_{\boldsymbol{k}}^{a} B_{\boldsymbol{q}}^{\dagger}|0\rangle = 0 \,, \quad N_{\boldsymbol{k}}^{a}|1\rangle = 0 \,, \ \text{etc.} \tag{149}$$

In short, $N_{\boldsymbol{k}}^{a}$ is a counting operator, reproducing the state when a is present with momentum $\boldsymbol{k}$, and zero otherwise, with the analogous statements for $N_{\boldsymbol{k}}^{b}$.

The model is fully defined by its Hamiltonian $H_0 + H_1$. If there were no reactions, the energy H_0 of the system would just be the kinetic energies and masses of the non-interacting constituents:

$$H_0 = E_0 \, \xi^{\dagger} \xi + \int d\boldsymbol{k} \left\{ e_a(k) N_{\boldsymbol{k}}^{a} + [e_b(k) + \mu] N_{\boldsymbol{k}}^{b} \right\} \,. \tag{150}$$

Here $e_{a,b}(k) = k^2/2m_{a,b}$, $\mu = m_b - m_a$. The energy scale has been chosen for convenience to have its zero at the state $|0\rangle$, i.e., when there is no particle and only the target $\mathcal{A}$. The operators $\xi^{\dagger}\xi$, $N_{\boldsymbol{k}}^{a}$ and $N_{\boldsymbol{k}}^{b}$ in H_0 simply count whether the object in question is present or not.

The interaction H_1 must produce the transitions in (143) and (144):

$$H_1 = \int d\boldsymbol{k} \left\{ [f(k) A_{\boldsymbol{k}} + g(k) B_{\boldsymbol{k}}] \xi^{\dagger} + [f^*(k) A_{\boldsymbol{k}}^{\dagger} + g^*(k) B_{\boldsymbol{k}}^{\dagger}] \xi \right\} \,. \tag{151}$$

Here $f(k)$ and $g(k)$ are functions only of $|\boldsymbol{k}|$, as required by rotational invariance because $a, b, \mathcal{A}$ and $\mathcal{A}^*$ all have spin 0. It is straightforward to generalize the model to nonzero spins; see Prob. 4. The Fourier transforms of $f(k)$ and $g(k)$ describe the spatial form of the interaction responsible for the transitions.

The physical content of H_1 can be best understood by first considering the case where it is sufficiently weak to make $\mathcal{A}^*$ very long lived, so that it is meaningful to talk of the decay processes $\mathcal{A}^* \to (a, b) + \mathcal{A}$ as if $\mathcal{A}^*$ were stable, and the energy released were precisely E_0. The total decay rates $\Gamma_{a,b}$ are then given by perturbation theory,

$$\Gamma_a = 4\pi \cdot 2\pi \, |\langle 0|A_{\boldsymbol{p}_a} H_1|1\rangle|^2 \, \rho_a \,, \quad \Gamma_b = 4\pi \cdot 2\pi \, |\langle 0|B_{\boldsymbol{p}_b} H_1|1\rangle|^2 \, \rho_b \,, \tag{152}$$

where the factor 4π comes from integrating over all directions of the isotropic distribution. The momenta are determined from energy conservation, i.e.,

$$e(p_a) = E_0 \,, \qquad e_b(p_b) + \mu = E_0 \,, \tag{153}$$

and the densities of states are $\rho_{a,b} = m_{a,b}p_{a,b}$. Only the term in H_1 that creates the final state can contribute to the matrix elements in (152), i.e., $A_{\boldsymbol{k}}^\dagger \xi$ and $B_{\boldsymbol{k}}^\dagger \xi$; furthermore,

$$A_{\boldsymbol{p}} A_{\boldsymbol{k}}^\dagger \xi |1\rangle = \delta(\boldsymbol{p}-\boldsymbol{k})|0\rangle \,. \tag{154}$$

Therefore the decay widths are

$$\Gamma_a = 8\pi^2 m_a p_a |f(p_a)|^2 \,, \qquad \Gamma_b = 8\pi^2 m_b p_b |g(p_b)|^2 \,. \tag{155}$$

The model has been designed to be exactly solvable so that effects that are either not present or hard to discern in perturbation theory can be examined. The most general state in the model's Hilbert space describing the reactions (137) to (139) has some amplitude for $\mathcal{A}^*$, and amplitudes for either a or b and the target $\mathcal{A}$:

$$|\Psi\rangle = C|1\rangle + \int d\boldsymbol{k} \left\{ \alpha(k) A_{\boldsymbol{k}}^\dagger + \beta(k) B_{\boldsymbol{k}}^\dagger \right\} |0\rangle \,. \tag{156}$$

The constant C and the functions α and β will be found by solving the Schrödinger equation. For that purpose we need the action of H on this state. H_0 just multiplies the separate terms by the energies in (150):

$$H_0|\Psi\rangle = E_0 C|1\rangle + \int d\boldsymbol{k} \left\{ e_a(k)\alpha(k) A_{\boldsymbol{k}}^\dagger + [e_b(k)+\mu]\beta(k) B_{\boldsymbol{k}}^\dagger \right\} |0\rangle \,. \tag{157}$$

When H_1 acts on $|1\rangle$, only the term which produces the transition to $|0\rangle$ gives a non-zero result because $\xi^\dagger|1\rangle = 0$:

$$H_1|1\rangle = \int d\boldsymbol{k} \, [f^*(k) A_{\boldsymbol{k}}^\dagger + g^*(k) B_{\boldsymbol{k}}^\dagger]|0\rangle \,. \tag{158}$$

By the same token, when H_1 acts on the terms in $|\Psi\rangle$ that have one or the other particle, only the operator that removes the particle and produces the intermediate state contributes. Eq. 154 and $B_{\boldsymbol{q}} A_{\boldsymbol{k}}^\dagger |0\rangle = 0$ then yield

$$H_1|\Psi\rangle = C \int d\boldsymbol{k} \, [f^*(k) A_{\boldsymbol{k}}^\dagger + g^*(k) B_{\boldsymbol{k}}^\dagger]|0\rangle + \int d\boldsymbol{k} \, [\alpha(k) f(k) + \beta(k) g(k)]|1\rangle \,. \tag{159}$$

The Schrödinger equation $(H-E)|\Psi\rangle = 0$ requires the coefficients of the linearly independent terms to vanish:

$$A_{\boldsymbol{k}}^\dagger|0\rangle : \quad [e_a(k) - E]\alpha(k) + C f^*(k) = 0 \,, \tag{160}$$

$$B_{\boldsymbol{k}}^\dagger|0\rangle : \quad [e_b(k) + \mu - E]\beta(k) + C g^*(k) = 0 \,, \tag{161}$$

$$|1\rangle : \quad (E_0 - E)C + \int d\boldsymbol{k} \, [\alpha(k) f(k) + \beta(k) g(k)] = 0 \,. \tag{162}$$

To describe the processes $a + \mathcal{A} \to (a,b) + \mathcal{A}$ we must construct a solution $|\Psi_{a,\boldsymbol{p}_a}^{(+)}\rangle$ in which a is incident with momentum $\boldsymbol{p}_a$ and the scattered waves are outgoing.

The amplitude $\alpha(k)$ pertains to particle a and therefore must be a solution of (160) satisfying these boundary conditions, i.e.,

$$\alpha(k) = \delta(\boldsymbol{k} - \boldsymbol{p}_a) + \frac{Cf^*(k)}{E - e_a(k) + i\epsilon} , \tag{163}$$

where $E = e_a(p_a)$. As b is not incident,

$$\beta(k) = \frac{Cg^*(k)}{E - e_b(k) - \mu + i\epsilon} . \tag{164}$$

These expressions are then substituted into (162),

$$(E_0 - E)C + f(p_a) + C \int d\boldsymbol{k} \left\{ \frac{|f(k)|^2}{E - e_a(k) + i\epsilon} + \frac{|g(k)|^2}{E - e_b(k) - \mu + i\epsilon} \right\} = 0 , \tag{165}$$

and therefore

$$C(E) = \frac{f(p_a)}{D(E)} , \tag{166}$$

where

$$D(E) = E - E_0 - \Sigma(E) , \tag{167}$$

$$\Sigma(E) = \int d\boldsymbol{k} \left\{ \frac{|f(k)|^2}{E - e_a(k) + i\epsilon} + \frac{|g(k)|^2}{E - e_b(k) - \mu + i\epsilon} \right\} . \tag{168}$$

Because $\Sigma(E)$ is complex, there is no need to add $i\epsilon$ to the denominator $D(E)$.

(b) Elastic and Inelastic Cross Sections

The scattering amplitudes can now be calculated from $T_{fi} = \langle \Phi_f | V | \Psi_i^{(+)} \rangle$. In this model, $V = H_1$, the initial state just constructed is $|\Psi_{a,\boldsymbol{p}_a}^{(+)}\rangle$ and, for example, in the elastic reaction the final state is $A_{\boldsymbol{q}}^\dagger |0\rangle$. Eq. 159 then yields the following results for the elastic and inelastic amplitudes:

$$T_{aa}(q|p_a) = \langle 0 | A_{\boldsymbol{q}} H_1 | \Psi_{a,\boldsymbol{p}_a}^{(+)} \rangle = C(E) f^*(q) \tag{169}$$

$$T_{ba}(q|p_a) = \langle 0 | B_{\boldsymbol{q}} H_1 | \Psi_{a,\boldsymbol{p}_a}^{(+)} \rangle = C(E) g^*(q) . \tag{170}$$

These expressions are valid off the energy shell, i.e., when q has any value; they only depend on the magnitudes of the momenta because in this model only s wave scattering is possible.

The elastic s-wave phase shift is related to the on-shell amplitude by the now familiar relations

$$\frac{1}{2ip}(e^{2i\delta} - 1) = f_{\text{el}}(p|p) = -4\pi^2 m T_{fi} . \tag{171}$$

The phase shift for the elastic reaction $a \to a$ is therefore

$$e^{2i\delta_a} = \frac{D(E) - i\, 8\pi^2 m_a p_a |f(p_a)|^2}{D(E)} , \tag{172}$$

where, recall, $E = e_a(p_a)$. This last expression is the diagonal element of the S matrix for the $j = 0$ component of the state $a + \mathcal{A}$:

$$S_{aa}(E) = e^{2i\delta_a(E)} \; . \tag{173}$$

Crucial features of the elastic and inelastic amplitudes arise from the imaginary part of $\Sigma(E)$. Using (103) we have

$$\mathrm{Im}\,\Sigma(E) = -\pi \int d\boldsymbol{k} \{ \delta(E - e_a(k))|f(k)|^2 + \delta(E - e_b(k) - \mu)\,|g(k)|^2 \} \, . \tag{174}$$

The δ-functions express energy conservation; the first one for elastic scattering is always satisfied, but the second for inelastic scattering requires E to be above the threshold at $(m_b - m_a) = \mu$. Thus

$$\mathrm{Im}\,\Sigma(E) = -4\pi^2 p_a m_a \,|f(p_a)|^2 - 4\pi^2 p_b m_b |g(p_b)|^2 \theta(E-\mu) \tag{175}$$

$$\equiv -\tfrac{1}{2}\Gamma_a(E) - \tfrac{1}{2}\Gamma_b(E)\theta(E-\mu) \, , \tag{176}$$

where p_b is the momentum of b in the $a \to b$ reaction, and the unit step function $\theta(E-\mu)$ expresses the threshold condition. The functions $\Gamma_{a,b}(E)$ are equal to the "naive" decay widths $\Gamma_{a,b}$ of Eq. 152 when $E = E_0$, i.e., when the incident (and therefore) total energy is that of the intermediate state $\mathcal{A}^*$ *in the absence of the interaction causing the decay.*

Using the expressions (176) for the widths, the elastic S-matrix element (173) reads

$$S_{aa}(E) = \frac{\mathrm{Re}\,D(E) - \frac{1}{2}i\Gamma_a(E) + \frac{1}{2}i\Gamma_b(E)\theta(E-\mu)}{\mathrm{Re}\,D(E) + \frac{1}{2}i\Gamma_a(E) + \frac{1}{2}i\Gamma_b(E)\theta(E-\mu)} \, . \tag{177}$$

Below the inelastic threshold S_{aa} is of the form X/X^*, i.e., it is just a phase. In other words, the phase shift δ_a is real below the inelastic threshold, which of course it must be whatever the dynamics may be. Above the threshold,

$$1 - |S_{aa}|^2 = \frac{\Gamma_a\Gamma_b}{(\mathrm{Re}\,D)^2 + \frac{1}{4}\Gamma_{\mathrm{tot}}^2} \, , \tag{178}$$

where the total width function is

$$\Gamma_{\mathrm{tot}}(E) = \Gamma_a(E) + \Gamma_b(E) \, . \tag{179}$$

In this simple model, the unitarity of the S matrix reduces to the identities

$$S_{aa}S_{aa}^* + S_{ba}S_{ba}^* = 1 \, , \tag{180}$$

$$S_{ab}S_{ab}^* + S_{bb}S_{bb}^* = 1 \, , \tag{181}$$

$$S_{aa}S_{ab}^* + S_{ba}S_{bb}^* = 0 \, . \tag{182}$$

It is important to bear in mind that $SS^\dagger = 1$ must be evaluated at some given energy, and therefore, in such unitarity relation, *all* matrix elements of S must be evaluated at the *same* energy E. This means, for example, that in (180) $E = e_b + \mu$ in S_{ba}^* and $E = e_a$ in S_{aa}.

Equation (180) is identical to (178), and therefore the latter's right-hand side is $|S_{ba}|^2$, or apart from a factor the cross section for the inelastic process $a \to b$. To confirm this, we compute the total cross section for inelastic scattering:

$$\sigma_{\text{in}} = 4\pi \cdot 2\pi |T_{ba}|^2 \rho_b \cdot (2\pi)^3 / v_a \tag{183}$$

$$= \frac{p_b}{p_a} 64\pi^5 m_a m_b \frac{|g(p_b)|^2 |f(p_a)|^2}{|D(E)|^2} \tag{184}$$

and therefore, recalling (167) and (176),

$$\sigma_{\text{in}}(E) = \frac{\pi}{p_a^2} \frac{\Gamma_b(E)\,\Gamma_a(E)}{[E - E_0 - \Sigma_R(E)]^2 + \frac{1}{4}[\Gamma_{\text{tot}}(E)]^2}\,, \tag{185}$$

where $\Sigma_R = \text{Re}\,\Sigma$. The relationship to the inelastic S-matrix element is thus of the expected form,

$$\sigma_{\text{in}}(E) = \frac{\pi}{p_a^2}\,|S_{ba}(E)|^2\,. \tag{186}$$

The total elastic cross section is $4\pi|f_{\text{el}}|^2$, so using (171),

$$\sigma_{\text{el}}(E) = \frac{\pi}{p_a^2} \frac{[\Gamma_a(E)]^2}{[E - E_0 - \Sigma_R(E)]^2 + \frac{1}{4}[\Gamma_{\text{tot}}(E)]^2}\,. \tag{187}$$

The partial wave elastic scattering amplitude is also a useful quantity. It is defined as in potential scattering, i.e.,

$$e^{i\delta_a} \sin\delta_a = \frac{1}{2i}(S_{aa} - 1) = \frac{\frac{1}{2}\Gamma_a}{E_0 - E - \Sigma_R - \frac{1}{2}i\Gamma_{\text{tot}}}\,. \tag{188}$$

As already emphasized, the phase shift is only real below the inelastic threshold.

These expressions for the cross sections are exact in this model. They have the same form, however, as s-wave elastic and inelastic cross sections dominated by a single intermediate state in more complex situations, though of course the functions $\Sigma, \Gamma_a, \Gamma_b$, are specific to the model.

If the interaction is weak, various approximations are possible. When $f \to 0$ and $g \to 0$, $|\Sigma_R| \ll E_0$ and the widths are very small. The cross sections are then very small except in the vicinity of the energy E_0 of the the intermediate state $\mathcal{A}^*$, where $\Gamma_{a,b}(E)$ can be replaced by $\Gamma_{a,b}(E_0)$, which are just the decay widths $\Gamma_{a,b}$ found from perturbation theory, as given by (155). When these approximations are valid, the cross sections are

$$\sigma_{\text{el}}(E) = \frac{\pi}{p_a^2} \frac{\Gamma_a^2}{(E - E_0)^2 + \frac{1}{4}\Gamma_{\text{tot}}^2}\,, \tag{189}$$

$$\sigma_{\text{in}}(E) = \frac{\pi}{p_a^2} \frac{\Gamma_b \Gamma_a}{(E - E_0)^2 + \frac{1}{4}\Gamma_{\text{tot}}^2}\,. \tag{190}$$

These are the *Breit-Wigner formulas* for elastic and inelastic s-wave scattering. At resonance

$$\sigma_{\text{el}}(E_0) = \frac{4\pi}{p_a^2} \frac{\Gamma_a^2}{(\Gamma_a + \Gamma_b)^2}\,; \tag{191}$$

if the resonance is below the inelastic threshold, this reaches the maximum value of $4\pi\lambda\!\!\!^{-2}$ allowed by probability conservation — or if you prefer, unitarity of the S matrix (recall §8.1).

The numerators in these Breit-Wigner formulas have a simple probabilistic meaning, which holds in much more complex situations if a set of reactions feed through one intermediate $j = 0$ states X. This is so because in such cases the cross sections are

$$\sigma_{fi}(E) = \frac{\pi}{p_i^2} \frac{\Gamma_{X\to f}\Gamma_{X\to i}}{(E-E_0)^2 + \frac{1}{4}\Gamma_X^2}, \tag{192}$$

where the partial widths are the total rates for the indicated processes, and Γ_X is the total width of X. In words, the cross section for $i \to f$ is proportional to the rate for feeding the intermediate state X from i times the rate for the decay $X \to f$.

The Breit-Wigner form survives intact if $\Sigma(E)$ is not negligible compared to E_0 but varies slowly near E_0. When that is so, the resonance is shifted to $E_* = E_0 + \Sigma_R(E_0)$, and the width functions can be replaced by the constants $\Gamma_{a,b}(E_*)$. For example, a better approximation than (190) is then provided by

$$\sigma_{\text{in}}(E) = \frac{\pi}{p_a^2} \frac{\Gamma_b(E_*)\,\Gamma_a(E_*)}{(E-E_*)^2 + \frac{1}{4}[\Gamma_{\text{tot}}(E_*)]^2}. \tag{193}$$

It is generally true that a discrete state suspended in a continuum undergoes a shift in energy if there is a coupling to the continuum that allows it to decay. We already encountered this in the one-dimensional example of §4.4, where the resonances are shifted from the positions of the perfectly enclosed states. In §10.3(b) and §10.8(d), we will learn that the interaction with the electromagnetic field responsible for the radiative decay of a state shifts the energy of that state.

Sometimes a better approximation is produced by keeping the first term in the Taylor expansion of $\Sigma_R(E)$ about E_0. However, as is evident from (185) and (187), if the couplings are strong, the energy dependence of the cross sections can depart significantly from simple Breit-Wigner forms.

Another interesting and generic property of inelastic collisions is displayed by the model: *the elastic scattering amplitude has a singularity at the inelastic threshold.* Call the threshold energy E_t, which is μ in the model. As $E \to E_t$, $\Gamma_b(E) \to 0$ like p_b, i.e., like $\sqrt{E - E_t}$. This square root, which results in a divergence of $d\sigma_{\text{el}}/dE$ at threshold, is the singularity in question.[1]

In (177) $\Sigma_R(E)$ and $\Gamma_a(E)$ are finite and well behaved at $E = E_t$, whereas $\Gamma_b \to 0$, so to leading order in Γ_b

$$e^{2i\delta_a(E)} \equiv \frac{Z + \frac{1}{2}i\Gamma_b}{Z^* + \frac{1}{2}i\Gamma_b} \simeq \frac{Z}{Z^*}\left(1 - \frac{\Gamma_b\Gamma_a}{2|Z|^2}\right), \tag{194}$$

where $Z = \text{Re}\, D(E_t) - \frac{1}{2}i\Gamma_a(E_t)$, $Z/Z^* = e^{2i\delta_t}$, and δ_t is the phase shift just below threshold, where there is only elastic scattering. Thus as $E \to E_t$ from above,

$$e^{2i\delta_a(E)} = e^{2i\delta_t}\left(1 - A\sqrt{E - E_t}\right), \tag{195}$$

[1] As we have already seen in the case of elastic scattering in Chapter 8, scattering amplitudes are analytic function in the complex energy plane with singularities stemming from eigenstates of the Hamiltonian — simple poles from bound states and branch cuts from scattering states. The singularity at threshold is due to a branch cut corresponding to the continuum of b states that starts there.

where A is real and positive. Let $f_{\rm el}(E)$ be the elastic s-wave amplitude, and p_t be p_a at threshold. Then

$$p_t f_{\rm el}(E) = \frac{1}{2i}(e^{2i\delta_a(E)} - 1) \to e^{i\delta_t}\sin\delta_t - \frac{1}{2i}e^{2i\delta_t}A\sqrt{E-E_t} \tag{196}$$

as $E \to E_t$ from above. The total elastic cross section is $4\pi|f_{\rm el}(E)|^2$, and therefore

$$\sigma_{\rm el}(E) = \sigma_{\rm el}(E_t)\left(1 - A\sqrt{E-E_t}\right) \tag{197}$$

for energies just above the inelastic threshold. *The energy dependence of the elastic cross section thus has a cusp at the inelastic threshold.*

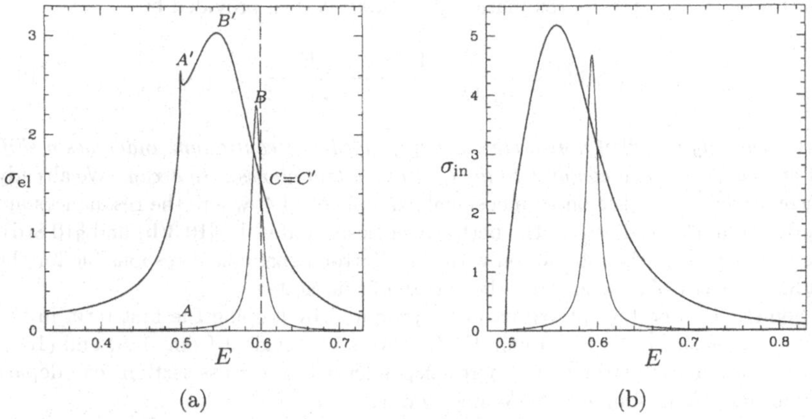

FIG. 9.1. Elastic and inelastic scattering according to the model of §9.3. The labels $A \cdots C'$ are defined in Fig. 9.2. The parameters are as follows: the particle masses are $m_a = \frac{1}{2}$, $m_b = 1$, so that the inelastic threshold is at $E_t = \frac{1}{2}$; the uncoupled excited state $\mathcal{A}^*$ has the energy $E_0 = 0.6$; and in the interaction (see Eq. 151) $f(k) = g(k) = \alpha/(k^2+1)$. (a) The elastic cross section as a function of $E = k^2$ for the coupling parameters $\alpha = 0.02$ (thin lines) and $\alpha = 0.05$ (thick lines). The narrow resonance near $E = E_0$ is for $\alpha = 0.02$; as the coupling grows to $\alpha = 0.05$ the resonance broadens, distorts and shifts to lower energy, and the threshold cusp becomes progressively more prominent. (b) The inelastic cross section for the same set of parameters.

Figure 9.1 shows the elastic and inelastic cross sections found from this solvable model for a selected set of parameters. As one can see, the simple Breit-Wigner forms only give an adequate rendition when the coupling is weak between the state $\mathcal{A}^*$ and the continuum in which it is suspended. As the strength of this coupling grows there are ever stronger and more striking departures from the familiar resonance form. The variation with energy of the elastic phase shift is best shown as an Argand plot of the function $\exp(2i\delta_a) \equiv x + iy$. This is provided by Fig. 9.2 for the two cases shown in Fig. 9.1, coupling sufficiently weak to make the Breit-Wigner form reliable, and when it is too strong for this. In the weak coupling case the Argand plot trajectory above the inelastic threshold remains close to circular, but when the coupling is strong enough to produce a cross section that is far from

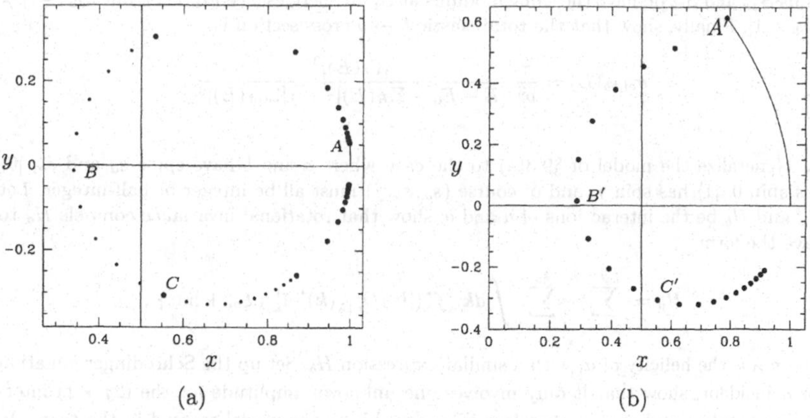

FIG. 9.2. Argand plots of $e^{2i\delta_a} \equiv x + iy$ as a function of the energy E. The parameters in (a) and (b) are the same as in the corresponding parts of Fig. 9.1. The aspect ratio of both plots is unity: circles would appear as circles. In both plots A is the inelastic threshold, and the continuous line terminating there is a segment of the unit circle; B is the maximum of the inelastic cross section (at $E = 0.595$ and $E \simeq 0.56$ respectively); and C denotes $E_0 = 0.6$, the position of the resonance when $\alpha \to 0$. In (a) the heavy dots denote $E = 0.50, 0.51, \ldots 0.70$, and the small dots are an enlargement about $E = 0.06$, namely 0.591, 0.592, ... 0.609. In (b) the dots denote $E = 0.50, 0.51, \ldots 0.70$. Note the discontinuity associated with the threshold cusp.

the Breit-Wigner shape, as in Fig. 9.1(b), the Argand plot is far from circular. A particularly striking and very general feature of scattering amplitudes is that the "speed" with which the point on the plot moves as a function of E is maximum at resonance; indeed, in §4.4(c) it was shown that this holds as well in one-dimensional elastic scattering — recall Fig. 4.10. This behavior of amplitudes is an important tool for extracting resonances from scattering data in highly inelastic collisions.

9.4 Problems

1. Compute and plot the elastic form factor for H; and for He using the wave function given in §6.2. Also compute and plot the inelastic form factor for the $1s \to 2p$ and $1s \to 3d$ transitions in H. Finally, calculate the elastic exchange amplitude for H, and compare it explicitly to the direct amplitude as a function of scattering angle for incident energies of 100 eV and 100 keV.

2. Prove Eq. 55, on the assumption that the atom's Hamiltonian is reflection invariant.

3. Show that time-reversal invariance and unitarity lead to $|S_{aa}|^2 = |S_{bb}|^2$ in the notation of §9.3(b). At first sight this seems wrong, because the elastic cross sections for $a \to a$ and $b \to b$ surely differ as they involve different interactions. Solve the scattering problem for b incident, and show that

$$S_{bb}(E) = \frac{\mathrm{Re}\, D(E) - \frac{1}{2} i\Gamma_b(E) + \frac{1}{2} i\Gamma_a(E)}{\mathrm{Re}\, D(E) + \frac{1}{2} i\Gamma_{\mathrm{tot}}(E)} .$$

Thus S_{aa} and S_{bb} do have the same modulus at equal energies (see Eq. 177), but $|S_{aa}-1| \neq |S_{bb}-1|$. Finally, show that the total elastic $b \to b$ cross section is

$$\sigma_{\mathrm{el},bb}(E) = \frac{\pi}{p_b^2} \frac{[\Gamma_b(E)]^2}{[E - E_0 - \Sigma_R(E)]^2 + \frac{1}{4}[\Gamma_{\mathrm{tot}}(E)]^2} .$$

4. Generalize the model of §9.3(a) to the case where a and b have spins s_a and s_b, $|0\rangle$ has spin 0, $|1\rangle$ has spin j, and of course (s_a, s_b, j) must all be integer or half-integer. Let H_a and H_b be the interactions of a and b; show that rotational invariance compels H_a to have the form

$$H_a = \sum_{m=-j}^{j} \sum_{\lambda=-s_a}^{s_a} \int d\boldsymbol{k}\, \{f_\lambda^*(|\boldsymbol{k}|) D_{m\lambda}^j(\boldsymbol{k})^* A_{\boldsymbol{k},\lambda}^\dagger \xi_m + \mathrm{h.c.}\} ,$$

where λ is the helicity of a, with a similar expression H_b. Set up the Schrödinger equation for a incident; show that it only involves one unknown amplitude for the $(2j+1)$ intermediate states; and that it therefore is as solvable as the model treated in the text. On the other hand, show that even when only elastic scattering is possible, the S matrix is non-trivial because it involves couplings between the various helicity states. Consider also the imposition and consequences of reflection invariance, especially for elastic scattering.

5. Consider the Hamiltonian $H = K + U + V$, where both U and V may cause scattering, but V is "strong" and must be treated exactly while U is "weak" and can be treated to leading order. This problem motivates *the distorted wave Born approximation*, which has many applications, e.g., in bremsstrahlung due to scattering in a strong Coulomb field, neutron capture accompanied by gamma emission, etc.

(a) Define T as the operator of Eq. 133 when the interaction is $V + U$, and T^0 this operator when $U = 0$. Using the identity Eq. 2.402, show that to first order in U

$$T = T^0 + \Lambda U \Lambda, \qquad \Lambda = 1 + \frac{1}{E - K - V + i\epsilon} V .$$

Let $|\chi^{(\pm)}\rangle$ be solutions of the problem when $U \equiv 0$ with the indicated boundary conditions, i.e.,

$$|\chi^{(\pm)}\rangle = \left(1 + \frac{1}{E - K - V \pm i\epsilon} V\right) |\Phi\rangle .$$

Show that the desired amplitude, to lowest order in U, is

$$T_{fi} = T_{fi}^0 + \langle \chi_f^{(-)} | U | \chi_i^{(+)} \rangle .$$

This is the distorted wave Born approximation. The "strong" interaction V distorts both the incoming and outgoing waves, but with time-reversed boundary conditions, so that the former is a plane wave in the past and the latter a plane wave in the future.

(b) Consider two phenomena: (i) bremsstrahlung; and (ii) the influence of spin-orbit coupling on scattering of an electron in a Coulomb field. Show that in (i), T^0 does not contribute. In (ii), show that T^0 contributes only to the amplitude for scattering without spin-flip, and that the leading contribution to spin-flip is due to interference between the two terms in T_{fi}. No calculations are required to answering this part of the problem!

6. This problem illustrates a standard analysis of resonant scattering, and is basically a numerical exercise. Consider the elastic amplitude as defined in Eq. 188; assume that the widths and Σ are energy independent and define the resonance energy as $E_* = E_0 + \Sigma_R$, or

$$a(E) \equiv e^{i\delta} \sin\delta = \frac{\frac{1}{2}\Gamma_a}{E_* - E - \frac{1}{2}i\Gamma_{\mathrm{tot}}} .$$

The resonance is assumed to be well above the inelastic threshold.

(a) Plot $a(E)$ as a function of E in the complex plane. Assume that Γ_{tot} is appreciably smaller than E_*, and do the calculation for several values of the branching fraction $\Gamma_a/\Gamma_{\text{tot}}$. Mark the points on $a(E)$ at constant intervals of E/E_*, and observe how the "speed" along this "trajectory" varies as E passes through the resonance.

(b) In general, a resonant elastic amplitude is not completely dominated by the resonance, but has a slowly varying "background" as well. Add a constant amplitude $b = |b|e^{i\alpha}$ to $a(E)$, chosen to saturate the unitarity bound. Plot the cross section for $\alpha = 0, \pm\frac{1}{4}\pi, \pm\frac{1}{2}\pi$. Note the dramatic effect this can have on the appearance of the resonance, and in particular that a resonance need not show itself as a prominent peak in the cross section, though it always shows up as a swiftly moving point in the the Argand plot. Confirm the last statement.

Endnotes

LLQM, Chapter XVIII, has a great deal of important information about inelastic collisions. The following are authoritative monographs: M.L. Goldberger and K.M. Watson, *Collision Theory,* Wiley (1964); N.F. Mott and H.S.W. Massey, *Theory of Atomic Collisions,* Oxford, 3rd ed. (1965); R.G. Newton, *Scattering Theory of Waves and Particles,* McGraw-Hill (1966); J.R. Taylor, *Scattering Theory,* Wiley (1972). Our treatment of the S matrix draws on M. Gell-Mann and M.L. Goldberger, *Phys. Rev.* **91**, 398 (1953), and G.C. Wick, *Rev. Mod. Phys.* **27**, 339 (1955).

10 Electrodynamics

10.1 Quantization of the Free Field

The quantum theory of the electromagnetic field and its interaction with matter was invented by Dirac about one and a half years after Heisenberg's discovery of quantum mechanics. Given the outstanding importance of this problem, the astonishing speed with which quantum mechanics was developed and how simple the solution looks in retrospect, one may ask why it took so long. The key for Dirac was his transformation theory, published just two months earlier, which we have been using in this book beginning with §2.1. It enabled him to write the Schrödinger equation for an arbitrary system in any representation, and for this purpose he used a representation in which the numbers of particles are the eigenvalues, and a Hamiltonian that is not diagonal in this very representation. That, of course, is precisely what is needed for the absorption and emission of photons, and allowed him to give the first derivation of spontaneous emission from first principles. The stumbling block had been how to describe processes in which particles are destroyed and created.[1] It was known already that the energy of the free classical electromagnetic field has the form of an infinite set of uncoupled harmonic oscillators when the fields are expanded into normal modes. Passage to the quantum theory of the free field is achieved by requiring these oscillators to obey the canonical quantization rules. In the case of empty space, the normal modes can be chosen as propagating plane waves. The excitations and de-excitations of these modes produce *discrete* increases and decreases of the linear momentum and energy of the field, i.e., the emission and absorption of photons.

The quantization of Maxwell's theory will seem innocent and unavoidable, but it constitutes one of the major watersheds in the history of theoretical physics. Canonical quantization was developed for nonrelativistic systems with a finite number of degrees of freedom, which is here extended to a Lorentz-invariant system with an infinite number of degrees of freedom. It was hardly obvious that this recipe would again succeed. Two tests must be passed. First, is the theory internally consistent? Second, is it empirically correct? In §10.2 we will present a powerful case for its internal consistency. Insofar as agreement with experiment is concerned, quantum electrodynamics is today the most successful theory in the history of physics. The evidence for this assertion is massive but involves a level of sophistication that goes far beyond this volume.

---

[1] Even Dirac found this hard to swallow: "When a light-quantum is absorbed it can be considered to jump into the zero [i.e., vacuum] state, and when one is emitted it can be considered to jump from the zero state to one in which it is in physical evidence, so that it appears to have been created." Cf., *Proc. Roy. Soc. A* **114**, 243 (1927).

(a) The Classical Theory

The first task is then to express the classical theory in terms of normal modes. It turns out that this can only be done in a simple manner if the electric and magnetic fields, $\boldsymbol{E}$ and $\boldsymbol{B}$, are expressed in terms of potentials. The potentials do not, however, have a unique physical significance because infinite families of potentials represent one and the same fields, and only the fields have a unique significance. Maxwell's equations, and all associated quantities, when expressed in terms of potentials, must therefore be invariant under gauge transformations — transformations between physically equivalent sets of potentials. Maxwell's theory has another basic symmetry: it is Lorentz invariant. It is not so simple a matter to demonstrate that the formulation that follows is both gauge and Lorentz invariant; indeed, such a demonstration involves both symmetries simultaneously. We will overlook this issue here, and choose the gauge which leads most directly to physically meaningful results.

This choice is called the radiation or transverse gauge; it has only a vector potential $\boldsymbol{A}$ and a vanishing scalar potential when the field has no sources. Furthermore, $\boldsymbol{A}$ is constrained to be divergence-free:

$$\boldsymbol{\nabla} \cdot \boldsymbol{A} = 0 \,. \tag{1}$$

The fields are determined from $\boldsymbol{A}$ by

$$\boldsymbol{B} = \boldsymbol{\nabla} \times \boldsymbol{A} \,, \qquad \boldsymbol{E} = -\frac{1}{c}\frac{\partial \boldsymbol{A}}{\partial t} \,, \tag{2}$$

and they will satisfy Maxwell's equations provided $\boldsymbol{A}$ satisfies the wave equation:

$$\nabla^2 \boldsymbol{A} - \frac{1}{c^2}\frac{\partial^2 \boldsymbol{A}}{\partial t^2} = 0 \,. \tag{3}$$

Any complete set of solutions of the wave equation satisfying the condition (1) can be used to define a set of normal modes. The appropriate choice is dictated by the boundary conditions that the fields are to satisfy. Here we deal with the fields in empty space, and will choose plane waves for this purpose. This choice makes linear momentum simple; if angular momentum is of interest, spherical waves should be chosen, as will be done in §10.1(c). There are many situations, however, such as in quantum optics, where the field is confined to a cavity containing a dispersive medium; in such cases the normal modes must satisfy the boundary conditions set by the cavity and the dispersion law appropriate to the medium.

In an infinite volume it is perhaps most natural to consider waves traveling in any and all directions with any frequency. But handling a continuous spectrum is somewhat tricky, and for that reason we start with an artificial boundary condition that renders the spectrum discrete, and leave the continuum limit for later. This is done, as in §3.4, by requiring the fields to be periodic on the surface of a large cube of volume V. Forgetting for a moment that we are dealing with a vector field, plane waves satisfying this boundary condition and the wave equation have the form

$$e^{i(\boldsymbol{k}\cdot\boldsymbol{r}-\omega_k t)} \equiv e^{ik\cdot x} \,, \qquad \omega_k = ck \,, \tag{4}$$

where the wave vectors $\boldsymbol{k}$ lie on the lattice

$$\boldsymbol{k} = (n_1, n_2, n_3) \cdot (2\pi/V^{\frac{1}{3}}) \,, \quad n_i = 0, \pm 1, \pm 2, \dots \,. \tag{5}$$

Eq. 4 introduces the shorthand

$$k \cdot x = \boldsymbol{k} \cdot \boldsymbol{r} - \omega_k t \,, \tag{6}$$

and the notation $x \equiv (\boldsymbol{r}, t)$ for space-time points. The waves (4) form a complete orthonormal set:

$$\frac{1}{V} \int_V d\boldsymbol{r}\, e^{i(\boldsymbol{k}-\boldsymbol{k}') \cdot \boldsymbol{r}} = \delta_{\boldsymbol{k}\boldsymbol{k}'} \,, \tag{7}$$

$$\frac{1}{V} \sum_{\boldsymbol{k}} e^{i\boldsymbol{k} \cdot (\boldsymbol{r}-\boldsymbol{r}')} = \delta(\boldsymbol{r}-\boldsymbol{r}') \tag{8}$$

Here the sum on k is actually a sum over all the three integers of Eq. 5, and the Kronecker delta in (7) requires equality of all the three pairs of integers that define $\boldsymbol{k}$ and $\boldsymbol{k}'$. The relation between periodic boundary conditions and a continuous spectrum of wave vectors is given by the rules

$$\frac{1}{V} \sum_{\boldsymbol{k}} \Leftrightarrow \int \frac{d\boldsymbol{k}}{(2\pi)^3} \,, \qquad \frac{V}{(2\pi)^3} \delta_{\boldsymbol{k}\boldsymbol{k}'} \Leftrightarrow \delta(\boldsymbol{k}-\boldsymbol{k}') \,, \quad V \to \infty \,. \tag{9}$$

The expansion of the vector potential in terms of plane waves is then

$$\boldsymbol{A}(x) = V^{-\frac{1}{2}} \sum_{\boldsymbol{k}} \left[e^{i\boldsymbol{k} \cdot \boldsymbol{r}} \boldsymbol{A}_k(t) + \text{c.c.} \right] , \tag{10}$$

where the second term is needed because $\boldsymbol{A}$ is real. To make $\boldsymbol{A}$ divergence-free; i.e., to satisfy (1), the coefficients in (10) must satisfy

$$\boldsymbol{k} \cdot \boldsymbol{A}_k(t) = 0 \,; \tag{11}$$

hence the vectors $\boldsymbol{A}_k(t)$ are perpendicular to the direction of propagation, which is the reason for the name transverse gauge. Because of the wave equation (3),

$$\boldsymbol{A}_k(t) = \boldsymbol{A}_k\, e^{-i\omega t} . \tag{12}$$

The mode expansions of the field strengths now follow immediately from (2) and (10):

$$\boldsymbol{E}(x) = \frac{i}{cV^{\frac{1}{2}}} \sum_{\boldsymbol{k}} \omega \left[e^{i\boldsymbol{k} \cdot \boldsymbol{r}} \boldsymbol{A}_k(t) - \text{c.c.} \right] , \tag{13}$$

$$\boldsymbol{B}(x) = \frac{i}{V^{\frac{1}{2}}} \sum_{\boldsymbol{k}} \left\{ e^{i\boldsymbol{k} \cdot \boldsymbol{r}} \left[\boldsymbol{k} \times \boldsymbol{A}_k(t) \right] - \text{c.c.} \right\} . \tag{14}$$

Condition (11) produces fields that are transverse, as required by Maxwell's equations.

It is often necessary to be more specific about the transverse character of the mode vectors $\boldsymbol{A}_k$. This is done by expanding them in terms of a pair of mutually orthogonal polarization vectors lying in the plane perpendicular to $\boldsymbol{k}$. One such pair are the real linear polarization vectors $\boldsymbol{\epsilon}_{k\alpha}$:

$$\boldsymbol{\epsilon}_{k1} \times \boldsymbol{\epsilon}_{k2} = \hat{\boldsymbol{k}} \,, \qquad \boldsymbol{\epsilon}_{k\alpha} \cdot \boldsymbol{\epsilon}_{k\beta} = \delta_{\alpha\beta} \quad (\alpha, \beta = 1, 2) \,. \tag{15}$$

Another are the complex circular polarization vectors

$$e_{k\pm1} = \mp\frac{1}{\sqrt{2}}(\epsilon_{k1} \pm i\epsilon_{k2}) , \tag{16}$$

where the phases are chose to agree with those for the spherical harmonics with $l = 1, m = \pm 1$; they satisfy

$$e^*_{k\lambda} \cdot e_{k\lambda'} = \delta_{\lambda\lambda'} , \quad e^*_{k\lambda} \times e_{k\lambda'} = i\lambda \hat{k}\, \delta_{\lambda\lambda'} , \quad i\hat{k} \times e_{k\lambda} = \lambda e_{k\lambda} , \quad (\lambda, \lambda' = \pm 1) . \tag{17}$$

It will turn out that in the quantum theory λ is the photon helicity.[1]

The next step is to express the field energy,

$$H_\gamma = \tfrac{1}{2} \int d\boldsymbol{r}\, (\boldsymbol{E}^2 + \boldsymbol{B}^2) , \tag{18}$$

in terms of the amplitudes $\boldsymbol{A}_k$. Here the notation H_γ is chosen because this object will, after quantization, become the Hamiltonian of the free electromagnetic field, i.e., of freely propagating photons.[2]

In calculating the field energy in terms of the $\boldsymbol{A}_k$, it is convenient to replace (13) and (14) by

$$\boldsymbol{E}(x) = \frac{i}{cV^{\frac{1}{2}}} \sum_k \omega\, e^{i\boldsymbol{k}\cdot\boldsymbol{r}} \left[A_k(t) - A^*_{-k}(t) \right] , \quad \text{etc.}; \tag{19}$$

then the orthogonality condition (7) gives

$$\int d\boldsymbol{r}\, \boldsymbol{E} \cdot \boldsymbol{E}^* = \sum_k k^2 |\boldsymbol{A}_k(t) - \text{c.c.}|^2 . \tag{20}$$

A similar calculation leads to

$$\int d\boldsymbol{r}\, \boldsymbol{B} \cdot \boldsymbol{B}^* = \sum_k k^2 |\boldsymbol{A}_k(t) + \text{c.c.}|^2 . \tag{21}$$

Hence the total energy of the field is a quadratic form in terms of the mode amplitudes $\boldsymbol{A}_k$:

$$H_\gamma = 2 \sum_k k^2 |\boldsymbol{A}_k(t)|^2 . \tag{22}$$

The time dependence of the amplitudes follows from substituting $\boldsymbol{A}(x)$ into the wave equation, which then requires

$$\left(\frac{d^2}{dt^2} - \omega^2 \right) \boldsymbol{A}_k(t) = 0 . \tag{23}$$

Hence the field energy H_γ is that of an infinite set of uncoupled simple harmonic oscillators — two for each and every possible wave vector $\boldsymbol{k}$. Of course t drops out in H_γ because it is a constant of motion.

---

[1]In the $\lambda = 1$ mode the electric field describes counter-clockwise or left-handed rotation *as seen by a receiver*, whereas for $\lambda = -1$ this rotation is clockwise.

[2]Note that we use rationalized units in defining H_γ, which simplify electrodynamics by avoiding many a 4π, but exacts the price that Coulomb's law is $e^2/4\pi r$.

All that is now needed to express Maxwell's theory in Hamiltonian form is to copy the theory of the simple harmonic oscillator: for each $\boldsymbol{k}$ define the real canonical variables

$$\boldsymbol{Q}_k(t) = \frac{1}{c}[\boldsymbol{A}_k(t) + \text{c.c.}] , \qquad \boldsymbol{P}_k(t) = -ik[\boldsymbol{A}_k(t) - \text{c.c.}] . \tag{24}$$

The energy then assumes the familiar form

$$H_\gamma = \tfrac{1}{2} \sum_k (\boldsymbol{P}_k^2 + \omega^2 \boldsymbol{Q}_k^2) . \tag{25}$$

Hamilton's equations,

$$\dot{\boldsymbol{P}}_k = -\frac{\partial H_\gamma}{\partial \boldsymbol{Q}_k} , \qquad \dot{\boldsymbol{Q}}_k = \frac{\partial H_\gamma}{\partial \boldsymbol{P}_k} , \tag{26}$$

which in this case form an infinite set of ordinary differential equations, translate Maxwell's partial differential equations into the canonical language of classical mechanics.

It is essential not to mistake these $\boldsymbol{P}_k$ and $\boldsymbol{Q}_k$ for the momenta and coordinates of any physical object — to not imagine, for example, that $\boldsymbol{P}_k$ is the momentum carried by the mode $\boldsymbol{k}$. Their purpose is to cast Maxwell's theory into the canonical format so that the familiar canonical quantization prescription can be applied to the electromagnetic field.

(b) Quantization

This prescription replaces the classical canonical variables by Hermitian operators satisfying the canonical commutation rules:[1]

$$[Q_{k\lambda}, P_{k'\lambda'}] = i\hbar \delta_{kk'} \delta_{\lambda\lambda'} , \tag{27}$$
$$[Q_{k\lambda}, Q_{k'\lambda'}] = 0 , \qquad [P_{k\lambda}, P_{k'\lambda'}] = 0 , \tag{28}$$

and assumes that the Hamiltonian of the quantized electromagnetic field is

$$H_\gamma = \tfrac{1}{2} \sum_{k\lambda} (P_{k\lambda}^2 + \omega^2 Q_{k\lambda}^2) . \tag{29}$$

The symbols of the classical theory have been used here for the Hermitian operators because the quantum and classical theories have exactly the same form, and also to avoid superfluous notation as we will have no further occasion to use the classical quantities.

In Eq. 27, $\hbar$ has not been set to one yet; in a first exposure to this topic it is best to see explicitly where $\hbar$ appears. Furthermore, the circular polarization description has been used, but this is purely a matter of taste; the commutation rules have precisely the same form in the linear polarization language after the replacements $Q_{k\lambda} \to Q_{k\alpha}, P_{k\lambda} \to P_{k\alpha}$, $(\lambda = \pm 1) \to (\alpha = 1, 2)$.

[1]In quantum optics, the Hermitian operators $Q_{k\lambda}$ and $P_{k\lambda}$ are called quadratures of this field mode, though with a variety of other normalizations and conventions.

As in the theory of the one-dimensional oscillator [§4.2(a)], the intuitive picture becomes clearer, and the equations simpler, if the Hermitian operators are replaced by the non-Hermitian operators

$$a_{k\lambda} = \frac{1}{\sqrt{2\hbar\omega}}(\omega Q_{k\lambda} + iP_{k\lambda}) , \tag{30}$$

and their adjoints $a^\dagger_{k\lambda}$. These operators are dimensionless and satisfy the commutation rules

$$[a_{k\lambda}, a^\dagger_{k'\lambda'}] = \delta_{kk'}\, \delta_{\lambda\lambda'} . \tag{31}$$

If the linear polarization description is desired, these operators would be written as $a_{k\alpha}$, with $\alpha = 1, 2$. It is often convenient to combine the two operators for a given value of $\boldsymbol{k}$ into a vector perpendicular to $\boldsymbol{k}$ by means of the circular or linear polarization vectors, respectively:

$$\boldsymbol{a}_k = \sum_{\lambda=\pm 1} a_{k\lambda}\boldsymbol{e}_{k\lambda} = \sum_{\alpha=1,2} a_{k\alpha}\boldsymbol{\epsilon}_{k\alpha} . \tag{32}$$

Beware that as a consequence of (32), the relationship between the operators in the linear and circular polarization bases differs from that between the polarization vector defined in (16), namely, for any given $\boldsymbol{k}$,

$$a_\lambda = \mp\frac{1}{\sqrt{2}}(a_1 \mp ia_2) \quad (\lambda = \pm 1) . \tag{33}$$

In terms of $\boldsymbol{a}_k$, the Hamiltonian is

$$H_\gamma = \sum_k \hbar\omega_k[\boldsymbol{a}^\dagger_k \cdot \boldsymbol{a}_k + \tfrac{1}{2}] = \sum_{k\lambda} \hbar\omega_k[N_{k\lambda} + \tfrac{1}{2}] , \tag{34}$$

where the number operators are

$$N_{k\lambda} = a^\dagger_{k\lambda}a_{k\lambda} , \tag{35}$$

and have the integer eigenvalues

$$n_{k\lambda} = 0, 1, 2, \ldots . \tag{36}$$

In the Heisenberg picture, the equations of motion for the $\boldsymbol{a}_k$ are

$$i\hbar\dot{\boldsymbol{a}}_k(t) = [\boldsymbol{a}_k(t), H_\gamma] = \hbar\omega_k\boldsymbol{a}_k(t) , \tag{37}$$

and have the solutions

$$\boldsymbol{a}_k(t) = \boldsymbol{a}_k\, e^{-i\omega_k t} . \tag{38}$$

Clearly, $\boldsymbol{a}_k(t)$ is the quantum analogue of the classical Fourier coefficient $\boldsymbol{A}_k(t)$. Indeed, on comparing (29) with (24), one sees that the the quantum fields are obtained from the classical fields by the replacement

$$\boldsymbol{A}_k(t) \to \sqrt{\hbar c/2k}\, \boldsymbol{a}_k(t) . \tag{39}$$

The *field operators* are thus

$$\boldsymbol{A}(x)=\sum_k\sqrt{\frac{\hbar c}{2Vk}}\,[e^{ik\cdot x}\boldsymbol{a}_k+\text{h.c.}]\,, \tag{40}$$

$$\boldsymbol{E}(x)=i\sum_k\sqrt{\frac{\hbar ck}{2V}}\,[e^{ik\cdot x}\boldsymbol{a}_k-\text{h.c.}]\,, \tag{41}$$

$$\boldsymbol{B}(x)=i\sum_k\sqrt{\frac{\hbar c}{2Vk}}\,[e^{ik\cdot x}(\boldsymbol{k}\times\boldsymbol{a}_k)-\text{h.c.}]\,. \tag{42}$$

It follows from (37) that *the Heisenberg equations of motion for the field operators* $\boldsymbol{E}(x)$ *and* $\boldsymbol{B}(x)$ *are Maxwell's equations.*

From our knowledge of the oscillator, we know that $\boldsymbol{a}_k$ lowers and $\boldsymbol{a}_k^\dagger$ raises the energy. What is new here is that these operators also raise and lower the momentum in the field. The operator for the linear momentum of the electromagnetic field is assumed to be the Hermitian counterpart of the classical expression, namely,

$$\boldsymbol{P}=\frac{1}{2c}\int d\boldsymbol{r}\,[\boldsymbol{E}\times\boldsymbol{B}-\boldsymbol{B}\times\boldsymbol{E}]\,. \tag{43}$$

Substituting (41) and (42), and using $\boldsymbol{k}\cdot\boldsymbol{a}_k=0$, then leads to

$$\boldsymbol{P}=\sum_{k\lambda}\hbar\boldsymbol{k}\,N_{k\lambda}\,, \tag{44}$$

where a term $\sum_k \boldsymbol{k}$ has been dropped as it vanishes by symmetry. If this operator $\boldsymbol{P}$ is indeed the total momentum, then the unitary operator

$$T(\boldsymbol{s})=e^{-i\boldsymbol{s}\cdot\boldsymbol{P}/\hbar} \tag{45}$$

must produce a spatial translation of the field operators through the distance $\boldsymbol{s}$, for example,

$$T^\dagger(\boldsymbol{s})\boldsymbol{E}(\boldsymbol{r},t)T(\boldsymbol{s})=\boldsymbol{E}(\boldsymbol{r}-\boldsymbol{s},t)\,. \tag{46}$$

The calculation showing that $\boldsymbol{P}$, as given by (44), actually passes this test is left for Prob. 2.

(c) Photons

The description of the electromagnetic field in terms of photons emerges directly from the expressions just derived for the energy and momentum of the field, because they show that in the absence of sources the number operators $\{N_{k\lambda}\}$ form a complete set of compatible observables.[1]

The stationary states of the free field can therefore be specified by the infinite set of integer quantum numbers $\{n_{k\lambda}\}$. These states, $|\{n_{k\lambda}\}\rangle$, can all be constructed from the *vacuum state,* $|0\rangle$, the state where all the $n_{k\lambda}$ vanish. The explicit form of

[1] "Absence of sources" is a more drastic simplification in quantum than in classical field theory. In the case of quantum electrodynamics, it means that vacuum polarization is ignored, and with it processes such as light-by-light scattering.

the arbitrary state is given by the straightforward generalization of the formula for the one-dimensional oscillator:

$$|\{n_{k\lambda}\}\rangle = \prod_{k\lambda} \frac{1}{\sqrt{n_{k\lambda}!}} (a^{\dagger}_{k\lambda})^{n_{k\lambda}} |0\rangle \equiv \prod_{k\lambda} \otimes |n_{k\lambda}\rangle \,. \tag{47}$$

The energy and momentum of this state are

$$E(\{n_{k\lambda}\}) = \sum_{k\lambda} \hbar\omega \, (n_{k\lambda} + \tfrac{1}{2}) \,, \tag{48}$$

$$\boldsymbol{P}(\{n_{k\lambda}\}) = \sum_{k\lambda} \hbar \boldsymbol{k} \, n_{k\lambda} \,. \tag{49}$$

Furthermore, from the theory of the oscillator we know that $a^{\dagger}_{k\lambda}$ increases $n_{k\lambda}$ by one, and that $a_{k\lambda}$ decreases it by one (unless $n_{k\lambda} = 0$), whereas both operators leave the quantum numbers for all other values of $\boldsymbol{k}$ and λ unchanged.

These results can be phrased succinctly in the photon language:

1. *the state* $a^{\dagger}_{k\lambda}|0\rangle$*, having the excitation energy* $\hbar\omega_k = \hbar kc$ *and the momentum* $\hbar\boldsymbol{k}$*, is a one-photon state;*

2. *the state* $|\{n_{k\lambda}\}\rangle$ *contains* $n_{k_1\lambda_1}, n_{k_2\lambda_2}, \ldots$ *photons in the modes* $\boldsymbol{k}_1\lambda_1$*,* $\boldsymbol{k}_2\lambda_2$*, ..., the integers* $\{n_{k\lambda}\}$ *being called occupation numbers;*

3. *when acting on* $|\{n_{k\lambda}\}\rangle$*, the destruction operator* $a_{k\lambda}$ *removes one photon of the indicated energy and momentum (if it is present), while the creation operator* $a^{\dagger}_{k\lambda}$ *adds one such photon;*

4. *photons obey Bose-Einstein statistics.*

This last, fundamental, statement follows from Eq. 47: as any pair of creation operators commute, this arbitrary state is symmetric under interchange of the quantum numbers of any pair of photons.

The names "creation" and "destruction" operators are not hyperbole, but have a concrete meaning. The interaction of an electromagnetic current $\boldsymbol{j}(r,t)$ with the electromagnetic field has an energy density proportional to $\boldsymbol{j}(x) \cdot \boldsymbol{A}(x)$, which is a linear form in the operators $\boldsymbol{a}^{\dagger}_k$ and $\boldsymbol{a}_k$. This term in the Hamiltonian can therefore raise or lower the number of photons — radiate or absorb light. All the radiative processes that occur in nature or the laboratory are described by these operators, and justify the terminology of creation and destruction.

A remark about the expression (48) for the energy is called for, because the term $\sum_{k\lambda} \frac{1}{2}$ is infinite, and thus assigns an infinite energy to the vacuum state! This is an artifact of the quantization procedure which would be avoided if the substitution (39) were made directly in the classical expression $\boldsymbol{A}^{*}_k \cdot \boldsymbol{A}_k$ for the energy; that is to say, one can make $O(\hbar)$ changes in the classical theory, which are by definition negligible in the classical limit, and which remove terms such as this in the quantum theory. Furthermore, only differences in energy ultimately matter.[1] In any event,

[1] Quite generally, in quantizing a classical field theory the freedom just mentioned is exploited to guarantee that the vacuum expectation values of observables vanish. This is done by "normal ordering" the products of operators so that all destruction operators stand on the right and annihilate the vacuum state.

henceforth we will use the expression

$$H_\gamma = \sum_{k\lambda} \hbar\omega_k \, \boldsymbol{a}_k^\dagger \cdot \boldsymbol{a}_k \tag{50}$$

for the Hamiltonian, and drop the term $\frac{1}{2}$ from (48).

The photon carries not only energy and linear momentum, but also an intrinsic angular momentum, or spin. But the photon spin does not conform to the general theory of angular momentum developed thus far. The reason is that this theory is non-relativistic, while the photon moves with the speed of light. The theory of angular momentum in a Lorentz covariant form is not needed, however, because Maxwell's classical theory is Lorentz invariant, and provides the correct prescription for the quantum theory. In the classical theory, the angular momentum carried by the field is

$$\boldsymbol{J}_{\text{cl}} = \frac{1}{c} \int d\boldsymbol{r} \, [\boldsymbol{r} \times (\boldsymbol{E} \times \boldsymbol{B})]_{\text{cl}} \, . \tag{51}$$

Two steps, both left for Prob. 3, lead from this classical expression to the operators for the orbital and spin angular momenta.[1] The first separates the total classical angular momentum into orbital and intrinsic parts; the second merely rewrites the classical forms in terms of Hermitian operators. The result is

$$\boldsymbol{J} = \boldsymbol{J}_{\text{sp}} + \boldsymbol{J}_{\text{orb}} \, , \tag{52}$$

where

$$\boldsymbol{J}_{\text{sp}} = \frac{1}{2c} \int d\boldsymbol{r} \, (\boldsymbol{E} \times \boldsymbol{A} - \boldsymbol{A} \times \boldsymbol{E}) \, , \tag{53}$$

$$\boldsymbol{J}_{\text{orb}} = \frac{1}{2c} \int d\boldsymbol{r} \sum_{i=1}^{3} [E_i (\boldsymbol{r} \times \boldsymbol{\nabla} A_i) + (\boldsymbol{r} \times \boldsymbol{\nabla} A_i) E_i] \, . \tag{54}$$

Observe that $\boldsymbol{J}_{\text{sp}}$ does not depend on a choice of origin, whereas $\boldsymbol{J}_{\text{orb}}$ does, as it should.

After the operators for $\boldsymbol{E}$ and $\boldsymbol{A}$ are substituted into $\boldsymbol{J}_{\text{sp}}$, the integration over space and (17) lead to

$$\boldsymbol{J}_{\text{sp}} = -i\hbar \sum_{k} \boldsymbol{a}_k^\dagger \times \boldsymbol{a}_k = \hbar \sum_{k\lambda} \lambda \hat{\boldsymbol{k}} \, n_{k\lambda} \, . \tag{55}$$

Hence the one-photon state of momentum $\hbar\boldsymbol{k}$ has angular momentum $\pm\hbar$ along $\boldsymbol{k}$, that is to say, the photon has helicity ± 1.

Note the striking difference from nonrelativistic angular momentum: the spin has no component along the direction of motion — there is no helicity 0 state (or in classical language, no longitudinally polarized electromagnetic waves). There is no logical inconsistency here, because there is no frame in which the photon is at rest and where the nonrelativistic theory of angular momentum would have to be valid. This is brought out very clearly by the Z^0 boson, a spin 1 object involved in the weak interaction, whose field theory is rather similar to Maxwell's, but has massive

[1] There is a general theory, based on Noether's theorem, for separating the angular momentum of a field into intrinsic and orbital parts.

field quanta that have all three helicity states, so that in the rest frame of a Z^0 the familiar theory of spin 1 obtains.

One photon states having both a definite helicity and total angular momentum can be constructed by the technique developed in §7.5(b). For this purpose states with a continuous momentum spectrum are needed because integrations over $\boldsymbol{k}$ are involved. Bearing (9) in mind, this is done by defining new creation and destruction operators and one-photon states that are δ-function normalized:

$$|\boldsymbol{k}\lambda\rangle = \sqrt{V/(2\pi)^3}\, a^\dagger_{k\lambda}|0\rangle \equiv a^\dagger_\lambda(\boldsymbol{k})|0\rangle \,. \tag{56}$$

so that

$$\langle \boldsymbol{k}\lambda|\boldsymbol{k}'\lambda'\rangle = \delta_{\lambda\lambda'}\,\delta(\boldsymbol{k}-\boldsymbol{k}')\,, \tag{57}$$

$$[a_\lambda(\boldsymbol{k}), a^\dagger_{\lambda'}(\boldsymbol{k}')] = \delta_{\lambda\lambda'}\,\delta(\boldsymbol{k}-\boldsymbol{k}')\,. \tag{58}$$

It is sometimes convenient to have expressions for the field operators in this description. Recalling (9), that for the vector potential is

$$\boldsymbol{A}(x) = \sqrt{\hbar c}\int \frac{d\boldsymbol{k}}{\sqrt{(2\pi)^3 2k}} \sum_\lambda \left[e^{ik\cdot x}\boldsymbol{e}_{k\lambda}\, a_\lambda(\boldsymbol{k}) + \text{h.c.}\right]. \tag{59}$$

The relation between helicity states of definite linear and angular momentum is then given by (7.199) and (7.206):

$$|kjm\lambda\rangle = \sqrt{\frac{2j+1}{4\pi}}\int d\boldsymbol{n}\; D^j_{m\lambda}(\boldsymbol{n})^*\, a^\dagger_\lambda(\boldsymbol{k})|0\rangle\,, \tag{60}$$

$$a^\dagger_\lambda(\boldsymbol{k})|0\rangle = \sum_{j=1}^{\infty}\sum_{m=-j}^{j}\sqrt{\frac{2j+1}{4\pi}}\, D^j_{m\lambda}(\boldsymbol{n})\,|kjm\lambda\rangle\,, \tag{61}$$

where $\boldsymbol{n}$ is the unit vector along $\boldsymbol{k}$. Because there is no one-photon state with zero helicity, the sum on j in Eq. 61 starts with $j=1$, and thereby establishes a crucial fact: *there is no one-photon state of total angular momentum zero.* As a consequence, *there can be no transitions between two states of any system that both have zero angular momentum due to the emission or absorption of one photon;* this is also known as the no zero-to-zero transition rule.

While one-photon states with a definite linear or angular momentum exist, as we have just seen, there is no such thing as a coordinate-space wave function associated with a one-photon state. That should come as no surprise, because in §1.1 we learned that a particle can only be localized to a distance of order its Compton wavelength — a distance that does not exist for a massless particle. Photon wave packets can, of course, be constructed by superposing the states $|\boldsymbol{k}\lambda\rangle$ about some mean value $\bar{k}$; in such states the energy density $\frac{1}{2}\left(E^2+B^2\right)$ will have an expectation value that is localized to within a distance of order $1/\bar{k}$, i.e., of the characteristic wave length.[1]

Many natural phenomena and experiments produce one-photon states that do not have a definite polarization, and must be described by a 2×2 density matrix,

[1] For a thorough discussion of photon localization, see L. Mandel and E. Wolf, *Optical Coherence and Quantum Optics,* Cambridge University Press (1995), §12.11.

which we define in the helicity basis, $\rho_{\lambda\lambda'}$. It is convenient to express ρ in terms of a set of Pauli matrices, $(\tau_1, \tau_2, \tau_3) \equiv \boldsymbol{\tau}$, which here are merely mathematical tools having nothing to do with rotations:

$$\rho = \tfrac{1}{2}(1 + \boldsymbol{\xi}\cdot\boldsymbol{\tau}) = \tfrac{1}{2}\begin{pmatrix} 1+\xi_3 & \xi_1 - i\xi_2 \\ \xi_1 + i\xi_2 & 1-\xi_3 \end{pmatrix} . \tag{62}$$

The three quantities ξ_i must be real to make ρ Hermitian. Furthermore, the condition

$$\mathrm{Tr}\,\rho^2 = \tfrac{1}{2}(1 + \boldsymbol{\xi}\cdot\boldsymbol{\xi}) \leq 1 \tag{63}$$

implies that

$$0 \leq (\xi_1^2 + \xi_2^2 + \xi_3^2) \leq 1 . \tag{64}$$

Hence the polarization of an arbitrary one-photon state of a given momentum is specified by a real three "vector" $\boldsymbol{\xi}$ which lies on or inside a unit sphere. The three quantities ξ_i are the Stokes parameters, and the sphere is the Poincaré sphere. Only when $\boldsymbol{\xi}$ reaches the surface is the state pure, and it has no polarization whatsoever if $\boldsymbol{\xi} = 0$.

Equation (62) has the same form as the density matrix for a spin $\frac{1}{2}$ object, with $\boldsymbol{\xi}$ playing a role that might appear to be that of the polarization vector $\boldsymbol{P}$. That is not so. Whereas $\boldsymbol{P}$ is truly a vector in physical space, $\boldsymbol{\xi}$ has a more subtle meaning, as was to be expected because the mathematical construct $\boldsymbol{\tau}$ is not the photon's intrinsic angular momentum.

In view of the basis that was used to write (62), when $\xi_1 = \xi_2 = 0$ the probability that the helicity is ± 1 is evidently $\frac{1}{2}(1 \pm \xi_3)$. Hence the north and south poles on the Poincaré sphere correspond to the pure states with helicity ± 1, and other points on the 3-axis are mixtures with the probabilities just stated. Here and afterwards "are" and "is" should be read as shorthands for "corresponds to."

What is the generalization to an arbitrary Stokes vector? The answer is expressed very elegantly in terms of the Poincaré sphere by setting $\boldsymbol{\xi} = \xi\boldsymbol{u}$, where $\boldsymbol{u}$ is a unit vector whose polar angles (ϑ, φ) are defined in the usual way (see Fig. 10.1). First, take the case of a pure state, $\xi = 1$, i.e., points on the sphere's surface. Then all points that lie on the equator are linearly polarized, and in particular the points $\varphi = 0$ and $\varphi = \pi$ are the linear polarizations $\boldsymbol{\epsilon}_x$ and $\boldsymbol{\epsilon}_y$, respectively (previously called $\boldsymbol{\epsilon}_{1,2}$). The linear polarization vector having the angle $\frac{1}{2}\varphi$ with respect to $\boldsymbol{\epsilon}_x$ is the point on the equator at φ. Any point on the surface that is neither on the equator nor at the poles corresponds to a pure elliptically polarized state, with the ratio of major to minor axis being

$$\frac{a}{b} = \left|\frac{1 + \cot\frac{1}{2}\vartheta}{1 - \cot\frac{1}{2}\vartheta}\right| . \tag{65}$$

Antipodal points are always orthogonal polarization states. Given that a photon is in the pure state corresponding to $\boldsymbol{u}$, the probability of finding it in the state corresponding to $\boldsymbol{v}$ has the very simple form

$$|\boldsymbol{e}_{\boldsymbol{u}}^* \cdot \boldsymbol{e}_{\boldsymbol{v}}|^2 = \tfrac{1}{2}(1 + \boldsymbol{u}\cdot\boldsymbol{v}) , \tag{66}$$

where $\boldsymbol{e}_{\boldsymbol{u}}$ and $\boldsymbol{e}_{\boldsymbol{v}}$ are the associated polarization vectors.

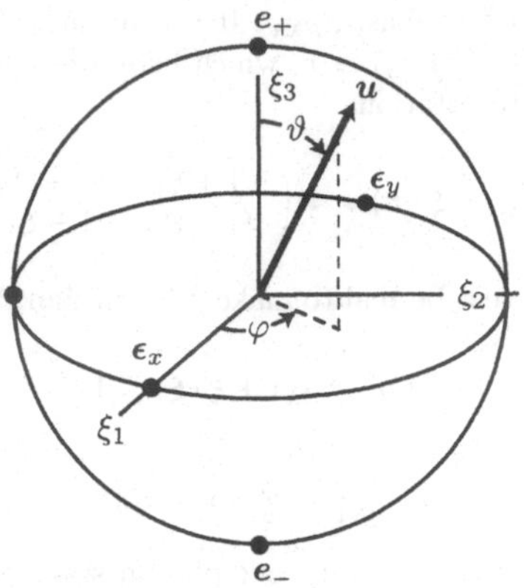

FIG. 10.1. The Poincaré sphere. The Stokes parameters are the projections of $\boldsymbol{\xi} = \xi \boldsymbol{u}$ onto the Cartesian axes which, it must be remembered, have no physical meaning *per se*. The circular polarization vectors $\boldsymbol{e}_\pm$ correspond to the North and South poles, while all linear polarization vectors correspond to points on the equator, oriented at an angle $\frac{1}{2}\varphi$ with respect to $\boldsymbol{\epsilon}_x$ and $\boldsymbol{\epsilon}_y$. All other points on the sphere correspond to elliptically polarized states. For an arbitrary Stokes vector with $\xi < 1$, the state is a mixture, and its density matrix is diagonal in the representation corresponding to the point $\boldsymbol{u}$ and the antipodal point.

When $\xi < 1$, the state is a mixture, and ρ is diagonal in the basis given by the point on the sphere's where $\boldsymbol{u}$ touches it, and the antipodal point. The probabilities for these two states in this basis are $\frac{1}{2}(1 \pm \xi)$.

The derivation of these result, left for Prob. 6, is a nice application of the theory of rotations of spin $\frac{1}{2}$. This, of course, is so because the basis here is also two-dimensional, which is why half-angles appear in the correspondence between points on the Poincaré sphere and the physical angles that relate the various polarization vectors.

(d) Space Reflection and Time Reversal

The electric field $\boldsymbol{E}$ is a force and therefore odd under space reflection and even under time reversal; Maxwell's equations then require $\boldsymbol{B}$ to be even under reflection and odd under time reversal. These symmetries are satisfied if the vector potential $\boldsymbol{A}$ is a polar vector odd under time reversal:

$$I_s \boldsymbol{A}(\boldsymbol{r}, t) I_s^{-1} = -\boldsymbol{A}(-\boldsymbol{r}, t) \,, \tag{67}$$

$$I_t \boldsymbol{A}(\boldsymbol{r}, t) I_t^{-1} = -\boldsymbol{A}(\boldsymbol{r}, -t) \,. \tag{68}$$

Time reversal, always the more subtle case, will be treated in detail. First note that it follows from (29) that the antiunitary transformation I_t turns a destruction operator into another destruction operator because $Q_{k\lambda}$ and $P_{k\lambda}$, being canonical coordinates and momenta, have opposite signature under time reversal which undoes $i \to -i$.

Applying the time reversal transformation to (59) and requiring (68) yields the condition

$$\int \frac{d\mathbf{k}}{\sqrt{k}} \sum_\lambda e^{-i\mathbf{k}\cdot\mathbf{r}}\, \mathbf{e}^*_\lambda(\mathbf{k}) I_t\, a_\lambda(\mathbf{k},t) I_t^{-1} = -\int \frac{d\mathbf{k}}{\sqrt{k}} \sum_\lambda e^{i\mathbf{k}\cdot\mathbf{r}}\, \mathbf{e}_\lambda(\mathbf{k}) a_\lambda(\mathbf{k},-t) \quad (69)$$

$$= -\int \frac{d\mathbf{k}}{\sqrt{k}} \sum_\lambda e^{-i\mathbf{k}\cdot\mathbf{r}}\, \mathbf{e}_\lambda(-\mathbf{k}) a_\lambda(-\mathbf{k},-t)\,, \quad (70)$$

where the last step changed the integration variable from $\mathbf{k}$ to $-\mathbf{k}$. To maintain the meaning of λ as the helicity, the polarization vectors must, for *all* $\mathbf{k}$, satisfy

$$\mathbf{e}^*_\lambda(\mathbf{k}) \times \mathbf{e}_\lambda(\mathbf{k}) = i\lambda\hat{\mathbf{k}}\,, \quad (71)$$

and therefore

$$\mathbf{e}_\lambda(-\mathbf{k}) = \mathbf{e}_{-\lambda}(\mathbf{k})\,. \quad (72)$$

Furthermore, $\mathbf{e}^*_\lambda(\mathbf{k}) = -\mathbf{e}_{-\lambda}(\mathbf{k})$. Hence the time-reversal transformation law for the destruction and creation operators is

$$I_t\, a_\lambda(\mathbf{k},t) I_t^{-1} = a_\lambda(-\mathbf{k},-t)\,, \quad I_t\, a^\dagger_\lambda(\mathbf{k},t) I_t^{-1} = a^\dagger_\lambda(-\mathbf{k},-t)\,. \quad (73)$$

This relation is consistent, as it must be, with the fact that the helicity does not change under time reversal.

A very similar calculation leads to the space reflection law for these photon operators:

$$I_s\, a_\lambda(\mathbf{k},t) I_s^{-1} = -a_{-\lambda}(-\mathbf{k},t)\,, \quad I_s\, a^\dagger_\lambda(\mathbf{k},t) I_s^{-1} = -a^\dagger_{-\lambda}(-\mathbf{k},t)\,. \quad (74)$$

The last minus sign is very important, for it shows that *the photon has odd intrinsic parity*, i.e., the one-photon state

$$a^\dagger_\lambda(\mathbf{k})|0\rangle \equiv |\mathbf{k}\lambda\rangle\,, \quad (75)$$

behaves as follows under space reflection,

$$I_s|\mathbf{k}\lambda\rangle = -|-\mathbf{k}-\lambda\rangle\,. \quad (76)$$

By that token, a state with n photons has an intrinsic parity $(-1)^n$.

That the momentum and helicity both change sign under reflection is obvious *a priori*, but the overall minus sign in (76) is not. Because of this property, the one-photon states (60) of definite total angular momentum have a parity opposite to what one would expect for such states for a particle which has even intrinsic parity. True enough, the odd intrinsic parity of the photon is just a consequence of the familiar fact that the electric field $\mathbf{E}$ is a polar vector. Nevertheless, that the photon has an intrinsic parity independent of all phase conventions, and that it is odd, leads to important selection rules in radiative processes, and can be used to determine the parity of states having no direct connection to so clearcut an observable as $\mathbf{E}$.

10.2 Causality and Uncertainty in Electrodynamics

The canonical quantization of classical electrodynamics assigns observables, such as the field operators $\boldsymbol{E}$ and $\boldsymbol{B}$, and the energy density $\frac{1}{2}(\boldsymbol{E}^2+\boldsymbol{B}^2)$, to *every* space-time point. Operators defined in this way are called *local observables.* Because these observables do not all commute with each other, there are uncertainty relations between them that impose lower limits on the accuracy with which certain pairs of such local observables can be measured.

As Maxwell's theory is Lorentz invariant, it should come as no surprise that a measurement of any one such local observable at $x = (\boldsymbol{r}, t)$ cannot be influenced in any way by a measurement of any other at $x' = (\boldsymbol{r}', t')$ if x and x' have a space-like separation, for no physical process can be transmitted between such points. This is called the *causality condition,* and *all* relativistic quantum field theories — not just electrodynamics — are required to obey it. It is astonishing that a number of powerful theorems are a consequence of the simple and plausible requirement that there cannot be any limit to the accuracy with which two local observables can be measured if nothing can communicate between them. From the perspective of nonrelativistic quantum mechanics, the most striking of these theorems is the connection between spin and statistics, which cannot be proven within nonrelativistic quantum mechanics but which is a theorem in Lorentz-invariant quantum field theory. The other exact consequences of the causality condition involve antiparticles, e.g., that the mass and lifetime of a particle and its antiparticle must be equal, and that a fermion and its antiparticle have opposite intrinsic parities.

In this section the commutation rules and uncertainty relations obeyed by the electromagnetic field operators will be derived in *two distinct* ways. As already mentioned in §1.1, because the momentum and position of a charged particle can be measured by means of its interaction with the electromagnetic field, there must be limits on the accuracy with which these fields can be measured, for if there were no such limits perfect knowledge of the field could be used to violate the uncertainty principle as it applies to a particle's position and momentum. We will show that the uncertainty relations regarding the fields, as required by the relations for the particle, are precisely the uncertainty relations that follow from the commutators of the field operators. This, then, is a gold-plated demonstration of the internal consistency of canonical quantization.

(a) Commutation Rules: Complementarity

All commutators between the local observables of the electromagnetic field can be computed once the commutation rule for the vector potential is know. Those for the field strengths are found by differentiation, and those for polynomials in the fields, such as the energy density, by straightforward algebra.

From (59) and (58), the commutator between the vector potentials at two distinct space-time points x_1 and x_2 is

$$[A_i(x_1), A_j(x_2)] = \hbar c \int \frac{d\boldsymbol{k}}{2k(2\pi)^3} [e^{ik\cdot(x_1-x_2)}\, t_{ij} - \text{c.c.}] \,, \tag{77}$$

where

$$t_{ij} = \sum_\lambda (e^i_{k\lambda})^* \, e^j_{k\lambda} \,, \tag{78}$$

$e^i_{k\lambda}$ being a Cartesian component of the polarization vector. This second rank tensor can only be a function of $\boldsymbol{k}$, so it must be of the form $a\delta_{ij} + bk_ik_j$, where a and b can be functions of k^2. Now $t_{ij} = t^*_{ji}$, and therefore a and b are real. Furthermore $t_{ii} = 2 = 3a + k^2b$, and because $\boldsymbol{e}_{k\lambda}$ is perpendicular to $\boldsymbol{k}$, $k_it_{ij} = 0 = k_j(a + bk^2)$. Hence

$$t_{ij} = \delta_{ij} - \frac{k_ik_j}{k^2} \, . \tag{79}$$

Thus

$$[A_i(x_1), A_j(x_2)] = -2i\hbar c \int \frac{d\boldsymbol{k}}{2k(2\pi)^3} \, t_{ij} \, e^{i\boldsymbol{k}\cdot(\boldsymbol{r}_1-\boldsymbol{r}_2)} \, \sin kc(t_1 - t_2) \, . \tag{80}$$

The commutators of physical interest are between the fields. For example, that for the electric fields is

$$[E_i(x_1), E_j(x_2)] = \frac{1}{c^2}\frac{\partial^2}{\partial t_1 \partial t_2}[A_i(x_1), A_j(x_2)] \, . \tag{81}$$

But

$$\frac{k_ik_j}{k^2}\frac{1}{c^2}\frac{\partial^2}{\partial t_1 \partial t_2}\left(e^{i\boldsymbol{k}\cdot\boldsymbol{r}} \sin kct\right) = \frac{\partial^2}{\partial r_{1i} \partial r_{2j}}\left(e^{i\boldsymbol{k}\cdot\boldsymbol{r}} \sin kct\right) , \tag{82}$$

where $\boldsymbol{r} = \boldsymbol{r}_1 - \boldsymbol{r}_2, t = t_1 - t_2$. Therefore the commutator for the electric field strengths is

$$[E_i(x_1), E_j(x_2)] = 2i\hbar c \left(\delta_{ij}\frac{1}{c^2}\frac{\partial^2}{\partial t_1 \partial t_2} - \frac{\partial^2}{\partial r_{1i} \partial r_{2j}} \right) D(r,t) \, , \tag{83}$$

where

$$D(r,t) = - \int \frac{d\boldsymbol{k}}{2k(2\pi)^3} e^{i\boldsymbol{k}\cdot\boldsymbol{r}} \sin kct \, . \tag{84}$$

The other commutators between the fields are calculated in the same way, the results being

$$[B_i(x_1), B_j(x_2)] = [E_i(x_1), E_j(x_2)] \, , \tag{85}$$

$$[E_i(x_1), B_i(x_2)] = 0 \, , \tag{86}$$

whereas non-parallel components of $\boldsymbol{E}$ and $\boldsymbol{B}$ do not commute:

$$[E_i(x_1), B_j(x_2)] = -2i\hbar \, \epsilon_{ijk} \frac{\partial^2}{\partial t_1 \partial r_{2k}} D(r,t) \, . \tag{87}$$

The function D is

$$D(r,t) = -\frac{1}{8\pi^2 r}\int_{-\infty}^{\infty} dk \sin kr \sin kct \, , \tag{88}$$

i.e.,

$$D(r,t) = \frac{1}{8\pi r}\,[\delta(r + ct) - \delta(r - ct)] \, . \tag{89}$$

Hence all the commutators between electromagnetic fields strengths at two space-time points vanish unless these points can be connected by a light signal; or put another way, if a light cone is drawn from one of the two points the other must be either on the forward or backward portion of this cone for the commutator not to vanish. This is the causality principle for the free electromagnetic field.

That the commutators all vanish for pairs of points that are not light-like is a consequence of the photon's zero mass, i.e., of the fact that Maxwell's equations only have solutions that propagate with the speed of light. Field theories with massive quanta, such as that for the aforementioned Z^0, will have commutators that are nonzero also for time-like separations, but not for space-like separations. And when the electromagnetic field is coupled to matter, the existence of antiparticles also produces non-vanishing commutators for time-like separations due to virtual electron-positron pairs, or any other states that have the quantum numbers of the photon — spin 1, charge 0 and odd intrinsic parity. But in no case will a properly formulated Lorentz-invariant field theory permit non-vanishing commutators for space-like separations.

The commutation rules between observables in the Heisenberg picture at equal times determine compatible observables, and as a consequence the representations that can be used to describe the system in question. For this purpose, we need the commutation rules at equal times, $t_1 = t_2$. Clearly $D(r,-t) = -D(r,t)$, and therefore

$$\lim_{t\to 0} D(r,t) = 0\,, \qquad \lim_{t\to 0} \frac{\partial^2}{\partial t^2} D(r,t) = 0\,. \tag{90}$$

Consequently

$$[E_i(\boldsymbol{r}_1,t), E_j(\boldsymbol{r}_2,t)] = [B_i(\boldsymbol{r}_1,t), B_j(\boldsymbol{r}_2,t)] = 0\,; \tag{91}$$

on the other hand, only first time derivatives appear in (87), so

$$[E_i(\boldsymbol{r}_1,t), B_j(\boldsymbol{r}_2,t)] \neq 0\,, \quad (i \neq j,\ \boldsymbol{r}_1 = \boldsymbol{r}_2)\,. \tag{92}$$

In short, *at any given instant, it is possible to specify either the electric field $\boldsymbol{E}(\boldsymbol{r},t)$* or *the magnetic field $\boldsymbol{B}(\boldsymbol{r},t)$ throughout all of space, but not both,* and therefore representations that specify one or the other field can be used to describe the system. Indeed, it can be shown that the electric (or magnetic) field forms a complete set of compatible observables.

Recall that the field operators are linear forms in the creation and destruction operators. Hence *states in which the electric or magnetic field strength is specified are not states with a definite number of photons*, but rather coherent superpositions of states with various (indeed, all) numbers of photons! This is a striking illustration of complementarity — of the principle that no one depiction of a system based on classical concepts provides a comprehensive rendition in the quantum domain, where in this case the complementary classical pictures are those of the continuous field and Newton's concept of light as a stream of particles. Indeed, it goes beyond that because only the electric or the magnetic field can be specified, but not both. It is, of course, not astonishing that complementarity would have a richer meaning in systems with an infinite number of degrees of freedom than in the systems we have encountered previously.

(b) Uncertainty Relations

The uncertainty relations for the field strengths follow from the general relationship between commutation rules and uncertainty relations of §2.3(c); for any two observables, and in any state,

$$\Delta\mathcal{O}_1\,\Delta\mathcal{O}_2 \geq \tfrac{1}{2}|\langle[\mathcal{O}_1,\mathcal{O}_2]\rangle| \,. \tag{93}$$

Consider then the average of a field operator over a small space-time volume Ω, e.g.,

$$\boldsymbol{E}(\Omega) = \frac{1}{\Omega}\int_\Omega d\Omega\, \boldsymbol{E}(x) \,. \tag{94}$$

According to (83), for non-parallel components of the averaged electric field and Ω_1 entirely in the future with respect to Ω_2, the uncertainty relation is

$$\Delta E_x(\Omega_1)\,\Delta E_y(\Omega_2) \geq \frac{\hbar c}{8\pi}\left|\int_{\Omega_1}\frac{d\Omega_1}{\Omega_1}\int_{\Omega_2}\frac{d\Omega_2}{\Omega_2}\frac{\partial^2}{\partial x_1 \partial y_2}\frac{\delta(r-ct)}{r}\right| \,. \tag{95}$$

Following Bohr and Rosenfeld, we now sketch how this uncertainty relation can also be derived from that for the position and momentum of a test body of charge Q whose spatial extent is a small volume V_1. If its momentum is measured at nearby times t_1' and t_1, that determines the electric field averaged over the space-time volume $\Omega_1 = V_1 T_1$, where $T_1 = t_1' - t_1$:

$$E_x(\Omega_1) \simeq \frac{p_x(t_1') - p_x(t_1)}{Q(t_1' - t_1)} \,. \tag{96}$$

Uncertainties in these momenta imply a corresponding uncertainty in the electric field,

$$\Delta E_x(\Omega_1) \gtrsim \frac{\hbar}{2QT_1\Delta x} \,. \tag{97}$$

Consider, now, the measurement of $\boldsymbol{E}$ in a small space-time region $\Omega_2 = V_2T_2$ which lies entirely in the future with respect to Ω_1. The uncertainty in the position and momentum of the test body implies an uncertain electromagnetic field that can propagate into Ω_2 and disturb a measurement of the field in that region. The uncertainty in the position Δx implies an uncertain dipole moment density in Ω_1 of order $Q\Delta x/V_1$ in the x-direction. At large distances, the leading contribution to the fields is the dipole field provided the test body's motion is nonrelativistic (as it must be as we are describing the test body with nonrelativistic quantum mechanics, and which can always be achieved by using a large enough mass). According to classical radiation theory, the retarded scalar potential is thus uncertain in the amount

$$\Delta\phi(\boldsymbol{r}_2, t_2) \simeq \frac{cQ\Delta x}{V_1}\int_{\Omega_1} d\boldsymbol{r}_1 dt_1 \frac{\partial}{\partial x_1}\frac{\delta(c(t_2-t_1) - |\boldsymbol{r}_2 - \boldsymbol{r}_1|)}{4\pi|\boldsymbol{r}_2 - \boldsymbol{r}_1|} \,. \tag{98}$$

There is also an uncertain current in the x-direction which leads to an uncertainty $\Delta A_x(\boldsymbol{r}_2, t_2)$ in the vector potential. Hence the uncertainty in the measurement of $E_x(\Omega_1)$ produces in Ω_2 an uncertainty in $\boldsymbol{E}$ in the amount

$$\Delta\boldsymbol{E}(\boldsymbol{r}_2, t_2) \simeq -\frac{\partial}{\partial \boldsymbol{r}_2}\Delta\phi(\boldsymbol{r}_2, t_2) - \frac{1}{c}\frac{\partial}{\partial t_2}\Delta\boldsymbol{A}(\boldsymbol{r}_2, t_2) \,. \tag{99}$$

If E_x is measured in Ω_1 there need be *no* uncertainty in A_y or A_z in Ω_2; the uncertainty in ΔA_x, because of $\boldsymbol{B} = \boldsymbol{\nabla}\times\boldsymbol{A}$, produces uncertainties in B_y and B_z, but *not* in B_x. Hence E_x and B_x can be measured simultaneously in regions that can be connected by light signals, or

$$\Delta E_x(\Omega_1)\Delta B_x(\Omega_2) \geq 0\,, \tag{100}$$

but perpendicular components of $\boldsymbol{E}$ and $\boldsymbol{B}$ have non-zero uncertainty products. These results agree with (85) to (86).

Finally, we calculate the uncertainty of E_y in Ω_2, bearing in mind that only ΔA_x is necessarily finite in Ω_2 when E_x is measured in Ω_1, and does not contribute to ΔE_y. According to (99), ΔE_y, averaged over Ω_2, is therefore

$$\Delta E_y(\Omega_2) \gtrsim -\int_{\Omega_2} \frac{d\Omega_2}{\Omega_2}\frac{\partial}{\partial y_2}\,\Delta\phi(\boldsymbol{r}_2,t_2) \tag{101}$$

$$\gtrsim -\frac{cQ\Delta x}{4\pi V_1}\int_{\Omega_2}\frac{d\Omega_2}{\Omega_2}\int_{\Omega_1} d\Omega_1\,\frac{\partial^2}{\partial x_1\partial y_2}\,\frac{\delta(r-ct)}{r}\,, \tag{102}$$

which when combined with (97) gives

$$\Delta E_x(\Omega_1)\Delta E_y(\Omega_2) \gtrsim \frac{\hbar c}{8\pi}\left|\int_{\Omega_2}\frac{d\Omega_2}{\Omega_2}\int_{\Omega_1}\frac{d\Omega_1}{\Omega_1}\,\frac{\partial^2}{\partial x_1\partial y_2}\,\frac{\delta(r-ct)}{r}\right|. \tag{103}$$

This result is in agreement with the one that follows from the commutation rule for the electric fields (Eq. 95).

That these distinct derivations give the same result constitutes one of the most impressive demonstrations of self-consistency in theoretical physics. It is, therefore, appropriate to summarize the argument again. It really starts by quantizing the sources of the field, which have a finite number of degrees of freedom, by the recipe of canonical quantization imposed on variables that were known to Newton — coordinates and momenta. Then it extends this recipe to the arcane and infinitely numerous canonical variables that represent the familiar electromagnetic field strengths. This second step produces commutation rules for the field operators, and these imply uncertainty relations for the fields strengths that guarantee that measurements on the fields cannot violate the uncertainty principle as it applies to the sources. This is a very welcome and indeed an indispensable outcome, but that it would all work out was not a foregone conclusion. In this regard, one last, crucial point: in the final result (Eq. 103), all reference to the test body (its charge and mass) has disappeared, as it had to — all that remains is a purely geometric quantity.

10.3 Vacuum Fluctuations

The uncertainty relations for the electromagnetic field operators derived in the previous section hold in any state of the free electromagnetic field, and in particular, in the vacuum state. This has a striking implication: *even in the vacuum the fields fluctuate.* From a formal point of view, this may not be surprising, because the quantization of the field is based on the quantization of the simple harmonic oscillator, and in the ground state of the oscillator the coordinate and momentum fluctuate about zero. And yet it is startling that the uncertainty principle compels the vacuum to be frenetic, which raises the question of whether its fluctuations are physically meaningful or a mathematical artifact. It is our purpose here to show that these vacuum fluctuations do have real, observable consequences.

The magnitude of the fluctuations follow from Eq. 103. The right-hand side is of order $\hbar c/L^4$, where L is the spatial dimension of the two space-time regions. Hence

$$\Delta E \sim \sqrt{\hbar c}/L^2 \sim \Delta B\,, \tag{104}$$

where the $\Delta E \sim \Delta B$ is a consequence of (85). Hence the energy of the fluctuations in a region of volume L^3 is

$$\Delta H \sim [(\Delta E)^2 + (\Delta B)^2]L^3 \sim \hbar c/L\,. \tag{105}$$

This is the energy carried by a photon of wavelength $\sim L$, and states that in a volume of dimension L the fluctuations in the field strengths are due to changes in the occupation numbers of order 1 for photon energies of order $\hbar c/L$.

Two observable consequences of vacuum fluctuations will be discussed here: the Casimir effect, which is a tiny force proportional to $\hbar$, exerted on conductors in vacuum; and the Lamb shift, a very small displacement of hydrogenic (and also other) energy levels. These effects involve the infamous infinities that beset quantum electrodynamics, and relativistic quantum fields theories quite generally. In the case of the Casimir effect, a complete calculation is possible with the means now at our disposal. The Lamb shift is a much more complex phenomenon, however, and requires the machinery of renormalization for a proper treatment; our discussion is crude, but will suffice for the purpose at hand.

(a) The Casimir Effect

Consider the three parallel, perfectly conducting plates shown in Fig. 10.2, sitting in empty space. We will learn that if the distance z between the two plates on the right is small, there is an attractive force between them of order $\hbar c/z^4$ per unit area.

The basic idea is that in a confined space the frequencies ω_i of the normal modes of the field differ from those in an unbounded region, and that, as a consequence, the zero point energies $\frac{1}{2}\sum_i \hbar\omega_i$ in these two situations differ. While the separate zero point energies have no direct physical meaning (indeed, are infinite), the difference between them does, and this produces a pressure on the enclosure of the finite region. This pressure is due to the difference in the spectrum of vacuum fluctuations in the enclosure as compared to the empty, unbounded space outside.

The normal modes in the space between the two perfectly conduction plates of size $L \times L$ separated by z, are standing waves with wave numbers

$$k_n(z) = \sqrt{(n\pi/z)^2 + k_x^2 + k_y^2}\,, \qquad n = 0, 1, \ldots\,. \tag{106}$$

The length L is taken to be sufficiently large to make the spectrum of wave numbers in the $x - y$ plane effectively continuous. For oscillations in the z-direction, there is just one (TE) mode when $n = 0$, while for $n \geq 1$ there are two.[1] The zero point

[1] Jackson, §8.4.

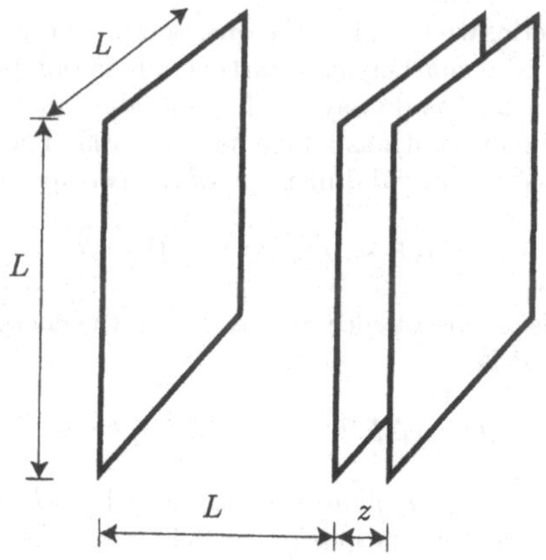

FIG. 10.2. Arrangement for derivation of the Casimir effect.

energy in this space is "therefore"

$$E(z) = \tfrac{1}{2}\hbar c \int_0^\infty \frac{L dk_x}{\pi} \int_0^\infty \frac{L dk_y}{\pi}\, 2 \sum_{n=0}^{\infty} \epsilon_n \, k_n(z), \tag{107}$$

where $\epsilon_0 = \frac{1}{2}$ and $\epsilon_n = 1$ for $n \geq 1$. The quotation marks on "therefore" warn that this is a highly divergent expression! As already indicated, however, the quantity that has physical meaning is the *difference* between this energy and the energy if the plates were absent. Furthermore, the notion of a perfect conductor has no meaning for arbitrary short wavelengths, because wavelengths much shorter than atomic dimensions cannot be confined.

In the light of these remarks two things are now done to save the day. First, a smooth cut-off function $f(k_n a)$ is put into the integrand of (107), which has the properties $f = 1$ for $ka \ll 1$ and $f \to 0$ for $ka \gg 1$, where a is a typical atomic dimension. Second, we consider the *difference* in energy between the cases (i) where z is small in the setup of Fig. 10.2, and (ii) where $z \equiv D$ is of order the large distance L:

$$\delta E(z) = E(z) + E_\infty(L - z) - E_\infty(D) - E_\infty(L - D)\,. \tag{108}$$

Here $E_\infty(D)$ means that the argument is a length large enough to make the spectrum of modes in the z-direction continuous, i.e.,

$$E_\infty(D) = \hbar c\, \frac{L^2}{\pi^2}\, \frac{\pi}{4} \int_0^\infty d\kappa^2 \int_0^\infty \frac{D dk_z}{\pi} \sqrt{k_z^2 + \kappa^2}\, f(a\sqrt{k_z^2 + \kappa^2})\,; \tag{109}$$

$\kappa^2 = k_x^2 + k_y^2$, and the fact that there is only one $n = 0$ mode is irrelevant when the spectrum is continuous. As we will soon show, the energy difference $\delta E(z)$ is finite and *independent* of the detailed form of the ad hoc function f. This is so because the difference between the spectra in the confined and open regions disappears as the frequencies increase, and as a consequence the actual form of the cutoff for $ka \sim 1$ is irrelevant to the difference δE.

The energies E_∞ in (108) are all extensive, i.e., proportional to the volume of integration, and therefore to the size of the gaps in the z directions. Hence

$$\delta E(z) = E(z) - E_\infty(z) . \tag{110}$$

The derivation is often presented with just two plates separated by z, and goes immediately to Eq. 110. The advantage of the arrangement in Fig. 10.2, which was used in the original derivation, is that it eliminates the concern that the energy of the region outside and near the two plates has been handled incorrectly.

In any event, we now have

$$\begin{aligned}\delta E(z) = \hbar c \, \frac{L^2}{4\pi} \int_0^\infty d\kappa^2 \Big\{ \sum_n \epsilon_n \sqrt{(n\pi/z)^2 + \kappa^2} \, f(a\sqrt{(n\pi/z)^2 + \kappa^2}) \\ - \frac{z}{\pi} \int_0^\infty dk_z \, \sqrt{k_z^2 + \kappa^2} \, f(a\sqrt{k_z^2 + \kappa^2}) \Big\} .\end{aligned} \tag{111}$$

To give the sum on n and the integral over k_z a similar appearance, define a continuous variable $n = zk_z/\pi$ in the latter, so that

$$\begin{aligned}\delta E(z) = \hbar c \, \frac{L^2}{4\pi} \left(\frac{\pi}{z}\right)^3 \int_0^\infty du \Big\{ \sum_n \epsilon_n \sqrt{n^2 + u} \, f((a\pi/z)\sqrt{n^2 + u}) \\ - \int_0^\infty dn \, \sqrt{n^2 + u} \, f((a\pi/z)\sqrt{n^2 + u}) \Big\} ,\end{aligned} \tag{112}$$

where $u = (z\kappa/\pi)^2$.

The difference between the sum and integral over n is evaluated with the Euler-Maclaurin formula. Assuming that the cut-off function f causes all derivatives of the integrand to vanish as $n \to \infty$, in our case this formula reads[1]

$$\sum_{n=0}^\infty \epsilon_n F(n) = \int_0^\infty dn \, F(n) - \sum_{r=1}^\infty \frac{B_{2r}}{(2r)!} \, F^{(2r-1)}(0) , \tag{113}$$

where $F^{(k)}(n) = d^k F(n)/dn^k$, and the B_{2r} are the Bernoulli numbers, $B_2 = \frac{1}{6}, B_4 = -\frac{1}{30}$, etc. From (112), with $w = n^2 + u$,

$$F(n) = \int_{n^2}^\infty w^{\frac{1}{2}} f(w^{\frac{1}{2}} \pi a/z) \, dw . \tag{114}$$

Thus

$$F^{(1)}(n) = -2n^2 f(n\pi a/z) , \tag{115}$$

and therefore $F^{(1)}(0) = 0$, $F^{(3)}(0) = -4$, while higher-order terms in (113) are down by ever higher powers of (a/z).

The final result for the energy difference per unit area is therefore

$$\frac{1}{L^2} \delta E(z) = -\hbar c \, \frac{\pi^2}{4} \, \frac{4}{4! \, 30} \, \frac{1}{z^3} . \tag{116}$$

---

[1] F.B. Hildebrand, *Introduction to Numerical Analysis*, McGraw-Hill, New York (1974), §5.8.

Note that this result does not depend on the cutoff function $f(k)$ provided, of course, that it does not alter the low-frequency modes, i.e., $f(0) = 1$.

Energy is gained as the plates recede from each other — an attractive force per unit area, i.e., a pressure P squeezing the plates together:

$$P(z) = -\hbar c \frac{\pi^2}{240} \frac{1}{z^4} = -\frac{0.013}{z^4} \text{ dyne/cm}^2 , \tag{117}$$

where in the last expression z is in units of μm.

The Casimir force between a sphere and a flat plate varies like $1/z^3$, and this version of the effect has been confirmed to good accuracy.[1]

(b) The Lamb Shift

A charge moving in empty space is not as devoid of interaction as one might naively suppose. This was already realized before the advent of quantum mechanics, because the interaction of the charge with its own electrostatic field produces a self-energy that is infinite if the charge is pointlike. This problem persists in quantum electrodynamics, though in a much less virulent form because of vacuum fluctuations of the electromagnetic field and the existence of anti-particles. The upshot is that the self-energy only tends to infinity like the logarithm of the source size, whereas the divergence is quadratic in classical electrodynamics. But this takes us well beyond the level of this volume, and here we will only deal with one important part of this topic, the influence on atomic energy levels due to vacuum fluctuations of the electromagnetic field. The treatment, due to Welton, will only be heuristic and qualitative.

The basic idea is that the vacuum fluctuation of the electric field causes a charged particle's position to fluctuate, so the potential it experiences is somewhat different from that when this fluctuation is ignored. Let $\delta \boldsymbol{r}$ be the fluctuation in position. Then if $V(\boldsymbol{r})$ is the electrostatic potential, the fluctuation produces a change in potential energy

$$\delta V(\boldsymbol{r}) = V(|\boldsymbol{r} + \delta \boldsymbol{r}|) - V(\boldsymbol{r}) = \delta \boldsymbol{r} \cdot \boldsymbol{\nabla} V + \tfrac{1}{2} \sum_{i,j} \delta x_i \delta x_j \frac{\partial^2 V}{\partial x_i \partial x_j} + \dots . \tag{118}$$

There is no preferred direction in the vacuum, and the field fluctuations are therefore isotropic in this state. Consequently the position fluctuation is also isotropic, and

$$\langle \delta V(\boldsymbol{r}) \rangle_0 = \tfrac{1}{6} \langle |\delta \boldsymbol{r}|^2 \rangle_0 \nabla^2 V , \tag{119}$$

where $\langle A \rangle_0$ is the expectation value of A in the vacuum state of the field.

If the vacuum fluctuations are disregarded, the electronic motion in an atom is known to be nonrelativistic as long as $\alpha Z \ll 1$, a condition we assume to hold. We now suppose that the fluctuation in position can also be estimated by treating the motion as nonrelativistic. If that is correct, the charge's interaction with the magnetic field fluctuations can be neglected, and the interaction with the electric

[1] S.K. Lamoreaux, *Phys. Rev. Lett.* **78**, 5 (1997). M. Bordag, U. Mohideen and V. Mostepanko, *Phys. Reports* **353**, 1(2001).

field can be handled with the dipole approximation, i.e., by ignoring the spatial variation of $\boldsymbol{E}$:

$$m\frac{d^2}{dt^2}\delta\boldsymbol{r}(t) = -e\boldsymbol{E}(\boldsymbol{r},t) \simeq -e\boldsymbol{E}(t)\,. \tag{120}$$

Decomposing the fluctuations into normal modes of frequency ω and polarization λ thus gives

$$\delta\boldsymbol{r}_{\omega\lambda} = \frac{e}{m\omega^2}\boldsymbol{E}_{\omega\lambda}\,, \tag{121}$$

and as $\delta\boldsymbol{r}$ has a vanishing expectation value,

$$(\Delta\boldsymbol{r}_{\omega\lambda})^2 = \langle|\delta\boldsymbol{r}_{\omega\lambda}|^2\rangle_0 - [\langle\delta\boldsymbol{r}_{\omega\lambda}\rangle_0]^2 = \frac{e^2}{m^2\omega^4}\langle|\boldsymbol{E}_{\omega\lambda}|^2\rangle_0\,. \tag{122}$$

The vacuum expectation value $\langle|\boldsymbol{E}_{\omega\lambda}|^2\rangle_0$ follows from our knowledge that the fluctuations of $\boldsymbol{E}$ and $\boldsymbol{B}$ are equal, and that the ground state energy of each mode is $\frac{1}{2}\hbar\omega$. Hence

$$\langle H\rangle_0 = \int_V d^3r \sum_{\boldsymbol{k},\lambda}\langle|\boldsymbol{E}_{\boldsymbol{k},\lambda}|^2\rangle_0 = \tfrac{1}{2}\sum_{\boldsymbol{k},\lambda}\hbar\omega\,, \tag{123}$$

and therefore $\langle|\boldsymbol{E}_{\boldsymbol{k},\lambda}|^2\rangle_0 = \hbar\omega/2V$. Consequently

$$(\Delta\boldsymbol{r})^2 = \sum_{\boldsymbol{k},\lambda}\frac{e^2}{m^2\omega^4}\frac{\hbar\omega}{2V} = \frac{e^2\hbar}{m^2}\int_0^\infty dk\,\frac{4\pi k^2}{(2\pi)^3\omega^3} = \frac{2\alpha}{\pi}\lambda\!\!\!^{-}_C{}^2\int_0^\infty\frac{d\omega}{\omega}\,, \tag{124}$$

where $\lambda\!\!\!^{-}_C = \hbar/mc$ is the electron's Compton wavelength.

This last integral diverges logarithmically at low and high frequencies — it has both an infrared and ultraviolet catastrophe. The infrared divergence is an artifact of this crude approximation because a bound state does not posses excitation energies of arbitrarily low energy, and therefore cannot respond to vacuum fluctuations of arbitrarily low frequencies. The ultraviolet divergence is more basic, and is cured in a proper calculation by two considerations. First, the nonrelativistic treatment does not hold for frequencies of order $mc/\hbar$ or higher; second, the vacuum fluctuations also alter the energy of an electron in free space, and the observable quantity is the difference in energy between the free and bound electron. Crudely speaking, therefore, the integral should be taken to run over the energy interval $(\bar{E}_b, mc^2)$, where $\bar{E}_b$ is some characteristic excitation energy of the bound state in question. If this is granted, we have

$$(\Delta\boldsymbol{r})^2 = \frac{2\alpha}{\pi}\lambda\!\!\!^{-}_C{}^2\,\ln\,(mc^2/\bar{E}_b)\,. \tag{125}$$

That the vacuum fluctuations appear in a logarithmic form is a hopeful sign, because logarithms are relatively insensitive to their argument, and therefore to the errors that have surely been committed.

We now apply these results to one electron in a point Coulomb field, $V = -Ze^2/4\pi r$. Combining (119) and (125) then results in

$$\delta V(\boldsymbol{r}) = \frac{\alpha Ze^2\lambda\!\!\!^{-}_C{}^2}{3\pi}\,\ln\left(\frac{mc^2}{\bar{E}_b}\right)\,\delta(\boldsymbol{r})\,. \tag{126}$$

This states that the modification of the electrostatic field is confined to the origin,[1] and therefore only alters s states. The Lamb shift, in this approximation, is just the expectation value of δV,

$$\Delta E_n = \frac{\alpha Z e^2 \lambda_C^2}{3\pi} \ln\left(\frac{mc^2}{\bar{E}_n}\right) |\psi_{n0}(0)|^2 , \tag{127}$$

where $\psi_{n0}(\boldsymbol{r})$ is the hydrogenic bound s state with principal quantum number n. As

$$|\psi_{n0}(0)|^2 = \frac{(Z/a_0)^3}{\pi n^3} , \tag{128}$$

where $a_0 = 4\pi\hbar^2/me^2$ is the Bohr radius in hydrogen, the final result is

$$\Delta E_n = \frac{8}{3\pi} \frac{\alpha^3 Z^4}{n^3} \ln\left(\frac{mc^2}{\bar{E}_n}\right) \text{Ry} , \tag{129}$$

where the Rydberg is $e^2/8\pi a_0$.

Note that this shift in energy is smaller than the unperturbed binding energy by a factor of order $\alpha(\alpha Z)^2 \sim \alpha(v/c)^2$, which is of order the fine structure splittings times α. Hence this "radiative correction" due to vacuum fluctuations of the field is of order α smaller than the kinematic relativistic corrections (recall §5.1). That it is so small is due to $(\Delta \boldsymbol{r})^2$ being tiny compared to a_0^2, namely, $\sim \alpha^3 a_0^2$ according to (125).

The exact calculation, to this same order in α, gives the following result:[2]

$$\Delta E_n = \frac{8}{3\pi} \frac{\alpha^3 Z^4}{n^3} \left[\ln\left(\frac{mc^2}{\bar{W}_n}\right) + \frac{19}{30}\right] \text{Ry} . \tag{130}$$

This is astonishingly close to the guesstimate (129), and shows that the heuristic argument has the basic effect well identified. The correct characteristic excitation energies are, however, far larger than a naive order-of-magnitude estimate, i.e., much larger than $\sim Z^2$ Ry; for the important case of the $n = 2$ level, $\bar{W}_2 \simeq 33\, Z^2$ Ry ! Equation (130) gives splendid agreement with experiment once much smaller corrections that are down by powers of α and αZ are taken into account.

10.4 Radiative Transitions

The interaction of the electromagnetic field with the constituents of matter accounts for a truly vast array of phenomena. These fall into two broad categories, divided by whether the characteristic velocities of the sources are small compared to the velocity of light or of comparable magnitude. In this volume we can only concern ourselves with the former. In the relativistic regime some of the key approximations

---

[1] In a more sophisticated calculation that treats the motion relativistically, $\delta V(\boldsymbol{r})$ has a spread of order λ_C, which vanishes when $c \to \infty$. For this and other reasons, states with $l \neq 0$ also have Lamb shifts, though far smaller than those for s states.

[2] Bethe and Salpeter, §21. Eq. 130 results from a proper relativistic and quantum mechanical treatment of the interaction between the electron and the vacuum fluctuations, as well as relatively small contributions from vacuum polarization and the electron's anomalous magnetic moment.

to be made below are invalid, and pair production and annihilation are also possible — processes that require a full-fledged relativistic treatment in which the sources are also described by quantum fields.

The term "radiative transition" is usually applied to the emission and absorption of light, whatever the wavelength may be, from microwave to gamma rays, and this section will be devoted to this topic. Scattering of light will be dealt with separately in §10.7.

(a) The Interaction Between Field and Sources

Due to gauge invariance, there are various equivalent but different formulations of the Hamiltonian. In the nonrelativistic regime there is a great advantage to the Coulomb gauge, which we adopt, for it separates the instantaneous Coulomb interaction responsible for the binding of slow charges from the interaction with the transverse radiation field.

Consider a set of charged point particles ($a = 1, \ldots, N$), with positions, momenta, charges, masses and magnetic moments $\boldsymbol{r}_a, \boldsymbol{p}_a, e_a, m_a, \boldsymbol{\mu}_a$. The complete Hamiltonian to leading order in v^2/c^2 is

$$H = H_\gamma + \sum_a \frac{1}{2m_a}\left(\boldsymbol{p}_a - \frac{e_a}{c}\boldsymbol{A}_a\right)^2 - \sum_a \boldsymbol{\mu}_a \cdot \boldsymbol{B}_a + \frac{1}{8\pi}\sum_{a \neq b} \frac{e_a e_b}{|\boldsymbol{r}_a - \boldsymbol{r}_b|} + V \,, \quad (131)$$

where H_γ is the energy of the free field (Eq. 34), and $\boldsymbol{A}_a \equiv \boldsymbol{A}(\boldsymbol{r}_a)$, etc. The magnetic moments are defined as in §5.3(a):

$$\boldsymbol{\mu}_a = \frac{e\hbar}{2m_a c} g_a \boldsymbol{s}_a \,, \quad (132)$$

where $\boldsymbol{s}_a$ is the spin of particle a and g_a its gyromagnetic ratio, or g-factor.

The last term V in (131) contains non-electromagnetic interactions, such as nuclear forces, should they play a role; in atomic problems it also contains the fine and hyperfine interactions, which to lowest order in v^2/c^2 do not involve coupling to the transverse field.

The Hamiltonian (131) has the following form:

$$H = H_\gamma + H_M + H_{\text{int}} \,, \quad (133)$$

where H_M is the energy of the sources (or matter) in the absence of the transverse field, and H_{int} is the interaction of matter with that field. The latter has terms linear and quadratic in the vector potential:

$$H_{\text{int}} = H_1 + H_2 \,. \quad (134)$$

The linear term H_1 is a sum of photon creation and destruction operators; it has matrix elements between states that differ by a single photon, and accounts for the emission and absorption of light. The quadratic term H_2 contains operators that create or destroy two photons, and terms that are products of a creation and a destruction operator; this last term in H_2 therefore contributes to scattering. The explicit forms of these two terms in the interaction with the radiation field are

$$H_1 = -\sum_a \frac{e_a}{m_a} \boldsymbol{p}_a \cdot \boldsymbol{A}_a - \sum_a \boldsymbol{\mu}_a \cdot \boldsymbol{B}_a , \tag{135}$$

$$H_2 = \sum_a \frac{e_a^2}{2m_a c^2} |\boldsymbol{A}_a|^2 . \tag{136}$$

In H_1 the order of operators does not matter because in the Coulomb gauge $\boldsymbol{\nabla} \cdot \boldsymbol{A} = 0$.

When the source is nonrelativistic, the various terms in the interaction have very different relative orders of magnitude. This is of capital importance in the analysis of radiative processes, because it leads to enormous simplifications as a consequence of the following facts:[1]

1. The first term in H_1, due to charge transport, is of order $\sqrt{\alpha} v^3 m$, where α is the fine structure constant and v the characteristic velocity.

2. The second term in H_1, due to intrinsic magnetic moments, is smaller than the first term by a factor of order v. (For a neutral particle with a magnetic moment, such as the neutron, there is only this term.)

3. Hence the ratio of H_1 to the typical energies $\sim mv^2$ of the system is at most $\sqrt{\alpha} v \ll 1$.

4. The radiation absorbed or emitted in transitions between the bound states of a nonrelativistic system has wavelengths long compared to the system's dimensions. As a consequence, *radiative transitions are dominated by the leading multipole allowed by the angular momentum and parity selection rules.*

5. The quadratic term H_2 is smaller than H_1 by the same factor $\sim \sqrt{\alpha} v \ll 1$, i.e., is comparable to what H_1 would yield in second-order perturbation theory.

6. *The emission and absorption of single photons (one photon transitions) are accounted for by H_1 in first-order perturbation theory,* as a consequence of No. 2, while the scattering of light is accounted for by a combination of H_2 in first order and H_1 in second order as a consequence of No. 5. Corrections to the leading order *amplitudes* are smaller by powers of $\sqrt{\alpha}$, i.e., powers of e when $\hbar = c = 1$.

These statements are consequences of simple order-of-magnitude estimates. The characteristic momenta of a particle confined to space of dimension a are $\sim 1/a$, and therefore $v \sim 1/ma$. The typical energy differences Δ in such a system are of order $mv^2 \sim 1/ma^2$, which is then also the characteristic frequency ω of emitted and absorbed radiation. The magnitude of the field is related to the photon energy by $\omega \sim B^2 \lambda^3$, where λ is the characteristic wavelength; therefore $B \sim \omega^2$ and $A \sim \omega$ because $B \sim A/\lambda$. Hence

$$e(p/m)A \sim emv^3 \sim \sqrt{\alpha} mv^3 , \quad \mu B \sim (e/m)\omega^2 \sim \sqrt{\alpha} mv^4 . \tag{137}$$

---

[1] In the remainder of this section we use units in which $\hbar = c = 1$.

This establishes statements 1 and 2. Furthermore $H_1/\Delta \sim \sqrt{\alpha}v$ as claimed in statement 3. The multipole expansion converges rapidly when $ka \ll 1$. But $k = \omega$ is the wave number, and as $\omega \sim 1/ma^2$ and $v \sim 1/ma$, hence

$$ka \sim v\,, \tag{138}$$

which is small as stated in No. 4. The quadratic term $H_2 \sim (e^2/m)\omega^2 \sim \alpha m v^4$, and therefore $H_2/H_1 \sim \sqrt{\alpha}v$, while the second-order contribution to transition amplitudes relative to first order is $\sim H_1/\Delta \sim \sqrt{\alpha}v$, as stated in No. 5, which then leads to statement 6.

(b) Transition Rates

In the remainder of this section we concern ourselves solely with the leading approximation to one-photon transitions, i.e., with H_1 to first order. It will be convenient to write this interaction as

$$H_1 = -e\sum_{k\lambda}\frac{1}{\sqrt{2Vk}}[J_{k\lambda}a_{k\lambda} + \text{h.c.}]\,, \tag{139}$$

where the operator $J_{k\lambda}$ only involves the degrees of freedom of the source:

$$J_{k\lambda} = \sum_a \left(\frac{q_a}{m_a}\boldsymbol{p}_a \cdot \boldsymbol{e}_{k\lambda} + i\boldsymbol{\mu}_a \cdot (\boldsymbol{k}\times \boldsymbol{e}_{k\lambda})\right) e^{i\boldsymbol{k}\cdot\boldsymbol{r}_a}\,. \tag{140}$$

Here q_a is the charge in units of e, so that $q = -1$ for electrons. Note that $J_{k\lambda}$ is dimensionless because $(Vk)^{-\frac{1}{2}}$ has the dimension of an inverse length, and therefore a mass or energy when $\hbar = c = 1$. All the rates to be evaluated here follow from the Golden Rule:

$$\Gamma_{i\to f} = 2\pi\,|\langle f|H_1|i\rangle|^2\,\delta(E_i - E_f)\,. \tag{141}$$

Consider first *spontaneous emission,* the case when there are no photons in the initial state. The initial and final states of the field are then the vacuum $|0\rangle$ and the state $|1_{k\lambda}\rangle \equiv a^\dagger_{k\lambda}|0\rangle$ with one photon of momentum $\boldsymbol{k}$ and helicity λ. The initial and final states of the source are $|i, f\rangle$, with energies $E_{i,f}$. The corresponding product states are written as

$$|i;0\rangle \equiv |i\rangle \otimes |0\rangle, \qquad |f;1_{k\lambda}\rangle \equiv |f\rangle \otimes a^\dagger_{k\lambda}|0\rangle\,. \tag{142}$$

The transition rate is given by the Golden Rule,

$$d\Gamma_{\text{sp};fi} = 2\pi|\langle f;1_{k\lambda}|H_1|i;0\rangle|^2\rho_f^{(\gamma)}\,. \tag{143}$$

The density of final photon states is

$$\rho_f^{(\gamma)} = \frac{V}{(2\pi)^3}\frac{d^3k}{d\omega} = \frac{V}{(2\pi)^3}\,k^2 d\Omega\,, \tag{144}$$

where $k = E_i - E_f$ and $d\Omega$ is the element of solid angle into which the photon propagates. Here the width symbol Γ has been used for the rate, instead of $\dot{P}$, as these quantities are identical when $\hbar = 1$.

Only one creation operator in H_1 can contribute to (143), because

$$\langle 0|a_{k\lambda}a^{\dagger}_{k'\lambda'}|0\rangle = \langle 0|[a_{k\lambda}, a^{\dagger}_{k'\lambda'}]|0\rangle = \delta_{kk'}\delta_{\lambda\lambda'} , \tag{145}$$

which takes care of the photon matrix element. The whole transition element is therefore

$$\langle f; 1_{k\lambda}|H_1|i; 0\rangle = -e\,\frac{1}{\sqrt{2Vk}}\,\langle f|J^{\dagger}_{k\lambda}|i\rangle , \tag{146}$$

where the last factor is a matrix element in the source subspace. The result for the spontaneous emission rate when the photon's direction and polarization are specified is thus

$$\frac{d\Gamma_{\text{sp};fi}}{d\Omega} = \frac{\alpha}{2\pi}\,k\,|\langle f|J^{\dagger}_{k\lambda}|i\rangle|^2 . \tag{147}$$

The *width* Γ_{fi} for the transition $i \to f$ is, by definition, the sum of the rates for all directions of emission and both polarizations, and over all degenerate final states should $|f\rangle$ be a member of a multiplet, as is often the case:

$$\Gamma_{\text{sp};fi} = \sum_{[f]}\sum_{\lambda}\int d\Omega\,\frac{d\Gamma_{fi}}{d\Omega} , \tag{148}$$

where $[f]$ indicates the members of the degenerate multiplet. The *total radiative width* Γ^{γ}_i is sum over all states $|f\rangle$ that can be reached via spontaneous photon emission, with each transition having its appropriate $k_{fi} = E_i - E_f$. If $|i\rangle$ can only decay by photon emission, the state's *lifetime* is

$$\tau_i = 1/\Gamma^{\gamma}_i . \tag{149}$$

When other decay modes are available (e.g., internal conversion accompanying γ-emission by nuclei), the total rates for all these modes sum to the inverse of the lifetime. Frequently $|i\rangle$ is also a member of a degenerate multiplet. If that is so, and the multiplet is not split by an external field (as in the Zeeman effect), the rate that is actually measured is the average over the members of the initial multiplet.

Stimulated emission occurs when the initial state has a finite photon population at the appropriate frequency. It stems from the basic oscillator matrix element

$$\langle n_{k\lambda}+1|a^{\dagger}_{k'\lambda'}|n_{k\lambda}\rangle = \delta_{kk'}\delta_{\lambda\lambda'}\sqrt{n_{k\lambda}+1} . \tag{150}$$

Hence if photons are present, (146) must be replaced by

$$\langle f; n_{k\lambda}+1|H_1|i; n_{k\lambda}\rangle = -e\,\frac{\sqrt{n_{k\lambda}+1}}{\sqrt{2Vk}}\,\langle f|J^{\dagger}_{k\lambda}|i\rangle . \tag{151}$$

The factor $\sqrt{n+1}$ is a hallmark of Bose statistics: the probability is enhanced by a pre-existing population of bosons. Were the photon a fermion, it would be impossible to radiate into an already occupied state.

The emission rate when photons are already present only differs from the spontaneous rate in that $|\langle f|H_1|i\rangle|^2$ has the extra factor $(n_{k\lambda}+1)$. To implement the Golden Rule, Eq. 141 must be integrated over an interval of final momenta, and this factor will only make a difference if every element in this interval has $n_{k\lambda} \neq 0$, i.e., if there is a mean number of photons $\bar{N}_{k\lambda}$ in the neighborhood of $\boldsymbol{k}$ with the requisite polarization. Thus a finite number of photons cannot produce stimulated emission unless the radiating system is in an enclosure, which would make the photon spectrum discrete. The general result for the rate of radiation into an infinite volume is therefore

$$\frac{d\Gamma_{\text{em};fi}}{d\Omega} = \frac{\alpha}{2\pi} k\, |\langle f|J^\dagger_{k\lambda}|i\rangle|^2 (\bar{N}_{k\lambda} + 1) , \tag{152}$$

which incorporates both spontaneous and stimulated emission.

Instead of the mean photon number $\bar{N}_{k\lambda}$, it is often more useful to characterize the initial state by its energy flux $I_{k\lambda}$ per unit area, per unit time in the interval $d\omega$. By definition, $(c/V)\hbar\omega\, \bar{N}\, V d^3k/(2\pi)^3 \equiv I\, d\omega\, d\Omega$, and therefore

$$I_{k\lambda} = \bar{N}_{k\lambda} k^3/(2\pi)^3 , \qquad (\hbar = c = 1) . \tag{153}$$

Finally, we come to *absorption*. When the final state of the absorber is discrete, the familiar form of the Golden Rule fails because there is no density of final states. But the initial photon states form a continuum, and the integral over any finite interval of incident energies will also remove the δ-function in (141). That is, for a set of initial states I, we define the rate by

$$\Gamma_{I\to f} = \int dE_i \left[(dN_i/dE_i)\, \Gamma_{i\to f}\right] = 2\pi\, |\langle f|H_1|i\rangle|^2 \rho_i , \tag{154}$$

where

$$\rho_i = \frac{V d^3k_i}{(2\pi)^3 dE_i} = \frac{V k^2 d\Omega}{(2\pi)^3} , \tag{155}$$

the latter holding only for a zero-mass incident particle, such as a photon. For absorption the transition matrix element is again given by the familiar oscillator result,

$$\langle f; n_{k\lambda} - 1|H_1|i; n_{k\lambda}\rangle = -e\frac{\sqrt{n_{k\lambda}}}{\sqrt{2Vk}} \langle f|J_{k\lambda}|i\rangle . \tag{156}$$

Hence the absorption rate is

$$\Gamma_{\text{abs};fi} = \frac{\alpha}{2\pi} k\, |\langle f|J_{k\lambda}|i\rangle|^2 \, \bar{N}_{k\lambda} . \tag{157}$$

The matrix elements that appear in the absorption and emission rates are very similar, but not identical: the former involves $J_{k\lambda}$, the latter $J^\dagger_{k\lambda}$. Unless something else is brought to bear, there is no generally valid relation between the emission and absorption rates. Time reversal is this "something else." This symmetry holds in both atomic and nuclear physics. The corresponding transformation I_t is antiunitary, and imposes the following relations on the term in (140) due to charge transport:

$$\langle f|\boldsymbol{p}\, e^{i\boldsymbol{k}\cdot\boldsymbol{r}}|i\rangle = \langle i|e^{-i\boldsymbol{k}\cdot\boldsymbol{r}}(-\boldsymbol{p})|f\rangle = -\langle f|\boldsymbol{p}\, e^{i\boldsymbol{k}\cdot\boldsymbol{r}}|i\rangle^* . \tag{158}$$

This generalizes to the whole operator $J_{k\lambda}$:

$$|\langle f|J_{k\lambda}|i\rangle|^2 = |\langle i|J^\dagger_{k\lambda}|f\rangle|^2 . \tag{159}$$

Therefore the emission and absorption rates satisfy a famous relation due to Einstein:

$$\frac{d\Gamma_{\text{abs}}}{d\Gamma_{\text{em}}} = \frac{\bar{N}_{k\lambda}}{\bar{N}_{k\lambda}+1} . \tag{160}$$

It holds separately for each polarization and element of solid angle, and for any single pair of initial and final states should they be in degenerate multiplets. Although the relation is often derived in the electric dipole approximation, it holds before any multipole expansion is performed, which is not surprising as Einstein derived it as a consequence of thermodynamic equilibrium.

(c) Dipole Transitions

We have already established that radiation emitted and absorbed by nonrelativistic systems has wavelengths long compared to the dimension a of the system. Hence the phase factors in $J_{k\lambda}$ can be expanded in a power series, the successive terms being of decreasing order in ka. The leading term is

$$J_{k\lambda} \simeq \sum_a q_a(\dot{\boldsymbol{r}}_a \cdot \boldsymbol{e}_{k\lambda}) \tag{161}$$

where we used $\boldsymbol{p}_a/m_a = \dot{\boldsymbol{r}}_a$ because in perturbation theory the field $\boldsymbol{A}$ is dropped from the source's equations of motion. Now

$$\langle f|\dot{\boldsymbol{r}}_a|i\rangle = -i\langle f|[\boldsymbol{r}_a, H_M]|i\rangle = i(E_f - E_i)\langle f|\boldsymbol{r}_a|i\rangle = \pm ik\,\langle f|\boldsymbol{r}_a|i\rangle \tag{162}$$

because $k = |E_i - E_f|$ in both emission and absorption.

In the long-wavelength limit, therefore, all radiative transitions result from the electric dipole moment of the source, i.e., from

$$\boldsymbol{D} = \sum_a q_a \boldsymbol{r}_a . \tag{163}$$

$\boldsymbol{D}$ is a polar vector. Consequently, in transitions of an isolated system between states $|i,f\rangle$ that are eigenstates of angular momentum and parity, the leading order transitions are those that satisfy *the electric dipole selection rule:*

$$|j_f - j_i| = 0,1\,; \quad |m_f - m_i| = 0,1\,; \quad \text{parity change} . \tag{164}$$

These are called *allowed* transitions. Returning to Eq. 147, the rate for electric dipole (E1) spontaneous emission is thus

$$\frac{d\Gamma_{\text{sp E1}}}{d\Omega} = \frac{\alpha}{2\pi}\,k^3\,|\langle f|\boldsymbol{D}\cdot\boldsymbol{e}^*_{k\lambda}|i\rangle|^2 . \tag{165}$$

Virtually all the emission and absorption lines seen in conventional atomic spectroscopy are allowed E1 transitions. States that do not satisfy the E1 selection rules can be connected by *"forbidden"* one-photon radiative transition. Higher multipoles are required, and will give rates slower by some power of $(ka)^2$. The preceding statement does not hold for one-photon transitions between $j = 0$ states, which are rigorously forbidden by angular momentum conservation because there are no one-photon states with zero helicity.

In many circumstances, neither the photon polarization nor the angular momentum projections of the participating states are known, and what is actually measured is the following average emission rate:

$$\frac{d\Gamma_{\text{sp E1}}}{d\Omega} = \frac{\alpha k^3}{2\pi}\frac{1}{2j_i+1}\sum_{m_i m_f \lambda} |\langle n_f j_f m_f | \boldsymbol{D}\cdot\boldsymbol{e}^*_{k\lambda} | n_i j_i m_i\rangle|^2 \,, \tag{166}$$

where $n_{i,f}$ are the quantum numbers aside from angular momentum. This quantity cannot depend on the direction of the emitted photon because the average over m_i produces a rotationally invariant ensemble. Hence the total rate is just the right-hand side times 4π. The Wigner-Eckart theorem can be used to carry out the remaining sum in (166), the final result being

$$\Gamma_{\text{sp E1}} = \frac{4\alpha k^3}{3}\frac{1}{2j_i+1}|\langle n_f j_f \,\|D\|\, n_i j_i\rangle|^2 \,. \tag{167}$$

Of course there are important situations in which the emitter is not an isolated system, and the states in an angular momentum multiplet are not degenerate. The most familiar example is the Zeeman effect, in which the transitions that are summed in (166) have different wavelengths, and each line has, in general, a distinct polarization that depends on the angle between the direction of the magnetic field and the photon momentum (see Prob. 10).

As already stated, when the E1 selection rule is not satisfied, a higher-order electric or magnetic multipole will, in general, produce a transition, though the rate may be unobservably low. The next-to-leading term is the magnetic dipole, M1. If the particle is charged, the M1 transition amplitude is as a coherent superposition of two terms: (i) the coupling of the intrinsic magnetic moment with the phase factor dropped from the spin term in (140); and (ii) the correction of order kr to the approximation (161). Both of these amplitudes are smaller than an E1 amplitude by one power of ka, the first because $\boldsymbol{B} = \boldsymbol{\nabla}\times\boldsymbol{A}$, the second explicitly so. They therefore give a rate slower by $(ka)^2$. The magnetic moment is an axial vector, and therefore *the M1 selection rule is*

$$|j_f - j_i| = 0, 1\,; \quad |m_f - m_i| = 0, 1\,; \quad \text{no parity change}\,. \tag{168}$$

The higher-order multipoles are tensor operators of rank $l = 2, 3, \dots$, and have a form essentially identical to that in classical electrodynamics.[1] The angular momentum selection rule follow from the "triangular inequality" of the Wigner-Eckart theorem: if l is the multipole order,

$$|j_i - j_f| \leq l \leq j_i + j_f \,. \tag{169}$$

Magnetic and electric multipoles of the same order have opposite reflection property, as exemplified by E1 and M1; this is so because $\boldsymbol{A}$ and $\boldsymbol{B}$ are polar and axial vector fields, respectively. The general result is

$$\text{E}l : \text{parity change} = (-1)^l \,, \qquad \text{M}l : \text{parity change} = -(-1)^l \,. \tag{170}$$

For example, if $J^\pm$ stands for the angular momentum and parity of a state, the transition $4^- \to 1^+$ is E3, whereas $4^+ \to 1^+$ is M3. Furthermore, the transition rates for El vary like k^{2l+1} (compare Eq. 167), and like k^{2l+3} for Ml.

---

[1]See Jackson, pp. 429–444.

10.5 Quantum Optics

Quantum optics is concerned with optical phenomena in which the quantum nature of the electromagnetic field plays an important role. This field has undergone dramatic growth in recent years, especially on the experimental front. Our attention will be narrowly confined to some aspects of quantum optics which have made possible many laboratory (not just *Gedanken*) experiments that touch on the foundations of quantum mechanics. Readers interested in a broader knowledge of quantum optics should consult the bibliography.

(a) The Beam Splitter

The beam splitter, among the simplest of optical devices, already demonstrates that classical electrodynamics is a poor guide when the optical state is one with a definite number of photons. This is to be expected, given that the photon concept is so foreign to classical wave optics. Indeed, the quantum theory of the beam splitter has some surprising and quite subtle features.

Let $\phi_{1,2,3}$ be classical fields entering and leaving the beam splitter, and let T and R be the complex transmission and reflection amplitudes. Then

$$\phi_2 = T\phi_1 , \qquad \phi_3 = R\phi_1 , \qquad |T|^2 + |R|^2 = 1 , \tag{171}$$

where we assume that the beam splitter is lossless. To describe how the beam splitter effects states with a definite number of photons, we must, at a minimum, introduce creation and destruction operators for the three channels. If they are related as in (171), that would lead to

$$[a_2, a_3^\dagger] = TR^*[a_1, a_1^\dagger] \neq 0 , \tag{172}$$

which is nonsense.

What went wrong? The error is that quantization requires the field to be resolved into a *complete* set of states, whose amplitudes are then replaced by destruction and creation operators. The "paradox" of Eq. 172 is due to our not having included a complete set in the description of the beam splitter, as one can surmise by noting that in the foiled attempt there are two outgoing states and just one ingoing state, which amounts to a change in dimension of the Hilbert space as a function of time.

The easiest way of divining the correct treatment is to consider reflection and transmission of light in one dimension by a dielectric plate that has a real index of refraction (Fig. 10.3a). Modes with different frequency are decoupled, and it suffices to consider waves of one frequency; for our purpose polarization can be ignored. In the regions outside the plate, four distinct wave forms can be present, two running toward the origin,

$$x \to -\infty : \quad u_1 \sim e^{ikx} , \qquad x \to +\infty : \quad u_0 \sim e^{-ikx} , \tag{173}$$

and two running away from it,

$$x \to -\infty : \quad u_3 \sim e^{-ikx} , \qquad x \to +\infty : \quad u_2 \sim e^{ikx} . \tag{174}$$

For an arbitrary incoming wave with amplitudes c_1 and c_0, the outgoing state is

$$\psi(x) = (Tc_1 + Rc_0)u_2(x) + (Rc_1 + Tc_0)u_3(x) . \tag{175}$$

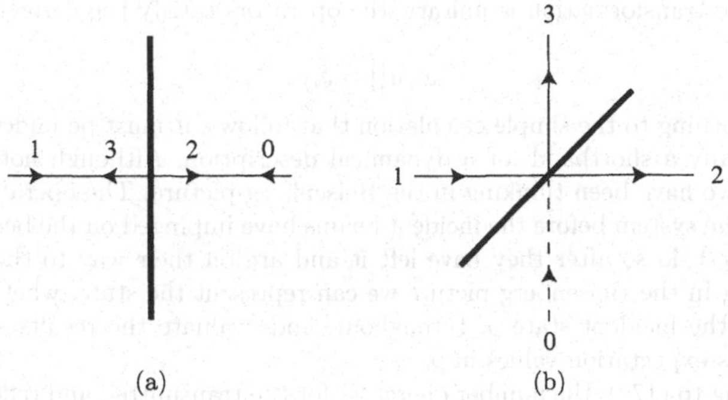

FIG. 10.3. Reflection and transmission by a plate normal to the incident state(a); and (b) a beam splitter at $\pi/4$.

Because there are no losses, conservation of energy requires

$$|c_1|^2 + |c_0|^2 = |Tc_1 + Rc_0|^2 + |Rc_1 + Tc_0|^2 , \tag{176}$$

and a bit of algebra shows that this, in turn, requires

$$|T|^2 + |R|^2 = 1 , \qquad \mathrm{Re}\, T^* R = 0 . \tag{177}$$

These conditions were already derived for one-dimensional potential scattering in §4.4(a) from the continuity equation, which is an equivalent consideration.

In short, the amplitudes of the outgoing and incoming waves are related to each other as follows:

$$\begin{pmatrix} c_2 \\ c_3 \end{pmatrix} = S \begin{pmatrix} c_1 \\ c_0 \end{pmatrix}, \qquad S = \begin{pmatrix} T & R \\ R & T \end{pmatrix}. \tag{178}$$

The matrix S is unitary in virtue of (177), though it is not the most general two-dimensional unitary matrix. It is nothing but the S matrix for this problem.

When the field is quantized, the amplitudes of all four waves must be included in the field operator, and (172) is nonsense because destruction and creation operators for *all four* channels must be introduced, and not just those that happen to be involved in a particular process. Should we be concerned with a *state* in which a particle is incident in channel 1, and no particle is incident in the other available incident channel 0, the *operators* pertaining to the empty mode cannot be blithely cast aside just because the mode happens to be unoccupied.

We now apply this lesson to the beam splitter. Its proper description requires four destruction operators a_i, $i = 0, 1, 2, 3$, corresponding to the channels shown in Fig. 10.3b. They are again related by a unitary matrix S if the device is lossless. The matrix has exactly the same form as in (178), though of course the functions $T(k)$ and $R(k)$ are not the same as in the one-dimensional exercise, but they must again satisfy the unitarity conditions (176). Hence

$$\begin{pmatrix} a_2 \\ a_3 \end{pmatrix} = \begin{pmatrix} Ta_1 + Ra_0 \\ Ra_1 + Ta_0 \end{pmatrix}, \qquad \begin{pmatrix} a_2^\dagger \\ a_3^\dagger \end{pmatrix} = \begin{pmatrix} T^* a_1^\dagger + R^* a_0^\dagger \\ R^* a_1^\dagger + T^* a_0^\dagger \end{pmatrix}. \tag{179}$$

Because the transformation is unitary, the operators satisfy the correct commutation rules:

$$[a_i, a_j^\dagger] = \delta_{ij} \; . \tag{180}$$

Before turning to the simple calculation that follows, it must be understood that (179) is really a shorthand for a dynamical description. Although not yet stated explicitly, we have been thinking in the Heisenberg picture. The operators (a_0, a_1) represent the system before the incident beams have impinged on the beam splitter, while (a_2, a_3) do so after they have left it and are on their way to the detectors. By working in the Heisenberg picture we can represent the state, whether pure or mixed, by the incident state ρ_i throughout, and evaluate the results seen by the detectors as expectation values in ρ_i.

According to (179), the number operators for the transmitted and reflected channels are

$$N_2 = |T|^2 N_1 + |R|^2 N_0 + X \; , \tag{181}$$

$$N_3 = |R|^2 N_1 + |T|^2 N_0 + Y \; , \tag{182}$$

$$X = T^* R \, a_1^\dagger a_0 + R^* T \, a_0^\dagger a_1 \; , \tag{183}$$

$$Y = R^* T \, a_1^\dagger a_0 + T^* R \, a_0^\dagger a_1 \; . \tag{184}$$

The total outgoing and incoming number operators are equal, $N_1 + N_0 = N_2 + N_3$, because

$$T^* R = ir \tag{185}$$

with r real. This, of course, is contingent on our assumption that the beam splitter is lossless, i.e., that the matrix S is unitary.

As an instructive check on what has just been done, consider a state with just one photon incident, which cannot display coincidences between the channels 2 and 3. In the quantum optics literature, eigenstates of the number operators, such as this state, are called *Fock states* (with dubious historical justification). Let $\doteq$ stand for equal in the state for which $n_1 = 1, n_0 = 0$; then

$$N_2 \doteq |T|^2 + X \; , \qquad N_3 \doteq |R|^2 + Y \; . \tag{186}$$

Note that X and Y have vanishing expectation values in any Fock state. Therefore

$$\langle N_2 N_3 \rangle_{10} = r^2 + \langle XY \rangle_{10} \; , \tag{187}$$

where $\langle \ldots \rangle_{10}$ is the expectation value in the incident state. Using the commutation rule (180),

$$XY = -r^2 (2 N_1 N_0 + N_1 + N_0) + Z \; , \tag{188}$$

where Z is has no diagonal matrix elements in Fock states. Therefore $\langle XY \rangle_{10} = -r^2$, and by combining this with (187) we have the required result,

$$\langle N_2 N_3 \rangle_{10} = 0 \; . \tag{189}$$

(b) Various States of the Field

The Fock states are a complete set, but they do not give the most convenient description of many important phenomena. Lasers are the light source in many experiments; their output can be modeled crudely by a pure state, but not one that is a linear combination of a finite number of Fock states. Most other sources have an output that is a statistical mixture of a broad spectrum of Fock states.

The quantum theory of the laser is a topic that is beyond the scope of this volume. A physical argument, due to Glauber, leads to a simple model of laser output that is adequate for our needs.[1] Namely, the field is coupled within the laser to the electric dipoles of the atoms and produces an oscillating polarization density of macroscopic proportions; the time derivative of a polarization density is a current distribution,[2] and this is of classical magnitude when the laser is operating well above its threshold. Furthermore, assuming the laser to have ideal stability, this current oscillates steadily in a perfectly predictable way. In other words, we may describe the current as a prescribed c-number current density.

Let $\boldsymbol{j}(\boldsymbol{r},t)$ be any such classical current density. To leading order in $\boldsymbol{A}$, the Hamiltonian governing its interaction with the field can be cast into the form[3]

$$H_1(t) = -\int d\boldsymbol{r}\, \boldsymbol{j}(\boldsymbol{r},t) \cdot \boldsymbol{A}(\boldsymbol{r}) \,. \tag{190}$$

This is written in the Schrödinger picture. On substituting the plane wave representation of the field operator (Eq. 135), this becomes

$$H_1(t) = \frac{1}{\sqrt{2}} \sum_{k\lambda} \{ f_{k\lambda}(t) a_{k\lambda} + f_{k\lambda}(t) a_{k\lambda}^{\dagger} \} , \tag{191}$$

$$f_{k\lambda}(t) = -\frac{1}{\sqrt{Vk}} \int d\boldsymbol{r}\, e^{i(\boldsymbol{k}\cdot\boldsymbol{r}-\omega t)}\, \boldsymbol{j}(\boldsymbol{r},t) \cdot \boldsymbol{e}_{k\lambda} \,. \tag{192}$$

The interaction H_1 produces no coupling between modes of different momentum or helicity, and each individual mode is simply a harmonic oscillator moving under the influence of a prescribed time-dependent force $f_{kl}(t)$. This is the problem already solved in §4.2(c), with the same definition of the force. Each mode is in a coherent state, specified by the complex parameter

$$z_k(t) = \frac{1}{i\sqrt{2}}\, e^{-i\omega t} \int_{-\infty}^{t} dt'\, e^{i\omega t'}\, f_k(t') \,, \tag{193}$$

where k stands for $(\boldsymbol{k}, \lambda)$. Therefore the state of the field produced by the classical current is the direct product of coherent states for all modes:

$$|t\rangle = e^{i\alpha(t)} \prod_k D_k(t)\, |0\rangle \equiv |\{z_k(t)\}\rangle \,, \tag{194}$$

where $|0\rangle$ is the vacuum state, $\alpha(t)$ is the (ignorable) sum of phases mentioned in §4.2(c), and D_k the unitary displacement operator

$$D_k(t) = \exp\left[z_k(t)\, a_k^{\dagger} - z_k^{*}(t)\, a_k \right] . \tag{195}$$

---

[1] R.J. Glauber in *Quantum Optics and Electronics*, C. DeWitt (ed.), Gordon and Breach (1965), p. 163.

[2] Jackson, pp. 255-6.

[3] This is (135) when the convection and magnetic dipole terms are replaced by continuous distributions, and in the latter the derivative of $\boldsymbol{A}$ is integrated by parts.

Because the coherent states are eigenstates of the annihilation and creation operators, the field operators $\boldsymbol{E}(\boldsymbol{r})$ and $\boldsymbol{B}(\boldsymbol{r})$ have expectation values in the state $|t\rangle$, and these expectation values agree exactly with the field strengths predicted for this phenomenon by classical electrodynamics.

Recall that in a coherent state the occupation numbers have a population that is a Poisson distribution. Therefore the state of the radiation field due to a classical current is a product of Poisson distributions,

$$p(\{n_k\},t) = \prod_k e^{-\bar{N}_k(t)} \frac{[\bar{N}_k(t)]^{n_k}}{n_k!} , \tag{196}$$

where

$$\bar{N}_k(t) = |z_k(t)|^2 \tag{197}$$

is the mean occupation number in this mode. Recall, furthermore, that in a Poisson distribution the dispersion or variance in occupation numbers is

$$\Delta N_k = \sqrt{\bar{N}_k}. \tag{198}$$

The Poisson distribution appears here because the source of the radiation is, by hypothesis, prescribed and does not change even though it is radiating. As a consequence, the probability of emission of any photon does not depend on what was radiated previously.

Of course, a system as complex as a real laser cannot be expected to be a perfectly stable system, and to radiate a field that is in a pure quantum state. For example, the phase of the oscillating polarization density may be chaotic, even if the amplitude is stable, and as a consequence produce a radiation field that is not in a pure state. Such thoughts motivate the Glauber-Sudarshan representation of the density matrix as a sum of projection operators onto coherent states,

$$\rho = \int d^2z\, \wp(z)|z\rangle\langle z| . \tag{199}$$

This is written here for one mode, with an obvious extension to all modes. As ρ is Hermitian, $\wp$ must be real. Other properties of $\wp$ follow from relating (199) to the n-basis. If $\wp$ is only a function of $|z|$, $\wp$ is also diagonal in the Fock basis:

$$\rho = \sum_n p_n\, |n\rangle\langle n| , \qquad p_n = \int d^2z\, \wp(|z|)\, |\langle z|n\rangle|^2 . \tag{200}$$

But $|\langle z|n\rangle|^2 = e^{-|z|^2}|z|^{2n}/n!$, and therefore

$$1 = \int d^2z\, \wp(z) , \tag{201}$$

$$p_n = \int d^2z\, \wp(z)\, e^{-|z|^2}\, |z|^{2n}/n! . \tag{202}$$

In short, if $\wp(z)$ only depends on the amplitude $|z|$ of the mode, and not on its phase arg z, the density matrix has no off-diagonal elements in the number basis, and as a consequence the field strengths must have vanishing expectation values in such states. One example of such a state is Eq. 203, which could model the radiation by a classical current of fixed amplitude but random phase.

The last expression for p_n shows that the function $\wp(z)$ is *not* a probability distribution in general, for were it to always satisfy the condition $\wp \geq 0$, it would follow that in all states $p_n \neq 0$ for all n. Hence any state with a finite number of photons does not have a positive semi-definite $\wp$, and for such states $\wp$ cannot be interpreted as a probability distribution. This is not too surprising, because in §4.2(d) we learned that the integration over the z-plane corresponds to an integration over phase space, and it is not possible, in general, to define a quantum mechanical phase space distribution. In fact, there is a relationship between $\wp$ and the Wigner phase space distribution (§2.2(f)); recall that the latter is also not positive definite in general.

There are, however, important states of the field for which $\wp$ is a probability distribution. For example, the field produced by a source whose amplitude A is definite but phase is random can be represented in the form of (199) with

$$\wp(z) = \frac{1}{2\pi A}\,\delta(A - |z|)\,. \tag{203}$$

Another very important example is *black body radiation*, for which the density matrix of the field will be derived in §11.3. The modes are uncoupled, and the complete density matrix is a direct product over all modes of

$$\rho_k = (1 - e^{-\beta\omega}) \sum_{n=0}^{\infty} e^{-n\beta\omega}\, |n\rangle\langle n|\,, \tag{204}$$

where $\omega = |\boldsymbol{k}|$, $\beta = 1/k_B T$, and k_B is Boltzmann's constant. The mean occupation number is given by the Planck distribution law,

$$\bar{N}_k = \frac{1}{e^{\beta\omega} - 1}, \tag{205}$$

and therefore the probability of finding n photons in the mode k is

$$p_n = \frac{(\bar{N}_k)^n}{(1 + \bar{N}_k)^{n+1}} \to \frac{1}{\bar{N}_k}\,, \qquad \bar{N}_k \gg 1\,. \tag{206}$$

The dispersion in N, whose evaluation is left as an exercise, is

$$\Delta N_k = \sqrt{\bar{N}_k(\bar{N}_k + 1)} \to \bar{N}_k\,, \qquad \bar{N}_k \gg 1\,. \tag{207}$$

This is to be compared with the result for the coherent state (Eq. 198); briefly put, thermal radiation has a far broader distribution than does the radiation from an ideal laser.

As a consequence of (202), thermal radiation has a Gaussian coherent state distribution,

$$\wp(z_k) = \frac{1}{\pi \bar{N}_k}\, e^{-|z_k|^2/\bar{N}_k}\,, \tag{208}$$

as can be confirmed by substituting into (202). Such a Gaussian distribution is characteristic of a chaotic system, and holds for a large variety of light sources.

(c) Photon Coincidences

We return to the beam splitter, and first consider the case where an *arbitrary beam* is incident in channel 1, channel 0 is empty, and coincidences between the exit channels 2 and 3 are counted. In the notation of (186),

$$N_2 \doteq |T|^2 N_1 + X \,, \quad N_3 \doteq |R|^2 N_1 + Y \,, \quad \langle XY \rangle = -r^2 \langle N_1 \rangle \,, \tag{209}$$

where the latter follows from (188), and $\langle \cdots \rangle$ is the expectation value in the incident state. Then the number of coincidences is

$$\langle N_2 N_3 \rangle = r^2 \{ \langle N_1^2 \rangle - \langle N_1 \rangle \} \,. \tag{210}$$

The fluctuation in coincidences is

$$\Delta N_2 \, \Delta N_3 \equiv \langle (N_2 - \langle N_2 \rangle)(N_3 - \langle N_3 \rangle) \rangle = \langle N_2 N_3 \rangle - \langle N_2 \rangle \langle N_3 \rangle \,. \tag{211}$$

From Eq. 209, $\langle N_2 \rangle \langle N_3 \rangle = r^2 \langle N_1 \rangle^2$, and therefore

$$\Delta N_2 \, \Delta N_3 = r^2 \{ \langle N_1^2 \rangle - \langle N_1 \rangle^2 - \langle N_1 \rangle \} \tag{212}$$

$$= r^2 \{ (\Delta N_1)^2 - \langle N_1 \rangle \} \tag{213}$$

$$\equiv r^2 Q \,. \tag{214}$$

This simple result is remarkable, for it shows that this correlations of fluctuations distinguishes between the most important categories of states:

1. When the state incident is chaotic, like black body radiation, $Q = \langle N_1 \rangle^2$, i.e., Q is positive. That Q is positive for an ordinary light beam is responsible for the famous Hanbury Brown-Twiss effect.

2. When it is a coherent state, like that produced by an ideal laser, $Q = 0$.

3. When it is a Fock state, which has no fluctuations in N_1, then Q is negative.

In the quantum optics literature states of the electromagnetic field for which Q is positive, and more generally for which $\wp(z)$ is non-negative, are called classical, which is a bit confusing as the Planck distribution is hardly classical.

The last state we will consider here has one photon incident in both channels 0 and 1, $n_0 = n_1 = 1$. According to (181), in this case $N_2 \doteq 1 + X, N_3 \doteq 1 + Y$, and $\langle XY \rangle = -4r^2$. Using $|T|^2 + |R|^2 = 1$ gives

$$\langle N_2 N_3 \rangle = \left(|T|^2 - |R|^2 \right)^2 \,. \tag{215}$$

Thus for a 50:50 beam splitter, there are *no coincidences — in all events one photon is reflected and the other is transmitted, so that both go to the same counter.*

This astonishing result actually illustrates, yet again, that when a quantum state has two paths for reaching an outcome, and no step is taken to ascertain the path, the amplitudes for these alternatives must be added coherently. To have a coincidence, both photons have to be either reflected or transmitted, and the amplitudes for these two options must be added coherently. Now $T = e^{i\varphi_1}/\sqrt{2}$, $R = e^{i\varphi_2}/\sqrt{2}$ when $|T|^2 - |R|^2 = 0$, and (185) requires $\varphi_1 - \varphi_2 = \frac{1}{2}\pi$. Apart from a common

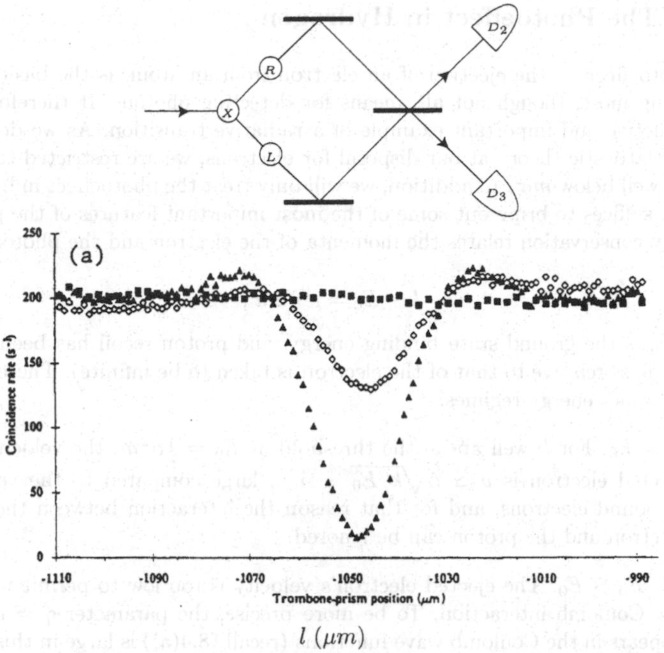

FIG. 10.4. Above, a cartoon version of the experiment by P.G. Kwiat, A.M. Steinberg and R.Y. Chiao, *Phys. Rev. A* **45**, 7729 (1992). The nonlinear crystal X produces two photons polarized in the plane of the page; R rotates the polarization through ϕ; and L is an optical "trombone" that provides a difference l in the length of the two paths. The coincidence rates as a function of l for $\phi = 0$ (▲), $\frac{1}{4}\pi$ (◇), $\frac{1}{2}\pi$ (■) are shown, demonstrating that the rate vanishes, within experimental errors, at $l = \phi = 0$ and grows as the distinguishability of the paths grows.

phase, the amplitude for the case where both photons are reflected or transmitted is therefore

$$\tfrac{1}{\sqrt{2}} \times \tfrac{1}{\sqrt{2}} + \tfrac{i}{\sqrt{2}} \times \tfrac{i}{\sqrt{2}} = 0\,, \tag{216}$$

whereas when one is reflected and the other transmitted the amplitude is

$$\tfrac{1}{\sqrt{2}} \times \tfrac{i}{\sqrt{2}} + \tfrac{i}{\sqrt{2}} \times \tfrac{1}{\sqrt{2}} = i\,. \tag{217}$$

An experiment confirming this effect is summarized in Fig. 10.4. It is, of course, no simple matter to produce a state in which one photon in each channel 0 and 1 is incident on the beam splitter in essence simultaneously. The trick is to use a nonlinear crystal, illuminated by a laser, which converts one photon into two having the *same* polarization. The experiment also allows the polarization of *one* of the incident photons to be rotated through an angle ϕ, so when $\phi = \frac{1}{2}\pi$ it becomes possible to *unambiguously* determine the path taken by the photons by measuring their polarization after passage through the beam splitter. In short, as ϕ grows from 0 to $\frac{1}{2}\pi$, the coherence of the amplitudes for the two paths shrinks to zero, and the coincidence rate is expected to increase correspondingly, which it does.

10.6 The Photoeffect in Hydrogen

The photoeffect — the ejection of an electron from an atom, is the basic process underlying most, though not all, means for detecting photons. It therefore offers an instructive and important example of a radiative transition. As we do not yet have a relativistic theory at our disposal for electrons, we are restricted to photon energies well below mc^2. In addition, we will only treat the photoeffect in hydrogen, but that suffices to bring out some of the most important features of the process.

Energy conservation relates the momenta of the electron and the photon, $\boldsymbol{p}$ and $\boldsymbol{k}$:

$$k - E_0 = p^2/2m \,, \tag{218}$$

where E_0 is the ground state binding energy, and proton recoil has been ignored (i.e., its mass relative to that of the electron is taken to be infinite). There are two rather distinct energy regimes:

1. $k \gg E_0$. For k well above the threshold at $E_0 = \frac{1}{2}\alpha^2 m$, the velocity of the ejected electron is $v \simeq \alpha\sqrt{k/E_0}$ — i.e., large compared to the velocity α of bound electrons, and for that reason the interaction between the ejected electron and the proton can be ignored.

2. $k - E_0 \lesssim E_0$. The ejected electron's velocity is too low to permit neglect of the Coulomb interaction. To be more precise, the parameter $\gamma = \alpha/v$ that appears in the Coulomb wave functions (recall §8.4(a)) is large in this regime, because
$$\frac{1}{\gamma} = pa_0 = \sqrt{\frac{k}{E_0} - 1} \,. \tag{219}$$
Nevertheless, the problem is tractable because in the threshold region the photon wavelength $\lambda\!\!\!^{-}$ is of order a_0/α, i.e., enormous compared to the atom's dimension, so only the lowest photon partial wave, with $j = 1$, interacts. That is, in this regime the photoelectron only emerges in a p wave. Furthermore, the low-energy calculation is valid up to $ka_0 \sim 1$, or $k \sim E_0/\alpha \gg E_0$, which overlaps the high-energy regime, and therefore the two calculations give a complete description of the process for all $k \ll m$.

In this whole nonrelativistic regime, the electron's spin can be ignored to leading order in v (see Eq. 137).

(a) High Energies

In the high-energy regime, the photon wavelength is not large compared to the size a_0 of the absorber, and the multipole expansion is not valid. Therefore the phase factor in the vector potential must be retained, and the matrix element for the process is to be taken from (156):

$$\langle \boldsymbol{p}; 0|H_1|1s; 1_{\boldsymbol{k}\lambda}\rangle = -\frac{e}{m}\frac{1}{\sqrt{2Vk}}\,(\boldsymbol{p}\cdot\boldsymbol{e}_{\boldsymbol{k}\lambda})\,\langle \boldsymbol{p}|e^{i\boldsymbol{k}\cdot\boldsymbol{r}}|1s\rangle \,, \tag{220}$$

where advantage was taken of the fact that the state $|\boldsymbol{p}\rangle$ of the photoelectron is a momentum eigenstate. The final factor in (220) is essentially the ground state momentum space wave function,

$$\langle \boldsymbol{p}|e^{i\boldsymbol{k}\cdot\boldsymbol{r}}|1s\rangle = \frac{1}{V^{\frac{1}{2}}}\int d^3r\, e^{i(\boldsymbol{k}-\boldsymbol{p})\cdot\boldsymbol{r}}\,\psi_0(r) \equiv \frac{1}{V^{\frac{1}{2}}}\,\varphi_0(q)\,, \tag{221}$$

evaluated at $\boldsymbol{q} = \boldsymbol{k} - \boldsymbol{p}$, the momentum transfer from photon to electron. This factor appears in the amplitude for the same reason as in other inelastic collisions in the Born approximation (see §9.1(c)). As $\psi_0(r) = (\pi a_0^3)^{-\frac{1}{2}}\, e^{-r/a_0}$,

$$\varphi_0(q) = 8\pi^{\frac{1}{2}}\,\frac{a_0^{3/2}}{(1+q^2a_0^2)^2}\,. \tag{222}$$

The differential cross section is then

$$d\sigma_\lambda = 2\pi|\langle f|H_1|i\rangle|^2\cdot\frac{Vd^3p}{(2\pi)^3dE_p}\cdot V\,, \tag{223}$$

where the factors are, from left to right, the rate, the density of final electron states, and the inverse of the photon flux. Hence

$$\frac{d\sigma_\lambda}{d\Omega} = \frac{\alpha}{2\pi}\,\frac{p}{km}\,|(\boldsymbol{p}\cdot\boldsymbol{e}_{k\lambda})\,\varphi_0(q)|^2\,, \tag{224}$$

that is, the area αa_0^2 times a kinematic factor. The angular distribution is given by the expression

$$\frac{|\boldsymbol{p}\cdot\boldsymbol{e}_{k\lambda}|^2}{[1+(\boldsymbol{p}-\boldsymbol{k})^2a_0^2]^4}\,. \tag{225}$$

The factor $[1+q^2a_0^2]^{-4}$ is proportional to the probability of finding an electron in the ground state with the momentum required by free space momentum conservation, and the numerator stems from the interaction between current and field, $\boldsymbol{p}\cdot\boldsymbol{A}$. The latter vanishes in the forward direction, i.e., when $\boldsymbol{p}$ is parallel to $\boldsymbol{k}$, which is required by angular momentum conservation. The initial state has $J_z = \pm 1$, where the z axis is taken along $\boldsymbol{k}$. Thus when the final electron state is expanded into angular momentum eigenstates, only $Y_{l,\pm 1}$ appears, and these all vanish at $\theta = \angle(\boldsymbol{p},\boldsymbol{k}) = 0$. Clearly, this must also hold when the final state is not free, as it will not be when we treat low energies.

When the photon beam is unpolarized, the cross section must be averaged over λ, which is done by using Eq. 79:

$$\tfrac{1}{2}\sum_\lambda |\boldsymbol{p}\cdot\boldsymbol{e}_{k\lambda}|^2 = \tfrac{1}{2}p^2\sin^2\theta\,, \qquad \theta - \angle(\boldsymbol{p},\boldsymbol{k})\,. \tag{226}$$

At high energies, furthermore, $pa_0 \gg 1$, and $p \gg k$, and therefore $1 + q^2a_0^2 \simeq (p^2 - 2pk\cos\theta)a_0^2$. The differential cross section for unpolarized photons in the high energy regime is therefore[1]

$$\frac{d\sigma}{d\Omega} = 32\alpha\, a_0^2\,(E_0/k)^{7/2}\,\sin^2\theta\,(1+4v\cos\theta)\,, \tag{227}$$

where v is again the velocity of the photoelectron. Hence the total cross section is

$$\sigma = \frac{256\pi}{3}\,\alpha\, a_0^2\left(\frac{E_0}{k}\right)^{7/2}\,. \tag{228}$$

[1] There is no discrepancy here with Bethe and Salpeter, Eq. 70.5; our cross section is an average over helicities, whereas theirs is not.

(b) The Cross Section Near Threshold

In this regime the wavelength is long compared to a_0, and the electric dipole approximation is valid. As already indicated, the influence of the Coulomb interaction on the ejected electron cannot be ignored, and the latter must therefore be described by a scattering state in the Coulomb field with asymptotic momentum $\boldsymbol{p}$.

At first sight one might think that this state has the continuum wave function $N\psi_{\boldsymbol{p}}(\boldsymbol{r}) = \langle \boldsymbol{r}|\boldsymbol{p}^{(+)}\rangle$ derived in §8.4, though here normalized to unity in the volume V. But that is not correct. The latter state has *outgoing scattered waves*, as required by causality when the *initially prepared state* is a free particle wave packet; in the photoeffect, by contrast, the detector selects a free particle wave packet *subsequent* to the collision, and as we learned in §8.2(c), such a state has *incoming scattered waves*. The idealized energy eigenstate that has this asymptotic form will here again be called $|\boldsymbol{p}^{(-)}\rangle$, and is related to the one with incoming scattered waves by time reversal, as shown in §8.2(d): $|\boldsymbol{p}^{(-)}\rangle = I_t|-\boldsymbol{p}^{(+)}\rangle$. Therefore the wave functions are related as follows:

$$\langle \boldsymbol{p}^{(-)}|\boldsymbol{r}\rangle = \langle \boldsymbol{r}|-\boldsymbol{p}^{(+)}\rangle = N\psi_{-\boldsymbol{p}}(\boldsymbol{r}) \,, \tag{229}$$

and the dipole matrix element is[1]

$$\langle \boldsymbol{p}^{(-)}|\boldsymbol{e}_{k\lambda}\cdot\boldsymbol{r}|1s\rangle = N\int d^3r\, \psi_{-\boldsymbol{p}}(\boldsymbol{r})\,(\boldsymbol{e}_{k\lambda}\cdot\boldsymbol{r})\,\psi_0(r)\,. \tag{230}$$

The factor N is $\sqrt{(2\pi)^3/V}$, because $\psi_{\boldsymbol{p}}$ was δ-function normalized in Eq. 233, and therefore

$$N\psi_{-\boldsymbol{p}}(\boldsymbol{r}) = \sqrt{4\pi/V}\sum_{l=0}^{\infty} i^l\sqrt{2l+1}\,C_l(p;r)Y_{l0}(\pi-\beta)\,, \tag{231}$$

where $\beta = \angle(\boldsymbol{r},\boldsymbol{p})$. Using $Y_{l0}(\pi-\beta) = (-1)^l Y_{l0}(\beta)$ and the addition theorem for spherical harmonics,

$$N\psi_{-\boldsymbol{p}}(\boldsymbol{r}) = 4\pi V^{-\frac{1}{2}}\sum_{lm} i^{-l}C_l(p;r)Y_{lm}(\theta\phi)Y^*_{lm}(\vartheta\varphi)\,, \tag{232}$$

where all angles are shown in Fig. 10.5. From this figure, and the definition (16) of the polarization vectors,

$$\boldsymbol{e}_{k\lambda}\cdot\boldsymbol{r} = r\sqrt{4\pi/3}\,Y_{1\lambda}(\vartheta\varphi)\,, \tag{233}$$

and therefore

$$\langle \boldsymbol{p}^{(-)}|\boldsymbol{e}_{k\lambda}\cdot\boldsymbol{r}|1s\rangle = -i\sqrt{\frac{64\pi^3}{3V}}Y_{1\lambda}(\theta\phi)\int_0^\infty C_1(p;r)\,r\,\psi_0(r)\,r^2dr\,. \tag{234}$$

As already anticipated, in the threshold region the only final states are p-waves with $m=\lambda$.

---

[1]It should be said that in this particular problem the use of the wrong final state $|\boldsymbol{p}^{(+)}\rangle$ leads to an amplitude with the wrong phase, but to the correct cross section. That does not mean that this is always an innocent error –e.g. in low-energy bremsstrahlung.

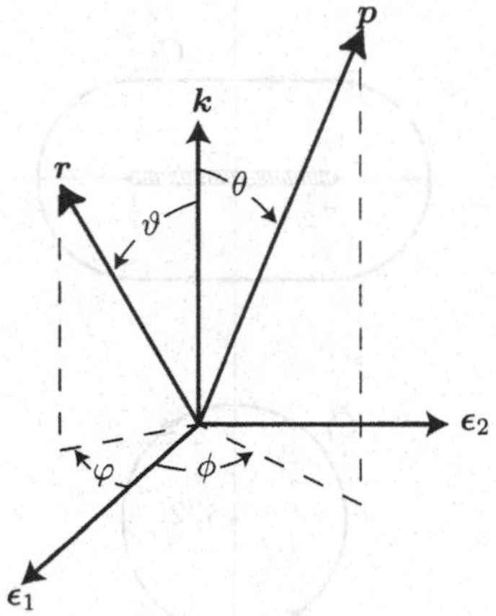

FIG. 10.5. Kinematic variables in the photoeffect.

The cross section is now found by the same steps as led to Eq. 224:

$$\frac{d\sigma_\lambda}{d\Omega} = 2\alpha\, a_0^2 \,\frac{kpa_0}{E_0}\, |I(p)|^2 \sin^2\theta , \tag{235}$$

where

$$I(p) = a_0^{-4} \int_0^\infty r^3 e^{-r/a_0}\, C_1(p;r)\, dr . \tag{236}$$

This result does not depend on the helicity of the incident photon.

An analytic expression for $I(p)$ can be found by using the integral representation for the radial wave function, as given by (8.241) and (8.242):

$$C_1(p;r) = (2pr)e^{ipr} e^{\frac{1}{2}\pi\gamma}\,\frac{\Gamma(2-i\gamma)}{2\pi i}\int_C e^s (s+2ipr)^{i\gamma-2}\, s^{-i\gamma-2}\, ds . \tag{237}$$

The trick is to modify this so as to make the r-integration simple, which is accomplished with the substitution $s = 2ipr(t-\frac{1}{2})$:

$$C_1(p;r) = \frac{e^{\frac{1}{2}\pi\gamma}\,\Gamma(2-i\gamma)}{2\pi(2pr)^2}\int_{C_t} e^{2iprt}\,(t-\tfrac{1}{2})^{-i\gamma-2}(t+\tfrac{1}{2})^{i\gamma-2}\, dt , \tag{238}$$

where C_t is the contour in the t-plane enclosing the cut from $t=\frac{1}{2}$ to $t=-\frac{1}{2}$, as shown in Fig. 10.6. The integration over r can now be done:

$$\int_0^\infty e^{2iprt}\, e^{-r/a_0}\, r dr = -\frac{a_0^2\gamma^2}{4(t+\frac{1}{2}i\gamma)^2} , \tag{239}$$

and therefore

$$I(p) = \frac{e^{\frac{1}{2}\pi\gamma}\,\Gamma(2-i\gamma)\,\gamma^4}{16i}\int_{C_t}\frac{dt}{2\pi i}\,\frac{(t-\frac{1}{2})^{-i\gamma-2}(t+\frac{1}{2})^{i\gamma-2}}{(t+\frac{1}{2}i\gamma)^2} . \tag{240}$$

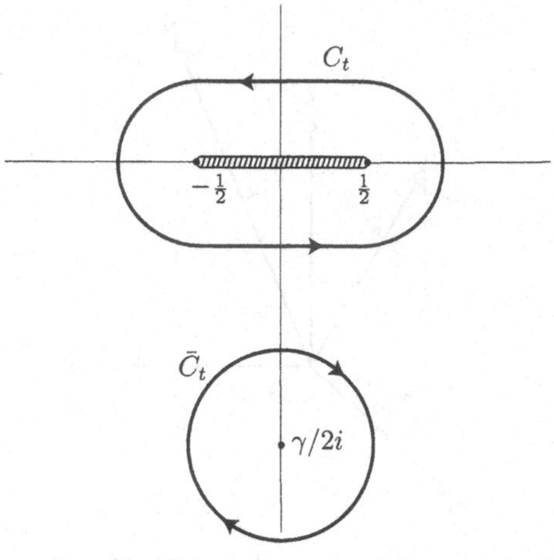

FIG. 10.6. Contours in the t-plane pertaining to the evaluation of $I(p)$ (Eq. 240).

The integrand has the aforementioned branch cut, and a second-order pole at $t = -\frac{1}{2}i\gamma$. As $|t| \to \infty$, the integrand falls off like $|t|^{-6}$, and therefore the contour can be deformed to $\bar{C}_t$, also shown in Fig. 10.6. Cauchy's theorem therefore gives the following value for the integral:

$$-\frac{d}{dt}\,(t-\tfrac{1}{2})^{-i\gamma-2}(t+\tfrac{1}{2})^{i\gamma-2}\,\Big|_{t=-\frac{1}{2}i\gamma} = \frac{64i\gamma}{(1+\gamma^2)^3}\left(\frac{i\gamma-1}{i\gamma+1}\right)^{i\gamma}. \tag{241}$$

The identities

$$|\Gamma(2-i\gamma)|^2 = \frac{(1+\gamma^2)\pi\gamma}{\sinh\pi\gamma}\,, \qquad \left|\left(\frac{i\gamma-1}{i\gamma+1}\right)^{i\gamma}\right|^2 = \exp(-4\gamma\cot^{-1}\gamma)\,,$$

then result in

$$|I(p)|^2 = \frac{32\pi\gamma}{(1+\gamma^{-2})^5}\,\frac{e^{-4\gamma\cot^{-1}\gamma}}{1-e^{-2\pi\gamma}}\,. \tag{242}$$

The final result for the low-energy cross section is therefore

$$\frac{d\sigma}{d\Omega} = 64\pi\,\alpha a_0^2\sin^2\theta\left(\frac{E_0}{k}\right)^4\frac{e^{-4\gamma\cot^{-1}\gamma}}{1-e^{-2\pi\gamma}}\,. \tag{243}$$

This expression takes the Coulomb interaction fully into account; it gives the exact result (to first order in α only, of course) for photon energies too low to eject the electron into higher partial waves than $l = 1$.

Eq. 243 joins smoothly to the high-energy result, as the qualitative argument already given would lead one to expect. In the intermediate energy regime, $k \sim O(E_0/\alpha)$ and $\gamma \ll 1$. Expanding (243) in γ gives

$$\frac{d\sigma_\lambda}{d\Omega} \simeq 32\alpha\,a_0^2\,(E_0/k)^{7/2}\,\sin^2\theta\,, \tag{244}$$

which agrees with (227) expect for the small term proportional to $v\cos\theta$. The latter is due to higher partial waves ignored in the low-energy calculation.

The behavior of the cross section near threshold is exceptionally interesting:[1]

$$\frac{d\sigma_\lambda}{d\Omega} = 64\pi\,\alpha\, a_0^2\, e^{-4}\sin^2\theta\left[1 - \frac{8}{3}\left(\frac{k}{E_0} - 1\right) + \cdots\right], \quad (k \to E_0)\,. \tag{245}$$

The remarkable feature of this result is that *the cross section is finite at threshold* (see Fig. 10.7). That is to say, the transition matrix element must diverge as $p \to 0$

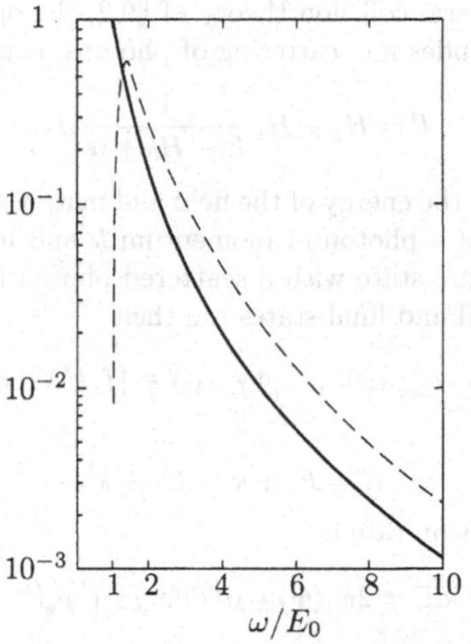

FIG. 10.7. The total cross section for the photoeffect in hydrogen near threshold. The quantity $\hat{\sigma}$ is the cross section divided by its threshold value, $(512\pi^2/3)\alpha a_0/e^4 = 6.31\times 10^{-18}$ cm^2. The solid curve is the exact *p*-wave calculation (Eq. 243), and the dashed curve the total cross section according to the plane wave approximation (Eq. 228), normalized in the same way. At energies higher than shown here the two curves converge only slowly because the factor $\exp(-4\gamma\cot^{-1}\gamma)$ tends to 1 very slowly.

to cancel the vanishing of the phase space factor, a divergence that is due to the infinite range of the Coulomb attraction. When the interaction is ignored, as it is in the high-energy calculation, the cross section vanishes at threshold, and it would do so as well if the Coulomb interaction were replaced by an interaction with a finite range.

Finite threshold cross sections also hold for electrons initially in bound states with $l \neq 0$, such as electrons in various shells of complex atoms. In such atoms the photon absorption probability as a function of frequency shows sharp absorption edges whose detailed features provide an important diagnostic tool.

[1] Note that in this formula e is the base of the natural logarithms, not the electronic charge!

10.7 Scattering of Photons

In classical electrodynamics, the scattering of light is a second-order process: the incident wave imposes an oscillatory motion on charged particles, and this acceleration causes the particles to radiate the scattered wave. In quantum theory the situation is similar. This is so because in scattering there is one photon in both the initial and final state, whereas the dominant term H_1 in the interaction H_{int} (recall (135)) between the field and its sources creates or destroys a single photon, and therefore can only connect the initial and final states in second order (or more generally, even orders). The other term H_2 in H_{int} is quadratic in the vector potential, and therefore contains products of one photon annihilation and one creation operator, and therefore connects the initial and final states in first order.

According to the general collision theory of §9.2, the operator T, whose matrix elements are the amplitudes for scattering of photons, is therefore in leading order

$$T = H_2 + H_1 \frac{1}{E - H_0 + i\epsilon} H_1 \,, \tag{246}$$

where $H_0 = H_\gamma + H_M$ is the energy of the field and matter without their interaction. Consider the collision of a photon of momentum $\boldsymbol{k}$ and helicity λ with the initial matter state $|i\rangle$, to a final state with a scattered photon $(\boldsymbol{k}', \lambda')$ and the target in the state $|f\rangle$. The initial and final states are then

$$|\Phi_{i,k\lambda}\rangle \equiv |i; k\lambda\rangle = a_{k\lambda}^\dagger |i; 0\rangle \,, \quad |\Phi_{f,k'\lambda'}\rangle \equiv |f; k'\lambda'\rangle = a_{k'\lambda'}^\dagger |f; 0\rangle \,, \tag{247}$$

with energies

$$E = E_i + k = E_f + k' \,. \tag{248}$$

The corresponding collision rate is

$$d\Gamma = 2\pi \, |\langle \Phi_{f,k'\lambda'} | T | \Phi_{i,k\lambda} \rangle|^2 \, \rho_{k'}^{(\gamma)} \,, \tag{249}$$

where $\rho_{k'}^{(\gamma)}$ is the density of final photon states (Eq. 144).

The matrix element of the quadratic interaction H_2 between these states is easily evaluated as only terms of the form $a_1^\dagger a_2$ and $a_2 a_1^\dagger$ can contribute. Using $\doteq$ to stand for "equal insofar as what contributes here is concerned," the operator $|\boldsymbol{A}(\boldsymbol{r})|^2$ is

$$|\boldsymbol{A}(\boldsymbol{r})|^2 \doteq \sum_{k_1 k_2} \frac{1}{\sqrt{2Vk_1 2Vk_2}} (\boldsymbol{e}_{k_1} \cdot \boldsymbol{e}_{k_2}^*) \, e^{i(\boldsymbol{k}_1 - \boldsymbol{k}_2)\cdot \boldsymbol{r}} \, (a_{k_1} a_{k_2}^\dagger + a_{k_2}^\dagger a_{k_1}) \tag{250}$$

$$= \frac{1}{2V} \sum_k \frac{1}{k} + \frac{1}{V} \sum_{k_1 k_2} \frac{1}{\sqrt{k_1 k_2}} \, (\boldsymbol{e}_{k_1} \cdot \boldsymbol{e}_{k_2}^*) \, e^{i(\boldsymbol{k}_1 - \boldsymbol{k}_2)\cdot \boldsymbol{r}} \, a_{k_2}^\dagger a_{k_1} \,, \tag{251}$$

where each k stands for both momentum and helicity. The first term in (251) is not an operator and cannot cause transitions.[1] The matrix element of (251) between the states defined in (247) is therefore

$$\langle 0 | a_{k'} a_{k_2}^\dagger a_{k_1} a_k^\dagger | 0 \rangle = \delta_{k'k_2} \delta_{kk_1} \,. \tag{252}$$

[1] This impotent (but infinite!) term could have been eliminated *ab initio* by defining the quantum theory as the expression obtained by canonical quantization of the classical theory with all destruction operators moved to the right ("normal ordered"), or equivalently, as the expression resulting from canonical quantization with the vacuum expectation value discarded.

From (136),

$$\langle f; k'\lambda'|H_2|i; k\lambda\rangle \equiv \frac{e^2}{2V\sqrt{kk'}}\,\mathcal{A}_2\,, \tag{253}$$

where

$$\mathcal{A}_2 = (\boldsymbol{e}^*_{k'\lambda'}\cdot\boldsymbol{e}_{k\lambda})\,\frac{1}{m}\sum_a \langle f|e^{i\boldsymbol{q}\cdot\boldsymbol{r}_a}|i\rangle = \frac{\boldsymbol{e}^*_f\cdot\boldsymbol{e}_i}{m}\,F_{fi}(\boldsymbol{q})\,, \tag{254}$$

and $\boldsymbol{q} = \boldsymbol{k} - \boldsymbol{k}'$ is the momentum transfer. Recall from §9.1 that $F_{fi}(\boldsymbol{q})$ is precisely the form factor that appears in scattering of a particle from this system.

The term of second order in H_1 is

$$T_2 = \sum_c \frac{\langle f; k'\lambda'|H_1|\Phi_c\rangle\langle\Phi_c|H_1|i; k\lambda\rangle}{E - E_c + i\epsilon}\,. \tag{255}$$

and involves two distinct types of intermediate states and energy denominators:

1. Those due to the annihilation operators in H_1 acting on $|i; k\lambda\rangle$. The only such operator has the same momentum and helicity $(\boldsymbol{k}, \lambda)$, and leaves no photon in the intermediate state, so in this case

$$|\Phi_c\rangle = |n; 0\rangle\,, \quad E - E_c = E_i + k - E_n\,, \tag{256}$$

where n designates some state of the target with energy E_n.

2. Those due to the creation operator in H_1 acting on $|i; k\lambda\rangle$, which produces an intermediate state $|\Phi_c\rangle$ with two photons, one being the incident photon, the other with quantum numbers $(\boldsymbol{k}_1, \lambda_1)$ which for the moment are unknown. Then in $\langle f; k'\lambda'|H_1|\Phi_c\rangle$ the destruction operators in H_1 must remove the incident photon from $|\Phi_c\rangle$, which requires $\boldsymbol{k}_1 = \boldsymbol{k}', \lambda_1 = \lambda'$, and thus

$$|\Phi_c\rangle = |n; k\lambda, k'\lambda'\rangle\,, \quad E - E_c = E_i + k - (E_n + k + k') = E_i - E_n - k'\,. \tag{257}$$

These two contributions to the amplitude can represented by the diagrams shown in Fig. 10.8.

The matrix elements of H_1 are in essence those of the operators $J_{k\lambda}$ defined in §10.4(b). The contribution of T_2 is then

$$\sum_c \frac{\langle f; k'\lambda'|H_1|\Phi_c\rangle\langle\Phi_c|H_1|i; k\lambda\rangle}{E - E_c + i\epsilon} \equiv \frac{e^2}{2V\sqrt{kk'}}\,\mathcal{A}_1\,, \tag{258}$$

$$\mathcal{A}_1 = \sum_n \left(\frac{\langle f|J^\dagger_{k'\lambda'}|n\rangle\langle n|J_{k\lambda}|i\rangle}{E_i - E_n + k + i\epsilon} + \frac{\langle f|J_{k\lambda}|n\rangle\langle n|J^\dagger_{k'\lambda'}|i\rangle}{E_i - E_n - k' + i\epsilon}\right)\,. \tag{259}$$

The collision rate now follows from (249) and (144):

$$d\Gamma = 2\pi\,\frac{e^4}{4V^2kk'}\,|\mathcal{A}_1 + \mathcal{A}_2|^2\,\frac{V\,d^3k'}{(2\pi)^3dk'} = \alpha^2\,\frac{k'}{k}\,|\mathcal{A}_1 + \mathcal{A}_2|^2\,\frac{d\Omega}{V}\,. \tag{260}$$

The differential cross section is this differential rate divided by the incident photon flux c/V, or $1/V$ in our units. Hence the final result for the angular distribution of

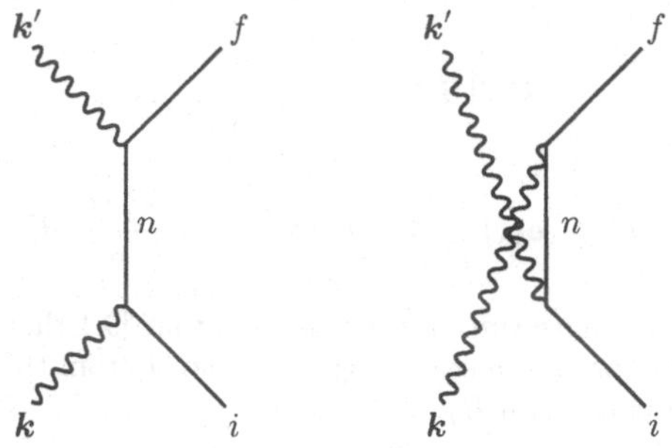

FIG. 10.8. Amplitudes that contribute coherently to photon scattering. On the left the intermediate state has no photons (Eq. 256), on the right both the incident and scattered photons (257). The diagrams are similar, but *not* identical to Feynman diagrams.

scattering with specified incident and final photon polarizations is

$$\frac{d\sigma}{d\Omega} = \frac{\alpha^2}{m^2}\frac{k'}{k}\Bigg|(\boldsymbol{e}_f^* \cdot \boldsymbol{e}_i)\, F_{fi}(\boldsymbol{q}) + m\sum_n \left(\frac{\langle f|J_{k'\lambda'}^\dagger|n\rangle\langle n|J_{k\lambda}|i\rangle}{E_i - E_n + k + i\epsilon} + \frac{\langle f|J_{k\lambda}|n\rangle\langle n|J_{k'\lambda'}^\dagger|i\rangle}{E_i - E_n - k' + i\epsilon}\right)\Bigg|^2 . \quad (261)$$

This is the Kramers-Heisenberg dispersion formula, derived (in the dipole approximation) shortly before Heisenberg's discovery of matrix mechanics, and in which it played a significant role. The cross section is seen to be of order r_0^2, where $r_0 = \alpha/m = e^2/4\pi mc^2$ is the classical electron radius, 2.82×10^{-13} cm.

The Kramers-Heisenberg formula has many interesting and important applications; it is the forerunner of more sophisticated dispersion relations in quantum field theory and the quantum theory of many particle systems. As it stands, it describes both elastic and inelastic scattering at all incident frequencies provided that the target can be described by nonrelativistic quantum mechanics. (In atomic physics inelastic scattering of light is called *Raman scattering.*)

Equation (261) also contains two famous results of classical electrodynamics as special cases: Thomson and Rayleigh scattering. The former is for scattering by a structureless charge; the latter the low-frequency limit for scattering by a spherically symmetric (or spin unaligned) polarizable system, such as an atom or molecule. Both cross sections have the form

$$\frac{d\sigma}{d\Omega} = r_0^2\, \frac{k'}{k} |\boldsymbol{e}_f^* \cdot \boldsymbol{e}_i|^2\, |\mathcal{A}|^2 . \quad (262)$$

In the case of Thomson scattering, $\mathcal{A} = 1$. In Rayleigh scattering

$$\mathcal{A}_R = 2m\omega^2\, \mathcal{D} , \quad (263)$$

where ω is the photon frequency and $\mathcal{D}$ the polarizability of the target in its ground state $|g\rangle$:

$$\mathcal{D} = \sum_{n\neq g} \frac{\langle g|X|n\rangle\langle n|X|g\rangle}{E_g - E_n} , \quad (264)$$

where $X = \sum_a x_a$, i.e., eX is a component of the electric dipole moment operator. Thomson and Rayleigh scattering are the subject of Prob. 14.

The Kramers-Heisenberg formula has one obvious defect: it is infinite when the photon energy equals one of the target's excitation energies. For such energies the scattering is resonant, a phenomenon know in optics as resonance fluorescence. It requires a more sophisticated approach than straightforward perturbation theory, which is the topic of the next section.

Some remarks about the dependence of the photon scattering amplitude on polarization and scattering angle are appropriate here. As in the case of scattering of particles with spin, symmetry provides a powerful tool. Note first that the dependence on polarization must, in both elastic and inelastic scattering, have the form

$$T_{fi} = \sum_{\alpha,\beta=1}^{3} (\boldsymbol{e}_f^*)_\alpha (\boldsymbol{e}_i)_\beta \, X_{\alpha\beta} \tag{265}$$

because the polarization vectors only enter via the final and initial photon states. The structure of the tensor $X_{\alpha\beta}$ is constrained by the usual symmetry requirement on T_{fi}. When the scattering is inelastic, these requirements are not very powerful, and the analysis is complicated. On the other hand, for elastic scattering symmetry is very restrictive, as can be surmised from (262). If the target has no spin, $X_{\alpha\beta}$ can only be a function of $\boldsymbol{k}_f$ and $\boldsymbol{k}_i$. Furthermore, if the wavelength is long compared to the target's size, as is true in scattering by atoms below the X-ray regime and by nuclei for photons having energies below ~ 100 MeV, the dipole approximation is valid for all the matrix elements in (261), and as a consequence the vectors $\boldsymbol{k}_{f,i}$ cannot appear in the elastic amplitude. Hence this amplitude must have the remarkably simple form

$$T_{fi}^{\text{el}} = (\boldsymbol{e}_f^* \cdot \boldsymbol{e}_i)\, A \,, \tag{266}$$

where A cannot depend on the scattering angle θ. Thomson and Rayleigh scattering exemplify this general result. If the target has spin, two further terms are possible:

$$(\boldsymbol{e}_f^* \times \boldsymbol{e}_i) \cdot \boldsymbol{S}\, B \,, \qquad (\boldsymbol{e}_f^*)_\alpha (\boldsymbol{e}_i)_\beta \, [S_\alpha S_\beta + S_\beta S_\alpha - \tfrac{2}{3}\delta_{\alpha\beta} S(S+1)]\, C \,, \tag{267}$$

where the second form requires $S \geq 1$, and B, C are again isotropic and independent of k. When the dipole approximation is invalid, the amplitude will be more complicated. In the case of spin 0 the general form of the amplitude is

$$T_{fi}^{\text{el}} = (\boldsymbol{e}_f^* \cdot \boldsymbol{e}_i)\, A_1(k,\theta) + (\boldsymbol{e}_f^* \cdot \boldsymbol{k}_i)(\boldsymbol{k}_f \cdot \boldsymbol{e}_i)\, A_2(k,\theta) \,, \tag{268}$$

and it of course grows in complexity with increasing spin.

10.8 Resonant Scattering and Spontaneous Decay

The energy spectrum of an atom changes profoundly when its interaction with the electromagnetic field is taken into account. When that interaction is ignored, the atom's discrete spectrum is suspended in the continuous spectrum of the free radiation field; when it is "turned on" the discrete levels disappear into the continuum, leaving as their trace resonances in the photon-atom scattering cross section. To be more concrete, consider the transition $i \to f$ in some atom. The noninteracting excited state is $|i\rangle$ and no photons; the noninteracting states into which it can decay are $|f;k\rangle$, the final atomic state and a photon of energy $k = \omega_0$, where ω_0 is the excitation energy. If the quantization volume V is finite, the spacing between the latter levels is of order $V^{-\frac{1}{3}}$. In any finite neighborhood ΔE about ω_0 there are thus $\sim V^{\frac{1}{3}}\Delta E$ states of the type $|f;k\rangle$. Were conventional perturbation theory to be attempted, then for any interaction strength that remains finite, no matter how small, a matrix whose dimension goes to infinity as $V \to \infty$ would have to be diagonalized. Surely something more powerful than perturbation theory is called for.[1]

This problem is actually analogous to that of a particle in the presence of a perfectly reflecting enclosure, i.e., a discrete spectrum of states trapped in the enclosure, and a continuum of states for scattering off the enclosure, with the latter appearing as sharp scattering resonances when weak transmission through the enclosure's walls becomes possible. A one-dimensional example of this type was discussed in detail in §4.5(c), and the analogy to the radiation phenomenon was pointed out. There we also learned that if the particle is initially in one of the trapped states, it thereafter decays with the time dependence $e^{-\Gamma t}$, where Γ is the width of the resonance displayed in scattering from the enclosure at the energy of the trapped state.

Essentially the same effect holds in the radiation case. If the photon frequency closely matches an excitation frequency of the atom, there is a resonance in the photon scattering amplitude of width Γ in the partial wave corresponding to the multipole of the radiative transition connection the ground state to that excited state, where Γ is the radiative transition rate; furthermore, if the excited state is prepared by a sudden process such as inelastic electron scattering or a spark discharge, it will then decay by photon emission with the exponential time dependence $e^{-\Gamma t}$. As we learned in §10.4(c), radiative widths are of order $\alpha\omega_0^3 a^2$, where a is the size of the system and ω_0 the transition energy, which is of order $1/ma^2$ (see §10.4(a)). Hence, the ratio of width to level spacing is $\Gamma/\omega_0 \sim \alpha\omega_0/mc^2$. This is very small in any nonrelativistic system, so these resonances are very sharp, and the lifetimes are very long compared to the characteristics periods of the system's internal motion. The dramatic phenomenon due to such resonances is called resonance fluorescence.

(a) Model Hamiltonian

Consider the resonance that is to be expected when the $1s$ ground state of H is excited to the $2p$ state, the Lyman α line. The terms in the Kramers-Heisenberg formula with matrix elements for $1s + \gamma \leftrightarrow 2p$ are then divided by a vanishing energy denominator because perturbation theory cannot handle such transitions. We therefore introduce a model Hamiltonian that only incorporates these transitions, but which leads to equations (in either the Heisenberg or Schrödinger pictures) that can be solved exactly for the states of interest. Readers of §9.3 will recognize that what follows is a variant of the model introduced there for the treatment of elastic

---

[1]The quantum theory of natural line width and resonance fluorescence is due to Weisskopf and Wigner.

and inelastic resonance scattering, when their widths are not small compared to level spacings, as is the case here.

It is convenient to introduce operators that produce the $1s \leftrightarrow 2p$ transitions,

$$b_m^\dagger = |2p, m\rangle\langle 1s| \,, \qquad b_m = |1s\rangle\langle 2p, m| \,, \tag{269}$$

where $m = \pm 1, 0$. The projection operators onto these states are

$$b_m^\dagger b_m = |2p, m\rangle\langle 2p, m| \,, \qquad b_m b_m^\dagger = |1s\rangle\langle 1s| \,. \tag{270}$$

The long-wavelength approximation is valid, so the amplitude for absorption of a photon with momentum $\boldsymbol{q}$ and helicity μ due to the true interaction H_1 is given by (156) and (162),

$$\langle 2p, m|H_1|1s; q\mu\rangle = \frac{-ie\omega_0}{\sqrt{2Vq}} \langle 2p, m|\boldsymbol{e}_{q\mu} \cdot \boldsymbol{r}|1s\rangle \,, \tag{271}$$

where $\omega_0 = E_{2p} - E_{1s}$ is the excitation energy. The dipole matrix elements can be written as

$$\langle 2p, m|\boldsymbol{r}|1s\rangle = D\, \boldsymbol{e}_m^* \,, \tag{272}$$

where

$$\boldsymbol{e}_0 = \hat{\boldsymbol{z}} \,, \qquad \boldsymbol{e}_{\pm 1} = \mp(\hat{\boldsymbol{x}} \pm i\hat{\boldsymbol{y}})/\sqrt{2} \,. \tag{273}$$

The model Hamiltonian that takes the resonant transitions fully into account is thus

$$H = H_\gamma + \omega_0 \sum_m b_m^\dagger b_m - ig \sum_{q\mu, m} \frac{1}{\sqrt{2Vq}} [(\boldsymbol{e}_{q\mu} \cdot \boldsymbol{e}_m^*) b_m^\dagger a_{q\mu} - (\boldsymbol{e}_{q\mu}^* \cdot \boldsymbol{e}_m) b_m a_{q\mu}^\dagger] \,, \tag{274}$$

where $g = e\omega_0 D$, and $1s$ is taken to be at the zero of the energy scale. In the model interaction H_I the terms $b_m^\dagger a_{q\mu}$ account for the absorption $\gamma + 1s \to 2p$, and their Hermitian conjugates for the emission $2p \to \gamma + 1s$.

Note that we have discarded the matrix elements for $1s \leftrightarrow 2p + \gamma$, i.e., for intermediate states of type (257) which give nonresonant terms in the Kramers-Heisenberg formula. This would seem to be an innocent approximation, but it is not, because both terms in the amplitude are required by nothing less than causality! The reason is that by discarding the nonresonant transitions in matrix elements of the vector potential, some of the creation and destruction operators in $\boldsymbol{A}$ have been discarded, and this butchered form of $\boldsymbol{A}$ does not lead to commutators for the field operators that vanish outside the light cone, as demanded by causality (recall §10.2). From a practical point of view the error being committed is negligible because the nonresonant terms in the correct formula only provide a smooth and very small background relative to the resonances.[1]

Because the nonresonant transitions have been discarded, the model has a constant of motion that the true Hamiltonian does not possess:

$$\mathcal{N} = \sum_{q\mu} a_{q\mu}^\dagger a_{q\mu} + \sum_m b_m^\dagger b_m \,. \tag{275}$$

[1]In quantum optics, the neglect of the nonresonant terms is often called the rotating wave approximation because of the analogous approximation in magnetic resonance (see §4.1(b)).

The amplitude for elastic scattering from the 1s state can be computed exactly because $[\mathcal{N}, H] = 0$, which breaks the complete Hilbert space of the photon-atom system into disconnected subspaces with differing eigenvalues of $\mathcal{N}$. The discarded terms in the true interaction for the transitions $\gamma + 2p \leftrightarrow 1s$ destroy conservation of $\mathcal{N}$, as is clear from the diagram on the right side of Fig. 10.8. The incident, intermediate and final states in the resonant diagram all have the eigenvalue $\mathcal{N}' = 1$, but the intermediate state of the nonresonant diagram has $\mathcal{N}' = 3$.

(b) The Elastic Scattering Cross Section

Let the incident and scattered momenta and helicities be $(\boldsymbol{k}\lambda, \boldsymbol{k}'\lambda')$ with $k = k'$, and the atom be in the ground state initially. The incident state has eigenvalue 1 of $\mathcal{N}$, and the whole scattering state must therefore belong to this sector. The most general such state is

$$|E\rangle = \sum_{q\mu} \chi_{q\mu} a_{q\mu}^\dagger |0\rangle + \sum_m C_m b_m^\dagger |0\rangle \ , \tag{276}$$

where $|0\rangle$ is the direct product of the electromagnetic vacuum and that of the atom in the 1s ground state. (Note that in this model, the inelastic process $\gamma + 1s \to \gamma + 2p$ is forbidden.)

Angular momentum conservation can also be used to advantage. Before the collision the projection of the whole system's angular momentum along the direction of the incident momentum $\boldsymbol{k}$ is the incident helicity λ. If we quantize the atom's angular momentum along this same direction, only the 2p state with $m = \lambda$ can be excited. Hence

$$C_m = \delta_{m\lambda} C \ , \tag{277}$$

and the angular wave function of the excited 2p state is $\boldsymbol{e}_{k\lambda}$. Rotational invariance also requires C to be independent of λ, as will be shown explicitly. The state (276) thus simplifies to

$$|E\rangle = \sum_{q\mu} \chi_{q\mu} a_{q\mu}^\dagger |0\rangle + C b_\lambda^\dagger |0\rangle \ . \tag{278}$$

The coefficients $\chi_{q\mu}$ and C are determined by requiring $|E\rangle$ to satisfy the Schrödinger equation. The action of the free part of H is

$$H_0 |E\rangle = \sum_{q\mu} q \chi_{q\mu} a_{q\mu}^\dagger |0\rangle + \omega_0 C b_\lambda^\dagger |0\rangle \ . \tag{279}$$

To evaluate the effect of the interaction H_I, we use (269), (271) and $(b_\lambda^\dagger)^2 = 0$:

$$H_I a_{q\mu}^\dagger |0\rangle = \frac{-ig}{\sqrt{2Vq}} \, (\boldsymbol{e}_{q\mu} \cdot \boldsymbol{e}_{k\lambda}^*) \, b_\lambda^\dagger |0\rangle \ , \tag{280}$$

$$H_I b_\lambda^\dagger |0\rangle = ig \sum_{q\mu} \frac{1}{\sqrt{2Vq}} \, (\boldsymbol{e}_{q\mu}^* \cdot \boldsymbol{e}_{k\lambda}) \, a_{q\mu}^\dagger |0\rangle \ . \tag{281}$$

Hence

$$H|E\rangle = \sum_{q\mu} \chi_{q\mu}\Big(q a_{q\mu}^\dagger - \frac{ig}{\sqrt{2Vq}}\,(\boldsymbol{e}_{q\mu}\cdot\boldsymbol{e}_{k\lambda}^*)\, b_\lambda^\dagger\Big)|0\rangle$$
$$+ C\Big(\omega_0 b_\lambda^\dagger + \sum_{q\mu} \frac{ig}{\sqrt{2Vq}}(\boldsymbol{e}_{q\mu}^*\cdot\boldsymbol{e}_{k\lambda})a_{q\mu}^\dagger\Big)|0\rangle\,. \quad (282)$$

The coefficients of the linearly independent terms in $(H-E)|E\rangle = 0$ must vanish separately, and therefore

$$(q-E)\chi_{q\mu} + \frac{igC}{\sqrt{2Vq}}\,(\boldsymbol{e}_{q\mu}^*\cdot\boldsymbol{e}_{k\lambda}) = 0\,, \quad (283)$$

$$(\omega_0 - E)C - ig\sum_{q\mu}\frac{\chi_{q\mu}}{\sqrt{2Vq}}\,(\boldsymbol{e}_{q\mu}\cdot\boldsymbol{e}_{k\lambda}^*) = 0\,. \quad (284)$$

These coupled equation must now be solved for the amplitudes $\chi_{q\mu}$ and C. From a mathematical standpoint, this problem is essentially identical to the separable non-local potential scattering problem treated in §8.2(b). Here too we first find an equation for C alone, whose solution will determine $\chi_{q\mu}$. To eliminate χ, we must invert $(q-E)$ in Eq. 283. In doing so the boundary condition of an incident plane wave with momentum $\boldsymbol{k}$ and helicity λ, and outgoing scattered waves, must be imposed in the now familiar manner:

$$\chi_{q\mu}^{(+)} = \delta_{qk}\delta_{\mu\lambda} + \frac{igC}{\sqrt{2Vq}}\,\frac{\boldsymbol{e}_{q\mu}^*\cdot\boldsymbol{e}_{k\lambda}}{k-q+i\epsilon}\,, \quad (285)$$

where $E = |\boldsymbol{k}|$ was used. Then (284) becomes

$$(\omega_0 - k)C - \frac{ig}{\sqrt{2Vk}} + g^2 C\sum_{q\mu}\frac{1}{2Vq}\,\frac{(\boldsymbol{e}_{k\lambda}^*\cdot\boldsymbol{e}_{q\mu})(\boldsymbol{e}_{q\mu}^*\cdot\boldsymbol{e}_{k\lambda})}{k-q+i\epsilon} = 0\,, \quad (286)$$

and therefore

$$C = -\frac{ig}{\sqrt{2Vk}}\,\frac{1}{k-\omega_0-\Phi(k+i\epsilon)}\,, \quad (287)$$

where

$$\Phi(k+i\epsilon) = g^2\sum_\mu\int\frac{d^3q}{2(2\pi)^3}\,\frac{1}{q}\,\frac{|\boldsymbol{e}_{k\lambda}^*\cdot\boldsymbol{e}_{q\mu}|^2}{k-q+i\epsilon}\,. \quad (288)$$

Consequently

$$\chi_{q\mu}^{(+)} = \delta_{qk}\delta_{\mu\lambda} + \frac{1}{k-q+i\epsilon}\,\frac{g^2}{2V\sqrt{kq}}\,\frac{\boldsymbol{e}_{q\mu}^*\cdot\boldsymbol{e}_{k\lambda}}{k-\omega_0-\Phi(k+i\epsilon)}\,. \quad (289)$$

According to Eq. 79,

$$\sum_\mu(\boldsymbol{e}^*\cdot\boldsymbol{e}_{q\mu})(\boldsymbol{e}_{q\mu}^*\cdot\boldsymbol{e}) = 1 - |\hat{\boldsymbol{q}}\cdot\boldsymbol{e}^*|^2\,; \quad (290)$$

therefore

$$\sum_\mu\int d\Omega_q\,|\boldsymbol{e}_{q\mu}\cdot\boldsymbol{e}|^2 = \frac{8\pi}{3}\,, \quad (291)$$

and

$$\Phi(k+i\epsilon) = \frac{g^2}{6\pi^2} \int_0^\infty \frac{q dq}{k - q + i\epsilon} . \tag{292}$$

This integral is linearly divergent because the integrand tends to a constant as $q \to \infty$. At first sight this is so because we have used the dipole approximation for large q, and it would converge if the Bessel function $j_0(qr)$ in the expansion of the plane wave had not been approximated by unity in the dipole matrix element (271). However, as we shall soon learn, there is a much deeper reason for the divergence. For the moment this is beside the point at issue, which can be understood by simply putting an upper limit K on the integral. A plausible choice is $K \sim 1/a_0 \sim \alpha m$, which is larger by a factor of order α than the resonance energy ω_0; furthermore, wavelengths short compared to the size of the system interact weakly.

We can now compute the scattering amplitude by using the standard formula

$$T_{fi} = \langle \Phi_f | H_I | \Psi_i^{(+)} \rangle .$$

In our case the noninteracting final state is $a^\dagger_{k'\lambda'}|0\rangle$, and the complete solution of the Schrödinger equation is $|\boldsymbol{k}\lambda^{(+)}\rangle$ whose amplitudes are C and $\chi^{(+)}_{k\lambda}$ as given by (287) and (289):

$$T_{fi} = \langle 0 | a_{k'\lambda'} H_I | \boldsymbol{k}\lambda^{(+)} \rangle . \tag{293}$$

Now $|\boldsymbol{k}\lambda^{(+)}\rangle$ has two types of terms: the one-photon states $a^\dagger_{q\mu}|0\rangle$ and the excited state $b^\dagger_\lambda|0\rangle$; when acted upon by H_I they turn into each other (cf. Eq. 280 and Eq. 281). Only $b^\dagger_\lambda|0\rangle$ can contribute to T_{fi}:

$$T_{fi} = C\langle 0 | a_{k'\lambda'} H_I b^\dagger_\lambda |0\rangle = \sum_{q\mu} \frac{igC}{\sqrt{2Vq}} (\boldsymbol{e}^*_{q\mu} \cdot \boldsymbol{e}_{k\lambda}) \langle 0 | a_{k'\lambda'} a^\dagger_{q\mu} |0\rangle , \tag{294}$$

or, using (287),

$$T_{fi} = \frac{g^2}{2Vk} \frac{\boldsymbol{e}^*_{k'\lambda'} \cdot \boldsymbol{e}_{k\lambda}}{k - \omega_0 - \Phi(k+i\epsilon)} . \tag{295}$$

The steps to the differential cross section are then the same as those from Eqs. 249–261 in the derivation of the Kramers-Heisenberg formula:

$$\frac{d\sigma}{d\Omega} = \alpha^2 D^4 \omega_0^4 \left| \frac{\boldsymbol{e}^*_{k'\lambda'} \cdot \boldsymbol{e}_{k\lambda}}{k - \omega_0 - \Phi(k+i\epsilon)} \right|^2 . \tag{296}$$

The function Φ is complex, and as a consequence the denominator does not vanish in Eq. 296, as it did in the Kramers-Heisenberg formula. The separation of Φ into its real and imaginary parts is done by applying Eq. 8.121 to Eq. 288:

$$\mathrm{Im}\, \Phi(k+i\epsilon) = -\pi g^2 \sum_\mu \int \frac{q dq\, d\Omega_q}{2(2\pi)^3} \delta(k-q) \, |\boldsymbol{e}^*_{k\lambda} \cdot \boldsymbol{e}_{q\mu}|^2 = -\frac{g^2 k}{6\pi} , \tag{297}$$

where (291) was used. But according to (165) the decay rate for the $2p \to 1s$ transition is

$$\Gamma = \frac{\alpha}{2\pi} \omega_0^3 \sum_\mu \int d\Omega_q |\langle 1s | \boldsymbol{e}^*_{q\mu} \cdot \boldsymbol{D} | 2p, m \rangle|^2 = \frac{4}{3} \alpha \omega_0^3 D^2 = \frac{g^2}{3\pi} \omega_0 , \tag{298}$$

and therefore

$$\mathrm{Im}\,\Phi(k+i\epsilon) = -\frac{k}{2\omega_0}\,\Gamma\,. \tag{299}$$

Furthermore,

$$\mathrm{Re}\,\Phi(k+i\epsilon) \equiv \Delta(k) = -\frac{\mathrm{P}}{\pi}\int dq\,\frac{\mathrm{Im}\,\Phi(q+i\epsilon)}{k-q}\,, \tag{300}$$

where P indicates the Cauchy principal value.

We already know that the radiative width Γ is tiny compared to the excitation energy ω_0, and the same is true for Δ once the integral (300) has been cut off. Then Φ only counts in the denominator of the scattering amplitude in the immediate neighborhood of $k=\omega_0$, and we can write (295) as

$$T_{fi} = \frac{g^2}{2Vk}\,\frac{\mathbf{e}^*_{k'\lambda'}\cdot\mathbf{e}_{k\lambda}}{k-\omega_0-\Delta+\frac{1}{2}i\Gamma} \tag{301}$$

where $\Delta=\Delta(\omega_0)$. The differential cross section is therefore

$$\frac{d\sigma}{d\Omega} = \alpha^2(D\omega_0)^4\left|\frac{\mathbf{e}^*_{k'\lambda'}\cdot\mathbf{e}_{k\lambda}}{k-\omega_0-\Delta+\frac{1}{2}i\Gamma}\right|^2\,, \tag{302}$$

that is, a Lorentzian resonance[1] slightly shifted from ω_0 and with a width Γ.

The total cross section is found by integrating over angles and summing on the final helicity, which by use of (291) gives

$$\sigma(k) = \frac{4\pi}{\omega_0^2}\,\frac{3}{8}\,\frac{\Gamma^2}{(k-\omega_0-\Delta)^2+\frac{1}{4}\Gamma^2}\,. \tag{303}$$

At resonance this cross section reaches its maximum value:

$$\sigma_{\mathrm{res}} = \frac{4\pi}{\omega_0^2}\cdot\frac{3}{2}\,. \tag{304}$$

This value is independent of all parameters in the problem, except the resonant frequency ω_0, and is huge in comparison to the off-resonance cross section. The latter is of order $\alpha^2\omega_0^2 D^4$, and therefore is smaller than σ_{res} by a factor $\sim\alpha^2(\omega_0 D)^4\sim\alpha^6$!

At this point we recall that the total cross section for particles without spin has the upper bound $4\pi(2l+1)/k^2$ for the partial wave of *orbital* angular momentum l. In the present problem only the $j=1$ photon partial wave scatters[2] because the only intermediate state is $2p$, and the resonant cross section might therefore be expected to be $3(4\pi/\omega_0^2)$. It is smaller than this because the photon has only two, not three, helicity states; the nonrelativistic theory does not apply as it stands.

---

[1]In nuclear physics this would be called a Breit-Wigner formula; recall §9.3.

[2]This statement only applies to the model Hamiltonian, or to be more general, to scattering in the resonance region. When the other terms in the Kramers-Heisenberg formula are kept, all photon partial waves scatter, not only because other atomic states are included, but also because of intermediate states with two photons.

(c) Decay of the Excited State

Assume that at $t = 0$ the atom is prepared in the excited state $|2p, m\rangle$. This can, for example, be achieved by a collision that takes place over a time interval short compared to $1/\Gamma$, such as through scattering of energetic electrons. Because $|2p, m\rangle$ is not an eigenstate of H, the probability $P(t)$ that the system will remain in this state changes with time. We will now show that to an excellent approximation $P(t) = e^{-\Gamma t}$. Here again there is a close analogy to the exponential decay of a state trapped inside a weakly penetrable enclosure (recall §4.5(d)).

At any time t subsequent to its preparation, the state is

$$|t\rangle = e^{-iHt}|2p, m\rangle \,, \tag{305}$$

where H is the model Hamiltonian (274). The initial state has eigenvalue $\mathcal{N}' = 1$, and $|t\rangle$ can therefore be expanded in the complete set of $\mathcal{N}' = 1$ eigenstates $|\boldsymbol{k}\lambda^{(+)}\rangle$ of energy k. The survival amplitude is

$$A(t) = \langle 2p, m|e^{-iHt}|2p, m\rangle = \sum_{k\lambda} |\langle 2p, m|\boldsymbol{k}\lambda^{(+)}\rangle|^2 e^{-ikt} \,. \tag{306}$$

The sought-after probability is

$$P(t) = |A(t)|^2 \,. \tag{307}$$

Only the term $Cb_\lambda^\dagger|0\rangle$ in $|\boldsymbol{k}\lambda^{(+)}\rangle$ contributes in (306). Using $\langle 2p, m|b_\lambda^\dagger|0\rangle = \boldsymbol{e}_m^* \cdot \boldsymbol{e}_{k\lambda}$ and (287),

$$A(t) = g^2 \sum_\lambda \int \frac{d^3k}{2k(2\pi)^3} \frac{|\boldsymbol{e}_m^* \cdot \boldsymbol{e}_{k\lambda}|^2}{|k - \omega_0 - \Phi(k + i\epsilon)|^2} \, e^{-ikt} \tag{308}$$

$$= \frac{g^2}{6\pi^2} \int_0^\infty kdk \, \frac{e^{-ikt}}{|k - \omega_0 - \Phi(k + i\epsilon)|^2} \,. \tag{309}$$

As before, we approximate Φ by its value at resonance, $\Delta - \frac{1}{2}i\Gamma$:

$$A(t) = \frac{g^2}{6\pi^2} \int_0^\infty kdk \, \frac{e^{-ikt}}{(k - \omega_0 - \Delta + \frac{1}{2}i\Gamma)(k - \omega_0 - \Delta - \frac{1}{2}i\Gamma)} \,. \tag{310}$$

The integral is dominated by the region about $k = \omega_0$. Therefore it can be extended to $-\infty$ without incurring a meaningful error, after which it is evaluated with Cauchy's theorem by closing the contour in the lower half-plane as dictated by the exponential when $t > 0$. Only the pole at $k = \omega_0 + \Delta - \frac{1}{2}i\Gamma$ contributes, and gives the result

$$A(t) = \frac{g^2}{6\pi^2} \cdot 2\pi i \cdot \frac{\omega_0}{i\Gamma} \, e^{-i(\omega_0 + \Delta)t} \, e^{-\frac{1}{2}\Gamma t}. \tag{311}$$

Using (298) then gives

$$A(t) = e^{-i(\omega_0 + \Delta)t} \, e^{-\frac{1}{2}\Gamma t} \,. \tag{312}$$

This demonstrates the oscillatory time dependence expected from the shift Δ, and the exponential decay law:

$$P(t) = e^{-\Gamma t} \,. \tag{313}$$

At first sight this would appear to be an impeccable derivation of the expected result, but not on second thought. The function $\Phi(z)$ of the complex variable $z = k + i\eta$ has a branch cut on the positive real axis, a fact that is ignored by the assumption that Φ is negligible except in the neighborhood of $z = \omega_0$ and replaceable there by the complex constant $\Delta - \frac{1}{2}i\Gamma$. For the same reason the evaluation of (309) was at best questionable: the integrand is not an analytic function whose only singularities are simple poles at $k = \omega_0 + \Delta \pm \frac{1}{2}i\Gamma$. Indeed, the multivalued nature of $\Phi(z)$ was already established in §8.2(b), where it was shown in Eq. 115 that the imaginary part of $\Phi(z)$ is discontinuous across real axis,

$$\lim_{\epsilon\to 0} [\Phi(k+i\epsilon) - \Phi(k-i\epsilon)] = 2i\,\mathrm{Im}\,\Phi(k+i\epsilon) = -i(k/\omega_0)\,\Gamma\,; \tag{314}$$

where (299) was used. To properly evaluate $A(t)$ while taking these properties of Φ into account, note that

$$\frac{2i\,\mathrm{Im}\,\Phi(k+i\epsilon)}{|k-\omega_0-\Phi(k+i\epsilon)|^2} = \frac{1}{k-\omega_0-\Phi(k+i\epsilon)} - \frac{1}{k-\omega_0-\Phi(k-i\epsilon)}\,. \tag{315}$$

Equation (309) can therefore be written as follows:

$$A(t) = \frac{i}{2\pi}\int_0^\infty dk\, e^{-ikt}\left(\frac{1}{G(k+i\epsilon)} - \frac{1}{G(k-i\epsilon)}\right), \tag{316}$$

where $G(z) = z - \omega_0 - \Phi(z)$. The two terms in (316) can then be combined into a single integral along the contour C that encloses the whole branch cut on the positive real axis:

$$A(t) = \frac{i}{2\pi}\int_C dz \frac{e^{-izt}}{G(z)}\,. \tag{317}$$

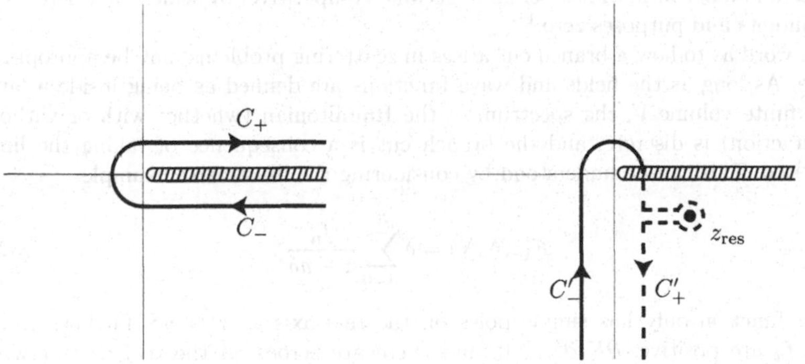

FIG. 10.9. Evaluation of Eq. 317 as $t \to \infty$. The contours $C_\pm$ and C'_- are on the sheet $\mathcal{R}_1$. The dashed contour C'_+ and the pole at z_{res} are on the sheet $\mathcal{R}_2$; they are reached by continuing from $\mathcal{R}_1$ through the cut from above.

The large t limit is best handled by converting the oscillatory factor into a decaying exponential, i.e., by converting this Fourier transform into a Laplace transform.

This is done by moving the contour (see Fig. 10.9). The lower portion C_- can be moved to C'_- on the negative imaginary axis without crossing the cut or any other singularities. The upper portion C_+ can be moved *through* the cut onto the next Riemann sheet $\mathcal{R}_2$, and then to C'_+ which is also along the negative imaginary on this sheet. But in the latter move some poles on the sheet $\mathcal{R}_2$ at z_n, with residues R_n, will be crossed. In particular, there is such a pole near $z_{\text{res}} = \omega_0 + \Delta - \frac{1}{2}i\Gamma$, as guaranteed by the analytic continuation of $\Phi(z)$ by means of the Taylor expansion about $z = \omega_0$,

$$\Phi(z) = \Phi(\omega_0) + (z - \omega_0)\Phi'(\omega_0) + \cdots \equiv \Phi_0 + (z - \omega_0)\Phi'_0 + \cdots , \tag{318}$$

so that

$$\frac{1}{G(z)} \simeq \frac{1 + \Phi'_0}{z - \omega_0 - \Phi_0(1 + \Phi'_0)} . \tag{319}$$

To order g^2, therefore, this pole is indeed at $\omega_0 + \Delta - \frac{1}{2}i\Gamma$, but its residue, which was 1 in the naive calculation is changed by a term of order g^2, i.e., to $1 + \Phi'_0$. As $z = k + i\eta$, we thus have

$$A(t) = \sum_n R_n e^{-iz_n t} + \frac{1}{2\pi} \int_{-\infty}^{0} d\eta\, e^{\eta t} \left(\frac{1}{G(i\eta)} - \frac{1}{\hat{G}(i\eta)} \right) , \tag{320}$$

where $\hat{G}(z)$ is the analytic continuation of G from the sheet $\mathcal{R}_1$ onto $\mathcal{R}_2$.

For long times, $A(t)$ will be dominated by the pole closest to the real axis, i.e., the resonance pole at z_{res} which the naive calculation that led to (312) took into account. The leading correction comes from the contributions to the integral near the origin, $\eta \to 0$. A detailed analysis shows, first, that this correction varies like $1/t^2$, and second, that the fact that this fall-off is slow compared to $e^{-\frac{1}{2}\Gamma t}$ does not mean that the exponential decay law has any significant error. The correction term has a prefactor that is exceedingly small so that it would be necessary to wait for a large number of lifetimes before it became competitive, by which time $P(t)$ is to all intents and purposes zero.[1]

A word as to how a branch cut arises in scattering problems may be appropriate here. As long as the fields and wave functions are defined as being inside a large but finite volume V, the spectrum of the Hamiltonian (whether with or without interaction) is discrete, and the branch cut is a consequence of taking the limit $V \to \infty$. This can be understood by considering the illustrative example

$$F(z; \delta, N) = \delta \sum_{n=0}^{N} \frac{f_n}{z - n\delta} . \tag{321}$$

This function only has simple poles on the real axis at $z = n\delta$; furthermore, if the f_n are positive, $\partial F/\partial z < 0$, and there are zeroes on the real axis between each singularity. As long as V is finite, $\Phi(z)$ is also a function of this type. The counterpart of $V \to \infty$ is the limit $N \to \infty, \delta \to 0$ such that $N\delta \equiv L$ remains finite. The function $F(z; \delta, N)$ behaves too wildly to be defined on the real axis in this

[1] The time dependence of this correction is process dependent. The $1/t^2$ behavior holds for radiative decay, but not for decay products that have mass, because the threshold is then not at $\eta = 0$. See M.L Goldberger and K. Watson, *Collision Theory,* Wiley (1964), pp. 445-450.

limit in the interval $0 \le x \le L$. But off from the real axis the limit is well behaved and is given by

$$\lim_{N\to\infty,\,\delta\to 0} F(z;\delta,N) = \int_0^L \frac{f(x')}{z-x'} . \qquad (322)$$

The simple poles have merged into the branch cut in this limit, and altered the character of the function fundamentally, making it multivalued. The so-called physical sheet of its Riemann surface is the one where the original function (321) was defined; by analytically continuing from this sheet through the cut from below or above one or more other sheets are reached. The character of the Riemann surface depends on the function $f(x)$. In the case of potential scattering, there are just two sheets, but in the present example, $\Phi(z)$ involves logarithms, and therefore an infinite number of sheets.

(d) The Connection Between Self-Energy and Resonance Width

By solving the Schrödinger equation exactly in the $\mathcal{N}' = 1$ subspace, an infinite subset of terms in the perturbation expansion has been summed, because the function Φ in the denominator of the scattering amplitude is of order g^2, which produces terms of all orders in g^2 when expanded into a power series:

$$\frac{1}{k-\omega_0-\Phi} = \frac{1}{k-\omega_0} + \frac{1}{k-\omega_0}\Phi\frac{1}{k-\omega_0} + \cdots . \qquad (323)$$

It is instructive to see explicitly how Φ emerges from the general formula for the fourth-order term of the scattering amplitude

$$\sum_{abc} \frac{\langle k';1s|H_I|a\rangle\langle a|H_I|b\rangle\langle b|H_I|c\rangle\langle c|H_I|k;1s\rangle}{(k-E_a+i\epsilon)(k-E_b+i\epsilon)(k-E_c+i\epsilon)} . \qquad (324)$$

Because of the selection rules on H_I, $|a\rangle = |2p\rangle$ and $k - E_a = k - \omega_0$; the same holds for $|c\rangle$. In contrast, $|b\rangle = a^\dagger_{q\mu}|1s\rangle$ and $k - E_b = k - q$, so that the only sum is on b, giving the following factor in (324):

$$\sum_{q\mu} \frac{|\langle 2p|H_I|1s;q\mu\rangle|^2}{k-q+i\epsilon} , \qquad (325)$$

which is indeed $\Phi(k+i\epsilon)$.

These corrections to the leading term of T_{fi} are *self-energy* effects, or in classical language, radiation damping effects. They take into account that the radiated field interacts with the source itself. To make this evident, let us calculate the energy of the $2p$ state using stationary state perturbation theory. Blindly applying the familiar formula gives

$$W_{2p} = \omega_0 + \langle 2p,\lambda|H_I|2p,\lambda\rangle + \sum_n \frac{|\langle 2p,\lambda|H_I|n\rangle|^2}{\omega_0 - E_n} + \cdots . \qquad (326)$$

The first-order term vanishes, and in the second-order term only the states of one photon and the $1s$ state can contribute:

$$W_{2p} = \omega_0 + \sum_{q\mu} \frac{|\langle 2p,\lambda|H_I a^\dagger_{q\mu}|0\rangle|^2}{\omega_0 - q} + \cdots = \omega_0 + \Phi(\omega_0) + \cdots . \qquad (327)$$

That is, the denominator in T_{fi} is modified because the excitation energy is not really the naive input spacing ω_0, but W_{2p}; the interaction of the atom's $2p$ state with "its own" radiation field has shifted its energy. This explanation does not quite hold water, however. The function $\Phi(\omega_0)$ does not exist for real ω_0 because of the singularity at $q = \omega_0$, and its correct contribution is selected in scattering theory by the outgoing wave condition $k \to k + i\epsilon$, but which has no justification in stationary state perturbation theory. The preceding analysis of decay has resolved this puzzle, however: when a state lies in the photon continuum, the interaction with the field produces a complex energy shift, the real part being in essence the shift in energy it would have were it a truly discrete level, and the imaginary part being half the width of the resonance in elastic photon scattering, or half the inverse lifetime of the state.

As we have seen, the width Γ is completely due to states of the same energy as the state in question, whereas the self-energy shift Δ has contributions from states of all energies — indeed, as we learned from (292), from states so distant in energy as to produce an infinite shift! Within the confines of the model being used here, this divergence can be taken to be an artifact due to taking the dipole approximation seriously at energies where it is inapplicable. But we can now understand that the problem is not avoided so easily, because if the self-energy has critical contributions from states at high energy, then the model itself cannot be used to compute the self-energy as it only has low-energy states of the atom in its Hilbert space. In fact, the interaction of the electron's bound states with the radiation field couples to states of such high energy as to render the binding essentially irrelevant to the self-energy problem. For that reason the crude approach to the Lamb shift in §10.3(b) offers a greatly superior first look at the self-energy problem than does the model of this section.

10.9 Problems

1. Verify that the Heisenberg operators $\boldsymbol{E}(x)$ and $\boldsymbol{B}(x)$ satisfy Maxwell's equations.

2. Show that the total momentum operator is the generator of spatial translations; i.e., prove Eq. 46.

3. Establish the decomposition of the angular momentum of the electromagnetic field into its orbital and intrinsic parts; for this purpose, show that

$$[\boldsymbol{r}\times(\boldsymbol{E}\times\boldsymbol{B})]_i = E_l(\boldsymbol{r}\times\boldsymbol{\nabla})_i A_l + \epsilon_{ilk}E_l A_k - \nabla_l(E_l\epsilon_{ijk}x_j A_k) .$$

4. Show that the *total* angular momentum of the electromagnetic field, when acting on a one-photon state, yields the eigenvalue equation:

$$\boldsymbol{J}\cdot\hat{\boldsymbol{k}}\,|\boldsymbol{k}\lambda\rangle = \hbar\lambda|\boldsymbol{k}\lambda\rangle .$$

This demonstrates that $\boldsymbol{J}_{\rm orb}$ is indeed the orbital angular momentum; why?

5. Using the theory developed in §7.5(b), show that the field operators can also be expressed directly in terms of operators that create and destroy photons in states of definite total angular momentum:

$$\boldsymbol{A}(x) = \frac{\sqrt{\hbar c}}{8\pi^2}\sum_{jm\lambda}\sqrt{2j+1}\int_0^\infty \frac{k^2dk}{\sqrt{k}}[e^{i\omega t}\boldsymbol{f}_{jm\lambda}(k,r)a^\dagger_{jm\lambda}(k) + \text{h.c.}] ,$$

where

$$f_{jm\lambda}(k,r) = \int dn\, e^*_{k\lambda} e^{-ik\cdot r}\, D^j_{m\lambda}(n)\,, \quad [a_{jm\lambda}(k), a^\dagger_{j'm'\lambda'}(k')] = \frac{\delta(k-k')}{kk'} \delta_{jj'}\, \delta_{mm'}\, \delta_{\lambda\lambda'}\,.$$

This expansion of A into states of total angular momentum about some fixed point leads to absorption and emission amplitudes of arbitrary multipole order.[1]

6. Find the basis of photon polarization states that diagonalize the arbitrary density matrix as given in Eq. 62 by carrying out an appropriate rotation in the space defined by the Poincaré sphere. An arbitrary polarization state is, in general, elliptically polarized, and has the form[2]

$$e = N\{e_+ - e^{2i\alpha}\, r\, e_-\}\,,$$

where α is the angle the major axis makes with respect to ϵ_x, and the major and minor axes are in the ratio $|(1+r)/(1-r)|$. Show that for the Stokes vector u, this ratio is given by Eq. 65, and that $\alpha = \varphi$. Finally, confirm Eq. 66.

7. It is a remarkable property of the Casimir force (Eq. 117), that it does not depend on any parameters pertaining to the confining plates if they are assumed to be perfect conductors. Show that this no longer true for the non-leading terms, which fall off faster that $1/z^4$ — that they depend on a and derivatives of $f(k)$ at $k = 0$.

8. Confirm that time reversal leads to Eq. 159, where $J_{k\lambda}$ is understood to contain both terms in Eq. 140, and the role of the photon polarization vectors is not to be ignored.

9. More than one multipole can often contribute to a transition. Illustrate this with the example given after Eq. 170, by showing that E3, M4 and E5 are all allowed and add *coherently* in the transition amplitude, and that the leading contribution is from E3 followed by the E3-M4 interference term.

10. Consider the Zeeman effect in an E1 transition $(j_i m_i) \to (j_f m_f)$. Show that the electric field of the emitted radiation is given by

$$E \propto \sum_{\lambda=\pm1} \sum_{\kappa=0,\pm1} e^*_{k\lambda}\, d^1_{-\lambda,\kappa}(\theta)\, \langle j_i m_i 1\kappa | j_f m_f\rangle,$$

where d^1 is the $j = 1$ rotation matrix, and θ the angle between the photon momentum k and the magnetic field. Note that the polarization, which is elliptical in general, is completely determined by rotational symmetry. Consider the special case $\theta = 0$, and show that only one circular polarization is emitted; why?

11. A neutral spin $\frac{1}{2}$ particle with magnetic moment $e\mu$ is bound by a combination of central and spin-orbit forces of comparable strengths. Show that the radiative transition rate is

$$\frac{d\Gamma}{d\Omega} = \frac{\alpha k \mu^2}{2\pi} |\langle f| e^{-ik\cdot r}\, [\sigma \cdot (k \times e^*_{k\lambda})] |i\rangle|^2\,.$$

Consider the $k \to 0$ limit. State all the selection rules, including the principal quantum number, and mark the allowed transitions on a hypothetical level scheme. Are the selection rules affected if the spin-orbit interaction is much weaker than the central force?

---

[1] Cf., e.g., K. Gottfried in *Preludes in Theoretical Physics*, A. de-Shalit, H. Feshbach and L. Van Hove (eds.), North-Holland (1966).

[2] Jackson, p. 300.

12. Show that the expectation values of the electromagnetic field operators in the state of Eq. 194 agree with the field strengths given by classical electrodynamics for the same current density.

13. Show that black body radiation has a dispersion ΔN given by Eq. 207.

14. (a) Consider scattering of light by a free electron. Evaluate the amplitudes depicted in Fig. 10.8, and show that this leads to the Thomson cross section.

(b) Show that Rayleigh's ω^4-law (1871!) for scattering of light by a neutral, polarizable atom (Eq. 263), is the low-frequency limit of the Kramers-Heisenberg formula. Compare the result for $\mathcal{D}$ with the dipole moment induced in the atom by an applied homogeneous electric field, which was the subject of Prob. 5.13. Hint: In carrying out this calculation, take advantage of the identity

$$\langle g|[\dot{X}, X]|g\rangle = -2i \sum_n (E_n - E_g)|\langle E_g|X|n\rangle|^2 ,$$

the so-called dipole sum rule, which was also used in the energy loss theory of §9.1(d). This sum rule played an important role in the Old Quantum Theory and the discovery of quantum mechanics.

15. A neutron n is radiatively captured by a spin $\frac{1}{2}$ nucleus A to form a total spin $S = 0$, s-wave bound state A^*, i.e., $n + A \to A^* + \gamma$. The nA interaction is $V_0 + V_1 \boldsymbol{\sigma}_n \cdot \boldsymbol{\sigma}_A$ where $V_{0,1}$ are central fields; A is considered to have infinite mass, and all degrees of freedom besides its spin are ignored. The binding energy and characteristic size of A^* are Δ and R.

(a) Show that the dipole approximation is valid if $v^2 m_n R \ll 1$, where v is the neutron's velocity. Under this approximation, show that capture can only occur from the continuum s-wave with $S = 1$ (i.e., ${}^3S_1$), and that the capture amplitude is

$$T = \frac{4\pi i \gamma}{\sqrt{2Vk}} \langle 0|(\boldsymbol{k} \times \boldsymbol{e}^*_{k\lambda}) \cdot \boldsymbol{\sigma}_n|h; m\rangle \, J , \qquad J = \int_0^\infty \varphi_p(r) u(r) \, r^2 dr .$$

Here φ_p is radial continuum wave function of the incident neutron whose momentum is $\boldsymbol{p}$; $\gamma\boldsymbol{\sigma}_n$ the neutron's magnetic moment; u and $|0\rangle$ the radial and spin states of A^*; $|h; m\rangle$ the spin states of the initial nA system, where h is the neutron's helicity and m the spin projection of A; and $(\boldsymbol{k}, \lambda)$ the momentum and helicity of the emitted photon.

(b) Assume that A is in a random spin state. Show that the angular distribution of the photon is $(1 + 2h\lambda \cos\theta)$, where $\theta = \angle(\boldsymbol{p}, \boldsymbol{k})$, and that this is consistent with angular momentum conservation.

(c) In general J cannot be evaluated without solving the Schrödinger equation for the nA system. Show, however, that in the limit of weak binding ($\Delta \to 0$), a model independent evaluation is possible, namely,

$$J = \sqrt{\frac{\beta}{2\pi V}} \, \frac{1}{p} \, \frac{p \cos\delta + \beta \sin\delta}{p^2 + \beta^2} \, e^{i\delta}$$

where $\delta(p)$ is the ${}^3S_1$ phase shift, and $\beta^2 = 2m_n\Delta$.

(d) Show that the differential cross section is

$$\frac{d\sigma}{d\Omega} = \frac{1}{2\pi v} \frac{\beta\gamma^2}{p^2} \left| \frac{p\cos\delta + \beta \sin\delta}{p^2 + \beta^2} \right|^2 k^3 (1 \pm \cos\theta) .$$

(e) At first sight the last expression appears to be horribly singular as $v \to 0$. Show, however, that as $v \to 0$,

$$\frac{d\sigma}{d\Omega} \to \frac{1}{2\pi v} \frac{\gamma^2}{\beta^3} (a_1\beta - 1)^2 \, k^3 (1 \pm \cos\theta) ,$$

where a_1 is the $^3S_1$ scattering length. Does this have a plausible dependence on the neutron velocity as $v \to 0$?

16. Consider radiative transitions in a system composed of a hypothetical spin $\frac{1}{2}$, electrically *neutral* particle n and its antiparticle $\bar{n}$; n and $\bar{n}$ have equal masses m, and magnetic moments $\pm eg/4m$; and $n\bar{n}$ annihilation is to be ignored. The $n\bar{n}$ interaction is $H_{n\bar{n}} = V_0(r) + V_1(r)(\boldsymbol{\sigma}_1 \cdot \boldsymbol{\sigma}_2)$. Assume that V_1 is repulsive and *very* weak in comparison to V_0, and call the splittings caused by V_1 fine structure (fs). The attraction V_0 can bind the levels shown in the Fig. 10.10, where fs splittings have been greatly exaggerated.[1] The characteristic size of these bound states is R.

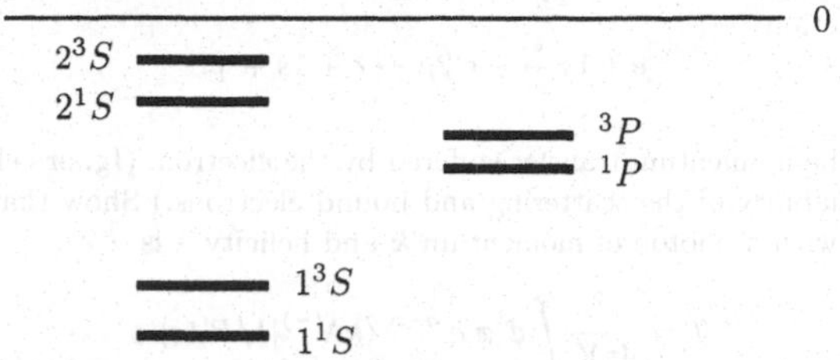

FIG. 10.10. Level scheme involved in Prob. 16.

(a) Cast the interaction between the electromagnetic field and the $n\bar{n}$ system into a form that fully exploits the symmetry under $n \leftrightarrow \bar{n}$.

(b) If $\boldsymbol{S}$ is the total $n\bar{n}$ spin, and $\boldsymbol{k}$ the photon momentum, show that the following selection rules hold in all one-photon processes for *all* values of kR: (i) $\Delta S = 1$ implies no parity change; (ii) $\Delta S = 0$ implies a parity change; (iii) there are no transitions between $S = 0$ states. Show that as a consequence one of the excited states cannot decay by one photon emission. (It can by three but not two-photon emission; can you show this ?!)

(c) Draw a detailed level diagram, including fs splittings, and for each transition indicate how the transition amplitude varies with k when $kR \ll 1$. In particular, show that the $2S \to 1S$ amplitudes vary like k^2.

(d) Calculate the total transition rate from one fs level to its partner, and show that all such rates are $(\alpha g^2/3m^2)\, k^3$. Verify that this expression has the correct dimensions, and write it in cgs units.

(e) When n and $\bar{n}$ collide, they can be captured into a bound state by emitting a photon of momentum k. Confine yourself to the case where the relative $n\bar{n}$ velocity v is very small — it is essential to take advantage of this circumstance! Using the results of the preceding parts, and assuming again that $kR \ll 1$, show that the dominant capture process leads to the bound ^3P state, and that the only other possibility is capture to the ^1S states, but with a rate smaller by a factor of order $(kR)^2$.

(f) Consider the elastic scattering of light by the $n\bar{n}$ ground state at a frequency sufficiently close to the ground state fs splitting to make this virtual transition dominant. Show that then the differential cross section is

$$\frac{d\sigma_{\lambda\lambda'}}{d\Omega} = \frac{\alpha^2 g^4 \Delta^2}{4m^4}\, \frac{|(\boldsymbol{k}' \times \boldsymbol{e}^*_{\lambda'}) \cdot (\boldsymbol{k} \times \boldsymbol{e}_\lambda)|^2}{(k^2 - \Delta^2)^2} .$$

Show that for an unpolarized photon beam the angular distribution is $(1 + \cos^2\theta)$.

---

[1] Note the traditional use of unambiguous spectroscopic notation: sometimes S means total spin, sometimes $L = 0$!

(g) The very strong selection rules of part (b) are characteristic of particle-antiparticle systems but do not apply as stated to positronium because the electron and positron have opposite intrinsic parities. How do the selection rules change if n and $\bar{n}$ have opposite intrinsic parity, i.e., if the S states have odd parity, etc.

17. The purpose of this problem is to show that resonance fluorescence produced by any mechanism will decay with the same resonance distribution as if it had been excited in photon scattering. (This statement applies in general to narrow resonances: how the narrow state is produced is irrelevant to its decay once transients that are short-lived compared to the lifetime have passed away.) Consider the excitation of the hydrogen $2p$ state by inelastic electron scattering within the framework of the model of Eq. 274, i.e., the two-step reaction

$$e + 1s \to e + 2p \to e + 1s + \gamma .$$

(a) Let $\boldsymbol{q}$ be the momentum transfer suffered by the electron. (Ignore electron spin and the indistinguishability of the scattering and bound electrons.) Show that the amplitude for a final state with a photon of momentum $\boldsymbol{k}$ and helicity λ is

$$T = \frac{e^2}{4\pi V} \int d^3x \, e^{i\boldsymbol{q}\cdot\boldsymbol{x}} \, \langle \boldsymbol{k}\lambda^{(-)} | 1/R | 1s \rangle \, ,$$

where $R = |\boldsymbol{x} - \boldsymbol{r}|$, with $\boldsymbol{x}$ and $\boldsymbol{r}$ the coordinates of the scattering and bound electrons. Here the final state is not quite the same as $|\boldsymbol{k}\lambda^{(+)}\rangle$ used in Eq. 293, etc., which has outgoing scattered waves, the reason being the same as in the photoeffect. (As in the latter, only the amplitude hinges on this point; show that use of the wrong state with outgoing boundary conditions gives the same cross section.)

(b) Show that

$$T = \frac{e^2}{V} \frac{1}{q^2} \frac{g^2}{\sqrt{2Vk}} F(q^2) \frac{\boldsymbol{q}\cdot\boldsymbol{e}^*_{k\lambda}}{k - \omega_0 - \Phi(k + i\epsilon)} \, ,$$

where $F(q^2)$ is an inelastic form factor defined by

$$\langle 2p, m | e^{i\boldsymbol{q}\cdot\boldsymbol{x}} | 1s \rangle = iF(q^2) \, (\boldsymbol{q}\cdot\boldsymbol{e}^*_m) \, ,$$

and which satisfies $F(0) = D$. Show that this is a valid representation of this matrix element, either by brute force, or preferably with a "what else can it be" argument based on symmetry.

(c) Show that the differential cross section is

$$\frac{d\sigma_\lambda}{dk d\Omega_\gamma d\Omega_e} = \frac{4}{\pi} \alpha^4 D^2 k m^2 \, \frac{1}{q^4} \frac{p_f}{p_i} \left| \frac{\omega_0 F(q^2) \, \boldsymbol{q}\cdot\boldsymbol{e}^*_{k\lambda}}{k - \omega_0 - \Phi(k + i\epsilon)} \right|^2 .$$

Here $p_{i,f}$ are the initial and final electron momenta, m the electron mass, and $d\Omega_{e,\gamma}$ the elements of solid angle for the scattered electron and radiated photon. This is a somewhat impenetrable result. Verify that it has the correct dimension. Then rewrite it in terms of the characteristic length in the problem, a_0, the Bohr radius, and for this purpose define $D = \bar{D} a_0$. Then show that in the resonance region

$$\frac{d\sigma_\lambda}{dk d\Omega_\gamma d\Omega_e} = \frac{3}{2\pi} a_0^3 \, \alpha^3 \, \bar{D}^2 \frac{1}{q^4 a_0^4} \frac{p_f}{p_i} \frac{|\omega_0 F(q^2) \boldsymbol{q}\cdot\boldsymbol{e}^*_{k\lambda}|^2}{(k - \omega_0 - \Delta)^2 + \frac{1}{4}\Gamma^2} \, .$$

(d) The angular distribution arises in two ways: from the electron momentum transfer, and from the photon's helicity and the angle Θ between $\boldsymbol{k}$ and $\boldsymbol{q}$. The dependence on electron direction via q^2 has a Rutherford-like $1/q^4$, as is always true in Coulomb scattering, but here there are complications. First, because the scattering is inelastic, $q \neq 0$ in the forward direction. Second, the form factor F introduces an angular dependence that depends in detail on the states involved. The distribution simplifies if the polarization is undetected; show that then

$$\frac{d\sigma}{dk d\Omega_\gamma d\Omega_e} = \frac{3}{2\pi} a_0^3 \alpha^3 \bar{D}^2 \frac{1}{q^2 a_0^2} \frac{p_f}{p_i} \frac{\omega_0^2 |F(q^2)/a_0|^2}{(k - \omega_0 - \Delta)^2 + \frac{1}{4}\Gamma^2} \sin^2 \Theta \, .$$

Endnotes

An extensive treatment of radiative processes can be found in V.B. Berestetskii, E.M. Lifshitz and L.P. Pitaevskii, *Quantum Electrodynamics,* 2nd. ed., Pergamon (1980), chapters V and VI.

On quantum optics, we have found the following instructive: R.J. Glauber, *Optical Coherence and Photon Statistics,* in *Quantum Optics and Electronics,* C. DeWitt *et al.* (eds.), Gordon and Breach (1965); R. Loudon, *The Quantum Theory of Light,* 2nd ed., Oxford University Press (1983); D.F. Walls and G.J. Milburn, *Quantum Optics,* Springer (1994); L. Mandel and E. Wolf, *Optical Coherence and Quantum Optics,* Cambridge University Press (1995); U. Leonhardt, *Measuring the Quantum State of Light,* Springer (1997); M.O. Scully and M.S. Zubarry, *Quantum Optics,* Cambridge University Press (1997).

11

Systems of Identical Particles

In classical physics, amazing consequences do not flow from the assumption that the constituents of a system are indistinguishable. If the constituents in question are particles with identical masses and charges, then the Hamiltonian is invariant under the interchange of the coordinates and momenta belonging to any pair of particles, which will lead to some algebraic simplifications in the solutions of the equations of motion. Call the set of all these solutions $\mathcal{S}$. If we imagine changing some of the charges and/or masses continuously, the solutions of the equations of motion will also change continuously and form a set $\mathcal{S}'$, each member of which corresponds to a member of $\mathcal{S}$ — a rather mundane story.

As we already saw in the simplest case of two particles in chapter 6, the story is very different in quantum mechanics. Indistinguishability, in addition to requiring the Hamiltonian, and all other observables, to be invariant under permutations, also imposes requirements on wave functions, which of course has no counterpart whatsoever in classical physics. Remarkable consequences of this requirement on the wave function were already encountered in the spectrum of helium (§6.2) and in elastic scattering (§8.7).

This chapter begins by examining the implications of indistinguishability for systems of more than two particles, which reveals that there are possibilities other than symmetric and antisymmetric wave functions, i.e., beyond Bose-Einstein and Fermi-Dirac statistics, though nature does not seem to have exploited them. This is followed by the formalism of second quantization, which is an extensions of the technique used in chapter 10 to quantize the electromagnetic field. Second quantization is the language in which relativistic quantum field theory is written, and it is also the language best suited to nonrelativistic many body theory. The chapter closes with the theory of free Bose and Fermi gases, and the approximate mean field theory of interacting Bose and Fermi systems.

11.1 Indistinguishability

Among the most striking conclusions of quantum theory is the knowledge that we can, with full confidence, describe any object at energies below its excitation threshold with just four degrees of freedom, as if it were a truly elementary particle. Call these degrees of freedom ξ_i for particle i; they could, for example, be the coordinate vector and a spin projection, or the momentum vector and the helicity, but any four commuting observables will do.

The Hamiltonian H of a system of N indistinguishable, or identical, particles is a function of the ξ_i that is invariant under all their interchanges, or permutations. Let P_{ij} be the operator that permutes ξ_i and ξ_j. Then, in an obvious shorthand, the identity of the particles requires

$$P_{12}H(1,2,3,\ldots,N)P_{12}^{-1} \equiv H(2,1,3,\ldots,N) = H(1,2,3,\ldots N)\,, \tag{1}$$

or in general

$$P_{ij}HP_{ij}^{-1} = H\,. \tag{2}$$

The same requirement holds for all the system's observables.

Because P_{ij} commutes with the total momentum, and must commute with spatial translations, it cannot be antiunitary. Therefore the arbitrary phases can be chosen to make $P_{ij}^2 = 1$, and the eigenvalues of every permutation operator are ± 1.

We now turn to the action of the P_{ij} on states. Let $\{|\xi_i'\rangle_i\}$ be a complete set for particle i, where ξ_i' are the eigenvalues of the observables ξ_i, and $|\ \rangle_i$ is a ket in the Hilbert space assigned to that particle. Then one complete set of N-particle states is the set of all products states,

$$|\xi_1'\xi_2'\ldots\xi_N'\rangle \equiv |\xi_1'\rangle_1 \otimes |\xi_2'\rangle_2 \cdots \otimes |\xi_N'\rangle_N\,. \tag{3}$$

The action of P_{12} on such a state is

$$\begin{aligned} P_{12}|\xi_1'\xi_2'\ldots\xi_N'\rangle &= |\xi_2'\rangle_1 \otimes |\xi_1'\rangle_2 \cdots \otimes |\xi_N'\rangle_N \\ &= |\xi_2'\xi_1'\ldots\xi_N'\rangle\,. \end{aligned} \tag{4}$$ 11-4A

Repeated use of (4) shows that different permutations do not, in general commute,

$$[P_{ij}, P_{kl}] \neq 0\,; \tag{5}$$

they will only commute if the labels (i,j) both differ from the labels (k,l). In general, therefore, only a limited set of permutation operators can be diagonalized simultaneously.

The two-particle case is trivial in this sense. There is only one permutation operator, P_{12}, whose eigenvalues $+1$ and -1 distinguish the symmetric from the antisymmetric 2-particle states already introduced in §6.1:

$$|\xi_1'\xi_2'\rangle_\pm = \tfrac{1}{\sqrt{2}}\big(\,|\xi_1'\xi_2'\rangle \pm |\xi_2'\xi_1'\rangle\,\big). \tag{6}$$

Note that neither the symmetric nor the antisymmetric states are, by themselves, complete; the simple product states $|\xi_1'\xi_2'\rangle$ which do form a complete set can only be recovered by combining both states in (6). This is illustrated by the example of two spin zero bosons, whose (symmetric) states only have even angular momenta — manifestly not a complete set.

The simplest nontrivial case is $N = 3$, in which there are three incompatible P_{ij}. Nevertheless, there is one symmetric and one antisymmetric state, wherein all three P_{ij} have the eigenvalues 1 and -1, respectively. Using the shorthand of (1), these states are

$$|123\rangle_\pm = \frac{1}{\sqrt{6}}\Big(\,|123\rangle \pm |213\rangle \pm |132\rangle \pm |321\rangle + |312\rangle + |231\rangle\,\Big)\,. \tag{7}$$

It should not come as a total surprise that all the P_{ij} can be simultaneously diagonal in some special states, for that is also true of $J = 0$ states in which all components of the angular momentum have the eigenvalue zero.

Altogether, there are 3! permutations of the product states $|123\rangle$, and thus there must be four other linear combinations beyond those of (7). Call these $|123\rangle_K$, with $K = 1, \dots 4$; in $|\ \rangle_K$ the P_{ij} are not all diagonal. But a further simplification exists in that the states $|\ \rangle_K$ can be chosen so that they fall into two sets of two such that the permutations do not mix members in one pair with those of the other pair. In the language of group theory, the permutations on three objects have four irreducible representations, the two one-dimensional representations of symmetric and antisymmetric states, and two distinct two-dimensional representations.[1]

For arbitrary N the situation generalizes in a similar fashion. There always exist totally symmetric and totally antisymmetric linear combinations, $|12\dots N\rangle_\pm$, in which all the P_{ij} have eigenvalues ± 1. In terms of the product states (3), they are

$$|12\dots N\rangle_\pm = \frac{1}{\sqrt{N!}} \left\{ \begin{matrix} \text{sym} \\ \text{det} \end{matrix} \right\} \begin{vmatrix} |1\rangle_1 & |1\rangle_2 & \cdots & |1\rangle_N \\ |2\rangle_1 & |2\rangle_2 & \cdots & |2\rangle_N \\ \vdots & \vdots & \ddots & \vdots \\ |N\rangle_1 & |N\rangle_2 & \cdots & |N\rangle_N \end{vmatrix} , \tag{8}$$

where, as always, det stands for determinant, and sym for the same linear combination but with all signs $+1$. In the antisymmetric case, (8) is called a Slater determinant.

There are, in addition, $(N! - 2)$ linear combinations that belong to various higher-dimensional representations of the permutation group on N objects. In consequence, the Hilbert space of N identical particles breaks up into subspaces $\mathfrak{H}_+$ and $\mathfrak{H}_-$ spanned by the totally symmetric and antisymmetric states, and a remainder $\mathfrak{H}_R$ that can be further broken down in terms of multi-dimensional irreducible representations. No symmetric observable, i.e., no conceivable observable for a system of identical particles, has matrix elements between any of these subspaces, and therefore no state that belongs at some instant to only one of these subspaces can, over time, acquire components in any of the other subspaces no matter what the interactions may be.

For better or worse, and to the best of our knowledge, nature does not take advantage of the spaces $\mathfrak{H}_R$. It is an empirically established fact that electrons and nucleons exist only in antisymmetric states, while photons and π mesons exist only in symmetric states. Direct (repeat, *direct*) experimental evidence is not in hand for this symmetry property of the other currently known basic constituents of matter, namely the electron neutrino, other charged leptons and their neutrinos, the hadrons that carry strangeness or other flavors, and the Z^0 and $W^\pm$ bosons of the electroweak interaction (nor of course for quarks and gluons). But a very powerful theoretical argument, and indirect empirical evidence, strongly support the statement that *all integer particles are bosons, and that all half-integer spin particles are fermions.* This is known as *the connection between spin and statistics*, which was already stated in §6.1.

Finally, a word regarding the theoretical argument just referred to. As an example of a violation of the spin-statistics connection, assume that the spin 1 photon is a fermion. The derivation of §10.2 then turns out to produce commutators between the various electromagnetic field operators that do not vanish outside the light

[1] For a comprehensive treatment of permutation symmetry, see A. Bohr and B.R. Mottelson, *Nuclear Structure*, W. A. Benjamin (1969), Vol. I, pp. 104–136.

cone, and therefore this assumption violates the principle of causality. This fact generalizes to the theorem that Lorentz covariant field theories for particles with half-integer spin only satisfy the causality principle if these particles are fermions, and for integer spin particles if they are bosons. In short, the connection between spin and statistics is a theorem of relativistic quantum field theory, and among its most important triumphs.

11.2 Second Quantization

In many circumstances the wave function is not a useful construct for treating systems of many interacting particles. It contains an enormous amount of information, and is therefore very difficult to compute with even questionable accuracy. Furthermore, most of this information is only indirectly relevant to the properties of the systems that are of greatest physical interest. To take an example, consider a system of N identical spin 0 bosons and the problem of computing the probability of finding one particle at $\boldsymbol{x}$. Because of indistinguishability, it suffices to find the probability of finding the particle nominally called 1 at $\boldsymbol{x}$. If we knew the wave function, $\Psi(\boldsymbol{x}_1\boldsymbol{x}_2\ldots\boldsymbol{x}_N)$, this probability would be

$$p_1(\boldsymbol{x}) = \int d\boldsymbol{x}_2\ldots d\boldsymbol{x}_N\,|\Psi(\boldsymbol{x}\boldsymbol{x}_2\ldots\boldsymbol{x}_N)|^2\,. \tag{9}$$

A similar formula, using the wave function in momentum space, holds for the probability of finding a particle with a given momentum. If the system has two-body interactions, the kinetic energy can be computed from this momentum distribution, and the energy due to the interactions from the probability of finding two particles at specific positions. The latter is found by integrating the N-particle density over all but two coordinates. In short, the information about higher order correlations is, so to say, thrown out by the integrations that lead to the quantities of physical interest.

It is also clear that if N is large, the probabilities of the preceding paragraph, which pertain to only one or a pair of particles, are insensitive to the value of N. In the terminology of statistical mechanics, these are intensive quantities, having a well-defined thermodynamic limit — the limit in which N and the volume tend to infinity with a finite ratio n, the particle density. And speaking of thermodynamics, for systems with very large N, the ground state (or any pure state) is often only an idealized notion, because most of these systems have excitation thresholds that tend to zero as $N \to \infty$, so that the zero temperature limit is rather artificial. Hence, such systems are best described by the density matrix for a system in thermal equilibrium.

What is needed, then, is a formalism that focuses on a small number of degrees of freedom, which does not refer explicitly to N, and which can handle mixtures on an equal footing with pure states. This list of desiderata is met by the technique of *second quantization.* [1] This name stems from the fact that in this formalism certain amplitudes that are c-numbers in conventional quantum mechanics are replaced by non-commuting operators, which results in *non-Hermitian field operators*

[1]Second quantization for bosons was invented by Dirac and for fermions by Jordan and Wigner.

$\psi(\boldsymbol{x}, t)$ whose equation of motion in the Heisenberg picture looks, superficially, like the ordinary Schrödinger equation (though its is actually nonlinear, whereas the Schrödinger equation is strictly linear).

Above and beyond its technical merits, the formalism of second quantization has a much deeper significance. It provides a vivid and direct mathematical expression of particle-wave duality. As we said, the operator $\psi(\boldsymbol{x}, t)$ and its relatives satisfy wavelike equations, and in addition, as we shall soon learn, these operators act on variables that are most corpuscular, because they create, destroy and count particles.

(a) Bose-Einstein Statistics

In §10.1 we saw that the free electromagnetic field can be thought of as a gas of non-interacting particles (photons) obeying Bose-Einstein statistics. A stationary state of the electromagnetic field was specified by giving the number of photons having various momenta and helicities. The basic operators that generated the whole Hilbert space, the creation and destruction operators, and the commutation rules between them guaranteed that all the state vectors were symmetric in the individual photon variables. Furthermore, this formulation of electrodynamics had all the desirable features referred to in the introduction to this section. Thus the expressions for the energy and momentum of the field, as well as the dynamical (Maxwell) equations, did not explicitly depend on the number of photons present. We therefore adapt the formalism of creation and destruction operators to our present problem.

The systems of interest to us now differ in certain fundamental ways from the radiation field, however. We are now concerned with particles having a rest mass, and confine ourselves to a nonrelativistic theory; and to systems where the number of particles is a constant of the motion. Thus we shall not deal with observables, such as the electromagnetic field strengths, which do not commute with the number operator.

Consider a system of spin zero bosons of mass m. The extension of the formalism to nonzero spin is straightforward, and will not be spelled out. As in the theory of the electromagnetic field, introduce creation and destruction operators, $a_{\boldsymbol{p}}^\dagger$ and $a_{\boldsymbol{p}}$, having the commutation rules

$$[a_{\boldsymbol{p}}, a_{\boldsymbol{p}'}^\dagger] = \delta_{\boldsymbol{p}\boldsymbol{p}'} , \qquad [a_{\boldsymbol{p}}, a_{\boldsymbol{p}'}] = 0 . \tag{10}$$

Here the label $\boldsymbol{p}$ specifies the momentum. We again employ periodic boundary conditions on a cube whose volume will now be called Ω. From these operators we then construct the number operators $N_{\boldsymbol{p}} = a_{\boldsymbol{p}}^\dagger a_{\boldsymbol{p}}$, with commutation rules

$$[a_{\boldsymbol{p}}, N_{\boldsymbol{p}}] = a_{\boldsymbol{p}} , \qquad [a_{\boldsymbol{p}}^\dagger, N_{\boldsymbol{p}}] = -a_{\boldsymbol{p}}^\dagger . \tag{11}$$

Define a normalized empty state, or vacuum, $|0\rangle$ by

$$a_{\boldsymbol{p}}|0\rangle = 0 \tag{12}$$

for all $\boldsymbol{p}$. The complete Hilbert space is then built up from $|0\rangle$. The state in which the commuting observables $\{N_{\boldsymbol{p}}\}$ have the eigenvalues $\{n_{\boldsymbol{p}}\}$ is given by

$$|\{n_{\boldsymbol{p}}\}\rangle = \prod_{\boldsymbol{p}} (n_{\boldsymbol{p}}!)^{-\frac{1}{2}} (a_{\boldsymbol{p}}^\dagger)^{n_{\boldsymbol{p}}} |0\rangle . \tag{13}$$

These states are obviously symmetric because the $\{a_{\boldsymbol{p}}^\dagger\}$ commute with each other. Thus the two-particle state $|\boldsymbol{p}_1\boldsymbol{p}_2\rangle \equiv a_{\boldsymbol{p}_1}^\dagger a_{\boldsymbol{p}_2}^\dagger|0\rangle$ satisfies $|\boldsymbol{p}_1\boldsymbol{p}_2\rangle = |\boldsymbol{p}_2\boldsymbol{p}_1\rangle$.

It is important to understand the difference between the space spanned by the vectors (13), and the Hilbert space of "conventional" quantum mechanics. In the conventional formulation, we specify the total number[1] N' of particles, and the Hilbert space $\mathfrak{H}_{N'}$ only contains state vectors describing N'-body systems. Operators that connect states with different numbers of particles do not arise. The space spanned by the states (13) is far larger, however, and one refers to it as *Fock space.* The Fock space $\mathfrak{F}$ is the sum of all the Hilbert spaces referred to above, i.e., $\mathfrak{F} = \mathfrak{H}_0 \oplus \mathfrak{H}_1 \oplus \mathfrak{H}_2 \oplus \ldots$. The operators $a_{\boldsymbol{p}}$ and $a_{\boldsymbol{p}}^\dagger$ are defined in $\mathfrak{F}$, though not in the conventional formulation, because they connect vectors in $\mathfrak{H}_{N'}$ and $\mathfrak{H}_{N''}$ with $N' = N'' \pm 1$. When dealing with systems (such as photons or pions) where the number of bosons is not conserved, a description in Fock space is indispensable, because observables (e.g., the field strengths) that are linear Hermitian forms in $a_{\boldsymbol{p}}$ and $a_{\boldsymbol{p}}^\dagger$ really exist. For systems such as liquid He, such observables do not occur, but the Fock space formulation is almost as indispensable.

All other observables may be constructed from the $a_{\boldsymbol{p}}$'s. As we are dealing with systems in which the total number of particles is conserved, we are only concerned with observables that are products of equal numbers of creation and destruction operators. The simplest example is the total momentum operator,

$$\boldsymbol{P} = \sum_{\boldsymbol{p}} \boldsymbol{p}\, a_{\boldsymbol{p}}^\dagger a_{\boldsymbol{p}} \,. \tag{14}$$

Equation (13) is an eigenstate of $\boldsymbol{P}$ with eigenvalue $\sum \boldsymbol{p}\, n_{\boldsymbol{p}}$, whereas $a_{\boldsymbol{p}}^\dagger$ increases the eigenvalue of any eigenstate of $\boldsymbol{P}$ in the amount $\boldsymbol{p}$, and vice versa for $a_{\boldsymbol{p}}$. If the particles in question do not interact, the Hamiltonian is

$$H = \sum_{\boldsymbol{p}} \frac{p^2}{2m}\, a_{\boldsymbol{p}}^\dagger a_{\boldsymbol{p}} \,, \tag{15}$$

and the states (13) are eigenstates of H with eigenvalue $\sum (p^2/2m) n_{\boldsymbol{p}}$.

The momentum representation operators are not sacrosanct. An equally important set of operators are defined by[2]

$$\psi(\boldsymbol{x}) = \Omega^{-\frac{1}{2}} \sum_{\boldsymbol{p}} e^{i\boldsymbol{p}\cdot\boldsymbol{x}} a_{\boldsymbol{p}} \,, \qquad \psi^\dagger(\boldsymbol{x}) = \Omega^{-\frac{1}{2}} \sum_{\boldsymbol{p}} e^{-i\boldsymbol{p}\cdot\boldsymbol{x}} a_{\boldsymbol{p}}^\dagger \,. \tag{16}$$

These destruction and creation operators are also called Bose *field operators.* Because

$$\Omega^{-1} \int_\Omega d\boldsymbol{x}\, e^{-i(\boldsymbol{p}-\boldsymbol{p}')\cdot\boldsymbol{x}} = \delta_{\boldsymbol{p},\boldsymbol{p}'} \,, \tag{17}$$

the inverse relations to (16) are

$$a_{\boldsymbol{p}} = \Omega^{-\frac{1}{2}} \int d\boldsymbol{x}\, e^{-i\boldsymbol{p}\cdot\boldsymbol{x}} \psi(\boldsymbol{x}) \,, \qquad a_{\boldsymbol{p}}^\dagger = \Omega^{-\frac{1}{2}} \int d\boldsymbol{x}\, e^{i\boldsymbol{p}\cdot\boldsymbol{x}} \psi^\dagger(\boldsymbol{x}) \,. \tag{18}$$

---

[1] We reserve the symbol N for the number operator [see (23)], and designate its eigenvalues by N' or n.

[2] In this chapter no confusion between the operators $\psi(\boldsymbol{x})$ and wave functions should occur, because we will always use the symbols φ and u for the latter.

The completeness relationship,

$$\Omega^{-1}\sum_{\boldsymbol{p}} e^{i(\boldsymbol{x}-\boldsymbol{x}')\cdot\boldsymbol{p}} = \delta(\boldsymbol{x}-\boldsymbol{x}'), \tag{19}$$

then leads to the *basic boson commutation rules*

$$[\psi(\boldsymbol{x}), \psi^\dagger(\boldsymbol{x}')] = \delta(\boldsymbol{x}-\boldsymbol{x}') . \tag{20}$$

$$[\psi(\boldsymbol{x}), \psi(\boldsymbol{x}')] = [\psi^\dagger(\boldsymbol{x}), \psi^\dagger(\boldsymbol{x}')] = 0 . \tag{21}$$

The operator for the total number of particles,

$$N = \sum_{\boldsymbol{p}} a_{\boldsymbol{p}}^\dagger a_{\boldsymbol{p}} , \tag{22}$$

can be expressed in terms of ψ and $\psi^\dagger$ by using (18) and (19):

$$N = \int d\boldsymbol{x}\ \psi^\dagger(\boldsymbol{x})\psi(\boldsymbol{x}) . \tag{23}$$

The commutation rules of N with the field operators then follow from (20) and (21):

$$[\psi(\boldsymbol{x}), N] = \psi(\boldsymbol{x}) , \qquad [\psi^\dagger(\boldsymbol{x}), N] = -\psi^\dagger(\boldsymbol{x}) . \tag{24}$$

Hence if $|N'\rangle$ is an eigenket of N with eigenvalue N', $\psi(\boldsymbol{x})|N'\rangle$ has the eigenvalue $N'-1$ of N, whereas $\psi^\dagger(\boldsymbol{x})|N'\rangle$ belongs to the eigenvalue $N'+1$. These facts also follow immediately from the definition of $\psi(\psi^\dagger)$ as a linear form in $a_{\boldsymbol{p}}(a_{\boldsymbol{p}}^\dagger)$.

We have already observed that $a_{\boldsymbol{p}}^\dagger|0\rangle$ is a one-particle state of momentum $\boldsymbol{p}$, and that the eigenvalues of $a_{\boldsymbol{p}}^\dagger a_{\boldsymbol{p}}$ are integers which were shown to be the number of particles of momentum $\boldsymbol{p}$. What are the analogous statements in the position representation? It is tempting, on the basis of (23), to interpret the positive Hermitian operator $\psi^\dagger(\boldsymbol{x})\psi(\boldsymbol{x})$ as the number density operator in position space— i.e., to assert that in any state $|\ \rangle$, $\langle\psi^\dagger(\boldsymbol{x})\psi(\boldsymbol{x})\rangle d\boldsymbol{x}$ is the mean number of particles in the volume element $d\boldsymbol{x}$— and to interpret $|\boldsymbol{x}\rangle \equiv \psi^\dagger(\boldsymbol{x})|0\rangle$ as the state in which "one particle is at the point $\boldsymbol{x}$." The latter is justified since it implies that the transformation function from $|\boldsymbol{x}\rangle$ to the one-particle momentum eigenstate $|\boldsymbol{p}\rangle \equiv a_{\boldsymbol{p}}^\dagger|0\rangle$ is

$$\langle\boldsymbol{x}|\boldsymbol{p}\rangle = \Omega^{-\frac{1}{2}}\langle 0|\psi(\boldsymbol{x})\int d\boldsymbol{x}' e^{i\boldsymbol{p}\cdot\boldsymbol{x}'}\psi^\dagger(\boldsymbol{x}')|0\rangle = \Omega^{-\frac{1}{2}}e^{i\boldsymbol{p}\cdot\boldsymbol{x}} , \tag{25}$$

which is correct. Furthermore, the basic commutation rule (14) leads to $\langle\boldsymbol{x}|\boldsymbol{x}'\rangle = \delta(\boldsymbol{x}-\boldsymbol{x}')$. To verify the interpretation of $\psi^\dagger(\boldsymbol{x})\psi(\boldsymbol{x})$ given above, we evaluate the result of applying the operator

$$N_{\Delta\Omega} \equiv \int_{\Delta\Omega} d\boldsymbol{x}\ \psi^\dagger(\boldsymbol{x})\psi(\boldsymbol{x}) \tag{26}$$

onto the ket $\psi^\dagger(\boldsymbol{y})|0\rangle$. If $\boldsymbol{y}$ does not lie in the volume $\Delta\Omega$, $N_{\Delta\Omega}\psi^\dagger(\boldsymbol{y})|0\rangle = 0$. On the other hand, when $\boldsymbol{y}$ does lie in $\Delta\Omega$,

$$N_{\Delta\Omega}\psi^\dagger(\boldsymbol{y})|0\rangle = \int_{\Delta\Omega} d\boldsymbol{x}\ \psi^\dagger(\boldsymbol{x})[\delta(\boldsymbol{x}-\boldsymbol{y}) + \psi^\dagger(\boldsymbol{y})\psi(\boldsymbol{x})]|0\rangle = \psi^\dagger(\boldsymbol{y})|0\rangle . \tag{27}$$

Thus $\psi^\dagger(\boldsymbol{y})|0\rangle$ is an eigenvector of $N_{\Delta\Omega}$ with eigenvalue one when $\boldsymbol{y}$ lies inside $\Delta\Omega$, and eigenvalue zero when $\boldsymbol{y}$ lies outside $\Delta\Omega$, no matter how small $\Delta\Omega$ may be. The argument is readily extended to an n-particle state $\prod_{i=1}^n \psi^\dagger(\boldsymbol{y}_i)|0\rangle$. Upon commuting the destruction operator through to the right, we obtain

$$\begin{aligned} N_{\Delta\Omega} \prod_{i=1}^n \psi^\dagger(\boldsymbol{y}_i)|0\rangle &= \int_{\Delta\Omega} d\boldsymbol{x}\, \psi^\dagger(\boldsymbol{x}) \sum_{i=1}^n \delta(\boldsymbol{x}-\boldsymbol{y}_i) \frac{\delta}{\delta\psi^\dagger(\boldsymbol{y}_i)} \prod_{j=1}^n \psi^\dagger(\boldsymbol{y}_j)|0\rangle \\ &= N'_{\Delta\Omega} \prod_{i=1}^n \psi^\dagger(\boldsymbol{y}_i)|0\rangle , \end{aligned} \tag{28}$$

where $N'_{\Delta\Omega}$ is the number of points in the set $(\boldsymbol{y}_1, \ldots, \boldsymbol{y}_n)$ that lie inside $\Delta\Omega$. Hence $N_{\Delta\Omega}$ is the operator whose eigenvalues specify the number of particles in the spatial volume $\Delta\Omega$, and the interpretation of the operator $\psi^\dagger(\boldsymbol{x})\psi(\boldsymbol{x})$ given above is therefore correct.

Having established that $\prod_{i=1}^n \psi^\dagger(\boldsymbol{x}_i)|0\rangle$ is an n-particle state in which the particles "are at" the indicated positions, we may now relate these states to the conventional wave functions. Let $\varphi_a(\boldsymbol{x}_1 \ldots \boldsymbol{x}_n)$ be a normalized and totally symmetric wave function for n-particles. Consider then the ket

$$|\varphi_a\rangle = (n!)^{-\frac{1}{2}} \int d\boldsymbol{x}_1 \ldots d\boldsymbol{x}_n\, \varphi_a(\boldsymbol{x}_1 \ldots \boldsymbol{x}_n)\, \psi^\dagger(\boldsymbol{x}_1) \ldots \psi^\dagger(\boldsymbol{x}_n)|0\rangle . \tag{29}$$

We shall show that the factor $(n!)^{-\frac{1}{2}}$ gives the conventional interpretation of $|\varphi_a\rangle$, i.e., that the transformation function to the coordinate representation is the wave function $\varphi_a(\boldsymbol{x}_1 \ldots \boldsymbol{x}_n)$. For this purpose it suffices to guarantee that

$$\langle\varphi_b|\varphi_a\rangle = \int d\boldsymbol{x}_1 \ldots d\boldsymbol{x}_n\, \varphi_b^*(\boldsymbol{x}_1 \ldots \boldsymbol{x}_n)\, \varphi_a(\boldsymbol{x}_1 \ldots \boldsymbol{x}_n) , \tag{30}$$

where $|\varphi_b\rangle$ is constructed in the same way as $|\varphi_a\rangle$. In order to evaluate $\langle\varphi_b|\varphi_a\rangle$ we require $\langle\psi(\boldsymbol{x}'_1) \ldots \psi(\boldsymbol{x}'_n)\psi^\dagger(\boldsymbol{x}_1) \ldots \psi^\dagger(\boldsymbol{x}_n)\rangle_0$, where $\langle \ldots \rangle_0$ is the expectation value in $|0\rangle$. Repeated use of the commutation rules leads to

$$\langle\psi(\boldsymbol{x}'_1) \ldots \psi(\boldsymbol{x}'_n)\psi^\dagger(\boldsymbol{x}_1) \ldots \psi^\dagger(\boldsymbol{x}_n)\rangle_0 = \sum_P \delta(\boldsymbol{x}_1 - \boldsymbol{x}'_1) \ldots \delta(\boldsymbol{x}_n - \boldsymbol{x}'_n) , \tag{31}$$

where $\sum_P$ is a sum over all $n!$ permutations P of the n vectors $(\boldsymbol{x}_1, \boldsymbol{x}_2, \ldots, \boldsymbol{x}_n)$; similarly, $\langle a_{\boldsymbol{k}_1} \ldots a_{\boldsymbol{k}_n} a^\dagger_{\boldsymbol{q}_1} \ldots a^\dagger_{\boldsymbol{q}_n}\rangle = \sum_P \delta_{\boldsymbol{k}_1\boldsymbol{q}_1} \delta_{\boldsymbol{k}_n\boldsymbol{q}_n}$. Hence

$$\begin{aligned} \langle\varphi_b|\varphi_a\rangle &= (n!)^{-1} \int d\boldsymbol{x}_1 \ldots d\boldsymbol{x}_n\, \varphi_b^*(\boldsymbol{x}_1 \ldots \boldsymbol{x}_n) \sum_P \varphi_a(\boldsymbol{x}_1 \ldots \boldsymbol{x}_n) \\ &= \int d\boldsymbol{x}_1 \ldots d\boldsymbol{x}_n\, \varphi_b^*(\boldsymbol{x}_1 \ldots \boldsymbol{x}_n)\varphi_a(\boldsymbol{x}_1 \ldots \boldsymbol{x}_n) , \end{aligned} \tag{32}$$

where the last equality is a consequence of the assumed symmetry of $\varphi_{a,b}(\boldsymbol{x}_1 \ldots \boldsymbol{x}_n)$; *QED*. In the same vein, define normalized and completely symmetric n-particle coordinate space kets by

$$|\boldsymbol{x}_1 \ldots \boldsymbol{x}_n; n\rangle = (n!)^{-\frac{1}{2}} \psi^\dagger(\boldsymbol{x}_1) \ldots \psi^\dagger(\boldsymbol{x}_n)|0\rangle . \tag{33}$$

The orthogonality relationship satisfied by these symmetric kets is therefore

$$\langle \boldsymbol{x}_1 \ldots \boldsymbol{x}_n; n | \boldsymbol{x}'_1 \ldots \boldsymbol{x}'_n; n \rangle = (n!)^{-1} \sum_P \delta(\boldsymbol{x}_1 \ldots \boldsymbol{x}'_1) \ldots \delta(\boldsymbol{x}_n - \boldsymbol{x}'_n) \,, \tag{34}$$

and the coordinate space wave function is indeed

$$\varphi_a(\boldsymbol{x}_1 \ldots \boldsymbol{x}_n) = \langle \boldsymbol{x}_1 \ldots \boldsymbol{x}_n; n | \varphi_a \rangle \,. \tag{35}$$

The effect of applying the field operators onto the kets (33) is readily determined. By inspection we have

$$\psi^\dagger(\boldsymbol{y}) | \boldsymbol{x}_1 \ldots \boldsymbol{x}_n; n \rangle = \sqrt{n+1}\, | \boldsymbol{y} \boldsymbol{x}_1 \ldots \boldsymbol{x}_n; n+1 \rangle \,. \tag{36}$$

Repeated use of the commutation rule leads to the somewhat more complicated result

$$\psi(\boldsymbol{y}) | \boldsymbol{x}_1 \ldots \boldsymbol{x}_n; n \rangle = \frac{1}{\sqrt{n}} \sum_{s=1}^{n} \delta(\boldsymbol{x}_s - \boldsymbol{y}) | \boldsymbol{x}_1 \ldots \boldsymbol{x}_{s-1} \boldsymbol{x}_{s+1} \ldots \boldsymbol{x}_n; n-1 \rangle \,. \tag{37}$$

Combining (36) and (37), and recalling that all these kets are symmetric functions of the coordinates, yields

$$\psi^\dagger(\boldsymbol{y}) \psi(\boldsymbol{y}') | \boldsymbol{x}_1 \ldots \boldsymbol{x}_n; n \rangle = \sum_{s=1}^{n} \delta(\boldsymbol{x}_s - \boldsymbol{y}') | \boldsymbol{x}_1 \ldots \boldsymbol{x}_{s-1} \boldsymbol{y} \boldsymbol{x}_{s+1} \ldots \boldsymbol{x}_n; n \rangle \,. \tag{38}$$

All the equipment required for the construction of arbitrary symmetric observables[1] in terms of the creation and destruction operators is now at hand. Consider the simplest observable, a symmetric sum of one-particle observables (e.g., the linear or angular momenta). In conventional quantum mechanics such an observable for an n-body system is written as

$$F = \sum_{i=1}^{n} f_i \,, \tag{39}$$

where f_i acts only on the coordinates of particle i. Let us first define the action of this operator in the Hilbert space $\mathfrak{H}_n$. If $|\boldsymbol{x}\rangle_i$ is a position eigenket of particle i, a symmetrized position ket for n particles is given by

$$|\boldsymbol{x}_1 \ldots \boldsymbol{x}_n\rangle_S = (n!)^{-\frac{1}{2}} \sum_P |\boldsymbol{x}_1\rangle_1 \ldots |\boldsymbol{x}_n\rangle_n \,. \tag{40}$$

The kets (40) span the space $\mathfrak{H}_n$. Applying F to (40) gives

$$\begin{aligned} F|\boldsymbol{x}_1 \ldots \boldsymbol{x}_n\rangle_S &= (n!)^{-\frac{1}{2}} \sum_P \int d\boldsymbol{y}\; |\boldsymbol{y}\rangle_1 |\boldsymbol{x}_2\rangle_2 \ldots |\boldsymbol{x}_n\rangle_n \; {}_1\langle \boldsymbol{y}|f_1|\boldsymbol{x}_1\rangle_1 \\ &+ (n!)^{-\frac{1}{2}} \sum_P \int d\boldsymbol{y}\; |\boldsymbol{x}_1\rangle |\boldsymbol{y}\rangle_2 \ldots |\boldsymbol{x}_n\rangle_n \; {}_2\langle \boldsymbol{y}|f_2|\boldsymbol{x}_2\rangle_2 \\ &+ \ldots \,. \end{aligned} \tag{41}$$

[1]Note that in the second quantization formalism, it is impossible to write down an operator which refers to one specific particle.

The matrix elements ${}_i\langle \boldsymbol{y}|f_i|\boldsymbol{x}\rangle_i$ are the same for all i, however, and is written simply as $\langle \boldsymbol{y}|f|\boldsymbol{x}\rangle$. Therefore (41) is

$$F|\boldsymbol{x}_1 \ldots \boldsymbol{x}_n\rangle_S = \sum_{s=1}^{n} |\boldsymbol{x}_1 \ldots \boldsymbol{x}_{s-1}\boldsymbol{y}\boldsymbol{x}_{s+1} \ldots \boldsymbol{x}_n\rangle\langle \boldsymbol{y}|f|\boldsymbol{x}_s\rangle \,. \tag{42}$$

Our problem is now solved: the Fock space representation of the observable F is

$$F = \int d\boldsymbol{x}d\boldsymbol{x}'\psi^\dagger(\boldsymbol{x})\langle \boldsymbol{x}|f|\boldsymbol{x}'\rangle\psi(\boldsymbol{x}') \,. \tag{43}$$

Equation (38) insures that (43) gives the correct result (42) in every n-particle subspace of $\mathfrak{F}$. An intuitive appreciation for expressions such as F should be acquired. As one sees, the integrand destroys a particle at $\boldsymbol{x}'$, and recreates it at $\boldsymbol{x}$, with amplitude $\langle \boldsymbol{x}|f|\boldsymbol{x}'\rangle$.

More complicated observables can be constructed in a similar fashion. The derivations are straightforward, though tedious. We shall only quote the result for a symmetric sum of two-body operators, such as a two-body interaction, i.e.,

$$G = \sum_{i<j} g_{ij} = \tfrac{1}{2}\sum_{i \neq j} g_{ij} \,.$$

Let $\langle \boldsymbol{x}_1\boldsymbol{x}_2|g|\boldsymbol{x}_1'\boldsymbol{x}_2'\rangle$ be the coordinate space matrix element of g between *unsymmetrized* two-particle position kets. Then G is expressed as follows:

$$G = \frac{1}{2}\int d\boldsymbol{x}_1 \ldots d\boldsymbol{x}_2' \; \psi^\dagger(\boldsymbol{x}_1)\psi^\dagger(\boldsymbol{x}_2)\langle \boldsymbol{x}_1\boldsymbol{x}_2|g|\boldsymbol{x}_1'\boldsymbol{x}_2'\rangle\psi(\boldsymbol{x}_2')\psi(\boldsymbol{x}_1') \,. \tag{44}$$

To give some examples of such operators, consider first the total momentum $\boldsymbol{P}$. Define the operator

$$\frac{\partial\psi(\boldsymbol{x})}{\partial\boldsymbol{x}} = i\frac{1}{\sqrt{\Omega}}\sum_{\boldsymbol{p}} \boldsymbol{p}\, e^{i\boldsymbol{p}\cdot\boldsymbol{x}} a_{\boldsymbol{p}} \,;$$

then

$$\boldsymbol{P} = \int d\boldsymbol{x}\; \psi^\dagger(\boldsymbol{x})\frac{1}{i}\frac{\partial\psi(\boldsymbol{x})}{\partial\boldsymbol{x}} \,. \tag{45}$$

Note the appearance of the familiar expression $(1/i)\partial/\partial\boldsymbol{x}$. The kinetic energy can, in a similar fashion, be written as

$$K = \frac{1}{2m}\int d\boldsymbol{x}\; \frac{\partial\psi^\dagger(\boldsymbol{x})}{\partial\boldsymbol{x}} \cdot \frac{\partial\psi(\boldsymbol{x})}{\partial\boldsymbol{x}} = -\frac{1}{2m}\int d\boldsymbol{x}\; \psi^\dagger(\boldsymbol{x})\nabla^2\psi(\boldsymbol{x}) \,. \tag{46}$$

The Coulomb energy in a system of particles carrying the charge e is an example of an operator of type G. In this case

$$\langle \boldsymbol{x}_1\boldsymbol{x}_2|g|\boldsymbol{x}_1'\boldsymbol{x}_2'\rangle = \frac{e^2\delta(\boldsymbol{x}_1 - \boldsymbol{x}_1')\delta(\boldsymbol{x}_2 - \boldsymbol{x}_2')}{4\pi|\boldsymbol{x}_1 - \boldsymbol{x}_2|} \tag{47}$$

and therefore the Coulomb energy is

$$V_C = \tfrac{1}{2}e^2\int \frac{\psi^\dagger(\boldsymbol{x}_1)\psi^\dagger(\boldsymbol{x}_2)\psi(\boldsymbol{x}_2)\psi(\boldsymbol{x}_1)}{4\pi|\boldsymbol{x}_1 - \boldsymbol{x}_2|}\, d\boldsymbol{x}_1 d\boldsymbol{x}_2 \,. \tag{48}$$

Finally, we point out that all of these relations can be written in an infinite variety of ways. Let $\{u_\nu(\boldsymbol{x})\}$ be any complete orthonormal set of one-particle wave functions, and define the creation and destruction operators $b_\nu^\dagger$ and b_ν by

$$b_\nu = \int d\boldsymbol{x}\; u_\nu^*(\boldsymbol{x})\psi(\boldsymbol{x}) \,, \quad b_\nu^\dagger = \int d\boldsymbol{x}\; u_\nu(\boldsymbol{x})\psi^\dagger(\boldsymbol{x}) \,. \tag{49}$$

The commutation rules (20) and (21) satisfied by the field operators then lead to

$$[b_\nu, b_{\nu'}^\dagger] = \delta_{\nu\nu'} \,, \qquad [b_\nu, b_{\nu'}] = 0 \,. \tag{50}$$

A unitary transformation from $\{u_\nu(\boldsymbol{x})\}$ to any other basis will lead to a new set of creation and destruction operators with precisely the same commutation rules. The operators F and G can be also written in terms of b_ν and $b_\nu^\dagger$:

$$F = \sum_{\nu\nu'} b_\nu^\dagger \langle\nu|f|\nu'\rangle b_{\nu'} \,, \tag{51}$$

$$G = \tfrac{1}{2}\sum_{[\nu]} b_{\nu_1}^\dagger b_{\nu_2}^\dagger \langle\nu_1\nu_2|g|\nu_1'\nu_2'\rangle b_{\nu_2'} b_{\nu_1'} \,. \tag{52}$$

(b) Fermi-Dirac Statistics

The second quantization formalism must undergo some fundamental changes if it is to apply to a system of indistinguishable fermions because the occupation numbers can only assume the values 0 and 1 if the exclusion principle is to be satisfied.

Consider a system of n fermions. Let $\{|\alpha_\nu\rangle_i\}$ be a complete set of states for particle i, where α_ν stands for any convenient set of four one-particle quantum numbers. These could be the position $\boldsymbol{x}$ and the projection s of the spin angular momentum on a given axis, or the four familiar quantum numbers used in spectroscopy. Recall from §11.1 that an antisymmetric n-particle state can be written as the determinant

$$|\alpha_{\nu_1}\dots\alpha_{\nu_N}\rangle_A \equiv \frac{1}{\sqrt{n!}}\begin{vmatrix} |\alpha_{\nu_1}\rangle_1 & \dots\dots & |\alpha_{\nu_1}\rangle_n \\ \cdot & & \cdot \\ \cdot & & \cdot \\ \cdot & & \cdot \\ |\alpha_{\nu_n}\rangle_1 & \dots\dots & |\alpha_{\nu_n}\rangle_n \end{vmatrix} \,. \tag{53}$$

The set of all such Slater determinants spans the Hilbert space $\mathfrak{H}_n$ for an n-body Fermi system. It is crucial to note that (53) is specified, except for phases, by stating whether the quantum number α_ν occurs or not. These are the only two possibilities, because (53) vanishes if α_ν appears more than once. Hence (53) is specified equally well by an infinite number of yes-no statements.

Let us introduce a vector space, the Fock space $\mathfrak{F}$, spanned by the states $|\{n_\nu\}\rangle$ specified by the occupation numbers $n_\nu = 0, 1$. There is then a correspondence between the state vectors (53) in $\mathfrak{H}_n$, and the vectors in $\mathfrak{F}$:

$$|\alpha_{\nu_1}\dots\alpha_{\nu_n}\rangle_A \Leftrightarrow \pm|n_1, n_2, \dots\rangle \,, \qquad \sum_i n_i = n \,. \tag{54}$$

The sign ambiguity appears because $|\alpha_{\nu_1}\alpha_{\nu_2}\alpha_{\nu_3}\dots\alpha_{\nu_n}\rangle$ and $|\alpha_{\nu_2}\alpha_{\nu_1}\alpha_{\nu_3}\dots\alpha_{\nu_n}\rangle$ have the same occupation numbers. This ambiguity is removed by the convention that $\nu_1 < \nu_2 < \dots < \nu_n$ in $|\alpha_{\nu_1}\dots\alpha_{\nu_n}\rangle_A$.

One should be quite clear on the meaning of (54). Consider a three-body system where one particle is in each of the states α_2, α_3, α_5. Then (54) reads $|\alpha_2\alpha_3\alpha_5\rangle_A \Leftrightarrow |0,1,1,0,1,0,0,\ldots\rangle$. That is, whereas the state $|\alpha_{\nu_1}\ldots\alpha_{\nu_n}\rangle_A$ is specified by n quantum numbers, the Fock space state requires an infinite set of quantum numbers for its specification. On the other hand, the quantum numbers in the $\mathfrak{F}$-description can only take on two values, whereas the α_ν will, in general, take on an infinite number of values.

We now introduce operators which work on the states $|\{n_\nu\}\rangle$. Define a pair of operators $a_\nu^\dagger$ and a_ν for each possible state $|\alpha_\nu\rangle$, with the algebraic properties

$$\{a_\nu^\dagger, a_{\nu'}\} = \delta_{\nu\nu'}\,, \qquad \{a_\nu, a_{\nu'}\} = 0\,, \tag{55}$$

where $\{A,B\} \equiv AB + BA$ is called the anticommutator of A and B. Due to (55), $(a_\nu^\dagger a_\nu)^2 = a_\nu^\dagger a_\nu(1 - a_\nu a_\nu^\dagger)$. But (55) requires

$$(a_\nu)^2 = (a_\nu^\dagger)^2 = 0\,, \tag{56}$$

and therefore

$$(a_\nu^\dagger a_\nu)^2 = a_\nu^\dagger a_\nu\,; \tag{57}$$

hence the eigenvalues of $a_\nu^\dagger a_\nu$ are zero and one. Define now the number operator N_ν by

$$N_\nu = a_\nu^\dagger a_\nu\,. \tag{58}$$

Then the eigenvalues of N_ν are the occupation number $n_\nu = 0, 1$ introduced earlier. Furthermore, a simple calculation shows that $[N_\nu, N_{\nu'}] = 0$. The kets $|\{n_\nu\}\rangle$ that span $\mathfrak{F}$ are therefore the simultaneous eigenvectors of the observables $\{N_\nu\}$.

The commutation rules of $a_\nu^\dagger$ and a_ν with N_ν are again important. Computations with Fermi operators are facilitated by the identities

$$\begin{aligned} [A, BC] &= \{A,B\}C - B\{A,C\}\,, \\ [AB, C] &= A\{B,C\} - \{A,C\}B\,. \end{aligned} \tag{59}$$

With their help we immediately find

$$[a_\nu, N_{\nu'}] = a_\nu \delta_{\nu\nu'}\,, \qquad [a_\nu^\dagger, N_{\nu'}] = -a_\nu^\dagger \delta_{\nu\nu'}\,. \tag{60}$$

These are precisely the same relations as in the Bose case [see (11)]. Hence $a_\nu^\dagger$ and a_ν are, respectively, creation and destruction operators. Conditions (56) and (57), which are built into the basic algebraic rules, assure that these operators cannot construct states that violate the exclusion principle.

As in the Bose case, a basis in $\mathfrak{F}$ is constructed from the empty of vacuum state in which $n_\nu = 0$ for all ν. This state $|0\rangle$ is again defined by $a_\nu|0\rangle = 0$ for all ν. Because

$$\langle a_{\nu_n}\ldots a_{\nu_1} a_{\nu_1}^\dagger \ldots a_{\nu_n}^\dagger\rangle_0 = 1\,,$$

the kets $a_{\nu_1}^\dagger\ldots a_{\nu_n}^\dagger|0\rangle$ are orthonormal. They are explicitly antisymmetric in virtue of (55). If $[a_{\nu_1}^\dagger\ldots a_{\nu_n}^\dagger]_C$ stands for the canonical order (where the smallest value of ν stands on the left, etc.), we have the correspondence

$$|\alpha_{\nu_1}\ldots\alpha_{\nu_n}\rangle_A \Leftrightarrow [a_{\nu_1}^\dagger\ldots a_{\nu_n}^\dagger]_C|0\rangle$$

between "conventional" states $|\ \rangle_A$ in $\mathfrak{H}_n$ and the vectors in $\mathfrak{F}$.

The creation and destruction operators in coordinate space may be introduced as in (49). Let $\langle \boldsymbol{x}s|\alpha_\nu\rangle$ be the one-particle coordinate space wave function for spin projection s, and define the field operators by

$$\psi_s(\boldsymbol{x}) = \sum_\nu a_\nu \langle \alpha_\nu|\boldsymbol{x}s\rangle \,, \qquad \psi_s^\dagger(\boldsymbol{x}) = \sum_\nu \langle \boldsymbol{x}s|\alpha_\nu\rangle a_\nu^\dagger \,. \tag{61}$$

Equation (55) then yield *the fermion anticommutation rules*

$$\{\psi_s(\boldsymbol{x}), \psi_{s'}^\dagger(\boldsymbol{x}')\} = \delta_{ss'}\delta(\boldsymbol{x}-\boldsymbol{x}') \,, \qquad \{\psi_s(\boldsymbol{x}), \psi_{s'}(\boldsymbol{x}')\} = 0 \,. \tag{62}$$

These rules imply the equivalent of (31):

$$\langle \psi_{s_n'}(\boldsymbol{x}_n')\ldots\psi_{s_1}(\boldsymbol{x}_1')\psi_{s_1}^\dagger(\boldsymbol{x}_1)\ldots\psi_{s_n}(\boldsymbol{x}_n)\rangle_0 = \sum_P \epsilon_P \prod_{i=1}^n \delta_{s_is_i'}\delta(\boldsymbol{x}_i - \boldsymbol{x}_i') \,, \tag{63}$$

where the unprimed (or primed) space *and* spin coordinates must be permuted as a unit, and ϵ_P is the signature of the permutation. Thus for $n = 2$ the right-hand side of (63) reads

$$\delta_{s_1s_1'}\delta(\boldsymbol{x}_1 - \boldsymbol{x}_1')\delta_{s_2s_2'}\delta(\boldsymbol{x}_2 - \boldsymbol{x}_2') - \delta_{s_1s_2'}\delta(\boldsymbol{x}_1 - \boldsymbol{x}_2')\delta_{s_2s_1'}\delta(\boldsymbol{x}_2 - \boldsymbol{x}_1') \,.$$

If $\varphi_a(\boldsymbol{x}_1s_1\ldots\boldsymbol{x}_ns_n)$ is an n-particle antisymmetric wave function, the ket in $\mathfrak{F}$ corresponding to it is

$$|\varphi_a\rangle = (n!)^{-\frac{1}{2}} \int d\boldsymbol{x}_1\ldots d\boldsymbol{x}_n \sum_{s_1\ldots s_n} \varphi_a(\boldsymbol{x}_1s_1\ldots\boldsymbol{x}_ns_n)\psi_{s_1}^\dagger(\boldsymbol{x}_1)\ldots\psi_{s_n}^\dagger(\boldsymbol{x}_n)|0\rangle \,. \tag{64}$$

This expression has the same structure as (29). The proof that (64) is the correct prescription will not be given for the present case.

The construction of expressions for symmetric observables proceeds in a manner that differs only in detail from the Bose case; they are left to the reader. The final expressions have precisely the same form as in the Bose case. Thus the generic one-body operator (type F) is

$$F = \sum_{ss'} \int d\boldsymbol{x}d\boldsymbol{x}'\psi_s^\dagger(\boldsymbol{x})\langle \boldsymbol{x}s|f|\boldsymbol{x}'s'\rangle\psi_{s'}(\boldsymbol{x}') \,, \tag{65}$$

and the two-body operator G is

$$G = \frac{1}{2}\sum_{s_1\ldots s_2'} \int d\boldsymbol{x}_1\ldots d^3x_2'\psi_{s_1}^\dagger(\boldsymbol{x}_1)\psi_{s_2}^\dagger(\boldsymbol{x}_2)\langle \boldsymbol{x}_1s_1\boldsymbol{x}_2s_2|g|\boldsymbol{x}_1's_1'\boldsymbol{x}_2's_2'\rangle\psi_{s_2'}(\boldsymbol{x}_2')\psi_{s_1'}(\boldsymbol{x}_1') \,. \tag{66}$$

The order of the operators $\psi_{s_2'}(\boldsymbol{x}_2')\psi_{s_1'}(\boldsymbol{x}_1')$ was immaterial in (44). Not so in the antisymmetric case; one *must* adhere to the order as given in (66). The most important observables are the number operator

$$N = \sum_s \int d\boldsymbol{x}\, \psi_s^\dagger(\boldsymbol{x})\psi_s(\boldsymbol{x}) \,; \tag{67}$$

the momentum

$$\boldsymbol{P} = \sum_s \int d\boldsymbol{x}\, \psi_s^\dagger(\boldsymbol{x})\frac{1}{i}\frac{\partial\psi_s(\boldsymbol{x})}{\partial\boldsymbol{x}} \,; \tag{68}$$

and the Coulomb energy

$$V_C = \tfrac{1}{2}e^2 \sum_{ss'} \int d\boldsymbol{x}d\boldsymbol{x}' \frac{\psi_s^\dagger(\boldsymbol{x})\psi_{s'}^\dagger(\boldsymbol{x}')\psi_{s'}(\boldsymbol{x}')\psi_s(\boldsymbol{x})}{4\pi|\boldsymbol{x}-\boldsymbol{x}'|} . \tag{69}$$

(c) The Equations of Motion

In the Fock space representation the Schrödinger picture is usually inappropriate, because it focuses attention on the intractable many-particle state vector. For this reason the Heisenberg picture will be used. Assuming that the Hamiltonian H is time-independent, we define the Heisenberg observables in the familiar way:

$$\psi(xt) = e^{iHt}\psi(x)e^{-iHt} . \tag{70}$$

The definition (70) applies equally well to the Fermi and Bose field operators, or to the creation and destruction operators in any representation. In (70) we have purposely written the argument of the field operator as x; this symbol will stand for $\boldsymbol{x}$ and the spin projection s, and instead of $\sum_s \int d\boldsymbol{x}$, we shall simply write $\int(dx)$ henceforth. Thus the number operator is $\int(dx)\psi^\dagger(x)\psi(x)$ for particles of any spin and either statistics.

The commutation rules between equal-time field operators is unaffected by the unitary transformation (70). For Fermi fields, for example, we still have

$$\{\psi^\dagger(xt), \psi(x't)\} = \delta(x-x') ,$$

where $\delta(x-x') \equiv \delta(\boldsymbol{x}-\boldsymbol{x}')\delta_{ss'}$. Knowledge of the equal-time algebraic rules therefore suffices to determine the commutator which appears in the equations of motion

$$i\dot{\psi}(xt) = [\psi(xt), H] . \tag{71}$$

Assume that H has the form

$$H = K + U + V , \tag{72}$$

where K is the kinetic energy

$$K = -\frac{1}{2m}\int (dx)\psi^\dagger(x)\nabla^2\psi(x) , \tag{73}$$

U is a one-body potential

$$U = \int (dx)(dx')\psi^\dagger(x)\langle x|u|x'\rangle\psi(x') \tag{74}$$

which describes the interaction of the system with any static external fields that may be present, and V is the two-body interaction

$$V = \frac{1}{2}\int (dx_1)\ldots(dx_2')\psi^\dagger(x_1)\psi^\dagger(x_2)\langle x_1x_2|v|x_1'x_2'\rangle\psi(x_2')\psi(x_1') . \tag{75}$$

Direct calculation gives

$$[\psi(y), V] = \frac{1}{2}\int (dx)(dx_1')(dx_2')\psi^\dagger(x)[\langle yx|v|x_1'x_2'\rangle \pm \langle xy|v|x_1'x_2'\rangle]\psi(x_2')\psi(x_1') , \tag{76}$$

where the plus sign holds for Bose statistics, the minus for Fermi statistics. The exchange term may be rewritten as follows:

$$\pm \int (dx_1')(dx_2')\langle xy|v|x_2'x_1'\rangle\psi(x_1')\psi(x_2') = \int (dx_1')(dx_2')\langle xy|v|x_2'x_1'\rangle\psi(x_2')\psi(x_1') .$$

But $\langle xy|v|x_2'x_1'\rangle = \langle yx|v|x_1'x_2'\rangle$, and therefore the two terms in (76) give the same contribution. In both statistics, the equations of motion are therefore

$$\begin{aligned} i\dot{\psi}(xt) = &-\frac{1}{2m}\nabla^2\psi(xt) + \int \langle x|u|x'\rangle\psi(x't)(dx') \\ &+ \int \psi^\dagger(x_2t)\langle xx_2|v|x_1'x_2'\rangle\psi(x_2't)\psi(x_1't)(dx_2)(dx_1')(dx_2') . \end{aligned} \tag{77}$$

This equation has the canonical form $i\dot{\psi}(xt) = \delta H/\delta\psi^\dagger(xt)$.

When u and v are both local and spin independent, the matrix elements are

$$\langle x|u|x'\rangle = \delta(x - x')U(\boldsymbol{x}) , \tag{78}$$

$$\langle x_1x_2|v|x_1'x_2'\rangle = \delta(x_1 - x_1')\delta(x_2 - x_2')v(\boldsymbol{x}_1 - \boldsymbol{x}_2) . \tag{79}$$

The equation of motion then simplifies to

$$i\dot{\psi}(xt) = -\frac{1}{2m}\nabla^2\psi(xt) + U(\boldsymbol{x})\psi(xt) + V_{\text{eff}}(\boldsymbol{x}t)\psi(xt) , \tag{80}$$

where

$$V_{\text{eff}}(\boldsymbol{x}t) = \int v(\boldsymbol{x} - \boldsymbol{x}')n(\boldsymbol{x}'t)d\boldsymbol{x}' , \tag{81}$$

and

$$n(\boldsymbol{x}t) = \sum_s \psi_s^\dagger(\boldsymbol{x}t)\psi_s(\boldsymbol{x}t) \tag{82}$$

is the number density operator.

There is a rather superficial resemblance between the equation of motion (80) and the Schrödinger equation. However, Schrödinger's equation is linear, whereas (80) is nonlinear because V_{eff} is a functional of ψ. Nevertheless, the field equation has an intuitively simple structure. If we momentarily ignore that $n(\boldsymbol{x}t)$ is an operator, we see that $V_{\text{eff}}(\boldsymbol{x}t)$ is the potential energy at $\boldsymbol{x}$ as computed in the classical manner from the two-body potential $v(\boldsymbol{x} - \boldsymbol{x}')$ and the density $n(\boldsymbol{x}t)$.

The solution of the operator equations of motion is equivalent to solving the Schrödinger equation in each and every subspace $\mathfrak{H}_{N'}$ of $\mathfrak{F}$. Clearly, that cannot be done if there are interactions between the particles. But the Heisenberg equations do provide equations of motion for the correlation functions. These equations cannot be solved exactly either, of course, but it is far easier to use our intuition in making approximations on correlation functions than on the many particle wave function. Furthermore, the second quantized method handles mixtures and pure states on an equal footing. This is indispensable when dealing with large systems at nonzero temperature.

(d) Distribution Functions

As in classical statistical mechanics, it usually suffices to restrict attention to the behavior of any one particle, or to any pair of particles, with all others being averaged over. This is done by introducing reduced density matrices and distribution functions. Once these are known it is possible to compute the expectation values of the most important symmetric observables.

For this purpose, define the *n-particle distribution matrices*

$$\langle x_1'|W_1|x_1\rangle = \operatorname{Tr}\rho\, \psi^\dagger(x_1)\psi(x_1') \,, \tag{83}$$

$$\langle x_1'x_2'|W_2|x_1x_2\rangle = \operatorname{Tr}\rho\, \psi^\dagger(x_1)\psi^\dagger(x_2)\psi(x_2')\psi(x_1') \,, \tag{84}$$

etc., where ρ is an arbitrary density matrix defined in $\mathfrak{F}$, the trace runs over any complete set in $\mathfrak{F}$, and as before, $x \equiv (\boldsymbol{x}, s)$. The diagonal elements of the distribution matrices, obtained by setting $x_1 = x_1'$, have a simple significance in coordinate space, and they merit a more compact notation:

$$D_n(x_1 \ldots x_n) \equiv \langle x_1 \ldots x_n|W_n|x_1 \ldots x_n\rangle \,. \tag{85}$$

Because $D_1(\boldsymbol{x}s) = \operatorname{Tr}\rho\, \psi_s^\dagger(\boldsymbol{x})\psi_s(\boldsymbol{x})$, it is clear that $D_1(\boldsymbol{x}s)$ is the density at $\boldsymbol{x}$ of particles having spin projection s. The off-diagonal elements of W_1 also have physical significance, because the momentum distribution, defined as the number of particles having momentum $\boldsymbol{p}$ and spin projection s, is

$$\operatorname{Tr}\rho\, a_{\boldsymbol{p}s}^\dagger a_{\boldsymbol{p}s} = \frac{1}{\Omega}\int d\boldsymbol{x}d\boldsymbol{x}' e^{-i\boldsymbol{p}\cdot(\boldsymbol{x}-\boldsymbol{x}')}\langle \boldsymbol{x}s|W_1|\boldsymbol{x}'s\rangle \,. \tag{86}$$

Thus W_1 incorporates both the coordinate space and momentum space densities. More generally, for an n-particle state W_n is the Wigner distribution, apart from normalization (recall §2.2(d)). As for W_1, it is normalized as follows:

$$\int d\boldsymbol{x} D_1(\boldsymbol{x}) = \langle N\rangle = \bar{N} \,. \tag{87}$$

The *pair distribution function*, $D_2(x_1, x_2)$, is of great importance. From the basic commutation rules it follows that

$$\psi^\dagger(x_1)\psi^\dagger(x_2)\psi(x_2')\psi(x_1') = \psi^\dagger(x_1)\psi(x_1')\psi^\dagger(x_2)\psi(x_2') - \delta(x_2 - x_2')\psi^\dagger(x_1)\psi(x_1') \tag{88}$$

for both Bose and Fermi statistics. Therefore

$$\int dx_1 dx_2 D_2(x_1, x_2) = \langle N^2\rangle - \langle N\rangle \,. \tag{89}$$

For a system with many particles, not to mention the thermodynamic limit, $\langle N\rangle$ is negligible, and $\langle N^2\rangle = \langle N\rangle^2$. In these circumstances, therefore

$$P(x_1, x_2) = \langle N\rangle^{-2} D_2(x_1, x_2) \tag{90}$$

is a joint probability distribution for any pair.

When the density matrix ρ is invariant under spatial symmetries, these are reflected in the distribution matrices. For example, neglecting gravity, a fluid in thermal equilibrium is spatially homogeneous and isotropic except in the vicinity of boundaries. Then $D_2(x_1x_2)$ depends only on the coordinate difference $|\boldsymbol{x}_1 - \boldsymbol{x}_2|$, unless $\boldsymbol{x}_1$ and/or $\boldsymbol{x}_2$ is near a boundary. Results of this type are consequences of symmetry assumptions about the ensemble of states. For instance, assume that ρ is translation invariant, i.e. $[\rho, \boldsymbol{P}] = 0$. The momentum operator generates translations:

$$\psi_s(\boldsymbol{x} - \boldsymbol{a}) = e^{i\boldsymbol{P}\cdot\boldsymbol{a}}\psi_s(\boldsymbol{x})e^{-i\boldsymbol{P}\cdot\boldsymbol{a}} . \tag{91}$$

The same transformation law applies to any operator $A(\boldsymbol{x})$, such as $\psi_s^\dagger(\boldsymbol{x})\psi_{s'}(\boldsymbol{x})$, constructed from the operators ψ and $\psi^\dagger$ at the point $\boldsymbol{x}$. Furthermore,

$$\operatorname{Tr}\rho\, A(\boldsymbol{x})B(\boldsymbol{x}') = \operatorname{Tr}\rho\, e^{i\boldsymbol{P}\cdot\boldsymbol{x}}A(0)e^{-i\boldsymbol{P}\cdot(\boldsymbol{x}-\boldsymbol{x}')}B(0)e^{-i\boldsymbol{P}\cdot\boldsymbol{x}'} . \tag{92}$$

Then $\operatorname{Tr} XY = \operatorname{Tr} YX$ and $[\boldsymbol{P}, \rho] = 0$ produce

$$\operatorname{Tr}\rho\, A(\boldsymbol{x})B(\boldsymbol{x}') = \operatorname{Tr}\rho\, A(\boldsymbol{x} - \boldsymbol{x}')B(0) . \tag{93}$$

As a consequence, the distribution matrices W_n are only functions of the coordinate differences if ρ is invariant under translations.

If ρ is rotationally invariant, there are further simplifications. Should spin-orbit forces be negligibly weak, ρ is invariant under orbital and spin rotations separately, and it then follows that $[\rho, \boldsymbol{L}] = [\rho, \boldsymbol{S}] = 0$. In this case,

$$\langle \boldsymbol{x}_1 s_1 \ldots \boldsymbol{x}_n s_n | W_n | \boldsymbol{x}'_1 s'_1 \ldots \boldsymbol{x}'_n s'_n \rangle$$

is a function of the absolute value of the coordinate differences, and vanishes unless

$$\sum_{m=1}^{n} s_m = \sum_{m=1}^{n} s'_m .$$

When ρ is invariant under translations, the momentum representation offers many advantages. Define

$$\begin{aligned}\langle p_1 \ldots p_n | \tilde{W}_n | p'_1 \ldots p'_n \rangle &\equiv \Omega^{-n} \int d\boldsymbol{x}_1 \ldots d\boldsymbol{x}'_n e^{-i(\boldsymbol{p}_1\cdot\boldsymbol{x}_1 + \ldots \boldsymbol{p}_n\cdot\boldsymbol{x}_n)} \\ &\qquad \times \langle x_1 \ldots x_n | W_n | x'_1 \ldots x'_n \rangle e^{i(\boldsymbol{p}'_1\cdot\boldsymbol{x}'_1 + \ldots \boldsymbol{p}'_n\cdot\boldsymbol{x}'_n)} \\ &= \operatorname{Tr}\rho\, a^\dagger_{p'_n} \ldots a^\dagger_{p'_1} a_{p_n} \ldots a_{p_1} ;\end{aligned} \tag{94}$$

then translation invariance implies that $\tilde{W}_n$ vanishes unless

$$\sum_{m=1}^{n} \boldsymbol{p}_m = \sum_{m=1}^{n} \boldsymbol{p}'_m .$$

11.3 Ideal Gases

Here we will derive the most important properties of ideal gases in thermal equilibrium. The language of second quantization will be used throughout.

(a) The Grand Canonical Ensemble

Because the total number of particles is not fixed in the formalism of second quantization, the state of a system is most appropriately described by the grand canonical ensemble. The density matrix for this ensemble is

$$\rho = \frac{1}{Z} e^{-\beta(H-\mu N)} , \tag{95}$$

where H and N are the energy and number operators, the temperature T is $1/k\beta$ with k being Boltzmann's constant, and μ is the chemical potential. Equation (95) applies both to interacting and non-interacting systems.

The grand partition function Z is determined by the requirement that $\mathrm{Tr}\,\rho = 1$:

$$Z = \mathrm{Tr}\, e^{-\beta(H-\mu N)} . \tag{96}$$

The thermodynamic potential Ξ, defined as

$$\Xi = -\frac{1}{\beta} \ln Z , \tag{97}$$

permits (95) to be written as

$$\rho = e^{\beta(\Xi+\mu N-H)} . \tag{98}$$

The chemical potential is determined by specifying the mean number of particles, i.e., by

$$\bar{N} = \mathrm{Tr}\, \rho N . \tag{99}$$

Then the thermodynamic properties of the system follow from (97) via

$$\Xi = E - TS - \mu\bar{N} , \tag{100}$$

where S is the entropy and E the mean energy,

$$S = -k\, \mathrm{Tr}\, \rho \ln\rho , \qquad E = \mathrm{Tr}\, \rho H . \tag{101}$$

In what follows, we shall use the abbreviations

$$\bar{A} = \langle A \rangle = \mathrm{Tr}\, \rho\, A$$

for the grand canonical average of an arbitrary observable A.

When there are no interactions between the particles, the Hamiltonian is a sum of commuting observables. In the momentum representation

$$H = \sum_{\mathbf{p}s} H_{\mathbf{p}s},$$

where

$$H_{\mathbf{p}s} = \frac{p^2}{2m}\, a^\dagger_{\mathbf{p}s} a_{\mathbf{p}s} \equiv \epsilon_p N_{\mathbf{p}s} , \tag{102}$$

s being the spin projection. The density matrix factorizes:

$$\rho = \prod_{\mathbf{p}s} \rho_{\mathbf{p}s} , \qquad \rho_{\mathbf{p}s} = e^{\beta\Xi_p} e^{-\beta(\epsilon_p-\mu)N_{\mathbf{p}s}} , \tag{103}$$

where

$$e^{-\beta\Xi_p} = \mathrm{Tr}_{ps}\, e^{-\beta(\epsilon_p-\mu)N_{ps}} . \tag{104}$$

Here Tr_{ps} designates the trace over the subspace spanned by the states belonging to the single mode with quantum numbers (p, s). Note that

$$\mathrm{Tr}_{ps}\, \rho_{ps} = 1 . \tag{105}$$

(b) The Ideal Fermi Gas

The eigenvalues n_{ps} of N_{ps} are 1 and 0 in this case. Hence

$$e^{-\beta\Xi_p} = \sum_{n=0}^{1} e^{-\beta(\epsilon_p-\mu)n} = 1 + e^{-\beta(\epsilon_p-\mu)} . \tag{106}$$

Therefore the mean number of particles with momentum p and spin projection s is

$$\langle N_{ps}\rangle = \mathrm{Tr}\,[\rho_{ps}N_{ps}\prod_{p's'\neq ps}\rho_{p's'}] = e^{\beta\Xi_p}\sum_{n=0}^{1} ne^{-\beta(\epsilon_p-\mu)n} , \tag{107}$$

or

$$\langle N_{ps}\rangle = \frac{1}{e^{\beta(\epsilon_p-\mu)}+1} . \tag{108}$$

This is the famous *Fermi distribution.*

The momentum distribution (108) has a very simple form in the zero temperature limit, $\beta\to\infty$:

$$\begin{aligned} \lim_{T\to 0}\langle N_{ps}\rangle &= 0 \qquad \epsilon_p > \mu \\ &= 1 \qquad \epsilon_p < \mu . \end{aligned} \tag{109}$$

Thus all single particle states with energy $\epsilon_p < \mu$ are occupied once, and all higher-lying single particle levels are empty. This is the ground-state distribution required by the exclusion principle. The chemical potential at $T = 0$ is called the *Fermi energy* ϵ_F, the momentum p_F for which $p_F^2/2m = \epsilon_F$ the *Fermi momentum*, the domain $|p| < p_F$ of momentum space the *Fermi sphere*, and the shell $|p| = p_F$ the *Fermi surface.*

The value of p_F is determined from the $T\to 0$ limit of the condition (99):

$$n = \frac{1}{\Omega}\sum_{ps}\langle N_{ps}\rangle = \frac{2J+1}{2\pi^2}\int_0^\infty \frac{p^2dp}{e^{\beta(\epsilon_p-\mu)}+1} , \tag{110}$$

where J is the spin of the particles and $n \equiv \bar{N}/\Omega$ is the mean number density. As $T\to 0$ (i.e., $\beta\to\infty$), the integral approaches the limiting value $\frac{1}{3}p_F^3$. The Fermi momentum is therefore

$$p_F = \left(\frac{6\pi^2}{2J+1}\right)^{\frac{1}{3}}\frac{1}{d} , \qquad n = 1/d^3 , \tag{111}$$

where d is the average separation between the particles.

When T is just above zero, excited states wherein particles slightly below the Fermi surface are raised above this surface become statistically significant. It is therefore clear that the momentum distribution must depart more and more from the form (109) as T increases. In fact, at high temperatures $\beta\mu \ll 1$, and (108) approaches the Maxwell distribution $e^{-\beta\epsilon_p}$. In metals the Fermi energy of the conduction electrons is of order 2-6 eV. On the other hand, 1 degree K is equivalent to 0.862×10^{-4}eV. Thus for most temperatures of interest $\beta\epsilon_F \gg 1$, and the momentum distribution departs only slightly from the step function (109). To be more precise, $\mu = \epsilon_F[1 + O(k^2T^2/\epsilon_F^2)]$ as $T \to 0$, and therefore $\langle N_{\boldsymbol{p}s}\rangle$ passes through the value $\frac{1}{2}$ very close to ϵ_F; furthermore $\langle N_{\boldsymbol{p}s}\rangle$ falls off exceedingly rapidly as ϵ_p increases beyond ϵ_F. A Fermi gas characterized by such a well-defined Fermi surface (or, equivalently, by $\epsilon_F \gg kT$) is said to be degenerate.

Consider the two simplest distributions in an ideal Fermi gas. We begin with the one-particle distribution matrix

$$\langle \boldsymbol{x}'s'|W_1|\boldsymbol{x}s\rangle = \langle \psi_s^\dagger(\boldsymbol{x})\psi_{s'}(\boldsymbol{x}')\rangle \ . \tag{112}$$

In §11.2(d) we learned that under the present circumstances (112) is diagonal in the spin coordinates, and only a function of $|\boldsymbol{x}-\boldsymbol{x}'| = R$. This can be seen directly by transforming (112) to momentum space; one then encounters the expression $\langle a_{\boldsymbol{p}s}^\dagger a_{\boldsymbol{p}'s'}\rangle$. Since all states in the trace can be characterized by a momentum and spin eigenvalue, this expectation value vanishes unless $\boldsymbol{p} = \boldsymbol{p}'$, $s = s'$. Therefore

$$\langle \boldsymbol{x}'s'|W_1|\boldsymbol{x}s\rangle = \delta_{ss'}\mathcal{G}(|\boldsymbol{x}-\boldsymbol{x}'|) \ ,$$

where

$$\mathcal{G}(R) = \frac{1}{\Omega}\sum_{\boldsymbol{p}} e^{-i\boldsymbol{p}\cdot\boldsymbol{R}}\langle N_{\boldsymbol{p}s}\rangle = \frac{1}{2\pi^2 R}\int_0^\infty p\,dp\frac{\sin pR}{e^{\beta(\epsilon_p-\mu)}+1} \ . \tag{113}$$

Analytic expressions are easily found in the low- and high-temperature limits. At $T = 0$,

$$\lim_{T\to 0}\mathcal{G}(R) = \frac{n}{2J+1}\frac{3j_1(p_F R)}{p_F R} \ , \tag{114}$$

where j_1 is a spherical Bessel function. The function (114) is peaked at $|\boldsymbol{x}-\boldsymbol{x}'| = 0$, and for $p_F|\boldsymbol{x}-\boldsymbol{x}'| \gg 1$ tends to 0,

$$\mathcal{G}(R) \sim -\frac{3n}{2J+1}\frac{\cos p_F R}{(p_F R)^2} \ , \qquad (p_F R \to \infty) \ . \tag{115}$$

In the high-temperature limit the integral is also elementary, and proportional to $\exp(-R^2 m/2\beta)$. The chemical potential is determined by requiring (113) to equal $n/(2J+1)$ when $R = 0$. Thus

$$\mathcal{G}(R) = \frac{n}{2J+1}e^{-\frac{1}{2}R^2/\lambda_T^2} \qquad (kT \gg \epsilon_F) \ , \tag{116}$$

where

$$\lambda_T = \frac{1}{\sqrt{mkT}} = 6.95\ (M\mathrm{K}^\circ)^{-\frac{1}{2}}\,\text{Å} \ , \tag{117}$$

is the *thermal wavelength*; in the second expression M is the mass in units of the proton's mass, and K° the temperature in degrees Kelvin. This is the de Broglie wavelength corresponding to the r.m.s. momentum of a Boltzmann gas at temperature T. As λ_T/d is proportional to $\sqrt{\epsilon_F/kT}$, (116) states that the non-locality of the one-particle distribution matrix is confined to a region very small compared to the mean interparticle separation in the Boltzmann limit ($kT \gg \epsilon_F$).

Next we turn to the two-particle distribution matrix $\langle x_1 x_2|W_2|x_1' x_2'\rangle$, where we again combine $\boldsymbol{x}$ and s into the single label x, and similarly for $\boldsymbol{p}$ and s. On transforming to the p-representation, we encounter $\langle a_{p_1}^\dagger a_{p_2}^\dagger a_{p_2'} a_{p_2'}\rangle$. Because the representation that diagonalizes ρ has definite occupation numbers $\{n_p\}$, the creation operators $a_{p_1}^\dagger$ and $a_{p_2}^\dagger$ must restore the particles destroyed by $a_{p_1'}$ and $a_{p_2'}$, i.e., either $p_1 = p_2'$, $p_2 = p_1'$, or $p_1 = p_1'$, $p_2 = p_2'$. Thus

$$\langle a_{p_1}^\dagger a_{p_2}^\dagger a_{p_2'} a_{p_1'}\rangle = (\delta_{p_1p_1'}\delta_{p_2p_2'} - \delta_{p_1p_2'}\delta_{p_2p_1'})\langle N_{p_1}\rangle\langle N_{p_2}\rangle\,, \tag{118}$$

because $\langle N_p N_{p'}\rangle = \langle N_p\rangle\langle N_{p'}\rangle$ when the system is non-interacting. Therefore

$$\begin{aligned}\langle x_1x_2|W_2|x_1'x_2'\rangle &= \delta_{s_1s_1'}\delta_{s_2s_2'}\mathcal{G}(|\boldsymbol{x}_1-\boldsymbol{x}_1'|)\mathcal{G}(|\boldsymbol{x}_2-\boldsymbol{x}_2'|)\\ &\quad - \delta_{s_1s_2'}\delta_{s_2s_1'}\mathcal{G}(|\boldsymbol{x}_1-\boldsymbol{x}_2'|)\mathcal{G}(|\boldsymbol{x}_2-\boldsymbol{x}_1'|)\,.\end{aligned} \tag{119}$$

The generalization of this result for the n-particle distribution matrix is a determinant whose elements are the functions $\mathcal{G}$. As might be expected, in a non-interacting system the higher order correlations are completely determined by the one-particle distribution matrix.

The pair distribution function $D_2(x_1x_2)$ is proportional to the probability for finding a particle at $\boldsymbol{x}_1$ with spin projection s_1, when another particle is at $\boldsymbol{x}_2$, s_2. According to (85), it is the diagonal element of (119):

$$D_2(x_1x_2) = \left\{\left(\frac{n}{2J+1}\right)^2 - \delta_{s_1s_2}[\mathcal{G}(|\boldsymbol{x}_1-\boldsymbol{x}_2|)]^2\right\}. \tag{120}$$

Equation (120) must be added to the list of peculiar results in the quantum theory of indistinguishable particles. In classical statistical mechanics, there are no correlations when there are no forces between the constituents of the system. In contrast to this, (120) states that the probability for finding two particles with equal spin projection is less than the random result for all separations, and is zero when the distance between them vanishes. This correlation is the expression of the Pauli principle in configuration space, and gives rise to the exchange energy already discussed in connection with the spectrum of He in §6.2.

Note that the anticommutation rules imply that $D_2(\boldsymbol{x}s;\boldsymbol{x}s) = 0$ no matter what the forces may be. On the other hand, $D_2(\boldsymbol{x}_1s;\boldsymbol{x}_2s) \leq (n/2J+1)^2$ does not hold irrespective of interactions.

In closing, note that (120) shows that there are no correlations in position between particles having different spin quantum numbers s. Furthermore, for $s_1 = s_2$, there is an important temperature variation of the spatial domain for which departures from the random result $(n/2J+1)^2$ are significant. When $kT \ll \epsilon_F$, we can use (114) to obtain

$$D_2(\boldsymbol{x}_1s;\boldsymbol{x}_2s) = \left(\frac{n}{2J+1}\right)^2\left\{1 - \left[\frac{3j_1(p_F|\boldsymbol{x}_1-\boldsymbol{x}_2|)}{p_F|\boldsymbol{x}_1-\boldsymbol{x}_2|}\right]^2\right\}. \tag{121}$$

Thus when $p_F|\boldsymbol{x}_1-\boldsymbol{x}_2| \gg 1$ (i.e., $|\boldsymbol{x}_1-\boldsymbol{x}_2|$ is large compared to the mean separation) the pair distribution function approaches the random result. In the Boltzmann limit ($kT \gg \epsilon_F$)

$$D_2(\boldsymbol{x}_1s;\boldsymbol{x}_2s) = \left(\frac{n}{2J+1}\right)^2\left\{1 - e^{-(\boldsymbol{x}_1-\boldsymbol{x}_2)^2/\lambda_T^2}\right\}, \tag{122}$$

and departures from the random result are confined to distances much smaller than the mean separation between particles. Hence classical statistics gives correct results in the high temperature limit (i.e., for distances large compared to λ_T).

(c) The Ideal Bose Gas

We only consider spin 0 particles; the extension to higher spins is straightforward. The occupation numbers range from zero to infinity in Bose statistics, so the partition function is

$$\sum_{\mathbf{p}} e^{-\beta\Xi_p} = \sum_{\mathbf{p}} \sum_{n=0}^{\infty} e^{-n\beta(\epsilon_p - \mu)} = \sum_{\mathbf{p}} \frac{1}{1 - e^{-\beta(\epsilon_{\mathbf{p}} - \mu)}} . \tag{123}$$

The expectation value of $N_{\mathbf{p}}$ is

$$\langle N_{\mathbf{p}} \rangle = e^{\beta\Xi_p} \sum_{n=0}^{\infty} n e^{-n\beta(\epsilon_p - \mu)} , \tag{124}$$

and therefore the mean number of particles of momentum $\mathbf{p}$ is

$$\langle N_{\mathbf{p}} \rangle = \frac{1}{e^{\beta(\epsilon_p - \mu)} - 1} . \tag{125}$$

This is the *Bose distribution.* The argument starting with (123) must hold for all $\mathbf{p}$, and therefore $\mu \leq 0$ is required in general.

In the important but special case of photons, where thermal equilibrium is established not through collisions but by emission and absorption, the number of particles is not a constant of motion and the chemical potential vanishes.[1] Furthermore, for a massless particle $\epsilon_p = p/c \equiv \omega$ when $\hbar = 1$. The result then is to turn (125) into the *Planck distribution law:*

$$\langle N_{\mathbf{p}\lambda} \rangle = \frac{1}{e^{\beta\omega} - 1} , \tag{126}$$

where λ designates any two orthogonal polarization modes. Because this distribution is isotropic and independent of polarization, the mean energy density is

$$\overline{W}(\omega)d\omega = 2\,\frac{p^2 dp\, 4\pi}{(2\pi)^3}\,\omega\langle N_{p\lambda}\rangle = \frac{1}{\pi^2}\frac{\omega^3 d\omega}{e^{\beta\omega} - 1} , \tag{127}$$

which is to be multiplied by $\hbar/c^3$ when $\hbar = c = 1$ units are not used.[2]

We now return to a gas of massive Bose particles. As in the case of fermions, the chemical potential is determined by

$$\bar{N} = \sum_{\mathbf{p}} \frac{1}{e^{\beta(\epsilon_p - \mu)} - 1} . \tag{128}$$

[1] See L.D. Landau and E.M. Lifshitz, *Statistical Physics*, 3rd Edition, Part I, Pergamon (1980), §63.

[2] Further properties of the Planck distribution can be found in §10.5.

In treating the Fermi gas, sums such as (128) were converted into integrals with the recipe

$$\sum_{\boldsymbol{p}} \to \frac{\Omega}{(2\pi)^3} \int d^3p \,. \tag{129}$$

Implicit in this step is the tacit assumption that all terms in the sum have the same dependence on Ω as $\Omega \to \infty$. This is valid for $\bar{N}$ in the Fermi case because there $\langle N_{\boldsymbol{p}} \rangle$ is independent of Ω for all $\boldsymbol{p}$, including $\boldsymbol{p} = 0$. In the Bose gas, on the other hand, this assumption is untenable, because the ground state is obtained by putting *all* the particles into the $\boldsymbol{p} = 0$ level. Hence $\langle N_0 \rangle = \bar{N} \equiv n\Omega$ when $T = 0$. As the temperature is raised, some particles are excited out of the $\boldsymbol{p} = 0$ level, but for sufficiently low temperatures *a finite fraction of all the bosons remains in this lowest level*, i.e., $\langle N_0 \rangle_T = n_0(T)\Omega$, with $n_0(T) < n$. The only question is whether $n_0(T)$ tends to zero asymptotically as $T \to \infty$, or whether $n_0(T)$ vanishes at a finite temperature T_c. This question will be answered shortly.

According to (125), when $\mu < 0$ the mean number of particles of momentum $\boldsymbol{p}$ is finite for all $\boldsymbol{p}$, including $\boldsymbol{p} = 0$. Hence the $\Omega \to \infty$ limit of $\langle N_{\boldsymbol{p}} \rangle$ is independent of Ω when μ is negative. The case where $\boldsymbol{p} = 0$ is macroscopically occupied, by which we mean $\langle N_0 \rangle \propto \Omega$, must therefore correspond to $\mu = 0$. Thus $\mu = 0$ for all $T < T_c$ (whether T_c is finite or infinite). For such temperatures, the $\boldsymbol{p} = 0$ term in all sums must be treated separately, and the step (129) only applies to the $\boldsymbol{p} \neq 0$ portions of such sums. In particular, the density is given by

$$n \equiv \lim_{\Omega \to \infty} (\bar{N}/\Omega) = n_0(T) + \frac{1}{(2\pi)^3} \int d^3p \frac{1}{e^{\beta\epsilon_p} - 1} \tag{130}$$

whenever $\mu = 0$. (Although $(e^{\beta\epsilon_p} - 1)^{-1} \to \infty$ as $p \to 0$, this singularity is completely suppressed by the vanishing volume of p-space.) If one prefers, the singular $\boldsymbol{p} = 0$ level can be combined with the others by writing the momentum distribution as

$$\lim_{\Omega \to \infty} \langle N_{\boldsymbol{p}} \rangle = (2\pi)^3 n_0(T) \delta(\boldsymbol{p}) + \frac{1}{e^{\beta\epsilon_p} - 1} \,, \tag{131}$$

with the δ-function keeping track of the peculiar character of the $\boldsymbol{p} = 0$ level.

The integral in (130) can be rewritten as

$$\frac{1}{\sqrt{2}\pi^2} (mkT)^{3/2} \int_0^\infty \frac{x^{\frac{1}{2}}\, dx}{e^x - 1} \,;$$

the last integral is $\Gamma(\frac{3}{2})\zeta(\frac{3}{2})$, where $\zeta(\frac{3}{2}) = 2.612$ is the indicated Riemann zeta function. Thus

$$n_0(T) = n - \zeta(\tfrac{3}{2}) \left(\frac{mkT}{2\pi} \right)^{3/2} . \tag{132}$$

The second term in (132) grows monotonically with T, and when it reaches the prescribed total density n, the density of particles in the $\boldsymbol{p} = 0$ state vanishes. Therefore there is a finite temperature T_c above which $n_0(T)$ vanishes (i.e., above which the number of particles $\langle N_0 \rangle$ in the $\boldsymbol{p} = 0$ level is not proportional to Ω). This temperature is seen to be

$$kT_c = 3.31\, n^{2/3}/m \,, \tag{133}$$

and the mean number of particles in the $\boldsymbol{p}=0$ mode is therefore $\bar{N}[1-(T/T_c)^{3/2}]$, or

$$n_0(T) = n[1-(T/T_c)^{3/2}] \,. \tag{134}$$

The phenomenon just described was discovered by Einstein, and it is called *Bose-Einstein condensation.* To repeat, below the critical temperature T_c, $\langle N_0 \rangle$ is an extensive parameter. Once T exceeds $T_c, \mu < 0$, and then $\langle N_0 \rangle$ is an intensive parameter like all the mean occupation numbers for $\boldsymbol{p} \neq 0$.

Finally we investigate the various distribution functions in the condensed phase ($T < T_c$). As in the Fermi gas, these are all functionals of

$$\mathcal{G}(R) \equiv \langle \psi^\dagger(\boldsymbol{R})\psi(0) \rangle = n_0(T) + \frac{1}{2\pi^2 R}\int_0^\infty pdp\, \frac{\sin(pR)}{e^{\beta\epsilon_p}-1} \,. \tag{135}$$

The qualitative features of this function are easier to see after the integration variable is changed to $\beta\epsilon_p$:

$$\mathcal{G}(R) = n_0(T) + \frac{1}{2\pi^2}(mkT)^{3/2}\frac{\lambda_T}{R}\int_0^\infty \sin\left(\frac{\sqrt{2x}R}{\lambda_T}\right)\frac{dx}{e^x-1} \,, \tag{136}$$

where λ_T is the thermal wavelength (117). The important contributions to the integral in (136) come from $x \lesssim 1$. Hence for $R \ll \lambda_T$, we can expand the sine to obtain

$$\mathcal{G}(R) \sim n - \frac{1}{2}\zeta(\tfrac{5}{2})\left(\frac{mkT}{2\pi}\right)^{3/2}\left(\frac{R}{\lambda_T}\right)^2 , \qquad (R \ll \lambda_T) \,. \tag{137}$$

When $R \gg \lambda_T$, the sine oscillates very rapidly except near $x=0$, and there we can replace $e^x - 1$ by x. Asymptotically, therefore,

$$\mathcal{G}(R) \sim n_0(T) + \frac{1}{2\pi}(mkT)^{3/2}\frac{\lambda_T}{R} \,, \qquad (R \gg \lambda_T) \,. \tag{138}$$

The pair distribution matrix is given by

$$\langle \boldsymbol{x}_1\boldsymbol{x}_2|W_2|\boldsymbol{x}_1'\boldsymbol{x}_2'\rangle = \langle \boldsymbol{x}_1|W_1|\boldsymbol{x}_1'\rangle\langle \boldsymbol{x}_2|W_1|\boldsymbol{x}_2'\rangle + \langle \boldsymbol{x}_1|W_1|\boldsymbol{x}_2'\rangle\langle \boldsymbol{x}_2|W_1|\boldsymbol{x}_1'\rangle \,. \tag{139}$$

in a free Bose gas (compare with (119)), and the pair distribution function is therefore

$$D_2(\boldsymbol{x}_1\boldsymbol{x}_2) = n^2 + [\mathcal{G}(|\boldsymbol{x}_1-\boldsymbol{x}_2|)]^2 \,. \tag{140}$$

As in the free Fermi case, there are correlations, but in the ideal Bose gas it is *more* probable to find two particles at some definite separation than we would expect classically, as is to be expected from the symmetry of Bose wave functions.

11.4 The Mean Field Approximation

The quantum mechanics of a system of identical interacting particles is obviously a very difficult subject. This volume is restricted to the simplest and most naive approximation, in which the force to which any one particle is exposed is taken to be due to the average force field created by the other particles. When stated in English, this would appear to ignore all correlations. But in quantum mechanics there are correlations if the particles are identical and the temperature is low. As we already know from the example of the ideal gas, this requirement imposes correlations even in the absence of forces.

It is intuitively clear that if the inter-particle forces are strong, there must be much stronger correlations than those imposed by symmetry or antisymmetry. For example, if the interaction has a very strong repulsion for separations smaller than r_0, the pair correlation function must nearly vanish for such separations. The repulsive correlations imposed by the exclusion principle in an ideal Fermi gas do not meet this requirement, while those in an ideal Bose gas have the opposite effect. Thus it is evident that the mean field approximation can only hope to be accurate if the interactions are, in some sense, weak; what is meant here by "in some sense" will be discussed presently.

The assumption that the force on a particle is to be approximated by the mean field implies that each particle is to move as if it were in an external potential. Therefore it has its "own" wave function, so to say, with the many-body wave function a product of these, though symmetrized or antisymmetrized.

Two many-body systems will be treated here: the dilute Bose-Einstein condensate, and the low-lying states of many electron atoms. By dilute we mean a density low enough so as to make the interaction in effect weak. In the case of atoms, the weak interaction restriction is not met, in fact, but the mean field (or Hartree-Fock) approximation is nevertheless important, for both pedagogical and practical reasons.

(a) The Dilute Bose-Einstein Condensate

Even an arbitrarily weak interaction is of crucial importance in a Bose gas. If the interaction is attractive, and has a range r_0, then the lowest state is one where all particles are collapsed into a microscopic volume of order r_0^3, giving a negative interaction energy proportional to N^2 that dominates the kinetic energy which is of order N.[1]

A weak repulsion also plays a crucial role, as shown by the following argument. Assume the gas is in a rigid enclosure, say a sphere of radius R, with $R^3 \propto N$. The ground-state wave function of one particle in this enclosure is

$$\varphi(r) = \frac{1}{\sqrt{2\pi R}}\,\frac{1}{r}\,\sin\left(\frac{\pi r}{R}\right). \tag{141}$$

The ground state wave function of the ideal gas is then

$$\Psi = \prod_{i=1}^{N} \varphi(\boldsymbol{r}_i)\,. \tag{142}$$

But this is a preposterous wave function: it states that the gas density is

$$n(r) = \frac{N}{2\pi R}\,\frac{1}{r^2}\,\sin^2\left(\frac{\pi r}{R}\right), \tag{143}$$

rather than being uniform except near the boundary.

---

[1]In this section N is the number of particles, and not the number operator.

The function Ψ is only correct when the interaction between particles *vanishes identically*. It minimizes the kinetic energy $K \propto \int \varphi\nabla^2\varphi$; according to (141), $K \sim N/R^2 \sim N^{\frac{1}{3}} \sim \Omega^{\frac{1}{3}}$, which is vastly smaller than an energy that is proportional to the system's volume. Properties proportional to the volume are called *extensive*, and most systems in nature have energies that are extensive. By contrast, a two body interaction of finite range will, in an extensive system, produce a potential energy proportional to $n^2\Omega$, which is enormous compared to K. Thus if there is any interaction between the particles, the wave function (142) ceases to bear any resemblance to reality; the correct wave function must produce a density that is uniform except near the walls of an enclosure.[1]

This enormous sensitivity to interactions of the free-particle wave function shown by the Bose gas does not occur in the Fermi case. Because of the exclusion principle, the free particle ground state is made up of wave function with wavelengths down to $1/k_F$, which is a distance of order the mean inter-particle spacing, and as a result the density produced by the naive many-body wave function is spatially uniform.

To find a realistic description of the Bose gas, the interaction must be taken into account *ab initio*. We do so for the case of a dilute system, such as the atomic Bose-Einstein gases that have been studied extensively in recent years. The important case of superfluid He^4 is far more difficult because in this system the interaction is far too strong to be handled by the mean field approximation.

In the condensate, the particles move very slowly, so the collisions between them can be described by the S-wave scattering length a. The true interaction can then be replaced by the pseudo-potential (Eq. 8.313),

$$v_{12} = \frac{4\pi a}{m}\, \delta(\boldsymbol{r}_1 - \boldsymbol{r}_2)\,, \tag{144}$$

bearing in mind that the reduced mass is $m/2$ in this case. The scattering length must be positive if the system is not to collapse.

The interaction is, in effect, weak if a is small compared to the mean interparticle distance d, where $d^3 = n$ is the mean density. Nevertheless, the true force can have both strong repulsive and attractive components as long as they result in a positive scattering length that is short compared to d. Thus the precise meaning of the term dilute is $d \gg a$. In the second-quantized form (Eq. 44), and the dilute gas approximation, the interaction between the particles is therefore

$$V = \frac{2\pi a}{m} \int d\boldsymbol{x}\; \psi^\dagger(\boldsymbol{x})\psi^\dagger(\boldsymbol{x})\psi(\boldsymbol{x})\psi(\boldsymbol{x})\,. \tag{145}$$

The mean field approximation assumes that the state of the system can be adequately described by assigning the constituents to one-body states moving in a mean field. In the case of a condensed Bose gas, a finite fraction of all particles occupy the lowest of these states, and at $T = 0$ all are in this state. Let $\phi(\boldsymbol{x})$ be this lowest one-body wave function; together with the functions $\{u_\nu(\boldsymbol{x})\}$ it forms a complete orthonormal set that vanishes on the walls of the enclosure. Further, let c and b_ν be the destruction operators for the corresponding one-body states, so that the whole Bose field operator is

$$\psi(\boldsymbol{x}) = \phi(\boldsymbol{x})\, c + \sum_\nu u_\nu(\boldsymbol{x})\, b_\nu\,. \tag{146}$$

[1] In this connection, see Problem 3.

By hypothesis, the ground state is

$$|G\rangle = \frac{1}{\sqrt{N!}}(c^\dagger)^N|0\rangle \ . \tag{147}$$

The objective is the expectation value of H in this state. The kinetic and potential energies involve the expressions $\psi(\boldsymbol{x})|G\rangle$ and $\psi(\boldsymbol{x})\psi(\boldsymbol{x})|G\rangle$, and in both only the term $\phi\, c$ of ψ gives a nonzero contribution because the operators b_ν acting on $|G\rangle$ give zero. When the destruction operator c acts on any function of $c^\dagger$ it can be represented by

$$c = \partial/\partial c^\dagger \ , \tag{148}$$

and therefore $\langle c^\dagger c\rangle_G = N$, and $\langle c^\dagger c^\dagger c c\rangle_G = N(N-1) \simeq N^2$ when $N \gg 1$. In view of this, we define what is called *the condensate wave function*

$$\Psi(\boldsymbol{x}) = \sqrt{N}\phi(\boldsymbol{x}) \ , \tag{149}$$

even though Ψ is not normalized to unity, but to N. In term of Ψ the ground-state energy is thus

$$\mathcal{E}[\Psi] = \int d\boldsymbol{x} \left\{ -\frac{1}{2m}\,\Psi^*(\boldsymbol{x})\nabla^2\Psi(\boldsymbol{x}) + \frac{2\pi a}{m}\,|\Psi(\boldsymbol{x})|^4 + U(\boldsymbol{x})|\Psi(\boldsymbol{x})|^2 \right\} \ , \tag{150}$$

where U is an external potential that represents the enclosure or an optical trap holding the gas.

The equation satisfied by Ψ is found by varying the functional $\mathcal{E}[\Psi]$. Two points must now be borne in mind: (i) the variation must keep $|\Psi|^2$ normalized to N, which is done by varying

$$\mathcal{E}[\Psi] - \epsilon \int d\boldsymbol{x}\, |\Psi(\boldsymbol{x})|^2 \ , \tag{151}$$

where ϵ is a Lagrange multiplier; and (ii) that in general Ψ is complex, and therefore (151) is to be stationary under variations of Ψ and Ψ^* independently. Varying with respect to Ψ^* then gives *the Gross-Pitaevskii equation*,

$$-\frac{1}{2m}\nabla^2\Psi + \frac{4\pi a}{m}|\Psi|^2\Psi + U\Psi = \epsilon\Psi \ . \tag{152}$$

This equation, and even more so its time-dependent generalization, has many interesting solutions beyond the ground state of the condensate.[1] We confine ourselves to the latter, and assume there is no applied force beyond the rigid walls of an enclosure, i.e., $U = 0$ in (152). As in the Schrödinger equation, this ground-state wave function is real. We anticipate, and will shortly confirm, that Ψ is a constant C except near the walls, and that the distance l in which Ψ rises from $\Psi = 0$ at the walls is microscopic. Hence the normalization integral is $C^2\Omega$, aside from a negligible surface term, and as the integral must equal N we have $C = \sqrt{n}$, where n is the number density. On the other hand, when Ψ is a constant and $U = 0$, (152) reduces to

$$|\Psi(\boldsymbol{x})|^2 = m\epsilon/4\pi a \ . \tag{153}$$

[1] F. Dalfovo, S. Giorgini, L.P. Pitaevskii and S. Stringari, *Rev. Mod. Phys.* **71**, 463 (1999).

When integrated over space, (153) becomes $N = (m\epsilon/4\pi a)\Omega$, and therefore

$$\epsilon = 4\pi an/m\,. \tag{154}$$

If the surface layer is microscopic, the behavior of Ψ near the walls cannot depend on the shape of the enclosure, and it suffices to consider a large box with a wall at the yz-plane, the interior being $x \geq 0$. The behavior of Ψ near this wall is then given by the one-dimensional Gross-Pitaevskii equation

$$\Psi'' + 8\pi a\Psi(n - \Psi^2) = 0\,. \tag{155}$$

The boundary conditions are

$$\Psi(0) = 0\,, \qquad \Psi(x) \to \sqrt{n}, \quad x \to \infty\,. \tag{156}$$

It is left to the reader to confirm that the desired solution of (155) is

$$\Psi(x) = \sqrt{n}\,\tanh(x/l)\,, \qquad l = 1/\sqrt{4\pi an} = d\sqrt{d/4\pi a}\,. \tag{157}$$

As expected the condensate density rises to its constant value in a microscopic distance that is large compared to the interparticle spacing when the system is dilute. Also as expected, this distance shrinks as the effect of the interaction increases, whether due to an increasing scattering length or density.

(b) The Hartree-Fock Equations

In the mean field approximation, the actual many body interaction is mimicked by a one-body "potential," and by the same token, the actual wave function of a system of identical fermions is approximated by an antisymmetrized product of N one particle wave functions. The approximation is made *self-consistent* by requiring the potential acting on a given particle to be the average field produced by all other particles. This averaging is done with the sought-for wave functions, and therefore the self-consistency requirement will lead to non-linear equations for the one-particle wave functions.

Let $\phi_1(1), \ldots, \phi_N(N)$ be these one-particle wave functions; in atomic physics they are often called spin-orbitals. They are assumed to be orthonormal. Here the suffix stands for a set of four quantum numbers, e.g., the $(n_1 l_1 j_1 m_1)$ of atomic spectroscopy, and the argument stands for the eigenvalues of four one-particle degrees of freedom, e.g., $(\boldsymbol{r}_1, s_1)$, where $s = \pm\frac{1}{2}$ is the spin projection. The ϕ_i are not a complete set, of course, but can be augmented with other functions that will play no more than a brief and purely formal role to make the set complete. Let $a_i^\dagger$ be the creation operator for the one-particle state whose wave function is ϕ_i. Then the N-body state we are working with is

$$|\Phi\rangle = \prod_{i=1}^{N} a_i^\dagger |0\rangle\,. \tag{158}$$

That is, the states $i = 1, \ldots, N$ have occupation numbers one, and all others are unoccupied.

The mean field approximation amounts to finding the best wave function of this type, where "best" is to be understood in the sense of the variational principle.

The equations that determine these functions are found by varying the expectation value of the Hamiltonian, viewed as a functional of Φ, with respect to the ϕ_i.

The Hamiltonian is assumed to have a one-body part, F, for the kinetic energy and any applied fields (e.g., that due to the nucleus in an atom); and a two-body interaction V between the particles. These operators have the same form as for the Bose case in (51) and (52):

$$F = \sum_{\mu\nu} a_\mu^\dagger \langle \mu|f|\nu\rangle a_\nu \,, \tag{159}$$

$$V = \tfrac{1}{2} \sum_{\mu\nu\lambda\sigma} a_\mu^\dagger a_\nu^\dagger \langle \mu\nu|v|\lambda\sigma\rangle a_\sigma a_\lambda \,. \tag{160}$$

In F and V, the sums run over the complete set, and not just the states in $|\Phi\rangle$. It turns out that the derivation is more transparent if these completely general forms for H are used, as compared to specializing at this point to specific interactions, such as those in an atom.

To evaluate the expectation value of F, consider first $\langle\Phi|a_\mu^\dagger a_\nu|\Phi\rangle$. Acting on $|\Phi\rangle$, a_ν gives zero unless ν is an occupied state; the same holds for $a_\mu^\dagger$ acting on the left, so μ must also be occupied. Using the anticommutation rule,

$$\langle\Phi|a_\mu^\dagger a_\nu|\Phi\rangle = \langle\Phi|\delta_{\mu\nu} - a_\nu a_\mu^\dagger|\Phi\rangle = \delta_{\mu\nu} \,, \qquad 1 \leq (\mu,\nu) \leq N \,, \tag{161}$$

because the creation operator $a_\mu^\dagger$ acting on $|\Phi\rangle$, with μ being an occupied state, gives zero as the occupation number cannot be greater than 1. Thus the expectation value of F is

$$\langle F\rangle_\Phi = \sum_{i=1}^{N} \langle i|f|i\rangle \tag{162}$$

$$= \sum_i \int d1d1'\, \phi_i^*(1)\langle 1|f|1'\rangle\, \phi_i(1') \,, \tag{163}$$

where the integrations include sums over spin projections. This is what one would expect for an independent particle system: each occupied state gives its own, additive contribution to the energy.

For the evaluation of $\langle V\rangle_\Phi$ we need

$$\langle\Phi|a_\mu^\dagger a_\nu^\dagger a_\sigma a_\lambda|\Phi\rangle = \delta_{\nu\sigma}\langle a_\mu^\dagger a_\lambda\rangle_\Phi - \langle a_\mu^\dagger a_\sigma a_\nu^\dagger a_\lambda\rangle_\Phi \tag{164}$$

$$= \delta_{\nu\sigma}\delta_{\mu\lambda} - \delta_{\nu\lambda}\delta_{\mu\sigma} \,, \tag{165}$$

where again all the quantum numbers refer to one-particle states occupied in $|\Phi\rangle$. Therefore

$$\langle V\rangle_\Phi = \tfrac{1}{2}\sum_{ij} \left[\langle ij|v|ij\rangle - \langle ij|v|ji\rangle\right] \equiv V_{\text{dir}} + V_{\text{ex}} \,, \tag{166}$$

where the so-called direct and exchange interaction energies are

$$V_{\text{dir}} = \tfrac{1}{2}\sum_{ij}\int d1d1'd2d2'\; \phi_i^*(1)\phi_j^*(2)\langle 12|v|1'2'\rangle\phi_i(1')\phi_j(2') \,, \tag{167}$$

$$V_{\text{ex}} = -\tfrac{1}{2}\sum_{ij}\int d1d1'd2d2'\; \phi_i^*(1)\phi_j^*(2)\langle 12|v|2'1'\rangle\phi_i(1')\phi_j(2') \,. \tag{168}$$

To summarize, the ground-state energy in the Hartree-Fock approximation is

$$\langle H\rangle_\Phi = \sum_{i=1}^{N} \langle i|f|i\rangle + \tfrac{1}{2} \sum_{i,j=1}^{N} \left[\langle ij|v|ij\rangle - \langle ij|v|ji\rangle\right] . \tag{169}$$

In varying the expectation value of the energy, as in the Bose case, the normalization of the ϕ_i is maintained by introducing a set of Lagrange multipliers $\epsilon_1 \dots \epsilon_N$. The functional that is to be stationary is thus

$$\mathcal{E}[\Phi] = \langle H\rangle_\Phi - \sum_{i=1}^{N} \epsilon_i \int d1\, |\phi_i(1)|^2 . \tag{170}$$

Because the functions ϕ_i are in general complex, $\mathcal{E}$ must be stationary with respect to both ϕ_i and ϕ_i^*. Therefore the equations that determine the ϕ_i are

$$\frac{\delta\,\langle H\rangle_\Phi}{\delta\,\phi_k^*(1)} = \epsilon_k \phi_k(1) ; \tag{171}$$

these are *the Hartree-Fock equations*, though for the moment in a form that is rather opaque.

The basic identity for implementing (171) is

$$\frac{\delta\,\phi_i(1)}{\delta\,\phi_j(1')} = \delta_{ij}\delta(1,1') \equiv \delta_{ij}\delta(\boldsymbol{r}-\boldsymbol{r}')\delta_{ss'} . \tag{172}$$

Then the variation of F, using (163), is straightforward:

$$\frac{\delta\,\langle F\rangle_\Phi}{\delta\,\phi_k^*(1)} = \int d1'\, \langle 1|f|1'\rangle \phi_k(i') . \tag{173}$$

That of $V_{\rm dir}$ requires more care:

$$\frac{\delta\,V_{\rm dir}}{\delta\,\phi_k^*(\bar{1})} = \tfrac{1}{2} \sum_j \int d1'd2d2'\; \phi_j^*(2)\langle \bar{1}2|v|1'2'\rangle \phi_k(1')\phi_j(2') \tag{174}$$

$$+ \tfrac{1}{2} \sum_i \int d1d1'd2'\; \phi_i^*(1)\langle 1\bar{1}|v|1'2'\rangle \phi_i(1')\phi_k(2') . \tag{175}$$

The second term equals the first, which is shown by taking the following steps on the former: replace i by j, use

$$\langle 12|v|1'2'\rangle = \langle 21|v|2'1'\rangle , \tag{176}$$

and then interchange $1'$ and $2'$. The final result is therefore

$$\frac{\delta\,V_{\rm dir}}{\delta\,\phi_k^*(1)} = \sum_i \int d1'd2d2'\; \phi_i^*(2)\langle 12|v|1'2'\rangle \phi_k(1')\phi_i(2') . \tag{177}$$

The same steps lead to

$$\frac{\delta\,V_{\rm ex}}{\delta\,\phi_k^*(1)} = -\sum_i \int d1'd2d2'\; \phi_i^*(2)\langle 12|v|2'1'\rangle \phi_k(1')\phi_i(2') . \tag{178}$$

The Hartree-Fock equations (Eq. 171), are thus

$$\epsilon_k \phi_k(1) = \int d1' \, \langle 1|f + U_{\rm dir} + U_{\rm ex}|1'\rangle \phi_k(1') \,, \tag{179}$$

where the direct and exchange potentials are

$$\langle 1|U_{\rm dir}|1'\rangle = \sum_i \int d2d2' \, \phi_i^*(2)\langle 12|v|1'2'\rangle \phi_i(2') \,, \tag{180}$$

$$\langle 1|U_{\rm ex}|1'\rangle = -\sum_i \int d2d2' \, \phi_i^*(2)\langle 12|v|2'1'\rangle \phi_i(2') \,. \tag{181}$$

These objects are not, in general, potentials in the conventional sense, but non-local, spin-dependent one-body operators, and functionals of the orbitals to boot. Nevertheless, the term potential is usually applied to them.

Now we specialize to the problem in which Hartree-Fock theory has seen the most service, the neutral Z-electron atom. The one- and two-body operators are then

$$f_1 = \tfrac{1}{2}p_1^2 - \frac{Z}{r_1} \,, \qquad v_{12} = \frac{1}{r_{12}} \,, \tag{182}$$

written in atomic units (§5.1), in which the unit of length is the Bohr radius in hydrogen and the unit of energy is 2 Ry $= \alpha^2 mc^2$. The electron-electron interaction is a "classical" potential — local and spin independent. Its matrix elements are

$$\langle 12|v|1'2'\rangle = \delta_{s_1 s_1'}\delta_{s_2 s_2'}\delta(\boldsymbol{r}_1 - \boldsymbol{r}_1')\delta(\boldsymbol{r}_2 - \boldsymbol{r}_2') \frac{1}{|\boldsymbol{r}_1 - \boldsymbol{r}_2|} \,. \tag{183}$$

The direct potential, using (180), is therefore

$$\langle \boldsymbol{r}s|U_{\rm dir}|\boldsymbol{r}'s'\rangle = \delta_{ss'}\delta(\boldsymbol{r} - \boldsymbol{r}') \sum_{j=1}^{Z}\sum_{s_2} \int d\boldsymbol{r}_2 \, \frac{|\phi_j(\boldsymbol{r}_2 s_2)|^2}{|\boldsymbol{r} - \boldsymbol{r}_2|} \,. \tag{184}$$

This is actually a local, spin-independent potential:

$$\langle \boldsymbol{r}s|U_{\rm dir}|\boldsymbol{r}'s'\rangle = \delta_{ss'}\delta(\boldsymbol{r} - \boldsymbol{r}') \, U_{\rm dir}(\boldsymbol{r}) \tag{185}$$

$$U_{\rm dir}(\boldsymbol{r}) = \int d\boldsymbol{r}' \, \frac{\rho(\boldsymbol{r}')}{|\boldsymbol{r} - \boldsymbol{r}'|} \,, \tag{186}$$

where $\rho(\boldsymbol{r})$ is the charge density of the electrons,

$$\rho(\boldsymbol{r}) = \sum_{j=1}^{Z}\sum_{s} |\phi_j(\boldsymbol{r}s)|^2 \,. \tag{187}$$

Thus $U_{\rm dir}$ is what would be anticipated from electrostatics, except that the charge distribution of the electron in the orbital ϕ_i should not be contributing to the force acting on this very electron, i.e., the sum in (187) should obey the restriction $j \neq i$. Looking back to (166) shows that this self-interaction is just an artifact created by the intuitively attractive division of the whole interaction into a direct and exchange part; $\langle V\rangle_\Phi$ has no $i = j$ term.

The exchange potential, as given by (181), is

$$\langle \boldsymbol{r}s|U_{\rm ex}|\boldsymbol{r}'s'\rangle = -\frac{1}{|\boldsymbol{r}-\boldsymbol{r}'|}\sum_{j=1}^{Z}\phi_j(\boldsymbol{r}s)\phi_j^*(\boldsymbol{r}'s') . \tag{188}$$

Clearly, it is non-local, and in general spin dependent.

The explicit form of the Hartree-Fock equations for a neutral atom is therefore

$$\Big(-\tfrac{1}{2}\nabla^2 - \frac{Z}{r} + U_{\rm dir}(\boldsymbol{r})\Big)\phi_i(\boldsymbol{r}s) + \sum_{s'}\int d\boldsymbol{r}'\, \langle \boldsymbol{r}s|U_{\rm ex}|\boldsymbol{r}'s'\rangle\, \phi_i(\boldsymbol{r}'s') = \epsilon_i\phi_i(\boldsymbol{r}s) . \tag{189}$$

These are non-linear integro-differential equations, which have been tackled with ever more powerful numerical methods since the 1930s.

The character of both potentials in (189) depends on the state. If we are dealing with a sufficiently light atom to make Russel-Saunders coupling a good starting point (recall §6.3), the orbitals would have the quantum numbers (nlm_ls). However, only if all values of m_l are occupied in Φ will the potential $U_{\rm dir}$ be spherically symmetric, and l a good quantum number! This remark should suffice to indicate that solving the Hartree-Fock equations is a challenging undertaking.

The exchange potential is a much more complicated creature. It is spin-independent (i.e., proportional to $\delta_{ss'}$) only if both values of s are occupied for every (lm_l). If there is an excess of, say, occupied $s = \frac{1}{2}$ spin-orbitals, it will have a term proportional to $\delta_{s,\frac{1}{2}}\delta_{s',\frac{1}{2}}$. This spin-dependent interaction has a purely electrostatic origin, and is of the type we encountered in §6.2 in the spectrum of He.

There is one important case in which the Hartree-Fock equations can be solved without doing any work: the ground state of a homogeneous Fermi gas of density n. This is a translationally invariant state, and therefore the "orbitals" are plane waves of momentum $\boldsymbol{k}$ with all levels inside the Fermi sphere doubly occupied. The calculations are left for Prob. 5, and lead to the following results. Because the wave functions are known, the ground-state energy is the same as in lowest-order perturbation theory, namely,

$$E_0 = 2\sum_{k\leq k_F}\frac{k^2}{2m} + \tfrac{1}{2}Nnu(0) - \frac{1}{\Omega}\sum_{k,k'\leq k_F} u(|\boldsymbol{k}-\boldsymbol{k}'|) , \tag{190}$$

where

$$u(k) = \int d\boldsymbol{r}\, e^{i\boldsymbol{k}\cdot\boldsymbol{r}}\, v(r) . \tag{191}$$

The terms in E_0 are, respectively, the kinetic energy, and the direct and exchange interaction energies.

Finally, we come to the energies ϵ_i appearing in the Hartree-Fock equations. It might seem that ϵ_i is the energy of the particle occupying the orbital ϕ_i, and that $\sum_i \epsilon_i$ is the ground state energy. This conjecture is roughly correct about the individual ϵ_i, but totally wrong about the total energy! According to (179),

$$\epsilon_i = \langle i|f + U_{\rm dir} + U_{\rm ex}|i\rangle \tag{192}$$

$$= \langle i|f|i\rangle + \sum_j \big[\langle ij|v|ij\rangle - \langle ij|v|ji\rangle\big] , \tag{193}$$

where (167) and (168) were used in the second step. Comparing with (169) therefore shows that

$$\sum_i \epsilon_i = \langle H\rangle_\Phi + \langle V\rangle_\Phi \,. \tag{194}$$

In short, adding up the putative one-particle energies double-counts the interactions between the particles.

Nevertheless, the energies ϵ_i do have a well-defined approximate meaning as the energy required to remove a particle from the level i by, for example, the photoeffect. The approximation assumes that N is sufficiently large so that the orbitals ϕ_i do not change appreciably if one particle is removed from the system. If this is granted, Eq. 169 leads directly to

$$\langle\Phi|H|\Phi\rangle - \langle\Phi^k|H|\Phi^k\rangle = \epsilon_k \,, \tag{195}$$

where $|\Phi^k\rangle$ is the $(N-1)$-particle state in which the orbital ϕ_k is missing. This identification of ϵ_k as the removal or separation energy is known as Koopman's theorem.

11.5 Problems

1. (a) Show that a "particle" (e.g., an atom, nucleus or nucleon) composed of N identical fermions is itself a fermion if N is odd, and a boson if N is even.

(b) Consider a nucleus of mass number A and atomic number Z. Show that it is a fermion or boson according to whether A is odd or even.

2. Let $\langle x_1 \ldots x_Z|\rho|x_1' \ldots x_Z'\rangle$ be the density matrix of a Z-electron state. The two-particle and one-particle reduced density matrices are defined by

$$\langle x_1x_2|\rho_2|x_1'x_2'\rangle = \int dx_3 \ldots dx_Z \langle x_1x_2x_3 \ldots x_Z|\rho|x_1'x_2'x_3 \ldots x_Z\rangle \,,$$

$$\langle x_1|\rho_1|x_1'\rangle = \int dx_2 \langle x_1x_2|\rho_2|x_1'x_2\rangle .$$

The ρ_n are normalized to 1, and are proportional to the distribution matrices W_n defined in the text.

(a) Consider the case where the state is adequately described by a single Slater determinant. Show that

$$\langle x|\rho_1|x'\rangle = \frac{1}{Z}\sum_{i=1}^{Z} \phi_i(x)\phi_i^*(x') \,,$$

where the ϕ_i are the spin-orbitals, and that

$$\langle x_1x_2|\rho_2|x_1'x_2'\rangle = \frac{Z}{Z-1} \det \begin{vmatrix} \langle x_1|\rho_1|x_1'\rangle & \langle x_1|\rho_1|x_2'\rangle \\ \langle x_2|\rho_1|x_1'\rangle & \langle x_2|\rho_1|x_2'\rangle \end{vmatrix} .$$

(b) Construct ρ_2 in the representation $|\boldsymbol{r}_1\boldsymbol{r}_2SM_S\rangle$ where (S, M_S) are the total spin quantum numbers. Assume that for each orbital state $u_\lambda(\boldsymbol{r})$, both spin states are occupied. Show that

$$\langle \boldsymbol{r}_1\boldsymbol{r}_2SM_S|\rho_2|\boldsymbol{r}_1'\boldsymbol{r}_2'S'M_S'\rangle = \delta_{SS'}\delta_{M_SM_S'}\frac{Z}{4(Z-1)} \times \left[\langle\boldsymbol{r}_1|\mathcal{D}|\boldsymbol{r}_1'\rangle\langle\boldsymbol{r}_2|\mathcal{D}|\boldsymbol{r}_2'\rangle + (-1)^S\langle\boldsymbol{r}_1|\mathcal{D}|\boldsymbol{r}_2'\rangle\langle\boldsymbol{r}_2|\mathcal{D}|\boldsymbol{r}_1'\rangle\right]$$

$$\langle\boldsymbol{r}|\mathcal{D}|\boldsymbol{r}'\rangle = \frac{2}{Z}\sum_{\lambda=1}^{Z/2} u_\lambda(\boldsymbol{r})u_\lambda^*(\boldsymbol{r}') \,.$$

What is the significance of the factor $(-1)^S$?

(c) Consider a Slater determinant consisting of a completely closed shell, i.e., both $j = l \pm \frac{1}{2}$ and all m occupied. Show that

$$\langle \boldsymbol{r} m_s | \rho_1 | \boldsymbol{r}' m_s' \rangle = \delta_{m_s m_s'} \frac{4\pi}{Z} \sum_{nl} (2l+1) R_{nl}(r) R_{nl}(r') \, P_l(\hat{\boldsymbol{r}} \cdot \hat{\boldsymbol{r}}') \,,$$

where the R_{nl} are the radial wave functions.

3. The wave function for a dilute Bose gas can be obtained by a variational calculation. This approach is not as flexible as that of the Gross-Pitaevskii equation, but it is instructive. Assume the gas is in a spherical enclosure of radius R and that interaction is given by (144). Take a trial function of the form (142), with

$$\varphi(\boldsymbol{r}) = C[1 - e^{-(R-r)/b}]$$

where b is to be varied.

Show that for small $\eta = b/R$ the kinetic and potential energies are

$$\langle K \rangle = \frac{\pi n R}{m} \left(\frac{1}{\eta} + \frac{7}{2} + \ldots \right) , \qquad \langle V \rangle = \frac{2\pi a n^2}{m} [1 + \tfrac{11}{4}\eta + \ldots] \,,$$

and that

$$b = \sqrt{3/22\pi \, an} \,.$$

Note that this confirms statements made in the text: the potential energy is extensive — proportional to $n^2\Omega$; and the kinetic energy is of order N/R^2 for the free-particle wave function (141), which is mimicked by b of order R (i.e., $\eta \sim 1$). Compare this result for b with (157).

4. *Stability of Bose and Fermi Systems.* A bound many-particle system is said to be extensive if its energy and volume are proportional to the number N of particles as $N \to \infty$. Ordinary solids and liquids are extensive, so are atomic nuclei, and so are most readers. For a system to be extensive, the inter-particle forces must meet certain conditions. It is intuitively not surprising that if the forces are purely attractive, a Bose system will collapse into a volume of order the force range and have a binding energy $\propto N^2$; i.e., it will not be extensive. On the other hand, one might suppose that the Pauli principle would suffice to make a many-fermion system extensive even for purely attractive forces. That is false, however, a fact that has played a central role in the development of nuclear physics. The purpose of this problem is to demonstrate that neither a Bose nor a Fermi system is extensive with purely attractive interactions.

(a) Consider a spin 0 Bose system, with the attractive interaction $V_{ij} = -V_0$ for $|\boldsymbol{r}_i - \boldsymbol{r}_j| \leq a$, and zero otherwise. Use the variational principle to prove that the state of lowest energy is bounded from above by $-CN^2$ as $N \to \infty$, where $C > 0$. (Hint: This does not require the explicit introduction of an actual trial function.)

(b) Use the same method to establish that this result also holds for a Fermi system with the same interaction, though of course with a different coefficient C.

5. *Energy of Homogeneous Fermi Gas.* Consider a spin $\frac{1}{2}$ fermion system in the limit $N \to \infty$ at constant density n; use periodic boundary conditions on the surface of a cube of volume Ω. Let the interaction be the local and spin independent potential be $v(|\boldsymbol{r}_i - \boldsymbol{r}_j|)$, whose Fourier transform $u(k)$ is defined in Eq. 191.

(a) Show that the ground-state energy to lowest order in v is given by Eq. 190.

(b) Show that this same result emerges from Hartree-Fock theory.

6. *Homogeneous Plasma.* This is the starting point for studying electronic plasmas, and also forms the crudest possible model of a conductor. The massive ions are replaced by a uniform *inert* charge distribution so that the whole system is electrically neutral, and only the light electrons flit about. Show that the electrons' interaction Hamiltonian is

$$V = \sum_{i<j}^{N} \frac{1}{\Omega} \sum_{q \neq 0} \frac{1}{q^2} e^{i\boldsymbol{q} \cdot (\boldsymbol{r}_i - \boldsymbol{r}_j)} ,$$

and that the Hartree-Fock separation energies are $\epsilon_k = \frac{1}{2}k^2 - U_{\text{ex}}(k)$, both in atomic units.

12
Interpretation

This chapter discusses the interpretation of quantum mechanics. That this basic topic has been postponed for so long, and that a whole chapter is being devoted to it, testifies to the unique character of quantum mechanics, for no text on a major topic in classical physics would have such a peculiar arrangement. The reason, of course, is that the mathematical structure of quantum mechanics is unfailingly successful in confronting an enormous range of phenomena, while the theory's epistemological framework does not have a clear link to the natural philosophy of classical physics and everyday life. As a consequence, the interpretation remains controversial.

Furthermore, in our view it is easier to present and absorb a discussion of the interpretation of quantum mechanics when the formalism can be used without fear. To some minds, however, the formalism is a superfluous barrier to understanding

"I'd just like to know what in hell is happening, that's all! I'd like to know what in hell is happening! Do you know what in hell is happening?"

FIG. 12.1. ©2002 The New Yorker Collection from cartoonbank.com. All Rights Reserved.

because ingenious expositions of much of the material in this chapter exist which assume no knowledge of the formalism — indeed, not of any physics or even calculus.[1] And because they rely on so little, these presentations offer stark demonstrations that profound and surprising conclusions follow from very general propositions, and do not require the elaborate machinery of quantum mechanics for their explication. It is our impression, however, that most physicists prefer to lean on the sturdy crutch of mathematics, rather than to test their critical acumen on subtle arguments that employ a sequence of seemingly simple steps, while constantly having to worry that they have overlooked some logical booby trap. Whatever your psychological disposition, the "elementary" discussions are very instructive whether studied before, during or after the presentation that follows.

This field has a rapidly growing literature due, primarily, to two developments: remarkable advances in experimental physics that have put many a *Gedanken* experiment onto a laboratory bench; and the connections between this topic and quantum computing.

12.1 The Critique of Einstein, Podolsky and Rosen

The issues to be discussed can be framed succinctly by recalling the famous 1935 paper "Can Quantum-Mechanical Description of Physical Reality Be Considered Complete?" by Einstein, Podolsky and Rosen (EPR). This section is largely a resumé of EPR, and in it all passages in italics are verbatim quotations from this paper.

EPR posed the following question: *Is the description given by quantum mechanics complete?* In raising this issue they explicitly acknowledged that quantum mechanics is correct.

To address their question, a definition of "complete" must be given. EPR did so in the form of two propositions, the second of which defines the concept "element of physical reality":

> 𝔸. *Every element of physical reality must have a counterpart in the physical theory.*

> 𝔹. *If, without in any way disturbing a system, we can predict with certainty (i.e., with probability equal to unity) the value of a physical quantity, then there exists an element of physical reality corresponding to this physical quantity.*

EPR invented a class of *Gedanken* experiments, each of which can can determine the eigenvalues of one member of a group of incompatible observables without disturbing the system to which the observable pertains. This they did by considering a system composed of two parts, I and II, which interacted in the past, but which subsequently separate to an extent where a measurement performed on one part is no longer able to produce an influence on the other. They illustrated their argument

---

[1] See, in particular, N.D. Mermin, *Boojums all the Way Through,* Cambridge University Press (1990), pp. 81–185. N.D. Mermin, *Am. J. Phys.* **38**, 38 (1985); **58**, 731 (1990). P.G. Kwiat and L. Hardy, ibid., **68**, 33 (2000).

with a particle that disintegrates into two particles moving in opposite directions along, say, the x-axis, so that a measurement of the momentum of I whose outcome displays the eigenvalue p_I determines that the momentum of the distant partner II is $-p_I$, whereas a measurement of position of I displaying the value x_I determines that of II as being $-x_I$. Although this example sufficed to illustrate the argument of EPR, it does not lend itself to most of the actual experiments that have been done, which have involved spins or photon polarizations. A more striking, practical and richer example, due to Bohm, is the spin singlet state $|0\rangle$ of two spin $\frac{1}{2}$ particles emanating from the decay of a $J = 0$ parent. Recall from §3.5(d) that this state is

$$|0\rangle = \tfrac{1}{\sqrt{2}} \left\{ |+\rangle_1 |-\rangle_2 - |-\rangle_1 |+\rangle_2 \right\} , \tag{1}$$

where $\pm$ is twice the spin projection along any direction — "any" because this state is spherically symmetric. It has perfect correlations between the two spin projections: if I is found to have one of the projections $v_1 = \pm$ along any direction $\boldsymbol{n}$, then the distant sibling will be found to have the opposite projection $v_2 = \mp$ along the same direction.

These examples make a fundamental point already emphasized in §1.1: the determination of a property of a system need not disturb the system. This insight, provided by EPR, contradicted the belief that such disturbances are inevitable, which played so large a role in the early development of quantum mechanics.

The possibility of determining a property of an object without coming into contact with it exists in classical physics: if the position, linear and angular momenta of a projectile flying through a force-free region are known at the moment when it disintegrates into two fragments, a subsequent measurement of these quantities belonging to one fragment determines those of the other no matter how far apart they may then be. Thus correlations between distant systems and phenomena are common in classical physics, and do not contradict the principle of causality. What causality requires is that there be no influence on an object due to any action taken in a region that is at a space-like position with respect to the object. It is the combination of this requirement with the existence of incompatible observables that creates the enigmatic situations that EPR pointed to.

We now turn to the EPR argument. Let A be an observable for system I, with eigenvalues a_α and a complete set of eigenfunctions $u_\alpha(x_1)$. Then if Ψ is any state of I and II, it can be written in the form [1]

$$\Psi(x_1, x_2) = \sum_\alpha \psi_\alpha(x_2)\, u_\alpha(x_1) . \tag{2}$$

If there is more than one term in this sum, Ψ is entangled, and that will be the case if the two systems interacted in the past. Furthermore, if B is another observable belonging to I that does not commute with A, and which has eigenvalues b_β and eigenfunctions $v_\beta(x_1)$, then Ψ can also be written in the form

$$\Psi(x_1, x_2) = \sum_\beta \varphi_\beta(x_2)\, v_\beta(x_1) . \tag{3}$$

[1] For arbitrary Ψ, the functions ψ_α do not form an orthogonal set, not do the φ_β, a complication that EPR overlooked, but which was resolved by Schrödinger in the papers that introduced the term "entanglement" and elucidated the significance of entangled states; *Proc. Cam. Phil. Soc.* **31**, 555 (1935); 446 (1936). This issue is the topic of Prob. 1.

In the example of the singlet state, A and B are the projection of the spin operator $\boldsymbol{\sigma}_1$ along any two distinct directions $\boldsymbol{n}_A$ and $\boldsymbol{n}_B$.

Expressions (2) and (3) hold at any time and for all states Ψ because the u_α and v_β are complete sets. Assume now that Ψ is a state in which I and II have acquired a space-like separation, so that *no real change can take place in system II in consequence of anything that may be done to system I.* From a measurement of the observable A belonging to system I, whose outcome finds it to be in the state u_ν, we know from (2) that the distant, undisturbed system II is in the state ψ_ν. Alternatively, by measuring B, and finding I to be in the state v_λ, we know from (3) that II is in the state φ_λ.

Suppose, furthermore, that the $\psi_\alpha(x_2)$ are the eigenfunctions of an observable P belonging to system II, with eigenvalues p_α, while the $\varphi_\beta(x_2)$ are the eigenfunctions of another observable Q of system II with eigenvalues q_β, and that P and Q do not commute. (In the spin singlet example, P and Q are the projections of $\boldsymbol{\sigma}_2$ along $\boldsymbol{n}_A$ and $\boldsymbol{n}_B$.) As no interaction between I and II is possible when they have a space-like separation, this has the following consequence:

> $\mathbb{C}$. *By measuring either A or B we are in a position to predict with certainty, and without in any way disturbing the second system, either the value of the quantity P (that is p_α) or the value of the quantity Q (that is q_β). In accordance with our criterion of reality, in the first case we must consider the quantity P as being an element of reality, [while] in the second case the quantity Q is an element of reality.*

On the other hand, when P and Q do not commute, quantum mechanics asserts that *precise knowledge of P precludes such a knowledge of Q; and furthermore, that any attempt to determine Q experimentally will alter the state of the system in such a way as to destroy the knowledge of P.* In short, according to quantum mechanics,

> $\mathbb{D}$. *When two operators corresponding to two physical quantities do not commute the two quantities cannot have simultaneous physical reality.*

But $\mathbb{D}$ contradicts $\mathbb{C}$. Given that the *Gedanken* experiments satisfy proposition $\mathbb{B}$, EPR concluded that quantum mechanics is a theory that fails to comply with proposition $\mathbb{A}$, i.e., *the quantum mechanical description of physical reality given by the wave function is not complete.*

Shortly after EPR appeared, Bohr argued (albeit, in his typically opaque style) that the EPR conception of reality was in conflict with how factual statements must be made in the quantum domain. An instructive illustration of Bohr's point is provided by the two-particle interferometer discussed in §1.2(c). Recall that, as in the EPR example, it also offers two complementary options: (i) the set of events in which the momentum of particle I along the direction perpendicular to the slits is determined, thereby selecting one of the two available paths for both particles, in which case the distant particles II build diffraction patterns due to just single slits; (ii) events in which the conjugate coordinate of particle I is determined, so that the two paths are not distinguished, and an interference pattern is formed between the coordinate of I and that of the distant particle II. But these are distinct experiments, measuring mutually incompatible observables of one particle; the same point applies to the spin singlet example, for distinct experimental arrangements must be used to measure the spin projections along distinct (and therefore incompatible) directions. And this was Bohr's basic point: EPR were being misled by their definition of an

FIG. 12.2. Niels Bohr and Albert Einstein
Photo by Paul Ehrenfest, courtesy of AIP Emilio Segrè Visual Archives

element of physical reality as it overlooks[1] "that the procedure of measurement has an essential influence on which the very definition of the physical quantities rests."

Many therefore take the position that EPR does not pose a paradox, because it is impossible to devise any experiment that simultaneously measures precise values of incompatible observables. Indeed, enigmas that stem from EPR usually entail counterfactual arguments: contradictions between an actual experiment that does measure P, and another that could, were it done instead, measures the incompatible observable Q. Nevertheless, entangled states have properties that appear to be due to non-local interventions, as is made strikingly clear by the spin singlet state (see §12.3). For if a long string of measurements of the distant spins $\boldsymbol{\sigma}_1$ and $\boldsymbol{\sigma}_2$ are made along the same direction, then in each and every individual case an observer who knows the outcomes $v_1 = \pm$ at end I can, without fail, tell the distant observer at II the outcome $v_2 = \mp$, and vice versa. This does not allow superluminal signaling between I and II because observer I cannot control which of the outcomes $\pm$ will appear in any one measurement at I, nor for that reason the correlated outcomes at II. The data have to be brought together before the "prediction" can be confirmed. Nevertheless, the correlations provoke the thought that a conspiracy is afoot, with the members of each pair carrying instructions that tell them what to do when they are questioned.

The EPR argument thus raises the question of whether there is a description of nature that is more refined than that of quantum mechanics, and which explains the "spooky action at a distance" (as Einstein put it) that quantum mechanics appears to imply. EPR did not suggest how quantum mechanics could be elaborated into a "local realistic" theory satisfying their definitions of realism and completeness, while leaving the established consequences of quantum mechanics intact. Here "local" refers to the requirement that a system cannot be affected by any action taken at a space-like separation from it. A natural elaboration would be to introduce

[1]N. Bohr, *Nature* **136**, 65 (1935)

additional "hidden variables" that provide a "complete" description of individual systems by assigning values to all the observables of such an individual prior to measurement, with an average over the hidden variables giving the quantum state as a description of ensembles of individuals.

12.2 Hidden Variables

Is the statistical interpretation of quantum mechanics inescapable, or is it a matter of taste because it is possible to devise logically consistent "completions" of quantum mechanics, in the sense of EPR? To understand the theorems and experiments that so drastically circumscribe such completions, it is necessary to have some understanding of what such hidden variable completions seek to accomplish.

The simplest non-trivial quantum systems inhabit the two-dimensional Hilbert space $\mathfrak{H}_2$, and a hidden variable theory exists in this case. Any Hermitian operator A in $\mathfrak{H}_2$ is some linear combination of 1 and $\boldsymbol{\sigma}$:

$$A = a_0 + \boldsymbol{a}\cdot\boldsymbol{\sigma} \,, \tag{4}$$

where a_0 and $\boldsymbol{a}$ are real. The eigenvalues of A are

$$v(A) = a_0 \pm a \,. \tag{5}$$

Quite aside from whether quantum mechanics is correct or complete, experimental physics tells us that these are the only values that are ever displayed by a two-level system in any single measurement of the arbitrary observable A. Therefore any theory must also have this feature, whether it be quantum mechanics, hidden variables, or some future theory that supersedes quantum mechanics.

Consider the pure quantum state with density matrix

$$\rho(\boldsymbol{n}) = \tfrac{1}{2}(1 + \boldsymbol{n}\cdot\boldsymbol{\sigma}) \,, \tag{6}$$

i.e., the state with "spin up" along $\boldsymbol{n}$. The expectation value of A in this state is

$$\langle A\rangle_{\boldsymbol{n}} = \operatorname{Tr} \rho(\boldsymbol{n})\, A = a_0 + \boldsymbol{n}\cdot\boldsymbol{a} = p_+(a_0 + a) + p_-(a_0 - a) \,, \tag{7}$$

and therefore the probabilities for finding the two values (5) are

$$p_\pm = \tfrac{1}{2}(1 \pm \hat{\boldsymbol{a}}\cdot\boldsymbol{n}) \,, \tag{8}$$

where $\hat{\boldsymbol{a}} = \boldsymbol{a}/a$.

A hidden variable description for $\mathfrak{H}_2$ was first given by Bell. In the version due to Mermin, the hidden variable is a real unit 3-vector $\boldsymbol{m}$ with a definite orientation. The connection between the "realistic" hidden variable description and the statistics of actually observable results is provided by (i) a function $v(A, \boldsymbol{n}; \boldsymbol{m})$ that unambiguously assigns one of the two eigenvalues (5) of the observable A, given the hidden variable $\boldsymbol{m}$ and the state $\rho(\boldsymbol{n})$; and (ii) a probability distribution that specifies the average over the hidden variable that reproduces the expectation value of A. In this example, the function (i) is

$$v(A, \boldsymbol{n}; \boldsymbol{m}) = \begin{cases} a + a_0 & \boldsymbol{m}\cdot\boldsymbol{a} < \boldsymbol{n}\cdot\boldsymbol{a} \\ a - a_0 & \boldsymbol{m}\cdot\boldsymbol{a} > \boldsymbol{n}\cdot\boldsymbol{a} \end{cases} \tag{9}$$

The recipe (ii) is that $\boldsymbol{m}$ has a uniform distribution, which then reproduces $\langle A\rangle_{\boldsymbol{n}}$:

$$\int \frac{d\boldsymbol{m}}{4\pi}\, v(A,\boldsymbol{n};\boldsymbol{m}) = \frac{1}{2}\left(\int_{-1}^{\phi}(a_0+a)dx + \int_{\phi}^{1}(a_0-a)dx\right) = a_0 + a\phi\,, \qquad (10)$$

where $\phi = \hat{\boldsymbol{a}}\cdot\boldsymbol{n}$.

To appreciate the gulf between the "realistic" and quantum mechanical conceptions, consider a succession of spin $\frac{1}{2}$ objects $\mathcal{S}_1, \mathcal{S}_2, \ldots$, prepared in the state $\rho(\boldsymbol{n})$ by the appropriate Stern-Gerlach apparatus $SG(\boldsymbol{n})$, that is to encounter another Stern-Gerlach magnet $SG(\boldsymbol{a})$, where the direction of $\boldsymbol{a}$ was chosen while the $\mathcal{S}_i$ are in free flight. In quantum orthodoxy, the state of $\mathcal{S}_i$ is exhaustively specified by $\boldsymbol{n}$, and all that can be said is that there is a known function of $\boldsymbol{n}\cdot\boldsymbol{a}$ that gives the probability that $\mathcal{S}_i$ will be deflected up or down along $\boldsymbol{a}$, but that its actual fate is beyond knowledge: *quantum mechanics assigns no value to the observable prior to measurement.*

In the hidden variable philosophy, each $\mathcal{S}_i$ is equipped with a vector $\boldsymbol{m}_i$ with a specific orientation, that in conjunction with $\boldsymbol{n}$ describes its state. When $\mathcal{S}_i$ encounters $SG(\boldsymbol{a})$, whether $(\boldsymbol{m}_i - \boldsymbol{n})\cdot\boldsymbol{a}$ is negative or positive settles uniquely whether $\mathcal{S}_i$ is deflected up or down. Furthermore, the assignments of $\boldsymbol{m}_i$ to the various specimens $\mathcal{S}_i$ are distributed so as to reproduce the quantum mechanical expectation value. Note that this hidden variable theory predicts with certainty the individual measurement outcomes for any and all $\boldsymbol{a}$, even though different directions of $\boldsymbol{a}$ correspond to incompatible observables. *Therefore, in this hidden variable theory all possible values of the observable exist whether or not a measurement is performed, with the measurement merely identifying one of these pre-existing values.*

A caution.[1] The hidden variable theory assigns the values ± 1 of, say, v_x and v_y to σ_x and σ_y, but these values cannot be added to assign values to the observable $\mu = \sigma_x + \sigma_y$, for doing so would say that v_μ can take on any of the values $(2, 0, -2)$ whereas the two eigenvalues of μ are $\pm\sqrt{2}$. This is not a failing of the hidden variable approach, but of understanding what is at stake. The hidden variables are to predict with certainty the outcome of any one measurement, and no such measurement corresponds to $v_x + v_y$. The SG apparatus that measures μ is specified by the parameters $a_0 = 0$ and $\boldsymbol{a} = \boldsymbol{u}_x + \boldsymbol{u}_y$, and then the rule (9) will give the correct results for v_μ.

The simplicity of the $\mathfrak{H}_2$ example might lead one to suppose that it should not be difficult to construct hidden variable descriptions for higher-dimensional Hilbert spaces, but that is not so. The Bell-Kochen-Specker (BKS) theorem states that it is not possible to consistently assign values to all the observables in any Hilbert space larger than $\mathfrak{H}_2$. The proof is difficult for $\mathfrak{H}_3$, but Mermin showed that it is easy for $\mathfrak{H}_4$.

The fact on which the BKS theorem relies is that if a set of observables $(A, B, \ldots)$ are compatible and satisfy the relation $f(A, B, \ldots) = 0$, and $(a, b, \ldots)$ are any set of their simultaneous eigenvalues, then $f(a, b, \ldots) = 0$. This is illustrated by the familiar identity $2\boldsymbol{L}\cdot\boldsymbol{S} + \boldsymbol{L}^2 + \boldsymbol{S}^2 - \boldsymbol{J}^2 = 0$. As a consequence, the *values* assigned by a hidden variable theory to the observables must also satisfy $f(v_A, v_B, \ldots) = 0$, and as we will now see it is on this rock that the attempt breaks apart. The problem

[1]This alludes to what turned out to be a misleading claim of von Neumann (which preceded the EPR paper) that hidden variable elaborations of quantum mechanics are impossible.

does not arise for $\mathfrak{H}_2$ because there is only one compatible observable, whereas there is more than one for all $\mathfrak{H}_d$ with $d \geq 3$, as is clear from angular momentum, because S_z and S_z^2 are compatible when $d \geq 3$.

All observables in $\mathfrak{H}_4$ can be represented by two *commuting* Pauli spin vectors $\boldsymbol{\sigma}_1$ and $\boldsymbol{\sigma}_2$. Consider then the following tableau:

$$\begin{matrix} \sigma_{1x} & \sigma_{2x} & \sigma_{1x}\sigma_{2x} \\ \sigma_{2y} & \sigma_{1y} & \sigma_{1y}\sigma_{2y} \\ \sigma_{1x}\sigma_{2y} & \sigma_{1y}\sigma_{2x} & \sigma_{1z}\sigma_{2z} \end{matrix} \quad \equiv \quad \begin{matrix} A_1 & B_1 & C_1 \\ A_2 & B_2 & C_2 \\ A_3 & B_3 & C_3 \end{matrix} \tag{11}$$

The operators in each row and each column are mutually compatible because $[\boldsymbol{\sigma}_1, \boldsymbol{\sigma}_2] = 0$ and the components of each separate spin anticommute. Furthermore,

$$1 = A_1A_2A_3 = B_1B_2B_3 = A_1B_1C_1 = A_2B_2C_2 = A_3B_3C_3 \,, \tag{12}$$

whereas

$$C_1C_2C_3 = -1 \,. \tag{13}$$

Assigning values to all these observables, and using (12), gives

$$1 = [v(A_1)v(A_2)v(A_3)v(B_1)v(B_2)v(B_3)]^2 \, v(C_1)v(C_2)v(C_3) \,, \tag{14}$$

which contradicts Eq. 13. *QED.*

This proof only uses *identities* between compatible observables, but it is based on the premise that it legitimate to assign *values* to incompatible observables, e.g., $\sigma_{1x} = A_1$ and $\sigma_{1y} = B_2$.

That physical states are correctly described by vectors in Hilbert space cannot be directly inferred from experiment, but if one accepts that the success of quantum mechanics is evidence "beyond all reasonable doubt" that this is a correct description, then the BKS theorem proves that quantum mechanics is not a statistical rendition of an underlying hidden variable theory. Bell's theorem is of a quite different nature; in contrast to the BKS theorem, in general it applies only to certain specific states, such as the one we are about to discuss, and it depends on imposing the EPR definition of locality, whereas locality does not play any role in the BKS argument. Furthermore, and of capital importance, the analysis based on Bell's inequality does not rely on quantum mechanics as such, and can be directly compared to experimental data.

12.3 Bell's Theorem

Bell's theorem and its generalizations establish that hidden variables theories can only reproduce the statistical predictions of quantum mechanics if the hidden variables in one region can be affected by measurements carried out in a space-like separated region (as they are in the de Broglie-Bohm theory). Such theories therefore violate the relativistic causality principle, surely an ironic outcome from the perspective of EPR ! The experiments motivated by Bell's theorem give results in excellent agreement with quantum mechanics, hence only *non-local* hidden variable theories are viable options, which at least in our view offers a cure that is worse than the malaise.[1]

Although Bell's theorem is motivated by the quantum mechanical description of a special class of states, it is not contingent on quantum mechanics as such. The theorem states a condition as to whether certain experimental data are compatible with the classical conceptions of reality and local causality, as expressed by EPR. The actual data violate this condition, but confirm the prediction of quantum mechanics. Unless the experiments have some undetected fatal flaw, a supposition that has become implausible, it is therefore established that the EPR concepts are invalid, whether or not quantum mechanics or some theory that might eventually supercede quantum mechanics proves to be the ultimate explanation for the data.

To prepare the ground for a proof of Bell's theorem, this section begins with a detailed description of the two-particle spin singlet state. The proof of Bell's original theorem is then given, followed by a generalization which, in contrast to the original version, is amenable to experimental tests. The first high confidence test is then described.

(a) The Spin Singlet State

Consider once more a particle of spin 0 that decays into two particles of spin $\frac{1}{2}$. In a frame in which the parent is localized in a packet of mean momentum zero with a momentum spread small compared to the momenta of the daughters, the final state is the product of a spatial wave function describing the daughters' motion in nearly opposite directions, multiplied by the entangled spin state $|0\rangle$, as given by (1). Only this spin factor will matter now. Because $|0\rangle$ is a state of zero angular momentum, it is rotationally invariant, as is most clear in the density matrix

$$\rho = \tfrac{1}{4}(1 - \boldsymbol{\sigma}_1 \cdot \boldsymbol{\sigma}_2) \,. \tag{15}$$

The expectation value of *any* component of *either* daughter's spin vanishes in $|0\rangle$ in virtue of rotational invariance. For example, measurement of σ_{2x} will yield the results ± 1 with equal probability. But there are correlations between the two spins, which can be displayed by two spatially separated experimental arrangements which, for every member of an ensemble of pairs in the state $|0\rangle$, measure the spin projection v_1 of particle 1 along and $\boldsymbol{n}_1$ and v_2 of 2 along $\boldsymbol{n}_2$. Then if $N(v_1, v_2)$ is the number of events observed with the indicated values, and N a sufficiently large total number of events, the data determines the empirical correlation function

$$C(\boldsymbol{n}_1, \boldsymbol{n}_2) = \frac{1}{N} \sum_{v_1, v_2 = \pm 1} N(v_1, v_2) v_1 v_2 \,. \tag{16}$$

For any rotationally invariant state, $C(\boldsymbol{n}_1, \boldsymbol{n}_2)$ can only be a function of $\boldsymbol{n}_1 \cdot \boldsymbol{n}_2$.

According to quantum mechanics, this correlation function is the expectation value of the operator $(\boldsymbol{\sigma}_1 \cdot \boldsymbol{n}_1)(\boldsymbol{\sigma}_2 . \boldsymbol{n}_2)$. In the case of the singlet state (5), it is

$$Q(\boldsymbol{n}_1, \boldsymbol{n}_2) = \operatorname{Tr} \rho \, (\boldsymbol{\sigma}_1 \cdot \boldsymbol{n}_1)(\boldsymbol{\sigma}_2 \cdot \boldsymbol{n}_2) = -\boldsymbol{n}_1 \cdot \boldsymbol{n}_2 \equiv -\cos\phi_{12} \,. \tag{17}$$

---

[1] It is unnecessary to invoke relativity here. In this context locality means that there are no influences between systems that do not interact. Further remarks concerning this will be made in §12.3(d).

For identical orientation of the two measurements, $\boldsymbol{n}_1 = \boldsymbol{n}_2$, this expectation value is -1, i.e., there is perfect anticorrelation. This is even clearer in the joint probability that $\boldsymbol{\sigma}_1$ is along $\boldsymbol{n}_1$ and $\boldsymbol{\sigma}_2$ along $\boldsymbol{n}_2$:

$$p_{12}(\boldsymbol{n}_1, \boldsymbol{n}_2) = \mathrm{Tr}\, \rho\, P_1(\boldsymbol{n}_1) P_2(\boldsymbol{n}_2) = \tfrac{1}{4}(1 - \boldsymbol{n}_1 \cdot \boldsymbol{n}_2) , \tag{18}$$

where $P_i(\boldsymbol{n}_i) = \frac{1}{2}(1 + \boldsymbol{n}_i \cdot \boldsymbol{\sigma}_i)$ is the spin projection operator for particle i. This probability vanishes when the spins are parallel, and equals $\frac{1}{2}$ when they are antiparallel because there are two possibilities in that case.

Hence when $\boldsymbol{n}_1 = -\boldsymbol{n}_2$ the state $|0\rangle$ satisfies the EPR condition $\mathbb{B}$, because by measuring the spin projection of one daughter along an arbitrary direction $\boldsymbol{n}$ "we can predict with certainty" that the distant undisturbed sibling has the opposite spin projection along this same direction. And this can be done with any other direction, which then produces the EPR conundrum ($\mathbb{C}$ vs. $\mathbb{D}$) because angular momenta along different directions do not commute.

The preceding argument was contingent on $|0\rangle$ having *perfect* anticorrelations for *any* direction. The separate terms $|\pm\rangle_1|\mp\rangle_2$ in $|0\rangle$ also have perfect anticorrelations, but only along the single direction that defines the eigenvalues as given. Perfect anticorrelation along any direction requires a coherent, entangled state.[1] To appreciate this, consider an *incoherent* spherically symmetric superposition ρ_R of states, each of which has perfect anticorrelations,

$$\rho_R = \int \frac{d\boldsymbol{n}}{4\pi} P_1(\boldsymbol{n}) P_2(-\boldsymbol{n}) . \tag{19}$$

This mixture has the imperfect anticorrelation $Q = -\frac{1}{3}\boldsymbol{n}_1 \cdot \boldsymbol{n}_2$, i.e., a probability of $\frac{1}{3}$ that the distant spin will have the opposite direction, which proves to be too weak a correlation to produce a violation of Bell's inequality.

(b) Bell's Theorem

Now to the hidden variable analysis of the two-particle spin singlet state. We will, for now, assume that the hidden variable description is deterministic, but this assumption will be dropped in the next subsection.

In a deterministic hidden variable theory, such as the $\mathfrak{H}_2$ example in §12.2, the hidden variables λ specify unambiguously the values of all "elements of reality" pertaining to a particular pair of particles, which by hypothesis is born in a dispersion free "h-state." These λ have values for any one pair such that given the directions $\boldsymbol{n}_1$ and $\boldsymbol{n}_2$ along which the spin of particles 1 and 2 are to be measured the outcomes $v_1(\lambda, \boldsymbol{n}_1)$ and $v_2(\lambda, \boldsymbol{n}_2)$ are *predetermined.* Furthermore, all such outcomes v_i are either 1 or -1.

If the EPR concepts are valid, correlations between outcomes v_1 and v_2 for an individual pair were imposed at their birth, and the statistical correlations observed in an experimental run are an average over the hidden variables that determine the individual outcomes. Under these assumptions, the observed correlations must have the form

$$R(\boldsymbol{n}_1, \boldsymbol{n}_2) = \int v_1(\lambda, \boldsymbol{n}_1)\, v_2(\lambda, \boldsymbol{n}_2)\, w(\lambda)\, d\lambda , \tag{20}$$

[1] The singlet state $|0\rangle$ is not the only one that has perfect correlations for a range of directions (see Prob. 3).

where $w(\lambda)\,d\lambda$ is the probability measure assigned to the h-states. Furthermore,

1. the values $v_i(\lambda, \boldsymbol{n})$ depend only on the setting $\boldsymbol{n}$ of the instrument with which the spin of daughter i is measured, but not on the setting of the other instrument;
2. the probability distribution $w(\lambda)$ does not depend on the settings $(\boldsymbol{n}_1, \boldsymbol{n}_2)$.

Equation (20) and these properties of v_i and w express the classical conceptions of local causality: the directions along which the spins are measured can be chosen long after the particles have separated, and therefore the value displayed by one daughter cannot depend on the direction chosen for the other, nor can the probability that the hidden variables have particular values at the daughters' birth depend on the as yet unchosen directions.

There are then two distinct questions: Do hidden variables exist such that $R(\boldsymbol{n}_1, \boldsymbol{n}_2)$ (i) agrees with the quantum mechanical prediction $Q(\boldsymbol{n}_1, \boldsymbol{n}_2)$ of Eq. 17; and/or (ii) agrees with the experimentally observed correlation $C(\boldsymbol{n}_1, \boldsymbol{n}_2)$? Here we focus on (i) and leave (ii) for later.

Bell's theorem, in its original version, is a consequence of the following inequality:

$$|R(\boldsymbol{a}, \boldsymbol{b}) - R(\boldsymbol{a}, \boldsymbol{c})| - R(\boldsymbol{b}, \boldsymbol{c}) \leq 1 , \tag{21}$$

where $(\boldsymbol{a}, \boldsymbol{b}, \boldsymbol{c})$ are arbitrary unit vectors. We postpone the short derivation of (21) for a moment, and go directly to the famous conclusion: *the correlations in the singlet state violate Bell's inequality.*

According to (21), if quantum mechanics is to satisfy Bell's inequality the expectation values (17) must satisfy

$$|Q(\boldsymbol{a}, \boldsymbol{b}) - Q(\boldsymbol{a}, \boldsymbol{c})| - Q(\boldsymbol{b}, \boldsymbol{c}) = |\cos\phi_{ab} - \cos\phi_{ac}| + \cos\phi_{bc} \leq 1 , \tag{22}$$

where $\cos\phi_{ab} = \boldsymbol{a}\cdot\boldsymbol{b}$, etc. As an example, let the three vectors be in a plane, and take the choice $\phi_{ab} = \phi_{bc} = \frac{1}{4}\pi$, so that $\phi_{ac} = \frac{1}{2}\pi$, which gives $\sqrt{2}$ for the left-hand side of Eq. 22, in gross violation of the inequality. *QED.*

The example just given is a special case of the following statement: Bell's inequality is violated by the singlet state unless $\boldsymbol{a} = \pm\boldsymbol{b}$ or $\boldsymbol{a} = \pm\boldsymbol{c}$ (see Prob. 2). When the inequality is satisfied, two of the three operators $(\boldsymbol{\sigma}_1\cdot\boldsymbol{a})(\boldsymbol{\sigma}_2\cdot\boldsymbol{b})$, $(\boldsymbol{\sigma}_1\cdot\boldsymbol{a})(\boldsymbol{\sigma}_2\cdot\boldsymbol{c})$, $(\boldsymbol{\sigma}_1\cdot\boldsymbol{b})(\boldsymbol{\sigma}_2\cdot\boldsymbol{c})$, whose expectation values appear in (22), are compatible, and the full power of incompatibility is veiled.

The proof is remarkably simple because in the singlet state the correlations are perfect when the widely separated observers use the same orientation, so that for all $\boldsymbol{n}$

$$R(\boldsymbol{n}, \boldsymbol{n}) = -1 . \tag{23}$$

Hence for all values of the hidden variables,

$$v_1(\lambda, \boldsymbol{n}) = -v_2(\lambda, \boldsymbol{n}) . \tag{24}$$

Therefore

$$R(\boldsymbol{n}_1, \boldsymbol{n}_2) = -\int v_1(\lambda, \boldsymbol{n}_1)\, v_1(\lambda, \boldsymbol{n}_2)\, w(\lambda)\, d\lambda . \tag{25}$$

As a consequence of (24),

$$R(\boldsymbol{a}, \boldsymbol{b}) - R(\boldsymbol{a}, \boldsymbol{c}) = -\int [v_1(\boldsymbol{a}, \lambda)v_1(\boldsymbol{b}, \lambda) - v_1(\boldsymbol{a}, \lambda)v_1(\boldsymbol{c}, \lambda)]\, w(\lambda)\, d\lambda ; \tag{26}$$

but $[v_1(\boldsymbol{b})]^2 = 1$, and therefore

$$R(\boldsymbol{a},\boldsymbol{b}) - R(\boldsymbol{a},\boldsymbol{c}) = -\int v_1(\boldsymbol{a},\lambda)v_1(\boldsymbol{b},\lambda)[1 - v_1(\boldsymbol{b},\lambda)v_1(\boldsymbol{c},\lambda)]\, w(\lambda)\, d\lambda\, . \tag{27}$$

Now $v_1(\boldsymbol{a})v_1(\boldsymbol{b}) = \pm 1$, and $[1 - v_1(\boldsymbol{b})v_1(\boldsymbol{c})] \geq 0$; therefore the absolute value of (27) gives the inequality

$$|R(\boldsymbol{a},\boldsymbol{b}) - R(\boldsymbol{a},\boldsymbol{c})| \leq \int [1 - v_1(\boldsymbol{b},\lambda)v_1(\boldsymbol{c},\lambda)]\, w(\lambda)\, d\lambda = 1 + R(\boldsymbol{b},\boldsymbol{c})\, , \tag{28}$$

which is Eq. 21; *QED.*

This conclusion cannot be drawn if the probability $w(\lambda)$ depends on either or both settings of the apparatus, or if the values v_i depend on the setting with which the distant particle is observed, i.e., if the hidden variable theory is non-local.

(c) The Clauser-Horne Inequality

Bell's original inequality (Eq. 21), cannot be tested as it stands because it relies on perfect anticorrelations (i.e., Eq. 23) when the settings of the two distant instruments agree, which cannot be attained in any actual experiment. However, inequalities of a similar character exist that do not depend on this unrealistic requirement, and also of importance, rely on weaker assumptions about the hidden variables than the derivation just given.

We continue to consider processes of the EPR type, and call $(a, a', \ldots)$ values of the observable A displayed by subsystem $\mathcal{S}_1$, $(b, b', \ldots)$ those of B belonging to subsystem $\mathcal{S}_2$. Here (a, b), etc., are generic symbols, with spin or photon polarization directions being only illustrative examples. Let N be the total number of EPR pairs introduced into the experiment, $N_1(a)$ and $N_2(b)$ the number of single counts measured at the detectors for systems $\mathcal{S}_1$ and $\mathcal{S}_2$, respectively, when set to the indicated values of the observables, and $N_{12}(a, b)$ the number of coincidence counts. All these counts may be small in number compared to N due to detection inefficiencies, attenuation in polarizers, and the like. Despite such unavoidable complications experiments can, however, accurately measure the probabilities

$$p_1(a) = N_1(a)/N\, , \quad p_2(b) = N_2(b)/N\, , \quad p_{12}(a,b) = N_{12}(a,b)/N\, . \tag{29}$$

Hidden variables λ are now introduced, but in contrast to Bell's derivation, it is no longer assumed that this elaboration of quantum mechanics is deterministic. That is, a particular set of parameters λ is not assumed to determine uniquely the measured value of say a, but appears in probability distributions $p_1(a;\lambda)$ for these values. The latter, when averaged over the probability distribution $w(\lambda)$ for the hidden variables, then yields the observed probability

$$p_1(a) = \int p_1(a;\lambda)\, w(\lambda) d\lambda\, . \tag{30}$$

Locality is imposed by requiring that given λ, the joint probability distribution for specific values of a and b does not have correlations between the distant subsystems $\mathcal{S}_1$ and $\mathcal{S}_2$,

$$p_{12}(a,b;\lambda) = p_1(a;\lambda)p_2(b;\lambda)\, , \tag{31}$$

and that the probability distribution $w(\lambda)$ for the hidden variables does not depend on the instrumental settings. No assumption is made about the special case $p_{12}(a, a; \lambda)$ when the distant settings are identical, i.e., there is no condition corresponding to Eq. 24. The probability of coincidences is the average

$$p_{12}(a,b) = \int p_1(a;\lambda)p_2(b;\lambda)\, w(\lambda)\, d\lambda \,. \tag{32}$$

Of course, the absence of correlations in $p_{12}(a, b; \lambda)$ does not preclude correlations in the measured probability $p_{12}(a, b)$.

An important generalization of Bell's inequality is a consequence of the following lemma: if the four real numbers (x, x', y, y') all lie in the interval (0,1), then

$$xy - xy' + x'y + x'y' - x' - y \leq 0 \,; \tag{33}$$

the proof is left to the reader. Now let $x = p_1(a; \lambda), \ldots, y' = p_2(b'; \lambda)$, and integrate (33) over λ with the weight $w(\lambda)$:

$$p_{12}(a,b) - p_{12}(a,b') + p_{12}(a',b) + p_{12}(a',b') - p_1(a') - p_2(b) \leq 0 \,, \tag{34}$$

or

$$\frac{p_{12}(a,b) - p_{12}(a,b') + p_{12}(a',b) + p_{12}(a',b')}{p_1(a') + p_2(b)} \leq 1 \,. \tag{35}$$

This is the Clauser-Horne inequality. It does not rely on flawless coincidence counters, or on the specific correlations of the spin singlet state, nor on knowledge of what fraction of the sample is detected because the total number N of EPR pairs drops out of the ratio so that only knowledge of the coincidence rate relative to the singles rates matters. As we will see shortly, there are two-photon states for which this inequality is violated by quantum mechanics. From a theoretical standpoint, Eq. 35 is also a stronger statement than Bell's original inequality because it does not assume that the elaboration of quantum mechanics is deterministic.

To repeat, if quantum mechanics were eventually shown to be false, while the experimental data violating the Bell inequalities withstand the test of time, as now appears close to certain, the new theory, whatever it might be, could not be a theory that embodies the classical conceptions of realism and locality as they are expressed by Eq. 20 and Eq. 31. In short, Bell's inequality and its generalizations provide an astonishingly powerful and yet remarkably simple tool for distinguishing the world as it really is from how our everyday experience prompts us to depict it.

(d) An Experimental Test of Bell's Inequality

A considerable number of experiments have been done to ascertain whether EPR states have correlations that violate an inequality of the Bell type; this continues to be an active field. As already stated, virtually all the experiments done thus far violate such inequalities and agree with quantum mechanics to varying but uniformly high degrees of precision. We confine our discussion to the historically critical two-photon experiment of Aspect and co-workers; it was the first high-accuracy test in which the decision as to what is to be measured is taken while the particles are traveling from their source to the detectors, which eliminated doubts that had been raised about earlier experiments.

In the Aspect experiment,[1] two photons traveling in opposite directions and having specified linear polarizations are detected. The parameters $a, \ldots$ in the Clauser-Horne inequality are then the angles $\theta_a, \ldots$ specifying these polarizations in the plane perpendicular to the direction along which the photons are flying apart. The setup is invariant under rotation about this direction, so $p_{12}(a, b)$ can depend only on $|\theta_a - \theta_b|$, and $p_1(a)$ does not depend on angle, provided that there are no experimental biases. If, furthermore, the four directions are chosen as in Fig. 12.3a, then (35) takes on a remarkably simple form:

$$\frac{3N_{12}(\phi) - N_{12}(3\phi)}{N_1 + N_2} - 1 \leq 0 \,. \tag{36}$$

Testing this inequality is the object of the experiment.

The two-photon states $|\gamma_1\gamma_2\rangle$ are produced in a sequence of atomic E1 transitions, $0^+ \to 1^- \to 0^+$, so that $|\gamma_1\gamma_2\rangle$ is also a 0^+ state (Fig. 12.3). The apparatus selects the component of $|\gamma_1\gamma_2\rangle$ in which the photons are propagating back-to-back along a direction $\boldsymbol{n}$. Conservation of angular momentum about $\boldsymbol{n}$ requires the two photons to have the same helicity μ. Let μ be the helicity of the photon emitted in the upper transition along the direction $\boldsymbol{n}$, and m the eigenvalue of the atom's angular momentum about $\boldsymbol{n}$ when it is in the intermediate 1^- state; then $m = -\mu$. If the atom is not in a magnetic field, the $m = \pm 1$ states are degenerate. Thus there are *two coherent but distinct paths* passing through $m = \mp 1$ and leading to final states in which both photons have helicity ± 1. Note that the term *path* has now assumed a more abstract meaning than paths through slits in a two-particle interferometer, but with the same consequence of producing an entangled state, as we will now see.

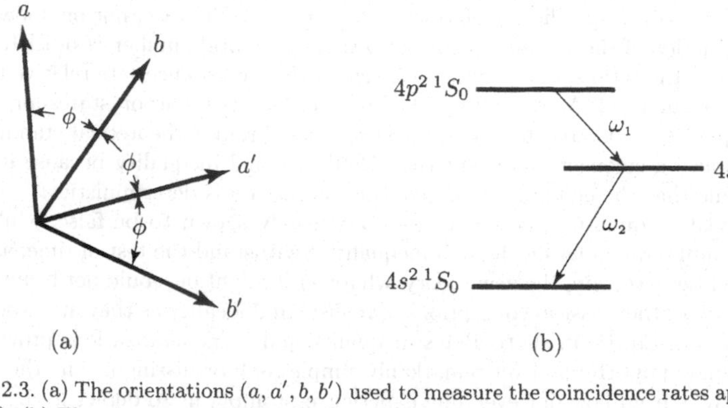

FIG. 12.3. (a) The orientations (a, a', b, b') used to measure the coincidence rates appearing in (35). (b) The photon cascade studied in the Aspect experiment; these are states of Ca with two valence electrons in the indicated configurations.

The two arms of the apparatus have filters such that the arm in the direction $\boldsymbol{n}$ only accepts photons with frequency near ω_1, while the opposite arm only accepts those near ω_2, where $\omega_{1,2}$ are the frequencies of the two E1 transitions. One state

[1] A. Aspect, J. Dalibard and G. Roger, *Phys. Rev. Lett.* **49**, 1804 (1982).

that can pass through the filters is

$$|\mu, \mu; \boldsymbol{n}\rangle = a_\mu^\dagger(\omega_1 \boldsymbol{n}) a_\mu^\dagger(-\omega_2 \boldsymbol{n})|0\rangle \ , \qquad (37)$$

with $\omega_1 \boldsymbol{n}$ and $-\omega_2 \boldsymbol{n}$ being the momenta of γ_1 and γ_2, respectively. This state does not have the correct symmetries, however: it does not have even parity, for under the reflection I_s, $\mu \to -\mu$, $\boldsymbol{n} \to -\boldsymbol{n}$; nor is it invariant under a rotation R_π through π about an axis perpendicular to $\boldsymbol{n}$, under which $\mu \to \mu$, $\boldsymbol{n} \to -\boldsymbol{n}$. Furthermore, the states produced by I_s or R_π acting on (37) have photons with the wrong frequencies to pass the filters, because the experimental arrangement is not invariant under I_s or R_π. But it is invariant under $I_s R_\pi$, which is also a symmetry of $|\gamma_1 \gamma_2\rangle$.

Hence the state selected by the experiment is

$$|\Phi\rangle = \tfrac{1}{\sqrt{2}}\,(1 + I_s R_\pi)|\mu, \mu; \boldsymbol{n}\rangle = \tfrac{1}{\sqrt{2}}\Big(|1, 1; \boldsymbol{n}\rangle + |-1, -1; \boldsymbol{n}\rangle\Big) \ , \qquad (38)$$

or in terms of the photon operators,

$$|\Phi\rangle = \frac{1}{\sqrt{2}}\Big(a_1^\dagger(\gamma_1) a_1^\dagger(\gamma_2) + a_{-1}^\dagger(\gamma_1) a_{-1}^\dagger(\gamma_2)\Big)|0\rangle \ , \qquad (39)$$

with the abbreviations $a_\mu^\dagger(\gamma_1) \equiv a_\mu^\dagger(\omega_1 \boldsymbol{n})$, $a_\mu^\dagger(\gamma_2) \equiv a_\mu^\dagger(-\omega_2 \boldsymbol{n})$. Note that (39) is an entangled state.

The experiment selects linear, not circular polarizations. The transformation to the linear basis is

$$a_{\pm 1}^\dagger(\boldsymbol{k}) = \mp\tfrac{1}{\sqrt{2}}\,[a_x^\dagger(\boldsymbol{k}) \pm i a_y^\dagger(\boldsymbol{k})] \ , \quad a_{\pm 1}^\dagger(-\boldsymbol{k}) = \mp\tfrac{1}{\sqrt{2}}\,[a_x^\dagger(-\boldsymbol{k}) \mp i a_y^\dagger(-\boldsymbol{k})] \ . \qquad (40)$$

Therefore

$$|\Phi\rangle = \frac{1}{\sqrt{2}}\Big(a_x^\dagger(\gamma_1) a_x^\dagger(\gamma_2) + a_y^\dagger(\gamma_1) a_y^\dagger(\gamma_2)\Big)|0\rangle = \frac{1}{\sqrt{2}}\, \boldsymbol{a}^\dagger(\gamma_1) \cdot \boldsymbol{a}^\dagger(\gamma_2)\, |0\rangle \ . \qquad (41)$$

The last expression shows explicitly that $|\Phi\rangle$ is a 0^+ state.

As is evident from (39) and (41), $|\Phi\rangle$ has perfect correlations like the two-particle spin singlet state. If both arms of the experiment are set to measure photons having the same linear polarization, then if one arm detects a photon polarized along that direction, the distant other arm is certain to detect a photon with the same linear polarization; the analogous statement holds for helicity. When the two arms are set to accept polarization modes that are neither identical nor orthogonal, there are only statistical correlations, as is the case in any experiment that tests Bell's inequality.

This is the appropriate point for examining the sequence of events in experiments of the EPR variety. If one of the two arms in the Aspect experiment is closer to the source than the other, and the two polarization detectors are parallel, does the closer event *determine* the polarization that will subsequently be seen by the more distant observer? In this and other experiments on EPR pairs, the detectors have a spacelike separation, so there is no invariant significance to the time sequence of the two events. The correlations have an invariant meaning, e.g., both arms have observed helicity -1, or neither did; it is meaningless to claim that one of these events was first and "caused" the other.

The quantity measured in this experiment is the number of coincidences, $N_{12}(\theta_1, \theta_2)$, in which γ_1 is linearly polarized along a direction making the angle

θ_1 with respect to the x-axis, and γ_2 is linearly polarized along the direction θ_2. It is given by the following expectation value of photon number operators:

$$N_{12}(\theta_1, \theta_2) = \langle\Phi|N_1(\theta_1)N_2(\theta_2)|\Phi\rangle\,, \tag{42}$$

where the number operator for γ_1 is

$$N_1(\theta_1) = a^\dagger_{\theta_1}(\gamma_1)a_{\theta_1}(\gamma_1) \tag{43}$$

$$a_{\theta_1}(\gamma_1) = a_x(\gamma_1)\,\cos\theta_1 + a_y(\gamma_1)\,\sin\theta_1\,, \tag{44}$$

and similarly for γ_2. The operators for the modes γ_1 and γ_2 commute because they have different momenta. Therefore

$$N_{12}(\theta_1, \theta_2) = \frac{1}{2}\sum_{ij} C_{ij}(\theta_1)C_{ij}(\theta_2) = \frac{1}{2}\,\mathrm{Tr}\,C(\theta_1)C^T(\theta_2)\,, \tag{45}$$

where $C(\theta_1)$ is the 2×2 matrix with elements

$$C_{ij}(\theta_1) = \langle 0|a_i N(\theta_1)a^\dagger_j|0\rangle \tag{46}$$

pertaining to γ_1, etc. A short calculation then gives the following result for the number of coincidences:

$$N_{12}(\theta_1, \theta_2) = \tfrac{1}{4}[1 + \cos 2(\theta_1 - \theta_2)]\,. \tag{47}$$

This depends only on the difference in angles, as it must. With the same normalization the single rates that appear in (36) are $N_1 = N_2 = \frac{1}{2}$.

When the quantum mechanical value is used for the coincident count, the inequality Eq. 36 reads

$$3\cos 2\phi - \cos 6\phi - 2 \le 0\,. \tag{48}$$

It is not obeyed, with the maximum violation occurring for $\phi = \pi/8$:

$$3\cos(\pi/4) - \cos(3\pi/4) - 2 = 2(\sqrt{2} - 1) = 0.828\,. \tag{49}$$

The Aspect experiment has the important "delayed choice" feature: each arm of the experiment switches its polarizer independently between the angles $\pi/8$ and $3\pi/8$ during an interval short compared to L/c, where L is the distance between the two detectors. The measured result is that Eq. 48 is violated by nearly 5 standard deviations. An earlier experiment that did not have the delayed choice feature had much higher counting rates and gave a violation of the inequality of 9 standard deviations. There were also runs over a range of angles ϕ which are in excellent agreement with the prediction from quantum mechanics, as are the numerical values of the violation of the inequality once acceptances, detection efficiencies, etc., are taken into account.

12.4 Locality

It is often said that quantum mechanics is a not a realistic local theory. Is this a verdict on quantum mechanics, or just the statement that theories based on classical definitions of reality and locality cannot reproduce the predictions of quantum mechanics? The Bell-Kochen-Specker theorem demonstrates that a "realistic" theory — one that assigns values to all observables — cannot mimic quantum mechanics. And we have also seen that a classical conception of locality, as formulated in Eq. 20 and Eq. 31 to derive the Bell and Clauser-Horne inequalities, is in conflict with both quantum mechanics and experiment. Hence the question is whether quantum mechanics itself is non-local. A definition of locality that does not entail any ingredients unknown to quantum mechanics is required for this purpose.

Let $\mathcal{S}$ be a system composed of two entangled components $\mathcal{S}_1$ and $\mathcal{S}_2$ which have separated to the point where they no longer interact with each other. At some later times $\mathcal{S}_1$ interacts with a measurement apparatus X, and $\mathcal{S}_2$ interacts with another apparatus Y, but there is *no coupling* between $\mathcal{S}_1$ and Y or between $\mathcal{S}_2$ and X, nor between the two apparatuses X and Y. We will first explore the consequence of defining quantum mechanical locality as the requirement that the measurement outcomes on any observable Q_1 of $\mathcal{S}_1$ be independent of all measurements performed on $\mathcal{S}_2$.[1]

The Hamiltonian of the entire system $\mathcal{S}_1 + X + \mathcal{S}_2 + Y$ therefore has the form

$$H = H_1 + H_2 \, , \qquad H_1 = K_1 + K_X + V_{1X} \, , \qquad H_2 = K_2 + K_Y + V_{2Y} \, , \tag{50}$$

where K_i is the Hamiltonian of $\mathcal{S}_i$, K_X that of X, V_{1X} the interaction between subsystem $\mathcal{S}_1$ and the apparatus X, etc. The crucial point is that H_1 and H_2 commute, because there are no interactions between $\Sigma_1 = \mathcal{S}_1 + X$ and $\Sigma_2 = \mathcal{S}_2 + Y$. Consequently the unitary operator describing the evolution of the whole system due to the interactions V_{1X} and V_{2Y} factorizes:

$$U = U_1 U_2 = U_2 U_1 \, , \qquad U_i = e^{-iH_i t} \, . \tag{51}$$

At the outset there are no correlations between the entangled systems $\mathcal{S}_{1,2}$ and the apparatus, nor between the two apparatuses, so the initial state of the complete system is

$$\rho^0 = \rho_S^0 \otimes \rho_X^0 \otimes \rho_Y^0 \, , \tag{52}$$

where ρ_S^0 is an entangled state of $\mathcal{S}_1 + \mathcal{S}_2$, and ρ_X is the state of apparatus X, etc. After the interactions V_{1X} and V_{2Y} come into play this state evolves as follows:

$$\rho^0 \to U(\rho_S^0 \otimes \rho_X^0 \otimes \rho_Y^0) U^\dagger \, . \tag{53}$$

Because ρ_S^0 is entangled, (53) does not factorize into the form $\rho(\Sigma_1) \otimes \rho(\Sigma_2)$.

Let Q_1 be any observable of the apparatus X, which is coupled only to $\mathcal{S}_1$. The expectation value $\langle Q_1 \rangle$ of Q_1 in the state (53) is then

$$\mathrm{Tr}\, Q_1 \rho = \mathrm{Tr}\, Q_1 U_1 U_2 \rho^0 U_2^\dagger U_1^\dagger = \mathrm{Tr}\, U_1^\dagger Q_1 U_1 \rho^0 \tag{54}$$

because $[Q_1, U_2] = 0$, and therefore

$$\langle Q_1 \rangle = \mathrm{Tr}\, Q_1 \, U_1 \rho^0 U_1^\dagger \, . \tag{55}$$

---

[1] G.C. Ghirardi, A. Rimini and T. Weber, *Lett. Nuovo Cimento* **27**, 293 (1980).

Here V_{2Y} has disappeared, so it does not matter whether the apparatus Y that can measure properties of the distant system $\mathcal{S}_2$ is turned on or off.[1] This result holds for any function of Q_1, so all moments of the distribution of Q_1 outcomes resulting from measurements on $\mathcal{S}_1$ are unaffected by any measurement on $\mathcal{S}_2$:

> *Quantum mechanics is a local theory in the sense that no measurement carried out on a system has outcomes whose statistical distributions depend on the choice of what to measure on another system with which it is entangled but with which it no longer interacts.*

This is not the whole story, however. If the initial state of $\mathcal{S}_1 + \mathcal{S}_2$ is entangled, there are correlations between the results of measurements on $\mathcal{S}_1$ and $\mathcal{S}_2$ after they have ceased to interact. The joint distribution $\langle Q_1 Q_2 \rangle$ depends on both V_{1X} and V_{2Y} and does not factorize into $\langle Q_1 \rangle \langle Q_2 \rangle$. And as we know from the spin singlet case, it is the correlations that have the non-local appearance.

Bell's theorem shows that such correlations cannot be accounted for by a local realistic theory. The perfect correlations displayed by the spin singlet state when both spins are measured along the same direction (or the counterpart in two photon states) do not, however, suffice to reach this conclusion, which requires an analysis over a range of relative orientations. It is significant, therefore, that there are multiparticle quantum states, discovered by Greenberger, Horne and Zeilinger (GHZ), which have perfect correlations, and that these perfect correlations are incompatible with a classical conception of locality.

The simplest GHZ example, in a version due to Mermin, is an object that disintegrates into three spin $\frac{1}{2}$ particles. Consider the following observables:

$$\Omega_1 = \sigma_{1x}\sigma_{2y}\sigma_{3y} \;, \quad \Omega_2 = \sigma_{1y}\sigma_{2x}\sigma_{3y} \;, \quad \Omega_3 = \sigma_{1y}\sigma_{2y}\sigma_{3x} \;. \tag{56}$$

It is assumed that the particles no longer interact. Because the components of $\boldsymbol{\sigma}_i$ belonging to particle i anticommute with each other and commute with those of the other particles, *all the* Ω_i *commute with each other*. Furthermore $\Omega_i^2 = 1$, so the eigenvalues ω_i of Ω_i are ± 1. The product of the Ω_i also plays a decisive role:

$$\Gamma \equiv \Omega_1 \Omega_2 \Omega_3 = -\sigma_{1x}\sigma_{2x}\sigma_{3x} \;. \tag{57}$$

Let $|\Psi\rangle$ be any one of the 2^3 simultaneous eigenstates of all three Ω_i, multiplied by a spatial wave function describing the outward propagation of the particles. Because $[\Gamma, \Omega_i] = 0$, it is also an eigenstate of Γ with eigenvalue

$$\gamma = \omega_1 \omega_2 \omega_3 = \pm 1 \;. \tag{58}$$

For example, take the state in which $\omega_{1,2,3} = 1$; in the notation of Eq. 1 the explicit expression is

$$|\Psi\rangle = \tfrac{1}{\sqrt{2}} \Big(|+\rangle_1 |+\rangle_2 |+\rangle_3 - |-\rangle_1 |-\rangle_2 |-\rangle_3 \Big) \;. \tag{59}$$

As in the two-particle singlet state, the expectation value of any single spin component vanishes in this state. Furthermore, the products of x and/or y components of

---

[1] This conclusion is only contingent on the absence of interactions between the systems Σ_1 and Σ_2. Their separation need not be spacelike. Putting the argument in terms of a spacelike separation is a shorthand for the requirement that there be no such interaction. Signaling is not possible between Σ_1 and Σ_2 if they no longer interact even if they are not far apart.

any pair (e.g., $\sigma_{1x}\sigma_{3y}$) also have vanishing expectation values. But there are correlations between the spins of all three particles that persist as they separate. These properties of the one, two and three particle observables hold for all the other seven simultaneous eigenstates of the Ω_i.

Consider any one of the GHZ states with simultaneous eigenvalues ω_i and γ. Because any Pauli matrix when squared is 1, and $|\omega_i|^2 = 1$, acting on such a state

$$\sigma_{1x} = \omega_1 \sigma_{2y} \sigma_{3y} \tag{60}$$

as a consequence of the definition of Ω_1. Hence a measurement of the compatible observables σ_{2y} and σ_{3y}, with outcomes v_{2y} and v_{3y}, establishes with certainty that a measurement of the distant untouched spin $\boldsymbol{\sigma}_1$, when measured along the x-direction, has the outcome

$$v_{1x} = \omega_1 v_{2y} v_{3y} \,. \tag{61}$$

According to (57), however, it is also true that $\sigma_{1x} = -\gamma\sigma_{2x}\sigma_{3x}$. The classical conception of locality would say that the outcome of a measurement on a system S_1 is not influenced by which measurement is chosen for the other systems S_2 and S_3 to which S_1 is not coupled. If that were so, we could equally well measure σ_{2x} and σ_{3x} instead, from which choice it would follow that

$$v_{1x} = -\gamma v_{2x} v_{3x} \,. \tag{62}$$

This logic therefore produces the tableau

$$\begin{array}{lll} & \Omega_i & \Gamma \\ v_{1x}: & \omega_1 v_{2y} v_{3y} & -\gamma v_{2x} v_{3x} \\ v_{2x}: & \omega_2 v_{1y} v_{3y} & -\gamma v_{1x} v_{3x} \\ v_{3x}: & \omega_3 v_{1y} v_{2y} & -\gamma v_{1x} v_{2x} \end{array} \tag{63}$$

As all the v's are ± 1, the product of Ω_i entries and the product of the Γ entries give $\omega_1\omega_2\omega_3 = \gamma = -\gamma^3$, or $1 = -1$!

This contradiction can only be established by combining results from measurements on incompatible observables belonging to the distant entangled co-conspirators: in the Ω_i choices measurements of y-components, in the Γ choices of x-components. From quantum mechanics one cannot infer that the numbers v_{1x} in (61) and (62) are the same; one would have to call them, say, $v_{1x}(Y)$ and $v_{1x}(X)$ to indicate which spin components of the other particles were measured in the two distinct and complementary (in Bohr's sense) situations.

All this being said, it is nonetheless so that here, as in the spin singlet example, the perfect correlations between distant observations have a non-local air about them. The observers of σ_{2y} and σ_{3y} cannot know what their next outcomes will be, but once they do, they know the outcome for the observer of σ_{1x} in that same event.[1] And if they switch to measuring σ_{2x} and σ_{3x}, their outcomes will tell them the outcome for σ_{1x} in each and every event. This does not produce a conflict with causality because the contradiction $1 = -1$ shows that prior values cannot be ascribed to the observables.

[1] The time ordering of these correlated outcomes does not have an invariant meaning, as already discussed.

Thus it is finally a matter of taste whether one calls quantum mechanics local or not. In the statistical distribution of measurement outcomes on separate systems in entangled states there is no hint of non-locality. Quantum theory does not offer any means for superluminal signaling. But quantum mechanics, and by that token nature itself, does display perfect correlations between distant outcomes, even though Bell's theorem establishes that pre-existing values cannot be assigned to such outcomes and it is impossible to predict which of the correlated outcome any particular event will reveal.

12.5 Measurement

Measurement plays a central role in the formulation of quantum mechanics and in its conceptual underpinnings, but how the measurement process is actually described remains to be considered. What follows is not a survey of the large, still growing and often controversial literature on this topic, but should provide sufficient background for studying other presentations.

Theoretical descriptions of measurement tend to have a stylized form, reminiscent of the theaters of classical Greece and Japan, which also have the purpose of illuminating deep truths, and sharpen that focus by setting aside the mundane details of real life. In that spirit, measurement theory is often phrased entirely in the abstract Hilbert space language, eschewing even cartoon versions of actual experiments. We will strive to be somewhat more concrete.

The issue at hand can be put as follows. Consider an object $\mathcal{O}$ which has the observable Λ with eigenvalues $\{\lambda_i\}$. Then ask: How are the λ_i displayed and measured; furthermore, if $\mathcal{O}$ is in a state $|\Psi\rangle$, how are the probabilities $|\langle\lambda_i|\Psi\rangle|^2$ determined? Note well that in responding to these questions we are not attempting to derive the statistical interpretation of quantum mechanics, which is assumed, but to examine whether this interpretation gives a consistent account of measurement.

(a) A Measurement Device

Since the early days of quantum mechanics, the Stern-Gerlach (SG) experiment has provided the most popular setting for discussing measurement. Here the object $\mathcal{O}$ is a particle that has a magnetic moment and therefore spin. The experiment "measures" the magnetic moment by passing a succession of specimens of $\mathcal{O}$, all prepared in an identical manner, through an inhomogeneous magnetic field that produces a deflection proportional to the projection of the magnetic moment along the direction of the field gradient. By design, the gradient is large along just one direction. However, from $\nabla \cdot \boldsymbol{B} = 0$ it follows that other components of the magnetic moment interact with the field, a fact that is often ignored in text-book descriptions. In a properly designed experiment this irremovable complication is under control,[1] but as this issue is irrelevant to our purpose we avoid it completely by constructing a soluble model that produces the same results as a good SG experiment.

---

[1]M. Hannout et al., *Am. J. Phys.* **66**, 377 (1998); A. Peres (1995), p. 15.

The model is defined by the Hamiltonian

$$H = \frac{p^2}{2m} + g\, p_z s_z\, f(x)\ , \qquad f(x) = \begin{cases} 1 & \text{if } 0 < x < a \\ 0 & \text{otherwise} \end{cases} \tag{64}$$

where s_z is the z component of the spin, which can have any particular magnitude $\frac{1}{2}$, 1, etc., and p_z is the momentum in the same direction; the interaction vanishes outside the slab $0 < x < a$. We take the liberty of calling the model defined by (64) an SG setup.

As we see from (64), in this model p_y, p_z and s_z are compatible constants of motion, but p_x is not because the interaction amounts to $2s+1$ potential steps $gp_z\mu$, where $\mu = s, s-1, \dots, -s$ are the eigenvalues of s_z. The particle trajectories can be anticipated from the solution of Hamilton's classical equation,

$$\dot{z}(t) = p_z/m + gs_z f(x)\ . \tag{65}$$

Consider a particle incident along the x-direction with $p_y = p_z = 0 =$ const. Then $\dot{z} = 0$ outside the slab; inside, however, $\dot{z} = gs_z$, producing a deflection in the z-direction proportional to s_z. Hence the classical trajectories will fan apart inside the slab with angles proportion to s_z, and be parallel to the x-direction on either side of the slab, as shown in Fig. 12.4

This model has features typical of experiments that assign values λ_i to an observable Λ (here $\hbar s_z$). The observable of interest is coupled to an observable Z of the "apparatus," and the interaction forces Z to display eigenvalues that fall into distinct *macroscopic* intervals z_i correlated to the *microscopic* eigenvalues λ_i. In this naive model the particle's position plays the role of Z, with z_i being a deflection. The model also elucidates how an operator in an abstract space is related to a concept (e.g., the magnetic moment) defined at the human scale of perception through its behavior as described by classical physics (here Eq. 65). Of course, not all observables that arise in quantum mechanics have so direct a linkage to macroscopic observation, e.g., the neutrino spin, or quark color which is still far more remote.

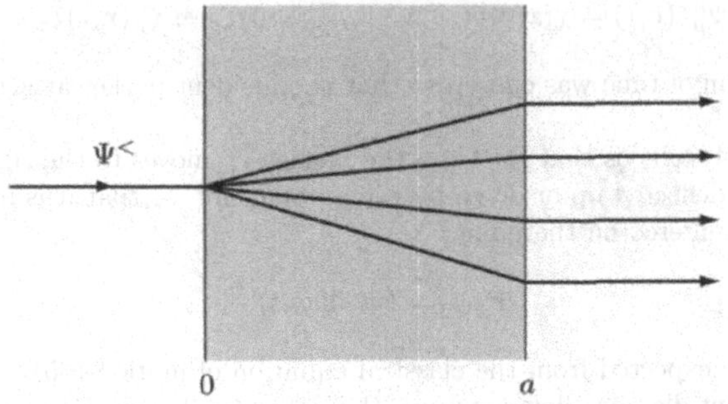

FIG. 12.4. An analogue to the Stern-Gerlach experiment as described by the Hamiltonian of Eq. 64, for the case of spin $\frac{3}{2}$.

Turning to quantum mechanics, both inside and outside the slab the solutions of the Schrödinger equation are plane waves multiplied by spin state ξ_μ:

$$\psi^{\text{out}} = e^{i\mathbf{k}\cdot\mathbf{r}}\, e^{-iEt}\, \xi_\mu \qquad (x < 0,\ x > a)\,, \tag{66}$$

$$\psi^{\text{ins}} = e^{ik'_x x}\, e^{i\mathbf{k}_\perp\cdot\mathbf{r}_\perp}\, e^{-iEt}\, \xi_\mu \qquad (0 < x < a)\,, \tag{67}$$

where

$$E = \frac{1}{2m}(k_x^2 + k_\perp^2) = \frac{1}{2m}[(k'_x)^2 + k_\perp^2] + g\mu k_z\,. \tag{68}$$

Let the state incident from the left be the free particle packet

$$\Psi_\mu^<(\mathbf{r},t) = \chi(\mathbf{r};t)\,\xi_\mu\,, \qquad (x < 0)\,, \tag{69}$$

$$\chi(\mathbf{r};t) = \int d\mathbf{k}\; a(\mathbf{k})\, e^{ik_x x}\, e^{i\mathbf{k}_\perp\cdot\mathbf{r}_\perp}\, e^{-ik^2t/2m}\,. \tag{70}$$

The momentum space amplitude $a(\mathbf{k})$ is peaked about the mean momentum $\bar{\mathbf{k}} = (K,0,0)$, with a spread $\kappa \lll K$, so that the packet's shape stays essentially unchanged during the interval needed to pass through the apparatus.

For sufficiently large K, reflection is negligible, so according to (67) the wave function inside the slab is simply

$$\Psi_\mu^{\text{ins}}(\mathbf{r},t) = \xi_\mu \int d\mathbf{k}\; a(\mathbf{k})\, e^{ik'_x x}\, e^{i\mathbf{k}_\perp\cdot\mathbf{r}_\perp}\, e^{-ik^2t/2m}\,. \tag{71}$$

Because the momentum integrals are dominated by the neighborhood of $\bar{\mathbf{k}}$, the momentum k'_x inside the slab can be approximated by

$$k'_x \simeq k_x - g\mu k_z m/k_x \simeq k_x - g\mu k_z/\bar{v};\,, \tag{72}$$

where $\bar{v} = K/m$ is the mean incident velocity. Hence

$$e^{ik'_x x}\, e^{ik_z z} \simeq e^{ik_x x}\, e^{ik_z(z - g\mu x/\bar{v})}\,, \tag{73}$$

and therefore

$$\Psi_\mu^{\text{ins}}(\mathbf{r},t) = \chi(x - \bar{v}t, y, z - g\mu x/\bar{v}; 0)\,\xi_\mu \equiv \chi_\mu(\mathbf{r},t)\,\xi_\mu\,, \tag{74}$$

where the origin of time was chosen so that the incident packet at arrives at $x \simeq 0$ at $t \simeq 0$.

Equation (74) tells us that for $t > 0$ the packet χ_μ moves to the right a distance $\bar{x} \simeq \bar{v}t$, and is deflected up or down by the amount $g\mu t$, so that it is in a region Ω_μ of size $\sim 1/\kappa$ centered on the point

$$\mathbf{r}_\mu(t) = (\bar{v}t, 0, g\mu t)\,. \tag{75}$$

This was to be expected from the classical equation of motion (65).

At a sufficient distance down-stream, the deflection becomes large compared to the size of the packet. After it leaves the slab the packet χ_μ will again move parallel to the x-axis at a distance $\bar{z}_\mu \simeq g\mu a/\bar{v}$. This deflection can be measured by having the particle strike a detector array with spatial resolution better than $ga/\bar{v}$, which thereby determines the eigenvalue μ of the observable s_z. As we see, the device described by the Hamiltonian (64) acts as an amplifier, which for two distinct input eigenvalues (μ_1, μ_2), produces two outputs that can be distinguished unambiguously, on an event-by-event basis, by techniques that do not require any knowledge of quantum mechanics, or for that matter physics altogether. We will call states such as the packets $\chi_\mu(\boldsymbol{r}, t)$ *macroscopically distinguishable states.*

Consider now an incident state which is some arbitrary combination of spin states,

$$\Psi^<(\boldsymbol{r}, t) = \chi(\boldsymbol{r}, t) \sum_{\mu=-s}^{s} c_\mu \, \xi_\mu \,. \tag{76}$$

In this state there are no correlations between position and spin, so no measurement of position can give information about spin, the observable of interest. According to the superposition principle, the output wave function to the right of the slab, which evolves out of $\Psi^<$, is the corresponding linear combination of the solutions (74),

$$\Psi^{\text{out}}(\boldsymbol{r}, t) = \sum_{\mu=-s}^{s} c_\mu \, \chi_\mu(\boldsymbol{r}, t) \, \xi_\mu \,, \qquad (x > a) \,. \tag{77}$$

Now there are correlations between position and spin. In each trial, one specimen in the state (76) is introduced into the setup; if a measurement of it's position finds it to be in the spatial region Ω_μ, we then have full confidence[1] that a measurement of the spin on that same specimen would yield the eigenvalue μ. By this we mean that if a specimen that *registers* in Ω_μ is subsequently passed into a second identical SG-apparatus, it is certain to emerge with the deflection appropriate to that particular value of μ.

When the experiment is repeated, the frequency with which the eigenvalue is "found" to have a particular value μ is the frequency with which specimens are found to be in the volume Ω_μ. According to the Born interpretation of the wave function, this frequency (or probability) is given by

$$\int_{\Omega_\mu} d\boldsymbol{r} \, |\Psi^{\text{out}}(\boldsymbol{r}, t)|^2 = |c_\mu|^2 = |\langle \mu | \Psi^{\text{out}} \rangle|^2 \,. \tag{78}$$

To repeat: *The "Stern-Gerlach" setup takes the unentangled input (76), in which there are no correlations between spin and position, and manufactures from it the entangled output (77) in which there is a one-to-one correlation between macroscopically distinct spatial regions Ω_μ and the eigenvalues μ of s_z, so that a determination of which region a specimen is in unambiguously determines the corresponding eigenvalue of the observable s_z for this same specimen; furthermore, the fraction of specimens "displaying" the eigenvalue μ determines $|c_\mu|^2$. In the output state the particle's location thus acts like a pointer of an instrument which, for each specimen, determines a value μ of s_z. Once made, this determination assigns the specimen to the normalized eigenstate $\chi_\mu \, \xi_\mu$.*

The last paragraph offers a first answer to the questions posed in the introduction to this section. However, the model is very simplistic in that the particle's own

[1] Full confidence holds because packets for different values μ no longer overlap and are therefore orthogonal, $\langle \chi_\mu | \Phi_\nu \rangle = \delta_{\mu\nu}$, which is required for the unambiguous distinction between states (cf. §2.2(e)).

position plays the role of apparatus, and one can ask whether a more realistic description would raise problems that this model hides.

In particular, the measurement procedure only determines the $|c_\mu|^2$, and not the complex amplitudes c_μ.[1] This seems to suggest that measurement "destroys" the relative phases between the various outputs in the superposition (77), which would make the process irreversible because the input state cannot be reconstructed from the output unless these phases are known. As we will see, this impression is not all that far from the truth, but when put so baldly it glosses over some important issues.

(b) Coherence and Entropy Following Measurement

Our model has a Hermitian Hamiltonian, and all its states evolve with a unitary transformation. Therefore the pure input state (76) must produce a pure output, and the density matrix of the output,

$$\langle \boldsymbol{r}\mu|\rho(t)|\boldsymbol{r}'\nu\rangle = c_\mu c_\nu^* \, \chi_\mu(\boldsymbol{r},t)\chi_\nu^*(\boldsymbol{r}',t) \,, \tag{79}$$

is that of a pure state, i.e., $\mathrm{Tr}\,\rho^2 = 1$. Furthermore, unitarity also guarantees that when the input is a mixture, the output will be a mixture having the same entropy.

The observations summarized by (78) affect only the "apparatus" coordinate $\boldsymbol{r}$, but not the spin itself. Measurements carried out on a subset of a system's degrees of freedom only "see" a reduced density matrix which, in general, will be the density matrix of a mixture even when the state of the whole system is pure. As a consequence output results can appear to have a larger entropy than the input, which leads to the impression of irreversibility. This important distinction between reduced and complete density matrices was already made in the discussion of the special case of a pure entangled state in §2.2(e), and it will now be restated in full generality.

Consider any complete set Q of compatible degrees of freedom of a system $\mathcal{S}$, and divide Q into two subsets Q_1 and Q_2 which operate in the subspaces $\mathfrak{H}_1$ and $\mathfrak{H}_2$ of the complete Hilbert space $\mathfrak{H}_1 \otimes \mathfrak{H}_2$. Let $\langle q_1 q_2|\rho|q_1' q_2'\rangle$ be the density matrix (pure or impure) of some state of $\mathcal{S}$, where q_1 are eigenvalues of Q_1, etc. An observable A that only acts in the subspace $\mathfrak{H}_1$ has the form $A_1 \cdot \mathbf{1}_2$ where $\mathbf{1}_2$ is the unit operator in $\mathfrak{H}_2$, i.e.,

$$\langle q_1 q_2|A|q_1' q_2'\rangle = \delta(q_2 - q_2')\,\langle q_1|A_1|q_1'\rangle \,. \tag{80}$$

Then the expectation value of A is

$$\mathrm{Tr}\,\rho A = \int dq_1 \,\langle q_1|\rho_1 A_1|q_1\rangle \,, \tag{81}$$

$$\langle q_1|\rho_1|q_1'\rangle = \int dq_2 \,\langle q_1 q_2|\rho|q_1' q_2\rangle \,, \tag{82}$$

where ρ_1 is the *reduced density matrix* in the subspace $\mathfrak{H}_1$, obtained from ρ by "tracing over" the subspace $\mathfrak{H}_2$. The analogous definition holds for the reduced

[1] In this experiment eigenvalues of just one observable (s_z) are measured, which does not suffice to determine the state in any Hilbert space. As shown in §2.2(d), in this example $[(2s+1)^2 - 1]$ observables must be measured to completely determine the state.

density matrix ρ_2 in the subspace $\mathfrak{H}_2$. The various density matrices have the following important properties, some of which were already proven in §2.2(e), with the remaining proofs left for Prob. 2 :

1. ρ_1 is only pure if the complete state is unentangled, i.e., has the form $\rho = \rho_1 \otimes \rho_2$, and ρ_1 itself is pure;

2. if, in particular, ρ is pure and entangled, then ρ_1 and ρ_2 are impure;

3. the entropy $S(\rho)$ of a state ρ of the whole system is smaller than the sum of the entropies of its reduced density matrices,[1]
$$S(\rho) \leq S(\rho_1) + S(\rho_2)\,, \tag{83}$$
where the equal sign holds only if $\rho = \rho_1 \otimes \rho_2$, i.e., if ρ is not entangled.

We now apply this lesson to our SG experiment, in which (for now) the whole system is the particle; the spatial coordinates form one subset of its degrees of freedom, and the spin component s_z the second subset. Any and all observations on the pure entangled state Ψ^{out} that are only sensitive to the coordinates, but are insensitive to the spin, can only "see" a reduced density matrix ρ_X, whose elements are

$$\langle \boldsymbol{r}|\rho_X(t)|\boldsymbol{r}'\rangle = \sum_\mu |c_\mu|^2 \chi_\mu(\boldsymbol{r},t)\chi_\mu^*(\boldsymbol{r}',t)\,. \tag{84}$$

This is *not* the density matrix of a pure state, because

$$\operatorname{Tr}\rho_X^2 = \int d\boldsymbol{r}\, \langle \boldsymbol{r}|\rho_X^2|\boldsymbol{r}\rangle = \sum_\mu |c_\mu|^4 \leq 1\,; \tag{85}$$

the sum is only 1 if all but one of the amplitudes c_μ vanish, i.e., if the output is not entangled. To all such partial observations, therefore, the pure entangled output appears to be a mixture, the phases of the amplitudes c_μ appear to have been "lost," and it appears as if the measurement has increased the entropy of the state. But is this familiar statement that phases are "destroyed" or "lost" in measurement not misleading? After all, the phases can be "rediscovered" by recombining the spatially separated output (77) on passing it through a second device described by a Hamiltonian (70) in which $g \to -g$. This will then have as its output the initial state (76), which would also seem to show that there really is no irreversibility.

This argument is not on the mark. The goal in measurement is to ascertain the state of an object by determining the state of a suitable apparatus, and the measurement concept is predicated on the strategy that knowledge of the object is to be gained by examining the apparatus. Thus far, however, the role of the apparatus has been played by the particle's own coordinate, and it is necessary to go beyond this simpleminded description to understand why "lost" phases and reversibility are not retrievable in any realistic measurement experiment.

---

[1]The division of degrees of freedom into commuting subsets does not require the system S to be composed of intuitively distinct subsystems; in fact, in the case of immediate interest all the degrees of freedom belong to a single particle. In multi-particle problems (for example) this theorem applies usefully to the entropies of physically distinct subsystem.

A more realistic model is obtained by adding counters – dynamical systems that undergo a change when a particle is deflected into one of the volumes Ω_μ. Call z_μ the degrees of freedom of the counter that detects a particle in Ω_μ, let $\varphi_0(z_\mu)$ be the unexcited state of this counter, and $\varphi_1(z_\mu)$ the state excited in the counter by the passage of the particle. For now assume the counters are perfectly efficient, so that the transition $\varphi_0 \to \varphi_1$ is *certain* to occur when a particle passes through Ω_μ, and not otherwise.[2] The initial state of the apparatus is then

$$\Phi = \prod_\mu \varphi_0(z_\mu) \,. \tag{86}$$

After a particle has passed though one of the regions Ω_μ, $\varphi_0(z_\mu)$ has evolved into $\varphi_1(z_\mu)$, but all the other counters remain in their initial state φ_0. The Hamiltonian must now be enlarged by adding an interaction that causes these excitations, say

$$H_c = \sum_\mu V_\mu(\boldsymbol{r})\{|\varphi_1(z_\mu)\rangle\langle\varphi_0(z_\mu)| + \text{h.c.}\} \,, \tag{87}$$

where $V_\mu(\boldsymbol{r})$ is only non-zero in the neighborhood Ω_μ of the indicated counter.

After the packet has travelled beyond the counters, the whole system – object plus apparatus – is in the pure entangled state

$$\Theta(Z, \boldsymbol{r}, t) = \sum_\mu c_\mu \xi_\mu \tilde{\chi}_\mu(\boldsymbol{r}, t)\, \varphi_1(z_\mu) \prod_{\nu \neq \mu} \varphi_0(z_\nu) \,, \tag{88}$$

where $Z \equiv z_s, z_{s-1}, \ldots$ denotes all the counter variables, and $\tilde{\chi}_\mu$ is the spatial wave function of a particle after it has passed counter μ. The complete density matrix, including the counters, is thus

$$\langle Z\boldsymbol{r}\mu|\rho(t)|Z'\boldsymbol{r}'\nu\rangle = c_\mu c_\nu^* \tilde{\chi}_\mu(\boldsymbol{r},t)\tilde{\chi}_\nu^*(\boldsymbol{r}',t)\varphi_1(z_\mu)\varphi_0(z_\nu)\varphi_1^*(z'_\nu)\varphi_0^*(z'_\mu) \prod_{\lambda \neq \mu,\nu} \varphi_0(z_\lambda)\varphi_0^*(z'_\lambda) \,. \tag{89}$$

Has coherence survived between states with different eigenvalues of s_z ? Consider, first, all possible ways of "looking at" the counters. The expectation value of any counter observable C is given by $\mathrm{Tr}\, \rho_c\, C$, where ρ_c is the counters' reduced density matrix, obtained by tracing (89) over all variables other than Z:

$$\begin{aligned} \langle Z|\rho_c(t)|Z'\rangle &= \sum_\mu \int d\boldsymbol{r}\, \langle Z\boldsymbol{r}\mu|\rho|Z'\boldsymbol{r}\mu\rangle \\ &= \sum_\mu |c_\mu|^2 \varphi_1(z_\mu)\varphi_1^*(z'_\mu) \prod_{\lambda\neq\mu} \varphi_0(z_\lambda)\varphi_0^*(z'_\lambda) \,. \end{aligned} \tag{90}$$

This expression shows that observations restricted to the counters cannot detect any coherence between states for different eigenvalues μ of the measured observable. Hence ρ_c gives the appearance that the state is no longer pure, which is the conclusion we drew from the naive model that led to (84).

---

[2]Hence H_c is an interaction that cannot be handled with perturbation theory, as that would leave the counter in the initial state with appreciable probability. Fig. 12.2 demonstrates that it is possible to devise interactions that have this desired unitary evolution.

Next, consider observations that "do not look" at the counters, and only "look at" the particle degrees of freedom $\boldsymbol{r}$ and μ. Such observations are completely described by the reduced density matrix $\rho_{\rm p}$ obtained by tracing over the counter variables, i.e., setting $Z = Z'$ and integrating over Z in (89). Assuming that the counters' states φ_0 and φ_1 are orthogonal, the density matrix that describes the particles only is

$$\langle \mu \boldsymbol{r}|\rho_{\rm p}(t)|\nu \boldsymbol{r}'\rangle = \delta_{\mu\nu}|c_\mu|^2 \, \tilde{\chi}_\mu(\boldsymbol{r},t)\tilde{\chi}^*_\mu(\boldsymbol{r}',t) \, . \tag{91}$$

Once again, there are no interference terms. (The last expression should be compared with (79), which describes the output when there are no counters.)

These conclusion regarding coherence hold whether or not one "looks at the counters." What matters is not whether one looks, but whether or not there is an interaction $H_{\rm c}$ between the particle (the object) and the counters (the apparatus) that guarantees that the latter become excited and entangled with the states of the object. If there is no such interaction, the counters' wave functions remains the same unentangled factor throughout, but nothing has been learned about the particle – no measurement has actually been performed on it.

We now return to the question of whether coherence between distinct outcomes, and reversibility, can be demonstrated. In contrast to the state (77) which describes a system that has no counters, the interference terms in the state (88) with counters cannot be recovered by simply passing the particles as described by $\rho_{\rm p}$ through a device that recombines the separated beams. Now reconstruction of the input would require an interaction which, with proper coordination, de-excites the counters and recombines the particle beams. That the output, absent such an unrealistic recombination, appears to be the result of an irreversible process is also demonstrated by applying (83), which tells us that the entropy of the particle's state $S(\rho_{\rm p})$ *plus* that of of the counters $S(\rho_{\rm c})$ is *larger* than that of Θ, and thus of the incident state.

To repeat, this apparent increase in entropy is due to interference terms in the complete density matrix (89) that do not survive in the object's reduced density matrix nor in that describing the apparatus. Short of a fantastic recombination of the output, how could these interference terms be made to show up in the *actual* output ? Consider one such term in (89). It has a factor $\tilde{\chi}_1\tilde{\chi}_2^*$ describing the particle, the product of two functions which have their support in macroscopically distinct regions, and therefore cannot contribute to the expectation value of any particle observables X unless it has non-vanishing matrix elements $\langle \boldsymbol{r}|X|\boldsymbol{r}'\rangle$ when $|\boldsymbol{r}-\boldsymbol{r}'|$ is macroscopic. Furthermore, the counters' factor is also the product of wave functions that have no overlap, so this interference term can only appear in the expectation value of a multi-body operator[1] that acts in the Hilbert spaces of both the particle and the counters, and is macroscopically nonlocal in both ! In short, these "lost" interference terms are visible in the output only to fantastic operators.[2] And if such an operator were to provide the coupling between the object and apparatus, this combined system could not serve as an experiment that determines the state of the object.

---

[1]In the notation of (80), such an operator must have off-diagonal matrix elements between simultaneous eigenstates of Q_1 and Q_2.

[2]If the analysis is stated entirely in the abstract Hilbert space language, and all Hermitian operators are considered to be observables, there is no distinction between physically meaningful operators and the physically fantastic mathematical constructs that are needed to display this coherence.

The conditions required for coherence to survive measurement are clarified by considering the more realistic case of counters that sometimes fail to respond. A simple way of modelling this is to assume that a passing particle causes the counter's state to undergo the transition

$$\varphi_0 \to \bar{\varphi}_1 = \alpha\varphi_1 + \sqrt{1 - |\alpha|^2}\,\varphi_0 \,. \tag{92}$$

This amounts to a counting inefficiency $(1 - |\alpha|^2)$, and a breakdown of orthogonality between the initial and final states of the counters, because

$$1 - |\alpha|^2 = |\langle\varphi_0|\bar{\varphi}_1\rangle|^2 \,. \tag{93}$$

When φ_1 is replaced by $\bar{\varphi}_1$ in (88), the result is still an entangled state for the whole system, but one in which the factors describing the subsystems are no longer orthogonal, with interesting consequences. One readily shows that the diagonal elements $(\mu = \nu)$ of the particle's reduced density matrix $\rho_{\rm p}$ are unaltered and given by (91), but that there are *off-diagonal* elements, namely

$$\langle\mu\boldsymbol{r}|\rho_{\rm p}(t)|\nu\boldsymbol{r}'\rangle = \left(1 - |\alpha|^2\right) c_\mu c_\nu^* \,\tilde{\chi}_\mu(\boldsymbol{r},t)\tilde{\chi}_\nu^*(\boldsymbol{r}',t) \qquad (\mu \neq \nu)\,. \tag{94}$$

Hence *coherence survives to the extent to which an experiment fails to distinguish between different eigenvalues of the observable being measured,*[1] and distinct final states of the object display interference if the final states of the apparatus overlap. A *Gedanken* experiment illustrating this point was discussed already in §2.2(e) (see Fig. 2.1). The conclusion drawn there can now be restated in more general terms.

When the apparatus ends up in one member of a set of orthogonal states, the experiment unambiguously determines whether or not the object is then in a particular eigenstate of the observable of interest, but the result becomes increasingly ambiguous with increasing overlap between the states of the apparatus; furthermore, states of the object with distinct eigenvalues interfere with a visibility that is a measure of the ambiguity with which the states assumed by the apparatus determine the states of the object.

(c) An Optical Analogue to the Stern-Gerlach Experiment

Quantum optics has produced sophisticated experiments which examine phenomena that are inaccessible to older techniques such as those of a conventional Stern-Gerlach (SG) experiment. A virtuoso example is an experiment by Mandel and collaborators, which can be interpreted as an optical analogue of an SG experiment, though it had a somewhat different purpose and for that reason was not interpreted in this way by its authors.

This experiment (see Fig. 12.5) examines a coherent superposition of one-photon states, $|1\rangle$ and $|2\rangle$. It can ascertain ("measure") which of these is occupied by any particular incident photon, the analogue being a superposition of spin up and down states of a spin $\frac{1}{2}$ particle and an SG experiment that ascertains in which of these any one particle "is in" after having passed through the apparatus. The last question was answered in the original Stern-Gerlach experiment by letting the

---

[1] Observations on the counters alone, as expressed by $\rho_{\rm c}$, do not display any coherence even when the counters are imperfect in the sense of (92).

deflected particles land on a photographic plate, to which the photon analogy would be to have $|1\rangle$ or $|2\rangle$ strike photomultipliers. But such primitive setups cannot be used to explore the coherence properties of the state subsequent to measurement. To accomplish this, in the optical experiment the one-photon states $|1\rangle$ and $|2\rangle$ are converted into two-photon states in two nonlinear crystals, so that detection of a *secondary* photon can determine whether the incident photon occupied $|1\rangle$ or $|2\rangle$. An ingenious feature of the experiment is that both secondary photons can be fed into the same mode of the electromagnetic field, which allows a detailed study of how coherence between apparatus states impacts measurement outcomes.

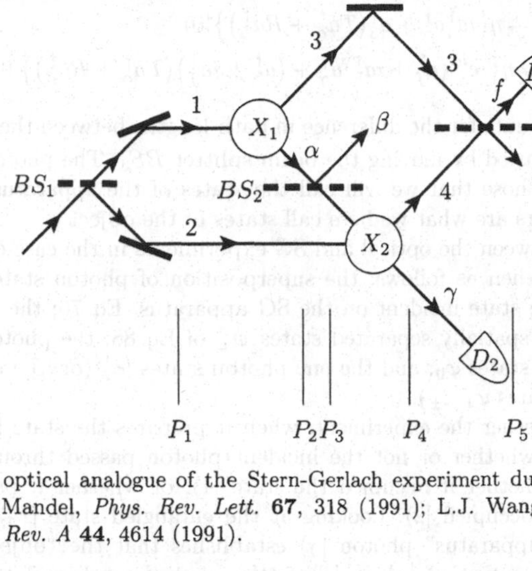

FIG. 12.5. An optical analogue of the Stern-Gerlach experiment due to X.Y. Zou, L.J. Wang and L. Mandel, *Phys. Rev. Lett.* **67**, 318 (1991); L.J. Wang, X.Y. Zou and L. Mandel, *Phys. Rev. A* **44**, 4614 (1991).

Now to a simplified description of the experiment. Photon states will be treated as if they were monochromatic and perfectly collimated modes, created by operators $a_1^\dagger$, etc., acting on the vacuum state $|0\rangle$, with polarization ignored as it plays no role. The key processes in the experiment occur at three beam splitters $BS_{1,2,3}$, and in two nonlinear crystals $X_{1,2}$. Recall from §10.5(a) that when $a_i^\dagger$ is incident at 45^0 on a lossless beam splitter, it produces the transition

$$a_i^\dagger \to T a_i^\dagger + R a_j^\dagger \; , \qquad (\, |T|^2 + |R|^2 = 1 \, , \; \mathrm{Re}\, T^* R = 0 \,) \, , \tag{95}$$

where $a_j^\dagger$ propagates at right-angle to $a_i^\dagger$, and both have the same frequency. The devices BS_1 and BS_3 are 50-50 splitters so for them $T = 1/\sqrt{2}$ and $R = i/\sqrt{2}$ (recall §10.5(c)), while BS_2 has a variable transmission amplitude T. In the nonlinear crystals an incident one-photon state $a_i^\dagger$ is transformed into a two-photon state,

$$a_i^\dagger \to \eta a_j^\dagger a_k^\dagger \, , \tag{96}$$

where η is very small – so small that processes in which both crystals convert one photon into two can be ignored. The actual two-photon states produced by $X_{1,2}$ are continua from which states with very narrow bands of momenta are extracted, a feature that is assumed to be already incorporated in (96). As a consequence, the states $a^\dagger_{\alpha,\gamma}$ are effectively collinear and of the same frequency, and $a^\dagger_{3,4}$ have in effect that same frequency.

The evolution of the incident one-photon state, as it passes through the planes $P_1 \dots P_5$ in the apparatus of Fig. 12.5, is therefore

$$|P_1\rangle = \tfrac{1}{\sqrt{2}}(a_1^\dagger + ia_2^\dagger)|0\rangle \tag{97}$$

$$|P_2\rangle = \tfrac{1}{\sqrt{2}}(ia_2^\dagger + \eta a_3^\dagger a_\alpha^\dagger)|0\rangle \tag{98}$$

$$|P_3\rangle = \tfrac{1}{\sqrt{2}}\{ia_2^\dagger + \eta a_3^\dagger(Ta_\alpha^\dagger + Ra_\beta^\dagger)\}|0\rangle \tag{99}$$

$$|P_4\rangle = \tfrac{1}{\sqrt{2}}\eta\{ia_4^\dagger a_\gamma^\dagger + a_3^\dagger(Ta_\alpha^\dagger + Ra_\beta^\dagger)\}|0\rangle \tag{100}$$

$$|P_5\rangle = \tfrac{1}{2}\eta\{ie^{i\phi}(a_f^\dagger + ia_5^\dagger)a_\gamma^\dagger + (a_5^\dagger + ia_f^\dagger)(Ta_\alpha^\dagger + Ra_\beta^\dagger)\}|0\rangle \,. \tag{101}$$

The phase $e^{i\phi}$ accounts for the difference in path lengths between the states $a_3^\dagger$ and $a_4^\dagger$ which can be varied by moving the beam splitter BS_3. The photon states with Greek names are those that we will call the states of the apparatus, while those labelled by numbers are what we here call states of the object.

The analogy between the optical and SG experiments in the case of $s = \frac{1}{2}$ (with eigenvalues $\pm$) is then as follows: the superposition of photon states $|1\rangle$ and $|2\rangle$ corresponds to the state incident on the SG apparatus, Eq. 76; the photon states $|3\rangle$ and $|4\rangle$ to the spatially separated states $\tilde{\psi}_\pm$ of Eq. 88; the photon vacuum to the initial counter states φ_0; and the one photon states $|\alpha\rangle$ (or $|\beta\rangle$) and $|\gamma\rangle$ to the excited counter states $\varphi_1(z_\pm)$.

We now ask whether the experiment, when it prepares the state $|P_5\rangle$, makes it possible to know whether or not the incident photon passed through the upper crystal X_1, i.e., whether it occupied the state $|1\rangle$, or whether it passed through X_2 and therefore occupied $|2\rangle$. Looking at the entangled state $|P_4\rangle$, we see that detection of an "apparatus" photon $|\gamma\rangle$ establishes that the "object" photon is in $|4\rangle$, whereas detection of $|\alpha\rangle$ or $|\beta\rangle$ settles that the "object" is in $|3\rangle$. As a consequence there should be no coherence between $|3\rangle$ and $|4\rangle$. Let $\mathcal{R}(i|j)$ be the coincidence rate between the channels i and j. Then apart from a common factor that accounts for experimental efficiencies and the like, these rates are

$$\mathcal{R}(f|\beta) = \tfrac{1}{4}\eta^2|R|^2\,, \qquad \mathcal{R}(f|\gamma) = \tfrac{1}{4}\eta^2\,. \tag{102}$$

As expected, these have no dependence on the phase ϕ, confirming that there is no coherence between the two states $|3\rangle$ and $|4\rangle$ that have been superposed into the output state $|f\rangle$ by the movable beam splitter BS_3.

As already mentioned, the experiment has the remarkable and crucial feature that the apparatus can be tuned so that the crystal X_1 feeds its secondary photon $|\alpha\rangle$ into the same mode $|\gamma\rangle$ as the other crystal's secondary photon. This amounts to the replacement $a_\alpha^\dagger \to e^{i\delta}\,a_\gamma^\dagger$ in $|P_3\rangle$ and in later stages in the evolution of the state. The phase δ depends on the details of the apparatus and needs not be known, as we will see. When $\alpha - \gamma$ identity is achieved, the state on reaching P_4 is

$$|\tilde{P}_4\rangle = \tfrac{1}{\sqrt{2}}\eta\{(ia_4^\dagger + Te^{i\delta}a_3^\dagger)a_\gamma^\dagger + Ra_3^\dagger a_\beta^\dagger\}|0\rangle\,, \tag{103}$$

and when it impinges on the detectors it is

$$|\tilde{P}_5\rangle = \tfrac{1}{2}\eta\{i(e^{i\phi} + e^{i\delta}T)a_f^\dagger a_\gamma^\dagger + (e^{i\delta}T - e^{i\phi})a_5^\dagger a_\gamma^\dagger + Ra_\beta^\dagger(a_5^\dagger + ia_f^\dagger)\}|0\rangle\,. \tag{104}$$

Now $f - \gamma$ coincidences between what we call the object and apparatus states have the rate

$$\mathcal{R}(f|\gamma) = \tfrac{1}{4}\eta^2\left[1 + 2T\cos(\phi - \delta) + T^2\right] . \tag{105}$$

This coincidence rate $\mathcal{R}(f|\gamma)$ displays coherence because identifying the states $|\alpha\rangle$ and $|\gamma\rangle$ affects the experiment's ability to distinguish whether an incident photon took a path through one crystal as compared to the other, i.e., whether the initial photon occupied $|1\rangle$ or $|2\rangle$. When $|T| = 1$ the two paths are equally likely, and the interference pattern has maximal visibility. On the other hand, if $T = 0$ the beam splitter BS_2 has been turned into a mirror and it becomes possible, by detecting $|\beta\rangle$, to know with full confidence which path was taken; hence there can be no interference pattern. In short, therefore, this experiment illustrates the general conclusions drawn at the end of the preceding subsection.

The singles rate at the detector D_1 also displays coherence when the two "apparatus" states are identified:

$$\mathcal{R}(f) = \mathcal{R}(f|\beta) + \mathcal{R}(f|\gamma) = \tfrac{1}{2}\eta^2\left[1 + T\cos(\phi - \delta)\right] . \tag{106}$$

That this is to be expected can be seen in (103). When $T = 1$, and thus $R = 0$, there is no entanglement between the apparatus state $|\gamma\rangle$ and object state, so that $|3\rangle$ and $|4\rangle$ are fully coherent and interfere maximally in forming the detected state $|f\rangle$. In this case, therefore, the experiment is in essence reversible – it has an output that is in essence identical with the input, but of course it does not produce any knowledge about the incident photon. As T decreases (and thus R increases) there is growing entanglement and growing knowledge as to whether a incident photon was $|1\rangle$ or $|2\rangle$, and decreasing coherence.

The experimental results are shown in Fig. 12.6, and are in excellent agreement with (106). The variation of the visibility with $|T|$ illustrates that the degree of confidence regarding the state of the apparatus is complementary to the coherence between states of the object.

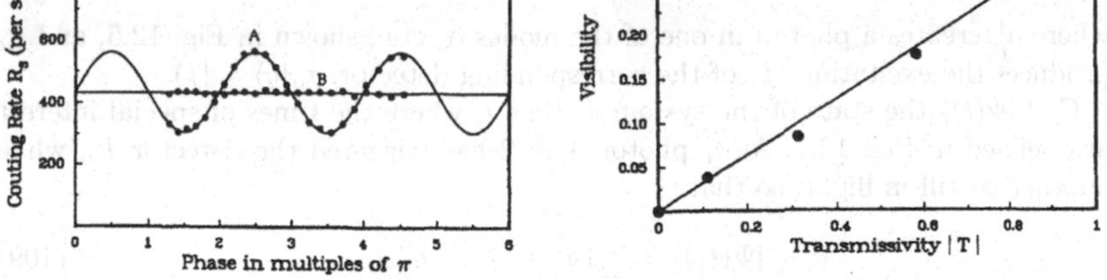

FIG. 12.6. The plot on the left shows the singles rate $\mathcal{R}(f)$ as a function of the phase ϕ, which is varied by moving the beam splitter BS_2; the curve A is for $|T| = 0.91$, and B for $|T| = 0$. The right-hand plot shows the visibility (i.e., amplitude) of the interference effect seen in $\mathcal{R}(f)$ as a function of the transmission amplitude $|T|$. In both plots the statistical errors are smaller than the dot size.

This experiment also bears on a famous dictum by Dirac,[1] which reads "[e]ach photon then interferes only with itself. Interference between two different photons never occurs." The interference between the photons in the channels f and γ demonstrates that this statement only applies to one-photon states, not to multiphoton states, and shows that the superposition principle, whose profound importance was first emphasized by Dirac, has consequences even he may not have fully appreciated.

(d) A Delayed Choice Experiment

Quantum mechanics does not assign values to observables prior to measurement, a feature of the theory that is confirmed by the experiments that violate the Bell-Clauser-Horne inequalities. That measurement does not reveal pre-existing values is also demonstrated by the possibility of leaving the decision "until the last moment" as to which one of two or more incompatible observables is to be measured. The Aspect experiment is such a delayed choice experiment. Another is the two-particle interferometer of §1.2, but that is only a *Gedanken* experiment. An elegant laboratory experiment that is essentially equivalent has been performed by Scully and collaborators. The delayed choice is between determining the path of a photon (in effect its momentum), or foregoing this path information in favor of displaying a two-photon interference pattern.

In this experiment a light beam impinges on a non-linear crystal, in which a single pair of photons is created coherently at one of two spots (see Fig. 12.7). One of these photons triggers the nearby detector D_0, whereas the other photon passes through an array of 50-50 beam splitters to one of four more distant detectors $D_{1\dots4}$. Nothing is altered while the photons are in flight, but as we will see this is, in effect, a delayed choice experiment.

Let $|g_i\rangle$ be the ground state of detector i, with $i = 0, \dots, 4$, and define

$$|\Xi\rangle = |0\rangle \otimes |g_0\rangle \otimes \cdots \otimes |g_4\rangle \,, \tag{107}$$

where $|0\rangle$ is the electromagnetic vacuum state. Schematically, the interaction of the photons with the detectors is

$$V = (a_1 + a_2)\tau_0 + \sum_{\mu=\alpha}^{\delta} a_\mu \, \tau_\mu \; + \; \text{h.c.} \,, \tag{108}$$

where $a_\mu^\dagger$ creates a photon in one of the modes α, etc., shown in Fig. 12.5, and τ_μ produces the excitation $|f\rangle$ of the corresponding detector: $\tau_\mu|g\rangle = |f\rangle$.

Call $|\Psi(t)\rangle$ the state of the system at time t, where the times of special interest are defined in Fig. 12.7. At t_1 photon 1 *or* 2 has triggered the detector D_0 while another is still in flight, so that

$$|\Psi(t_1)\rangle = \tfrac{1}{\sqrt{2}}(a_3^\dagger + e^{i\phi}a_5^\dagger)\tau_0|\Xi\rangle \,, \tag{109}$$

where the phase ϕ accounts for the path difference due to the adjustable position of D_0. A calculation like the one that led to (101) shows that after the second photon

[1] The most recent appearance is on p. 9 of P.A.M. Dirac, *The Principles of Quantum Mechanics*, 4th ed., Oxford (1958).

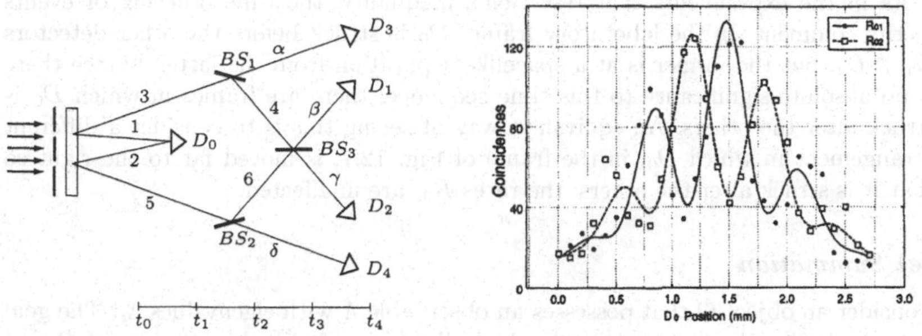

FIG. 12.7. The arrangement of the experiment by Y-Ho. Kim, R. Yu, S.P. Kulik, Y. Shih and M.O. Scully, *Phys. Rev. Lett*, **84**, 1 (2000) is sketched on the left. The rates R_{01} and R_{02} for detecting one photon in D_0 and another in D_1 or D_2 are shown on the right.

has disappeared the state is

$$|\Psi(t_4)\rangle = \tfrac{1}{2}\left\{\tau_3 + e^{i\phi}\tau_4 + \tfrac{1}{\sqrt{2}}\, i(1+ie^{i\phi})\tau_2 - \tfrac{1}{\sqrt{2}}(1-ie^{i\phi})\tau_1\right\}\tau_0|\Xi\rangle\,. \tag{110}$$

Thus the rates for events in which one photon triggers D_0, and the other photon subsequently triggers either D_1 or D_2, are proportional to $|1 \pm ie^{i\phi}|^2$, i.e.,

$$R_{01} = \tfrac{1}{4}(1+\sin\phi)\,, \quad R_{02} = \tfrac{1}{4}(1-\sin\phi)\,. \tag{111}$$

These are interference patterns between points separated in both space and time. The other two rates show no such interference: $R_{03} = R_{04} = \frac{1}{4}$. The data in Fig. 12.5(b) is consistent with Eq. 111, bearing in mind that the sensitivity of D_0 depends on ϕ, a complication that is ignored in the derivation of (110).

The absence of interference in R_{03} (R_{04}) reflects the fact that in such events it is unambiguous from which spot in the crystal the photon originated that subsequently struck the detector D_3 (D_4). On the other hand, both spots can be the source when D_1 or D_2 is struck, so the two possible paths add coherently.

That the setup in Fig. 12.7 is equivalent to a delayed choice experiment can be seen by considering a somewhat different arrangement.[1] Namely, replace the beam splitters BS_1 and BS_2 by mirrors, and let BS_3 be removed at random. Of course D_3 and D_4 are then decoupled and irrelevant. If "in" and "out" designate the cases where BS_3 is in or out of place, the final states in these two circumstances are

$$|\Psi_{\rm in}(t_4)\rangle = \tfrac{1}{2}\left\{(1+ie^{i\phi})\tau_2 + (i+e^{i\phi})\tau_1\right\}\tau_0|\Xi\rangle\,, \tag{112}$$

$$|\Psi_{\rm out}(t_4)\rangle = \tfrac{1}{\sqrt{2}}(\tau_2 + e^{i\phi}\tau_1)\tau_0|\Xi\rangle\,. \tag{113}$$

Thus when BS_3 is in place, the amplitudes for triggering D_1 and D_2 are, apart from a common factor, the same as in (110), and therefore the rates R_{01} and R_{02} show the same interference patterns as those in (111). On the other hand, when BS_3 is removed, the rates R_{01} and R_{02} show no intereference because the paths are now determined.

[1] This is the delayed choice experiment originally proposed by Wheeler (*WZ*, p. 183).

As in the experiments that test Bell's inequality, the time ordering of events merits comment. In the laboratory frame, D_0 is struck before the other detectors $D_1 \dots D_4$, but the former is at a spacelike separation from the latter. Hence there is no absolute significance to this time sequence; there are frames in which D_0 is struck after the others. An equivalent way of seeing this is to consider a different arrangement in which D_0, in the frame of Fig. 12.7, is moved far to the right so that it is struck after the others; the rates R_{0i} are unaffected.

(e) Summation

Consider an object $\mathcal{O}$ that possesses an observable Λ with eigenvalues λ_i. The goal is to determine the frequency with which distinct, individual specimens of $\mathcal{O}$ in a state $|\Psi\rangle$ display the eigenvalue λ_i, that is, are "found" to be in the state $|\lambda_i\rangle$. For this purpose there must be an interaction H_c between $\mathcal{O}$ and an apparatus $\mathcal{A}$ that forces the eigenvalues λ_i to be uniquely correlated to the final states of $\mathcal{A}$. In the input $\mathcal{O}$ and $\mathcal{A}$ are isolated from each other and H_c is inoperative, so that the whole system $\mathcal{O} + \mathcal{A}$ is in an uncorrelated state,

$$|\Theta_{\text{in}}\rangle = |\Phi_{\text{in}}\rangle \otimes |\Psi\rangle \,, \quad |\Psi\rangle = \sum_i c_i \, |\lambda_i\rangle \,, \tag{114}$$

where $|\Phi_{\text{in}}\rangle$ is a state of $\mathcal{A}$. The unitary transformation generated by such an interaction H_c produces the entangled output state

$$|\Theta_{\text{out}}\rangle = \sum_i c_i \, |\Phi_i\rangle \otimes |\lambda_i\rangle \,, \tag{115}$$

where the $|\Phi_i\rangle$ are orthogonal states of $\mathcal{A}$.

An ideal measurement experiment is one whose output provides an unambiguous macroscopic distinction between these apparatus states $|\Phi_i\rangle$, so that the determination that $\mathcal{A}$ is in the state $|\Phi_i\rangle$ in a particular trial guarantees that the object $\mathcal{O}$ is in the state $|\lambda_i\rangle$ in this trial. Consequently, such a specimen of $\mathcal{O}$ is assigned to the normalized post-measurement state $|\lambda_i\rangle$.

All results of observations on the apparatus can be expressed in terms of its reduced density matrix $\rho(\mathcal{A})$,

$$\rho(\mathcal{A}) = \sum_i \langle \lambda_i | \Theta_{\text{out}} \rangle \langle \Theta_{\text{out}} | \lambda_i \rangle = \sum_i |c_i|^2 \, P(\Phi_i) \,, \tag{116}$$

where $P(\Phi_i)$ is the projection operator onto $|\Phi_i\rangle$. Although $|\Theta_{\text{out}}\rangle$ is a pure state, $\rho(\mathcal{A})$ is not – it has positive entropy (unless there is just one term in Eq. 114). This is true of *reduced* density matrices quite generally. However, a "good" measurement setup manufactures an output whose *complete* (repeat, complete) density matrix ρ_{out} is, to all intents and purposes, indistinguishable from an effective density matrix ρ_{eff} which gives the impression that the entropy has increased even though the evolution is unitary.

To see how this comes about, write the complete density matrix in the form

$$\rho_{\text{out}} = \sum_i |c_i|^2 \, P(\Phi_i) \otimes P(\lambda_i) + \rho_{\text{int}} \,, \tag{117}$$

$$\rho_{\text{int}} = \sum_{i \neq j} c_i c_j^* \left(|\Phi_i\rangle\langle\Phi_j| \right) \otimes \left(|\lambda_i\rangle\langle\lambda_j| \right) . \tag{118}$$

The operator $\rho_{\rm int}$ due to interference between the separate terms in (115) is essentially unobservable in a "good" experiment. If this indeed so, the *complete* density matrix is, to all intents and purposes,

$$\rho_{\rm eff} = \sum_i |c_i|^2 \, P(\Phi_i) \otimes P(\lambda_i) \, , \tag{119}$$

which is not pure; in fact, it has the same entropy as the reduced density matrix $\rho(\mathcal{A})$.

To understand why $\rho_{\rm int}$ is effectively zero, it is instructive to compare the experiments of present concern with an EPR experiment. Assume there are only two terms in (115), with $c_1 = -c_2 = 1/\sqrt{2}$, so as to correspond to the EPR spin singlet example. Then

$$\rho_{\rm int} = -\tfrac{1}{2}\{(|\Phi_1\rangle\langle\Phi_2|) \otimes (|\lambda_1\rangle\langle\lambda_2|) + \text{h.c.}\} \, . \tag{120}$$

In the EPR experiment, $|\Phi_{1,2}\rangle$ are orthogonal states of a system with *one* degree of freedom, a single spin (say $\boldsymbol{\sigma}_1$), and $|\lambda_{1,2}\rangle$ those of another single spin ($\boldsymbol{\sigma}_2$). The interference term $\langle\Phi_1|\boldsymbol{\sigma}_1|\Phi_2\rangle$ is of order one. Such intereference terms are crucial in this case; were they forgotten the correlations between $\boldsymbol{\sigma}_1$ and $\boldsymbol{\sigma}_2$ would come out wrong and the analysis based on Bell's theorem would fail.

In any real measurement experiment, on the other hand, the apparatus is a system with a huge number of degrees of freedom, and thus vastly more complex than the two-state counters used in the model of §5(b). Furthermore, the states $|\Phi_{1,2}\rangle$ are to be macroscopically distinguishable. For these reasons such states are so fiercely orthogonal that only fantastic operators $\mathcal{F}$ have detectable off-diagonal matrix elements $\langle\Phi_1|\mathcal{F}|\Phi_2\rangle$.

Schrödinger's famous cat parody both explains and illustrates the last statement. In this skit $\mathcal{O}$ is a radioactive nucleus and $\mathcal{A}$ the cat, with the decay triggering a mechanism that poisons the cat. Does observing the cat's death give irrefutable evidence that the nucleus has decayed, bearing in mind that the probability distribution has interference terms in which the cat's state is "alive" while that of the nucleus is "decayed," and vice versa? Such interference terms can, however, only be displayed by an operator $\mathcal{F}$ acting on the cat's variables such that $\langle\text{alive}|\mathcal{F}|\text{dead}\rangle \neq 0$. In one of these states blood is flowing, in the other it is coagulated. Thus $\mathcal{F}$ must connect two states in which a vast number K of constituents have very different momenta. To give a matrix element that is not immeasurably small, $\mathcal{F}$ must act *simultaneously* on virtually all of these, for if N of these constituents are not acted upon by $\mathcal{F}$ they will, in the matrix element, produce a factor that is, figuratively speaking, some very small number raised to the power N. In the language of §11.2, $\mathcal{F}$ must be a K-body operator, and must be coupled by the interaction H_c to the object $\mathcal{O}$ on which the measurement is being attempted – which is why we call such operators "fantastic." This argument applies to any realistic detector (cat, photographic emulsion, scintillator array, etc.) able to amplify the microscopic distinctions between states of $\mathcal{O}$ to the point where they are susceptible to detection.[1]

Coherence between distinct outcomes becomes ever more irretrievable as one moves from the idealized realm of simple models to increasingly realistic descrip-

[1] Schrödinger himself cautioned that the superposition of a dead and living cat is "quite ridiculous" (cf. *WZ*, p. 157)

tions of real experiments. For example, consider again the optical Stern-Gerlach experiment of Fig. 12.3. When the beam splitter BS_2 is perfectly reflecting, knowing which of the one-photon states $|f\rangle$ or $|\gamma\rangle$ is occupied determines which of the states $|1\rangle$ or $|2\rangle$ is occupied, i.e., in the spin analogy whether the spin is up or down. As long as one sticks to this stripped-down description, it is possible to set up mirrors, and the like, to demonstrate coherence between $|f\rangle$ and $|\gamma\rangle$. But if a realistic description of the detectors $D_{1,2}$ is included, one is in the cat dead vs. alive realm, because displaying coherence between distinct outcomes then requires coherence to be established between the states of the two detectors after one or the other has unleashed a cascade of complex excitations.

In sum, the detectable outcomes of a successful measurement experiment form a probability distribution in which there are one-to-one correlations between states of the apparatus and the object, and the quantum mechanical interference terms are effectively invisible, so that this distribution has the same character as those of everyday statistics, and therefore has an unambiguous interpretation. Because we ascribe the quantum state to individual systems, this probability distribution refers to potentialities, only one of which is actually realized in any one trial.[1]

12.6 Problems

1. Consider the following incoherent, rotationally invariant superposition of projection operators onto states of two spins with opposite orientation:

$$\rho_R = \int \frac{d\boldsymbol{n}}{16\pi}\,(1+\boldsymbol{n}\cdot\boldsymbol{\sigma}_1)(1-\boldsymbol{n}\cdot\boldsymbol{\sigma}_2)\,.$$

Show that this is the density matrix of a mixture; that

$$\mathrm{Tr}\,\rho_R(\boldsymbol{n}_1\cdot\boldsymbol{\sigma}_1)(\boldsymbol{n}_2\cdot\boldsymbol{\sigma}_2) = -\tfrac{1}{3}\boldsymbol{n}_1\cdot\boldsymbol{n}_2\,,$$

in contrast to Eq. 17; and that the argument leading to Bell's inequality cannot be made in terms of this state.

2. Show that in addition to Eq. 28, there is also the inequality

$$|R(\boldsymbol{a},\boldsymbol{b}) + R(\boldsymbol{a},\boldsymbol{c})| \leq 1 - R(\boldsymbol{b},\boldsymbol{c})\,,$$

and that as a consequence the quantum mechanical correlations only satisfy the Bell inequalities if $\boldsymbol{a} = \pm\boldsymbol{b}$ or $\boldsymbol{a} = \pm\boldsymbol{c}$.

3. One might suppose that the EPR conclusion is a peculiarity of the spin singlet state because its eigenvalue $J = 0$ compels the anticorrelation in any direction as a consequence of angular momentum conservation. That this not so is clear from the Clauser-Horne inequality, but can be seen explicitly from the $J = 2$ state

$$|2\rangle = \tfrac{1}{\sqrt{2}}\{|+\rangle_1|+\rangle_2 + |-\rangle_1|-\rangle_2\}\,,$$

[1]Here there is a distinct advantage to ascribing the quantum state to individuals as compared to ensembles. In the latter philosophy, there is no indication that the probability distribution describes actual phenomena, i.e., individual, solitary events. For that reason von Neumann, in his classic 1932 treatment, stated the dynamical axioms as follows: (i) between measurements, states evolve with the unitary transformation generated by the conventional Hamiltonian; (ii) if a measurement occurs, the state *collapses* at that instant to the state describing the outcome, and evolves thereafter in accordance with (i) until the next measurement.

where the spin projections are along the z-axis. Show that this state has perfect correlations along any direction in the zx-plane.

4. Show the statement concerning expectation values of spin operators following (59) for the state with eigenvalues $\omega_1 = \omega_2 = \omega_3 = 1$ holds for all the simultaneous eigenstates of the operators $\Omega_{1,2,3}$.

5. Confirm Eq. 94 regarding coherence in the final state of the object when there is imperfect counting efficiency in measurement. Show that the final state of the apparatus has no coherence effects when the whole system is described by (88), and the final and initial counter states are not orthogonal (i.e., that Eq. 92 holds).

6. From the definition of the entropy $S = -k \,\mathrm{Tr}\, \rho \ln \rho$, derive the inequality Eq. 83 concerning the apparent increase in entropy.

7. The figure shows an interesting *Gedanken* experiment due to L. Hardy. An electron-positron pair in the state $|e^+\rangle \otimes |e^-\rangle \equiv |e^+e^-\rangle$ is injected into two interferometers, where the beam splitters $BS^{\pm}$ can either be in place or not, but the other two beam splitters are immovable. If the pair meets at P it is certain to annihilate into photons, i.e., $|e^+e^-\rangle \to |\gamma\rangle$.

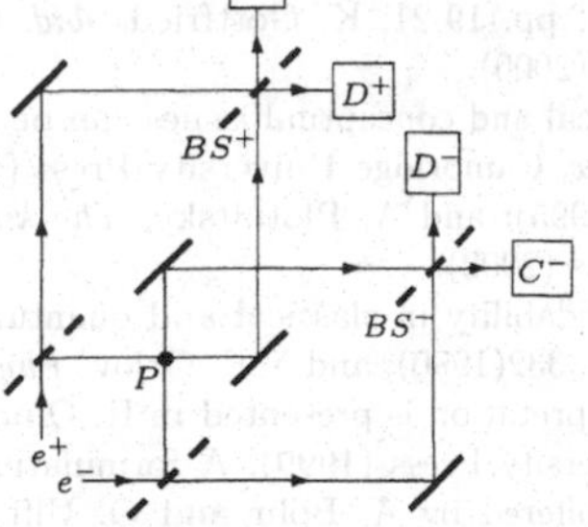

(a) First consider the case where there is *no* annihilation interaction. Show that the state remains unentangled as it moves through the apparatus, and that the output always produces coincidences between the detectors C^+ and C^-, and no counts in D^+ or D^-.

(b) When the annihilation process is operative, show that the state is entangled whether either or both beam splitters $BS^{\pm}$ are in place. Further, if both are in place show that there are D^+D^- coincidences in 1/16 events; if neither is in place, that there are such coincidences in 1/4 of the events; and if only one is in place that there are no such coincidences. In the latter case, with only one of the two beam splitters in place, show that there are C^+C^- coincidences in 1/8 events.

Endnotes

An excellent and technically less demanding account of much of the material in this chapter is provided by G. Greenstein and A.J. Zajonc, *The Quantum Challenge*, Jones and Bartlett, New York (1997) For a sophisticated and often original treatment that examines many issues ignored here and in most accounts, see A. Peres, *Quantum Theory: Concepts and Methods*, Kluwer (1995). Most of the classic early papers on the interpretation of quantum mechanics are in the collection by Wheeler and Zurek (*WZ*). For a comprehensive survey with a good bibliography, see F. Laloë, *Am. J. Phys.* **69**, 655 (2001).

Bell's publications are collected in J.S. Bell, *Speakable and unspeakable in quantum mechanics*, Cambridge University Press (1987); and in *John S. Bell on the Foundations of Quantum Mechanics*, M. Bell, K. Gottfried and M. Veltman (eds.), World Scientific (2001).

Authoritative treatments of EPR, Bell's theorem, the Bell-Kochen-Specker theorem and related issues, can be found in J.F. Clauser and A. Shimony, *Rep. Prog. Phys.* **41**, 1881 (1978); N.D. Mermin, *Rev. Mod. Phys.* **65**, 803 (1993); *Am. J. Phys.* **66**, 920 (1998); *Found. Phys.* **29**, 571 (1999). For GHZ states and related topics, see especially D.M. Greenberger, M.A. Horne, A. Shimony and A. Zeilinger, *Am. J. Phys.* **58**, 1131 (1990). An interesting discussion of coherence related to that of §12.5 is in A. Stern, Y. Aharanov and Y. Imry, *Phys. Rev. A* **41**, 3436 (1990)

Successive measurements are treated by Y. Aharanov, P.G. Bergman and J.L. Lebowitz, *Phys. Rev. B* **134**, 1310 (1964). It is possible to measure canonically conjugate variables simultaneously, though of course with non-vanishing dispersions for both; see E. Arthurs and J.L. Kelly, *Bell Syst. Tech.J.* **44**, 725 (1964); W.K. Wooters and W.H. Zurek, *Phys. Rev. D* **19**, 473 (1979) (*WZ*); S. Stenholm, *Ann. Phys.* **218**, 233 (1992); U. Leonhardt, *Measuring the Quantum State of Light,* Cambridge University Press (1997), Chapter 6.

Bell was sardonically critical of the prominent role given to measurement in the Copenhagen formulation of quantum mechanics and all its descendents, such as this book, and also of those that rely on coupling to the environment. See J. Bell, *Physics World,* August 1990, pp. 33-40, reprinted in the 2001 collections cited above. For rebuttals, see R.E. Peierls, *ibid.*, January 1991, pp. 19-21; K. Gottfried, *ibid.* October 1991, pp. 34-40; K. Gottfried, *Nature* **405**, 533 (2000).

Treatments of philosophical and conceptual issues can be found in A. Shimony, *Search for a naturalistic world view*, Cambridge University Press (1993); B. d'Espagnat, *Veiled Reality,* Addison Wesley (1995); and A. Plotnitsky, *The knowable and the unknowable*, University of Michigan Press (2002).

The role of logical undecidability in classical and quantum physics are explored by I. Kanter, *Phys. Rev. Lett.* **64**, 332(1990); and Y.F. Orlov, *Phys. Rev. A* **65**, 042106 (2002). The consistent history interpretation is presented in R. Omnès, *Understanding Quantum Mechanics,* Princeton University Press (1999). A formulation in which symmetry is the fundamental ingredient is offered by A. Bohr and O. Ulfbeck, *Rev. Mod. Phys.* **67**, 1 (1997).

The most famous hidden variable theory is the pilot wave theory due to L. de Broglie and D. Bohm; see D. Bohm and B.J. Hiley, *The Undivided Universe*, Routledge (1993). The de Broglie-Bohm theory introduces a position $\boldsymbol{x}_i(t)$ for every particle. The motion of $\boldsymbol{x}_i(t)$ is dictated by the probability current as computed from the corresponding Schrödinger equation, and as a consequence the theory is dramatically nonlocal – the movement of $\boldsymbol{x}_i(t)$ depends on the instantaneous value of all the others positions. In contrast to orthodox quantum mechanics, the pilot wave theory could, at least in principle, be understood by our fictitious Maxwell (see opening page of Chapter 2).

There is a large literature on the implications for measurement of coupling to the environment. See H.D. Zeh in *Decoherence and the Appearance of the Classical World,* D. Giulini *et al* (eds.) Springer (1996); and in *New Developments on Fundamental Problems in Quantum Physics*, M. Ferrero and A. van der Merve (eds.), Kluwer (1997). W.H. Zurek, *Physics Today,* October 1991, p. 36; *Philos. Trans. R. Soc. London A* **356** (1998). A.O. Caldeira and A.J. Legget, *Phys. Rev. Lett.* **46**, 211 (1981) A.J. Legget, S. Chakravery, A.T. Dorsey, M.P.A. Fisher, A.Garg and W. Zwerger, *Rev. Mod. Phys.* **59**, 1 (1987). T.A. Brun, *Am. J. Phys.* **70**, 719 (2002). For experimental work on this topic see S. Haroche, *Physics Today,* July 1998, p. 36; J.R. Friedman *et al, Nature* **406**, 43 (2000) C. van der Wal *et al, Science* **290**, 773 (2000).

13
Relativistic Quantum Mechanics

13.1 Introduction

The Special Theory of Relativity was already firmly established when quantum mechanics was discovered in 1925, so it was understood that if a fundamental theory purported to supercede classical mechanics it must be Lorentz covariant. At that time, the only known "elementary" particles were the photon, electron and proton. Construction of a relativistic theory of photons was accomplished rather quickly by Dirac — in effect, by canonical quantization of Maxwell's expression for the energy of the electromagnetic field, as described in §10.1. In atomic and molecular phenomena, protons move so slowly that no one was concerned with their high velocity behavior until much later. On the other hand, that relativistic corrections to the Bohr model of hydrogen can account for the fine structure of hydrogenic spectral lines had been demonstrated by Sommerfeld in 1916. For that reason, the search for a relativistic, quantum mechanical description of electrons began immediately. This, however, proved to be a far more difficult task than the quantization of the electromagnetic field.

Dirac derived his relativistic equation for the electron in 1928 from assumptions that were seen as compelling at that time, but which turned out to be based on misconceptions that were only revealed by the difficult search for a correct interpretation of his equation. The elusiveness of that interpretation can be gauged by noting that it took about two years (1925–27) to largely unravel the meaning of the nonrelativistic formalism, whereas the same cast needed some six years (1927–34) to come to an understanding of the Dirac equation, and of relativistic quantum mechanics more generally.

The basic assumptions Dirac inferred from the spectacular success of the Schrödinger equation were:

1. the marriage between relativity and quantum mechanics is to be embodied in a one-particle wave equation of the Schrödinger type;

2. a quantum mechanical equations of motion must be first order in time.

To understand the salience of the second assumption, consider the most obvious relativistic wave equation for a particle of mass m. It emerges from the relation among energy, momentum and mass,

$$E^2 = p^2 + m^2 \,, \tag{1}$$

via Schrödinger's recipe

$$\boldsymbol{p} \to \frac{1}{i}\boldsymbol{\nabla}, \quad E \to i\frac{\partial}{\partial t} \,. \tag{2}$$

The result is the *Klein-Gordon equation*,

$$(\partial^2 + m^2)\psi = 0 \,, \tag{3}$$

where

$$\partial^2 = \frac{\partial^2}{\partial t^2} - \nabla^2 \tag{4}$$

is the wave operator, and $1/m$ (or $\hbar/mc$) is the Compton wavelength of the particle in question.

It is a consequence of the Klein-Gordon equation that the wave function satisfies the continuity equation

$$\frac{\partial \rho}{\partial t} + \nabla \cdot j = 0 \,, \tag{5}$$

where

$$\rho = \psi^* \frac{\partial}{\partial t}\psi - \text{c.c.}, \quad j = \psi \nabla \psi^* - \text{c.c.} \,. \tag{6}$$

But this density ρ is not positive definite, so it is impossible to interpret ψ as a probability amplitude.

Dirac was not primarily motivated by this problem of the Klein-Gordon equation, but by his belief that evolution must, whatever the system, be described by a unitary transformation performed on some wave function as the initial condition, whence his quest for a first-order equation in time.[1] Once it is accepted that the equation is to be of first-order in time, Lorentz covariance requires that it also be first-order in space. Dirac derived an equation of this character by introducing a 4×4 generalization of the Pauli matrices (the matrices called γ_μ below). The nonrelativistic approximation to his equation was the Schrödinger equation for a particle of spin $\frac{1}{2}$ with the known g factor of 2, and its exact solution in a Coulomb field was in splendid agreement with the hydrogen fine structure. Dirac (and his contemporaries) therefore believed, at first, that he had *proven* that the electron's spin was a direct consequence of compelling a wedding of relativity with quantum mechanics.

The Dirac equation suffered from one disease that had also afflicted the Klein-Gordon equation: it possessed solutions of negative energy, $-(p^2 + m^2)^{\frac{1}{2}}$. Attempts to discard these solutions failed until Dirac introduced a profound re-interpretation in which he assigned the negative energy solutions to a positive energy states of a hitherto unknown particle, *the positron*, having the same mass but opposite charge to the electron. This spectacular prediction was confirmed in 1932.

By 1934 it was understood that while Dirac's original argument was somewhat misleading, his equation, considered as a one-particle theory, nevertheless describes relativistic electrons correctly provided effects due to the existence of positrons are ignorable. Furthermore, we now know that Lorentz-covariant theories for free particles of any spin and any magnetic moment can be constructed. The most profound aspect of this successful interpretation is that the symbols ψ appearing

[1]In this regard he was correct, but he was mislead (as were most) insofar as the Klein-Gordon equation is concerned. Pauli and Weisskopf subsequently showed that this is a correct equation for a charged scalar quantum *field operator*, and that it does *not* govern the time evolution of the wave function (i.e., *state vector*). Indeed, in quantum field theory, the Dirac equation also governs the evolution of the field operator, and not the state vector. The latter, in all quantum field theories, evolves with a unitary transformation, as Dirac had assumed.

in the Dirac equation cannot be interpreted consistently as wave functions of the Schrödinger type, but are *field operators* similar to the field operators that arise in the quantum theory of the electromagnetic field, though anticommuting and thus describing particles that are fermions (see §11.2(b)).

The bulk of this chapter is confined to phenomena involving electrons and other spin $\frac{1}{2}$ particles (such as the muon) in which effects due to the existence of their antiparticles is either negligible or so small as to make a discussion in which they are ignored an excellent first approximation. Hence we shall treat the Dirac equation as if it were the theory of a one-particle system, and largely ignore the equation's negative energy solutions.

Furthermore, and in contrast with most of this volume, we shall adopt a wave-mechanical formulation, as compared to the Hilbert space language. This is not just a matter of style. The true role of Dirac's equation is that of a "classical" field equation, like Maxwell's; both are elevated to equations for operators in quantum field theory, and only when this is done can the theory be stated correctly in the Hilbert space language.

13.2 The Dirac Equation

The Dirac equation for a free spin $\frac{1}{2}$ particle is, as we shall see, the prescription of how states of spin $\frac{1}{2}$ transform from one inertial frame to another. The set of all such transformations forms the *Lorentz group* $\mathcal{L}$, and some understanding of this group is thus necessary.

To begin with, the contravariant coordinate 4-vector is

$$x^\mu = (x^0, x^1, x^2, x^3) = (t, x, y, z) = (t, \boldsymbol{r}) \; ; \tag{7}$$

the covariant counterpart is $x_\mu = (t, -\boldsymbol{r})$. The group $\mathcal{L}$ is the set of all homogeneous linear transformations that leave the quadratic form

$$\mathcal{Q} = t^2 - x^2 - y^2 - z^2 \tag{8}$$

invariant. The 3-dimensional rotation group is the subgroup of $\mathcal{L}$ formed by those transformations that also leave t invariant. A Lorentz transformation therefore has the form

$$x^\mu \to \Lambda^\mu{}_\nu \, x^\nu \, , \tag{9}$$

where the 4×4 matrix Λ depends on the relative orientation and velocity of the two frames. By definition, *any* set of four real quantities $V^\mu \equiv (V^0, \boldsymbol{V})$ that transform like x^μ constitutes a 4-vector. Hence if V^μ and W^μ are any two 4-vectors, $V_\mu W^\mu = V^0 W^0 - \boldsymbol{V} \cdot \boldsymbol{W}$ is an invariant.

The elements of the Lorentz group are parametrized by the 4×4 matrix Λ, and the group separates into two subgroups according to whether $\det \Lambda$ equals 1 or -1. The formers is called the *proper*, the latter the *improper* Lorentz group. The improper subgroup includes space reflection and time reversal, but not the combination $I_s I_t$ (i.e., reflection through the space-time origin), which has unit determinant. Of course, the distinction between proper rotations and improper reflections already arose in nonrelativistic quantum mechanics, but only when it came to analyzing the consequences of reflection-invariant *interactions*. In striking contrast, reflection invariance is of fundamental importance in relativistic quantum mechanics because it already imposes restrictions on the theory of *free* particles.

(a) Lorentz Transformations of Spinors

Recall from §7.4 that spin $\frac{1}{2}$, and other half-integral angular momenta, arise because it is possible to represent the rotations of a Euclidean 3-vector by homogeneous linear transformations of 2×2 matrices. It is remarkable that a very simple generalization describes Lorentz transformations of 4-vectors by linear transformations of such 2×2 matrices. Namely, to the *real* 4-vector x^μ we assign the Hermitian matrix

$$X_+ = \begin{pmatrix} t-z & -x+iy \\ -x-iy & t+z \end{pmatrix} = t - \boldsymbol{r}\cdot\boldsymbol{\sigma} \; ; \tag{10}$$

the need for the subscript "+" is related to reflections, as shall soon be explained. Then the basic invariant $\mathcal{Q}$ is simply the determinant of X_+:

$$\det X_+ = \mathcal{Q} \, . \tag{11}$$

Hence a Lorentz transformation of the 4-vector x^μ can be represented as a homogeneous linear transformation on X_+ that leaves it Hermitian and does not alter its determinant. The most general homogeneous linear transformation of the matrix X_+ has the form AX_+B, where A and B are arbitrary matrices that are not functions of x^μ. Hermiticity will be preserved if $B = A^\dagger$, and therefore the required transformation is

$$X'_+ = \Omega_+ X_+ \Omega_+^\dagger \, . \tag{12}$$

From $\det AB = (\det A)(\det B)$, and $\det A^\dagger = (\det A)^*$, we have

$$\det X'_+ = \det X_+ |\det \Omega_+|^2 \, . \tag{13}$$

Hence (12) will be a Lorentz transformation if

$$|\det \Omega_+| = 1 \, . \tag{14}$$

As rotations form a subgroup, this must reduce to the familiar unitary transformation law for rotations in the special case where the time-component is left unchanged. But

$$\mathrm{Tr}\, X_+ = 2t = \mathrm{Tr}\, X_+ \Omega_+^\dagger \Omega_+ \, , \tag{15}$$

because $\mathrm{Tr}\, AB = \mathrm{Tr}\, BA$. Therefore Ω_+ is indeed unitary for a rotation. By the same token, Ω_+ *is not unitary if the time-component of 4-vectors are changed by the Lorentz transformation.*

Two-component *Weyl spinors,*

$$\chi_+ = \begin{pmatrix} \chi_{+1} \\ \chi_{+2} \end{pmatrix} \, , \tag{16}$$

are then defined in a similar manner to that for Pauli spinors.[1] However Ω_+ is not unitary, and this is a significant new feature in the transformation law. Define the transformed spinor χ'_+ in the usual way:

$$X_+\chi_+ \to \Omega_+(X_+\chi_+) \equiv X'_+\chi'_+ = (\Omega_+ X_+ \Omega_+^\dagger)\chi'_+ \, . \tag{17}$$

[1] A word about notation. The following definition of a Weyl spinor holds in all representations, and in particular, in both coordinate and momentum space. Later, when writing the Dirac equation in momentum space the symbol χ will be used, and φ in coordinate space.

Therefore

$$\chi'_+ = (\Omega_+^\dagger)^{-1} \chi_+ . \tag{18}$$

When Ω_+ is a rotation it is unitary and $\chi'_+ = \Omega_+ \chi_+$, which is the familiar form, but it does not hold for a Lorentz transformation.

We now turn to the spatial reflection I_s, which will introduce a surprising feature with no counterpart in the nonrelativistic theory. The reflection of the matrix X_+ representing x^μ produces the following transformation:

$$I_s: \quad X_+ \to \begin{pmatrix} t+z & x-iy \\ x+iy & t-z \end{pmatrix} = t + \boldsymbol{r}\cdot\boldsymbol{\sigma} \equiv X_- . \tag{19}$$

This *new* matrix X_- also has the determinant $t^2 - |\boldsymbol{r}|^2 = \mathcal{Q}$. For the reflection to be represented by (12), it is therefore necessary that

$$\Omega_+\Omega_+^\dagger = 1, \qquad \Omega_+\boldsymbol{\sigma}\Omega_+^\dagger = -\boldsymbol{\sigma}, \tag{20}$$

or $\Omega_+\boldsymbol{\sigma} = -\boldsymbol{\sigma}\Omega_+$. But there is no 2×2 matrix that anticommutes with all three Pauli matrices! Therefore two 2×2 matrices, X_+ and X_-, are needed to describe the 4-vector x^μ if space reflection is to represented as a transformation on spinors.

This seemingly innocent fact has an astonishing implication of great consequence: *a reflection-invariant relativistic theory of spin $\frac{1}{2}$ particles cannot be based on 2-component Weyl spinors, but requires the introduction of 4-component Dirac spinors.* That is to say, if the reflection I_s is to take a state represented by χ_+ into another state contained in the theory, it is necessary to introduce a second, distinct type of Weyl spinor, χ_-, which is the image of χ_+ under I_s. Note that this requirement goes beyond that of reflection invariance, for it only demands that a reflected state exists, and not that it has the same dynamics — indeed, dynamics has not yet been mentioned.

The definition of the second type of Weyl spinor χ_- follows the pattern of that for χ_+. A proper Lorentz transformation on the second matrix X_- representing x^μ is performed by a 2×2 matrix Ω_-,

$$X_- \to \Omega_- X_- \Omega_-^\dagger , \tag{21}$$

and for this second spinor,

$$\chi_- \to (\Omega_-^\dagger)^{-1} \chi_- . \tag{22}$$

The 4-component Dirac spinor is defined as

$$\psi = \begin{pmatrix} \chi_+ \\ \chi_- \end{pmatrix} . \tag{23}$$

Proper Lorentz transformations do not mix χ_+ with χ_-, so under such transformations the Dirac spinor transforms with the block-diagonal 4×4 matrix Ω defined as

$$\Omega = \begin{pmatrix} \Omega_+ & 0 \\ 0 & \Omega_- \end{pmatrix} . \tag{24}$$

Under space reflection the Weyl spinors are interchanged, $\chi_+ \leftrightarrow \chi_-$; hence the Dirac spinor undergoes[1]

$$I_s: \quad \psi \to \gamma^0\psi, \quad \gamma^0 = \begin{pmatrix} 0 & 1 \\ 1 & 0 \end{pmatrix} . \tag{25}$$

---

[1]The rationale for the peculiar notation γ^0 will become clear presently.

This intricate relationship between proper and improper Lorentz transformations for particles with spin arises already in the familiar context of classical electrodynamics, or in quantum language, for particles of spin 1. From the electric and magnetic fields we can define the two complex field strengths

$$\boldsymbol{F}^{\pm} = \boldsymbol{E} \pm i\boldsymbol{B} , \tag{26}$$

in terms of which Maxwell's equations have the decoupled form

$$\partial_t \boldsymbol{F}^{\pm} \pm \boldsymbol{\nabla} \times \boldsymbol{F}^{\pm} = 0 . \tag{27}$$

Under proper Lorentz transformations the three components of $\boldsymbol{F}^+$ become linear combinations of only themselves, and the same holds for $\boldsymbol{F}^-$. But the electric and magnetic fields have opposite signature under space reflection, and therefore I_s produces $\boldsymbol{F}^+ \leftrightarrow \boldsymbol{F}^-$. This is but one of the analogies between the Dirac and Maxwell theories that will be mentioned from time to time, and which are explored in Prob. 1.

We now turn to constructing the matrix Ω_+. Like any 2×2 matrix it can be written as

$$\Omega_+ = (a + i\boldsymbol{b}\cdot\boldsymbol{\sigma}) , \tag{28}$$

where a and $\boldsymbol{b}$ are in general complex. According to (14), if Ω_+ is to represent a Lorentz transformation it must satisfy

$$\det \Omega_+ = a^2 + \boldsymbol{b}\cdot\boldsymbol{b} = e^{2i\phi} , \tag{29}$$

where ϕ is real. This phase can be eliminated by the change of variables $a \to e^{i\phi}a, \boldsymbol{b} \to e^{i\phi}\boldsymbol{b}$, which is of no consequence since the phase factor drops out from (12). Then $a^2 + \boldsymbol{b}\cdot\boldsymbol{b} = 1$, and the choice of ϕ can be taken such that

$$a = +(1 - \boldsymbol{b}\cdot\boldsymbol{b})^{\frac{1}{2}} . \tag{30}$$

Hence a (proper) Lorentz transformation can be fully parametrized by the complex 3-vector $\boldsymbol{b}$, i.e., by six real parameters. This is as it must be: three angles are needed to define the relative orientation of two frames, and a 3-vector of velocity for their relative motion.

The special case of a rotation is already familiar from §7.4(b). In this case $\boldsymbol{b}$ is the real vector

$$\boldsymbol{b} = -\hat{\boldsymbol{n}} \sin \tfrac{1}{2}\theta , \tag{31}$$

where $\hat{\boldsymbol{n}}$ is the axis and θ the angle of rotation. Hence $a = \cos \frac{1}{2}\theta$, and

$$\Omega_+ = \cos \tfrac{1}{2}\theta - i\hat{\boldsymbol{n}}\cdot\boldsymbol{\sigma} \sin \tfrac{1}{2}\theta = \exp(-\tfrac{1}{2} i\theta \hat{\boldsymbol{n}}\cdot\boldsymbol{\sigma}) . \tag{32}$$

To construct Ω_+ for Lorentz transformations it suffices to consider a transformation along one direction, say that of z, in which case the other spatial coordinates do not change and can be set to zero. The quadratic form of concern is then

$$Q = t^2 - z^2 = (t - z)(t + z) . \tag{33}$$

Relative velocity v is the physically obvious parametrization of the transformation, but as the rather complicated Einstein addition law for successive transformations shows, v is not a happy choice. A better choice is one that is additive, like angles of rotations about a definite axis. To that end, note that $\mathcal{Q}$ is invariant under

$$(t-z) \to (t-z)e^{-\zeta} \equiv t'-z', \quad (t+z) \to (t+z)e^{\zeta} \equiv t'+z' , \tag{34}$$

and that two successive transformations of this type, with parameters ζ_1 and ζ_2, are equivalent to a single transformation with $\zeta_1+\zeta_2$. This desirable, additive parameter ζ is called *rapidity.*[1] Its relation to relative velocity follows from (34), which implies that[2]

$$t' = t\cosh\zeta + z\sinh\zeta , \quad z' = z\cosh\zeta + t\sinh\zeta . \tag{35}$$

This is an active Lorentz transformation: a point at rest in a frame K moves in the positive z'-direction in the frame K' reached by the transformation. The sought-after relation between rapidity and relative velocity v is therefore

$$v = \tanh\zeta, \quad \cosh\zeta = \frac{1}{\sqrt{1-v^2}} \equiv \gamma, \quad \sinh\zeta = v\gamma . \tag{36}$$

The matrix Ω_+ for this special Lorentz transformation can now be constructed, because (34) states that the matrix X_+ transforms as follows:

$$X_+ = t - z\sigma_z = \begin{pmatrix} t-z & 0 \\ 0 & t+z \end{pmatrix} \to X'_+ = \begin{pmatrix} (t-z)e^{-\zeta} & 0 \\ 0 & (t+z)e^{\zeta} \end{pmatrix} . \tag{37}$$

This is accomplished by the non-unitary matrix

$$\Omega_+ = e^{-\frac{1}{2}\zeta\sigma_z} , \tag{38}$$

which is easily verified:

$$\begin{aligned} X'_+ &= e^{-\frac{1}{2}\zeta\sigma_z}(t - z\sigma_z)\exp e^{-\frac{1}{2}\zeta\sigma_z} && (39)\\ &= e^{-\zeta\sigma_z}(t-z\sigma_z) = (\cosh\zeta - \sigma_z\sinh\zeta)(t - z\sigma_z) \\ &= (t\cosh\zeta + z\sinh\zeta) - \sigma_z(z\cosh\zeta + t\sinh\zeta) , && (40) \end{aligned}$$

which agrees with (35).

What about the matrices Ω_- that perform Lorentz transformations on the other matrix, X_-, representing a 4-vector? A pure Lorentz transformation along z must reproduce the result (35) when applied to $t + z\sigma_z$, and a short calculation shows that this is performed by

$$\Omega_- = e^{\frac{1}{2}\zeta\sigma_z} . \tag{41}$$

The choice of the z direction in the foregoing is merely a convention, of course. More generally, for a Lorentz transformation with rapidity ζ along the arbitrary direction $\hat{\boldsymbol{n}}$ the two matrices are

$$\Omega_+ = \exp(-\tfrac{1}{2}\zeta\hat{\boldsymbol{n}}\cdot\boldsymbol{\sigma}), \qquad \Omega_- = \exp(\tfrac{1}{2}\zeta\hat{\boldsymbol{n}}\cdot\boldsymbol{\sigma}) . \tag{42}$$

---

[1]Rapidities *do not add* for Lorentz transformations along different axes, just as angles of rotation do not add for rotations about different axes. Put more learnedly, the Lorentz group is non-Abelian, which is hardly astonishing since rotations form a subgroup that is not Abelian.

[2]The appearance of hyperbolic functions reflects the fact that $\mathcal{Q}$ is a hyperboloid.

These matrices perform a "pure" Lorentz transformation, one in which the two frames have the same orientation; such a transformation is often called a *boost*. The parameter $\boldsymbol{b}$, defined in (28), is thus pure imaginary for a boost,

$$\boldsymbol{b}_{\pm} = \pm i\hat{\boldsymbol{n}} \sinh \tfrac{1}{2}\zeta \ , \tag{43}$$

and real for a pure rotation. For a pure rotation, furthermore, there is no distinction between Ω_+ and Ω_-, because t can, in this case, be removed by a shift in the origin of time; thus for a rotation about $\hat{\boldsymbol{n}}$,

$$\Omega_{\pm} = \exp(-\tfrac{1}{2}i\theta\hat{\boldsymbol{n}} \cdot \boldsymbol{\sigma}) \ . \tag{44}$$

To summarize, for an arbitrary Lorentz transformation parametrized by the complex 3-vector $\boldsymbol{b}$, the 2×2 matrices are $\Omega_+ = a + i\boldsymbol{b} \cdot \boldsymbol{\sigma}$ and $\Omega_- = a + i\boldsymbol{b}^* \cdot \boldsymbol{\sigma}$, where $a^2 + \boldsymbol{b} \cdot \boldsymbol{b} = 1$. For pure rotations, $\boldsymbol{b}$ is real, $\Omega_+ = \Omega_-$ and unitary. In general, when $\boldsymbol{b}$ is complex, the inverse transformation is given by $\boldsymbol{b} \to -\boldsymbol{b}$, as is readily verified. Consequently,

$$\left(\Omega_+^\dagger\right)^{-1} = \Omega_- \ , \qquad \left(\Omega_-^\dagger\right)^{-1} = \Omega_+ \ , \tag{45}$$

and therefore the transformation laws (18) and (22) are

$$\chi_+ \to \Omega_- \chi_+ \ , \qquad \chi_- \to \Omega_+ \chi_- \ . \tag{46}$$

(b) The Free-Particle Dirac Equation

We are now ready to derive the Dirac equation for a free spin $\frac{1}{2}$ particle of mass m. Consider, first, an arbitrary pure state at rest in the frame K, where nonrelativistic quantum mechanics holds, and which can therefore be represented by a 2-component Pauli spinor ξ. As seen in the frame K' defined by the transformation (35), such a state has the momentum

$$p_z/m = v_z\gamma = \sinh \zeta \ . \tag{47}$$

The crucial point is that this transformation can be carried out in two ways: by use of Ω_+ or Ω_-, which define two distinct 2-component Weyl spinors for the *same* momentum. According to (46), the required transformations are

$$\chi_+(p_z) = e^{\frac{1}{2}\zeta\sigma_z}\xi, \quad \chi_-(p_z) = e^{-\frac{1}{2}\zeta\sigma_z}\xi \ . \tag{48}$$

On eliminating ξ, we have

$$\chi_-(p_z) = e^{-\zeta\sigma_z}\chi_+(p_z) \ . \tag{49}$$

Believe it or not, apart from some embellishments, this is the Dirac equation for a free particle!

To cast this cryptic form into the standard notation, note first that

$$me^{-\zeta\sigma_z} = m(\cosh\zeta - \sigma_z \sinh\zeta) = E - \sigma_z p_z \ . \tag{50}$$

Hence (49) implies

$$(E - \sigma_z p_z)\chi_+(p_z) = m\chi_-(p_z) \ , \tag{51}$$

$$(E + \sigma_z p_z)\chi_-(p_z) = m\chi_+(p_z) \ . \tag{52}$$

But the z-direction enjoys no special privilege, or put more learnedly, the equations must be rotationally invariant. For arbitrary $\boldsymbol{p}$ they therefore read

$$(E - \boldsymbol{\sigma}\cdot\boldsymbol{p})\chi_+(\boldsymbol{p}) = m\chi_-(\boldsymbol{p}) , \tag{53}$$

$$(E + \boldsymbol{\sigma}\cdot\boldsymbol{p})\chi_-(\boldsymbol{p}) = m\chi_+(\boldsymbol{p}) . \tag{54}$$

This is the Dirac equation for a free particle in momentum space.

A consequence of importance now follows from the fact that for a zero mass particle there is no coupling between the two Weyl spinors $\chi_\pm$:

$$(E - \boldsymbol{p}\cdot\boldsymbol{\sigma})\chi_+ = 0, \qquad (E + \boldsymbol{p}\cdot\boldsymbol{\sigma})\chi_- = 0 . \tag{55}$$

These are the Weyl equations in momentum space. When $m = 0, |\boldsymbol{p}| = E$, and these equations therefore are

$$(1 - \hat{\boldsymbol{p}}\cdot\boldsymbol{\sigma})\chi_+(\boldsymbol{p}) = 0 , \tag{56}$$

$$(1 + \hat{\boldsymbol{p}}\cdot\boldsymbol{\sigma})\chi_-(\boldsymbol{p}) = 0 . \tag{57}$$

They express a very simple but important message, because $\frac{1}{2}\hat{\boldsymbol{p}}\cdot\boldsymbol{\sigma}$ is the helicity operator: *the Weyl spinors $\chi_\pm$ are eigenstates of helicity $\pm\frac{1}{2}$ in the zero mass limit.*

As we shall see in the next section, the interaction of a charged spin $\frac{1}{2}$ particle, such as the electron, with the electromagnetic field does not introduce any mixing between the two Weyl spinors $\chi_\pm$; the only coupling is due to the mass. As a consequence, in electromagnetic phenomena involving momenta large compared to m the Dirac equation separates into two uncoupled equations for the two helicity states, so that in this high-energy regime the helicity is a constant of motion.

If spin $\frac{1}{2}$ particles of zero mass exist, they could be described by just one of they two Weyl spinors, that is, exist in only one helicity state. Here there is again an analogy with electrodynamics, where it is possible to at least conceive of a theory in which only one of the two fields $\boldsymbol{E} \pm i\boldsymbol{B}$ exists, corresponding to just one of the two modes of circular polarization. Such a theory would, of course, be very different from Maxwell's because it would violate reflection invariance with a vengeance. The weak interaction does, however, violate reflection invariance maximally, in the sense that this interaction only couples to one of the two Weyl spinors in the 4-component Dirac spinor even for particles which have mass, such as electrons and quarks.

The substitution of $\boldsymbol{p} \to -i\boldsymbol{\nabla}, E \to i\partial_t$ in (53) and (54) produces *the Dirac equation in coordinate space*:

$$(\partial_t + \boldsymbol{\sigma}\cdot\boldsymbol{\nabla})\varphi_+(\boldsymbol{r},t) + im\varphi_-(\boldsymbol{r},t) = 0 , \tag{58}$$

$$(\partial_t - \boldsymbol{\sigma}\cdot\boldsymbol{\nabla})\varphi_-(\boldsymbol{r},t) + im\varphi_+(\boldsymbol{r},t) = 0 , \tag{59}$$

where $\varphi_\pm(\boldsymbol{r},t)$ are the Fourier transforms of the momentum space spinors,

$$\varphi_\pm(\boldsymbol{r},t) = \int d^3p\, e^{-iEt}e^{i\boldsymbol{p}\cdot\boldsymbol{r}}\chi_\pm(\boldsymbol{p}) , \tag{60}$$

with $E = \sqrt{p^2 + m^2}$. Both of these Weyl spinors satisfy the Klein-Gordon equation,

$$(\partial^2 + m^2)\chi_\pm = 0 , \tag{61}$$

which follows on multiplying (58) and (59) by $(\partial_t \mp \boldsymbol{\sigma}\cdot\boldsymbol{\nabla})$ and using $(\boldsymbol{\sigma}\cdot\boldsymbol{\nabla})^2 = \nabla^2$. This confirms that the Klein-Gordon equation merely encodes the kinematic relationship $E^2 = p^2 + m^2$, and has nothing to say about spin and the related issue of Lorentz transformations. The electromagnetic analogy is that Maxwell's first-order equations imply the wave equation, but not vice versa, and that the wave equation has nothing to say about polarization or the transformation properties of the electric and magnetic field strengths.

That the Dirac equation is covariant does not require proof because the foregoing derivation was predicated on the requirement that states for arbitrary momenta were to be manufactured by Lorentz transformations from rest frame states. Other derivations, and in particular Dirac's original chain of argument, which is reproduced in many texts,[1] require a proof of covariance. While the covariance of (58) and (59) is not at issue, it is also true that in this form the equation does not "look" covariant, just as the covariance of Maxwell's equations is anything but manifest in their standard form.

For that reason, and for use in more sophisticated contexts, it is often useful to cast Dirac's equation into a form in which its "looks" covariant. This is done in terms of Dirac spinors and matrices. The spinor was already defined, but is written out again for clarity's sake:

$$\psi = \begin{pmatrix} \varphi_{+1} \\ \varphi_{+2} \\ \varphi_{-1} \\ \varphi_{-2} \end{pmatrix} \equiv \begin{pmatrix} \psi_1 \\ \psi_2 \\ \psi_3 \\ \psi_4 \end{pmatrix} . \tag{62}$$

The matrix γ^0 defined in (25), and the 4×4 matrices

$$\Sigma^i = \begin{pmatrix} \sigma_i & 0 \\ 0 & -\sigma_i \end{pmatrix} , \quad i = 1, 2, 3, \tag{63}$$

where each entry is the indicated 2×2 matrix, allow one to write a single equation for ψ:

$$\left(\frac{\partial}{\partial t} + \Sigma^i \frac{\partial}{\partial x^i} + im\gamma^0 \right) \psi = 0 . \tag{64}$$

Because m is an invariant, the matrix multiplying it is best eliminated, which can be done by multiplying from the left by γ^0 and use of

$$(\gamma^0)^2 = 1 . \tag{65}$$

The other three Dirac matrices are defined as $\gamma^0 \Sigma^i$, and are thus

$$\gamma^i = \begin{pmatrix} 0 & -\sigma_i \\ \sigma_i & 0 \end{pmatrix} \equiv -\gamma_i . \tag{66}$$

The differential operator in (64) therefore becomes

$$\gamma^0 \frac{\partial}{\partial t} + \gamma^i \frac{\partial}{\partial x^i} = \gamma^0 \partial_0 + \gamma^i \partial_i = \gamma^0 \partial_0 + \boldsymbol{\gamma} \cdot \boldsymbol{\nabla} = \gamma^\mu \partial_\mu , \tag{67}$$

because[2]

$$\frac{\partial}{\partial x^\mu} \equiv \partial_\mu = (\partial_t, \boldsymbol{\nabla}), \quad \frac{\partial}{\partial x_\mu} \equiv \partial^\mu = (\partial_t, -\boldsymbol{\nabla}) . \tag{68}$$

---

[1] For example, Bjorken and Drell, §2.2.

[2] It is very easy to make sign mistakes at this point! Note that the definition $z = x^3$ implies that $\partial/\partial z = \partial/\partial x^3 = \nabla_3 = -\nabla^3$.

With these definitions, *the Dirac equation assumes the standard covariant-looking form*

$$(i\gamma^\mu \partial_\mu - m)\psi = 0 \ . \tag{69}$$

There is nothing sacred about any particular representation for the Dirac matrices and spinors. The freedom to choose representations is merely the fact that the Dirac equation is invariant under the transformation

$$\psi \to \Xi\psi, \quad \gamma^\mu \to \Xi\gamma^\mu\Xi^{-1}, \tag{70}$$

where Ξ is any 4×4 matrix with constant elements. The representation (66) arose naturally in the development given here, and is most convenient in phenomena where the energy is high compared to the rest mass, because then the decoupling between the two Weyl spinors that constitute the Dirac spinor can produce great simplifications. At the other extreme, when relativistic corrections are small, it is not convenient, but another representation is then more suitable. It is this latter representation which is most frequently displayed in textbooks.

The four Dirac matrices form a Minkowski 4-vector in the same sense as the three Pauli matrices form a Euclidean 3-vector. That is, if under a particular Lorentz transformation the coordinate 4-vector undergoes the transformation $x^\mu \to \Lambda^\mu{}_\nu x^\nu$, the corresponding 4×4 matrix Ω has the result

$$\Omega\gamma^\mu\Omega^{-1} = \Lambda^\mu{}_\nu\gamma^\nu \ . \tag{71}$$

In this notation, covariance is established by multiplying the Dirac equation from the left by Ω, using (71), the fact that ∂^μ is a 4-vector, and the spinor transformation law $\psi \to \Omega\psi$.

(c) Charge and Current Densities

The last question about the free Dirac theory that needs to be addressed is whether there is a counterpart to the probability conservation law of Schrödinger's equation. The latter involves bilinear combinations of the wave function and its complex conjugate. Given that hint, we rewrite the Dirac equation as follows:

$$i\partial_t\varphi_+ = -i\boldsymbol{\sigma}\cdot\boldsymbol{\nabla}\varphi_+ + m\varphi_-, \quad i\partial_t\varphi_- = i\boldsymbol{\sigma}\cdot\boldsymbol{\nabla}\varphi_- + m\varphi_+ \ . \tag{72}$$

We need the complex conjugate of these equations, and a convenient notation for this and for bilinear forms built from spinors. For that purpose, consider any n-dimensional complex vector Φ with components ϕ_α, $\alpha = 1, \ldots, n$, and M any $n \times n$ matrix. Then

$$(M\Phi)^*_\alpha \equiv M^*_{\alpha\beta}\phi^*_\beta = \phi^*_\beta(M^\dagger)_{\beta\alpha} \equiv (\Phi^* M^\dagger)_\alpha \ . \tag{73}$$

In our case, $M = \boldsymbol{\sigma} = \boldsymbol{\sigma}^\dagger$, $\alpha, \beta = 1, 2$, and therefore

$$(\boldsymbol{\sigma}\varphi)^*_\alpha = (\varphi^*\boldsymbol{\sigma})_\alpha \ . \tag{74}$$

Hence

$$-i\partial_t\varphi^*_{+\alpha} = i\boldsymbol{\nabla}\cdot(\varphi^*_+\boldsymbol{\sigma})_\alpha + m\varphi^*_{-\alpha} \ , \tag{75}$$

$$-i\partial_t\varphi^*_{-\alpha} = -i\boldsymbol{\nabla}\cdot(\varphi^*_-\boldsymbol{\sigma})_\alpha + m\varphi^*_{+\alpha} \ . \tag{76}$$

Therefore

$$\partial_t(\varphi_+^*\varphi_+) = -\nabla\cdot(\varphi_+^*\boldsymbol{\sigma}\varphi_+) + im(\varphi_-^*\varphi_+ - \text{c.c.}) \,, \tag{77}$$
$$\partial_t(\varphi_-^*\varphi_-) = \nabla\cdot(\varphi_-^*\boldsymbol{\sigma}\varphi_-) - im(\varphi_-^*\varphi_+ - \text{c.c.}) \,. \tag{78}$$

Here, and often in what follows, we use the shorthand

$$\Theta M\Phi \equiv \theta_\alpha M_{\alpha\beta}\phi_\beta \,, \tag{79}$$

where Θ and Φ are any n-component vectors. Then (77) and (78) can be combined into the continuity equation

$$\frac{\partial\rho}{\partial t} + \nabla\cdot\boldsymbol{j} = 0 \,, \tag{80}$$

with the density and current being

$$\rho = \varphi_+^*\varphi_+ + \varphi_-^*\varphi_- \,, \tag{81}$$
$$\boldsymbol{j} = \varphi_+^*\boldsymbol{\sigma}\varphi_+ - \varphi_-^*\boldsymbol{\sigma}\varphi_- \,. \tag{82}$$

The density ρ is positive definite, in conformity with Dirac's original stipulation.

This density and current have two remarkable properties. First, they do not contain any terms mixing the two Weyl spinors, which is of capital importance in the electromagnetic interaction of Dirac particles (and of even greater import in the weak interaction). Second, the current $\boldsymbol{j}$ has no spatial derivative, in contrast to the Schrödinger current. On second thought, the latter is not surprising because in a relativistic theory ρ and $\boldsymbol{j}$ must form a 4-vector, which is only possible if they have closely related forms.

As the Dirac equation is covariant, and the derivatives in the continuity equation form a 4-vector, it must be that

$$j^\mu = (\rho, \boldsymbol{j}) \tag{83}$$

is also a 4-vector, and that the continuity equation has the Lorentz invariant form

$$\partial_\mu j^\mu = 0 \,. \tag{84}$$

An explicit proof that $(\rho, \boldsymbol{j})$ forms a 4-vector is, nevertheless, instructive because it involves the important issue of how bilinear combinations of spinors behave under Lorentz transformations. First, since φ_+ and φ_- have the same behavior under rotations, the bilinear expressions $\varphi_a^*\varphi_b$, where $a, b = \pm$, are rotationally invariant for any and all combinations of a and b (where the shorthand of (79) is used again). But not all these combinations $\varphi_a^*\varphi_b$ are Lorentz invariant! This is a consequence of the non-unitary nature of the matrices $\Omega_\pm$:

$$\varphi_\pm \to \exp(\pm\tfrac{1}{2}\zeta\hat{\boldsymbol{n}}\cdot\boldsymbol{\sigma})\varphi_\pm, \quad \varphi_\pm^* \to \varphi_\pm^*\exp(\pm\tfrac{1}{2}\zeta\hat{\boldsymbol{n}}\cdot\boldsymbol{\sigma}) \,, \tag{85}$$

from which it follows that only the combinations

$$\varphi_+^*\varphi_-, \qquad \varphi_-^*\varphi_+ \,, \tag{86}$$

are Lorentz invariant. Neither is invariant under reflection because that interchanges φ_+ and φ_-. Hence the even and odd combinations S and P are, respectively, a scalar and pseudoscalar density:

$$S = \varphi_+^*\varphi_- + \varphi_-^*\varphi_+ \,, \tag{87}$$
$$P = \varphi_+^*\varphi_- - \varphi_-^*\varphi_+ \,. \tag{88}$$

Armed with this information we can now determine how the combinations that appear in the density ρ behave under Lorentz transformation:

$$\begin{aligned}\varphi_\pm^* \varphi_\pm &\rightarrow \varphi_\pm^* \exp(\pm\zeta \hat{\boldsymbol{n}} \cdot \boldsymbol{\sigma})\varphi_\pm \\ &= (\varphi_\pm^* \varphi_\pm) \cosh\zeta \pm [\hat{\boldsymbol{n}} \cdot (\varphi_\pm^* \boldsymbol{\sigma} \varphi_\pm)] \sinh\zeta \,. \end{aligned} \tag{89}$$

Therefore the density undergoes the Lorentz transformation

$$\rho \rightarrow \rho \cosh\zeta + \hat{\boldsymbol{n}} \cdot \boldsymbol{j} \sinh\zeta \,, \tag{90}$$

as required for the time component of a 4-vector.

The covariant-looking form for the Dirac equation is convenient, and appears often in the literature. A covariant-looking form for the 4-current is also useful. To construct it, note that the invariant S can be written as

$$S = \psi_\beta^* (\gamma^0)_{\beta\alpha} \psi_\alpha \equiv \psi^* \gamma^0 \psi \,, \tag{91}$$

where now $\alpha, \beta = 1, \ldots, 4$. This prompts the introduction of the so-called Pauli adjoint

$$\bar{\psi}_\alpha \equiv \psi_\beta^* (\gamma^0)_{\beta\alpha} \,, \tag{92}$$

in terms of which the invariant is simply

$$S = \bar{\psi}\psi \,. \tag{93}$$

Turning to j^μ, recall that $(\gamma^0)^2 = 1$, so that

$$\rho = \psi^* \psi = \psi^* \gamma^0 \gamma^0 \psi = \bar{\psi} \gamma^0 \psi \,; \tag{94}$$

furthermore

$$\gamma^0 \gamma^i = \begin{pmatrix} \sigma_i & 0 \\ 0 & -\sigma_i \end{pmatrix}, \tag{95}$$

and therefore

$$j^\mu = \bar{\psi} \gamma^\mu \psi \,, \tag{96}$$

which is the desired form that displays the 4-vector nature of the current.

13.3 Electromagnetic Interaction of a Dirac Particle

The theory of a free particle must pass tests posed by theory itself, such as Lorentz covariance, but experimental tests are contingent on interactions. In the microscopic realm the most pervasive and well understood interaction is that of electromagnetism. The next task, therefore, is to extend the Dirac equation to include that interaction.

(a) The Dirac Equation in the Presence of a Field

The natural thing to try first is the familiar recipe of classical electrodynamics[1] for including the electromagnetic interaction of a particle carrying the charge e. In terms of canonical momenta, this recipe is the substitution

$$p_\mu \longrightarrow p_\mu - \frac{e}{c} A_\mu \,, \tag{97}$$

[1] Cf., e.g., Jackson, §12.1.

where the 4-vector A_μ is composed of the scalar and vector potentials, $A_\mu \equiv (V, -\boldsymbol{A})$, which may, in general, be time dependent. In quantum mechanics, this becomes the recipe of Schrödinger:

$$i\partial_\mu \longrightarrow i\partial_\mu - eA_\mu \,, \tag{98}$$

where now $c = 1$ in addition to $\hbar = 1$. Assuming that this is correct implies that in the presence of an electromagnetic field the Dirac equation for a particle of spin $\frac{1}{2}$ and charge e reads

$$\gamma^\mu \left(i\frac{\partial}{\partial x^\mu} - eA_\mu \right) \psi = m\psi \,. \tag{99}$$

Equation (99) can be inferred from an argument which has had a remarkably sweeping success physics: the requirement of invariance under the *gauge transformation*

$$\psi \rightarrow e^{-ie\lambda(x)}\, \psi \,, \tag{100}$$

where $\lambda(x)$ is an arbitrary function. The identity

$$(i\partial_\mu - eA_\mu)e^{-ie\lambda}\psi = e^{-ie\lambda}(i\partial_\mu - eA_\mu + e\partial_\mu\lambda)\psi \tag{101}$$

implies that the Dirac equation (99) will be invariant if, in addition to the transformation (100), the potentials undergo the gauge transformation familiar from classical electrodynamics,

$$A_\mu \longrightarrow A_\mu + \partial_\mu\lambda \,. \tag{102}$$

The current of the free Dirac particle contains no derivatives. Hence, if the recipe (98) is correct, *in the Dirac theory the 4-current of a charged particle does not depend explicitly on the potentials,* in contrast to Schrödinger current whose spatial components depend explicitly on the vector potential.[1] There is, of course, an implicit dependence via the wave function on the fields to which the particle is exposed.

The covariant expressions are elegant, but tend to hide what is "really" going on. That is the case with Eq. 99, which becomes more revealing when written out in terms of the 2-component Weyl spinors. To that end, define

$$(P^0, \boldsymbol{P}) \equiv i\partial^\mu - eA^\mu = \left(i\partial_t - eV, \frac{1}{i}\boldsymbol{\nabla} - e\boldsymbol{A} \right) \,. \tag{103}$$

Then

$$(P^0 - \boldsymbol{P}\cdot\boldsymbol{\sigma})\varphi_+ = m\varphi_- \,, \tag{104}$$

$$(P^0 + \boldsymbol{P}\cdot\boldsymbol{\sigma})\varphi_- = m\varphi_+ \,, \tag{105}$$

which have the same structure as the free particle equations in momentum space, Eqs. (53) and (54).

If the energy of the phenomenon of interest is sufficiently high compared to the mass, the right-hand side of these equations can be dropped, and there is no

---

[1] It should be said that (98) is the simplest recipe consistent with gauge invariance, and is therefore called *minimal coupling*. Gauge invariance by itself permits an additional "Pauli" coupling which can be used to give the particle an arbitrary magnetic moment; see Prob. 2.

coupling between the two Weyl spinors $\varphi_\pm$, or put in physical terms, *the helicity is a constant of motion for an ultrarelativistic Dirac particle moving in an arbitrary electromagnetic field.*

To underscore the import of this conclusion, assume that there exists a Lorentz-scalar field $U(x)$ with which the particle can interact. Such a field transforms like the mass, and would be included by the substitution $m \to m + U$. In this hypothetical situation the interaction with the external field U would produce changes in helicity even if the mass can be neglected in comparison to the energy.

(b) The Magnetic Moment

In nonrelativistic quantum mechanics there is no restriction on the magnetic moment of a charged particle; it can have any g-factor. But the Dirac equation (99) has only two free parameters, m and e, and must, therefore, predict a value for g! That it does so correctly for the electron was shown by Dirac when he first proposed his equation.[1]

This conclusion is reached by constructing the Schrödinger equation as the nonrelativistic limit of Dirac's equation. Recall that for a particle at rest, there is no distinction between the two Weyl spinors. In the presence of a field a particle cannot remain at rest, but if the field is sufficiently weak, it can be in a state with momenta small compared to m. In this limit, therefore, it is to be expected that $\varphi_+ \approx \varphi_-$. For that reason, define the two-component spinors,

$$\tilde{\Psi} = \frac{1}{\sqrt{2}}(\varphi_+ + \varphi_-)\,, \quad \tilde{\Phi} = \frac{1}{\sqrt{2}}(\varphi_+ - \varphi_-)\,, \tag{106}$$

with the expectation that $\tilde{\Phi} \to 0$ in the nonrelativistic limit. The Dirac equation, without approximation, then reads

$$P^0\tilde{\Psi} - (\boldsymbol{P}\cdot\boldsymbol{\sigma})\tilde{\Phi} = m\tilde{\Psi}, \quad P^0\tilde{\Phi} - (\boldsymbol{P}\cdot\boldsymbol{\sigma})\tilde{\Psi} = -m\tilde{\Phi}\,. \tag{107}$$

In the nonrelativistic context energies are measured relative to m, and this is accomplished by defining new two-component spinors

$$\tilde{\Psi} = e^{-imt}\Psi, \quad \tilde{\Phi} = e^{-imt}\Phi\,, \tag{108}$$

in terms of which the Dirac equation is

$$P^0\Psi - (\boldsymbol{P}\cdot\boldsymbol{\sigma})\Phi = 0\,, \tag{109}$$

$$P^0\Phi - (\boldsymbol{P}\cdot\boldsymbol{\sigma})\Psi = -2m\Phi\,, \tag{110}$$

still without approximation.

The patient is now ready for amputation. The expression $P^0\Phi$ is of order (energy) $\times\,\Phi$, which by hypothesis is small compared to $m\Phi$, and therefore

$$\Phi \approx \frac{1}{2m}(\boldsymbol{P}\cdot\boldsymbol{\sigma})\Psi\,. \tag{111}$$

[1] At that time the magnetic moment of the proton was not known, and so it came as a big surprise when it was found to differ greatly from the Dirac value. Today the proton (and neutron) magnetic moments are understood in terms of their quark structure.

This shows how Φ "disappears" in the nonrelativistic limit:

$$\frac{|\Phi|}{|\Psi|} \sim \frac{\bar{p}}{m}, \quad m \to \infty, \tag{112}$$

where $\bar{p}$ is a momentum characteristic of the state. Eliminating Φ from (109) then yields

$$\left(P^0 - \frac{1}{2m}(\boldsymbol{P}\cdot\boldsymbol{\sigma})^2\right)\Psi = 0 . \tag{113}$$

But

$$\begin{aligned}(\boldsymbol{P}\cdot\boldsymbol{\sigma})^2 &= \boldsymbol{P}^2 + \tfrac{1}{2}i\epsilon_{ijk}\sigma_i[P^j, P^k] \\ &= \boldsymbol{P}^2 - \tfrac{1}{2}e\,\epsilon_{ijk}\sigma_i(\nabla^j A^k - \nabla^k A^j) \\ &= (\frac{1}{i}\boldsymbol{\nabla} - e\boldsymbol{A})^2 - e\boldsymbol{\sigma}\cdot\boldsymbol{B} ,\end{aligned} \tag{114}$$

where $\boldsymbol{B} = \boldsymbol{\nabla}\times\boldsymbol{A}$ is the magnetic field.

The non-relativistic limit of Dirac's equation is therefore

$$i\frac{\partial\Psi}{\partial t} = \left(\frac{1}{2m}(\frac{1}{i}\boldsymbol{\nabla} - e\boldsymbol{A})^2 - \frac{e}{2m}\boldsymbol{\sigma}\cdot\boldsymbol{B} + eV\right)\Psi . \tag{115}$$

This is the Schrödinger equation[1] for a particle of charge e and magnetic moment

$$\mu = \frac{e}{2m}, \tag{116}$$

or $\mu = e\hbar/2mc$ when $\hbar$ and c are restored. As promised, the g-factor is 2.

The electron's actually measured g-factor is not quite 2:

$$g_e - 2 \approx 0.00232 . \tag{117}$$

This discrepancy is known experimentally to far higher precision than indicated by (117) — to about 1 part per 10^8, and is fully accounted for by quantum electrodynamics. The same statement applies to the muon, to slightly lower accuracy. If the Dirac equation is used to describe neutrons or protons, it must be understood that the equation is then not a first, good approximation to the correct description, but a phenomenological equation that is only valid if the particle in question is subjected to probes having wavelengths long compared to its size, which is of order 10^{-13} cm for nucleons. In particular, the magnetic moment must be inserted "by hand," as is obvious because the neutron, were it really a Dirac particle, would have zero magnetic moment. How this is done in a covariant manner is outlined in Prob. 2.

The charge density and current associated with the nonrelativistic approximation can also be recovered from the Dirac 4-current. From (81), (106) and (111), the density is

$$\rho = |\varphi_+|^2 + |\varphi_-|^2 = |\Psi|^2 + |\Phi|^2 \tag{118}$$

$$\approx |\Psi|^2 + \frac{1}{4m^2}|\boldsymbol{\sigma}\cdot\boldsymbol{P}\Psi|^2 . \tag{119}$$

---

[1] To be historically accurate, this is the Pauli equation, and the two-component wave function that appears in it the Pauli spinor. To add emphasis to what is the nonrelativistic approximation to the Dirac theory we use the *appellation controlée Schrödinger* in this section.

The last correction term is of order $(\bar{p}/m)^2 \sim (v/c)^2$, and is negligible if the nonrelativistic approximation is valid. In terms of the spinors Ψ and Φ, the Dirac current (Eq. 82) is

$$\boldsymbol{j} = \Psi^* \boldsymbol{\sigma} \Phi + \Phi^* \boldsymbol{\sigma} \Psi \,. \tag{120}$$

Using the approximation (111) for Φ gives

$$\begin{aligned} j^k &= \frac{1}{2m}\left(\Psi^* \sigma_k \sigma_i P^i \Psi + (P^i \Psi)^* \sigma_i \sigma_k \Psi\right) \\ &= \frac{1}{2m}\left(\Psi^* P^k \Psi + (P^k \Psi)^* \Psi + \epsilon_{ikj}[\Psi^* \sigma_i \nabla^j \Psi + (\nabla^j \Psi)^* \sigma_i \Psi]\right) \,, \end{aligned} \tag{121}$$

and therefore

$$\boldsymbol{j} = \frac{1}{2m}[\Psi^* \boldsymbol{P} \Psi + (\boldsymbol{P}\Psi)^* \Psi] + \frac{1}{2m} \boldsymbol{\nabla} \times (\Psi^* \boldsymbol{\sigma} \Psi) \,. \tag{122}$$

Thus $e\boldsymbol{j}$ is the Schrödinger expression for the electromagnetic current of a particle with charge e and magnetic moment $e\boldsymbol{\sigma}/2m$.

(c) The Fine Structure Hamiltonian

The Schrödinger approximation to Dirac's equation, (115), is the leading term of a low-velocity expansion, and in the case of a hydrogenic atom it results in the familiar Bohr spectrum. The next term in this expansion is responsible for the leading-order fine structure, already evaluated in §5.3, from a Hamiltonian that shall now be derived. We shall not, however, assume that the external field is a Coulomb field, but take it to be an arbitrary *static* electromagnetic field. The characteristic small expansion parameter is $\bar{p}/m$, and this means that an expansion in inverse powers of the mass is required. As we shall see, this is not an entirely straightforward matter, and a powerful technique — the Foldy-Wouthuysen transformation — for carrying it out systematically exists.[1] The dominant terms can be found without such sophistication, however, and such a derivation is instructive quite aside from the importance of the result.

The first point to be noted is somewhat astonishing: the large component Ψ of the Dirac spinor cannot be the Schrödinger wave function after the small component Φ is eliminated from the Dirac equation! The reason is that the Schrödinger wave function ψ is normalized to one, whereas in the Dirac theory it is $|\Psi|^2 + |\Phi|^2$ which is normalized to one. That is, ψ must satisfy

$$\sum \int |\psi|^2 \, d^3r = \sum \int (|\Psi|^2 + |\Phi|^2) \, d^3r \approx \sum \int \left(|\Psi|^2 + \frac{1}{4m^2}\Psi^* \Lambda^2 \Psi\right) d^3r \,, \tag{123}$$

where the sum is over the two spin components, and Λ is the Hermitian operator

$$\Lambda = \boldsymbol{\sigma} \cdot \boldsymbol{P} = \boldsymbol{\sigma} \cdot (\boldsymbol{p} - e\boldsymbol{A}) \,, \quad \boldsymbol{p} = -i\boldsymbol{\nabla} \,. \tag{124}$$

Therefore the connection between the Dirac and Schrödinger spinors is

$$\psi \approx \left(1 + \frac{\Lambda^2}{8m^2}\right)\Psi, \quad \Psi \approx \left(1 - \frac{\Lambda^2}{8m^2}\right)\psi, \tag{125}$$

[1] See Bjorken and Drell, §4.

to order $(\bar{p}/m)^2$, because $\Lambda\psi$ has a magnitude of order $\bar{p}\psi$. (Identification of Ψ with the Schrödinger wave function in the preceding subsection was permissible because (100) is only valid to order $\bar{p}/m$.)

The approximate relation (111) between Φ and Ψ must be improved to the same order. To that end, set

$$\Phi = \frac{\Lambda}{2m}\Psi + \Phi' \tag{126}$$

in (110), and solve for Φ'. This produces

$$\Phi \approx \frac{1}{2m}\Lambda\Psi - \frac{1}{4m^2}P^0\Lambda\Psi \,. \tag{127}$$

The goal is to recast the Dirac equation (109) into the Schrödinger form

$$H_{\text{eff}}\psi = E\psi \,, \tag{128}$$

for that would identify the effective Hamiltonian. For that purpose, (109) is approximated by

$$P^0\Psi = \frac{1}{2m}\Lambda^2\Psi - \frac{1}{4m^2}\Lambda P^0\Lambda\Psi \,, \tag{129}$$

where terms of order $(\bar{p}/m)^3$ and higher have been dropped. Finally, the spinor Ψ is replaced by the Schrödinger wave function by use of (125) on the left-hand side and the leading term on the right-hand side; in the order $1/m^2$ term $\Psi \approx \psi$ is adequate, however, and therefore

$$P^0\left(1 - \frac{\Lambda^2}{8m^2}\right)\psi \approx \frac{\Lambda^2}{2m}\psi - \frac{\Lambda^4}{16m^3}\psi - \frac{1}{4m^2}\Lambda P^0\Lambda\psi \,. \tag{130}$$

Furthermore, in the static case we are considering, the replacement

$$P^0 \longrightarrow (E - eV) \tag{131}$$

can be made. As a consequence, the effective Hamiltonian can be written as

$$H_{\text{eff}} = H_0 + \delta H \,, \tag{132}$$

where

$$H_0 = \frac{1}{2m}\Lambda^2 + eV \tag{133}$$

is the nonrelativistic Hamiltonian of Eq. 113, and the sought-after correction to it is

$$\delta H = \frac{1}{8m^2}P^0\Lambda^2 - \frac{1}{16m^3}\Lambda^4 - \frac{1}{4m^2}\Lambda P^0\Lambda \,. \tag{134}$$

In this correction term, the eigenvalue E can be replaced by its nonrelativistic approximant E_{nr}, which satisfies

$$(E_{\text{nr}} - eV)\psi = \frac{1}{2m}\Lambda^2\psi \,. \tag{135}$$

To exploit this, P^0 must be moved to the right, which leads to the following expression for the correction to the Hamiltonian:

$$\delta H = -\frac{1}{8m^3}\Lambda^4 + \frac{e}{4m^2}\left(\Lambda[V,\Lambda] - \tfrac{1}{2}[V,\Lambda^2]\right) \,. \tag{136}$$

This expression is evaluated with the help of

$$\Lambda[V,\Lambda] - \tfrac{1}{2}[V,\Lambda^2] = \tfrac{1}{2}[\Lambda,[V,\Lambda]], \tag{137}$$

$$[V,\Lambda] = -i\boldsymbol{\sigma}\cdot\boldsymbol{E}, \tag{138}$$

where $\boldsymbol{E} = -\boldsymbol{\nabla}V$ is the electric field. Hence

$$\begin{aligned}\tfrac{1}{2}[\Lambda,[V,\Lambda]] &= -\tfrac{1}{2}i[\boldsymbol{\sigma}\cdot\boldsymbol{P},\boldsymbol{\sigma}\cdot\boldsymbol{E}] \\ &= -\tfrac{1}{2}\boldsymbol{\nabla}\cdot\boldsymbol{E} - \tfrac{1}{2}i\boldsymbol{\sigma}\cdot(\boldsymbol{\nabla}\times\boldsymbol{E}) - \boldsymbol{\sigma}\cdot(\boldsymbol{E}\times\boldsymbol{P}),\end{aligned} \tag{139}$$

because

$$\boldsymbol{\sigma}\cdot[(\boldsymbol{P}\times\boldsymbol{E}) - (\boldsymbol{E}\times\boldsymbol{P})] = -i\boldsymbol{\sigma}\cdot(\boldsymbol{\nabla}\times\boldsymbol{E}) - 2\boldsymbol{\sigma}\cdot(\boldsymbol{E}\times\boldsymbol{P}). \tag{140}$$

In a static situation, Maxwell's equations tell us that $\boldsymbol{\nabla}\times\boldsymbol{E} = 0$.

The final result for the order $(\bar{p}/m)^2$ correction to the Hamiltonian is therefore

$$\delta H = -\frac{1}{8m^3}(\boldsymbol{\sigma}\cdot\boldsymbol{P})^4 - \frac{e}{8m^2}[\boldsymbol{\nabla}\cdot\boldsymbol{E} + 2\boldsymbol{\sigma}\cdot(\boldsymbol{E}\times\boldsymbol{P})]. \tag{141}$$

The first term is just the $(v/c)^2$ correction to the energy:

$$\sqrt{\Lambda^2 + m^2} = m + \frac{1}{2m}\Lambda^2 - \frac{1}{8m^3}\Lambda^4 + \ldots\,. \tag{142}$$

The term proportional to $\boldsymbol{\nabla}\cdot\boldsymbol{E} = \rho_{\text{ext}}$ is the Darwin term; it is only nonzero within the source distribution ρ_{ext} for the externally applied electric field. The last term, in the case of a central electrostatic field, becomes the familiar spin-orbit coupling, because then $\boldsymbol{P} = \boldsymbol{p}$, and

$$\boldsymbol{E}\times\boldsymbol{p} = -\frac{dV}{dr}\hat{\boldsymbol{r}}\times\boldsymbol{p} = -\frac{1}{r}\frac{dV}{dr}\boldsymbol{L}\,. \tag{143}$$

The fine structure Hamiltonian in an atom is thus

$$\delta H = -\frac{1}{8m^3}\boldsymbol{p}^4 - \frac{Ze^2}{8m^2}\delta^3(\boldsymbol{r}) + \frac{e}{4m^2}\frac{1}{r}\frac{dV}{dr}\boldsymbol{\sigma}\cdot\boldsymbol{L}\,. \tag{144}$$

Here the nuclear charge distribution has been treated as if it were a point charge; this is very accurate in all but the heaviest electronic atoms, but not in μ-mesic atoms because their Bohr radii are 200 times smaller.

(d) Antiparticles and Charge Conjugation

A detailed treatment of how antiparticles arise from the theory is beyond the level of this volume, but a brief discussion is in order here. [1] This will be done by showing that the Dirac equation for a particle of charge e in a static electromagnetic field has solutions of negative energy $-E$ that are simply related to solutions of positive energy E but for the charge $-e$.

---

[1] This argument is due to Dirac, *The Principles of Quantum Mechanics,* Oxford University Press (1930), pp. 255–256.

The Dirac equation for a stationary state is obtained from (99) by assuming a periodic time dependence for the wave function:

$$\psi(\boldsymbol{r},t) = e^{-iEt}\,\psi(E,\boldsymbol{r}) \,, \tag{145}$$

i.e.,

$$[\gamma^0(E - eV) + \boldsymbol{\gamma}\cdot(i\boldsymbol{\nabla} + e\boldsymbol{A})]\psi(E) = m\psi(E) \,. \tag{146}$$

The wave function for a particle with the opposite charge, and energy E', is designated by $\hat{\psi}(E')$, and satisfies

$$[\gamma^0(E' + eV) + \boldsymbol{\gamma}\cdot(i\boldsymbol{\nabla} - e\boldsymbol{A})]\hat{\psi}(E') = m\hat{\psi}(E') \,. \tag{147}$$

Both E and E' are assumed to be physically reasonable eigenvalues, which in the relativistic context means non-negative, though smaller than m for bound states.

That these equations actually have solutions for arbitrarily large energies of either sign is clear because for applied fields with bounded variation in space, there are solutions of sufficiently high momentum to render the fields irrelevant, and in this regime the equations reduce to the Dirac equation for a free particle in momentum space, which has solutions with energies $\pm\sqrt{p^2 + m^2}$.

Our goal is to transform the equation for $\psi(E)$ into the equation for $\hat{\psi}(E)$, i.e., with the same eigenvalue. The existence of solutions with negative eigenvalues of E allows us to consider $E \to -E$ in (146), which would be impossible in a Schrödinger equation for any Hamiltonian with a lower bound to its spectrum. That does not yet put (146) into the form of (147), but further progress in the desired direction is made by also taking the complex conjugate, which yields

$$-[(\gamma^0)^*(E + eV) + \boldsymbol{\gamma}^*\cdot(i\boldsymbol{\nabla} - e\boldsymbol{A})]\psi^*(-E) = m\psi^* \,. \tag{148}$$

The minus sign on the left-hand side, and the complex conjugates of the Dirac matrices, must still be removed somehow. For that purpose, assume for a moment that there exists a 4×4 matrix Γ such that

$$\Gamma(\gamma^\mu)^*\Gamma^{-1} = -\gamma^\mu \,. \tag{149}$$

If that is the case, $\Gamma\psi^*(-E)$ *satisfies the Dirac equation for a particle of mass m but opposite charge $-e$, and with positive energy E*, i.e.,

$$\Gamma\psi^*(-E) = \hat{\psi}(E) \,. \tag{150}$$

This is the promised demonstration that the solutions for one sign of the charge and negative energy are the solutions for positive energy and the opposite charge.

It only remains to show that such a matrix Γ does exist. With our conventions, γ^2 is pure imaginary, and the other Dirac matrices are real. Hence what is needed is[1]

$$\Gamma\gamma^2 = \gamma^2\Gamma, \quad \Gamma\gamma^{0,1,3} = -\gamma^{0,1,3}\Gamma \,. \tag{151}$$

A short calculation shows that

$$\gamma^\mu\gamma^\nu = -\gamma^\nu\gamma^\mu \,, \qquad \mu \neq \nu \,; \tag{152}$$

[1]Beware that Γ is not quite the matrix C that appears in the treatment of charge conjugation in most texts (e.g., Bjorken and Drell, §5.2).

hence apart from a constant factor $\Gamma = \gamma^2$.

The symmetry of the Dirac equation that has just been established is called *charge conjugation.* For a charged spin $\frac{1}{2}$ particle like the electron, this symmetry states that the theory is invariant under the interchange of electrons and positrons, with momenta and helicities unchanged.[2] This symmetry continues to hold in quantum electrodynamics, that is, when the electromagnetic field is itself quantized, whereas the field was treated as a specified numerical function in the foregoing. Charge conjugation is not, however, a universal symmetry. It is violated dramatically by the weak interaction, but it is a symmetry of the strong interaction.

13.4 Scattering of Ultra-Relativistic Electrons

As a second application of the Dirac equation, consider high-energy scattering of electrons or muons by an electrostatic field V, where by high is meant an incident momentum large compared to the mass m. As electrons are so very light, this ultra-relativistic (UR) limit is readily reached, and applies to many phenomena involving them. By today's standards, even muons of modest energy are in the UR regime. To the extent that nucleons can be described as a collection of free quarks, in multi-GeV collisions involving nucleons, quarks can also be treated as zero mass Dirac particles. What follows therefore describes, to varying degrees of accuracy, a far wider set of situations than one might suspect at first sight. The case of momenta comparable to m poses no basic difficulties but requires rather more complicated calculations and offers too little insight in comparison to the effort entailed to warrant a treatment here.

When the mass is negligible, the Dirac equations (104) and (105) decouple into equations for each of the two helicity states:

$$(i\partial_t - eV - \boldsymbol{\sigma}\cdot\boldsymbol{p})\varphi_+ = 0 \, , \tag{153}$$

$$(i\partial_t - eV + \boldsymbol{\sigma}\cdot\boldsymbol{p})\varphi_- = 0 \, , \tag{154}$$

where $\varphi_\pm$ are the two-component Weyl spinors and $\boldsymbol{p} = -i\nabla$. The equation for φ_+ turns into that for φ_- under space reflection, and therefore it suffices to consider one of them, say, that for the positive helicity state φ_+.

We are interested in a stationary state of energy k with a particle of momentum $\boldsymbol{k} = k\hat{\boldsymbol{z}}$ incident along the z, and therefore put

$$\varphi_+(\boldsymbol{r},t) = e^{-ikt}\,\phi_+(\boldsymbol{r}) \, . \tag{155}$$

Then ϕ_+ satisfies

$$(k - eV - \boldsymbol{\sigma}\cdot\boldsymbol{p})\phi_+ = 0 \, . \tag{156}$$

Multiplying by $(k + \boldsymbol{\sigma}\cdot\boldsymbol{p})$ gives

$$(k^2 + \nabla^2)\phi_+ = (k + \boldsymbol{\sigma}\cdot\boldsymbol{p})eV\phi_+ \, . \tag{157}$$

This equation has the same form as the Schrödinger equation in nonrelativistic (NR) potential scattering, and can, therefore, be converted into an integral equation that

[2] Hence charge conjugation should not be confused with time reversal, even though it involves a complex conjugation. Under time reversal momenta are reversed, but charge is unaffected.

specifies the boundary condition by the technique of §8.2(a):

$$\phi_+(\boldsymbol{r}) = \phi_{\rm inc}(\boldsymbol{r}) - \frac{e}{4\pi}\int \frac{e^{ikR}}{R}(k + \boldsymbol{\sigma}\cdot\boldsymbol{p}')V(\boldsymbol{r}')\phi_+(\boldsymbol{r}')\,d^3r' \,, \tag{158}$$

where $R = |\boldsymbol{r} - \boldsymbol{r}'|$. The incident state of helicity $+\frac{1}{2}$ is

$$\phi_{\rm inc} = e^{ikz}\chi_+(\boldsymbol{k}) \,. \tag{159}$$

The last factor is the momentum space spinor

$$\chi_+(\boldsymbol{k}) = Ne^{\frac{1}{2}\zeta\sigma_z}\xi_\uparrow \,, \qquad \sinh\zeta = k/m \,, \tag{160}$$

where $\xi_\uparrow$ is the Pauli spinor for spin up along z in the rest frame, and N is a normalizing factor.

As $r \to \infty$, the Green's function has the asymptotic form

$$\frac{e^{ikR}}{R} \sim \frac{e^{ikr}}{r}\exp(-i\boldsymbol{k}'\cdot\boldsymbol{r}') \,, \tag{161}$$

where $\boldsymbol{k}' = k\hat{\boldsymbol{r}}$ is the scattered momentum; by energy conservation $|\boldsymbol{k}'| = k$. The asymptotic spinor is thus

$$\phi_+(\boldsymbol{r}) \sim e^{ikz}\chi_+(\boldsymbol{k}) + \frac{e^{ikr}}{r}F^+(\boldsymbol{k}', \boldsymbol{k}) \,, \tag{162}$$

where

$$F^+_\alpha = -\frac{e}{4\pi}\sum_{\beta=1}^{2}\int d^3r\, e^{-i\boldsymbol{k}'\cdot\boldsymbol{r}}(k + \boldsymbol{\sigma}\cdot\boldsymbol{p})_{\alpha\beta}V(\boldsymbol{r})\phi_{+\beta}(\boldsymbol{r}) \,, \tag{163}$$

and the fact that F^+ is a spinor has been made explicit. On integrating by parts,

$$F^+ = -\frac{e}{4\pi}(k + \boldsymbol{\sigma}\cdot\boldsymbol{k}')\int d^3r\, e^{-i\boldsymbol{k}'\cdot\boldsymbol{r}}\, V(\boldsymbol{r})\phi_+(\boldsymbol{r}) \,. \tag{164}$$

The scattering amplitude is, as always, the coefficient of e^{ikr}/r. That for helicity $\frac{1}{2}$ is F^+, while that for helicity $-\frac{1}{2}$ it is

$$F^- = -\frac{e}{4\pi}(k - \boldsymbol{\sigma}\cdot\boldsymbol{k}')\int d^3r\, e^{-i\boldsymbol{k}'\cdot\boldsymbol{r}}\, V(\boldsymbol{r})\phi_-(\boldsymbol{r}) \,. \tag{165}$$

We seek the amplitude for finding a scattered particle with momentum $\boldsymbol{k}'$ and helicity λ. This amplitude is the scalar product of F^+ with the momentum space spinor $\chi_\pm(\boldsymbol{k}')$ representing the observed scattered state, where $\pm$ stands for $\lambda = \pm\frac{1}{2}$. According to the Weyl equation (55), the first factor in F^+ is the projection operator for helicity $+\frac{1}{2}$ for an ultrarelativistic particle moving along $\boldsymbol{k}'$:

$$\chi_{+\beta}(\boldsymbol{k}')^*(k + \boldsymbol{\sigma}\cdot\boldsymbol{k}')_{\beta\alpha} = 2k\chi_{+\alpha}(\boldsymbol{k}')^* \tag{166}$$

$$\chi_{-\beta}(\boldsymbol{k}')^*(k + \boldsymbol{\sigma}\cdot\boldsymbol{k}')_{\beta\alpha} = 0 \,. \tag{167}$$

This, then, is a reminder that in §13.3(a) we learned that helicity is conserved in an ultrarelativistic collision from an arbitrary electromagnetic field, and in particular, therefore, from the electrostatic field V. This conclusion does not hinge on perturbation theory; apart from dropping the mass terms from the Dirac equation, no approximations have been made.

The amplitude for ultrarelativistic scattering from the potential V is thus

$$f(\boldsymbol{k}',\boldsymbol{k}) = -\frac{ek}{2\pi}\int d^3r\, e^{-i\boldsymbol{k}'\cdot\boldsymbol{r}}\, V(\boldsymbol{r})\, \left(\chi_+(\boldsymbol{k}'), \phi_+(\boldsymbol{r})\right) , \tag{168}$$

where the scalar products between two spinors χ and ϕ is defined as

$$(\chi,\phi) \equiv \sum_{\alpha=1}^{2} \chi_\alpha^* \phi_\alpha . \tag{169}$$

The amplitude is written purposely as f, and not f^+, because *from reflection invariance we already know that the amplitudes for scattering of the two helicity states must be equal.*

To go further one must either solve the integral equation (158) or make some approximation. We confine ourselves to the Born approximation, and in (168) put

$$\phi_+ \approx e^{ikz}\chi_+(\boldsymbol{k}) . \tag{170}$$

Then

$$f(k\theta) = -\frac{k}{2\pi} h(k\theta)\Phi(\boldsymbol{q}) , \tag{171}$$

where

$$h(k\theta) = \left(\chi_+(\boldsymbol{k}'), \chi_+(\boldsymbol{k})\right) \tag{172}$$

$$\Phi(\boldsymbol{q}) = e\int d^3r\, e^{i\boldsymbol{q}\cdot\boldsymbol{r}}\, V(\boldsymbol{r}) , \tag{173}$$

and $\boldsymbol{q} = \boldsymbol{k} - \boldsymbol{k}'$ is the momentum transfer, and $q = 2k\sin\frac{1}{2}\theta$. In the case of a Coulomb field due to a point charge Ze

$$\Phi(\boldsymbol{q}) = -\frac{4\pi Z\alpha}{q^2} . \tag{174}$$

The quantity Φ, aside from a factor, is the Born approximation amplitude in the nonrelativistic limit. The novel features in the UR amplitude f are the factor of k coming from the spin-projection operator, and the factor h involving the spinors for the incident and scattered states.

To calculate h we need

$$\chi_+(k) = Ne^{\frac{1}{2}\zeta\sigma_z}\xi_\uparrow = Ne^{\frac{1}{2}\zeta}\xi_\uparrow . \tag{175}$$

If we choose to normalize the spinor to unity, $N = e^{-\frac{1}{2}\zeta}$. The spinor for the scattered electron is obtained by rotating the incident spinor through the scattering angle θ about any axis perpendicular to z, say, y:

$$\chi_+(\boldsymbol{k}') = e^{-\frac{1}{2}i\theta\sigma_y}\chi_+(\boldsymbol{k}) = e^{-\frac{1}{2}i\theta\sigma_y}\xi_\uparrow . \tag{176}$$

Therefore

$$h = \left(\xi_\uparrow, e^{\frac{1}{2}i\theta\sigma_y}\xi_\uparrow\right) = \cos\tfrac{1}{2}\theta . \tag{177}$$

The differential cross section is then $|f^2|$. Collecting the various factors yields the cross section for high energy scattering in either helicity state by a Coulomb field,

$$\frac{d\sigma}{d\Omega} = \frac{(Z\alpha)^2 \cos^2 \frac{1}{2}\theta}{4k^2 \sin^4 \frac{1}{2}\theta} \qquad (k \gg m) \ . \tag{178}$$

The cross section for all values of k, when the incident beam has no spin polarization and there is no spin detection in the scattered state, is given by the Mott formula:[1]

$$\frac{d\sigma}{d\Omega} = \frac{(Z\alpha)^2(1 - v^2 \sin^2 \frac{1}{2}\theta)}{4k^2 v^2 \sin^4 \frac{1}{2}\theta} \ , \tag{179}$$

where v is the velocity (in units of c); this reduces to the Rutherford cross section when $v \ll 1$, i.e., when $v = k/m$. In comparison to the Rutherford cross section, the UR cross section has a faster decrease with angle, and at fixed angle a slower fall-off with momentum, because the Rutherford factor m^2/k^4 is replaced by $k^{-2}v^{-2} \to k^{-2}$.

That the spin dependence of scattering by an electrostatic field has a rather complicated energy dependence is now evident. For $v \ll 1$, the spin is oblivious to the interaction; for $v \to 1$ helicity is conserved. This situation is best described in terms of the 2×2 matrix $\mathcal{M}$ that keeps track of all possible transitions in the spin space, as it was used in §8.5. In the UR regime, $\mathcal{M}$ is the unit matrix times the expression (171) *provided* the helicity basis is used for both the initial and final states, i.e., spin projections along the incident and final momentum, respectively. In the NR regime, by contrast, $\mathcal{M}$ is the unit matrix times the Rutherford amplitude *provided* spin states along the *same* direction for both the initial and final states are used to define the basis for $\mathcal{M}$. If the helicity basis is used in the NR regime, $\mathcal{M}$ must have off-diagonal terms to account for the amount of helicity flip required to maintain the incident spin direction. For velocities at neither extreme, $\mathcal{M}$ is not diagonal in either basis.[1]

In closing it should be said that the spin dependence in the relativistic regime reflects the fact that there is a magnetic field in the electron's rest frame, the same effect as leads to spin-orbit coupling. As $v \to 1$ this magnetic effect is of the same magnitude as that due to the electric field.

13.5 Bound States in a Coulomb Field

The relativistic bound state problem has important applications in atomic and nuclear physics. In hydrogenic atoms, the relativistic corrections to the energy eigenvalues are of order $Z^2\alpha^2$, and do not depend on the mass of the bound particle. They are small, therefore, as long as $Z \ll 137$. Spectroscopy yields results of extremely high accuracy, however, and therefore provides incisive tests of relativistic quantum

---

[1] This general case requires a rather more complicated calculation than the UR or NR limits. For the derivation see Bjorken and Drell, §7.1.

[1] A complete treatment of this spin dependence for all values of v is in Berestetskii et al., p. 320.

mechanics even in hydrogen, positronium, and muonium, the bound $e^-\mu^+$ system. In heavy atoms, phenomena involving K-shell electrons such as X-ray spectra, internal conversion and the photoelectric effect have substantial relativistic effects which must be incorporated by use of the Dirac equation.

According to the Dirac equation, the bound-state energy eigenvalues in the Coulomb field of a point charge Ze are[1]

$$E_{nj} = m\left(1+\frac{Z^2\alpha^2}{\left[n-(j+\frac{1}{2})+\sqrt{(j+\frac{1}{2})^2-Z^2\alpha^2}\,\right]^2}\right)^{-\frac{1}{2}}, \tag{180}$$

where n is the principle quantum number and j the total angular momentum. This formula is remarkable in that it does not depend on the orbital angular momentum, and therefore certain pairs of levels for a given n, such as $2s_{1/2}$ and $2p_{1/2}$, and $3p_{3/2}$ and $3d_{3/2}$, remain degenerate to all orders in $Z^2\alpha^2 \sim (v/c)^2$ even though no obvious symmetry requires this.[2] The lowest-order fine structure, which emerges from (180) by expanding in powers of $Z^2\alpha^2$, is

$$E = m - m\frac{Z^2\alpha^2}{2n^2}\left[1+\frac{Z^2\alpha^2}{n}\left(\frac{1}{j+\frac{1}{2}}-\frac{3}{4n}\right)+O(Z^4\alpha^4)\right] \tag{181}$$

and was already derived in §5.3; as stated there, the observed spectrum shows that there is a small splitting (the Lamb shift) between these degenerate pairs, which is completely explained by quantum electrodynamics.

It is natural to label states in the same manner as in the nonrelativistic situation, and for that reason to use the basis in which one two-component spinor reduces to the Schrödinger wave function in that limit. This is the basis of Eq. 106. Setting[1]

$$\tilde{\Psi}(\boldsymbol{r},t) = e^{-iEt}\psi(\boldsymbol{r})\ , \quad \tilde{\Phi}(\boldsymbol{r},t) = e^{-iEt}\phi(\boldsymbol{r})\ , \tag{182}$$

and the vector potential to zero, puts the Dirac equation (107) into the form

$$(E-m-eV)\psi = (\boldsymbol{p}\cdot\boldsymbol{\sigma})\phi\,, \tag{183}$$

$$(E+m-eV)\phi = (\boldsymbol{p}\cdot\boldsymbol{\sigma})\psi\,, \tag{184}$$

where $\boldsymbol{p} = -i\nabla$. For a central field V, these equations are invariant under simultaneous rotation of the coordinate and spin, but not when only one or the other is rotated alone. [3] Therefore the total angular momentum,

$$\boldsymbol{J} = \boldsymbol{L} + \tfrac{1}{2}\boldsymbol{\sigma}\,, \tag{185}$$

---

[1]This very formula was first derived by Sommerfeld in 1916 from the "old" quantum theory of Bohr by treating the classical problem relativistically before imposing the Bohr-Sommerfeld quantization condition. Spin had not yet been invented, so it is puzzling that Sommerfeld found the correct result for the Dirac and not the Klein-Gordon equation. This issue is analyzed in L.C. Biedenharn, *Found. of Phys.* **13**, 13 (1983).

[2]The Lenz vector is not a constant of motion in the relativistic situation, but there is a conserved operator in addition to angular momentum that has no classical analogue and which accounts for the degeneracy; see L.C. Biedenharn, *Phys. Rev.* **126**, 845 (1962); P.C. Martin and R.J. Glauber, *ibid.* **109**, 1307 (1958).

[1]Beware that in what follows ψ is not the Schrödinger wave function used in connection with the approximate treatment of fine structure in §13.3(a); it is defined by the exact Dirac equation.

[3]This statement in English requires no computational confirmation, as many treatments seem to suggest. As shown in §13.2 , the Weyl spinors, and hence their relatives such as ψ and ϕ, have the same transformation law under rotations as wave functions for a spin $\frac{1}{2}$ particle in nonrelativistic quantum mechanics.

is a constant of motion, but the orbital angular momentum $\boldsymbol{L}$ is not. As for space reflection, $\boldsymbol{p}\cdot\boldsymbol{\sigma}$ is a pseudoscalar, and therefore ψ and ϕ must have opposite signature under I_s. This conforms with the the fact that the original Weyl spinors $\varphi_\pm$ are interchanged under reflection. Hence

$$I_s: \qquad \psi \to \eta\psi, \quad \phi \to -\eta\phi\,, \tag{186}$$

where η is the parity of the so-called "large" spinor ψ — the one that survives if relativistic effects are ignored.

The preceding conclusions regarding angular momentum and parity suffice to determine the angular and spin dependence of both ψ and ϕ, leaving the Dirac equation to only determine their radial dependence. This is so because, for a single spin $\frac{1}{2}$ particle in a central field, once j and m are specified there are just two 2-component spinors with these quantum numbers, having $l = j \pm \frac{1}{2}$ and parity $(-1)^l$. If one of these is chosen to be the "large" spinor ψ, the other must be the "small" one ϕ. In view of this we *define* the "large" spinor ψ to have the conventional angular momentum quantum numbers (jm) and parity η of the state in question, which then fixes l via $\eta = (-1)^l$. These quantum numbers are then used to *designate* the complete 4-component spinor. This definition does not involve any approximation. Exact solutions of the Dirac equation will be found, i.e., valid when the adjectives "large" and "small" are not justified.

With this understanding, the components of ψ are thus

$$\begin{aligned}\psi_{nljm}(\boldsymbol{r}, s) &= \langle \boldsymbol{r}s|nljm\rangle = \sum_{m_l m_s} \langle \boldsymbol{r}s|nlm_l m_s\rangle\langle lm_l \tfrac{1}{2}m_s|jm\rangle \\ &\equiv \frac{i}{r}F_{nlj}(r)\langle l\,m{-}s\;\tfrac{1}{2}s|jm\rangle Y_{l\,m-s}(\Omega)\,,\end{aligned} \tag{187}$$

where $s = \pm\frac{1}{2}$, $Y_{lm}(\Omega)$ is a spherical harmonic with Ω standing for the angles (ϑ, φ), and F the radial wave function that remains to be found. This spinor, as a whole, has the form

$$\psi_{nljm} = \frac{i}{r}F_{nlj}(r)\; u_{ljm}(\Omega)\,, \tag{188}$$

with

$$u_{ljm} = \begin{pmatrix} \langle l\,m{-}\frac{1}{2}\;\frac{1}{2}\frac{1}{2}|jm\rangle\; Y_{l\,m-\frac{1}{2}} \\ \langle l\,m{+}\frac{1}{2}\;\frac{1}{2}{-}\frac{1}{2}|jm\rangle\; Y_{l\,m+\frac{1}{2}} \end{pmatrix}. \tag{189}$$

The "small" spinor ϕ has the same total angular momentum quantum numbers (jm) as its partner ψ, but the opposite parity. In terms of the conventional spectroscopic notation, this gives the following designations for the spinors:

$$\begin{array}{ccccccc} \psi & s_{\frac{1}{2}} & p_{\frac{1}{2}} & p_{\frac{3}{2}} & d_{\frac{3}{2}} & d_{\frac{5}{2}} & \cdots \\ \phi & p_{\frac{1}{2}} & s_{\frac{1}{2}} & d_{\frac{3}{2}} & p_{\frac{3}{2}} & f_{\frac{5}{2}} & \cdots \\ \kappa & -1 & 1 & -2 & 2 & -3 & \cdots \end{array} \tag{190}$$

The important quantum number κ will be defined shortly.

The spinor ϕ could be constructed in the same manner as ψ, but a more painless route is provided by the *Ansatz*

$$\phi_{nljm}(\boldsymbol{r}) = \frac{1}{r}G_{nlj}(r)\;(\hat{\boldsymbol{r}}\cdot\boldsymbol{\sigma})u_{ljm}(\Omega)\,. \tag{191}$$

The operator $\hat{\boldsymbol{r}}\cdot\boldsymbol{\sigma}$ is a pseudoscalar, which guarantees that ϕ has the same total angular momentum as ψ but opposite parity. Although $\hat{\boldsymbol{r}}\cdot\boldsymbol{\sigma}$ does not commute with $\boldsymbol{L}$, for a spin $\frac{1}{2}$ particle total angular momentum and parity define a unique value of l, so ϕ is an eigenstate of $\boldsymbol{L}^2$ with its eigenvalue given by (190).

Differential equations for the radial functions F and G can now be found. For this purpose, the following identity, valid for any function $f(r)$, comes in handy:

$$\begin{aligned}(\boldsymbol{\sigma}\cdot\boldsymbol{p})f(r)u(\Omega) = (\boldsymbol{\sigma}\cdot\hat{\mathbf{r}})^2(\boldsymbol{\sigma}\cdot\boldsymbol{p})fu &= (\boldsymbol{\sigma}\cdot\hat{\mathbf{r}})[\hat{\boldsymbol{r}}\cdot\boldsymbol{p}+i\boldsymbol{\sigma}\cdot\hat{\boldsymbol{r}}\times\boldsymbol{p}]fu \\ &= (\boldsymbol{\sigma}\cdot\hat{\mathbf{r}})\left(\frac{1}{i}\frac{\partial}{\partial r}+\frac{i}{r}\boldsymbol{\sigma}\cdot\boldsymbol{L}\right)f(r)u(\Omega)\,. \qquad (192)\end{aligned}$$

Now

$$\boldsymbol{\sigma}\cdot\boldsymbol{L} = j(j+1)-l(l+1)-\tfrac{3}{4} \equiv -(\kappa+1)\,, \qquad (193)$$

where

$$\kappa = \mp(j+\tfrac{1}{2}), \quad l = j\mp\tfrac{1}{2}\,. \qquad (194)$$

Hence, for a given j, wave functions of opposite parity have opposite κ, and for that reason, ψ and ϕ have opposite κ, where κ is the value appropriate to the quantum numbers of ψ, as listed in (190). The identity (192) thus leads to

$$(\boldsymbol{\sigma}\cdot\boldsymbol{p})\psi = (\boldsymbol{\sigma}\cdot\boldsymbol{r})u(\Omega)\left(\frac{d}{dr}+\frac{1+\kappa}{r}\right)\frac{F}{r}\,, \qquad (195)$$

$$(\boldsymbol{\sigma}\cdot\boldsymbol{p})\phi = -iu(\Omega)\left(\frac{d}{dr}+\frac{1-\kappa}{r}\right)\frac{G}{r}\,. \qquad (196)$$

The ordinary differential equations that determine F and G, and the energy spectrum, are therefore

$$(E-m-eV)F = -\frac{dG}{dr}+\frac{\kappa}{r}G\,, \qquad (197)$$

$$(E+m-eV)G = \frac{dF}{dr}+\frac{\kappa}{r}F\,. \qquad (198)$$

Observe that the only quantum number aside from the energy that appears here is κ. This does not explain the degeneracy in the Coulomb case, however, because the preceding statement holds for any central field.

These coupled first-order equations can be turned into uncoupled second-order equations of the hypergeometric type. The solutions are a decaying exponential multiplied by a confluent hypergeometric function, and the latter must be a polynomial if the solution is to be normalizable. This requirement produces the eigenvalues (Eq. 180).

Only a cursory discussion of how wave functions and eigenvalues can be found will be given here.[1] Consider, first, the asymptotic form for a field that drops off

---

[1]The solution of these equations is given in detail by Berestetskii et al., §36. A good deal of information about the hydrogenic bound states, and references to yet more complete treatments, is to be found in Bethe and Salpeter, §14, and M.E. Rose, *Relativistic Electron Theory,* Wiley (1961). All these authors also discuss solutions of the Dirac equation for scattering in a Coulomb field.

more rapidly than a Coulomb field, i.e., $\lim_{r\to\infty} rV = 0$. The equations for large r are then

$$(E-m)F = -G', \quad (E+m)G = F' , \tag{199}$$

and therefore the asymptotic forms are

$$F \sim Ne^{-\lambda r}, \quad G \sim -N\sqrt{\frac{m-E}{m+E}}e^{-\lambda r}, \quad \lambda = \sqrt{m^2 - E^2} , \tag{200}$$

with N the normalization constant. Note that λ satisfies $E = [m^2 + (i\lambda)^2]^{\frac{1}{2}}$, as required by the Klein-Gordon equation for the momentum $i\lambda$. This is the analogue of the nonrelativistic expression $E = (i\lambda)^2/2m$. This relativistic asymptotic form reduces to the nonrelativistic one when $|E-m| \ll m$. Also noteworthy is the ratio

$$\frac{|G|^2}{|F|^2} = \frac{m-E}{m+E} \sim \frac{B}{2m} , \tag{201}$$

where the last expression is for small binding energy B, and shows explicitly how "small" ϕ is in a nonrelativistic bound state. This order-of-magnitude expression is consistent with (112), by the way, because $B \sim \bar{p}^2/m$.

For a Coulomb field, $eV = -Z\alpha/r$, and the wave functions fall off far more slowly as $r \to \infty$, as one would suspect from the behavior of the nonrelativistic wave functions. This is exemplified by the ground state, $1s_{1/2}$. Because this state has no radial nodes,[1] it is easily determined by setting

$$F = Nr^{\gamma}e^{-\lambda r}, \quad G = N'F , \tag{202}$$

with the power γ to be determined by the differential equations (197) and (198) when $\kappa = -1$. Equating powers of r then yields

$$N'/N = -\sqrt{(m-E)/(m+E)} , \tag{203}$$

as in (200), and

$$\lambda(\gamma+1) = Z\alpha(E+m), \quad (\gamma-1)(E+m) = -Z\alpha\lambda . \tag{204}$$

Hence

$$\lambda = mZ\alpha = Z/a_0 , \tag{205}$$

where a_0 is the Bohr radius; the ground-state energy is

$$E_0 = m\sqrt{1 - Z^2\alpha^2} , \tag{206}$$

in agreement with (180), and the power γ is

$$\gamma = \sqrt{1 - Z^2\alpha^2} . \tag{207}$$

The latter reduces to the nonrelativistic limit $\gamma = 1$ when $Z^2\alpha^2 \ll 1$. That is to say, whereas the nonrelativistic $1s$ state has no singularity at $r = 0$, the relativistic state has the weak, integrable, singularity

$$r^{\gamma-1} \approx r^{Z^2\alpha^2} . \tag{208}$$

---

[1]This simple technique yields the wave functions and eigenvalues for all nodeless states, those of the highest j for a given value of n. See Prob. 3.

Finally, combining this result with (188), leads to the normalized $1s_{1/2}$ wave function

$$\psi(1s_{\frac{1}{2}};\boldsymbol{r}) = \frac{1}{\sqrt{4\pi}}\left(\frac{2Z}{a_0}\right)^{3/2}\sqrt{\frac{1+\gamma}{2\Gamma(1+2\gamma)}}\,(2Zr/a_0)^{\gamma-1}\,e^{-Zr/a_0}\,\xi\,, \tag{209}$$

where $\Gamma(z)$ is the gamma function and ξ a two component Pauli spinor. The other two components of the Dirac wave function for this state are then found from (191). If (209) is written as $g(r)\xi$, then

$$\phi_\uparrow(1s;\boldsymbol{r}) = g(r)\frac{1-\gamma}{Z\alpha}\begin{pmatrix}\cos\vartheta \\ e^{i\varphi}\sin\vartheta\end{pmatrix}, \tag{210}$$

$$\phi_\downarrow(1s;\boldsymbol{r}) = g(r)\frac{1-\gamma}{Z\alpha}\begin{pmatrix}e^{-i\varphi}\sin\vartheta \\ -\cos\vartheta\end{pmatrix}. \tag{211}$$

13.6 Problems

1. The analogy between Maxwell's and Dirac's equations is briefly mentioned in §13.2, and is explored here at greater length. At a descriptive level, note that in both cases, the first-order equations specify the behavior of the fields ($\boldsymbol{E}$ and $\boldsymbol{B}$, or ψ) under rotations and reflections, and also their orientation (i.e., polarization or spin), whereas the second-order (wave or Klein-Gordon) equations do neither, admitting fields of any rank under rotations and any character under reflection, and only give the relation between energy and momentum, i.e., whether it is $\omega = k$ or $\omega^2 = k^2 + m^2$. To go beyond this, define the column vector

$$\mathsf{F}^\pm = \begin{pmatrix}F_x^\pm \\ F_y^\pm \\ F_z^\pm\end{pmatrix}, \qquad \boldsymbol{F}^\pm = \boldsymbol{E} \pm i\boldsymbol{B}\,.$$

Show that Maxwell's equations can be written in the form

$$(\partial_t \pm \boldsymbol{S}\cdot\boldsymbol{\nabla})\mathsf{F}^\pm = 0\,,$$

where $\boldsymbol{S}$ is the 3×3 angular momentum matrix in the Cartesian basis. Show that for a plane wave of amplitude $\phi_\pm$ propagating along $\boldsymbol{n}$ this reduces to

$$(1 \mp \boldsymbol{S}\cdot\boldsymbol{n})\phi_\pm = 0\,,$$

and give the physical interpretation thereof.

2. The Dirac equation for a particle in an electromagnetic field (Eq. 99) can be modified by adding a so-called Pauli term,

$$\left(i\gamma^\mu\partial_\mu - e\gamma^\mu A_\mu - \frac{e\kappa}{4m}\sigma_{\mu\nu}F^{\mu\nu} - m\right)\psi = 0\,,$$

where

$$F^{\mu\nu} = \partial^\mu A^\nu - \partial^\nu A^\mu\,, \qquad \sigma_{\mu\nu} = \tfrac{1}{2}i(\gamma^\mu\gamma^\nu - \gamma^\nu\gamma^\mu)\,.$$

Show first that this equation is also covariant and gauge invariant. More importantly, use the technique that led to Eq. 115 shows that this modified equation ascribes a so-called anomalous magnetic moment to the electron, i.e., $eg/4m$ where the g-factor is not 2, as in the unblemished Dirac equation (cf. Eq. 116), but $2+\kappa$.

3. In nonrelativistic quantum mechanics, the lowest-lying bound-state radial wave function for a given value of l is nodeless. While it is not obvious what the counterpart of this statement is for the Dirac equation, it is at least plausible to guess that it holds for the lowest-lying state of a given j. Show that this is correct; that for these states the ratio F/G is still given by (203), that the power defined by (202) is $\gamma = \sqrt{n^2 - Z^2\alpha^2}$, and that, in agreement with (180), the energy eigenvalues are

$$E = m\sqrt{1 - (Z^2\alpha^2/n^2)}\,, \qquad (j + \tfrac{1}{2} = n)\,.$$

4. Verify the expressions in the text for "small" components $\phi(1s_{\frac{1}{2}})$ of the ground-state wave function.

Endnotes

Extensive treatments of the Dirac equation can be found in J.D. Bjorken and S.D. Drell, *Relativistic Quantum Mechanics*, McGraw-Hill (1964); V.B. Berestetskii, E.M. Lifshitz and L.P. Pitaevskii, *Quantum Electrodynamics*, Pergamon (1980).

For historical accounts of relativistic quantum mechanics and quantum electrodynamics see A. Pais, *Inward Bound*, Oxford University Press (1986), ch. 13, 15; S.S. Schweber, *QED and the Men Who Made it*, Princeton University Press (1994), ch. 1; S. Weinberg, *Quantum Field Theory, Vol.I*, Cambridge University Press (1995), ch. 1; A.I. Miller, *Early Quantum Electrodynamics*, Cambridge University Press (1994).

Appendix

Physical Constants[1]

Quantity	Symbol	Value
electron mass	m_e	0.510999 MeV$/c^2$
muon mass	m_μ	105.658 MeV$/c^2$
proton mass	m_p	938.272 MeV$/c^2$
nucleon mass difference	$m_n - m_p$	1.2933 MeV$/c^2$
fine structure constant	$\alpha = e^2/4\pi\hbar c$	$7.29735\times10^{-3} = 1/137.036$
electron Compton wavelength	$\lambda\!\!\!^{-}_e = \hbar/m_e c$	3.86159×10^{-11} cm
Bohr radius	$4\pi\hbar^2/m_e e^2$	0.529177×10^{-8} cm
Rydberg	$\frac{1}{2}\alpha^2 m_e c^2$	13.6057 eV
Bohr magneton	$\mu_B = e\hbar/2m_e$	5.78838×10^{-5} eV/T
nuclear magneton	$\mu_N = e\hbar/2m_p$	3.15245×10^{-8} eV/T
electron magnetic moment		$1.00116\,\mu_B$
proton magnetic moment		$2.79285\,\mu_N$
neutron magnetic moment		$-1.91304\,\mu_N$
Boltzmann constant	k_B	8.6173×10^{-5} eV/K
energy-length conversion		$1 = 197.326$ MeV fm
energy-area unit, electron	$\hbar^2/2m_e$	3.810×10^{-16} cm^2 eV
energy-area unit, proton	$\hbar^2/2m_p$	20.75×10^{-26} cm^2 MeV
de Broglie wavelength, electron	$\hbar/p$	$1.952\,[E(\text{eV})]^{-\frac{1}{2}}\times10^{-8}$ cm
de Broglie wavelength, proton	$\hbar/p$	$4.555\,[E(\text{MeV})]^{-\frac{1}{2}}$ fm

Operator Identities

Let $[q_i, p_j] = i\delta_{ij}$, then

$$[q_i, G(p)] = i\partial G/\partial p_i\,, \qquad [p_i, F(q)] = -i\partial F/\partial q_i\,.$$

For arbitrary operators $A, B\ldots$

$$[A, BC] = [A,B]C + B[A,C] = \{A,B\}C - B\{A,C\}\,,$$

$$[AB, CD] = A[B,C]D + AC[B,D] + [A,C]DB + C[A,D]B\,,$$

$$[AB, CD] = A\{B,C\}D - \{A,C\}BD + CA\{B,D\} - C\{A,D\}B\,.$$

Identities relevant to propagators, etc.,

$$\frac{1}{A-B} = \frac{1}{A} + \frac{1}{A}B\frac{1}{A-B} = \frac{1}{A} + \frac{1}{A-B}B\frac{1}{A}\,.$$

[1]For accurate and up-to-date values see www.pdg.lbl.gov/

For vector operators $\boldsymbol{A}, \boldsymbol{B}, \boldsymbol{C}$,

$$[\boldsymbol{A}\times(\boldsymbol{B}\times\boldsymbol{C})]_i = A_j B_i C_j - (\boldsymbol{A}\cdot\boldsymbol{B})C_i\,, \qquad [(\boldsymbol{A}\times\boldsymbol{B})\times\boldsymbol{C}]_i = A_j B_i C_j - A_i(\boldsymbol{B}\cdot\boldsymbol{C})\,.$$

The BCH theorem for arbitrary A and B,

$$e^A e^B = \exp\left(A + B + \frac{1}{2}[A,B] + \frac{1}{12}\left([A,[A,B]) - [B,[A,B]\right)\ldots\right)\,;$$

if $[A,B]$ commutes with A and B,

$$e^A e^B = e^{A+B}\, e^{\frac{1}{2}[A,B]}\,.$$

Angular Momentum

Let $[J_i, J_k] = i\epsilon_{ijk}J_k$ and $J_\pm = J_1 \pm iJ_2$, then

$$[J_+, J_-] = 2J_3\,, \quad [J_+, J_3] = -J_+\,, \quad [J_-, J_3] = J_-\,,$$

$$J_3|jm\rangle = m|jm\rangle\,, \quad J_\pm|jm\rangle = \sqrt{j(j+1) - m(m\pm 1)}\,|jm\pm 1\rangle\,.$$

$$\sigma_i\sigma_j = i\epsilon_{ijk}\sigma_k + \delta_{ij}$$

$$(\boldsymbol{\sigma}\cdot\boldsymbol{A})(\boldsymbol{\sigma}\cdot\boldsymbol{B}) = \boldsymbol{A}\cdot\boldsymbol{B} + i\boldsymbol{\sigma}\cdot(\boldsymbol{A}\times\boldsymbol{B})\,,$$

$$\exp(-\tfrac{1}{2}i\theta\,\boldsymbol{n}\cdot\boldsymbol{\sigma}) = \cos\tfrac{1}{2}\theta - i\boldsymbol{n}\cdot\boldsymbol{\sigma}\sin\tfrac{1}{2}\theta\,.$$

$$D^j_{mm'}(R) = e^{-i\alpha m}\, d^j_{mm'}(\beta)\, e^{-i\gamma m'}\,.$$

$$D^{\frac{1}{2}}(R) = \begin{pmatrix} e^{-\frac{1}{2}i(\alpha+\gamma)}\cos\frac{1}{2}\beta & -e^{-\frac{1}{2}i(\alpha-\gamma)}\sin\frac{1}{2}\beta \\ e^{\frac{1}{2}i(\alpha-\gamma)}\sin\frac{1}{2}\beta & e^{\frac{1}{2}i(\alpha+\gamma)}\cos\frac{1}{2}\beta \end{pmatrix}\,.$$

$$D^l_{m0}(\alpha\beta\gamma) = \sqrt{\frac{4\pi}{2l+1}}\, Y^*_{lm}(\beta\alpha)\,.$$

$$d^j_{mm'}(\beta) = d^j_{m'm}(-\beta) = d^j_{-m',-m}(\beta)\,,$$

$$d^j_{mm'}(\beta) = (-1)^{m-m'} d^j_{mm'}(-\beta) = (-1)^{m-m'} d^j_{m'm}(\beta) = (-1)^{m-m'} d^j_{-m,-m'}(\beta)\,.$$

For integer $l_{1,2,3}$, and if $2L = l_1 + l_2 + l_3$ is even,

$$\langle l_1 0 l_2 0 | l_3 0\rangle = (-1)^{l_1+L}\frac{L!\sqrt{2l_1+1}}{(L-l_1)!(L-l_2)!(L-l_3)!}\sqrt{\frac{(l_1+l_2-l_3)!(l_2+l_3-l_1)!(l_3+l_1-l_2)!}{(2L+1)!}}\,,$$

but vanishes if $2L$ is odd.

CLEBSCH-GORDAN COEFFICIENTS, SPHERICAL HARMONICS, AND d FUNCTIONS

Note: A square-root sign is to be understood over *every* coefficient, *e.g.*, for $-8/15$ read $-\sqrt{8/15}$.

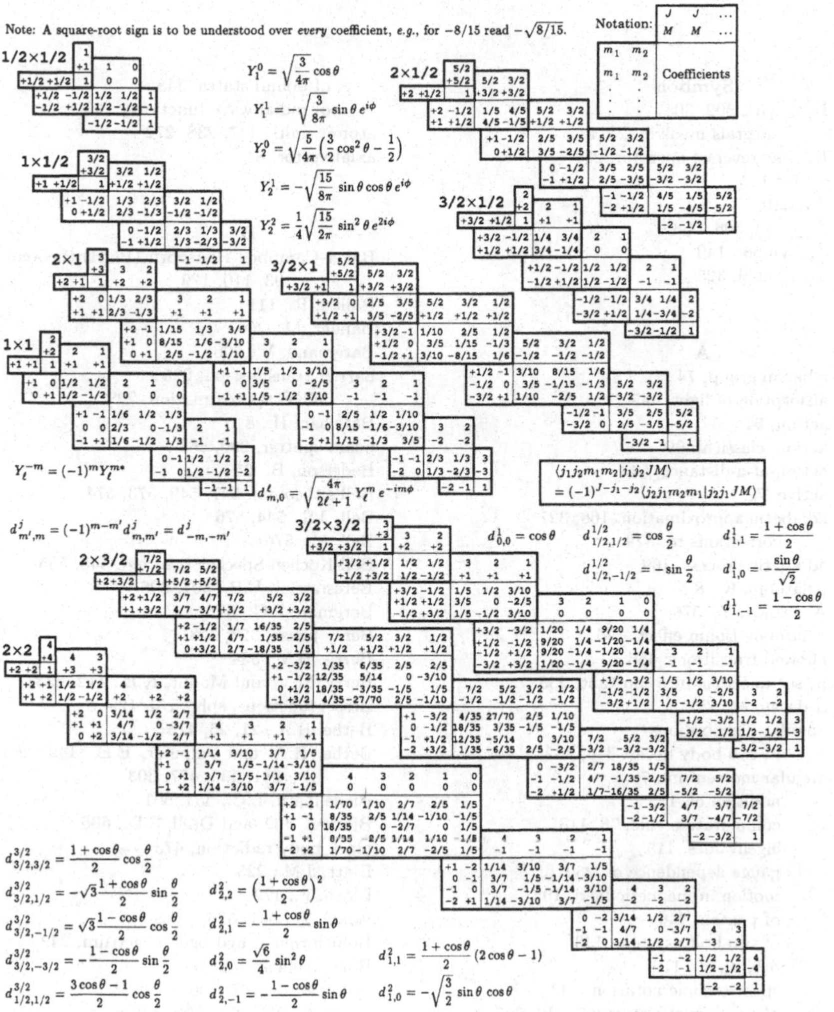

$$Y_1^0 = \sqrt{\frac{3}{4\pi}}\cos\theta$$

$$Y_1^1 = -\sqrt{\frac{3}{8\pi}}\sin\theta\, e^{i\phi}$$

$$Y_2^0 = \sqrt{\frac{5}{4\pi}}\left(\frac{3}{2}\cos^2\theta - \frac{1}{2}\right)$$

$$Y_2^1 = -\sqrt{\frac{15}{8\pi}}\sin\theta\cos\theta\, e^{i\phi}$$

$$Y_2^2 = \frac{1}{4}\sqrt{\frac{15}{2\pi}}\sin^2\theta\, e^{2i\phi}$$

$$Y_\ell^{-m} = (-1)^m Y_\ell^{m*} \qquad d^\ell_{m,0} = \sqrt{\frac{4\pi}{2\ell+1}}\, Y_\ell^m\, e^{-im\phi}$$

$$\langle j_1 j_2 m_1 m_2 | j_1 j_2 J M\rangle = (-1)^{J-j_1-j_2}\langle j_2 j_1 m_2 m_1 | j_2 j_1 J M\rangle$$

$$d^j_{m',m} = (-1)^{m-m'} d^j_{m,m'} = d^j_{-m,-m'}$$

$$d^1_{0,0} = \cos\theta \qquad d^{1/2}_{1/2,1/2} = \cos\frac{\theta}{2} \qquad d^{1/2}_{1/2,-1/2} = -\sin\frac{\theta}{2}$$

$$d^1_{1,1} = \frac{1+\cos\theta}{2} \qquad d^1_{1,0} = -\frac{\sin\theta}{\sqrt{2}} \qquad d^1_{1,-1} = \frac{1-\cos\theta}{2}$$

$$d^{3/2}_{3/2,3/2} = \frac{1+\cos\theta}{2}\cos\frac{\theta}{2} \qquad d^{3/2}_{3/2,1/2} = -\sqrt{3}\frac{1+\cos\theta}{2}\sin\frac{\theta}{2}$$

$$d^{3/2}_{3/2,-1/2} = \sqrt{3}\frac{1-\cos\theta}{2}\cos\frac{\theta}{2} \qquad d^{3/2}_{3/2,-3/2} = -\frac{1-\cos\theta}{2}\sin\frac{\theta}{2}$$

$$d^{3/2}_{1/2,1/2} = \frac{3\cos\theta-1}{2}\cos\frac{\theta}{2} \qquad d^{3/2}_{1/2,-1/2} = -\frac{3\cos\theta+1}{2}\sin\frac{\theta}{2}$$

$$d^2_{2,2} = \left(\frac{1+\cos\theta}{2}\right)^2 \qquad d^2_{2,1} = -\frac{1+\cos\theta}{2}\sin\theta \qquad d^2_{2,0} = \frac{\sqrt{6}}{4}\sin^2\theta$$

$$d^2_{2,-1} = -\frac{1-\cos\theta}{2}\sin\theta \qquad d^2_{2,-2} = \left(\frac{1-\cos\theta}{2}\right)^2$$

$$d^2_{1,1} = \frac{1+\cos\theta}{2}(2\cos\theta-1) \qquad d^2_{1,0} = -\sqrt{\frac{3}{2}}\sin\theta\cos\theta$$

$$d^2_{1,-1} = \frac{1-\cos\theta}{2}(2\cos\theta+1) \qquad d^2_{0,0} = \left(\frac{3}{2}\cos^2\theta - \frac{1}{2}\right)$$

Figure 31.1: The sign convention is that of Wigner (*Group Theory*, Academic Press, New York, 1959), also used by Condon and Shortley (*The Theory of Atomic Spectra*, Cambridge Univ. Press, New York, 1953), Rose (*Elementary Theory of Angular Momentum*, Wiley, New York, 1957), and Cohen (*Tables of the Clebsch-Gordan Coefficients*, North American Rockwell Science Center, Thousand Oaks, Calif., 1974). The coefficients here have been calculated using computer programs written independently by Cohen and at LBNL.

FIG. 13.1. Courtesy, The Particle Data Group.

Index

T

Z

Graduate Texts in Contemporary Physics

B.M. Smirnov: **Physics of Atoms and Ions**

M. Stone: **The Physics of Quantum Fields**

F.T. Vasko and A.V. Kuznetsov: **Electronic States and Optical Transitions in Semiconductor Heterostructures**

A.M. Zagoskin: **Quantum Theory of Many-Body Systems: Techniques and Applications**